Lecture Notes in Computer Science 13842

The series Lecture Notes in Computer Science (LNCS), including its subseries Lecture Notes in Artificial Intelligence (LNAI) and Lecture Notes in Bioinformatics (LNBI), has established itself as a medium for the publication of new developments in computer science and information technology research, teaching, and education.

LNCS enjoys close cooperation with the computer science R & D community, the series counts many renowned academics among its volume editors and paper authors, and collaborates with prestigious societies. Its mission is to serve this international community by providing an invaluable service, mainly focused on the publication of conference and workshop proceedings and postproceedings. LNCS commenced publication in 1973.

Lei Wang · Juergen Gall · Tat-Jun Chin ·
Imari Sato · Rama Chellappa
Editors

Computer Vision – ACCV 2022

16th Asian Conference on Computer Vision
Macao, China, December 4–8, 2022
Proceedings, Part II

Springer

Editors
Lei Wang
University of Wollongong
Wollongong, NSW, Australia

Juergen Gall
University of Bonn
Bonn, Germany

Tat-Jun Chin
University of Adelaide
Adelaide, SA, Australia

Imari Sato
National Institute of Informatics
Tokyo, Japan

Rama Chellappa
Johns Hopkins University
Baltimore, MD, USA

ISSN 0302-9743 ISSN 1611-3349 (electronic)
Lecture Notes in Computer Science
ISBN 978-3-031-26283-8 ISBN 978-3-031-26284-5 (eBook)
https://doi.org/10.1007/978-3-031-26284-5

This Springer imprint is published by the registered company Springer Nature Switzerland AG
The registered company address is: Gewerbestrasse 11, 6330 Cham, Switzerland

Preface

The 16th Asian Conference on Computer Vision (ACCV) 2022 was held in a hybrid mode in Macau SAR, China during December 4–8, 2022. The conference featured novel research contributions from almost all sub-areas of computer vision.

For the main conference, 836 valid submissions entered the review stage after desk rejection. Sixty-three area chairs and 959 reviewers made great efforts to ensure that every submission received thorough and high-quality reviews. As in previous editions of ACCV, this conference adopted a double-blind review process. The identities of authors were not visible to the reviewers or area chairs; nor were the identities of the assigned reviewers and area chairs known to the authors. The program chairs did not submit papers to the conference.

After receiving the reviews, the authors had the option of submitting a rebuttal. Following that, the area chairs led the discussions and final recommendations were then made by the reviewers. Taking conflicts of interest into account, the area chairs formed 21 AC triplets to finalize the paper recommendations. With the confirmation of three area chairs for each paper, 277 papers were accepted. ACCV 2022 also included eight workshops, eight tutorials, and one grand challenge, covering various cutting-edge research topics related to computer vision. The proceedings of ACCV 2022 are open access at the Computer Vision Foundation website, by courtesy of Springer. The quality of the papers presented at ACCV 2022 demonstrates the research excellence of the international computer vision communities.

This conference is fortunate to receive support from many organizations and individuals. We would like to express our gratitude for the continued support of the Asian Federation of Computer Vision and our sponsors, the University of Macau, Springer, the Artificial Intelligence Journal, and OPPO. ACCV 2022 used the Conference Management Toolkit sponsored by Microsoft Research and received much help from its support team.

All the organizers, area chairs, reviewers, and authors made great contributions to ensure a successful ACCV 2022. For this, we owe them deep gratitude. Last but not least, we would like to thank the online and in-person attendees of ACCV 2022. Their presence showed strong commitment and appreciation towards this conference.

December 2022

Lei Wang
Juergen Gall
Tat-Jun Chin
Imari Sato
Rama Chellappa

Organization

General Chairs

Gérard Medioni	University of Southern California, USA
Shiguang Shan	Chinese Academy of Sciences, China
Bohyung Han	Seoul National University, South Korea
Hongdong Li	Australian National University, Australia

Program Chairs

Rama Chellappa	Johns Hopkins University, USA
Juergen Gall	University of Bonn, Germany
Imari Sato	National Institute of Informatics, Japan
Tat-Jun Chin	University of Adelaide, Australia
Lei Wang	University of Wollongong, Australia

Publication Chairs

Wenbin Li	Nanjing University, China
Wanqi Yang	Nanjing Normal University, China

Local Arrangements Chairs

Liming Zhang	University of Macau, China
Jianjia Zhang	Sun Yat-sen University, China

Web Chairs

Zongyuan Ge	Monash University, Australia
Deval Mehta	Monash University, Australia
Zhongyan Zhang	University of Wollongong, Australia

AC Meeting Chair

Chee Seng Chan	University of Malaya, Malaysia

Area Chairs

Aljosa Osep	Technical University of Munich, Germany
Angela Yao	National University of Singapore, Singapore
Anh T. Tran	VinAI Research, Vietnam
Anurag Mittal	Indian Institute of Technology Madras, India
Binh-Son Hua	VinAI Research, Vietnam
C. V. Jawahar	International Institute of Information Technology, Hyderabad, India
Dan Xu	The Hong Kong University of Science and Technology, China
Du Tran	Meta AI, USA
Frederic Jurie	University of Caen and Safran, France
Guangcan Liu	Southeast University, China
Guorong Li	University of Chinese Academy of Sciences, China
Guosheng Lin	Nanyang Technological University, Singapore
Gustavo Carneiro	University of Surrey, UK
Hyun Soo Park	University of Minnesota, USA
Hyunjung Shim	Korea Advanced Institute of Science and Technology, South Korea
Jiaying Liu	Peking University, China
Jun Zhou	Griffith University, Australia
Junseok Kwon	Chung-Ang University, South Korea
Kota Yamaguchi	CyberAgent, Japan
Li Liu	National University of Defense Technology, China
Liang Zheng	Australian National University, Australia
Mathieu Aubry	Ecole des Ponts ParisTech, France
Mehrtash Harandi	Monash University, Australia
Miaomiao Liu	Australian National University, Australia
Ming-Hsuan Yang	University of California at Merced, USA
Palaiahnakote Shivakumara	University of Malaya, Malaysia
Pau-Choo Chung	National Cheng Kung University, Taiwan

Qianqian Xu	Key Laboratory of Intelligent Information Processing, Institute of Computing Technology, Chinese Academy of Sciences, China
Qiuhong Ke	Monash University, Australia
Radu Timofte	University of Würzburg, Germany and ETH Zurich, Switzerland
Rajagopalan N. Ambasamudram	Indian Institute of Technology Madras, India
Risheng Liu	Dalian University of Technology, China
Ruiping Wang	Institute of Computing Technology, Chinese Academy of Sciences, China
Sajid Javed	Khalifa University of Science and Technology, Abu Dhabi, UAE
Seunghoon Hong	Korea Advanced Institute of Science and Technology, South Korea
Shang-Hong Lai	National Tsing Hua University, Taiwan
Shanshan Zhang	Nanjing University of Science and Technology, China
Sharon Xiaolei Huang	Pennsylvania State University, USA
Shin'ichi Satoh	National Institute of Informatics, Japan
Si Liu	Beihang University, China
Suha Kwak	Pohang University of Science and Technology, South Korea
Tae Hyun Kim	Hanyang Univeristy, South Korea
Takayuki Okatani	Tohoku University, Japan/RIKEN Center for Advanced Intelligence Project, Japan
Tatsuya Harada	University of Tokyo/RIKEN, Japan
Vicky Kalogeiton	Ecole Polytechnique, France
Vincent Lepetit	Ecole des Ponts ParisTech, France
Vineeth N. Balasubramanian	Indian Institute of Technology, Hyderabad, India
Wei Shen	Shanghai Jiao Tong University, China
Wei-Shi Zheng	Sun Yat-sen University, China
Xiang Bai	Huazhong University of Science and Technology, China
Xiaowei Zhou	Zhejiang University, China
Xin Yu	University of Technology Sydney, Australia
Yasutaka Furukawa	Simon Fraser University, Canada
Yasuyuki Matsushita	Osaka University, Japan
Yedid Hoshen	Hebrew University of Jerusalem, Israel
Ying Fu	Beijing Institute of Technology, China
Yong Jae Lee	University of Wisconsin-Madison, USA
Yu-Chiang Frank Wang	National Taiwan University, Taiwan
Yumin Suh	NEC Laboratories America, USA

Yung-Yu Chuang	National Taiwan University, Taiwan
Zhaoxiang Zhang	Chinese Academy of Sciences, China
Ziad Al-Halah	University of Texas at Austin, USA
Zuzana Kukelova	Czech Technical University, Czech Republic

Additional Reviewers

Abanob E. N. Soliman
Abdelbadie Belmouhcine
Adrian Barbu
Agnibh Dasgupta
Akihiro Sugimoto
Akkarit Sangpetch
Akrem Sellami
Aleksandr Kim
Alexander Andreopoulos
Alexander Fix
Alexander Kugele
Alexandre Morgand
Alexis Lechervy
Alina E. Marcu
Alper Yilmaz
Alvaro Parra
Amogh Subbakrishna Adishesha
Andrea Giachetti
Andrea Lagorio
Andreu Girbau Xalabarder
Andrey Kuehlkamp
Anh Nguyen
Anh T. Tran
Ankush Gupta
Anoop Cherian
Anton Mitrokhin
Antonio Agudo
Antonio Robles-Kelly
Ara Abigail Ambita
Ardhendu Behera
Arjan Kuijper
Arren Matthew C. Antioquia
Arjun Ashok
Atsushi Hashimoto
Atsushi Shimada
Attila Szabo
Aurelie Bugeau
Avatharam Ganivada
Ayan Kumar Bhunia
Azade Farshad
B. V. K. Vijaya Kumar
Bach Tran
Bailin Yang
Baojiang Zhong
Baoquan Zhang
Baoyao Yang
Basit O. Alawode
Beibei Lin
Benoit Guillard
Beomgu Kang
Bin He
Bin Li
Bin Liu
Bin Ren
Bin Yang
Bin-Cheng Yang
BingLiang Jiao
Bo Liu
Bohan Li
Boyao Zhou
Boyu Wang
Caoyun Fan
Carlo Tomasi
Carlos Torres
Carvalho Micael
Cees Snoek
Chang Kong
Changick Kim
Changkun Ye
Changsheng Lu
Chao Liu
Chao Shi
Chaowei Tan
Chaoyi Li
Chaoyu Dong
Chaoyu Zhao
Chen He
Chen Liu
Chen Yang
Chen Zhang
Cheng Deng
Cheng Guo
Cheng Yu
Cheng-Kun Yang
Chenglong Li
Chengmei Yang
Chengxin Liu
Chengyao Qian
Chen-Kuo Chiang
Chenxu Luo
Che-Rung Lee
Che-Tsung Lin
Chi Xu
Chi Nhan Duong
Chia-Ching Lin
Chien-Cheng Lee
Chien-Yi Wang
Chih-Chung Hsu
Chih-Wei Lin
Ching-Chun Huang
Chiou-Ting Hsu
Chippy M. Manu
Chong Wang
Chongyang Wang
Christian Siagian
Christine Allen-Blanchette

Christoph Schorn
Christos Matsoukas
Chuan Guo
Chuang Yang
Chuanyi Zhang
Chunfeng Song
Chunhui Zhang
Chun-Rong Huang
Ci Lin
Ci-Siang Lin
Cong Fang
Cui Wang
Cui Yuan
Cyrill Stachniss
Dahai Yu
Daiki Ikami
Daisuke Miyazaki
Dandan Zhu
Daniel Barath
Daniel Lichy
Daniel Reich
Danyang Tu
David Picard
Davide Silvestri
Defang Chen
Dehuan Zhang
Deunsol Jung
Difei Gao
Dim P. Papadopoulos
Ding-Jie Chen
Dong Gong
Dong Hao
Dong Wook Shu
Dongdong Chen
Donghun Lee
Donghyeon Kwon
Donghyun Yoo
Dongkeun Kim
Dongliang Luo
Dongseob Kim
Dongsuk Kim
Dongwan Kim
Dongwon Kim
DongWook Yang
Dongze Lian
Dubing Chen
Edoardo Remelli
Emanuele Trucco
Erhan Gundogdu
Erh-Chung Chen
Rickson R. Nascimento
Erkang Chen
Eunbyung Park
Eunpil Park
Eun-Sol Kim
Fabio Cuzzolin
Fan Yang
Fan Zhang
Fangyu Zhou
Fani Deligianni
Fatemeh Karimi Nejadasl
Fei Liu
Feiyue Ni
Feng Su
Feng Xue
Fengchao Xiong
Fengji Ma
Fernando Díaz-del-Rio
Florian Bernard
Florian Kleber
Florin-Alexandru Vasluianu
Fok Hing Chi Tivive
Frank Neumann
Fu-En Yang
Fumio Okura
Gang Chen
Gang Liu
Gao Haoyuan
Gaoshuai Wang
Gaoyun An
Gen Li
Georgy Ponimatkin
Gianfranco Doretto
Gil Levi
Guang Yang
Guangfa Wang
Guangfeng Lin
Guillaume Jeanneret
Guisik Kim
Gunhee Kim
Guodong Wang
Ha Young Kim
Hadi Mohaghegh Dolatabadi
Haibo Ye
Haili Ye
Haithem Boussaid
Haixia Wang
Han Chen
Han Zou
Hang Cheng
Hang Du
Hang Guo
Hanlin Gu
Hannah H. Kim
Hao He
Hao Huang
Hao Quan
Hao Ren
Hao Tang
Hao Zeng
Hao Zhao
Haoji Hu
Haopeng Li
Haoqing Wang
Haoran Wen
Haoshuo Huang
Haotian Liu
Haozhao Ma
Hari Chandana K.
Haripriya Harikumar
Hehe Fan
Helder Araujo
Henok Ghebrechristos
Heunseung Lim
Hezhi Cao
Hideo Saito
Hieu Le
Hiroaki Santo
Hirokatsu Kataoka
Hiroshi Omori
Hitika Tiwari
Hojung Lee
Hong Cheng

Hong Liu
Hu Zhang
Huadong Tang
Huajie Jiang
Huang Ziqi
Huangying Zhan
Hui Kong
Hui Nie
Huiyu Duan
Huyen Thi Thanh Tran
Hyung-Jeong Yang
Hyunjin Park
Hyunsoo Kim
HyunWook Park
I-Chao Shen
Idil Esen Zulfikar
Ikuhisa Mitsugami
Inseop Chung
Ioannis Pavlidis
Isinsu Katircioglu
Jaeil Kim
Jaeyoon Park
Jae-Young Sim
James Clark
James Elder
James Pritts
Jan Zdenek
Janghoon Choi
Jeany Son
Jenny Seidenschwarz
Jesse Scott
Jia Wan
Jiadai Sun
JiaHuan Ji
Jiajiong Cao
Jian Zhang
Jianbo Jiao
Jianhui Wu
Jianjia Wang
Jianjia Zhang
Jianqiao Wangni
JiaQi Wang
Jiaqin Lin
Jiarui Liu
Jiawei Wang
Jiaxin Gu
Jiaxin Wei
Jiaxin Zhang
Jiaying Zhang
Jiayu Yang
Jidong Tian
Jie Hong
Jie Lin
Jie Liu
Jie Song
Jie Yang
Jiebo Luo
Jiejie Xu
Jin Fang
Jin Gao
Jin Tian
Jinbin Bai
Jing Bai
Jing Huo
Jing Tian
Jing Wu
Jing Zhang
Jingchen Xu
Jingchun Cheng
Jingjing Fu
Jingshuai Liu
JingWei Huang
Jingzhou Chen
JinHan Cui
Jinjie Song
Jinqiao Wang
Jinsun Park
Jinwoo Kim
Jinyu Chen
Jipeng Qiang
Jiri Sedlar
Jiseob Kim
Jiuxiang Gu
Jiwei Xiao
Jiyang Zheng
Jiyoung Lee
John Paisley
Joonki Paik
Joonseok Lee
Julien Mille
Julio C. Zamora
Jun Sato
Jun Tan
Jun Tang
Jun Xiao
Jun Xu
Junbao Zhuo
Jun-Cheng Chen
Junfen Chen
Jungeun Kim
Junhwa Hur
Junli Tao
Junlin Han
Junsik Kim
Junting Dong
Junwei Zhou
Junyu Gao
Kai Han
Kai Huang
Kai Katsumata
Kai Zhao
Kailun Yang
Kai-Po Chang
Kaixiang Wang
Kamal Nasrollahi
Kamil Kowol
Kan Chang
Kang-Jun Liu
Kanchana Vaishnavi Gandikota
Kanoksak Wattanachote
Karan Sikka
Kaushik Roy
Ke Xian
Keiji Yanai
Kha Gia Quach
Kibok Lee
Kira Maag
Kirill Gavrilyuk
Kohei Suenaga
Koichi Ito
Komei Sugiura
Kong Dehui
Konstantinos Batsos
Kotaro Kikuchi

Kouzou Ohara
Kuan-Wen Chen
Kun He
Kun Hu
Kun Zhan
Kunhee Kim
Kwan-Yee K. Wong
Kyong Hwan Jin
Kyuhong Shim
Kyung Ho Park
Kyungmin Kim
Kyungsu Lee
Lam Phan
Lanlan Liu
Le Hui
Lei Ke
Lei Qi
Lei Yang
Lei Yu
Lei Zhu
Leila Mahmoodi
Li Jiao
Li Su
Lianyu Hu
Licheng Jiao
Lichi Zhang
Lihong Zheng
Lijun Zhao
Like Xin
Lin Gu
Lin Xuhong
Lincheng Li
Linghua Tang
Lingzhi Kong
Linlin Yang
Linsen Li
Litao Yu
Liu Liu
Liujie Hua
Li-Yun Wang
Loren Schwiebert
Lujia Jin
Lujun Li
Luping Zhou
Luting Wang
Mansi Sharma
Mantini Pranav
Mahmoud Zidan Khairallah
Manuel Günther
Marcella Astrid
Marco Piccirilli
Martin Kampel
Marwan Torki
Masaaki Iiyama
Masanori Suganuma
Masayuki Tanaka
Matan Jacoby
Md Alimoor Reza
Md. Zasim Uddin
Meghshyam Prasad
Mei-Chen Yeh
Meng Tang
Mengde Xu
Mengyang Pu
Mevan B. Ekanayake
Michael Bi Mi
Michael Wray
Michaël Clément
Michel Antunes
Michele Sasdelli
Mikhail Sizintsev
Min Peng
Min Zhang
Minchul Shin
Minesh Mathew
Ming Li
Ming Meng
Ming Yin
Ming-Ching Chang
Mingfei Cheng
Minghui Wang
Mingjun Hu
MingKun Yang
Mingxing Tan
Mingzhi Yuan
Min-Hung Chen
Minhyun Lee
Minjung Kim
Min-Kook Suh
Minkyo Seo
Minyi Zhao
Mo Zhou
Mohammad Amin A. Shabani
Moein Sorkhei
Mohit Agarwal
Monish K. Keswani
Muhammad Sarmad
Muhammad Kashif Ali
Myung-Woo Woo
Naeemullah Khan
Naman Solanki
Namyup Kim
Nan Gao
Nan Xue
Naoki Chiba
Naoto Inoue
Naresh P. Cuntoor
Nati Daniel
Neelanjan Bhowmik
Niaz Ahmad
Nicholas I. Kuo
Nicholas E. Rosa
Nicola Fioraio
Nicolas Dufour
Nicolas Papadakis
Ning Liu
Nishan Khatri
Ole Johannsen
P. Real Jurado
Parikshit V. Sakurikar
Patrick Peursum
Pavan Turaga
Peijie Chen
Peizhi Yan
Peng Wang
Pengfei Fang
Penghui Du
Pengpeng Liu
Phi Le Nguyen
Philippe Chiberre
Pierre Gleize
Pinaki Nath Chowdhury
Ping Hu

Ping Li
Ping Zhao
Pingping Zhang
Pradyumna Narayana
Pritish Sahu
Qi Li
Qi Wang
Qi Zhang
Qian Li
Qian Wang
Qiang Fu
Qiang Wu
Qiangxi Zhu
Qianying Liu
Qiaosi Yi
Qier Meng
Qin Liu
Qing Liu
Qing Wang
Qingheng Zhang
Qingjie Liu
Qinglin Liu
Qingsen Yan
Qingwei Tang
Qingyao Wu
Qingzheng Wang
Qizao Wang
Quang Hieu Pham
Rabab Abdelfattah
Rabab Ward
Radu Tudor Ionescu
Rahul Mitra
Raül Pérez i Gonzalo
Raymond A. Yeh
Ren Li
Renán Rojas-Gómez
Renjie Wan
Renuka Sharma
Reyer Zwiggelaar
Robin Chan
Robin Courant
Rohit Saluja
Rongkai Ma
Ronny Hänsch
Rui Liu
Rui Wang
Rui Zhu
Ruibing Hou
Ruikui Wang
Ruiqi Zhao
Ruixing Wang
Ryo Furukawa
Ryusuke Sagawa
Saimunur Rahman
Samet Akcay
Samitha Herath
Sanath Narayan
Sandesh Kamath
Sanghoon Jeon
Sanghyun Son
Satoshi Suzuki
Saumik Bhattacharya
Sauradip Nag
Scott Wehrwein
Sebastien Lefevre
Sehyun Hwang
Seiya Ito
Selen Pehlivan
Sena Kiciroglu
Seok Bong Yoo
Seokjun Park
Seongwoong Cho
Seoungyoon Kang
Seth Nixon
Seunghwan Lee
Seung-Ik Lee
Seungyong Lee
Shaifali Parashar
Shan Cao
Shan Zhang
Shangfei Wang
Shaojian Qiu
Shaoru Wang
Shao-Yuan Lo
Shengjin Wang
Shengqi Huang
Shenjian Gong
Shi Qiu
Shiguang Liu
Shih-Yao Lin
Shin-Jye Lee
Shishi Qiao
Shivam Chandhok
Shohei Nobuhara
Shreya Ghosh
Shuai Yuan
Shuang Yang
Shuangping Huang
Shuigeng Zhou
Shuiwang Li
Shunli Zhang
Shuo Gu
Shuoxin Lin
Shuzhi Yu
Sida Peng
Siddhartha Chandra
Simon S. Woo
Siwei Wang
Sixiang Chen
Siyu Xia
Sohyun Lee
Song Guo
Soochahn Lee
Soumava Kumar Roy
Srinjay Soumitra Sarkar
Stanislav Pidhorskyi
Stefan Gumhold
Stefan Matcovici
Stefano Berretti
Stylianos Moschoglou
Sudhir Yarram
Sudong Cai
Suho Yang
Sumitra S. Malagi
Sungeun Hong
Sunggu Lee
Sunghyun Cho
Sunghyun Myung
Sungmin Cho
Sungyeon Kim
Suzhen Wang
Sven Sickert
Syed Zulqarnain Gilani
Tackgeun You
Taehun Kim

Takao Yamanaka
Takashi Shibata
Takayoshi Yamashita
Takeshi Endo
Takeshi Ikenaga
Tanvir Alam
Tao Hong
Tarun Kalluri
Tat-Jen Cham
Tatsuya Yatagawa
Teck Yian Lim
Tejas Indulal Dhamecha
Tengfei Shi
Thanh-Dat Truong
Thomas Probst
Thuan Hoang Nguyen
Tian Ye
Tianlei Jin
Tianwei Cao
Tianyi Shi
Tianyu Song
Tianyu Wang
Tien-Ju Yang
Tingting Fang
Tobias Baumgartner
Toby P. Breckon
Torsten Sattler
Trung Tuan Dao
Trung Le
Tsung-Hsuan Wu
Tuan-Anh Vu
Utkarsh Ojha
Utku Ozbulak
Vaasudev Narayanan
Venkata Siva Kumar Margapuri
Vandit J. Gajjar
Vi Thi Tuong Vo
Victor Fragoso
Vikas Desai
Vincent Lepetit
Vinh Tran
Viresh Ranjan
Wai-Kin Adams Kong
Wallace Michel Pinto Lira
Walter Liao
Wang Yan
Wang Yong
Wataru Shimoda
Wei Feng
Wei Mao
Wei Xu
Weibo Liu
Weichen Xu
Weide Liu
Weidong Chen
Weihong Deng
Wei-Jong Yang
Weikai Chen
Weishi Zhang
Weiwei Fang
Weixin Lu
Weixin Luo
Weiyao Wang
Wenbin Wang
Wenguan Wang
Wenhan Luo
Wenju Wang
Wenlei Liu
Wenqing Chen
Wenwen Yu
Wenxing Bao
Wenyu Liu
Wenzhao Zheng
Whie Jung
Williem Williem
Won Hwa Kim
Woohwan Jung
Wu Yirui
Wu Yufeng
Wu Yunjie
Wugen Zhou
Wujie Sun
Wuman Luo
Xi Wang
Xianfang Sun
Xiang Chen
Xiang Li
Xiangbo Shu
Xiangcheng Liu
Xiangyu Wang
Xiao Wang
Xiao Yan
Xiaobing Wang
Xiaodong Wang
Xiaofeng Wang
Xiaofeng Yang
Xiaogang Xu
Xiaogen Zhou
Xiaohan Yu
Xiaoheng Jiang
Xiaohua Huang
Xiaoke Shen
Xiaolong Liu
Xiaoqin Zhang
Xiaoqing Liu
Xiaosong Wang
Xiaowen Ma
Xiaoyi Zhang
Xiaoyu Wu
Xieyuanli Chen
Xin Chen
Xin Jin
Xin Wang
Xin Zhao
Xindong Zhang
Xingjian He
Xingqun Qi
Xinjie Li
Xinqi Fan
Xinwei He
Xinyan Liu
Xinyu He
Xinyue Zhang
Xiyuan Hu
Xu Cao
Xu Jia
Xu Yang
Xuan Luo
Xubo Yang
Xudong Lin
Xudong Xie
Xuefeng Liang
Xuehui Wang
Xuequan Lu

Xuesong Yang
Xueyan Zou
XuHu Lin
Xun Zhou
Xupeng Wang
Yali Zhang
Ya-Li Li
Yalin Zheng
Yan Di
Yan Luo
Yan Xu
Yang Cao
Yang Hu
Yang Song
Yang Zhang
Yang Zhao
Yangyang Shu
Yani A. Ioannou
Yaniv Nemcovsky
Yanjun Zhu
Yanling Hao
Yanling Tian
Yao Guo
Yao Lu
Yao Zhou
Yaping Zhao
Yasser Benigmim
Yasunori Ishii
Yasushi Yagi
Yawei Li
Ye Ding
Ye Zhu
Yeongnam Chae
Yeying Jin
Yi Cao
Yi Liu
Yi Rong
Yi Tang
Yi Wei
Yi Xu
Yichun Shi
Yifan Zhang
Yikai Wang
Yikang Ding
Yiming Liu
Yiming Qian
Yin Li
Yinghuan Shi
Yingjian Li
Yingkun Xu
Yingshu Chen
Yingwei Pan
Yiping Tang
Yiqing Shen
Yisheng Zhu
Yitian Li
Yizhou Yu
Yoichi Sato
Yong A.
Yongcai Wang
Yongheng Ren
Yonghuai Liu
Yongjun Zhang
Yongkang Luo
Yongkang Wong
Yongpei Zhu
Yongqiang Zhang
Yongrui Ma
Yoshimitsu Aoki
Yoshinori Konishi
Young Jun Heo
Young Min Shin
Youngmoon Lee
Youpeng Zhao
Yu Ding
Yu Feng
Yu Zhang
Yuanbin Wang
Yuang Wang
Yuanhong Chen
Yuanyuan Qiao
Yucong Shen
Yuda Song
Yue Huang
Yufan Liu
Yuguang Yan
Yuhan Xie
Yu-Hsuan Chen
Yu-Hui Wen
Yujiao Shi
Yujin Ren
Yuki Tatsunami
Yukuan Jia
Yukun Su
Yu-Lun Liu
Yun Liu
Yunan Liu
Yunce Zhao
Yun-Chun Chen
Yunhao Li
Yunlong Liu
Yunlong Meng
Yunlu Chen
Yunqian He
Yunzhong Hou
Yuqiu Kong
Yusuke Hosoya
Yusuke Matsui
Yusuke Morishita
Yusuke Sugano
Yuta Kudo
Yu-Ting Wu
Yutong Dai
Yuxi Hu
Yuxi Yang
Yuxuan Li
Yuxuan Zhang
Yuzhen Lin
Yuzhi Zhao
Yvain Queau
Zanwei Zhou
Zebin Guo
Ze-Feng Gao
Zejia Fan
Zekun Yang
Zelin Peng
Zelong Zeng
Zenglin Xu
Zewei Wu
Zhan Li
Zhan Shi
Zhe Li
Zhe Liu
Zhe Zhang
Zhedong Zheng

Zhenbo Xu
Zheng Gu
Zhenhua Tang
Zhenkun Wang
Zhenyu Weng
Zhi Zeng
Zhiguo Cao
Zhijie Rao
Zhijie Wang
Zhijun Zhang
Zhimin Gao
Zhipeng Yu
Zhiqiang Hu
Zhisong Liu
Zhiwei Hong
Zhiwei Xu
Zhiwu Lu
Zhixiang Wang
Zhixin Li
Zhiyong Dai
Zhiyong Huang
Zhiyuan Zhang
Zhonghua Wu
Zhongyan Zhang
Zhongzheng Yuan
Zhu Hu
Zhu Meng
Zhujun Li
Zhulun Yang
Zhuojun Zou
Ziang Cheng
Zichuan Liu
Zihan Ding
Zihao Zhang
Zijiang Song
Zijin Yin
Ziqiang Zheng
Zitian Wang
Ziwei Yao
Zixun Zhang
Ziyang Luo
Ziyi Bai
Ziyi Wang
Zongheng Tang
Zongsheng Cao
Zongwei Wu
Zoran Duric

Contents – Part II

Applications of Computer Vision, Vision for X

Applications of Computer Vision, Vision for X

Skin Tone Diagnosis in the Wild: Towards More Robust and Inclusive User Experience Using Oriented Aleatoric Uncertainty

Emmanuel Malherbe[1], Michel Remise[2], Shuai Zhang[1], and Matthieu Perrot[1](✉)

[1] L'Oréal AI Research, New York, USA
{emmanuel.malherbe,shuai.zhang,matthieu.perrot}@rd.loreal.com
[2] ANEO, Boulogne-Billancourt, France
mremise@aneo.fr

Abstract. The past decade has seen major advances in deep learning models that are trained to predict a supervised label. However, estimating the uncertainty for a predicted value might provide great information beyond the prediction itself. To address this goal, using a probabilistic loss was proven efficient for aleatoric uncertainty, which aims at capturing noise originating from the observations. For multidimensional predictions, this estimated noise is generally a multivariate normal variable, characterized by a mean value and covariance matrix. While most of literature have focused on isotropic uncertainty, with diagonal covariance matrix, estimating full covariance brings additional information, such as the noise orientation in the output space.

We propose in this paper a specific decomposition of the covariance matrix that can be efficiently estimated by the neural network. From our experimental comparison to the existing approaches, our model offers the best trade-off between uncertainty orientation likeliness, model accuracy and computation costs. Our industrial application is skin color estimation based on a selfie picture, which is at the core of an online make-up assistant but is a sensitive topic due to ethics and fairness considerations. Thanks to oriented uncertainty, we can reduce this risk by detecting uncertain cases and proposing a simplified color correction bar, thus making user experience more robust and inclusive.

1 Introduction

Even if they are still likely to make wrong predictions, Deep Learning models are now state of the art for many problems [28]. More and more industrial applications are now based on such models, such as face verification for security by [43], autonomous driving by [39], or cancer detection by [6], with a certain degree of risk in case of wrong prediction. The risk can be also in terms of financial costs, for instance when [36] estimate construction prices or when [38] predicts user retention. Cosmetics and Beauty Tech industries also suffers from imperfect deep learning models, which provide mass personalization via smartphone

L. Wang et al. (Eds.): ACCV 2022, LNCS 13842, pp. 3–21, 2023.
https://doi.org/10.1007/978-3-031-26284-5_1

applications [1,20]. Within this context, we currently develop a model that estimates facial skin color from selfie pictures taken in the wild. This model is the core diagnosis for an online make-up assistant service, which for instance recommends foundation shades to the user. Such application highly suffers from model errors, since a poor estimation may lead to a degraded personalization and a disappointed customer. Moreover, the skin color is an ethically sensitive feature to predict, because it is linked to the ethnicity. Such a model is thus at the core of AI fairness and inclusivity issue, as explained by [17]. The risk is very high for the provider of such service, both legally and for brand reputation.

These models being trained to minimize *on average* the errors on samples (e.g., MSE, cross-entropy, ...), errors are always likely to occur, even for an ideal training with no overfitting. This is true for training samples but even more for unseen data in-the-wild. Among the possible causes to such errors, a good part find roots during the training of the model. For instance, the quality of the training data-set can explain noisy predictions, typically when the coverage is not good enough and there are underrepresented zones in the train data. Having such zones is difficult to fully avoid, but may lead to fairness and ethic issues, for instance when these zones relate to ethnicity. Besides, some ground-truth labels can be imperfect, due to wrong manual annotation or noisy measurement device, so that the model might learn to reproduce this noise. The model capacity can also limit its capability to learn enough patterns on the training data-set. Last, some prediction errors find their only cause at inference time, typically because of poor input quality. For instance, in the case of pictures, estimation can be impacted by bad lighting conditions, blurry picture or improper framing.

Since it is impossible or very costly to avoid such errors, we would like to spot potentially wrong predictions by estimating their uncertainties. Modeling this uncertainty would provide tools to reduce or at least control the risks associated with high errors, for example by requesting a human validation in uncertain cases. Such validation would improve the overall user experience and make it more robust, fair and inclusive. For neural networks, the most common approaches for uncertainty modeling are aleatoric and epistemic, as explained by [23]. In practice, epistemic uncertainty relies on sampling multiple predictions by leveraging dropout randomness, while for aleatoric uncertainty, the model learns to predict from the input patterns a distribution of the prediction. We focus in this paper on the second approach, since [23] found it to be more relevant for real-time applications and large datasets with few outliers. Ideally, such uncertainty shall express the *orientation* in which the ground truth stands from the actual prediction, thus reducing the cost of manual labelling [41]. In the case of a smartphone application, such orientation enables the user to easily refine a poor prediction. More generally, in the context of Active Learning [14] or Active Acquisition [47], uncertainty's orientation helps to better target additional data annotation or acquisition.

To accurately estimate this oriented uncertainty in real-time, we propose a parameterization of a full covariance matrix. When compared to the state-of-the-art uncertainty methods, it offers the best trade-off between performance, color accuracy and uncertainty orientation. We propose in this paper the following contributions. First, we present an uncertainty model based on a specific decomposition of the covariance matrix efficiently predictable by a neural network. Second, we propose to extract relevant information from this covariance, such as the uncertainty magnitude and orientation. Last, we performed experiments on the task of in-the-wild skin color estimation from selfie pictures, with several applications of uncertainty for the scenario of online make-up assistant.

2 Related Work

Errors Detection in Machine Learning. Detecting and understanding errors has long been a hot topic for machine learning, partly due to the cost or risk induced by wrong predictions [21]. For error detection in classification, probabilistic models are natively providing a probability that indicates how sure the model is for the predicted class, as [13] described. Similarly, Gaussian Processes are popular for regression problems and natively provide a variance for each prediction, as depicted by [45]. On the other hand, non-probabilistic classification models only predict raw scores, which take values of any magnitude and are thus hard to interpret for error detection. This score can however be converted to a probability, as [34] proposed for binary Support Vector Machines classifier, that was extended for multi-class models by [46]. [30] proposes a posterior probability estimation for the best outcome of a ranking system, with industrial applications in Natural Language Processing. These approaches all rely on a cross-validation performed during the training, in order to calibrate the scores conversion on unseen samples. Despite their efficacy, these posterior probability estimations remain conversion of scores in a discrete output space - classes, recommendation objects - and do not apply to regression tasks.

Uncertainty in Deep Learning. More recently, [28] described how Deep Learning introduced models with higher representation capabilities, which can be leveraged to estimate an uncertainty of the prediction. As [14] explains, a first approach is to consider the model as Bayesian, whose weights follow a random distribution obtained after the training. This uncertainty is denoted epistemic, as by [8,22]. In practice, [15] proposed to approximate it with one model by performing multiple predictions with stochastic drop-out, without changing the training procedure. Given an input, it simulates thus a Monte Carlo sampling among the possible models induced by dropout. This iteration leads to much higher inference time, that [35] proposes to reduce by approximating the sampling with analytical formulas, starting from dropout underlying Bernoulli distribution. They get thus reduced run times, while results tend to be similar to slightly worse than with sampling. [8] describes another approach denoted as aleatoric uncertainty, where the model learns to predict the uncertainty from the input. To do so, the model is trained to predict a probability distribution instead of the ground truth only, as formalized by [22]. In practice, this distribution is generally Gaussian, which captures the most likely output value and the covariance around it. According to [23], aleatoric is more relevant than epistemic in the case of large datasets as well as real-time applications.

Oriented Aleatoric Uncertainty Methods. While previous cited approaches focused on a single-valued aleatoric uncertainty, [10,33] proposed to predict the full covariance matrix using Cholesky decomposition, proposed by [5]. The matrix is built as $\hat{L}^T\hat{L}$ where $\hat{L}$ is predicted as a lower triangular matrix with positive elements in its diagonal. However, they do not extract nor leverage any *orientation* information induced by the covariance, that we focus on in this paper. One reason is that their output spaces are high dimensional, where the orientation is hard to interpret and exploit. In 3D space, [37] obtained promising results by combining aleatoric uncertainty and Kalman filter for tracking object location in a video. However, the covariance decomposition they propose for pure aleatoric uncertainty model is only valid for 2 dimensional output space. They only rely on their final activation functions to reduce risk of non-positive definite matrix, which would not work on any data-set. More recently, [29] proposes a full rank aleatoric uncertainty to visualize detected keypoints areas in the 2D image. This interesting usage of uncertainty is limited to 2D in practice since each keypoint has its own uncertainty area.

3 Problem

3.1 Color Estimation: A Continuous 3D Output

We consider as our real world use-case the problem of skin color estimation from a selfie picture. In this scenario, the user takes a natural picture, from any smartphone and under unknown lighting condition. This picture's pixels are represented in standard RGB, the output color space for most smartphones. We want to estimate the user's skin color as a real color measured by a device, and not a self-declared skin type nor a-posteriori manual annotation. To do so, we consider the skin color measured by a spectrocolorimeter, whose spectrum is converted into the L*a*b* color space as defined by the CIE (Commission Internationale de l'Eclairage), as done by [44]. Compared to the hardware-oriented standard RGB space, this 3 dimensional space is built so that the Euclidean distance between two colors approaches the human perceived difference. The ground truth y is thus represented as 3 continuous values, and the mean squared error $\mathcal{L}_{\text{MSE}}$ approximates the perceived difference between colors y and $\hat{y}$:

$$y = (L^*, a^*, b^*)^T \in \mathbb{R}^3, \quad \mathcal{L}_{\text{MSE}}(y, \hat{y}) = \|y - \hat{y}\|^2 \tag{1}$$

where $\hat{y}$ is the model prediction for input picture x and $\|.\|$ denotes the $L2$ norm.

Color and skin tone can be efficiently estimated by regression Convolutional Neural Network, as done by [3,7,26,27,31]. Beyond predicting y, we focus in this paper in estimating the oriented uncertainty, as described in the next part.

3.2 Oriented Uncertainty

Following the prediction space described above, we now discuss what form of uncertainty could be predicted. The simplest form would be a simple real value estimating the *magnitude* of the prediction error, measuring thus our level of uncertainty. This enables to apply a threshold on this estimated value for filtering unsure cases and has been widely studied in literature (see Sect. 2). However, for the case of multi-dimensional predictions, a single-value uncertainty treats each output dimension equally, in an *isotropic* manner. We focus instead on a full rank uncertainty, which is *oriented* since it expresses the most likely orientation of prediction errors.

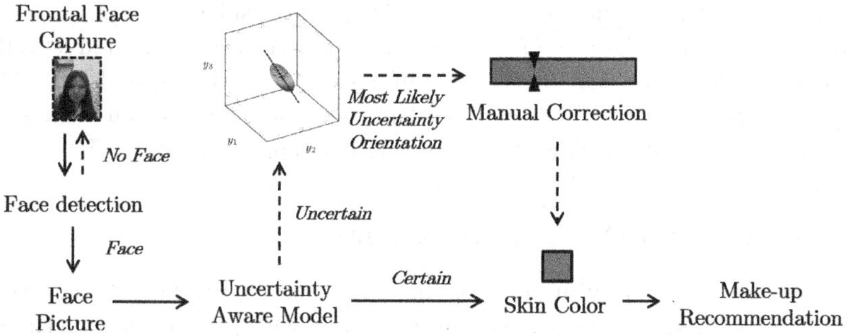

Fig. 1. Pipeline for our color estimation use-case, where the uncertainty is used for filtering uncertain predictions and defining 1-D color bar correction.

The epistemic uncertainty can also provide orientation information, by providing points cloud in the prediction space (see Sect. 2). However, we focus in this paper on aleatoric uncertainty, for the following reasons. First, [23] advised it for large datasets, and when there is a need for *real-time* application. Second, epistemic uncertainty is mostly advised for detecting inputs out of training data distribution, because model presents higher variability for such data. This appears useless in the case of selfie pictures as input, since face detectors (as the one of [24]) easily ensure to detect outliers before feeding the neural network. In the following, we detail how our model estimates oriented aleatoric uncertainty.

4 Model

4.1 Aleatoric Uncertainty Principle

The aleatoric uncertainty approach is to consider that the model no longer predicts a single value $\hat{y}$ but instead a distribution of random value. In practice, we assume this distribution to be a multivariate normal law, meaning $y \sim \mathcal{N}(\hat{\mu}, \hat{\Sigma})$ where $\hat{\mu}$ and $\hat{\Sigma}$ are typically estimated by a neural network from the input. The loss to be optimized by our model then relates to the likelihood of the ground truth y with respect to this distribution, written $p_{\hat{\mu},\hat{\Sigma}}(y)$. For a multi-dimensional output space, $y \in \mathbb{R}^d$, $\hat{\mu} \in \mathbb{R}^d$, $\hat{\Sigma} \in \mathbb{R}^{d \times d}$, and the likelihood is written:

$$p_{\hat{\mu},\hat{\Sigma}}(y) = \frac{1}{(2\pi)^{d/2}|\hat{\Sigma}|^{1/2}} e^{-\frac{1}{2}(y-\hat{\mu})^T \hat{\Sigma}^{-1}(y-\hat{\mu})} \tag{2}$$

where $|\hat{\Sigma}|$ is the determinant of matrix $\hat{\Sigma}$. Practically we minimize an affine transformation of the log-likelihood $log(p_{\hat{\mu},\hat{\Sigma}}(y))$, discarding the constant terms:

$$\mathcal{L}_p(y, \hat{\mu}, \hat{\Sigma}) = (y - \hat{\mu})^T \hat{\Sigma}^{-1} (y - \hat{\mu}) + \log\left(|\hat{\Sigma}|\right) \tag{3}$$

This loss is similar to the MSE (Eq. 1) where the model still learns to predict the most likely value $\hat{\mu}$, which is equivalent to $\hat{y}$. Besides, it learns to predict a matrix $\hat{\Sigma}$ that represents a rich form of uncertainty.

A straightforward choice for covariance $\hat{\Sigma}$ is to consider isotropic noise, like in [23]. In this case, $\hat{\Sigma} = \hat{\sigma}^2 I_d$, where $\hat{\sigma} \in \mathbb{R}+$ is the estimated standard deviation in every direction of the output space. This corresponds to assuming $y_j \sim \mathcal{N}(\hat{\mu}_j, \hat{\sigma}^2)$ $\forall j = 1..d$ and only provides information on uncertainty magnitude. In the following, we focus on a richer representation of $\hat{\Sigma}$.

4.2 Covariance for Oriented Uncertainty

To capture the uncertainty in any direction, we need to estimate the covariance matrix $\hat{\Sigma}$ as a symmetric positive definite matrix (SPD), meaning it verifies:

$$\hat{\Sigma}^T = \hat{\Sigma} \text{ and } v^T \hat{\Sigma} v > 0 \;\; \forall v \neq \vec{0} \in \mathbb{R}^d$$

While the symmetry of $\hat{\Sigma}^T$ can be easily ensured by construction, the positive semi-definiteness property is not straightforward to satisfy when $\hat{\Sigma}$ is the output of an uncontrolled neural network. To build $\hat{\Sigma}$, [33] and [10] propose to use Cholesky decomposition or equivalently LDL decomposition $\hat{\Sigma} = \hat{L}\hat{D}\hat{L}^T$, where $\hat{L}$ is a lower unit triangular matrix and $\hat{D}$ a diagonal matrix (as formalized by [19]). The independent components $\hat{D}_{i,i} > 0$ and $\hat{L}_{i,j}$ $(i < j)$ are produced by a regression layer. However, we propose not to use such decomposition. Indeed, we observed an difficult optimization of such model, and poor results for the main prediction task. We explain this phenomenon by the intuition that $\hat{L}_{i,j}$, are *low lever features* in the sense that they are hard to interpret, contrary to standard deviations $\hat{\sigma}_j$ for instance. Our main proof remains experimental and further theoretical explanation goes beyond the scope of this paper.

4.3 Euler Angles Decomposition

To build $\hat{\Sigma}$, we instead rely on the following decomposition of SPD:

$$\hat{\Sigma} = \hat{U}\hat{D}\hat{U}^T \text{ with } \hat{D} = Diag(\hat{\sigma}_1^2, \ldots, \hat{\sigma}_d^2)$$

where $\hat{U} \in \mathbb{R}^{d \times d}$ is a unitary matrix with columns being eigenvectors of $\hat{\Sigma}$ and $\hat{\sigma}_1^2, \ldots, \hat{\sigma}_d^2 \in \mathbb{R}^+$ are the corresponding eigenvalues. Each eigenvector $\hat{U}_j \in \mathbb{R}^d$ is an uncertainty orientation associated with a standard deviation σ_j, so that $\hat{\Sigma}$ geometrically corresponds ellipsoidal level sets of the distribution.

To simplify the notations and place ourselves in the color output space, we now consider the 3D space for y ($d = 3$). We propose to express $\hat{U}$ as the multiplication of the rotation matrices around each canonical axis [42]:

$$\hat{U} = R_{y_1}(\hat{\theta}_1)R_{y_2}(\hat{\theta}_2)R_{y_3}(\hat{\theta}_3) \in \mathbb{R}^{3\times 3}$$

where $R_{y_j}(\hat{\theta}_j)$ are the rotation matrices around axis y_j and $\hat{\theta}_1$, $\hat{\theta}_2$, $\hat{\theta}_3 \in [-\pi/4, \pi/4]$ are the rotation angles respectively around y_1, y_2 and y_3 axes, also named Euler angles (see appendix for explicit matrices expressions). This choice of representation enables the model to predict high-level interpretable features, meaning the Euler angles $\hat{\theta}_i$. Euler angles representation usually suffer from periodicity and discontinuity problems, as [9] points out. However, in our problem of covariance matrix $\hat{\Sigma}$ construction, we can enforce a narrow range for the $\hat{\theta}_i$. Indeed, having $\hat{\theta}_i \in [-\pi/4, \pi/4]$ ensures that $\hat{\Sigma}$ covers the whole SPD space, while keeping each $\hat{\sigma}_i$ closely associated to the canonical axe y_i, thus easing the optimization. Without such boundaries, one notes that $\hat{\theta} = (\pi/2, 0, 0)$, $\hat{\sigma} = (1, 1, 2)$ and $\hat{\theta} = (0, 0, 0)$, $\hat{\sigma} = (1, 2, 1)$ would give the same $\hat{\Sigma}$, which makes optimization difficult since $\hat{\sigma}_i$ values are shifted. While we focus on the 3 dimensional case, this parameterization could be extended to higher dimensions [40].

In practice, to compute the loss of Eq. 3 we build $|\hat{\Sigma}|$ and $\hat{\Sigma}^{-1}$ as:

$$|\hat{\Sigma}| = \prod_{j=1}^{d} \hat{\sigma}_j^2, \quad \hat{\Sigma}^{-1} = \hat{U} Diag(\hat{\sigma}_1^{-2}, \ldots, \hat{\sigma}_d^{-2})\hat{U}^T$$

which are differentiable, so that the loss is minimizable by gradient descent.

To enforce the boundaries on each $\hat{\theta}_i$, we preferred not to use strict clipping of neurons values. Indeed, doing so was leading to numerical issues with vanishing gradient for the $\hat{\theta}_i$ regressions. Instead, we let $\hat{\theta}_i$ as raw outputs of linear regressions, and included a penalization term in the loss:

$$\mathcal{L}_\theta(\hat{\theta}) = \lambda_\theta \sum_{j=1}^{d} \max(0, \hat{\theta}_i - \pi/4) + \max(0, -\hat{\theta}_i - \pi/4) \quad (4)$$

where $\lambda_\theta > 0$ is a hyperparameter. This penalization corresponds to soft boundary constraints, meaning $\hat{\theta}_i$ can take values beyond $\pi/4$ or below $-\pi/4$, but then the loss is increased. The total loss function takes the form:

$$\mathcal{L}(y, \hat{\mu}, \hat{\Sigma}, \hat{\theta}) = \mathcal{L}_p(y, \hat{\mu}, \hat{\Sigma}) + \mathcal{L}_\theta(\hat{\theta})$$

4.4 Oriented Uncertainty Benefits

We now propose to extract information from the matrix $\hat{\Sigma}$ predicted on a new picture. First, we can interpret an uncertainty magnitude. In case of isotropic uncertainty $\hat{\Sigma} = \hat{\sigma}^2 I_d$, the magnitude is directly given by $\hat{\sigma}$. [14] showed that by applying a threshold on $\hat{\sigma}$, we filter out the most unsure cases. We extend this to oriented uncertainty by considering the determinant of the covariance matrix:

$$|\hat{\Sigma}| = \prod_j \hat{\sigma}_j^2 \in \mathbb{R} \tag{5}$$

We can also extract the most likely orientation of y with respect to predicted $\hat{y}$:

$$\hat{v} = \hat{U}_{j*} \in \mathbb{R}^d \text{ where } j^* = \underset{j}{\text{argmax}}\ \hat{\sigma}_j \tag{6}$$

which is of norm 1. One notes that error is equally likely to stand in the orientation $\hat{v}$ and $-\hat{v}$ due to the symmetry of the normal distribution.

4.5 Probabilities Re-scaling for Models Comparison

We now describe an optional step in the training, that do not serve the general purpose of the model and has no impact on $|\hat{\Sigma}|$ and $\hat{v}$ computation. We use it to get unbiased likelihood in order to compare models in our experiments.

When looking at $p_{\hat{\mu},\hat{\Sigma}}(y)$ values (Eq. 2) on the test samples during the optimization process, we observed that uncertainty models tend to show *overconfidence*, meaning they predict $\hat{\Sigma}$ with lower volume through the epochs. This phenomenon of overconfidence of neural network has regularly be observed [4], for instance with probabilities inferred after softmax always close to 0% or 100% (see Fig. 8 in Appendix for an illustration). To avoid this, we propose to multiply every predicted covariance $\hat{\Sigma}$ by a unique factor $\beta \in \mathbb{R}^+$, in order to adapt the magnitudes of the distributions $\mathcal{N}(\hat{\mu}, \hat{\Sigma})$ while keeping their shape and orientation. Such factor does not change the relative order between the magnitudes $|\hat{\Sigma}|$ among samples, typically when filtering most uncertain cases. We propose to consider these optimal β^* value as:

$$\beta^* = \text{argmax}_\beta \frac{1}{N} \sum_i^N \log\left(p_{\hat{\mu}_i,\beta\hat{\Sigma}_i}(y_i)\right) \tag{7}$$

where the sum covers a *subpart* of the train data, while we estimate all $\hat{\mu}_i$ and $\hat{\Sigma}_i$ by a model trained on the remaining of the train data. This sub-training is necessary for computing $p_{\hat{\mu}_i,\beta\hat{\Sigma}_i}(y_i)$ from unbiased inputs. Such re-scaling is similar to scores conversion for probabilistic SVM [34], where a cross-validation is performed on the training data to learn the conversion. Due to the computation costs, we preferred to split the training data into 2 equal parts reflecting the validation strategy such as group-out or stratification. We can then re-scale the estimated covariances as $\beta^*\hat{\Sigma}$ when computing metrics and comparing models.

5 Experiments

5.1 Dataset

Our dataset is composed of various selfie pictures taken by different people with their own smartphone in indoor and outdoor environments. Besides, they had their skin color measured in a controlled environment using a spectrophotometer under a specific protocol to reduce measurement noise. In order to face real-world skins diversity and include various race groups, we performed several acquisitions in different countries (see Table 1). For standardization purpose, we pre-processed all pictures by detecting facial landmarks using the face detector of [24] and then placing eyes in a standardized location. Resulting images are thus centered on the face and resized to 128×128. This pre-processing step is identically done at inference time in our real-world application (see Fig. 1).

5.2 Evaluation and Implementation Details

To compute all results of this sections, we performed a 5-fold cross validation on our dataset, and evaluated the predictions on the successive test folds. To avoid biased predictions, each volunteer pictures are grouped in the same fold. Furthermore, we stratified the folds with respect to y_1 value, that corresponds to L^* in the color space, that can be interpreted as skin intensity.

Using this process, we compared the following models:

- Regression w/o Uncertainty: pure regression CNN (Sect. 3.1, [27])
- Aleatoric Isotropic: aleatoric single-valued uncertainty, as proposed by [23]
- Aleatoric (Cholesky): the aleatoric uncertainty model using Cholesky decomposition, as proposed by [10]
- Epistemic via Dropout: epistemic uncertainty model proposed by [15] for 100 sampled predictions
- Epistemic Sampling-free: epistemic uncertainty model with direct covariance estimation proposed by [35]
- Aleatoric (Ours): the uncertainty model described in Sect. 4.3

For each model, we used the same architecture for the convolutional network, 4 convolutions blocks with skip connections followed by a dropout layer (similar to ResNet [18]). Only the loss and final dense layers have different architecture between models. We used ReLu activations except for oriented uncertainty in which $\hat{\theta}_j$ are linear output of regression. The corresponding soft constraint term (Eq. 4) is scaled by $\lambda_\theta = 10$. For aleatoric uncertainty models, the lower bound for $\hat{\sigma}_j$ is set as 10^{-2} to avoid numerical overflow. Contrary to [23], we prefer our model to directly estimate $\hat{\sigma}_j$ instead of $\log(\hat{\sigma}_j)$, for avoiding weights updates to produce too high variation on $\hat{\sigma}_j$. For all models, we used the optimization procedure of [25] on batches of size 16 and a learning rate 10^{-4} during 400 epochs. We implemented our networks and layers using TensorFlow [2].

Table 1. Description of the data used for our experiments. The full data-set is still under collection and will cover more countries.

Country	China	France	India	Japan	USA	Total
# People	53	249	22	32	832	1188
# Pictures	936	3967	368	144	10713	16128

5.3 Metrics for Regression and Uncertainty

In order to evaluate the core regression task, for each sample of the test folds we computed the ΔE^* [44] between the prediction and the ground-truth y:

$$\Delta E^* = \|\hat{\mu} - y\| \tag{8}$$

where $\hat{\mu}$ is replaced by $\hat{y}$ for the regression model.

For uncertainty models, we evaluated the quality of predicted distributions $\mathcal{N}(\hat{\mu}, \hat{\Sigma})$ when compared to the ground truth y. For epistemic uncertainty via dropout, we considered a normal distribution of covariance $\hat{\Sigma}$ computed from sampled predictions. For each model, we computed the scaling factor β^* (Eq. 7, Sect. 4.5) and considered the average likelihood after scaling:

$$\langle p^*_\beta \rangle = \frac{1}{N}\sum_i^N p_{\hat{\mu}_i, \beta^* \hat{\Sigma}_i}(y_i), \quad \langle \log(p^*_\beta) \rangle = \frac{1}{N}\sum_i^N \log\left(p_{\hat{\mu}_i, \beta^* \hat{\Sigma}_i}(y_i)\right) \tag{9}$$

For orientation-aware models, we computed the angle error α between the orientation $\hat{v}$ (Eq. 6) and the actual ground truth y orientation from $\hat{y}$:

$$\alpha = cos^{-1}\left(\hat{v}.\frac{(y - \hat{\mu})}{\|y - \hat{\mu}\|}\right)$$

To assess the quality of estimated errors distributions, we also considered the probability of having an error ΔE below a certain threshold E:

$$\mathbb{P}_{\hat{\mu},\hat{\Sigma}}(\|\hat{\mu} - y\| < E) = \iiint_{\|e\|^2 < E^2} p_{\hat{\mu},\hat{\Sigma}}(\hat{\mu} + e)\, d^3e \tag{10}$$

where the integral does not depend on $\hat{\mu}$ (see Eq. 2). This quantity differs from $|\hat{\Sigma}|$ by having a clear probabilistic interpretation, but is slower to compute. We compared these numerically computed values on test samples to the actual

Table 2. Comparisons of the models. For each metrics, ↑ indicates when higher is better, and ↓ indicates when lower is better. For accuracy metrics, best result is highlighted in italics bold while second best is in bold.

Model	Performance	Regression task		Uncertainty metrics		
	Inference time (ms) ↓	$\hat{\mu}$ accuracy $\langle \Delta E^* \rangle$ ↓	vs baseline ↓	Likelihood $\langle log\,(p_{\beta^*}) \rangle$ ↑	ROC-AUC $\Delta E < 1$ ↑	Angle $\langle \alpha \rangle$ ↓
No uncertainty	68.1 ± 6.1	***3.54***	*baseline*	–	–	–
Aleat. isotropic	70.2 ± 7.3	3.59	+1.5%	−7.86	58.47	–
Aleat. Cholesky	73.2 ± 4.2	4.49	+26.8%	***−7.02***	***62.74***	***36.85°***
Epist. w dropout	7820 ± 19	***3.54***	+0%	−9.07	58.15	40.88°
Epist sample-free	81.2 ± 5.3	***3.54***	+0%	−9.08	57.70	41.55°
Aleat. (ours)	71.4 ± 4.7	3.65	+3.1%	**−7.33**	**59.09**	**39.43°**

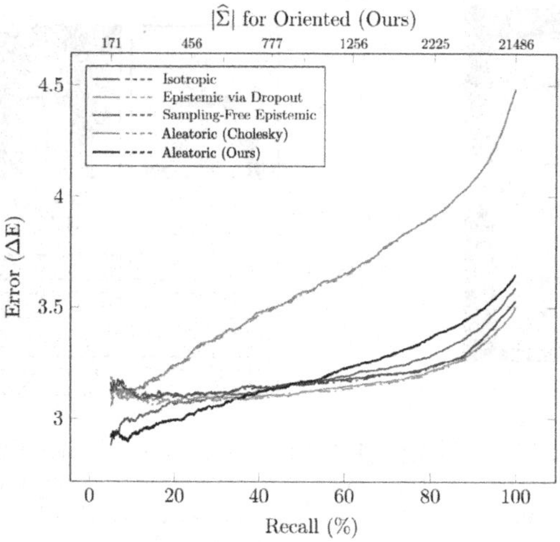

Fig. 2. ΔE with respect to recall for samples with $|\hat{\Sigma}|$ below a threshold (Eq. 5). For plain curves, successive thresholds are applied on samples $|\hat{\Sigma}|$ values, for dashed curves on $\mathbb{P}_{\hat{\mu},\hat{\Sigma}}(||\hat{\mu} - y|| < E)$ values. On top, indicative $|\hat{\Sigma}|$ values for our model are shown. Points with recall below 5%, noiser, are omitted.

realization of $||\hat{\mu} - y|| < E$ to get the ROC-AUC [11]. We choose as threshold $E = 1$ which corresponds to human perception of color dissimilarities [44].

Table 2 shows the computed metrics. As expected, the color estimation accuracy is unchanged for epistemic uncertainty models. The difference is very little for our model, while the Cholesky uncertainty shows 26% *higher average error.* According to our intuition described in Sect. 4.1, this degradation seems to be due to the complex optimization process, since $\hat{\Sigma}$ representation relies on low level features (see Fig. 7 in Appendix for an illustration). Besides, estimated distributions appears better for the Cholesky, while second best metrics are for our oriented uncertainty model. Using our model, the angle error is reduced by 2° compared to the sampling-free epistemic, and the log-likelihood is significantly higher. Compared to other models, our method gives thus the best trade-off with maintained accuracy, fast computation and accurate uncertainty distribution.

5.4 Uncertainty for Samples Selection

A first benefit of uncertainty-aware models is to detect samples whose error ΔE is likely to be high by applying a threshold on predicted $|\hat{\Sigma}|$ (Eq. 5). In our pipeline shown in Fig. 1, this conditions the manual color correction step for the uncertain cases. In order to evaluate how efficient this condition is, we applied for each model successive thresholds on test samples $|\hat{\Sigma}|$. In practice, for every threshold T we compute the recall for selected samples as $\frac{1}{N}\sum_i^N \mathbb{1}(|\hat{\Sigma}_i| < T)$, where $\mathbb{1}$ denotes the indicator function and $\hat{\Sigma}_i$ is the i-th test sample estimated covariance. The average ΔE error is similarly computed for samples verifying

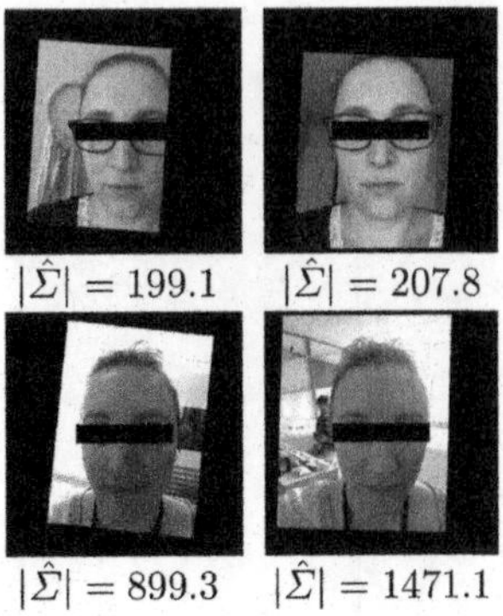

Fig. 3. Pictures with lowest $|\hat{\Sigma}|$ value (top) and highest $|\hat{\Sigma}|$ value (bottom) from same person (/16 pictures).

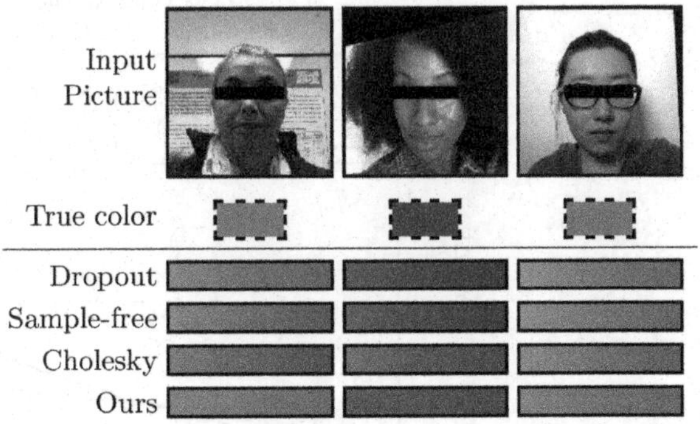

Fig. 4. Examples of input pictures with the 1D correction color bars estimated by each uncertainty model. The bar from Cholesky model suffers from the higher error of the estimated $\hat{\mu}$.

$|\hat{\Sigma}_i| < T$. Plain curves in Fig. 2 show both quantities when computed for every threshold T values. This is similar to a precision-recall curves for binary classification, as described by [12], that we extend to regression tasks using estimated $|\hat{\Sigma}|$ (as done by [16,30]). Equivalently, we computed the dashed line curves by selecting samples verifying $\mathbb{P}_{\hat{\mu},\hat{\Sigma}}(\|\hat{\mu} - y\| < E) > T$. Selection based on this second quantity requires heavier computation (Eq. 10 versus Eq. 5) and is thus not convenient for real-time application.

We see in Fig. 2 that most curves are very close for all models, except for the Cholesky model. The accuracy error for Cholesky uncertainty is indeed much higher under all thresholds, which was expected from overall accuracy (Table 2). Besides, we see that for the most certain test samples (for recall $\leq 20\%$), our model indeed selects the samples with *actual* lowest errors. Last, dashed and plain curves are hard to distinguish, which means that thresholding on $|\hat{\Sigma}|$ is a good approximation for thresholding on $\mathbb{P}_{\hat{\mu},\hat{\Sigma}}(\|\hat{\mu} - y\| < E)$.

For a given person, we also looked at rejected and valid pictures based on the $|\hat{\Sigma}|$ criteria (see Fig. 3). In general, uncertainty is higher when lighting conditions are bad, such as yellow light, strong back light and shadowed face. These patterns seems to be leveraged by the model for estimating the uncertainty, and are indeed strongly impacting the skin color estimation task.

5.5 Color Control Using Uncertainty Orientation

For our use case, another benefit of oriented uncertainty is to enable a better user experience for manual color correction. This correction is requested for users with $|\hat{\Sigma}|$ falling above the operational threshold. Requesting a non-expert user to re-define a color in 3D is practically impossible. Based on the low angle error α obtained, we propose to use the most likely orientation $\hat{v}$ (Eq. 6) as the only

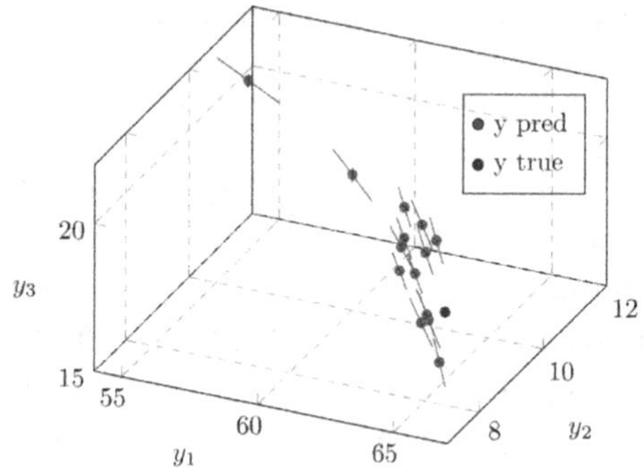

Fig. 5. Predictions and uncertainties for all pictures of a given panelist. We see that uncertainties are generally oriented towards the ground truth (in blue), and that further predictions have higher uncertainty in the main axis (in red). (Color figure online)

degree of freedom for the color control. To confirm this intuition, Fig. 5 shows an example of $\hat{v}$ orientations estimated for all pictures of a single volunteer: $\hat{v}$ is generally oriented towards the real color. Figure 4 shows practical examples of color bars. The bars are centered on estimated color $\hat{\mu}$, directed towards $\hat{v}$. The bar length was fixed at $\Delta E^* = 5$ in the color space, after discussion with user experience experts in order to have a standard control bar expressiveness.

5.6 Uncertainty Orientation for Foundation Recommendation

We present here preliminary results that illustrate additional benefits of our oriented uncertainty for online make-up assistant service. For the sub-set of our data-set from acquisition in Japan (see Table 1), a make-up artist have assessed the most suitable foundation shade chosen in a range of 25 shades. We also measured the color f in L*a*b* space for each shade of the range, using same spectrophotometer and comparable protocol than for the skin color. For each participant, we note $f^* \in \mathbb{R}^3$ the color of the best foundation shade chosen by make-up artist. Based on our make-up knowledge, this best foundation should be the closest from the skin tone, meaning the predicted $\hat{\mu}$ color should be close from this best shade color f^*. Based on this, we computed for each picture the difference between the estimated skin color and this foundation color and as a $\Delta E^*_{f^*} = \|\hat{\mu} - f^*\|^2$, as well as the probability distribution value $p_{\hat{\mu},\hat{\Sigma}}(f)$ (Eq. 2) for every shade color f of the range. Using those 25 values, we could thus rank the shades in a scenario of product recommendation. For the regression model, we similarly ranked the shades using ΔE^* which can be considered the best products recommendation when not using uncertainty. We display in Table 3 the average rank of the best foundation shade according to make-up artist, as well as the average ΔE^* and $log(p_{\hat{\mu},\beta^*\hat{\Sigma}}(f^*))$ values for this shade color:

$$\langle \log(p^*_\beta(f^*)) \rangle = \frac{1}{N} \sum_i^N \log\left(p_{\hat{\mu}_i,\beta^*\hat{\Sigma}_i}(f_i{}^*)\right)$$

Table 3. Metrics on Japan data for foundation recommendation using uncertainty orientation. Reference is ideal foundation color (instead of panelist skintone for Table 2). Best result is highlighted in italics bold, second best in bold.

Model	$\langle \Delta E^*_{f^*} \rangle$	$\langle \log(p^*_\beta(f^*)) \rangle$	Shade rank
Regression	**3.20**	–	$2.43^{th}/25$
Aleat. Isotropic	3.22	−7.63	$2.54^{th}/25$
Aleat. Cholesky	4.34	**−7.43**	$3.34^{th}/25$
Epist. w Dropout	**3.20**	−12.25	$\mathbf{2.30^{th}}/25$
Epist. Sample-Free	**3.20**	−11.27	$2.82^{th}/25$
Aleat. Ours	3.37	***−7.26***	$\boldsymbol{2.20^{th}}/25$

where scaling factor β^* do not impact the products ranking but helps for models comparison. We get the best rank for our model, even if the raw skin to foundation distance $\Delta E^*_{f^*}$ is higher than with some other models. We emphasize that these results do not include any manual color correction as described in Sect. 5.5. Details for the products recommendation go beyond the scope of this paper.

6 Conclusion

In this paper, we proposed to estimate multivariate aleatoric uncertainty by using a different parameterization of the covariance matrix based on Euler angles. We experimented our model on a real-world data-set, which addresses skin color estimation from selfie pictures. This use-case is at the core of a make-up online assistant but is very sensitive in terms of ethics and AI fairness. The uncertainty estimation is an answer to reduce ethic risks, among other benefits it brings. Our oriented uncertainty model showed a similar accuracy to pure regression model, contrary to the model using Cholesky decomposition which got 26.8% higher errors on the core diagnosis task. Furthermore, when comparing to other approaches, our model obtained the best metrics about the estimated distributions, such as the angle error for the most likely error orientation. This shows its ability to infer the scale and orientation of the actual prediction error.

The proposed model can be used for real-time color adjustment. Users with uncertain predictions are requested to make manual correction via a simplified UX with 1D color bar whose orientation is given by the uncertainty. Besides, we experimented another benefit of our model in the case of foundation recommendation, where the orientation helps to recommend the best product. In our future work, we will evaluate the benefits of the uncertainty bar for the end user. To do so, we are currently conducting a study were panelist are asked to correct their diagnosed skin tone using the uncertainty-aware color bar. Beyond, we plan to use our uncertainty model as domain discriminator of a conditional generative adversarial network [32], in order to less penalize generated pictures whose predicted label lies in the most likely orientation of uncertainty.

A Rotation Matrices

$$R_{y1}(\hat{\theta}_1) = \begin{bmatrix} 1 & 0 & 0 \\ 0 & \cos\hat{\theta}_1 & -\sin\hat{\theta}_1 \\ 0 & \sin\hat{\theta}_1 & \cos\hat{\theta}_1 \end{bmatrix}$$

$$R_{y2}(\hat{\theta}_2) = \begin{bmatrix} \cos\hat{\theta}_2 & 0 & \sin\hat{\theta}_2 \\ 0 & 1 & 0 \\ -\sin\hat{\theta}_2 & 0 & \cos\hat{\theta}_1 \end{bmatrix}$$

$$R_{y3}(\hat{\theta}_3) = \begin{bmatrix} \cos\hat{\theta}_3 & -\sin\hat{\theta}_3 & 0 \\ \sin\hat{\theta}_3 & \cos\hat{\theta}_1 & 0 \\ 0 & 0 & 1 \end{bmatrix}$$

B Neural Network Architecture

See Fig. 6.

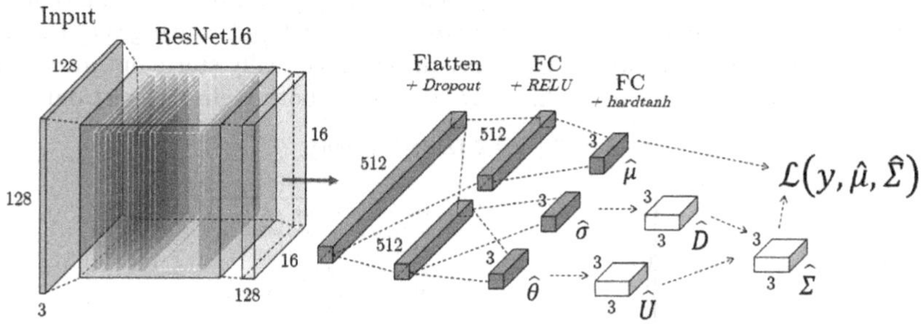

Fig. 6. Architecture for our neural network. The ResNet-16 is a simplified version of convolution blocks of ResNet [18] and the same convolution architecture was used for all compared models.

C Behavior During Training

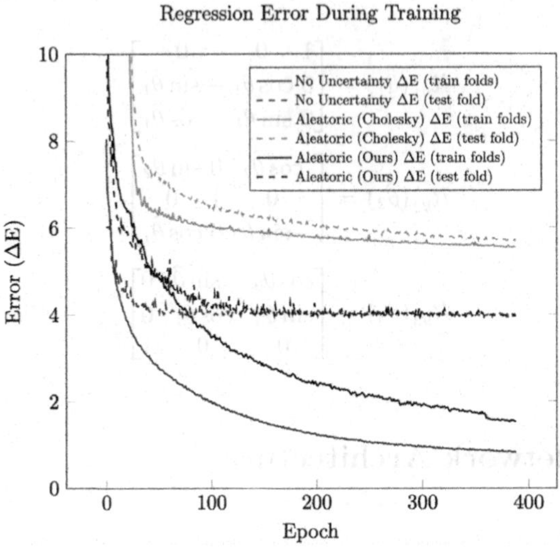

Fig. 7. Evolution of ΔE (Eq. 8) through epochs of the 1st fold of our evaluation. Our model learns slower compared to the pure regression model, but converges to the same value on test fold. On the contrary, Cholesky uncertainty really starts optimization at around 30 epochs, and saturates to high ΔE values for both train and test data.

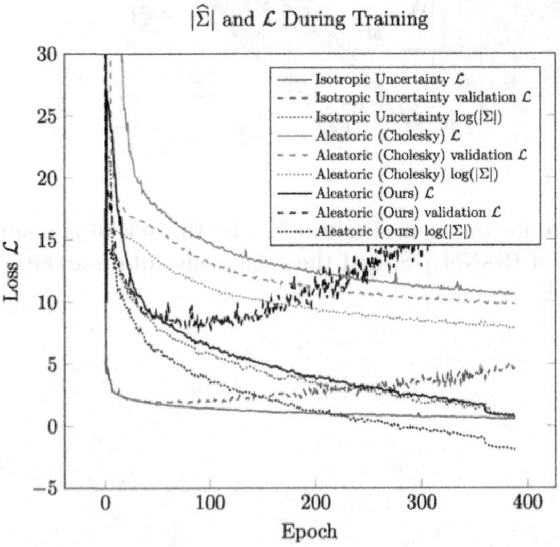

Fig. 8. Evolution of log-likelihood loss $\mathcal{L}$ (Eq. 3) through epochs of the 1st fold of our evaluation. Similar overfitting behavior for $\mathcal{L}$ occurs for isotropic and our uncertainty model, that is explained by the decreasing volume of covariance matrix $|\Sigma|$ to very low values during epochs. Besides, Cholesky model keeps a larger volume for matrix Σ with no overfiting, but we interpret the higher values for $\mathcal{L}$ as a more difficult gradient descent, which seems also visible from Table 1 and Fig. 5. Those behaviors motivates our re-scaling of Sect. 4.5 for comparing $\mathcal{L}$ between models.

References

1. Aarabi, P.: Method, system and computer program product for generating recommendations for products and treatments, 12 September 2017, uS Patent 9,760,935
2. Abadi, M., et al.: TensorFlow: a system for large-scale machine learning. In: 12th USENIX Symposium on Operating Systems Design and Implementation (OSDI 2016), pp. 265–283 (2016)
3. Bokaris, P.A., Malherbe, E., Wasserman, T., Haddad, M., Perrot, M.: Hair tone estimation at roots via imaging device with embedded deep learning. Electron. Imaging **2019**(6), 483-1–483-11 (2019)
4. Bulatov, K.B., Polevoy, D.V., Mladenov, V., Spasov, G., Georgieva, P., Petrova, G.: Reducing overconfidence in neural networks by dynamic variation of recognizer relevance. In: ECMS, pp. 488–491 (2015)
5. Cholesky, A.L.: Sur la résolution numérique des systèmes d'équations linéaires. Bulletin de la Sabix. Société des amis de la Bibliothèque et de l'Histoire de l'École polytechnique (39), 81–95 (2005)
6. Coudray, N., et al.: Classification and mutation prediction from non-small cell lung cancer histopathology images using deep learning. Nat. Med. **24**(10), 1559 (2018)
7. Das, A., Dantcheva, A., Bremond, F.: Mitigating bias in gender, age and ethnicity classification: a multi-task convolution neural network approach. In: Leal-Taixé, L., Roth, S. (eds.) ECCV 2018. LNCS, vol. 11129, pp. 573–585. Springer, Cham (2019). https://doi.org/10.1007/978-3-030-11009-3_35
8. Der Kiureghian, A., Ditlevsen, O.: Aleatory or epistemic? Does it matter? Struct. Saf. **31**(2), 105–112 (2009)
9. Diebel, J.: Representing attitude: Euler angles, unit quaternions, and rotation vectors. Matrix **58**(15–16), 1–35 (2006)
10. Dorta, G., Vicente, S., Agapito, L., Campbell, N.D., Simpson, I.: Structured uncertainty prediction networks. In: Proceedings of the IEEE Conference on Computer Vision and Pattern Recognition, pp. 5477–5485 (2018)
11. Fawcett, T.: An introduction to roc analysis. Pattern Recogn. Lett. **27**(8), 861–874 (2006)
12. Flach, P., Kull, M.: Precision-recall-gain curves: PR analysis done right. In: Advances in Neural Information Processing Systems, pp. 838–846 (2015)
13. Hastie, T., Tibshirani, R., Friedman, J.: The Elements of Statistical Learning. SSS, Springer, New York (2009). https://doi.org/10.1007/978-0-387-84858-7
14. Gal, Y.: Uncertainty in deep learning. Ph.D. thesis, University of Cambridge (2016)
15. Gal, Y., Ghahramani, Z.: Dropout as a Bayesian approximation: representing model uncertainty in deep learning. In: International Conference on Machine Learning, pp. 1050–1059 (2016)
16. Gurevich, P., Stuke, H.: Learning uncertainty in regression tasks by artificial neural networks. arXiv preprint arXiv:1707.07287 (2017)
17. Hazirbas, C., Bitton, J., Dolhansky, B., Pan, J., Gordo, A., Canton Ferrer, C.: Towards measuring fairness in AI: the casual conversations dataset. arXiv e-prints arXiv:2104.02821, April 2021
18. He, K., Zhang, X., Ren, S., Sun, J.: Deep residual learning for image recognition. In: 2016 IEEE Conference on Computer Vision and Pattern Recognition, CVPR 2016, Las Vegas, NV, USA, 27–30 June 2016, pp. 770–778. IEEE Computer Society (2016). https://doi.org/10.1109/CVPR.2016.90
19. Higham, N.J.: Analysis of the Cholesky Decomposition of a Semi-Definite Matrix. Oxford University Press, Oxford (1990)

20. Holder, C.J., Obara, B., Ricketts, S.: Visual Siamese clustering for cosmetic product recommendation. In: Carneiro, G., You, S. (eds.) ACCV 2018. LNCS, vol. 11367, pp. 510–522. Springer, Cham (2019). https://doi.org/10.1007/978-3-030-21074-8_40
21. Holzinger, A.: From machine learning to explainable AI. In: 2018 World Symposium on Digital Intelligence for Systems and Machines (DISA), pp. 55–66. IEEE (2018)
22. Kendall, A., Cipolla, R.: Modelling uncertainty in deep learning for camera relocalization. In: 2016 IEEE International Conference on Robotics and Automation (ICRA), pp. 4762–4769. IEEE (2016)
23. Kendall, A., Gal, Y.: What uncertainties do we need in Bayesian deep learning for computer vision? In: Advances in Neural Information Processing Systems, pp. 5574–5584 (2017)
24. King, D.E.: Dlib-ml: a machine learning toolkit. J. Mach. Learn. Res. **10**, 1755–1758 (2009)
25. Kingma, D.P., Ba, J.: Adam: a method for stochastic optimization. arXiv preprint arXiv:1412.6980 (2014)
26. Kips, R., Tran, L., Shakhmametova, A., Malherbe, E., Askenazi, B., Perrot, M.: Toward online foundation shade recommendation: skin color estimation from smartphone images through deep learning. In: Proceedings of the 31st IFSCC Congress, Yokohama, Japan (2020)
27. Kips, R., Tran, L., Malherbe, E., Perrot, M.: Beyond color correction: skin color estimation in the wild through deep learning. Electron. Imaging **2020**, 82-1–82-8 (2020)
28. LeCun, Y., Bengio, Y., Hinton, G.: Deep learning. Nature **521**(7553), 436 (2015)
29. Lu, C., Koniusz, P.: Few-shot keypoint detection with uncertainty learning for unseen species. In: Proceedings of the IEEE/CVF Conference on Computer Vision and Pattern Recognition, pp. 19416–19426 (2022)
30. Malherbe, E., Vanrompay, Y., Aufaure, M.A.: From a ranking system to a confidence aware semi-automatic classifier. Procedia Comput. Sci. **60**, 73–82 (2015)
31. Masood, S., Gupta, S., Wajid, A., Gupta, S., Ahmed, M.: Prediction of human ethnicity from facial images using neural networks. In: Satapathy, S.C., Bhateja, V., Raju, K.S., Janakiramaiah, B. (eds.) Data Engineering and Intelligent Computing. AISC, vol. 542, pp. 217–226. Springer, Singapore (2018). https://doi.org/10.1007/978-981-10-3223-3_20
32. Mirza, M., Osindero, S.: Conditional generative adversarial nets. arXiv preprint arXiv:1411.1784 (2014)
33. Peretroukhin, V., Wagstaff, B., Kelly, J.: Deep probabilistic regression of elements of SO(3) using quaternion averaging and uncertainty injection. In: Proceedings of the IEEE Conference on Computer Vision and Pattern Recognition Workshops, pp. 83–86 (2019)
34. Platt, J., et al.: Probabilistic outputs for support vector machines and comparisons to regularized likelihood methods. In: Advances in Large Margin Classifiers, vol. 10, no. 3, pp. 61–74 (1999)
35. Postels, J., Ferroni, F., Coskun, H., Navab, N., Tombari, F.: Sampling-free epistemic uncertainty estimation using approximated variance propagation. arXiv preprint arXiv:1908.00598 (2019)
36. Rafiei, M.H., Adeli, H.: Novel machine-learning model for estimating construction costs considering economic variables and indexes. J. Constr. Eng. Manag. **144**(12), 04018106 (2018)
37. Russell, R.L., Reale, C.: Multivariate uncertainty in deep learning. IEEE Trans. Neural Netw. Learn. Syst. **33**, 7937–7943 (2021)

38. Sabbeh, S.F.: Machine-learning techniques for customer retention: a comparative study. Int. J. Adv. Comput. Sci. Appl. **9**(2) (2018)
39. Sallab, A.E., Abdou, M., Perot, E., Yogamani, S.: Deep reinforcement learning framework for autonomous driving. Electron. Imaging **2017**(19), 70–76 (2017)
40. Schaub, H., Tsiotras, P., Junkins, J.L.: Principal rotation representations of proper n× n orthogonal matrices. Int. J. Eng. Sci. **33**(15), 2277–2295 (1995)
41. Settles, B.: Active learning literature survey. Technical report, University of Wisconsin-Madison Department of Computer Sciences (2009)
42. Stuelpnagel, J.: On the parametrization of the three-dimensional rotation group. SIAM Rev. **6**(4), 422–430 (1964)
43. Taigman, Y., Yang, M., Ranzato, M., Wolf, L.: DeepFace: closing the gap to human-level performance in face verification. In: Proceedings of the IEEE Conference on Computer Vision and Pattern Recognition, pp. 1701–1708 (2014)
44. Tkalcic, M., Tasic, J.F.: Colour spaces: perceptual, historical and applicational background, vol. 1. IEEE (2003)
45. Williams, C.K., Rasmussen, C.E.: Gaussian Processes for Machine Learning, vol. 2. MIT Press Cambridge, Cambridge (2006)
46. Zadrozny, B., Elkan, C.: Transforming classifier scores into accurate multiclass probability estimates. In: Proceedings of the Eighth ACM SIGKDD International Conference on Knowledge Discovery and Data Mining, pp. 694–699. ACM (2002)
47. Zhang, Z., Romero, A., Muckley, M.J., Vincent, P., Yang, L., Drozdzal, M.: Reducing uncertainty in undersampled MRI reconstruction with active acquisition. In: Proceedings of the IEEE Conference on Computer Vision and Pattern Recognition, pp. 2049–2058 (2019)

Spatio-Channel Attention Blocks for Cross-modal Crowd Counting

Youjia Zhang, Soyun Choi, and Sungeun Hong(✉)

Department of Electrical and Computer Engineering, Inha University, Incheon, South Korea

{zhangyoujia,sychoi}@inha.edu, csehong@inha.ac.kr

Abstract. Crowd counting research has made significant advancements in real-world applications, but it remains a formidable challenge in cross-modal settings. Most existing methods rely solely on the optical features of RGB images, ignoring the feasibility of other modalities such as thermal and depth images. The inherently significant differences between the different modalities and the diversity of design choices for model architectures make cross-modal crowd counting more challenging. In this paper, we propose Cross-modal Spatio-Channel Attention (CSCA) blocks, which can be easily integrated into any modality-specific architecture. The CSCA blocks first spatially capture global functional correlations among multi-modality with less overhead through spatial-wise cross-modal attention. Cross-modal features with spatial attention are subsequently refined through adaptive channel-wise feature aggregation. In our experiments, the proposed block consistently shows significant performance improvement across various backbone networks, resulting in state-of-the-art results in RGB-T and RGB-D crowd counting.

Keywords: Crowd counting · Cross-modal · Attention

1 Introduction

Crowd counting is a core and challenging task in computer vision, which aims to automatically estimate the number of pedestrians in unconstrained scenes without any prior knowledge. Driven by real-world applications, including traffic monitoring [1], social distancing monitoring [16], and other security-related scenarios [56], crowd counting plays an indispensable role in various fields. Over the last few decades, an increasing number of researchers have focused on this topic, achieving significant progress [50,51,60,63] in counting tasks.

In general, the various approaches for crowd counting can be roughly divided into two categories: detection-based methods and regression-based methods. Early works [8,10,12,22,46,54] on crowd counting mainly use detection-based approaches, which detect human heads in an image as can be seen in drone object detection tasks [9,18]. In fact, most detection-based methods assume that the crowd in the image is made up of individuals that can be detected

L. Wang et al. (Eds.): ACCV 2022, LNCS 13842, pp. 22–40, 2023.
https://doi.org/10.1007/978-3-031-26284-5_2

(a) RGB-T (RGB<T) (b) RGB-T (RGB>T) (c) RGB-D (RGB<D) (d) RGB-D (RGB>D)

Fig. 1. Visualization of RGB-T and RGB-D pairs. (a) shows the positive effects of thermal images in extremely poor illumination. (b) shows the negative effects caused by additional heating objects. In case (c), the depth image provides additional information about the position and size of the head, but in scene (d), part of the head information in the depth image is corrupted by cluttered background noise.

by the given detectors. There are still some significant limitations for tiny heads in higher density scenarios with overlapping crowds and serious occlusions. To reduce the above problems, some researchers [6,7,19,40] introduce regression-based methods where they view crowd counting as a mapping problem from an image to the count or the crowd-density map. In this way, the detection problem can be circumvented. Recent works [26,44,59,64] have shown the significant success of density map estimation in RGB crowd counting, and it has become the mainstream method for the counting task in complex crowd scenes.

Despite substantial progress, crowd counting remains a long-standing challenging task for various situations such as occlusion, high clutter, non-uniform lighting, non-uniform distribution, etc. Due to the limitations of methods or insufficient data sources, most previous works have focused on the RGB crowd counting and failed to obtain high-quality density maps in unconstrained scenarios. RGB images can provide strong evidence in bright illumination conditions while they are almost invisible in the dark environment. Recently, some researchers [31,33] indicate that the thermal feature helps better distinguish various targets in poor illumination conditions. Therefore, as shown in Fig. 1-(a), thermal images can be used to improve the performance of crowd counting tasks. However, from Fig. 1-(b), we can find that additional heating objects seriously interfere with the thermal image.

Furthermore, besides exploiting various contextual information from the optical cues, the depth feature [3,27,55] has recently been utilized as supplementary information of the RGB feature to generate improved crowd density maps. From Fig. 1-(c), we observe that crowd movement leads to aggregation behavior and large variation in head size. Thus, it is insufficient to generate pixel-wise density maps using only RGB data. Depth data naturally complements RGB information by providing the 3D geometry to 2D visual information, which is robust to varying illumination and provides additional information about the localization and size of heads. However, as shown in Fig. 1-(d),the depth information would be critically affected by noise under the condition of cluttered background.

Despite the many advantages of using additional modalities besides RGB in crowd counting, there are few related studies in crowd counting. In this study, we focus on two main issues that need to be addressed for a more robust cross-modal crowd counting as follows:

i. Crowd counting data is obtained from the different modalities, which means that they vary significantly in distribution and properties. Since there are substantial variations between multimodal data, how to effectively identify their differences and unify multiple types of information is still an open problem in cross-modal crowd counting. Additionally, since additional modality other than RGB is not fixed (either thermal or depth), it is very critical to adaptively capture intra-modal and inter-modal correlation.
ii. For cross-modal crowd counting, various types can be given as additional modalities besides RGB. At this time, constructing a cross-modal model architecture from scratch usually entails a lot of trial and error. We have seen the well-explored model architectures show promising results in various fields through transfer learning or model tuning. In this respect, if the widely used RGB-based unimodal architecture can be extended to multimodal architecture without bells and whistles, we can relieve the effort of design choice of the model architecture.

In this paper, we propose a novel plug-and-play module called Cross-modal Spatio-Channel Attention (CSCA) block consisting of two main modules. First, Spatial-wise Cross-modal Attention (SCA) module utilizes an attention mechanism based on the triplet of 'Query', 'Key', and 'Value' widely used in non-local-based models [53,58,66]. Notably, conventional non-local blocks are unsuitable for cross-modal tasks due to their high computational complexity and the fact that they are designed to exploit self-attention within a single modality. To address these issues, SCA module is deliberately designed to understand the correlation between cross-modal features, and reduces computational overhead through the proposed feature re-assembling scheme. Given an input feature map with size C×H×W, the computational complexity of a typical non-local block is $O(CH^2W^2)$ while our SCA block can reduce the complexity by a factor of G_l to $O(CH^2W^2/G_l)$. Here, the G_l refers to the spatial re-assembling factor in the lth block of the backbone network.

Importantly, we note that non-local blocks widely used in various vision tasks focus on spatial correlation of features but do not directly perform feature recalibration at the channel level. To supplement the characteristics of non-local-based blocks considering only spatial correlation, our Channel-wise Feature Aggregation (CFA) module adaptively adjusts the spatially-correlated features at the channel level. Unlike previous works, which use pyramid pooling feature differences to measure multimodal feature variation [31] or fine-tune the density map [23], our SCA aims to spatially capture global feature correlations between multimodal features to alleviate misalignment in multimodal data. Moreover, instead of adding auxiliary tasks for other modal information [3,27,55], our method dynamically aggregate complementary features along the channel dimension by the CFA block. We show that the proposed two main modules complement each other on two challenging RGB-T [31] and RGB-D [27]. Additionally, we conduct various ablation studies and comparative experiments in cross-modal settings that have not been well explored. In summary, the major contributions of this work are as follows:

- We propose a new plug-and-play module that can significantly improve performance when integrated with any RGB-based crowd counting model.
- Our cross-modal attention with feature re-assembling is computationally efficient and has higher performance than conventional non-local attention. Moreover, our channel-level recalibration further improves the performance.
- The proposed CSCA block consistently demonstrates significant performance enhancements across various backbone networks and shows state-of-the-art performance in RGB-T and RGB-D crowd counting.
- We extensively conduct and analyze comparative experiments between multi-modality and single-modality, which have not been well explored in existing crowd counting, and this can provide baselines for subsequent studies.

2 Related Works

2.1 Detection-Based Methods

Early works of crowd counting focus on detection-based approaches. The initial method [28,49] mainly employs detectors based on hand-crafted features. The typical traditional methods [8,10,22,24] train classifiers to detect pedestrians using wavelet, HOG, edge, and other features extracted from the whole body, which degrades seriously for those very crowded scenes. With the increase of crowd density, the occlusion between people becomes more and more serious. Therefore, methods [12,54] based on partial body detection (such as head, shoulders, etc.) are proposed to deal with the crowd counting problem. Compared with the overall detection, this method has a slight improvement in effect. Recently, detection-based [2,27,30,47] methods have also been further improved with the introduction of more advanced object detectors. For example, some works [27,30] design specific architectures such as Faster RCNN [39] or RetinaNet [29] to promote the performance of crowd counting by improving the robustness of detectors for tiny head detection. But, they still present unsatisfactory results on challenging images with very high densities, significant occlusions, and background clutter in extremely dense crowds, which are quite common in crowd counting tasks.

2.2 Regression–Based Methods

Some of the recent approaches [4,7,15,26,32,43,52,64] attempt to tackle this problem by bypassing completely the detection problem. They directly address it as a mapping problem from images to counts or crowd density maps. Recently, numerous models utilize basic CNN layers [15,52], multi-column based models

[4,64], or single-column based models [5,26,34] to extract rich information for density estimation and crowd counting. To capture multi-scale information at different resolutions, MCNN [64] proposes a three branches architecture. And, CrowdNet [4] combines low-level features and high-level semantic information. To reduce the information redundancy caused by the multi-column method, SANet [5] is built on the Inception architecture. And, CSRNet [26] adopts dilated convolution layers to expand the receptive field. Regression-based methods capture low-level features and high-level semantic information through multi-scale operations or dilated convolution, which show remarkable success in crowd counting tasks. However, most methods only focus on the RGB images and may fail to accurately recognize the semantic objects in unconstrained scenarios.

2.3 Multi-modal Learning for Crowd Counting

Recently, multimodal feature fusion has been widely studied across various fields [17,36,45]. The cross-modal feature fusion aims to learn better feature representations by exploiting the complementarity between multiple modalities. Some cross-modal fusion methods [14,45] fuse the RGB source and the depth source in the "Early Fusion" or "Late Fusion" way. Besides, inspired by the success of multi-task learning in various computer vision tasks, some researchers [42,65] utilize the depth feature as the auxiliary information to boost the performance on the RGB crowd counting task. Similarly, the perspective information [41] during the density estimation procedure is also treated as auxiliary data to improve the performance of RGB crowd counting. However, these methods rely on hand-crafted features instead of fusing multimodal data with predicting weights from different modalities. Moreover, some researchers [13,27] combine the depth information into the detection procedure for the crowd counting task, which would be limited by the performance of the detectors. The IADM [31] proposes to collaboratively represent the RGB and thermal images. However, it does not fully consider the different characteristics and feature space misalignment of RGB data and thermal imaging data. Different from previous work, in this work, we propose a novel cross-modal spatio-channel attention block, which adaptively aggregates cross-modal features based on both spatial-wise and channel-wise attention. Furthermore, our SCA block is robust to cross-modal feature variation at lower complexity compared to the conventional non-local block.

3 Proposed Method

3.1 Overview

We proposed a unified framework to extend the existing baseline models from unimodal crowd counting to the multimodal scene. As shown in Fig. 2, the framework for cross-modal crowd counting consists of two parts: modality-specific branches and the Cross-modal Spatio-Channel Attention (CSCA) block. Given pairs of multimodal data like RGB-T or RGB-D pairs, we first utilize the backbone network (*e.g.*, BL [34], CSRNet [26], etc.) to extract the modality-specific

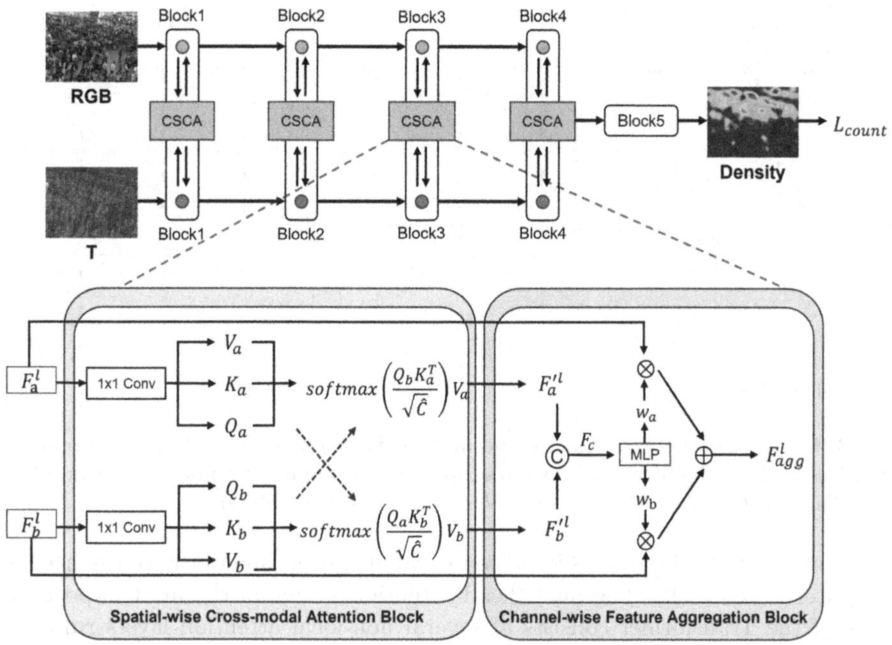

Fig. 2. The architecture of the proposed unified framework for extending existing baseline models from unimodal crowd counting to multimodal scenes. Our CSCA module is taken as the cross-modal solution to fully exploit the multimodal complementarities. Specifically, the CSCA consists of SCA to model global feature correlations among multimodal data, and CFA to dynamically aggregate complementary features.

features of RGB images and thermal images. To fully exploit the multimodal complementarities, we embed CSCA after different convolutional blocks of the two modality-specific branches to further update the feature maps and get the complementary fusion information.

To ensure informative feature interaction between modalities, our CSCA consists of two operations: Spatial-wise Cross-modal Attention (SCA) block and Channel-wise Feature Aggregation (CFA) block, as illustrated in the bottom part of Fig. 2. The SCA block utilizes an improved cross-modal attention block to model the global feature correlations between RGB images and thermal/depth images, which has a larger receptive field than the general CNN modules and facilitates complementary aggregation operation. And, the CFA block adaptively extracts and fuses cross-modal information based on channel correlations. Consequently, the framework can capture the non-local dependencies and aggregate the complementary information of cross-modal data to expand from unimodal crowd counting to multimodal crowd counting.

3.2 Spatial-Wise Cross-modal Attention (SCA)

SCA module aims to capture global feature correlations among multimodal data with less overhead based on cross-modal attention. Recently, the Transformer-

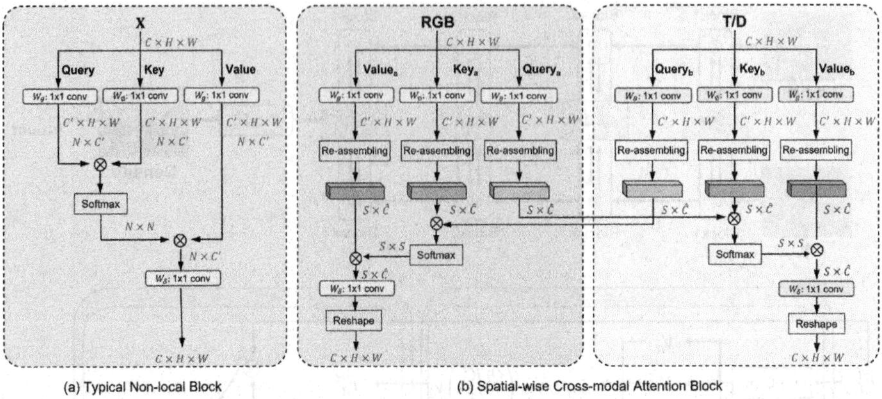

Fig. 3. Comparison between block structures (a) typical non-local block (b) proposed SCA block where the blue font indicates dimension transformation of SCA block, $\hat{C} = C' G_l$ and $S \times \hat{C} = N \times C'$ while $S = N/G_l$. (Color figure online)

based architecture [48] has achieved comparable or even superior performance on computer vision tasks. The multi-head attention as one of the most important parts of the Transformer consists of several non-local attention layers running in parallel. A typical non-local block [53] is shown in Fig. 3-(a), which aims to capture the global relationships among all locations of the input image. Consider an input feature map $X \in \mathcal{R}^{C \times H \times W}$, where C, H, and W indicate the channel number, spatial height, and width, respectively. The feature map is first projected and flattened into three vectors: Query $Q \in \mathcal{R}^{N \times C'}$, Key $K \in \mathcal{R}^{N \times C'}$ and Value $V \in \mathcal{R}^{N \times C'}$ by three different 1×1 convolutions, where $N = H \times W$, and C' refers to the new embedding channel, $C' = C/2$. Then, the triplet of Query, Key, and Value are used to encode the relationship at any position in the feature map, thereby capturing the global context information. The output can be computed as a weighted sum of values where the weight is assigned to each value by the compatibility of the query with the corresponding key. This procedure can be computed efficiently with matrix multiplications as:

$$Z = f(QK^T)V, \tag{1}$$

where the $f(\cdot)$ is a normalization function, such as softmax, rescaling, etc. Finally, the non-local block recovers the channel dimension from C' to C by an additional 1×1 convolution layer.

However, the typical non-local block can only work on unimodal data, and the high complexity introduced by matrix multiplication hinders its use. To alleviate these problems, we propose the SCA block. Different from the typical non-local block, the input of our SCA Block is a pair of images with the different modalities. And it's Key and Value are from the same modality, while the Query is from another modality. For example, the Key and Value are from the RGB image modality, and the Query is generated from thermal image or depth image modality, and vice versa.

$$Z_a = softmax(\frac{Q_b K_a^T}{\sqrt{\hat{C}}})V_a, \tag{2}$$

where the a means the RGB modality and b means another modality (*i.e.*, thermal or depth modality). The output of the SCA block (Fig. 3-(b)), is calculated by multiplying Value with attention weights, which presents the similarity between the corresponding Query in modality b and all the Keys in another modality a. In other words, the SCA block aggregates and aligns the information from two modalities.

Thanks to the great power of cross-attention, many works [25,35,57,66] have utilized it to solve the misalignment. However, the matrix multiplication operation is very time and memory-consuming compared to convolutions and activation blocks. Some researchers [31] alleviate the problem by reducing the dimensions of the Key and Value through the pooling operation. But, it is hard to ensure that the performance would not degenerate too much by losing a lot of global information. Instead of the pooling function, we perform re-assembling operations in the SCA block to randomly divide the feature map into G_l patches in the spatial dimension, where l means the lth block of the backbone network. Thus, the Key, Value, and Query vectors in our SCA block become $K, V, Q \in \mathcal{R}^{S \times \hat{C}}$, where $S = HW/G_l$, $\hat{C} = C' G_l$. Finally, we recover the channel dimension and reshape the result to $C \times H \times W$ size. Therefore, the computational cost of our SCA block is $O(CH^2W^2/G_l)$, which is G_l times less than the computational complexity $O(CH^2W^2)$ of a typical non-local block.

3.3 Channel-Wise Feature Aggregation (CFA)

The SCA block can capture global features at all locations in multimodal images by spatial-wise attention. To make full use of the complementarity of multimodal data, we need to complementarily aggregate cross-modal features according to their characterization capabilities. To achieve this, we propose a CFA block to fusion the cross-modal features and capture channel-wise interactions. As shown in the bottom part of Fig. 2, we first concatenate the outputs of the SCA block to get the F_c. We then use an MLP layer and softmax function to learn the weight vectors $w_a, w_b \in \mathcal{R}^{C \times H \times W}$, which are used to re-weight the RGB feature and thermal/depth feature across the channels.

$$[w_a, w_b] = softmax(MLP(F_c)). \tag{3}$$

To fully exploit the aggregated information and suppress feature noise and redundancy simultaneously, we fusion the cross-modal features using the two channel-wise adaptive weights. The final aggregation feature map F_{agg}^l at l^{th} block can be obtained by the following formula:

$$F_{agg}^l = w_a * F_a^l + w_b * F_b^l. \tag{4}$$

Through the SCA block, the network can increase sensitivity to informative features on the important channels and enhance the final feature map. In practice, for each layer l, the updated feature maps $f_a^l = (F_{agg}^l + F_a^l)/2$ and $f_b^l = (F_{agg}^l + F_b^l)/2$ are propagated to the next blocks except the last layer. Hence, the module can adaptively provide more discriminative and richer characteristics to increase cross-modal crowd counting performance.

4 Experiments

To evaluate our method, we carry out detailed experiments on two cross-modal crowd counting datasets: RGBT-CC benchmark [31] and ShanghaiTechRGBD [27]. And we also provide a thorough ablation study to validate the key components of our approach. Experimental results show that the proposed method considerably improves the performance of the cross-modal crowd counting.

4.1 Implementation Details

In this work, our code is based on PyTorch [38]. Following [31], our model is implemented with three backbone networks: BL [34], MCNN [64], and CSRNet [26]. To make fair comparisons, we maintain a similar number of parameters to the original backbone models, and the same channel setting with the work [31], which takes 70%, 60%, and 60% of the original values.

For the input size, we feed the 640×480 RGB-T pairs and 1920×1080 RGB-D pairs, respectively. Notably, the original loss function of the adopted backbone network is used to train our framework. During training, our framework is optimized by Adam [21] with a 1e-5 learning rate. All the experiments are conducted using $1\times$ RTX A6000 GPU.

Datasets: The popular datasets, RGBT-CC [31] and ShanghaiTechRGBD [27], in the cross-modal crowd counting field are used in this study. The RGBT-CC dataset is a new challenging benchmark for RGB-T crowd counting, including 2,030 RGB-thermal pairs with resolution 640 × 480. All the images are divided into three sets: training set with 1030 pairs, validation set with 200 pairs, and testing set with 800 pairs, respectively. The ShanghaiTechRGBD dataset is a large-scale RGB-D dataset, which consists of 1193 training RGB-D image pairs and 1000 test RGB-D image pairs with a fixed resolution of 1920 × 1080.

Evaluation Metrics: As commonly used in previous works, we adopt the Root Mean Square Error (RMSE) and the Grid Average Mean absolute Error (GAME) as the evaluation metrics.

$$RMSE = \sqrt{\frac{1}{N}\sum_{i=1}^{N}(P_i - \hat{P}_i)^2}$$
$$GAME(L) = \frac{1}{N}\sum_{i=1}^{N}\sum_{l=1}^{4^L}|P_i^l - \hat{P}_i^l|, \tag{5}$$

Table 1. The performance of our model implemented with different backbones on RGBT-CC.

Backbone	GAME(0)↓	GAME(1)↓	GAME(2)↓	GAME(3)↓	RMSE↓
MCNN [64]	21.89	25.70	**30.22**	**37.19**	37.44
MCNN+CSCA	**17.37**	**24.21**	30.37	37.57	**26.84**
CSRNet [26]	20.40	23.58	**28.03**	**35.51**	35.26
CSRNet+CSCA	**17.02**	**23.30**	29.87	38.39	**31.09**
BL [34]	18.70	22.55	26.83	34.62	32.64
BL+CSCA	**14.32**	**18.91**	**23.81**	**32.47**	**26.01**

Table 2. The performance of our model implemented with different backbones on ShanghaiTechRGBD.

Backbone	GAME(0)↓	GAME(1)↓	GAME(2)↓	GAME(3)↓	RMSE↓
BL [34]	8.94	11.57	15.68	22.49	12.49
BL+CSCA	**5.68**	**7.70**	**10.45**	**15.88**	**8.66**
CSRNet [26]	4.92	6.78	9.47	13.06	7.41
Ours	**4.39**	**6.47**	**8.82**	**11.76**	**6.39**

where N means numbers of testing images, P_i and $\hat{P}_i$ refer to the total count of the estimated density map and corresponding ground truth, respectively. Different from the RMSE, the GAME divides each image into 4^L non-overlapping regions to measure the counting error in each equal area, where P_i^l and $\hat{P}_i^l$ are the estimated count and ground truth in a region l of image i. Note that $GAME(0)$ is equivalent to Mean Absolute Error (MAE).

4.2 Effectiveness of Plug-and-Play CSCA Blocks

To evaluate the effectiveness of the proposed plug-and-play CSCA module, we easily incorporate CSCA as a plug-and-play module into backbone networks (*e.g.*, MCNN, CSRNet, and BL), and extend existing models from unimodal to multimodal on the RGBT-CC dataset. We report the result in Table 1. Note that for all baseline methods in the table, we estimate crowd counts by feeding the concatenation of RGB and thermal images. As can be observed, almost all instances of CSCA based method significantly outperform the corresponding backbone networks. Especially, our method far outperforms MCNN by 4.52 in MAE and 10.60 in RMSE. Compared to the CSRNet backbone, our CSCA provides a 3.88 and 4.17 improvement in MAE and RMSE, respectively. For the BL backbone, our method surpasses the baseline on all evaluation metrics.

Similarly, we also apply CSCA to the RGB-D crowd counting task. Furthermore, previous research [31] showed that MCNN is inefficient on the ShanghaiTechRGBD dataset with high-resolution images due to its time-consuming multi-column structure. So here we only compare the two backbone networks:

Table 3. Counting performance evaluation on RGBT-CC.

Backbone	GAME(0)↓	GAME(1)↓	GAME(2)↓	GAME(3)↓	RMSE↓
UCNet [61]	33.96	42.42	53.06	65.07	56.31
HDFNet [37]	22.36	27.79	33.68	42.48	33.93
BBSNet [11]	19.56	25.07	31.25	39.24	32.48
MVMS [62]	19.97	25.10	31.02	38.91	33.97
MCNN+IADM [31]	19.77	23.80	28.58	35.11	30.34
CSRNet+IADM [31]	17.94	21.44	26.17	33.33	30.91
BL+IADM [31]	15.61	19.95	24.69	32.89	28.18
CmCaF [23]	15.87	19.92	24.65	**28.01**	29.31
Ours	**14.32**	**18.91**	**23.81**	32.47	**26.01**

Table 4. Counting performance evaluation on ShanghaiTechRGBD.

Backbone	GAME(0)↓	GAME(1)↓	GAME(2)↓	GAME(3)↓	RMSE↓
UCNet [61]	10.81	15.24	22.04	32.98	15.70
HDFNet [37]	8.32	13.93	17.97	22.62	13.01
BBSNet [11]	6.26	8.53	11.80	16.46	9.26
DetNet [30]	9.74	–	–	–	13.14
CL [20]	7.32	–	–	–	10.48
RDNet [44]	4.96	–	–	–	7.22
MCNN+IADM [31]	9.61	11.89	15.44	20.69	14.52
BL+IADM [31]	7.13	9.28	13.00	19.53	10.27
CSRNet+IADM [31]	**4.38**	**5.95**	**8.02**	**11.02**	7.06
Ours	4.39	6.47	8.82	11.76	**6.39**

BL and CSRNet. As shown in Table 2, our framework consistently outperforms their corresponding backbone networks on all evaluation metrics. In conclusion, all experimental results confirm that CCIM is universal and effective for cross-modal crowd counting.

4.3 Comparison with Previous Methods

In addition, to evaluate that our proposed model can be more effective and efficient, we compare our method with state-of-the-art methods on RGB-T and RGB-D crowd counting tasks. For fair comparisons, we directly use the optimal results of the comparison methods in the original papers. As shown in Table 3, compared with multimodal models, including UCNet [61], HDFNet [37], and BBSNet [62], our method outperforms these models by a wide margin on all evaluation metrics. Moreover, our model also achieves better performance on the RGBT-CC dataset compared to specially designed crowd counting models,

Table 5. Ablation study using the BL backbone on RGBT-CC.

Method	GAME(0)↓	GAME(1)↓	GAME(2)↓	GAME(3)↓	RMSE↓
RGB	32.89	38.81	44.49	53.44	59.49
T	17.80	22.88	28.50	37.30	30.34
Early fusion	18.25	22.35	26.78	34.96	37.71
Late fusion	16.63	20.33	24.81	32.59	32.66
NL	17.91	21.12	25.06	32.61	32.60
SCA	17.85	21.13	25.07	32.56	30.60
CFA	17.78	21.36	25.69	33.14	32.37
NL+CFA	16.69	20.75	25.63	33.92	29.39
CSCA (SCA+CFA)	**14.32**	**18.91**	**23.81**	**32.47**	**26.01**

Table 6. Ablation study using the BL backbone on ShanghaiTechRGBD. Since methods based on non-local blocks run out of memory, we do not report those results.

Method	GAME(0)↓	GAME(1)↓	GAME(2)↓	GAME(3)↓	RMSE↓
RGB	6.88	8.62	11.86	18.31	10.44
D	15.17	17.92	22.59	30.76	23.65
Early fusion	7.04	8.74	11.82	18.33	10.62
Late fusion	6.55	8.01	10.45	15.43	9.93
SCA	6.94	8.19	10.48	15.91	10.61
CFA	6.43	8.07	10.80	16.12	9.86
CSCA (SCA+CFA)	**5.68**	**7.70**	**10.45**	**15.88**	**8.66**

including IADM [31], MVMS [11], and CmCaF [23]. Especially, our CSCA implemented with the BL backbone achieves state-of-the-art performance on almost all evaluation metrics.

For the RGB-D crowd counting task, we also compare two categories of methods including multimodal methods and specially-designed models for cross-modal crowd counting. For fair comparisons, we use them to estimate the crowd counts by following the work [31]. The performance of all comparison methods is shown in Table 4. The results of the proposed method on the ShanghaiTechRGBD dataset are encouraging, as we used the same parameters as our experiments on the RGBT-CC dataset and did not perform fine-tuning. Moreover, our model implemented with the CSRNet backbone outperforms most advanced methods and achieves a new state-of-the-art result of 6.39 in RMSE, the main evaluation metric for crowd counting.

4.4 Ablation Studies

To verify the effectiveness of the proposed universal framework for cross-modal crowd counting, we investigate how different inputs affect the model. First, we

Table 7. The performance of our model implemented with BL backbone under different illumination conditions on RGBT-CC.

Illumination	Input	GAME(0)↓	GAME(1)↓	GAME(2)↓	GAME(3)↓	RMSE↓
Brightness	RGB	26.96	31.82	37.07	46.00	48.42
	T	19.19	24.07	30.28	39.93	32.07
	RGB-T	**14.41**	**18.85**	**24.71**	**34.20**	**24.74**
Darkness	RGB	45.78	51.03	54.65	62.51	87.29
	T	16.38	21.65	26.67	34.59	28.45
	RGB-T	**14.22**	**18.97**	**22.89**	**30.69**	**27.25**

feed RGB images and thermal (or depth) images into the network, respectively, following the unimodal crowd counting setting of the backbone network. Then we adopt the early fusion strategy, feeding the concatenation of RGB and thermal images into BL network. We also apply late fusion where the different modal features are extracted separately with the backbone network and then combine multimodal features in the last layer to estimate the crowd density map. From Table 5 and Table 6, we can observe an interesting finding that thermal images are more discriminative than RGB images for the RGBT-CC dataset. However, RGB images are more efficient than depth images on ShanghaiTechRGBD. Interestingly, naive feature fusion strategies such as early fusion and late fusion even damage the original accuracy.

To explore the effectiveness of each CSCA component, we remove the SCA block and CFA block, respectively. As shown in Table 5 and Table 6, regardless of which component is removed, the performance of our model would drop severely. These results suggest that the main components of CSCA work complementary. Additionally, we perform additional experiments replacing the SCA block with the typical non-local block. Remarkably, our CSCA is far superior to non-local-based methods in terms of various evaluation metrics. Since the non-local-based methods were out of memory on the ShanghaiTechRGBD dataset where image resolution is 1920×1080, we could not report the results using non-local blocks. These results suggest that the SCA blocks have more practical potential than the typical non-local blocks.

To further validate the complementarity of the multimodal data for crowd counting, we infer our model using BL backbone under different illumination conditions on the RGBT-CC dataset. As shown in Table 7, we conduct experiments under bright and dark scenes, with three different input settings, including unimodal RGB images, unimodal thermal images, and RGB-T multimodal images, respectively. Notably, thermal images can provide more discriminative features than RGB images, especially in lower illumination cases. Moreover, comparing the single-modal results, it is easy to conclude that our CSCA can complementarily fuse thermal information and RGB optical information.

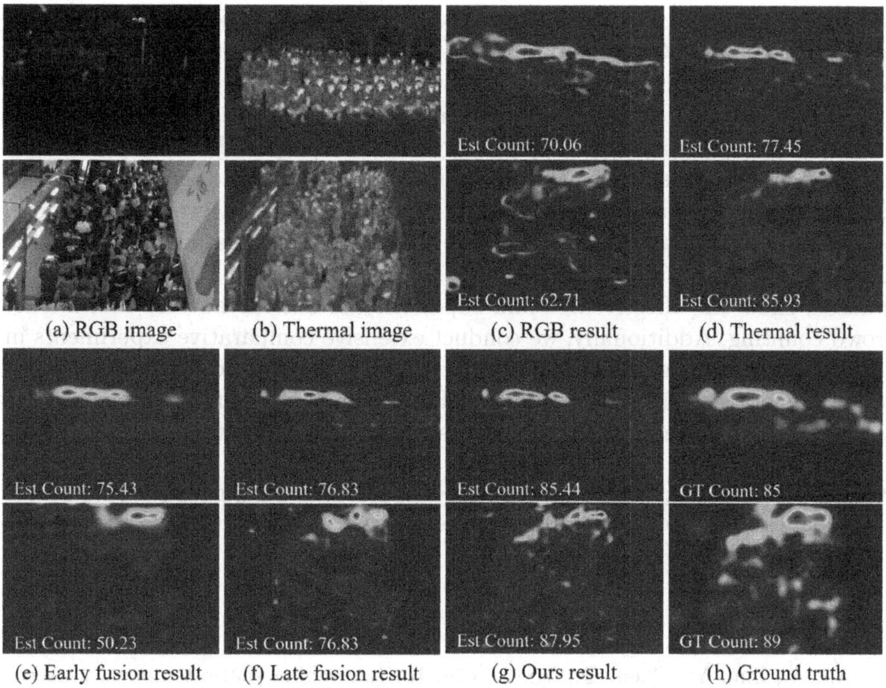

Fig. 4. Qualitative Results. (a) and (b) show the input RGB images and thermal images in different illumination conditions. (c)-(d) are the results of the RGB-based and thermal-based network. (e)-(f) are the results of the early fusion and late fusion. (g) and (h) refers to the results of our CSCA and the ground truth.

5 Qualitative Results

From the visualization results in cases (a) to (d) of Fig. 4, we can easily find that thermal images can facilitate the crowd counting task better than RGB images. Moreover, as shown in Fig. 4-(b), thermal images can also provide additional crowd location information to a certain extent, especially in dark scenes or bright scenes with less light noise. As we discussed earlier, inappropriate fusions fail to exploit the potential complementarity of multimodal data and even degrade the performance, such as the early fusion and late fusion shown in Fig. 4-(e) and Fig. 4-(f). Our proposed CSCA, a plug-and-play module, achieve significant improvements for cross-modal crowd counting by simply integrating into the backbone network as shown in Fig. 4-(g). This result shows the effectiveness of CSCA's complementary multi-modal fusion.

6 Conclusion

Whereas previous crowd counting approaches have mainly focused on RGB-based frameworks, we focus on a cross-modal representation that can additionally utilize thermal or depth images. Specifically, we propose a novel plug-and-play module named Cross-modal Spatio-Channel Attention (CSCA), which can be integrated with any RGB-based model architecture. Our CSCA blocks consist of two main modules. First, SCA is designed to spatially capture global feature correlations among multimodal data. The proposed SCA to deal with cross-modal relationships is computationally more efficient than existing non-local blocks by re-assembling the feature maps at the spatial level. Furthermore, our CFA module adaptively recalibrates the spatially-correlated features at the channel level. Extensive experiments on the RGBT-CC and ShanghaiTechRGBD datasets show the effectiveness and superiority of our method for multimodal crowd counting. Additionally, we conduct extensive comparative experiments in various scenarios using unimodal or multimodal representations, which have not been well addressed in crowd counting. These results are expected to be used as baselines for subsequent studies for cross-modal crowd counting.

Acknowledgements. This work was partly supported by Samsung Research Funding & Incubation Center of Samsung Electronics under Project Number SRFC-TD2103-01 and INHA UNIVERSITY Research Grant.

References

1. Ali, A., Zhu, Y., Chen, Q., Yu, J., Cai, H.: Leveraging spatio-temporal patterns for predicting citywide traffic crowd flows using deep hybrid neural networks. In: IEEE International Conference on Parallel and Distributed Systems (ICPADS), pp. 125–132. IEEE (2019)
2. Bathija, A., Sharma, G.: Visual object detection and tracking using yolo and sort. Int. J. Eng. Res. Technol. **8**(11), 705–708 (2019)
3. Bondi, E., Seidenari, L., Bagdanov, A.D., Del Bimbo, A.: Real-time people counting from depth imagery of crowded environments. In: Proceedings of International Conference on Advanced Video and Signal Based Surveillance (AVSS), pp. 337–342. IEEE (2014)
4. Boominathan, L., Kruthiventi, S.S., Babu, R.V.: Crowdnet: a deep convolutional network for dense crowd counting. In: ACM International Conference on Multimedia (ACMMM), pp. 640–644 (2016)
5. Cao, X., Wang, Z., Zhao, Y., Su, F.: Scale aggregation network for accurate and efficient crowd counting. In: Proc. of European Conf. on Computer Vision (ECCV). pp. 734–750 (2018)
6. Chan, A.B., Liang, Z.S.J., Vasconcelos, N.: Privacy preserving crowd monitoring: counting people without people models or tracking. In: Proceedings of Computer Vision and Pattern Recognition (CVPR), pp. 1–7. IEEE (2008)
7. Chan, A.B., Vasconcelos, N.: Bayesian poisson regression for crowd counting. In: Proceedings of International Conference on Computer Vision (ICCV), pp. 545–551. IEEE (2009)
8. Dalal, N., Triggs, B.: Histograms of oriented gradients for human detection. In: Proceedings of Computer Vision and Pattern Recognition (CVPR), vol. 1, pp. 886–893. IEEE (2005)
9. Du, D., et al.: Visdrone-det2019: the vision meets drone object detection in image challenge results. In: Proceedings of of International Conference on Computer Vision Workshops (ICCVW) (2019)

10. Enzweiler, M., Gavrila, D.M.: Monocular pedestrian detection: survey and experiments. IEEE Trans. Pattern Anal. Mach. Intell. (TPAMI) **31**(12), 2179–2195 (2008)
11. Fan, D.-P., Zhai, Y., Borji, A., Yang, J., Shao, L.: BBS-Net: RGB-D salient object detection with a bifurcated backbone strategy network. In: Vedaldi, A., Bischof, H., Brox, T., Frahm, J.-M. (eds.) ECCV 2020. LNCS, vol. 12357, pp. 275–292. Springer, Cham (2020). https://doi.org/10.1007/978-3-030-58610-2_17
12. Felzenszwalb, P.F., Girshick, R.B., McAllester, D., Ramanan, D.: Object detection with discriminatively trained part-based models. IEEE Trans. Pattern Anal. Mach. Intell. (TPAMI) **32**(9), 1627–1645 (2010)
13. Fu, H., Ma, H., Xiao, H.: Real-time accurate crowd counting based on rgb-d information. In: IEEE International Conference on Image Processing (ICIP), pp. 2685–2688. IEEE (2012)
14. Fu, K., Fan, D.P., Ji, G.P., Zhao, Q.: JL-DCF: joint learning and densely-cooperative fusion framework for rgb-d salient object detection. In: Proceedings of Computer Vision and Pattern Recognition (CVPR), pp. 3052–3062 (2020)
15. Fu, M., Xu, P., Li, X., Liu, Q., Ye, M., Zhu, C.: Fast crowd density estimation with convolutional neural networks. Eng. Appl. Artif. Intell. **43**, 81–88 (2015)
16. Ghodgaonkar, I., et al.: Analyzing worldwide social distancing through large-scale computer vision. arXiv preprint arXiv:2008.12363 (2020)
17. Hong, S., Im, W., Yang, H.S.: Cbvmr: content-based video-music retrieval using soft intra-modal structure constraint. In: International Conference on Multimedia Retrieval (ICMR), pp. 353–361 (2018)
18. Hong, S., Kang, S., Cho, D.: Patch-level augmentation for object detection in aerial images. In: Proceedings of International Conference on Computer Vision Workshops (ICCVW) (2019)
19. Idrees, H., Saleemi, I., Seibert, C., Shah, M.: Multi-source multi-scale counting in extremely dense crowd images. In: Proceedings of Computer Vision and Pattern Recognition (CVPR), pp. 2547–2554 (2013)
20. Idrees, H., et al.: Composition loss for counting, density map estimation and localization in dense crowds. In: Proceedings of European Conference on Computer Vision (ECCV), pp. 532–546 (2018)
21. Kingma, D.P., Ba, J.: Adam: a method for stochastic optimization. arXiv preprint arXiv:1412.6980 (2014)
22. Leibe, B., Seemann, E., Schiele, B.: Pedestrian detection in crowded scenes. In: Proceedings of Computer Vision and Pattern Recognition (CVPR), vol. 1, pp. 878–885. IEEE (2005)
23. Li, H., Zhang, S., Kong, W.: Rgb-d crowd counting with cross-modal cycle-attention fusion and fine-coarse supervision. IEEE Trans. Ind. Inf. **19**, 306–316 (2022)
24. Li, M., Zhang, Z., Huang, K., Tan, T.: Estimating the number of people in crowded scenes by mid based foreground segmentation and head-shoulder detection. In: Proceedings of International Conference on Pattern Recognition (ICPR), pp. 1–4. IEEE (2008)
25. Li, X., Hou, Y., Wang, P., Gao, Z., Xu, M., Li, W.: Trear: transformer-based rgb-d egocentric action recognition. IEEE Trans. Cogn. Dev. Syst. **14**, 246–252 (2021)
26. Li, Y., Zhang, X., Chen, D.: Csrnet: dilated convolutional neural networks for understanding the highly congested scenes. In: Proceedings of Computer Vision and Pattern Recognition (CVPR), pp. 1091–1100 (2018)

27. Lian, D., Li, J., Zheng, J., Luo, W., Gao, S.: Density map regression guided detection network for rgb-d crowd counting and localization. In: Proceedings of Computer Vision and Pattern Recognition (CVPR), pp. 1821–1830 (2019)
28. Lin, S.F., Chen, J.Y., Chao, H.X.: Estimation of number of people in crowded scenes using perspective transformation. IEEE Trans. Syst. Man Cybern.-Part A: Syst. Hum. **31**(6), 645–654 (2001)
29. Lin, T.Y., Goyal, P., Girshick, R., He, K., Dollár, P.: Focal loss for dense object detection. In: Proceedings of International Conference on Computer Vision (ICCV), pp. 2980–2988 (2017)
30. Liu, J., Gao, C., Meng, D., Hauptmann, A.G.: Decidenet: counting varying density crowds through attention guided detection and density estimation. In: Proceedings of Computer Vision and Pattern Recognition (CVPR), pp. 5197–5206 (2018)
31. Liu, L., Chen, J., Wu, H., Li, G., Li, C., Lin, L.: Cross-modal collaborative representation learning and a large-scale rgbt benchmark for crowd counting. In: Proceedings of Computer Vision and Pattern Recognition (CVPR), pp. 4823–4833 (2021)
32. Liu, X., Van De Weijer, J., Bagdanov, A.D.: Leveraging unlabeled data for crowd counting by learning to rank. In: Proceedings of Computer Vision and Pattern Recognition (CVPR), pp. 7661–7669 (2018)
33. Liu, Z., et al.: Visdrone-cc2021: the vision meets drone crowd counting challenge results. In: Proceedings of International Conference on Computer Vision (ICCV), pp. 2830–2838 (2021)
34. Ma, Z., Wei, X., Hong, X., Gong, Y.: Bayesian loss for crowd count estimation with point supervision. In: Proceedings of International Conference on Computer Vision (ICCV), pp. 6142–6151 (2019)
35. Mehta, S., Rastegari, M.: Mobilevit: light-weight, general-purpose, and mobile-friendly vision transformer. arXiv preprint arXiv:2110.02178 (2021)
36. . Nagrani, A., Yang, S., Arnab, A., Jansen, A., Schmid, C., Sun, C.: Attention bottlenecks for multimodal fusion. In: Proceedings of Neural Information Processing Systems (NeurIPS), vol. 34, pp14200–14213 (2021)
37. Pang, Y., Zhang, L., Zhao, X., Lu, H.: Hierarchical dynamic filtering network for RGB-D salient object detection. In: Vedaldi, A., Bischof, H., Brox, T., Frahm, J.-M. (eds.) ECCV 2020. LNCS, vol. 12370, pp. 235–252. Springer, Cham (2020). https://doi.org/10.1007/978-3-030-58595-2_15
38. Paszke, A., et al.: Pytorch: an imperative style, high-performance deep learning library. In: Proceedings of Neural Information Processing Systems (NeurIPS), vol. 32 (2019)
39. Ren, S., He, K., Girshick, R., Sun, J.: Faster r-cnn: towards real-time object detection with region proposal networks. In: Proceedings of Neural Information Processing Systems (NeurIPS), vol. 28 (2015)
40. Shao, J., Loy, C.C., Kang, K., Wang, X.: Slicing convolutional neural network for crowd video understanding. In: Proceedings of Computer Vision and Pattern Recognition (CVPR), pp. 5620–5628 (2016)
41. Shi, M., Yang, Z., Xu, C., Chen, Q.: Revisiting perspective information for efficient crowd counting. In: Proceedings of Computer Vision and Pattern Recognition (CVPR), pp. 7279–7288 (2019)
42. Shi, Z., Mettes, P., Snoek, C.G.: Counting with focus for free. In: Proceedings of International Conference on Computer Vision (ICCV), pp. 4200–4209 (2019)
43. Sindagi, V.A., Patel, V.M.: Cnn-based cascaded multi-task learning of high-level prior and density estimation for crowd counting. In: Proceedings of International

Conference on Advanced Video and Signal Based Surveillance (AVSS), pp. 1–6. IEEE (2017)
44. Sindagi, V.A., Patel, V.M.: Generating high-quality crowd density maps using contextual pyramid cnns. In: Proceedings of International Conference on Computer Vision (ICCV), pp. 1861–1870 (2017)
45. Sun, T., Di, Z., Che, P., Liu, C., Wang, Y.: Leveraging crowdsourced gps data for road extraction from aerial imagery. In: Proceedings of Computer Vision and Pattern Recognition (CVPR), pp. 7509–7518 (2019)
46. Tuzel, O., Porikli, F., Meer, P.: Pedestrian detection via classification on riemannian manifolds. IEEE Trans. Pattern Anal. Mach. Intell. (TPAMI) **30**(10), 1713–1727 (2008)
47. Valencia, I.J.C., et al.: Vision-based crowd counting and social distancing monitoring using tiny-yolov4 and deepsort. In: 2021 IEEE International Smart Cities Conference (ISC2), pp. 1–7. IEEE (2021)
48. Vaswani, A., et al.: Attention is all you need. In: Proceedings of Neural Information Processing Systems (NeurIPS), vol. 30 (2017)
49. Viola, P., Jones, M.J., Snow, D.: Detecting pedestrians using patterns of motion and appearance. Int. J. Comput. Vision (IJCV) **63**(2), 153–161 (2005)
50. Walach, E., Wolf, L.: Learning to count with CNN boosting. In: Leibe, B., Matas, J., Sebe, N., Welling, M. (eds.) ECCV 2016. LNCS, vol. 9906, pp. 660–676. Springer, Cham (2016). https://doi.org/10.1007/978-3-319-46475-6_41
51. Wan, J., Liu, Z., Chan, A.B.: A generalized loss function for crowd counting and localization. In: Proceedings of Computer Vision and Pattern Recognition (CVPR), pp. 1974–1983 (2021)
52. Wang, C., Zhang, H., Yang, L., Liu, S., Cao, X.: Deep people counting in extremely dense crowds. In: ACM International Conference on Multimedia (ACMMM), pp. 1299–1302 (2015)
53. Wang, X., Girshick, R., Gupta, A., He, K.: Non-local neural networks. In: Proceedings of Computer Vision and Pattern Recognition (CVPR), pp. 7794–7803 (2018)
54. Wu, B., Nevatia, R.: Detection and tracking of multiple, partially occluded humans by bayesian combination of edgelet based part detectors. Int. J. Comput. Vision (IJCV) **75**(2), 247–266 (2007)
55. Xu, M., et al.: Depth information guided crowd counting for complex crowd scenes. Pattern Recogn. Lett. **125**, 563–569 (2019)
56. Xu, M., Li, C., Lv, P., Lin, N., Hou, R., Zhou, B.: An efficient method of crowd aggregation computation in public areas. IEEE Trans. Circ. Syst. Video Technol. (TCSVT) **28**(10), 2814–2825 (2017)
57. Xu, T., Chen, W., Wang, P., Wang, F., Li, H., Jin, R.: Cdtrans: cross-domain transformer for unsupervised domain adaptation. arXiv preprint arXiv:2109.06165 (2021)
58. Yang, Y., Li, G., Wu, Z., Su, L., Huang, Q., Sebe, N.: Weakly-supervised crowd counting learns from sorting rather than locations. In: Vedaldi, A., Bischof, H., Brox, T., Frahm, J.-M. (eds.) ECCV 2020. LNCS, vol. 12353, pp. 1–17. Springer, Cham (2020). https://doi.org/10.1007/978-3-030-58598-3_1
59. Zhang, A., et al.: Relational attention network for crowd counting. In: Proceedings of International Conference on Computer Vision (ICCV), pp. 6788–6797 (2019)
60. Zhang, H., Kyaw, Z., Chang, S.F., Chua, T.S.: Visual translation embedding network for visual relation detection. In: Proceedings of Computer Vision and Pattern Recognition (CVPR), pp. 5532–5540 (2017)

61. Zhang, J., et al.: Uc-net: uncertainty inspired rgb-d saliency detection via conditional variational autoencoders. In: Proceedings of Computer Vision and Pattern Recognition (CVPR), pp. 8582–8591 (2020)
62. Zhang, Q., Chan, A.B.: Wide-area crowd counting via ground-plane density maps and multi-view fusion cnns. In: Proceedings of Computer Vision and Pattern Recognition (CVPR), pp. 8297–8306 (2019)
63. Zhang, S., Wu, G., Costeira, J.P., Moura, J.M.: Fcn-rlstm: deep spatio-temporal neural networks for vehicle counting in city cameras. In: Proceedings of International Conference on Computer Vision (ICCV), pp. 3667–3676 (2017)
64. Zhang, Y., Zhou, D., Chen, S., Gao, S., Ma, Y.: Single-image crowd counting via multi-column convolutional neural network. In: Proceedings of Computer Vision and Pattern Recognition (CVPR), pp. 589–597 (2016)
65. Zhao, M., Zhang, J., Zhang, C., Zhang, W.: Leveraging heterogeneous auxiliary tasks to assist crowd counting. In: Proceedings of Computer Vision and Pattern Recognition (CVPR), pp. 12736–12745 (2019)
66. Zhu, Z., Xu, M., Bai, S., Huang, T., Bai, X.: Asymmetric non-local neural networks for semantic segmentation. In: Proceedings of International Conference on Computer Vision (ICCV), pp. 593–602 (2019)

Domain Generalized RPPG Network: Disentangled Feature Learning with Domain Permutation and Domain Augmentation

Wei-Hao Chung, Cheng-Ju Hsieh, Sheng-Hung Liu, and Chiou-Ting Hsu(✉)

National Tsing Hua University, Hsinchu, Taiwan
cthsu@cs.nthu.edu.tw

Abstract. Remote photoplethysmography (rPPG) offers a contactless method for monitoring physiological signals from facial videos. Existing learning-based methods, although work effectively on intra-dataset scenarios, degrade severely on cross-dataset testing. In this paper, we address the cross-dataset testing as a domain generalization problem and propose a novel DG-rPPGNet to learn a domain generalized rPPG estimator. To this end, we develop a feature disentangled learning framework to disentangle rPPG, identity, and domain features from input facial videos. Next, we propose a domain permutation strategy to further constrain the disentangled rPPG features to be invariant to different domains. Finally, we design a novel adversarial domain augmentation strategy to enlarge the domain sphere of DG-rPPGNet. Our experimental results show that DG-rPPGNet outperforms other rPPG estimation methods in many cross-domain settings on UBFC-rPPG, PURE, COHFACE, and VIPL-HR datasets.

1 Introduction

Since the outbreak of new epidemics, remote estimation of human physiological states has attracted enormous attention. Remote Photoplethysmography (rPPG), which analyzes the blood volume changes in optical information of facial videos, is particularly useful in remote heart rate (HR) estimation. Earlier methods [1–7] usually adopted different prior assumptions to directly analyze the chromaticity of faces. Recent deep learning-based methods [8–18], through either multi-stage or end-to-end training networks, have achieved significant breakthroughs in rPPG estimation.

Although existing learning-based methods performed satisfactorily in intra-dataset testing, their cross-dataset testing performance tends to degrade severely. This cross-dataset or cross-domain issue is especially critical in rPPG estimation, because different rPPG datasets were recorded using their own equipment, under

Supplementary Information The online version contains supplementary material available at https://doi.org/10.1007/978-3-031-26284-5_3.

L. Wang et al. (Eds.): ACCV 2022, LNCS 13842, pp. 41–57, 2023.
https://doi.org/10.1007/978-3-031-26284-5_3

different environments or lighting conditions and thus exhibit a broad diversity. For example, the videos in UBFC-rPPG dataset [19] were recorded at 30 fps in a well-lighted environment; whereas the videos in COHFACE dataset [20] were recorded at 20 fps under two illumination conditions. The PURE dataset [21] even includes different motion settings when recording the videos. Therefore, if the training and testing data are from different datasets, the model trained in one dataset usually fails to generalize to another one.

We address the cross-dataset testing issue as a domain generalization (DG) problem and assume different "domains" refer to different characteristics (e.g., illumination conditions or photographic equipment) in the rPPG benchmarks. Domain generalization has been developed to facilitate the model to unseen domains at the inference time. Previous methods [22–31] have shown the effectiveness of DG on many classification tasks. However, many of these DG mechanisms, such as contrastive loss or triplet loss, are designed for classification problems and are inapplicable to the regression problem of rPPG estimation. Moreover, because rPPG signals are extremely vulnerable in comparison with general video content, any transformation across different domains (e.g., video-to-video translation, illumination modification, and noise perturbation) will substantially diminish the delicate rPPG signals.

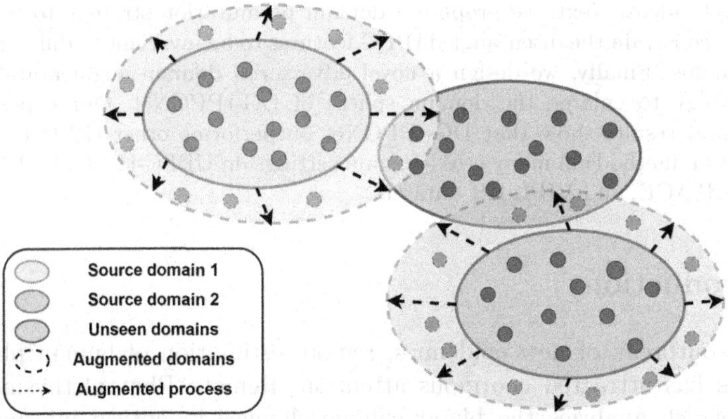

Fig. 1. Illustration of the proposed domain augmentation.

In this paper, we propose a novel Domain Generalized rPPG Network (DG-rPPGNet) via disentangled feature learning to address the domain generalization problem in rPPG estimation. Through the feature disentanglement framework, we first disentangle the rPPG, identity (ID), and domain features from the input data. Next, we develop two novel strategies, including domain permutation and domain augmentation, to cope with the disentangled feature learning. We devise a domain permutation strategy to ensure that the disentangled rPPG features are also invariant to different source domains. In addition, to generalize the model towards unseen domains, we propose a learnable domain augmentation strategy to enlarge the domain sphere during the model training. As illustrated in Fig. 1, the proposed domain augmentation aims to generate "adversarial domains" which maximally degrade the prediction accuracy of the rPPG estimator, to offer the model with information outside the source domain boundaries.

Our contributions are summarized below:

1) We propose a novel end-to-end training network DG-rPPGNet for rPPG estimation. To the best of our knowledge, this is the first work focusing on domain generalization issue in rPPG estimation.
2) We devise a disentangled feature learning framework, cooperated with domain permutation and domain augmentation, to significantly increase the generalization capability of rPPG estimation on unseen domains.
3) Experimental results on UBFC-rPPG, PURE, COHFACE, and VIPL-HR datasets show that the proposed DG-rPPGNet outperforms other rPPG estimation methods in cross-domain testing.

2 Related Work

2.1 Remote Photoplethysmography Estimation

Earlier methods [1–7] adopted different assumptions to design hand-crafted methods for rPPG estimations and usually do not generalize well to videos recorded in less-controlled scenarios. The learning-based methods [8–18], either through multi-stage processing or end-to-end training, benefit from the labeled data and largely improve the estimation performance over traditional methods. For example, in [13], a Dual-GAN framework is proposed to learn a noise-resistant mapping from the pre-processed spatial-temporal maps into the ground truth blood volume pulse (BVP) signals. In [17], a network is proposed to enhance highly compressed videos and to recover rPPG signals from the enhanced videos. In [9], a multi-task framework is developed to augment the rPPG dataset and to predict rPPG signals simultaneously. In [18], a video transformer is proposed to adaptively aggregate both local and global spatio-temporal features to enhance rPPG representation. Nevertheless, these learning-based methods mostly focus on improving intra-dataset performance but rarely concern the domain generalization issue.

2.2 Feature Disentanglement

Disentangled feature learning aims to separate the informative (or explanatory) variations from multifactorial data and has been extensively studied for learning task-specific feature representation in many computer vision tasks. For example, disentangled representation learning has been included in detecting face presentation attacks [32] or in unsupervised cross-domain adaptation [33]. In [34], a cross-verified feature disentangling strategy is proposed for remote physiological measurement. By disentangling physiological features from non-physiological features, the authors in [34] improved the robustness of physiological measurements from disentangled features.

2.3 Domain Generalization

Domain generalization aims to learn a representation or a model from multiple source domains and to generalize to unseen target domains. Because the target domains are unseen in the training stage, there is no way to match the source-to-target distributions or to minimize the cross-domain shift when developing the model. Therefore, most domain generalization methods either focus on learning domain-invariant representation or design different augmentation strategies to enlarge the source domains. For example, to address the DG issue in object detection, the authors in [26] designed a disentangled network to learn domain-invariant representation on both the image and instance levels. In [27], to tackle the semantic segmentation issue in real-world autonomous driving scenarios, the authors proposed using domain randomization and pyramid consistency to learn generalized representation. In [29], the authors proposed to augment perturbed images to enable the image classifier to generalize to unseen domains.

3 Proposed Method

3.1 Problem Statement and Overview of DG-rPPGNet

In the domain generalization setting, we are given a set of M source domains $\mathcal{S} = \{\mathcal{S}_1, ..., \mathcal{S}_M\}$ but have no access to the target domain $\mathcal{T}$ during the training stage. Let $\mathcal{S}_i = \{(x_j, s_j, y_j^{id}, y_j^{domain})\}_{j=1}^{N_i}$ denote the ith source domain, and x_j, s_j, y_j^{id}, and y_j^{domain} denote the facial video, the ground truth PPG signal, the subject ID label, and the domain label, respectively. Assuming that the unseen target domain $\mathcal{T}$ has very different distribution with the source domains, our goal is to learn a robust and well-generalized rPPG estimator to correctly predict the rPPG signals for the facial videos from any unseen target domain.

In this paper, we propose a novel DG-rPPGNet to tackle the domain generalization problem in rPPG estimation from three aspects. First, we develop a disentangled feature learning framework to disentangle rPPG-relevant features from other domain-dependent variations. Second, we devise a domain permutation strategy to ensure that the disentangled rPPG features are invariant to different source domains. Finally, we design a learnable domain augmentation strategy to augment the source domains to enable DG-rPPGNet to generalize to unseen domains.

Figure 2 illustrates the proposed DG-rPPGNet, which includes a global feature encoder F, three extractors (i.e., S_{rPPG}, S_{id}, and S_{domain}) for feature disentanglement, two decoders (i.e., $D_{feature}$ and D_{video}), two rPPG estimators (i.e., E_{rPPG}^{global} and E_{rPPG}^{disent}), one ID classifier C_{id}, and one domain classifier C_{domain}. All these components in DG-rPPGNet are jointly trained using the total loss defined in Sect. 3.5.

3.2 Disentangled Feature Learning

In this subsection, we describe the disentangled feature learning in DG-rPPGNet and the corresponding loss terms. For each input facial video x, we first obtain

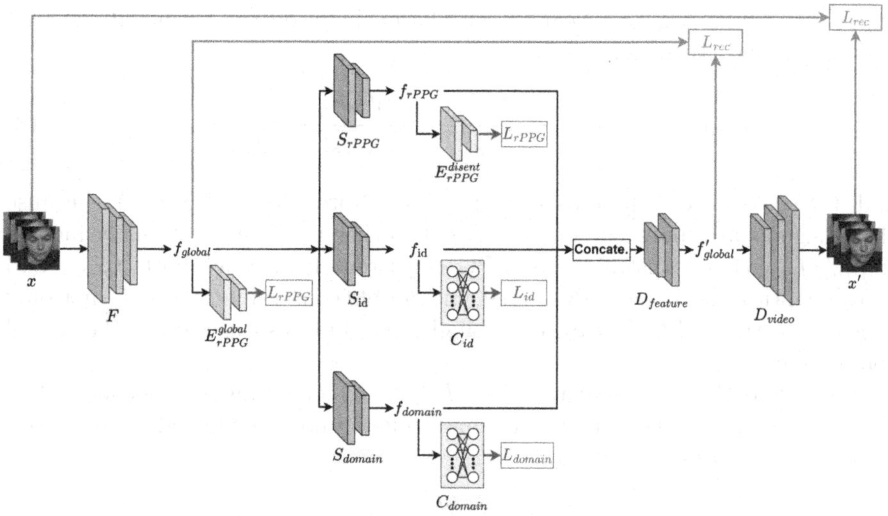

Fig. 2. The proposed DG-rPPGNet, consisting of a global feature encoder F, a rPPG extractor S_{rPPG}, an ID extractor S_{id}, a domain extractor S_{domain}, a feature decoder $D_{feature}$, a video decoder D_{video}, two rPPG estimators E_{rPPG}^{global} and E_{rPPG}^{disent}, an ID classifier C_{id}, and a domain classifier C_{domain}.

its global feature f_{global} using the global feature encoder F by,

$$f_{global} = F(x). \tag{1}$$

Next, we use three extractors S_{rPPG}, S_{id}, and S_{domain} to disentangle the rPPG feature f_{rPPG}, the ID feature f_{id}, and the domain feature f_{domain} from f_{global}, respectively:

$$f_{rPPG} = S_{rPPG}(f_{global}), \tag{2}$$

$$f_{id} = S_{id}(f_{global}), \text{ and} \tag{3}$$

$$f_{domain} = S_{domain}(f_{global}). \tag{4}$$

To constrain the disentangled feature learning in DG-rPPGNet, we define three prediction consistent losses for each of the features f_{rPPG}, f_{id}, and f_{domain}, and one reconstruction loss to jointly train the model.

The prediction consistent losses are defined by minimizing the prediction losses of the corresponding labels by,

$$L_{rPPG}^{disent} = L_{np}(E_{rPPG}^{global}(f_{global}), s) + L_{np}(E_{rPPG}^{disent}(f_{rPPG}), s), \tag{5}$$

$$L_{id}^{disent} = CE(C_{id}(f_{id}), y^{id}), \text{ and} \tag{6}$$

$$L_{domain}^{disent} = CE(C_{domain}(f_{domain}), y^{domain}), \tag{7}$$

where L_{np} is the negative Pearson correlation between the predicted signal s' and the ground truth signal s:

$$L_{np}(s', s) = 1 - \frac{(s - \bar{s})^t (s' - \bar{s'})}{\sqrt{(s - \bar{s})^t (s - \bar{s})} \sqrt{(s' - \bar{s'})^t (s' - \bar{s'})}}, \tag{8}$$

and $CE(\cdot)$ denotes the cross-entropy loss. Note that, in Eq. (5), because rPPG signals are more vulnerable than the other two, we additionally include $E^{global}_{rPPG}(f_{global})$ to constrain the rPPG consistent loss. Since both f_{global} and f_{rPPG} capture the same rPPG signal, the double constraints in Eq. (5) not only consolidate the rPPG feature disentanglement but also accelerate the model convergence.

We define the reconstruction loss L^{disent}_{rec} by enforcing the decoder $D = D_{video} \circ D_{feature}$ to reconstruct the input video in both the global feature space f_{global} and the color space x by,

$$L^{disent}_{rec} = ||f_{global} - f'_{global}||_1 + ||x - x'||_1, \tag{9}$$

where

$$f'_{global} = D_{feature}(f_{rPPG}, f_{id}, f_{domain}), \text{ and} \tag{10}$$
$$x' = D_{video}(f'_{global}). \tag{11}$$

3.3 Domain Permutation for Domain-Invariant Feature Learning

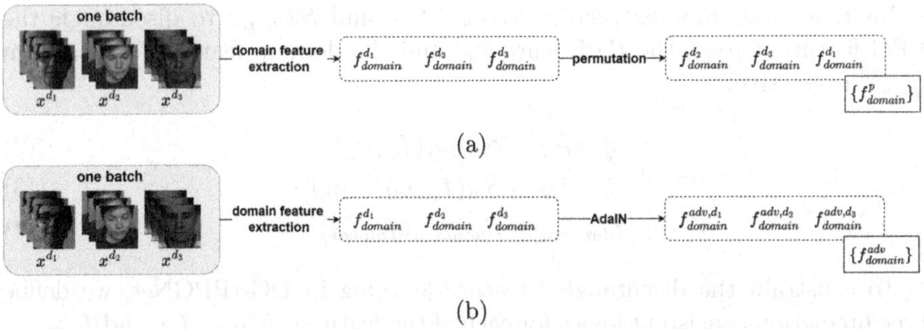

Fig. 3. Illustrations of the (a) domain permutation; and (b) domain augmentation.

In Eqs. (5) (6) and (7), although we constrain DG-rPPGNet to extract f_{rPPG}, f_{id}, and f_{domain} from f_{global}, there is no guarantee that these features are successfully disentangled. Specifically, our major concern is that the extracted rPPG feature f_{rPPG} should capture not only rPPG-relevant information but also be invariant to different domains. In other words, we expect that f_{rPPG} should contain little or no domain-dependent variations. Therefore, in this subsection, we devise a novel domain permutation strategy to consolidate the feature disentanglement and further encourage S_{rPPG} to focus on extracting domain-invariant rPPG features.

Given one batch of input facial videos, we first extract their global features by Eq. (1) and then extract the rPPG features, ID features, and domain features by Eqs. (2) (3) and (4). Next, we randomly permute the locations of domain features $\{f_{domain}\}$ within this batch and have the permuted domain features $\{f^p_{domain}\}$ by,

$$\{f^p_{domain}\} = Permute(\{f_{domain}\}), \tag{12}$$

where $Permute(\cdot)$ is a random permutation operation.

An example is given in Fig. 3(a), where the input batch consists of three videos x^{d_1}, x^{d_2}, and x^{d_3} sampled from different domains and their original domain features are $f^{d_1}_{domain}$, $f^{d_2}_{domain}$, and $f^{d_3}_{domain}$, respectively. After random permutation, the three videos have their new domain features as $f^{d_2}_{domain}$, $f^{d_3}_{domain}$, and $f^{d_1}_{domain}$, respectively. Our rationale is that, if the disentangled rPPG feature f_{rPPG} and ID feature f_{id} are indeed invariant to different domains, then the global feature f^p_{global} reconstructed using the permuted domain features, i.e.,

$$f^p_{global} = D_{feature}(f_{rPPG}, f_{id}, f^p_{domain}), \tag{13}$$

should carry the same rPPG feature and ID feature as the original one f_{global}.

Next, we reconstruct a video by decoding f^p_{global} and then encode this reconstructed video to obtain its global feature $f^{'p}_{global}$ by,

$$f^{'p}_{global} = F(D_{video}(f^p_{global})). \tag{14}$$

Finally, we extract the three features f^p_{rPPG}, f^p_{id}, and $f^{'p}_{domain}$ from $f^{'p}_{global}$ by,

$$f^p_{rPPG} = S_{rPPG}(f^{'p}_{global}), \tag{15}$$

$$f^p_{id} = S_{id}(f^{'p}_{global}), \text{ and} \tag{16}$$

$$f^{'p}_{domain} = S_{domain}(f^{'p}_{global}). \tag{17}$$

Similar to Eqs. (5) (6) and (7), we define the prediction consistent losses to constrain f^p_{rPPG}, f^p_{id}, and $f^{'p}_{domain}$ by,

$$\begin{aligned} L^p_{rPPG} &= L_{np}(E^{global}_{rPPG}(f^{'p}_{global}), s) \\ &+ L_{np}(E^{disent}_{rPPG}(f^p_{rPPG}), s), \end{aligned} \tag{18}$$

$$L^p_{id} = CE(C_{id}(f^p_{id}), y^{id}), \text{ and} \tag{19}$$

$$L^p_{domain} = CE(C_{domain}(f^{'p}_{domain}), Permute(y^{domain})), \tag{20}$$

where $Permute(y^{domain})$ is the ground truth domain label of f'^{p}_{domain}.

We also define a reconstruction loss L^{p}_{rec} to constrain that (1) the rPPG and ID features should remain unchanged, before and after the domain permutation; and (2) the permuted domain features and the global features should remain the same after the decoding and re-encoding steps. We thus formulate the reconstruction loss L^{p}_{rec} in the feature spaces by,

$$L^{p}_{rec} = ||f_{rPPG} - f'^{p}_{rPPG}||_1 + ||f_{id} - f'^{p}_{id}||_1 + ||f^{p}_{domain} - f'^{p}_{domain}||_1 + ||f^{p}_{global} - f'^{p}_{global}||_1. \tag{21}$$

3.4 Domain Augmentation via AdaIN

In Sect. 3.2 and Sect. 3.3, the disentangled feature learning and domain permutation involve only the set of source domains S in the model training but are oblivious to any external domains. Because data augmentation is widely adopted to alleviate the data shortage, we adopt the idea and design a domain augmentation strategy to enlarge the sphere of source domains.

Unlike most generic augmentation methods, the proposed domain augmentation has two specific goals. First, the augmented domains should well preserve discriminative information in the original source domains S; and second, they should offer diverse characteristics different from those in S so as to simulate the unseen domains. To balance the two competing goals, we propose (1) using AdaIN [35] to generate the augmented domains by transforming the style of S without changing their discriminative content, and then (2) enforcing the augmented domains to act as the adversaries of S so as to expand the sphere of S. In the second part, we adopt the idea of adversary examples [36] and define "adversarial domains" as the domains generated to mislead the rPPG estimators and to offer unseen information to the model.

We formulate the proposed data augmentation as an adversary domain learning problem in the parameter space of AdaIN. Given one batch of input facial videos, we extract their global features by Eq. (1) and then extract the rPPG features f_{rPPG}, ID features f_{id}, and domain features f_{domain} by Eqs. (2) (3) and (4). respectively. Next, we use AdaIN [35] to transform the domain feature from f_{domain} to f^{adv}_{domain} by,

$$f^{adv}_{domain} = AdaIN(f_{domain}, \alpha, \beta), \tag{22}$$

where

$$AdaIN(f, \alpha, \beta) = \alpha \cdot \frac{f - \mu_f}{\sigma_f} + \beta, \tag{23}$$

α and β are two learnable parameters; μ_f and σ_f are the mean and standard deviation of the feature map f, respectively. An example is shown in Fig. 3(b), where the domain features $f^{d_i}_{domain} (i = 1, 2, 3)$ of x^{d_i} are transformed by AdaIN into f^{adv,d_i}_{domain}.

To ensure f^{adv}_{domain} behaves like adversarial domains, we constrain the two parameters α and β by "maximizing" the prediction losses L_{np} of the two rPPG estimators E^{global}_{rPPG} and E^{disent}_{rPPG} by,

$$(\alpha, \beta) = \underset{\alpha,\beta}{argmax}\ L_{np}(E^{global}_{rPPG}(f^{'adv}_{global}), s) + L_{np}(E^{disent}_{rPPG}(f^{adv}_{rPPG}), s), \tag{24}$$

where $f^{'adv}_{global}$ and f^{adv}_{rPPG} are the re-encoded global feature and the extracted rPPG feature of $f^{'adv}_{global}$ obtained by:

$$f^{'adv}_{global} = F(D_{video}(f^{adv}_{global})), \text{ and} \tag{25}$$

$$f^{adv}_{rPPG} = S_{rPPG}(f^{'adv}_{global}), \tag{26}$$

and $f^{'adv}_{global}$ is re-encoded from the reconstructed global feature f^{adv}_{global} by,

$$f^{adv}_{global} = D_{feature}(f_{rPPG}, f_{id}, f^{adv}_{domain}). \tag{27}$$

To facilitate the implementation of Eq. (24), we use the gradient reversal layer (GRL) [37] to flip the gradients (1) between F and E^{global}_{rPPG}, and (2) between S_{rPPG} and E^{disent}_{rPPG}. Hence, we reformulate Eq. (24) by,

$$(\alpha, \beta) = \underset{\alpha,\beta}{argmin}\ L_{np}(E^{global}_{rPPG}(GRL(f^{'adv}_{global})), s) + L_{np}(E^{disent}_{rPPG}(GRL(f^{adv}_{rPPG})), s). \tag{28}$$

Finally, we extract the other two features f^{adv}_{id} and f^{adv}_{domain} from $f^{'adv}_{global}$ by,

$$f^{adv}_{id} = S_{id}(f^{'adv}_{global}), \text{ and} \tag{29}$$

$$f^{'adv}_{domain} = S_{domain}(f^{'adv}_{global}). \tag{30}$$

Similar to Eqs. (18) (19) and (21), we again impose the prediction consistent losses L^{adv}_{rPPG}, L^{adv}_{id} and the reconstruction loss L^{adv}_{rec} to constrain the model learning by,

$$L^{adv}_{rPPG} = L_{np}(E^{global}_{rPPG}(f^{'adv}_{global}), s) + L_{np}(E^{disent}_{rPPG}(f^{adv}_{rPPG}), s), \tag{31}$$

$$L^{adv}_{id} = CE(C_{id}(f^{adv}_{id}), y^{id}), \text{ and} \tag{32}$$

$$L^{adv}_{rec} = ||f_{rPPG} - f^{adv}_{rPPG}||_1 + ||f_{id} - f^{adv}_{id}||_1 + ||f^{adv}_{domain} - f^{'adv}_{domain}||_1 + ||f^{adv}_{global} - f^{'adv}_{global}||_1. \tag{33}$$

Here, we do not include the domain prediction consistent loss, because there exist no ground truth labels for the adversarial domains.

3.5 Loss Function

Finally, we include the feature disentanglement loss L_{total}^{disent}, the domain permutation loss L_{total}^{p}, and the domain augmentation loss L_{total}^{adv} to define the total loss L_{total}:

$$L_{total} = L_{total}^{disent} + L_{total}^{p} + L_{total}^{adv}, \tag{34}$$

where

$$L_{total}^{disent} = \lambda_1 L_{rPPG}^{disent} + \lambda_2 L_{id}^{disent} + \lambda_3 L_{domain}^{disent} + L_{rec}^{disent}, \tag{35}$$

$$L_{total}^{p} = \lambda_1 L_{rPPG}^{p} + \lambda_2 L_{id}^{p} + \lambda_3 L_{domain}^{p} + L_{rec}^{p}, \text{ and} \tag{36}$$

$$L_{total}^{adv} = \lambda_1 L_{rPPG}^{adv} + \lambda_2 L_{id}^{adv} + L_{rec}^{adv}, \tag{37}$$

and λ_1, λ_2, and λ_3 are hyper-parameters and are empirically set as 0.01 in all our experiments.

3.6 Inference Stage

In the inference stage, we include only the global feature encoder F and the global rPPG estimator E_{rPPG}^{global} to predict the rPPG signals of the test facial videos. We do not include the rPPG extractor S_{rPPG} and the rPPG estimator E_{rPPG}^{disent} during the inference stage, because they are trained only on source domains and may not well disentangle the features on unseen domains.

4 Experiments

4.1 Datasets and Cross-Domain Setting

The UBFC-rPPG dataset [19] consists of 42 RGB videos from 42 subjects; i.e., each subject contributes one single video. The videos were recorded by Logitech C920 HD Pro at 30 fps with resolution of 640 × 480 pixels in uncompressed 8-bit format. The PPG signals and corresponding heart rates were collected by CMS50E transmissive pulse oximeter. We follow the setting in [13] to split the dataset into the training and testing sets with videos from 30 and 12 subjects, respectively.

The PURE dataset [21] consists of 60 RGB videos from 10 subjects. Each subject performs 6 different activities, including (1) sitting still and looking directly at the camera, (2) talking, (3) slowly moving the head parallel to the camera, (4) quickly moving the head, (5) rotating the head with 20° angles, and (6) rotating the head with 35° angles. All videos were recorded using an eco274CVGE camera at 30 fps and with resolution of 640 × 480 pixels. The PPG signals were captured by using Pulox CMS50E finger clip pulse oximeter with sampling rate 60 Hz. To align with the videos, the PPG signals are reduced 30 Hz. We follow the setting in [9] to split the dataset into the training and testing sets with videos from 7 and 3 subjects, respectively.

The COHFACE dataset [20] consists of 160 one-minute-long sequence RGB videos from 40 subjects. The videos were recorded under two illumination conditions, including (1) a well-lighted environment, and (2) a natural light environment. All videos were recorded using Logitech HD C525 at 20 fps with resolution of 640×480 pixels. The PPG signals were taken by a contact blood volume pulse sensor model SA9308M. We follow the setting in [9] to split the dataset into the training and testing sets with videos from 24 and 16 subjects, respectively.

The VIPL-HR dataset [38] contains 2378 RGB videos from 107 subjects. The dataset was recorded using 3 different devices under 9 scenarios. We follow the setting in [15] and use a subject-exclusive 5-fold cross-validation protocol on VIPL-HR. In addition, because the facial videos and PPG signals have different sampling rates, we resample the PPG signals to match the corresponding video frames by linear interpolation.

Cross-Domain Setting. When experimenting on these datasets, we consider each dataset refers to one domain, except COHFACE, which is considered as two domains and each one refers to either the well-lighted or natural light settings. In Sect. 4.4 and 4.5, we adopt two experimental settings by randomly choosing two datasets from UBFC-rPPG, PURE, and COHFACE to form the set of source domains and then testing on (1) the remaining one (that is, we have three cross-domain settings: "P+C→U", "U+C→P", and "U+P→C") in Sect. 4.4 and 4.5; and (2) the VIPL-HR dataset in Sect. 4.5.

4.2 Implementation Details

The architectures of the global feature encoder F, the extractor S, the decoder D, the rPPG estimator E, and the classifier C in DG-rPPGNet are given in the supplementary file. We train DG-rPPGNet in two stages. We first train the disentangled feature learning model with domain permutation for 300 epochs and then fine-tune the model with domain augmentation for 100 epochs. We train the model using Nvidia RTX 2080 and RTX 3080 with one sample from each domain in one batch, and use Adam optimizer with the learning rate of 0.0002. For all the facial videos, we use [39] to detect face landmarks, crop the coarse face area and resize the cropped areas into 80×80 pixels. In each epoch, we randomly sample 60 consecutive frames from the training videos of each domain to train DG-rPPGNet.

4.3 Evaluation Metrics

To assess the performance of DG-rPPGNet on rPPG estimation, we follow [9] to derive heart rate (HR) from the predicted rPPG signals and then evaluate the results in terms of the following metrics: (1) Mean absolute error (MAE), (2) Root mean square error (RMSE), and (3) Pearson correlation coefficient (R).

4.4 Ablation Study

We conduct ablation studies on three cross-domain settings, including "P+C→U", "U+C→P", and "U+P→C". In Table 1, L_{total}^{disent}, L_{total}^{p}, and L_{total}^{adv} indicate that we include the corresponding losses as defined in Eqs. (35) (36) and (37), respectively, to train DG-rPPGNet.

We first evaluate the effectiveness of domain permutation. When including $L_{total}^{disent} + L_{total}^{p}$ in the model training, we significantly improve the performance over using L_{total}^{disent} alone by reducing MAE about 85% and RMSE about 86% in "P+C→U". However, because the proposed domain permutation focuses on learning domain-invariant features within the source domains, the model still lacks the ability to generalize to unseen domains. Nevertheless, although we achieve no improvement in "U+C→P" and "U+P→C" with $L_{total}^{disent} + L_{total}^{p}$, we see that the setting $L_{total}^{disent} + L_{total}^{p} + L_{total}^{adv}$ significantly outperforms $L_{total}^{disent} + L_{total}^{adv}$ when we further include domain augmentation in DG-rPPGNet. These results show that the proposed domain permutation works cooperatively with domain augmentation to support the model to learn domain-invariant and rPPG-discriminative features in the augmented domains. Finally, when including L_{total}^{adv} in DG-rPPGNet, we see significant performance improvement with reduced MAE (about 81%, 28%, and 3%; and about 92%, 50% and 42%) and RMSE (about 80%, 29%, and 6%; and about 89%, 54% and 41%), without and with L_{total}^{p}, respectively. These results verify that the proposed domain augmentation substantially enlarges the domain sphere and enables the model to generalize to unseen domains.

Table 1. Ablation study

Loss terms			P+C→U			U+C→P			U+P→C		
L_{total}^{disent}	L_{total}^{p}	L_{total}^{adv}	MAE↓	RMSE↓	R↑	MAE↓	RMSE↓	R↑	MAE↓	RMSE↓	R↑
✓			7.74	12.34	0.40	6.14	10.26	0.56	12.54	15.30	0.07
✓	✓		1.17	1.71	0.83	6.53	11.96	0.47	14.33	16.93	0.08
✓		✓	1.46	2.46	0.78	4.36	7.19	0.70	12.18	14.35	**0.39**
✓	✓	✓	**0.63**	**1.35**	**0.88**	**3.02**	**4.69**	**0.88**	**7.19**	**8.99**	0.30

In Table 2, we evaluate the proposed domain augmentation by comparing with random domain augmentation. We simulate the random domain augmentation by replacing the parameters α and β in Eq. (28) with values randomly sampled from standard Gaussian distribution. The results in Table 2 show that the proposed method outperforms the random augmentation with reduced MAE (by about 38%, 50% and 33%) and RMSE (by about 18%, 56% and 29%) and verify its effectiveness on domain generalization.

Table 2. Evaluation of domain augmentation

Augmentation	P+C→U			U+C→P			U+P→C		
	MAE↓	RMSE↓	$R\uparrow$	MAE↓	RMSE↓	$R\uparrow$	MAE↓	RMSE↓	$R\uparrow$
Random	1.02	1.65	0.82	6.08	10.54	0.56	10.80	12.71	0.25
Adversarial	**0.63**	**1.35**	**0.88**	**3.02**	**4.69**	**0.88**	**7.19**	**8.99**	**0.30**

4.5 Results and Comparison

We compare our results on the three settings: "P+C→U", "U+C→P", and "U+P→C", with previous methods [1–5,9,13] in Tables 3, 4, and 5, respectively. Note that, the methods [1–5] (marked with †) are not learning-based methods and thus have neither training data nor cross-domain issue. The other learning-based methods (marked with $*$) all adopt different cross-domain settings from ours. Although there exist no results reported under the same cross-domain settings as ours for a fair comparison, we include their results here to assess the relative testing performance on these rPPG datasets.

Table 3. Cross-domain test on "P+C→U"

Method	MAE↓	RMSE↓
GREEN† [2]	8.29	15.82
ICA† [3]	4.39	11.60
POS† [4]	3.52	8.38
CHROM† [1]	3.10	6.84
Multi-task* [9]	1.06	2.70
Dual-GAN* [13]	0.74	**1.02**
DG-rPPGNet	**0.63**	1.35

Table 4. Cross-domain test on "U+C→P"

Method	MAE↓	RMSE↓
LiCVPR† [5]	28.22	30.96
POS† [4]	22.25	30.20
ICA† [3]	15.23	21.25
GREE† [2]	9.03	13.92
CHROM† [1]	3.82	6.8
Multi-task* [9]	4.24	6.44
DG-rPPGNet	**3.02**	**4.69**

In Table 3, we show our results on "P+C→U" and compare with previous methods with testing results on UBFC-rPPG. We cite the performance of the four non-learning-based methods, i.e., GREEN [2], ICA [3], POS [4], and CHROM [1], from [8]. The two methods, Multi-task [9] and Dual-GAN [13], are trained on PURE dataset; and their cross-domain setting is considered as "P→U". Table 3 shows that DG-rPPGNet achieves the best performance with MAE 0.63 even without involving UBFC-rPPG in the training stage.

In Table 4, we show our results on "U+C→P" and compare with previous methods with testing results on PURE. We cite the result of LiCVPR [5] from [15], and use an open source toolbox [40] to obtain the results of GREEN [2], POS [4], and ICA [3] on PURE. The model of Multi-task [9] is trained on UBFC-rPPG alone, i.e., its cross-domain setting here is "U→P". We again show that DG-rPPGNet outperforms the other methods with MAE 3.02 and RMSE 4.69.

In Table 5, we show our results on "U+P→C" and compare with previous methods with testing results on COHFACE. Again, we use the open source toolbox [40] to obtain the results of POS [4], ICA [3], and GREEN [2] on COHFACE. The results once again show that DG-rPPGNet outperforms the other methods with MAE 7.19 and RMSE 8.99.

Table 5. Cross-domain test on "U+P→C"

Method	MAE↓	RMSE↓
POS† [4]	19.86	24.57
LiCVPR† [5]	19.98	25.59
ICA† [3]	14.27	19.28
GREEN† [2]	10.94	16.72
CHROM† [1]	7.8	12.45
DG-rPPGNet	**7.19**	**8.99**

Table 6. Cross-domain test on VIPL-HR

Method	MAE↓	RMSE↓
Averaged GT	22.21	26.70
DG-rPPGNet (U+P)	18.38	18.86
DG-rPPGNet (U+C)	18.23	18.81
DG-rPPGNet (P+C)	15.95	17.47

Finally, in Table 6, we evaluate the generalization capability of the proposed DG-rPPGNet. We use different combinations of the three small-scale datasets PURE, COHFACE, and UBFC-rPPG as the source domains and then test on the large-scale dataset VIPL-HR [38]. Because there exist no similar experimental results, we show the averaged ground truth signals (marked by "Averaged GT") of VIPL-HR as the baseline results for comparison. The results show that, even trained on small-scale datasets, the proposed DG-rPPGNet substantially exceeds the baseline results and reduces MAE (by about 17%, 18%, and 28%) and RMSE (by about 29%, 30%, and 35%).

5 Conclusion

In this paper, we propose a DG-rPPGNet to address the domain generalization issue in rPPG estimation. The proposed DG-rPPGNet includes (1) a feature disentangled learning framework to extract rPPG, ID, and domain features from facial videos; (2) a novel domain permutation strategy to constrain the domain invariant property of rPPG features; and (3) an adversarial domain augmentation strategy to increase the domain generalization capability. Experimental results on UBFC-rPPG, PURE, COHFACE, and VIPL-HR datasets show that the proposed DG-rPPGNet outperforms other rPPG estimation methods in many cross-domain testings.

References

1. De Haan, G., Jeanne, V.: Robust pulse rate from chrominance-based rppg. IEEE Trans. Biomed. Eng. **60**, 2878–2886 (2013)

2. Verkruysse, W., Svaasand, L.O., Nelson, J.S.: Remote plethysmographic imaging using ambient light. Opt. Express **16**, 21434–21445 (2008)
3. Poh, M.Z., McDuff, D.J., Picard, R.W.: Non-contact, automated cardiac pulse measurements using video imaging and blind source separation. Opt. Express **18**, 10762–10774 (2010)
4. Wang, W., Den Brinker, A.C., Stuijk, S., De Haan, G.: Algorithmic principles of remote PPG. IEEE Trans. Biomed. Eng. **64**, 1479–1491 (2016)
5. Li, X., Chen, J., Zhao, G., Pietikainen, M.: Remote heart rate measurement from face videos under realistic situations. In: Proceedings of the IEEE Conference on Computer Vision and Pattern Recognition, pp. 4264–4271 (2014)
6. Wang, W., Stuijk, S., De Haan, G.: A novel algorithm for remote photoplethysmography: spatial subspace rotation. IEEE Trans. Biomed. Eng. **63**, 1974–1984 (2015)
7. Poh, M.Z., McDuff, D.J., Picard, R.W.: Advancements in noncontact, multiparameter physiological measurements using a webcam. IEEE Trans. Biomed. Eng. **58**, 7–11 (2010)
8. Song, R., Chen, H., Cheng, J., Li, C., Liu, Y., Chen, X.: Pulsegan: learning to generate realistic pulse waveforms in remote photoplethysmography. IEEE J. Biomed. Health Inf. **25**, 1373–1384 (2021)
9. Tsou, Y.Y., Lee, Y.A., Hsu, C.T.: Multi-task learning for simultaneous video generation and remote photoplethysmography estimation. In: Proceedings of the Asian Conference on Computer Vision (2020)
10. Bousefsaf, F., Pruski, A., Maaoui, C.: 3D convolutional neural networks for remote pulse rate measurement and mapping from facial video. Appl. Sci. **9**, 4364 (2019)
11. Chen, W., McDuff, D.: Deepphys: video-based physiological measurement using convolutional attention networks. In: Proceedings of the European Conference on Computer Vision (ECCV), pp. 349–365 (2018)
12. Lee, E., Chen, E., Lee, C.-Y.: Meta-rPPG: remote heart rate estimation using a transductive meta-learner. In: Vedaldi, A., Bischof, H., Brox, T., Frahm, J.-M. (eds.) ECCV 2020. LNCS, vol. 12372, pp. 392–409. Springer, Cham (2020). https://doi.org/10.1007/978-3-030-58583-9_24
13. Lu, H., Han, H., Zhou, S.K.: Dual-gan: joint bvp and noise modeling for remote physiological measurement. In: Proceedings of the IEEE/CVF Conference on Computer Vision and Pattern Recognition, pp. 12404–12413 (2021)
14. Niu, X., Han, H., Shan, S., Chen, X.: Synrhythm: learning a deep heart rate estimator from general to specific. In: 2018 24th International Conference on Pattern Recognition (ICPR), pp. 3580–3585. IEEE (2018)
15. Špetlík, R., Franc, V., Matas, J.: Visual heart rate estimation with convolutional neural network. In: Proceedings of the British Machine Vision Conference, Newcastle, UK, pp. 3–6 (2018)
16. Tsou, Y.Y., Lee, Y.A., Hsu, C.T., Chang, S.H.: Siamese-rppg network: Remote photoplethysmography signal estimation from face videos. In: Proceedings of the 35th Annual ACM Symposium on Applied Computing, pp. 2066–2073 (2020)
17. Yu, Z., Peng, W., Li, X., Hong, X., Zhao, G.: Remote heart rate measurement from highly compressed facial videos: an end-to-end deep learning solution with video enhancement. In: Proceedings of the IEEE/CVF International Conference on Computer Vision, pp. 151–160 (2019)
18. Yu, Z., Shen, Y., Shi, J., Zhao, H., Torr, P.H., Zhao, G.: Physformer: facial video-based physiological measurement with temporal difference transformer. In: Proceedings of the IEEE/CVF Conference on Computer Vision and Pattern Recognition, pp. 4186–4196 (2022)

19. Bobbia, S., Macwan, R., Benezeth, Y., Mansouri, A., Dubois, J.: Unsupervised skin tissue segmentation for remote photoplethysmography. Pattern Recogn. Lett. **124**, 82–90 (2019)
20. Heusch, G., Anjos, A., Marcel, S.: A reproducible study on remote heart rate measurement. arXiv preprint arXiv:1709.00962 (2017)
21. Stricker, R., Müller, S., Gross, H.M.: Non-contact video-based pulse rate measurement on a mobile service robot. In: The 23rd IEEE International Symposium on Robot and Human Interactive Communication, pp. 1056–1062. IEEE (2014)
22. Balaji, Y., Sankaranarayanan, S., Chellappa, R.: MetaReg: towards domain generalization using meta-regularization. Adv. Neural Inf. Process. Syst. **31**, 1–11 (2018)
23. Li, Y., Yang, Y., Zhou, W., Hospedales, T.: Feature-critic networks for heterogeneous domain generalization. In: International Conference on Machine Learning, pp. 3915–3924. PMLR (2019)
24. Wang, Z., Luo, Y., Qiu, R., Huang, Z., Baktashmotlagh, M.: Learning to diversify for single domain generalization. In: Proceedings of the IEEE/CVF International Conference on Computer Vision, pp. 834–843 (2021)
25. Li, D., Yang, Y., Song, Y.Z., Hospedales, T.M.: Deeper, broader and artier domain generalization. In: Proceedings of the IEEE International Conference on Computer Vision, pp. 5542–5550 (2017)
26. Lin, C., Yuan, Z., Zhao, S., Sun, P., Wang, C., Cai, J.: Domain-invariant disentangled network for generalizable object detection. In: Proceedings of the IEEE/CVF International Conference on Computer Vision, pp. 8771–8780 (2021)
27. Yue, X., Zhang, Y., Zhao, S., Sangiovanni-Vincentelli, A., Keutzer, K., Gong, B.: Domain randomization and pyramid consistency: simulation-to-real generalization without accessing target domain data. In: Proceedings of the IEEE/CVF International Conference on Computer Vision, pp. 2100–2110 (2019)
28. Li, L., et al.: Progressive domain expansion network for single domain generalization. In: Proceedings of the IEEE/CVF Conference on Computer Vision and Pattern Recognition, pp. 224–233 (2021)
29. Zhou, K., Yang, Y., Hospedales, T., Xiang, T.: Deep domain-adversarial image generation for domain generalisation. In: Proceedings of the AAAI Conference on Artificial Intelligence, vol. 34, pp. 13025–13032 (2020)
30. Shankar, S., Piratla, V., Chakrabarti, S., Chaudhuri, S., Jyothi, P., Sarawagi, S.: Generalizing across domains via cross-gradient training. arXiv preprint arXiv:1804.10745 (2018)
31. Kim, D., Yoo, Y., Park, S., Kim, J., Lee, J.: Selfreg: self-supervised contrastive regularization for domain generalization. In: Proceedings of the IEEE/CVF International Conference on Computer Vision, pp. 9619–9628 (2021)
32. Wang, G., Han, H., Shan, S., Chen, X.: Cross-domain face presentation attack detection via multi-domain disentangled representation learning. In: Proceedings of the IEEE/CVF Conference on Computer Vision and Pattern Recognition, pp. 6678–6687 (2020)
33. Lee, S., Cho, S., Im, S.: Dranet: disentangling representation and adaptation networks for unsupervised cross-domain adaptation. In: Proceedings of the IEEE/CVF Conference on Computer Vision and Pattern Recognition, pp. 15252–15261 (2021)
34. Niu, X., Yu, Z., Han, H., Li, X., Shan, S., Zhao, G.: Video-based remote physiological measurement via cross-verified feature disentangling. In: Vedaldi, A., Bischof, H., Brox, T., Frahm, J.-M. (eds.) ECCV 2020. LNCS, vol. 12347, pp. 295–310. Springer, Cham (2020). https://doi.org/10.1007/978-3-030-58536-5_18

35. Karras, T., Laine, S., Aila, T.: A style-based generator architecture for generative adversarial networks. In: Proceedings of the IEEE/CVF Conference on Computer Vision and Pattern Recognition, pp. 4401–4410 (2019)
36. Szegedy, C., et al.: Intriguing properties of neural networks. arXiv preprint arXiv:1312.6199 (2013)
37. Ganin, Y., Lempitsky, V.: Unsupervised domain adaptation by backpropagation. In: International Conference on Machine Learning, pp. 1180–1189. PMLR (2015)
38. Niu, X., Han, H., Shan, S., Chen, X.: VIPL-HR: a multi-modal database for pulse estimation from less-constrained face video. In: Jawahar, C.V., Li, H., Mori, G., Schindler, K. (eds.) ACCV 2018. LNCS, vol. 11365, pp. 562–576. Springer, Cham (2019). https://doi.org/10.1007/978-3-030-20873-8_36
39. Bulat, A., Tzimiropoulos, G.: How far are we from solving the 2d & 3d face alignment problem? (and a dataset of 230,000 3d facial landmarks). In: International Conference on Computer Vision (2017)
40. McDuff, D., Blackford, E.: IPHYS: an open non-contact imaging-based physiological measurement toolbox. In: 41st Annual International Conference of the IEEE Engineering in Medicine and Biology Society (EMBC), pp. 6521–6524. IEEE (2019)

Complex Handwriting Trajectory Recovery: Evaluation Metrics and Algorithm

Zhounan Chen[1], Daihui Yang[1], Jinglin Liang[1], Xinwu Liu[4], Yuyi Wang[3,4], Zhenghua Peng[1], and Shuangping Huang[1,2](✉)

[1] South China University of Technology, Guangzhou, China
eehsp@scut.edu.cn
[2] Pazhou Laboratory, Guangzhou, China
[3] Swiss Federal Institute of Technology, Zurich, Switzerland
[4] CRRC Institute, Zhuzhou, China

Abstract. Many important tasks such as forensic signature verification, calligraphy synthesis, etc., rely on handwriting trajectory recovery of which, however, even an appropriate evaluation metric is still missing. Indeed, existing metrics only focus on the writing orders but overlook the fidelity of glyphs. Taking both facets into account, we come up with two new metrics, the adaptive intersection on union (AIoU) which eliminates the influence of various stroke widths, and the length-independent dynamic time warping (LDTW) which solves the trajectory-point alignment problem. After that, we then propose a novel handwriting trajectory recovery model named Parsing-and-tracing ENcoder-decoder Network (PEN-Net), in particular for characters with both complex glyph and long trajectory, which was believed very challenging. In the PEN-Net, a carefully designed double-stream parsing encoder parses the glyph structure, and a global tracing decoder overcomes the memory difficulty of long trajectory prediction. Our experiments demonstrate that the two new metrics AIoU and LDTW together can truly assess the quality of handwriting trajectory recovery and the proposed PEN-Net exhibits satisfactory performance in various complex-glyph languages including Chinese, Japanese and Indic. The source code is available at https://github.com/ChenZhounan/PEN-Net.

Keywords: Trajectory recovery · Handwriting · Evaluation metrics

1 Introduction

Trajectory recovery from static handwriting images reveals the natural writing order while ensuring the glyph fidelity. There are a lot of applications including forensic signature verification [17,21], calligraphy synthesis and imitation [33,34],

---

Supplementary Information The online version contains supplementary material available at https://doi.org/10.1007/978-3-031-26284-5_4.

L. Wang et al. (Eds.): ACCV 2022, LNCS 13842, pp. 58–74, 2023.
https://doi.org/10.1007/978-3-031-26284-5_4

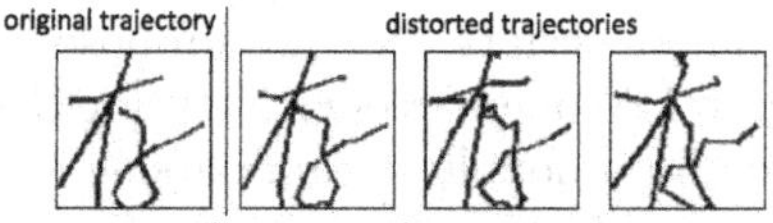

RMSE↓	2.33	2.33(+0%)	2.33(+0%)
DTW↓	80.1	84.0(+4.9%)	84.0(+4.9%)
AIoU↑	0.43	0.31(**-28%**)	0.28(**-35%**)

Fig. 1. Evaluation scores of distorted trajectories. Three distorted trajectories are obtained by moving half of the points a fixed distance from the original trajectory. The first distorted trajectory shows small glyph structure distortion and its evaluation scores are 2.33, 80.1 and 0.43. The moving angles of points in the other two distorted trajectories are quite different from the first one, hence they show varied degrees of glyph distortion, but this obvious visual differences do not affect the RMSE value and only slightly worsens the DTW value (+4.9 % and +4.9 %, respectively). In contrast, the AIoU value exhibits a significant difference (–28 % and –35 %, respectively).

handwritten character recognition [22,27], handwriting robot [30,31], etc. This paper deals with two main challenges the task of complex handwriting trajectory recovery faces.

On the one hand, surprisingly, a proper task-targeted evaluation metric for handwriting trajectory recovery is still missing. The existing approaches are in three classes, but indeed, they are all with non-negligible drawbacks:

1. Human vision is usually used [4,7,13,20,24,25]. This non-quantitative and human-involved method is expensive, time-consuming and inconsistent.
2. Some work quantified the recovery quality indirectly through the accuracy of a handwriting recognition model [1,8,14], but the result inevitably depends on the recognition model and hence brings unfairness to evaluation.
3. Direct quantitative metrics have been borrowed from other fields [10,15,19], without task-targeted adaption. These metrics overlook the glyph fidelity, and some of them even ignore either the image-level fidelity (which is insufficient for our task, compared with glyph fidelity) or the writing order. For example, Fig. 1 presents the effects of different metrics on glyph fidelity. As it shows, three distorted trajectories show varied degrees of glyph distortion, however, metrics on writing order such as the root mean squared error (RMSE) and the dynamic time warping (DTW) cannot effectively and sensitively reflect their varied degrees of glyph degradation, since trajectory points jitter in the same distance.

We propose two evaluation metrics, the adaptive intersection on union (AIoU) and the length-independent dynamic time warping (LDTW). AIoU assesses the glyph fidelity and eliminates the influence of various stroke widths. LDTW is robust to the number of sequence points and overcomes the evaluation bias of classical DTW [10].

On the other hand, the existing trajectory recovery algorithms are not good in dealing with characters with both complex glyph and long trajectory, such as Chinese and Japanese scripts, thus we propose a novel learning model named Parsing-and-tracing ENcoder-decoder Network (PEN-Net). In the character encoding phase, we add a double-stream parsing encoder in the PEN-Net by creating two branches to analyze stroke context information along two orthogonal dimensions. In the decoding phase, we construct a global tracing decoder to alleviate the drifting problem of long trajectory writing order prediction.

Our contributions are threefold:

- We propose two trajectory recovery evaluation metrics, AIoU to assess glyph correctness which was overlooked by most, if not all, existing quantitative metrics, and LDTW to overcome the evaluation bias of the classical DTW.
- We propose a new trajectory recovery model called PEN-Net by constructing a double-stream parsing encoder and a global tracing decoder to solve the difficulties in the case of complex glyph and long trajectory.
- Our experiments demonstrate that AIoU and LDTW can truly assess the quality of handwritten trajectory recovery and the proposed PEN-Net exhibits satisfactory performance in various complex-glyph datasets including Chinese [16], Japanese [18] and Indic [3].

2 Related Work

2.1 Trajectory Recovery Evaluation Metrics

There is not much work on evaluation metrics for trajectory recovery. Many techniques rely on the human vision to qualitatively assess the recovery quality [4,7,13,20,24,25]. However, these non-quantitative human evaluations are expensive, time-consuming and inconsistent.

Trajectory recovery has been proved beneficial to handwriting recognition. As a byproduct, one may use the final recognition accuracy to compare the recovery efficacy [1,8,14]. Instead of directly assessing the quality of different recovery trajectories, they compare the accuracy of the recognition models among which the only difference is the intermediate recovery trajectory. Though, to some extent, this method reflects the recovery quality, it is usually disturbed by the recognition model, and only provides a relative evaluation.

Most of direct quantitative metrics only focus on the evaluation of the writing order. Lau *et al.* [15] designed a ranking model to assess the order of stroke endpoints, and Nel *et al.* [19] designed a hidden Markov model to assess the writing order of local trajectories such as stroke loops. However, these methods are unable to assess the sequence points' deviation to the groundtruth. Hence, these two evaluations are seldom adopted in the subsequent studies.

Metrics borrowed from other fields have also been used, such as the RMSE borrowed from signal processing and the DTW from speech recognition [10]. RMSE directly calculates distances between two sequences and strictly limits the number of trajectory points, which makes it hard to use in practice. It is too strict to require the recovered trajectory to recall exactly all the groundtruth points one-by-one, since trajectory recovery is an ill-posed problem that the unique recovery solution cannot be obtained without constraints [26]. DTW introduces an elastic matching mechanism to obtain the most possible alignment between two trajectories. However, DTW is not robust to the number of trajectory points, and prefers trajectories with fewer points. Actually, the number of points is irrelevant to the writing order or glyph fidelity, and shouldn't affect the judgement of the recovery quality.

Aforementioned quantitative metrics only focus on the evaluation of the writing order, and neither involves the glyph fidelity. Note that the glyph correctness is also an essential aspect of trajectory recovery, since glyphs reflect the content of characters and the writing styles of a specific writer. Only one of the latest trajectory recovery work [2] borrows the metric LPIPS [32] which compares the deep features of images. However, LPIPS is not suitable for such a fine task of trajectory recovery, as we observed that the deep features are not informative enough to distinguish images with only a few pixel-level glyph differences.

2.2 Trajectory Recovery Algorithms

Early studies in the 1990s s relied on heuristic rules, often including two major modules: local area analysis and global trajectory recovery [4,12,13,25,26]. These algorithms are difficult to devise, as they rely on delicate hand-engineered features. Rule-based methods are sophisticated, and not flexible enough to handle most of the practical cases, hence these methods are considered not robust, in particular for characters with both complex glyph and long trajectory.

Inspired by the remarkable progress in deep-learning-based image analysis and sequence generation over the last few years, deep neural networks are used for trajectory recovery. Sumi *et al.* [29] applied variational autoencoders to mutually convert online trajectories and offline handwritten character images, and their method can handle single-stroke English letters with simple glyph structures. Nevertheless, instead of predicting the entire trajectory from a plain encoding result, we can consider employing a selection mechanism (e.g., attention mechanism) to analyze the glyph structure between the prediction of two successive points, since the relative position of continuous points can be quite variable. Zhao *et al.* [35,36] proposed a CNN model to iteratively generate stroke point sequence. However, besides CNNs, we also need to consider applying RNNs to analyze the context in handwriting trajectories, which may contribute to the recovery of long trajectories.

Bhunia *et al.* [3] introduced an encoder-decoder model to recover the trajectory of single-stroke Indic scripts. This algorithm employs a one-dimensional feature map to encode characters, however, it needs more spatial information to tackle complex handwritings (on a two-dimensional plane). Nguyen *et al.* [20] improved the encoder-decoder model by introducing a Gaussian mixture model (GMM), and tried to recover multi-stroke trajectories, in particular Japanese scripts. However, since the prediction difficulty of long trajectories remains unsolved, this method does not perform well in the case of complex characters. Archibald *et al.* [2] adapted the encoder-decoder model to English text with arbitrary width, which attends to text-line-level trajectory recovery, but not designed specifically for complex glyph and long trajectory sequences, either.

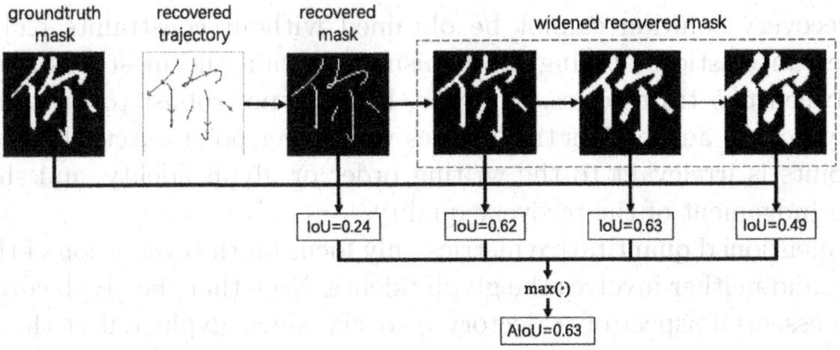

Fig. 2. Illustration of the adaptively-dilating mechanism of AIoU.

3 Glyph-and-Trajectory Dual-modality Evaluation

Let I represent a handwritten image, typically in the form of grayscale. The trajectory recovery system takes I as input and predicts the trajectory which can be mathematically represented as a time series $p = (p_1, \ldots, p_N)$, where N is the trajectory length, and $p_i = \left(x_i, y_i, s_i^1, s_i^2, s_i^3\right)$, x_i and y_i are coordinates of p_i, s_i^1, s_i^2 and s_i^3 are pen tip states which are described in detail in Sect. 4.3. And the corresponding groundtruth trajectory is $q = (q_1, \ldots, q_M)$ of length M.

3.1 Adaptive Intersection on Union

We propose the first glyph fidelity metric, Adaptive Intersection on Union (AIoU). It firstly performs a binarization process on the input image I using a thresholding algorithm, e.g., OTSU [23], to obtain the ground-truth binary mask, denoted G, which indicates whether a pixel belongs to a character stroke. Meanwhile, the predicted trajectory p is rendered into a bitmap (predicted mask) of width 1 pixel, by drawing lines between neighboring points if they belong to a stroke, denoted P. We define the IoU (Intersection over Union) between G and P as follow, which is similar to the mask IoU [5] $IoU(G, P) = |G \cap P| / |G \cup P|$.

An input handwritten character usually has various stroke widths while the predicted stroke widths are fixed, nevertheless, the stroke width shouldn't influence the assessment of the glyph similarity. To reduce the impacts of stroke width, we propose a dynamic dilation algorithm to adjust the stroke width adaptively. Concretely, as shown in Fig. 2, we adopt a dilation algorithm [9] with a kernel of 3×3 to widen the stroke along until the IoU score reaches the maximum, denoted $AIoU(G, P)$. Since the image I is extracted as the binary mask G, the ground-truth trajectory of I is not involved in the calculation of the AIoU, making the criteria still effective even without the ground-truth trajectory.

3.2 Length-Independent Dynamic Time Warping

Variable lengths make it hard to align handwriting trajectories. As shown in Fig. 3, when comparing two handwriting trajectories with different lengths, the

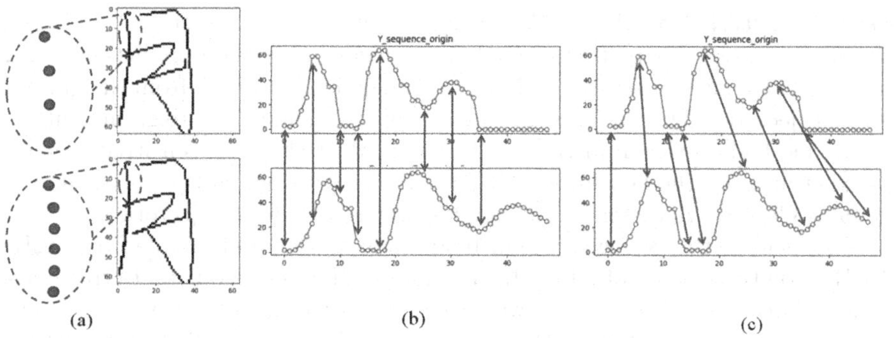

Fig. 3. Comparison between the one-to-one and the elastic matching. (a) Original and upsampling handwriting trajectories of a same character. (b) One-to-one, (c) elastic matching of two trajectories. Above and Below waveforms are Y coordinate sequences of the original-sampling and upsampling handwriting trajectory, respectively. Partial correspondence pairs are illustrated as red connection lines. (Color figure online)

direct one-to-one stroke-point correspondence cannot represent the correct alignment of strokes. We modify the well-known DTW [10] to compare two trajectories whose lengths are allowed to be different, which uses an elastic matching mechanism to obtain the most possible alignment.

The original DTW relies on the concept of alignment paths, $DTW(q, p) = \min_\phi \left\{\sum_{t=1}^{T} d\left(q_{i_t}, p_{j_t}\right)\right\}$, where the minimization is taken over all possible alignment paths ϕ (which is solved by a dynamic programming), $T \leq M + N$ is the alignment length and $d\left(q_{i_t}, p_{j_t}\right)$ refers to the (Euclidean) distance between the two potentially matched (determined by ϕ) points q_{i_t} and p_{j_t}.

We observe that the original DTW empirically behaves like a monotonic function of T that is usually proportional to N, so it in general prefers short strokes and even gives good score to incomplete strokes, which is what we want to get rid of. Intuitively, this phenomenon is interpretable: DTW is the minimization of a sum of T terms, and T depends on N. We suggest a normalized version of DTW, called the length-independent DTW (LDTW)

$$LDTW(q, p) = \frac{1}{T} DTW(q, p). \tag{1}$$

It is worth noting that, since the alignment problem also exists during the training process, we use a soft dynamic time warping loss (SDTW Loss) [6] to realize a global-alignment optimization, see Sect. 4.3.

3.3 Analysis of AIoU and LDTW

In this part, we firstly investigate how the values of our proposed metrics(AIoU and LDTW) and other recently used metrics change in response to the errors in different magnitudes. Secondly, we analyze the impacts of the changes in the number of trajectory points and stroke width to LDTW and AIoU respectively.

Error-Sensitivity Analysis. We simulate a series of common trajectory recovery errors across different magnitude by generating pseudo-predictions with errors such as point or stroke level insertion, deletion, and drift from the ground truth trajectories. Specific implementations of error simulation (e.g., magnitude setting method) is shown in Appendix. We conduct the error-sensitivity analysis experiment on the benchmark OLHWDB1.1 (described in Sect. 5.1), since it contains Chinese characters with complex glyphs and long trajectories.

We calculate the average score in 1000 randomly-selected sample from OLHWDB1.1 on the metrics of AIoU, LDTW and LPIPS across different error magnitudes. For better visualization, we normalize the values of the three metrics to [0, 1]. As Fig. 4 illustrates, firstly, the values of AIoU and LPIPS, two metrics on the glyph and image level respectively, decrease as the magnitude of the four error types increases. Secondly, the value of LDTW, the proposed quantitative metric for sequence similarity comparison, increases along with the magnitude of the four error types. These two results prove that the three metrics are sensitive to the four errors. Furthermore, as the changing trend of AIoU is faster than LPIPS, the former is more sensitive to the errors than the latter.

Invariance Analysis. In terms of metric invariance, stroke width change and trajectory points number change are two critical factors. The former highlights different handwriting brush strokes (e.g., brushes, pencils, or water pens), which only affects stroke widths and keep the original glyph of the characters. The latter regards to the change of the total number of points in a character to simulate different handwriting speeds.

This analysis is also based on OLHWDB1.1 and the data preprocessing is the same with the error sensitivity analysis mentioned above. As shown in Fig. 4(e), on the overall, the trend of our proposed AIoU is more stable compared with LPIPS as the stroke width of the character increases, indicating that AIoU is more robust to the changes of stroke widths so that it can truly reflect the glyph fidelity of a character. In terms of the trajectory points number change, as shown in Fig. 4(f), the value of DTW rises with the increase of the number of points in the character trajectory while our proposed LDTW, on the other hand, shows a smooth and steady trend. This is because LDTW applies length normalization techniques but DTW does not. The results proves that our proposed LDTW is more robust to the changes in the number of points in a character trajectory.

4 Parsing-and-Tracing ENcoder-Decoder Network

As shown in Fig. 5, PEN-Net is composed of a double-stream parsing encoder and a global tracing decoder. Taking a static handwriting image as input, the double-stream parsing encoder analyzes the stroke context and parses the glyph structure, obtaining the features that will be used by the global tracing decoder to predict trajectory points.

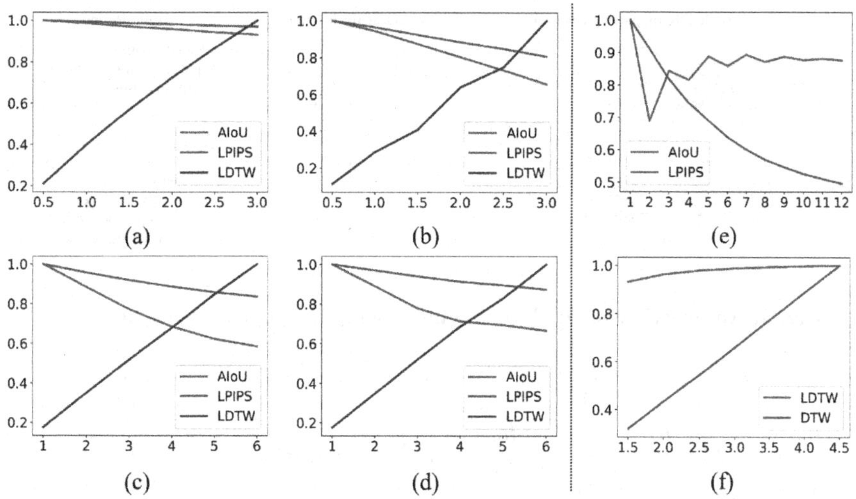

Fig. 4. Left: Sensitivity curves across error magnitudes: AIoU, LPIPS (Instead of $LPIPS$, we show $1 - LPIPS$, for a better visual comparison), LDTW results of 4 error types: (a) Stroke insertion error. (b) Stroke deletion error. (c) Trajectory point drift error. (d) Stroke drift error. X-axes of (a) and (b) are the number of inserted and deleted strokes, respectively. X-axes of (c) and (d) are the drifted pixel distance of point and stroke, respectively. Right: Sensitivity curves across change magnitudes: (e) LPIPS (Instead of $LPIPS$, we show $1 - LPIPS$, for a better visual comparison) and AIoU results of the change of stroke widths (X-axis). (f) DTW and LDTW results of the change of sample rates (X-axis). Y-axes refer to the normalized metric value for all sub-figures.

4.1 Double-Stream Parsing Encoder

Existing methods (e.g., DED-Net [3], Cross-VAE [29]) compress the feature to only one dimension vector, which are not informative enough to maintain the complex two-dimensional information of characters. Actually, every two-dimensional stroke can be projected to horizontal and vertical axes. In the double-stream parsing encoder, we construct two CRNN [28] branches denoted as $CRNN_X$ and $CRNN_Y$ to decouple handwriting images to horizontal and vertical features V_x and V_y, which are complementary in two perpendicular directions for the parsing of glyph structure. Each branch is composed of a CNN to extract the vertical or horizontal stroke features, and a 3-layer BiLSTM to analyze the relationship between strokes, e.g., which stroke should be drawn earlier, what is the relative position between strokes. To extract stroke features of single direction in the CNN of each stream, we use asymmetric poolings, which is found to be effective experimentally. Details of proposed CNNs are shown in Fig. 5.

The stroke region is always sparse in a handwriting image, and the blank background disturbs the stroke feature extraction. To this end, we use an attention mechanism to attend to the stroke foreground. The attention mechanism fuses V_x and V_y, and obtains the attention score s_i of each feature v_i to let the glyph parsing feature Z focus on the stroke foreground:

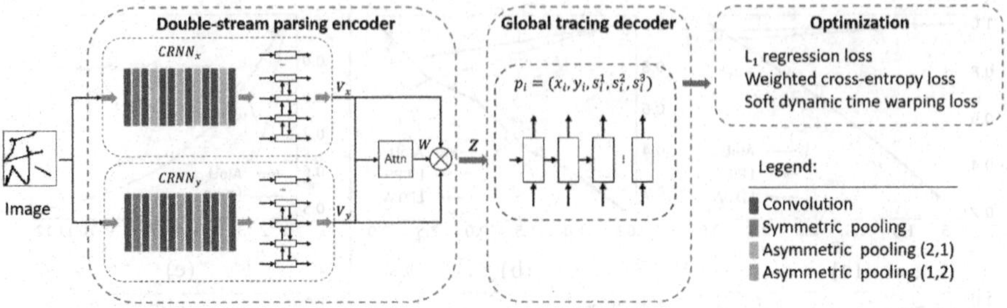

Fig. 5. An overview of the Parsing-and-tracing Encoder-decoder Network.

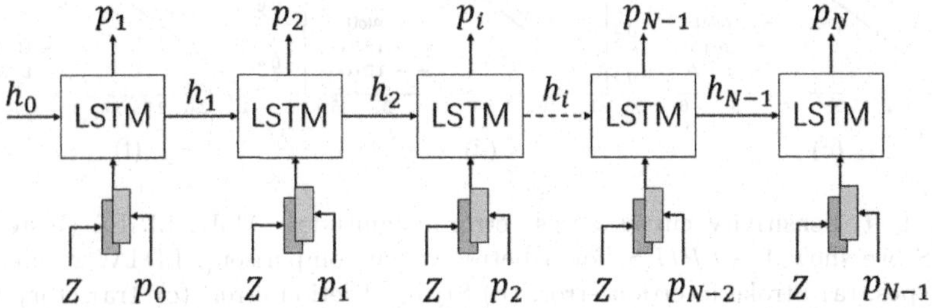

Fig. 6. The architecture of global tracing decoder. Z is the glyph parsing feature, p_i represent the trajectory point at time i, h_i is the hidden state of the LSTM at time i. Z is concatenated with p_i. We initialize the hidden state of LSTM decoder with the hidden state outputs of BiLSTMs encoder, which is similar to the [3].

$$s_i = f(v_i) = Uv_i, \tag{2}$$

$$w_i = \frac{e^{s_i}}{\sum_{j=1}^{|V|} e^{s_j}}, \tag{3}$$

$$Z = \sum_{i=1}^{|V|} v_i * w_i, \tag{4}$$

where V is obtained by concatenating V_x and V_y, v_i is the component of V, $|V|$ the length of V, U is learnable parameters of a fully-connected layer. We apply a simplified attention strategy to acquire the attention score s_i of the feature v_i.

4.2 Global Tracing Decoder

We adopt a 3-layer LSTM as the decoder to predict the trajectory points sequentially. In particular, the decoder uses the position and the pen tip state at time step $i-1$ to predict those at time step i, similar to [3,20].

During decoding, previous trajectory recovery methods [3,20] only utilize the initial character coding. As a result, the forgetting phenomenon of RNN [11] causes the so-called trajectory-point position drifting problem during the

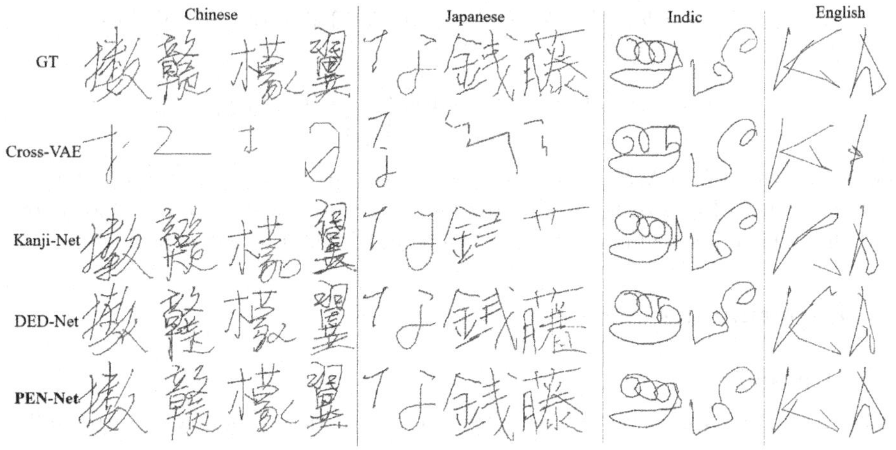

Fig. 7. Sample visualization of recovered trajectories of our proposed PEN-Net, Cross-VAE [29], Kanji-Net [20] and DED-Net [3]. Each color represents a stroke, and colors of strokes from starting to ending is represented from blue to red. (Color figure online)

subsequent decoding steps, especially for characters with long trajectories. To alleviate this drifting problem, we propose a global tracing mechanism by using the glyph parsing feature Z at each decoding step. The whole decoding process is shown in Fig. 6.

4.3 Optimization

Similar to [3,20], we use the L_1 regression loss and the cross-entropy loss to optimize the coordinates and the pen tip states of the trajectory points, respectively. Similar to [33], during the process of optimizing pen tip states, we define three states "pen-down", "pen-up" and "end-of-sequence" respectively, which are denoted as s_i^1, s_i^2, s_i^3 of p_i. It is obvious that "pen-down" data points are much more than the other two classes. To solve the biased dataset issue, we add weights ("pen-down" is set to 1, "pen-up" 5, and "end-of-sequence" 1, respectively) to the cross-entropy loss.

These hard-losses are insufficient because they require a one-to-one stroke-point correspondence, which is too strict for handwriting trajectories of variable lengths. We borrow the soft dynamic time warping loss (SDTW Loss) [6], which has never been used for trajectory recovery, to supplement the global-alignment goal of the whole trajectory and to alleviate the alignment learning problem.

The DTW algorithm can solve the alignment issue during optimization using an elastic matching mechanism. However, since containing the hard minimization operation which is not differentiable, DTW cannot be used as an loss function directly. Hence, we place the minimization operation by a soft-minimization

$\min^{\gamma}\{a_1, \ldots, a_n\} = \gamma \log \sum_{i=1}^{n} e^{-a_i/\gamma}, \gamma > 0$. We define the SDTW loss

$$L_{sdtw} = \mathrm{SDTW}(q, p) = \min_{\phi}^{\gamma} \left\{ \sum_{t=1}^{T} d\left(q_{i_t}, p_{j_t}\right) \right\}.$$

The total loss is $L = \lambda_1 L_1 + \lambda_2 L_{wce} + \lambda_3 L_{sdtw}$, where $\lambda_1, \lambda_2, \lambda_3$ are parameters to balance the effects of the L_1 regression loss, the weighted cross-entropy loss and the SDTW loss, which are set to $0.5, 1$ and $1/6000$ in our experiments.

5 Experiments

5.1 Datasets

Information of datasets is given as follows, and statistics of them are in Appendix.

Chinese Script. CASIA-OLHWDB(1.0–1.2) [16] is a million-level online handwritten character dataset. We conduct experiments on all of the Chinese characters from OLHWDB1.1 which covers the most frequently used characters of GB2312-80. The largest amounts of trajectory points and strokes reach 283 (with an average of 61) and 29 (average of 6), respectively.

English Script. We collect all of the English samples from the symbol part of CASIA-OLHWDB (1.0–1.2), covering 52 classes of English letters.

Japanese Script. Referring to [20], we conduct Japanese handwriting recovery experiments on two datasets including Nakayosi_t-98-09 for training and Kuchibue_d-96-02 for testing. The largest amount of trajectory points and strokes reach 3544 (with an average of 111) and 35 (average of 6), respectively.

Indic Script. Tamil dataset [3] contains samples of 156 character classes. The largest amount of trajectory points reach 1832 (average of 146).

5.2 Experimental Setting

Implementation Details. We normalize the online trajectories to [0, 64) range. In addition, in terms of the Japanese and Indic datasets, because their points densities are so high that points may overlap each other after the rescaling process, we remove the redundant points in the overlapping areas and then down-sample remaining trajectory points by half. We convert the online data to its offline equivalent by rendering the image using the online coordinate points. Although the rendered images are not real offline patterns, they are useful to evaluate the performance of trajectory recovery [3,20]. In addition, we train our model 500,000 iterations on the Chinese and Japanese datasets, and 200,000 iterations on the English and Indic dataset, with a RTX3090 GPU. The batch size is set to 512. The optimizer is Adam with the learning rate of 0.001.

Table 1. Comparisons with state-of-the-art methods on four different language datasets. ↓ / ↑ denote the smaller/larger, the better.

Datasets	Method	Evaluation metric				
		AIoU ↑	LPIPS ↓	**LDTW**↓	DTW ↓	RMSE ↓
Chinese	Cross-VAE [29]	0.146	0.402	13.64	1037.9	20.32
	Kanji-Net [20]	0.326	0.186	5.51	442.7	15.43
	DED-Net [3]	0.397	0.136	4.08	302.6	15.01
	PEN-Net	**0.450**	**0.113**	**3.11**	**233.8**	**14.39**
English	Cross-VAE [29]	0.238	0.206	7.43	176.7	20.39
	Kanji-Net [20]	0.356	0.121	5.98	149.1	18.66
	DED-Net [3]	0.421	0.089	4.70	109.4	16.05
	PEN-Net	**0.461**	**0.074**	**3.21**	**77.35**	**15.12**
Indic	Cross-VAE [29]	0.235	0.228	4.89	347.3	16.01
	Kanji-Net [20]	0.340	0.163	3.04	234.0	15.65
	DED-Net [3]	0.519	0.084	2.00	130.5	14.52
	PEN-Net	**0.546**	**0.074**	**1.62**	**104.9**	**12.92**
Japanese	Cross-VAE [29]	0.164	0.346	22.7	1652.2	38.79
	Kanji-Net [20]	0.290	0.236	6.92	395.0	19.47
	DED-Net [3]	0.413	0.150	4.70	214.0	18.88
	PEN-Net	**0.476**	**0.125**	**3.39**	**144.5**	**17.08**
Chinese (complex)	Cross-VAE [29]	0.159	0.445	16.15	1816.4	26.24
	Kanji-Net [20]	0.311	0.218	5.56	668.4	15.26
	DED-Net [3]	0.363	0.168	4.34	483.4	16.08
	PEN-Net	**0.411**	**0.143**	**3.58**	**402.9**	**15.61**
Japanese (complex)	Cross-VAE [29]	0.154	0.489	41.84	6230.3	60.28
	Kanji-Net [20]	0.190	0.435	9.68	1264.5	20.32
	DED-Net [3]	0.341	0.250	4.35	536.1	18.31
	PEN-Net	**0.445**	**0.186**	**2.95**	**344.4**	**16.00**

5.3 Comparison with State-of-the-Art Approaches

In this section, we quantitatively evaluate the quality of trajectory, recovered by our PEN-Net and existing state-of-the-art methods including DED-Net [3], Cross-VAE [29] and Kanji-Net [20], on the above-mentioned four datasets via five different evaluation metrics of which AIoU and LDTW are proposed by us.

As Table 1 shows, our PEN-Net expresses satisfactory and superior performance compared to other approaches, with an average of 13% to 20% gap away from the second-best in all of the five evaluation criteria on the first four datasets. Moreover, to further validate the models' effects for complex handwritings, we build two subsets by extracting 5% of samples with the most strokes from the Japanese and Chinese testing set independently, where the number of strokes of each sample is over 15 and 10 corresponding to the two languages. According to the data(Chinese/Japanese complex in the table), PEN-Net still performs better than SOTA methods. Particularly, on Japanese complex set, PEN-Net expresses superior performance compared to other approaches, with an average of 27.3% gap away from the second-best in all of the five evaluation criteria.

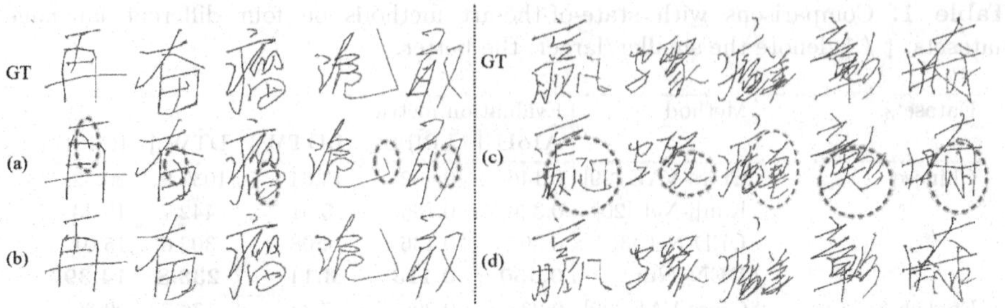

Fig. 8. Left: Sample visualization of recovered trajectories of models (a) without and (b) with double-stream mechanism. Right: Sample visualization of models (c) without and (d) with global tracing mechanism. Stroke errors are circled in red. (Color figure online)

Table 2. Ablation study on each component of PEN-Net.

DS	GLT	ATT	SDTW	**AIoU**↑	LPIPS ↓	**LDTW**↓	DTW↓	RMSE↓
√	√	√	√	**0.451**	**0.113**	**3.11**	**233.8**	14.39
√	√	√		0.433	0.118	3.53	261.1	**14.06**
√	√			0.426	0.123	3.62	262.4	15.05
√				0.417	0.130	3.89	291.3	14.40
				0.406	0.140	4.07	301.6	15.17

As the visualization results in Fig. 7, Cross-VAE [29], Kanji-Net [20] and DED-Net [3] can recover simple characters' trajectories (English, Indic, and part of Japanese characters). However, their methods exhibit error phenomena, such as stroke duplication and trajectory deviation, in complex situations. Cross-VAE [29] may fail at recovering trajectories of complex characters (Chinese and Japanese), and Kanji-Net [20] cannot recover the whole trajectory of complex Japanese characters. In contrast, our PEN-Net makes accurate and reliable recovery prediction on both simple and complex characters, demonstrating an outstanding performance in terms of both visualization and quantitative metrics compared with the three prior SOTA works.

5.4 Ablation Study of PEN-Net

In this section, we conduct ablation experiments on the effectiveness of PEN-Net's core components, including double-stream (DS) mechanism, global tracing (GLT) mechanism, attention (ATT) mechanism and SDTW loss. We use Chinese dataset to evaluate PEN-Net's performance for complex handwriting trajectory recovery. The evaluation metrics are the same as in Sect. 5.3. The experiment results are reported in Table 2 in which the first row relates to the full model with all components, and we gradually ablate each component one-by-one down to the plain baseline model at the bottom row.

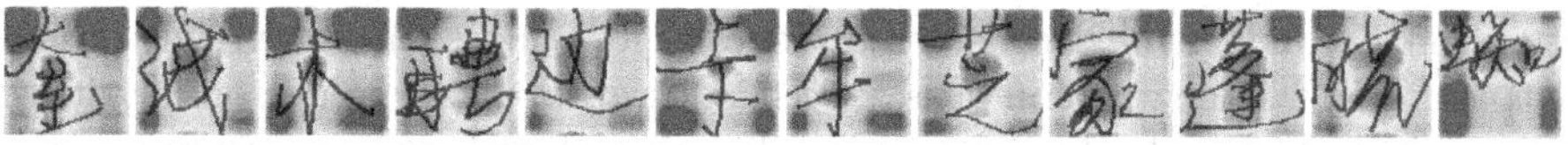

Fig. 9. Sample visualization of attention scores maps. The maps are obtained by extracting and multiplying two attention-weighted vectors corresponding to V_x and V_y mentioned in Sect. 4.1. (Color figure online)

Double-Stream Mechanism. In this test, we remove CRNNy from the backbone of the double-stream encoder and remains the CRNNx. As the 4th and 5th rows in Table 2 show, CRNNy contributes 2.7% and 7.14% improvement on glyph fidelity metrics (AIoU and LPIPS), and 4.4%, 3.41%, 5.1% improvement on writing order metrics (LNDTW, DTW, RMSE). Additionally, as Fig. 8 reveals, the model without CRNNy cannot make an accurate prediction on vertical strokes of Chinese characters.

Global Tracing Mechanism. As the 3rd and 4th row in Table 2 show, GLT further improves the performance based on all the metrics except RMSE. The value rise in RMSE, from 14.40 to 15.05, is because this generic metric overemphasizes the point-by-point absolute deviation, which negatively affects the overall quality evaluation of the handwriting trajectory matching. In addition, as Fig. 8 shows, drifting phenomenon occurs in the recovered trajectories if GLT is removed, while, in contrast, the phenomenon disappears vise versa.

Attention Mechanism. As the 2nd and 3rd rows showed in Table 2, ATT also improves the performance of the model. Furthermore, as the attention heat-map visualization showed in Fig. 9, the stroke region always attracts more attention(in red color) than the background area(in blue color) in a character image.

SDTW Loss. As the 1st and the 2nd rows showed in Table 2, the SDTW loss also contributes to the performance enhancement of the model.

Finally, based on these ablation studies, the PEN-Net dramatically boost the trajectory recovery performance over the baseline by 10.8% on AIoU, 23.6% on LDTW, 22.5% on DTW, 5.1% on RMSE, 19.3% on LPIPS. Consequently, we claim that the four components of PEN-Net: double-stream mechanism, global tracing, attention mechanism and SDTW loss, all play pivotal roles w.r.t. the final performance of trajectory recovery.

6 Conclusion

We have proposed two evaluation metrics AIoU and LDTW specific for trajectory recovery, and have proposed the PEN-Net for complex character recovery.

There are several possible future directions. First, local details such as loops play an important role in some writing systems, to which we will pay more attention. Second, we have considered recovering the most natural writing order, but, as far as we know, no one has succeeded in recovering the personal writing order, which should also be a promising direction. Third, one can try to replace the decoder part by some trendy methods, e.g., transformer. Besides, we can go beyond the encoder-decoder framework, and treat this task as, for example, a decision-making problem and then use the techniques of reinforcement learning.

Acknowledgements. This work has been supported by the National Natural Science Foundation of China (No.62176093, 61673182), the Key Realm Research and Development Program of Guangzhou (No.202206030001), and the GuangDong Basic and Applied Basic Rescarch Foundation (No.2021A1515012282).

References

1. Al-Ohali, Y., Cheriet, M., Suen, C.Y.: Dynamic observations and dynamic state termination for off-line handwritten word recognition using hmm. In: Proceedings Eighth International Workshop on Frontiers in Handwriting Recognition, pp. 314–319. IEEE (2002)
2. Archibald, T., Poggemann, M., Chan, A., Martinez, T.: Trace: a differentiable approach to line-level stroke recovery for offline handwritten text. arXiv preprint arXiv:2105.11559 (2021)
3. Bhunia, A.K., et al.: Handwriting trajectory recovery using end-to-end deep encoder-decoder network. In: 2018 24th International Conference on Pattern Recognition (ICPR), pp. 3639–3644. IEEE (2018)
4. Boccignone, G., Chianese, A., Cordella, L.P., Marcelli, A.: Recovering dynamic information from static handwriting. Pattern Recogn. **26**(3), 409–418 (1993)
5. Cheng, B., Girshick, R., Dollár, P., Berg, A.C., Kirillov, A.: Boundary iou: improving object-centric image segmentation evaluation. In: Proceedings of the IEEE/CVF Conference on Computer Vision and Pattern Recognition, pp. 15334–15342 (2021)
6. Cuturi, M., Blondel, M.: Soft-dtw: a differentiable loss function for time-series. In: International Conference on Machine Learning, pp. 894–903. PMLR (2017)
7. Doermann, D.S., Rosenfeld, A.: Recovery of temporal information from static images of handwriting. Int. J. Comput. Vision **15**(1–2), 143–164 (1995)
8. Guo, J.K., Doermann, D., Rosenfeld, A.: Forgery detection by local correspondence. Int. J. Pattern Recogn. Artif. Intell. **15**(04), 579–641 (2001)
9. Haralick, R.M., Sternberg, S.R., Zhuang, X.: Image analysis using mathematical morphology. IEEE Trans. Pattern Anal. Mach. Intell. **4**, 532–550 (1987)
10. Hassaïne, A., Al Maadeed, S., Bouridane, A.: Icdar 2013 competition on handwriting stroke recovery from offline data. In: 2013 12th International Conference on Document Analysis and Recognition, pp. 1412–1416. IEEE (2013)
11. Hochreiter, S., Schmidhuber, J.: Long short-term memory. Neural Comput. **9**(8), 1735–1780 (1997)
12. Jager, S.: Recovering writing traces in off-line handwriting recognition: using a global optimization technique. In: Proceedings of 13th International Conference on Pattern Recognition, vol. 3, pp. 150–154. IEEE (1996)
13. Kato, Y., Yasuhara, M.: Recovery of drawing order from single-stroke handwriting images. IEEE Trans. Pattern Anal. Mach. Intell. **22**(9), 938–949 (2000)

14. Lallican, P.M., Viard-Gaudin, C., Knerr, S.: From off-line to on-line handwriting recognition. In: Proceedings of the Seventh International Workshop on Frontiers in Handwriting Recognition, pp. 303–312 (2000)
15. Lau, K.K., Yuen, P.C., Tang, Y.Y.: Directed connection measurement for evaluating reconstructed stroke sequence in handwriting images. Pattern Recogn. **38**(3), 323–339 (2005)
16. Liu, C.L., Yin, F., Wang, D.H., Wang, Q.F.: Casia online and offline Chinese handwriting databases. In: 2011 International Conference on Document Analysis and Recognition, pp. 37–41. IEEE (2011)
17. Munich, M.E., Perona, P.: Visual identification by signature tracking. IEEE Trans. Pattern Anal. Mach. Intell. **25**(2), 200–217 (2003)
18. Nakagawa, M., Matsumoto, K.: Collection of on-line handwritten Japanese character pattern databases and their analyses. Doc. Anal. Recogn. **7**(1), 69–81 (2004)
19. Nel, E.M., du Preez, J.A., Herbst, B.M.: Verification of dynamic curves extracted from static handwritten scripts. Pattern Recogn. **41**(12), 3773–3785 (2008)
20. Nguyen, H.T., Nakamura, T., Nguyen, C.T., Nakawaga, M.: Online trajectory recovery from offline handwritten Japanese kanji characters of multiple strokes. In: 2020 25th International Conference on Pattern Recognition (ICPR), pp. 8320–8327. IEEE (2021)
21. Niels, R., Vuurpijl, L.: Automatic trajectory extraction and validation of scanned handwritten characters. In: Tenth International Workshop on Frontiers in Handwriting Recognition. Suvisoft (2006)
22. Noubigh, Z., Kherallah, M.: A survey on handwriting recognition based on the trajectory recovery technique. In: 2017 1st International Workshop on Arabic Script Analysis and Recognition (ASAR), pp. 69–73. IEEE (2017)
23. Otsu, N.: A threshold selection method from gray-level histograms. IEEE Trans. Syst. Man Cybern. **9**(1), 62–66 (1979)
24. Pan, J.C., Lee, S.: Offline tracing and representation of signatures. In: Proceedings. 1991 IEEE Computer Society Conference on Computer Vision and Pattern Recognition, pp. 679–680. IEEE (1991)
25. Plamondon, R., Privitera, C.M.: The segmentation of cursive handwriting: an approach based on off-line recovery of the motor-temporal information. IEEE Trans. Image Process. **8**(1), 80–91 (1999)
26. Qiao, Y., Nishiara, M., Yasuhara, M.: A framework toward restoration of writing order from single-stroked handwriting image. IEEE Trans. Pattern Anal. Mach. Intell. **28**(11), 1724–1737 (2006)
27. Rabhi, B., Elbaati, A., Hamdi, Y., Alimi, A.M.: Handwriting recognition based on temporal order restored by the end-to-end system. In: 2019 International Conference on Document Analysis and Recognition (ICDAR), pp. 1231–1236. IEEE (2019)
28. Shi, B., Bai, X., Yao, C.: An end-to-end trainable neural network for image-based sequence recognition and its application to scene text recognition. IEEE Trans. Pattern Anal. Mach. Intell. **39**(11), 2298–2304 (2016)
29. Sumi, T., Iwana, B.K., Hayashi, H., Uchida, S.: Modality conversion of handwritten patterns by cross variational autoencoders. In: 2019 International Conference on Document Analysis and Recognition (ICDAR), pp. 407–412. IEEE (2019)
30. Yao, F., Shao, G., Yi, J.: Trajectory generation of the writing-brush for a robot arm to inherit block-style Chinese character calligraphy techniques. Adv. Rob. **18**(3), 331–356 (2004)

31. Yin, H., Alves-Oliveira, P., Melo, F.S., Billard, A., Paiva, A., et al.: Synthesizing robotic handwriting motion by learning from human demonstrations. In: Twenty-Fifth International Joint Conference on Artificial Intelligence (IJCAI 2016), pp. 3530–3537 (2016)
32. Zhang, R., Isola, P., Efros, A.A., Shechtman, E., Wang, O.: The unreasonable effectiveness of deep features as a perceptual metric. In: Proceedings of the IEEE Conference on Computer Vision and Pattern Recognition, pp. 586–595 (2018)
33. Zhang, X.Y., Yin, F., Zhang, Y.M., Liu, C.L., Bengio, Y.: Drawing and recognizing Chinese characters with recurrent neural network. IEEE Trans. Pattern Anal. Mach. Intell. **40**(4), 849–862 (2018). https://doi.org/10.1109/TPAMI.2017.2695539
34. Zhao, B., Tao, J., Yang, M., Tian, Z., Fan, C., Bai, Y.: Deep imitator: handwriting calligraphy imitation via deep attention networks. Pattern Recogn. **104**, 107080 (2020)
35. Zhao, B., Yang, M., Tao, J.: Pen tip motion prediction for handwriting drawing order recovery using deep neural network. In: 2018 24th International Conference on Pattern Recognition (ICPR), pp. 704–709. IEEE (2018)
36. Zhao, B., Yang, M., Tao, J.: Drawing order recovery for handwriting Chinese characters. In: ICASSP 2019–2019 IEEE International Conference on Acoustics, Speech and Signal Processing (ICASSP), pp. 3227–3231. IEEE (2019)

Fully Transformer Network for Change Detection of Remote Sensing Images

Tianyu Yan, Zifu Wan, and Pingping Zhang(✉)

School of Artificial Intelligence, Dalian University of Technology, Dalian, China
zhpp@dlut.edu.cn

Abstract. Recently, change detection (CD) of remote sensing images have achieved great progress with the advances of deep learning. However, current methods generally deliver incomplete CD regions and irregular CD boundaries due to the limited representation ability of the extracted visual features. To relieve these issues, in this work we propose a novel learning framework named Fully Transformer Network (FTN) for remote sensing image CD, which improves the feature extraction from a global view and combines multi-level visual features in a pyramid manner. More specifically, the proposed framework first utilizes the advantages of Transformers in long-range dependency modeling. It can help to learn more discriminative global-level features and obtain complete CD regions. Then, we introduce a pyramid structure to aggregate multi-level visual features from Transformers for feature enhancement. The pyramid structure grafted with a Progressive Attention Module (PAM) can improve the feature representation ability with additional interdependencies through channel attentions. Finally, to better train the framework, we utilize the deeply-supervised learning with multiple boundary-aware loss functions. Extensive experiments demonstrate that our proposed method achieves a new state-of-the-art performance on four public CD benchmarks. For model reproduction, the source code is released at https://github.com/AI-Zhpp/FTN.

Keywords: Fully Transformer Network · Change detection · Remote sensing image

1 Introduction

Change Detection (CD) plays an important role in the field of remote sensing. It aims to detect the key change regions in dual-phase remote sensing images captured at different times but over the same area. Remote sensing image CD has been used in many real-world applications, such as land-use planning, urban expansion management, geological disaster monitoring, ecological environment protection. However, due to change regions can be any shapes in complex

Supplementary Information The online version contains supplementary material available at https://doi.org/10.1007/978-3-031-26284-5_5.

L. Wang et al. (Eds.): ACCV 2022, LNCS 13842, pp. 75–92, 2023.
https://doi.org/10.1007/978-3-031-26284-5_5

scenarios, there are still many challenges for high-accuracy CD. In addition, remote sensing image CD by handcrafted methods is time-consuming and labor-intensive, thus there is a great need for fully-automatic and highly-efficient CD.

In recent years, deep learning has been widely used in remote sensing image processing due to its powerful feature representation capabilities, and has shown great potential in CD. With deep Convolutional Neural Networks (CNN) [12,15,17], many CD methods extract discriminative features and have demonstrated good CD performances. However, previous methods still have the following shortcomings: 1) With the resolution improvement of remote sensing images, rich semantic information contained in high-resolution images is not fully utilized. As a result, current CD methods are unable to distinguish pseudo changes such as shadow, vegetation and sunshine in sensitive areas. 2) Boundary information in complex remote sensing images is often missing. In previous methods, the extracted changed areas often have regional holes and their boundaries can be very irregular, resulting in a poor visual effect [28]. 3) The temporal information contained in dual-phase remote sensing images is not fully utilized, which is also one of the reasons for the low performance of current CD methods.

To tackle above issues, in this work we propose a novel learning framework named Fully Transformer Network (FTN) for remote sensing image CD, which improves the feature extraction from a global view and combines multi-level visual features in a pyramid manner. More specifically, the proposed framework is a three-branch structure whose input is a dual-phase remote sensing image pair. We first utilize the advantages of Transformers [9,29,42] in long-range dependency modeling to learn more discriminative global-level features. Then, to highlight the change regions, the summation features and difference features are generated by directly comparing the temporal features of dual-phase remote sensing images. Thus, one can obtain complete CD regions. To improve the boundary perception ability, we further introduce a pyramid structure to aggregate multi-level visual features from Transformers. The pyramid structure grafted with a Progressive Attention Module (PAM) can improve the feature representation ability with additional interdependencies through channel attentions. Finally, to better train the framework, we utilize the deeply-supervised learning with multiple boundary-aware loss functions. Extensive experiments show that our method achieves a new state-of-the-art performance on four public CD benchmarks. In summary, the main contributions of this work are as follow:

- We propose a novel learning framework (*i.e.*, FTN) for remote sensing image CD, which can improve the feature extraction from a global view and combine multi-level visual features in a pyramid manner.
- We propose a pyramid structure grafted with a Progressive Attention Module (PAM) to further improve the feature representation ability with additional interdependencies through channel attentions.
- We introduce the deeply-supervised learning with multiple boundary-aware loss functions, to address the irregular boundary problem in CD.

- Extensive experiments on four public CD benchmarks demonstrate that our framework attains better performances than most state-of-the-art methods.

2 Related Work

2.1 Change Detection of Remote Sensing Images

Technically, the task of change detection takes dual-phase remote sensing images as inputs, and predicts the change regions of the same area. Before deep learning, direct classification based methods witness the great progress in CD. For example, Change Vector Analysis (CVA) [16,48] is powerful in extracting pixel-level features and is widely utilized in CD. With the rapid improvement in image resolution, more details of objects have been recorded in remote sensing images. Therefore, many object-aware methods are proposed to improve the CD performance. For example, Tang *et al.* [41] propose an object-oriented CD method based on the Kolmogorov–Smirnov test. Li *et al.* [23] propose the object-oriented CVA to reduce the number of pseudo detection pixels. With multiple classifiers and multi-scale uncertainty analysis, Tan *et al.* [40] build an object-based approach for complex scene CD. Although above methods can generate CD maps from dual-phase remote sensing images, they generally deliver incomplete CD regions and irregular CD boundaries due to the limited representation ability of the extracted visual features.

With the advances of deep learning, many works improve the CD performance by extracting more discriminative features. For example, Zhang *et al.* [52] utilize a Deep Belief Network (DBN) to extract deep features and represent the change regions by patch differences. Saha *et al.* [36] combine a pre-trained deep CNN and traditional CVA to generate certain change regions. Hou *et al.* [14] take the advantages of deep features and introduce the low rank analysis to improve the CD results. Peng *et al.* [33] utilize saliency detection analysis and pre-trained deep networks to achieve unsupervised CD. Since change regions may appear any places, Lei *et al.* [22] integrate Stacked Denoising AutoEncoders (SDAE) with the multi-scale superpixel segmentation to realize superpixel-based CD. Similarly, Lv *et al.* [31] utilize a Stacked Contractive AutoEncoder (SCAE) to extract temporal change features from superpixels, then adopt a clustering method to produce CD maps. Meanwhile, some methods formulate the CD task into a binary image segmentation task. Thus, CD can be finished in a supervised manner. For example, Alcantarilla *et al.* [1] first concatenate dual-phase images as one image with six channels. Then, the six-channel image is fed into a Fully Convolutional Network (FCN) to realize the CD. Similarly, Peng *et al.* [34] combine bi-temporal remote sensing images as one input, which is then fed into a modified U-Net++ [57] for CD. Daudt *et al.* [7] utilize Siamese networks to extract features for each remote sensing image, then predict the CD maps with fused features. The experimental results prove the efficiency of Siamese networks. Further more, Guo *et al.* [11] use a fully convolutional Siamese network with a contrastive loss to measure the change regions. Zhang *et al.* [49] propose a deeply-supervised image fusion network for CD. There are also some works focused on specific object CD. For example, Liu *et al.* [28] propose a dual-task constrained deep Siamese convolutional network for building CD. Jiang *et al.* [19] propose a pyramid feature-based attention-guided Siamese network for building CD. Lei *et al.* [21] propose a hierarchical paired channel fusion network for street scene CD. The aforementioned methods have shown great success in feature learning for CD. However, these methods have limited global representation capabilities and usually focus on local regions of changed objects. We find that Transformers have strong characteristics in extracting global features. Thus, different from previous works, we take the advantages of Transformers, and propose a new learning framework for more discriminative feature representations.

2.2 Vision Transformers for Change Detection

Recently, Transformers [42] have been applied to many computer vision tasks, such as image classification [9,29], object detection [4], semantic segmentation [44], person re-identification [27,51] and so on. Inspired by that, Zhang *et al.* [50] deploy a Swin Transformer [29] with a U-Net [35] structure for remote sensing image CD. Zheng *et al.* [56] design a deep Multi-task Encoder-Transformer-Decoder (METD) architecture for semantic CD. Wang *et al.* [45] incorporate a Siamese Vision Transformer (SViT) into a feature difference framework for CD. To take the advantages of both Transformers and CNNs, Wang *et al.* [43] propose to combine a Transformer and a CNN for remote sensing image CD. Li *et al.* [24] propose an encoding-decoding hybrid framework for CD, which has the advantages of both Transformers and U-Net. Bandara *et al.* [3] unify hierarchically structured Transformer encoders with Multi-Layer Perception (MLP) decoders in a Siamese network to efficiently render multi-scale long-range details for accurate CD. Chen *et al.* [5] propose a Bitemporal Image Transformer (BIT) to efficiently and effectively model contexts within the spatial-temporal domain for CD. Ke *et al.* [20] propose a hybrid Transformer with token aggregation for remote sensing image CD. Song *et al.* [39] combine the multi-scale Swin Transformer and a deeply-supervised network for CD. All these methods have shown that Transformers can model the inter-patch relations for strong feature representations. However, these methods do not take the full abilities of Transformers in multi-level feature learning. Different from existing Transformer-based CD methods, our proposed approach improves the feature extraction from a global view and combines multi-level visual features in a pyramid manner.

3 Proposed Approach

As shown in Fig. 1, the proposed framework includes three key components, *i.e.*, Siamese Feature Extraction (SFE), Deep Feature Enhancement (DFE) and Progressive Change Prediction (PCP). By taking dual-phase remote sensing images as inputs, SFE first extracts multi-level visual features through two shared Swin Transformers. Then, DFE utilizes the multi-level visual features to generate summation features and difference features, which highlight the change regions with temporal information. Finally, by integrating all above features, PCP introduces

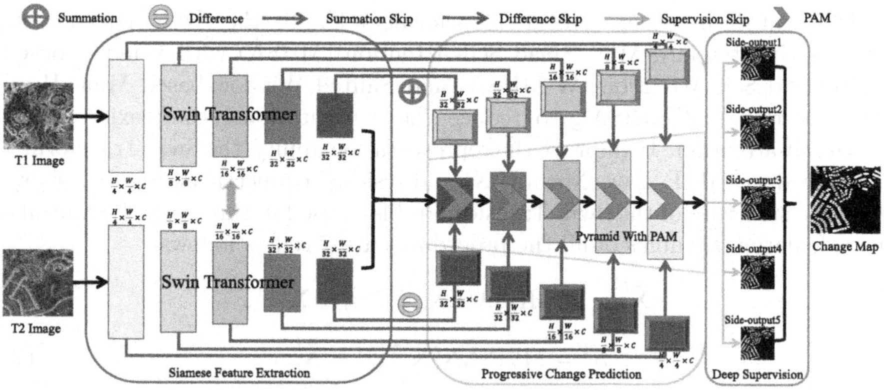

Fig. 1. The overall structure of our proposed framework.

a pyramid structure grafted with a Progressive Attention Module (PAM) for the final CD prediction. To train our framework, we introduce the deeply-supervised learning with multiple boundary-aware loss functions for each feature level. We will elaborate these key modules in the following subsections.

3.1 Siamese Feature Extraction

Following previous works, we introduce a Siamese structure to extract multi-level features from the dual-phase remote sensing images. More specifically, the Siamese structure contains two encoder branches, which share learnable weights and are used for the multi-level feature extraction of images at temporal phase 1 (T1) and temporal phase 2 (T2), respectively. As shown in the left part of Fig. 1, we take the Swin Transformer [29] as the basic backbone of the Siamese structure, which involves five stages in total.

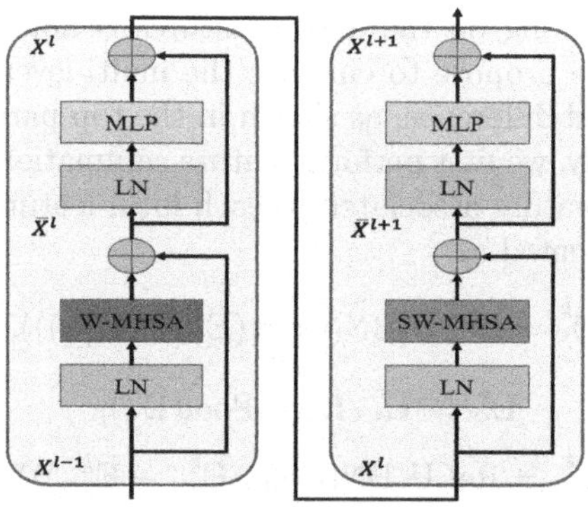

Fig. 2. The basic structure of the used Swin Transformer block.

Different from other typical Transformers [9,42], the Swin Transformer replaces the standard Multi-Head Self-Attention (MHSA) with Window-based Multi-Head Self-Attention (W-MHSA) and Shifted Window-based Multi-Head Self-Attention (SW-MHSA), to reduce the computational complexity of the global self-attention. To improve the representation ability, the Swin Transformer also introduces MLP, LayerNorm (LN) and residual connections. Figure 2 shows the basic structure of the Swin Transformer block used in this work. Technically, the calculation formulas of all the procedures are given as follows:

$$\bar{\mathbf{X}}^l = \text{W-MHSA}(\text{LN}(\mathbf{X}^{l-1})) + \mathbf{X}^{l-1}, \tag{1}$$

$$\mathbf{X}^l = \text{MLP}(\text{LN}(\bar{\mathbf{X}}^{l-1})) + \bar{\mathbf{X}}^l, \tag{2}$$

$$\bar{\mathbf{X}}^{l+1} = \text{SW-MHSA}(\text{LN}(\mathbf{X}^l)) + \mathbf{X}^l, \tag{3}$$

$$\mathbf{X}^{l+1} = \text{MLP}(\text{LN}(\bar{\mathbf{X}}^{l+1})) + \bar{\mathbf{X}}^{l+1}, \tag{4}$$

where $\bar{\mathbf{X}}$ is the output of the W-MHSA or SW-MHSA module and $\mathbf{X}$ is the output of the MLP module. At each stage of the original Swin Transformers, the feature resolution is halved, while the channel dimension is doubled. More specifically, the feature resolution is reduced from $(H/4) \times (W/4)$ to $(H/32) \times (W/32)$, and the channel dimension is increased from C to $8C$. In order to take advantages of global-level information, we introduce an additional Swin Transformer block to enlarge the receptive field of the feature maps. Besides, to reduce the computation, we uniformly reduce the channel dimension to C, and generate encoded features $[\mathbf{E}_{T1}^1, \mathbf{E}_{T1}^2, ..., \mathbf{E}_{T1}^5]$ and $[\mathbf{E}_{T2}^1, \mathbf{E}_{T2}^2, ..., \mathbf{E}_{T2}^5]$ for the T1 and T2 images, respectively. Based on the shared Swin Transformers, the multi-level visual features can be extracted. In general, features in the high-level capture more global semantic information, while features in the low-level retain more local detail information. Both of them help the detection of change regions.

3.2 Deep Feature Enhancement

In complex scenarios, there are many visual challenges for remote sensing image CD. Thus, only depending on the above features is not enough. To highlight the change regions, we propose to enhance the multi-level visual features with feature summation and difference, as shown in the top part and bottom part of Fig. 1. More specifically, we first perform feature summation and difference, then introduce a contrast feature associated to each local feature [30]. The enhanced features can be represented as:

$$\bar{\mathbf{E}}_S^k = \text{ReLU}(\text{BN}(\text{Conv}(\mathbf{E}_{T1}^k + \mathbf{E}_{T2}^k))), \tag{5}$$

$$\mathbf{E}_S^k = [\bar{\mathbf{E}}_S^k, \bar{\mathbf{E}}_S^k - \text{Pool}(\bar{\mathbf{E}}_S^k)], \tag{6}$$

$$\bar{\mathbf{E}}_D^k = \text{ReLU}(\text{BN}(\text{Conv}(\mathbf{E}_{T1}^k - \mathbf{E}_{T2}^k))), \tag{7}$$

$$\mathbf{E}_D^k = [\bar{\mathbf{E}}_D^k, \bar{\mathbf{E}}_D^k - \text{Pool}(\bar{\mathbf{E}}_D^k)], \tag{8}$$

where $\mathbf{E}_S^k$ and $\mathbf{E}_D^k$ $(k = 1, 2, ..., 5)$ are the enhanced features with point-wise summation and difference, respectively. ReLU is the rectified linear unit, BN is the batch normalization, Conv is a 1×1 convolution, and Pool is a 3×3 average pooling with padding=1 and stride=1. [,] is the concatenation operation in channel. Through the proposed DFE, change regions and boundaries are highlighted with temporal information. Thus, the framework can make the extracted features more discriminative and obtain better CD results.

3.3 Progressive Change Prediction

Since change regions can be any shapes and appear in any scales, we should consider the CD predictions at various cases. Inspired by the feature pyramid [25], we propose a progressive change prediction, as shown in the middle part of Fig. 1. To improve the representation ability, a pyramid structure with a Progressive Attention Module (PAM) is utilized with additional interdependencies through channel attentions. The structure of the proposed PAM is illustrated in Fig. 3. It first takes the summation features and difference features as inputs, then a channel-level attention is applied to enhance the features related to change regions. Besides, we also introduce a residual connection to improve the learning ability. The final feature map can be obtained by a 1×1 convolution. Formally, the PAM can be represented as:

$$\mathbf{F}^k = \mathrm{ReLU}(\mathrm{BN}(\mathrm{Conv}([\mathbf{E}_S^k, \mathbf{E}_D^k]))), \tag{9}$$

$$\mathbf{F}_A^k = \mathbf{F}^k * \sigma(\mathrm{Conv}(\mathrm{GAP}(\mathbf{F}^k))) + \mathbf{F}^k, \tag{10}$$

where σ is the Sigmoid function and GAP is the global average pooling.

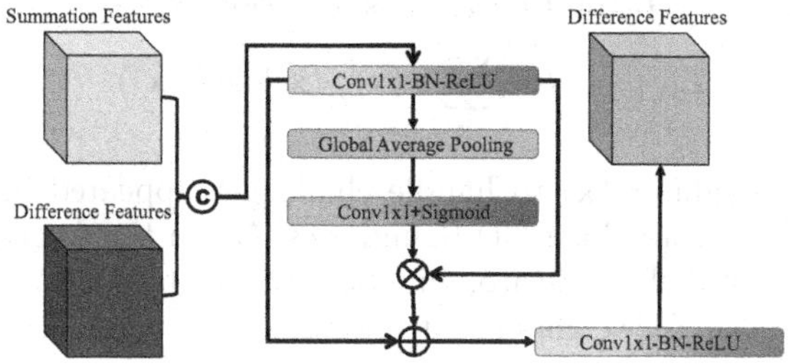

Fig. 3. The structure of our proposed Progressive Attention Module (PAM).

To achieve the progressive change prediction, we build the decoder pyramid grafted with a PAM as follows:

$$\mathbf{F}_P^k = \begin{cases} \mathbf{F}_A^k, & k = 5, \\ \mathrm{UM}(\mathrm{SwinBlock}^n(\mathbf{F}_P^{k+1}))) + \mathbf{F}_A^k, & 1 \leq k < 5. \end{cases} \tag{11}$$

where UM is the patch unmerging block used in Swin Transformers for upsampling, and $\mathrm{SwinBlock}^n$ is the Swin Transformer block with n layers. From the above formula, one can see that our PCP can make full use of the interdependencies within channels, and can progressively aggregate multi-level visual features to improve the perception ability of the change regions.

3.4 Loss Function

To optimize our framework, we adopt the deeply-supervised learning [53–55] with multiple boundary-aware loss functions for each feature level. The overall loss is defined as the summation over all side-outputs and the final fusion prediction:

$$\mathcal{L} = \mathcal{L}^f + \sum_{s=1}^{S} \alpha_s \mathcal{L}^s \tag{12}$$

where $\mathcal{L}^f$ is the loss of the final fusion prediction and $\mathcal{L}^s$ is the loss of the s-th side-output, respectively. S denotes the total number of the side-outputs and α_s is the weight for each level loss. In this work, our method includes five side-outputs, *i.e.*, $S = 5$.

To obtain complete CD regions and regular CD boundaries, we define $\mathcal{L}^f$ or $\mathcal{L}^s$ as a combined loss with three terms:

$$\mathcal{L}^{f/s} = \mathcal{L}_{WBCE} + \mathcal{L}_{SSIM} + \mathcal{L}_{SIoU}, \tag{13}$$

where $\mathcal{L}_{WBCE}$ is the weighted binary cross-entropy loss, $\mathcal{L}_{SSIM}$ is the structural similarity loss and $\mathcal{L}_{SIoU}$ is the soft intersection over union loss. The $\mathcal{L}_{WBCE}$ provides a probabilistic measure of similarity between the prediction and ground truth from a pixel-level view. The $\mathcal{L}_{SSIM}$ captures the structural information of change regions in patch-level. The $\mathcal{L}_{SIoU}$ is inspired by measuring the similarity of two sets, and yields a global similarity in map-level. More specifically, given the ground truth probability $g_l(\mathbf{x})$ and the estimated probability $p_l(\mathbf{x})$ at pixel $\mathbf{x}$ to belong to the class l, the $\mathcal{L}_{WBCE}$ loss function is

$$\mathcal{L}_{WBCE} = -\sum_{\mathbf{x}} w(\mathbf{x}) g_l(\mathbf{x}) \log(p_l(\mathbf{x})). \tag{14}$$

Here, we utilize weights $w(\mathbf{x})$ to handle challenges appeared in CD: the class imbalance and the errors along CD boundaries. Given the frequency f_l of class l in the training data, the indicator function I, the training prediction P, and the gradient operator ∇, weights are defined as:

$$w(\mathbf{x}) = \sum_{l} I(P(\mathbf{x} == l)) \frac{median(\mathbf{f})}{f_l} + w_0 I(|\nabla P(\mathbf{x})| > 0), \tag{15}$$

where $\mathbf{f} = [f_1, ..., f_L]$ is the vector of all class frequencies. The first term models median frequency balancing [2] to handle the class imbalance problem by highlighting classes with low probability. The second term assigns higher weights on the CD boundaries to emphasize on the correct prediction of boundaries.

The $\mathcal{L}_{SSIM}$ loss considers a local neighborhood of each pixel [46]. Let $\hat{\mathbf{x}} = \{\mathbf{x}_j : j = 1, ..., N^2\}$ and $\hat{\mathbf{y}} = \{\mathbf{y}_j : j = 1, ..., N^2\}$ be the pixel values of two corresponding patches (size: $N \times N$) cropped from the prediction P and the ground truth G respectively, the $\mathcal{L}_{SSIM}$ loss is defined as:

$$\mathcal{L}_{SSIM} = 1 - \frac{(2\mu_{\mathbf{x}}\mu_{\mathbf{x}} + \epsilon)(2\sigma_{\mathbf{xy}} + \epsilon)}{(\mu_{\mathbf{x}}^2 + \mu_{\mathbf{y}}^2 + \epsilon)(\sigma_{\mathbf{x}}^2 + \sigma_{\mathbf{y}}^2 + \epsilon)}, \tag{16}$$

where $\mu_{\mathbf{x}}$, $\mu_{\mathbf{y}}$ and $\sigma_{\mathbf{x}}$, $\sigma_{\mathbf{y}}$ are the mean and standard deviations of $\hat{\mathbf{x}}$ and $\hat{\mathbf{y}}$ respectively. $\sigma_{\mathbf{xy}}$ is their covariance. $\epsilon = 10^{-4}$ is used to avoid dividing by zero.

In this work, one metric of interest at test time is the Intersection over Union (IoU). Thus, we also introduce the soft IoU loss [32], which is differentiable for learning. The $\mathcal{L}_{SIoU}$ is defined as:

$$\mathcal{L}_{SIoU} = 1 - \frac{\sum_{\mathbf{x}} p_l(\mathbf{x}) g_l(\mathbf{x})}{\sum_{\mathbf{x}} [p_l(\mathbf{x}) + g_l(\mathbf{x}) - p_l(\mathbf{x}) g_l(\mathbf{x})]}. \tag{17}$$

When utilizing all above losses, the $\mathcal{L}_{WBCE}$ loss can relieve the imbalance problem for change pixels, the $\mathcal{L}_{SSIM}$ loss highlights the local structure of change boundaries, and the $\mathcal{L}_{SIoU}$ loss gives more focus on the change regions. Thus, we can obtain better CD results and make the framework easier to optimize.

4 Experiments

4.1 Datasets

LEVIR-CD [6] is a public large-scale CD dataset. It contains 637 remote sensing image pairs with a 1024×1024 resolution (0.5 m). We follow its default dataset split, and crop original images into small patches of size 256×256 with no overlapping. Therefore, we obtain 7120/1024/2048 pairs of image patches for training/validation/test, respectively.

WHU-CD [18] is a public building CD dataset. It contains one pair of high-resolution (0.075m) aerial images of size 32507×15354. As no definite data split is widely-used, we crop the original image into small patches of size 256×256 with no overlap and randomly split it into three parts: 6096/762/762 for training/validation/test, respectively.

SYSU-CD [37] is also a public building CD dataset. It contains 20000 pairs of high-resolution (0.5 m) images of size 256×256. We follow its default dataset split for experiments. There are 12000/4000/4000 pairs of image patches for training/validation/test, respectively.

Google-CD [26] is a very recent and public CD dataset. It contains 19 image pairs, originating from Google Earth Map. The image resolutions are ranging from 1006×1168 pixels to 4936×5224 pixels. We crop the images into small patches of size 256×256 with no overlap and randomly split it into three parts: 2504/313/313 for training/validation/test, respectively.

4.2 Evaluation Metrics

To verify the performance, we follow previous works [3,49] and mainly utilize F1 and Intersection over Union (IoU) scores with regard to the change-class as the primary evaluation metrics. Additionally, we also report the precision and recall of the change category and overall accuracy (OA).

4.3 Implementation Details

We perform experiments with the public MindSpore toolbox and one NVIDIA A30 GPU. We used the mini-batch SGD algorithm to train our framework with an initial learning rate 10^{-3}, moment 0.9 and weight decay 0.0005. The batch size is set to 6. For the Siamese feature extraction backbone, we adopt the Swin Transformer pre-trained on ImageNet-22k classification task [8]. To fit the input size of the pre-trained Swin Transformer, we uniformly resize image patches to 384×384. For other layers, we randomly initialize them and set the learning rate with 10 times than the initial learning rate. We train the framework with 100 epochs. The learning rate decreases to the 1/10 of the initial learning rate at every 20 epoch. To improve the robustness, data augmentation is performed by random rotation and flipping of the input images. For the loss function in the model training, the weight parameters of each level are set equally. For model reproduction, the source code is released at https://github.com/AI-Zhpp/FTN.

4.4 Comparisons with Sate-of-the-Arts

In this section, we compare the proposed method with other outstanding methods on four public CD datasets. The experimental results fully verify the effectiveness of our proposed method.

Table 1. Quantitative comparisons on LEVIR-CD and WHU-CD datasets.

Methods	LEVIR-CD					WHU-CD				
	Pre.	Rec.	F1	IoU	OA	Pre.	Rec.	F1	IoU	OA
FC-EF [7]	86.91	80.17	83.40	71.53	98.39	71.63	67.25	69.37	53.11	97.61
FC-Siam-Diff [7]	89.53	83.31	86.31	75.92	98.67	47.33	77.66	58.81	41.66	95.63
FC-Siam-Conc [7]	91.99	76.77	83.69	71.96	98.49	60.88	73.58	66.63	49.95	97.04
BiDateNet [28]	85.65	89.98	87.76	78.19	98.52	78.28	71.59	74.79	59.73	81.92
U-Net++MSOF [34]	90.33	81.82	85.86	75.24	98.41	91.96	89.40	90.66	82.92	96.98
DTCDSCN [28]	88.53	86.83	87.67	78.05	98.77	63.92	82.30	71.95	56.19	97.42
DASNet [28]	80.76	79.53	79.91	74.65	94.32	68.14	73.03	70.50	54.41	97.29
STANet [6]	83.81	**91.00**	87.26	77.40	98.66	79.37	85.50	82.32	69.95	98.52
MSTDSNet [39]	85.52	90.84	88.10	78.73	98.56	—	—	—	—	—
IFNet [49]	**94.02**	82.93	88.13	78.77	98.87	**96.91**	73.19	83.40	71.52	98.83
SNUNet [10]	89.18	87.17	88.16	78.83	98.82	85.60	81.49	83.50	71.67	98.71
BIT [5]	89.24	89.37	89.31	80.68	98.92	86.64	81.48	83.98	72.39	98.75
H-TransCD [20]	91.45	88.72	90.06	81.92	99.00	93.85	88.73	91.22	83.85	99.24
ChangeFormer [3]	92.05	88.80	90.40	82.48	99.04	91.83	88.02	89.88	81.63	99.12
Ours	92.71	89.37	**91.01**	**83.51**	**99.06**	93.09	**91.24**	**92.16**	**85.45**	**99.37**

Quantitative Comparisons. We present the comparative results in Table 1 and Table 2. The results show that our method delivers excellent performance. More specifically, our method achieves the best F1 and IoU values of 91.01% and 83.51% on the LEVIR-CD dataset, respectively. They are much better than previous methods. Besides, compared with other Transformer-based methods, such as BIT [5], H-TransCD [20] and ChangeFormer [3], our method shows consistent improvements in terms of all evaluation metrics. On the WHU-CD dataset, our method shows significant improvement with the F1 and IoU values of 92.16% and 85.45%, respectively. Compared with the second-best method, our method improves the F1 and IoU values by 0.9% and 1.6%, respectively. On the SYSU-CD dataset, our method achieves the F1 and IoU values of 81.53% and 68.82%, respectively. The SYSU-CD dataset includes more large-scale change regions. We think the improvements are mainly based on the proposed DFE. On the Google-CD dataset, our method shows much better results than compared methods. In fact, our method achieves the F1 and IoU values of 85.58% and 74.79%, respectively. We note that the Google-CD dataset is recently proposed and it is much challenging than other three datasets. We also note that the performance of precision, recall and OA is not consistent in all methods. Our method generally achieve better recall values than most methods. The main reason may be that our method gives higher confidences to the change regions.

Table 2. Quantitative comparisons on SYSU-CD and Google-CD datasets.

Methods	SYSU-CD					Google-CD				
	Pre.	Rec.	F1	IoU	OA	Pre.	Rec.	F1	IoU	OA
FC-EF [7]	74.32	75.84	75.07	60.09	86.02	80.81	64.39	71.67	55.85	85.85
FC-Siam-Diff [7]	**89.13**	61.21	72.57	56.96	82.11	85.44	63.28	72.71	57.12	87.27
FC-Siam-Conc [7]	82.54	71.03	76.35	61.75	86.17	82.07	64.73	72.38	56.71	84.56
BiDateNet [28]	81.84	72.60	76.94	62.52	89.74	78.28	71.59	74.79	59.73	81.92
U-Net++MSOF [34]	81.36	75.39	78.26	62.14	86.39	91.21	57.60	70.61	54.57	95.21
DASNet [28]	68.14	70.01	69.14	60.65	80.14	71.01	44.85	54.98	37.91	90.87
STANet [6]	70.76	**85.33**	77.37	63.09	87.96	**89.37**	65.02	75.27	60.35	82.58
DSAMNet [49]	74.81	81.86	78.18	64.18	89.22	72.12	80.37	76.02	61.32	94.93
MSTDSNet [39]	79.91	80.76	80.33	67.13	90.67	—	—	—	—	—
SRCDNet [26]	75.54	81.06	78.20	64.21	89.34	83.74	71.49	77.13	62.77	83.18
BIT [5]	82.18	74.49	78.15	64.13	90.18	92.04	72.03	80.82	67.81	96.59
H-TransCD [20]	83.05	77.40	80.13	66.84	90.95	85.93	81.73	83.78	72.08	97.64
Ours	86.86	76.82	**81.53**	**68.82**	**91.79**	86.99	**84.21**	**85.58**	**74.79**	**97.92**

Qualitative Comparisons. To illustrate the visual effect, we display some typical CD results on the four datasets, as shown in Fig. 4. From the results, we can see that our method generally shows best CD results. For example, when change regions have multiple scales, our method can correctly identify most of them, as shown in the first row. When change objects cover most of the image regions, most of current methods can not detect them. However, our method can still detect them with clear boundaries, as shown in the second row. In addition, when change regions appear in complex scenes, our method can maintain the contour shape. While most of compared methods fail, as shown in the third row. When distractors appear, our method can reduce the effect and correctly detect change regions, as shown in the fourth row. From these visual results, we can see that our method shows superior performance than most methods.

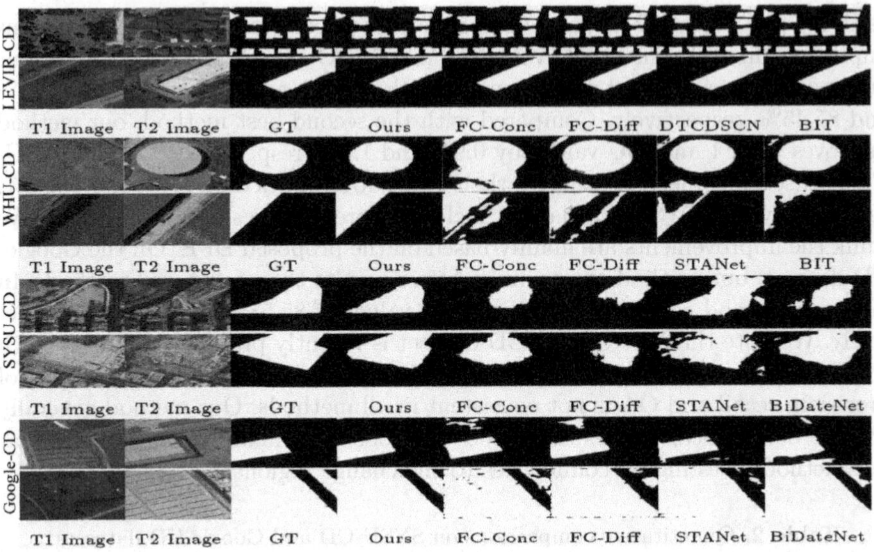

Fig. 4. Comparison of typical change detection results on four CD datasets.

To further verify the effectiveness, we provide more hard samples in Fig. 5. As can be seen, our method performs better than most methods (1st row). Most of current methods can not detect the two small change regions in the center, while our method can accurately localize them. Besides, we also show failed examples in the second row of Fig. 5. As can be seen, all compared methods can not detect all the change regions. However, our method shows more reasonable results.

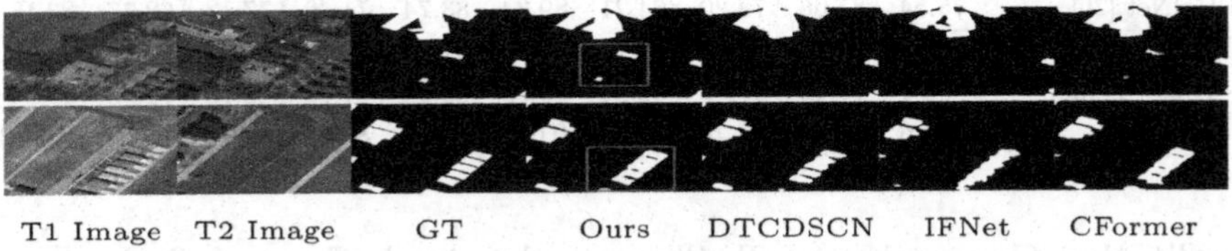

Fig. 5. Comparison of typical change detection results on hard and failed samples.

4.5 Ablation Study

In this subsection, we perform extensive ablation studies to verify the effect of key components in our framework. The experiments are conducted on LEVIR-CD dataset. However, other datasets have similar performance trends.

Effects of Different Siamese Backbones. As shown in the 2–3 rows of Table 3, we introduce the VGGNet-16 [38] and Swin Transformer as Siamese backbones. To ensure a fair comparison, we utilize the basic Feature Pyramid (FP) structure [25]. From the results, one can see that the performance with the Swin Transformer can be consistently improved in terms of Recall, F1, IoU and OA. The main reason is that the Swin Transformer has a better ability of modeling long-range dependency than VGGNet-16.

Table 3. Performance comparisons with different model variants on LEVIR-CD.

Models	Pre.	Rec.	F1	IoU	OA
(a) VGGNet-16+FP	91.98	82.65	87.06	77.09	98.75
(b) SwinT+FP	91.12	87.42	89.23	80.56	98.91
(c) SwinT+DFE+FP	91.73	88.43	90.05	81.89	99.00
(d) SwinT+DFE+PCP	92.71	89.37	91.01	83.51	99.06

Effects of DFE. The fourth row of Table 3 shows the effect of our proposed DFE. When compared with the $Model(b)\ SwinT + FP$, DFE improves the F1 value from 89.23% to 90.05%, and the IoU value from 80.56% to 81.89%, respectively. The main reason is that our DFE considers the temporal information with feature summation and difference, which highlight change regions.

Effects of PCP. In order to better detect multi-scale change regions, we introduce the PCP, which is a pyramid structure grafted with a PAM. We compare it with FP. From the results in the last row of Table 3, one can see that our PCP achieves a significant improvement in all metrics. Furthermore, adding the PCP also achieves a better visual effect, in which the extracted change regions are complete and the boundaries are regular, as shown in Fig. 6.

Fig. 6. Visual comparisons of predicted change maps with different models.

In addition, we also introduce the Swin Transformer blocks in the PCP as shown in Eq. 11. To verify the effect of different layers, we report the results in Table 4. From the results, we can see that the models show better results with equal layers. The best results can be achieved with $n = 4$. With more layers, the computation is larger and the performance decreases in our framework.

Effects of Different Losses. In this work, we introduce multiple loss functions to improve the CD results. Table 5 shows the effects of these losses. It can be seen that using the WBCE loss can improve the F1 value from 88.75% to 90.01% and the IoU from 79.78% to 81.83%. Using the SSIM loss achieves the F1 value

Table 4. Performance comparisons with different decoder layers on LEVIR-CD.

Layers	Pre.	Rec.	F1	IoU	OA
(2,2,2,2)	91.18	87.00	89.04	80.24	98.90
(4,4,4,4)	91.65	88.42	90.01	81.83	99.00
(6,6,6,6)	91.70	88.30	89.96	81.76	98.99
(8,8,8,8)	91.55	88.47	89.98	81.79	98.99
(2,4,6,8)	92.13	85.71	88.80	79.86	98.89

Table 5. Performance comparisons with different losses on LEVIR-CD.

Losses	Pre.	Rec.	F1	IoU	OA
BCE	90.68	86.91	88.75	79.78	98.88
WBCE	91.65	88.42	90.01	81.83	99.00
WBCE+SSIM	91.71	88.57	90.11	82.27	99.01
WBCE+SSIM+SIoU	92.71	89.37	91.01	83.51	99.06

of 90.11% and the IoU of 82.27%. Using the SIoU loss achieves the F1 value of 91.01% and the IoU of 83.51%. In fact, combining all of them can achieve the best results, which prove the effectiveness of all loss terms.

More Structure Discussions. There are some key differences between our work and previous fully Transformer structures: The works in [13,47] are taking single images as inputs and using an encoder-decoder structure. However, our framework utilizes a Siamese structure to process dual-phase images. In order to fuse features from two encoder streams, we propose a pyramid structure grafted with a PAM for the final CD prediction. Thus, apart from the input difference, our work progressively aggregates multi-level features for feature enhancement.

5 Conclusion

In this work, we propose a new learning framework named FTN for change detection of dual-phase remote sensing images. Technically, we first utilizes a Siamese network with the pre-trained Swin Transformers to extract long-range dependency information. Then, we introduce a pyramid structure to aggregate multi-level visual features, improving the feature representation ability. Finally, we utilize the deeply-supervised learning with multiple loss functions for model training. Extensive experiments on four public CD benchmarks demonstrate that our proposed framework shows better performances than most state-of-the-art methods. In future works, we will explore more efficient structures of Transformers to reduce the computation and develop unsupervised or weakly-supervised methods to relieve the burden of remote sensing image labeling.

Acknowledgements. This work is partly sponsored by CAAI-Huawei MindSpore Open Fund (No.CAAIXSJLJJ-2021-067A), the College Students' Innovative Entrepreneurial Training Plan Program (No. 20221014141123) and the Fundamental Research Funds for the Central Universities (No. DUT20RC(3)083).

References

1. Alcantarilla, P.F., Stent, S., Ros, G., Arroyo, R., Gherardi, R.: Street-view change detection with deconvolutional networks. Auton. Robots **42**(7), 1301–1322 (2018). https://doi.org/10.1007/s10514-018-9734-5
2. Badrinarayanan, V., Kendall, A., Cipolla, R.: Segnet: a deep convolutional encoder-decoder architecture for image segmentation. IEEE Trans. Pattern Anal. Mach. Intell. **39**(12), 2481–2495 (2017)
3. Bandara, W.G.C., Patel, V.M.: A transformer-based siamese network for change detection. arXiv:2201.01293 (2022)
4. Carion, N., Massa, F., Synnaeve, G., Usunier, N., Kirillov, A., Zagoruyko, S.: End-to-end object detection with transformers. In: Vedaldi, A., Bischof, H., Brox, T., Frahm, J.-M. (eds.) ECCV 2020. LNCS, vol. 12346, pp. 213–229. Springer, Cham (2020). https://doi.org/10.1007/978-3-030-58452-8_13
5. Chen, H., Qi, Z., Shi, Z.: Remote sensing image change detection with transformers. IEEE Trans. Geosci. Remote Sens. **60**, 1–14 (2021)
6. Chen, H., Shi, Z.: A spatial-temporal attention-based method and a new dataset for remote sensing image change detection. Remote Sens. **12**(10), 1662 (2020)
7. Daudt, R.C., Le Saux, B., Boulch, A.: Fully convolutional siamese networks for change detection. In: IEEE International Conference on Image Processing, pp. 4063–4067. IEEE (2018)
8. Deng, J., Dong, W., Socher, R., Li, L.J., Li, K., Fei-Fei, L.: Imagenet: a large-scale hierarchical image database. In: IEEE Conference on Computer Vision and Pattern Recognition, pp. 248–255 (2009)
9. Dosovitskiy, A., et al.: An image is worth 16×16 words: transformers for image recognition at scale. In: International Conference on Learning Representations, pp. 1–13 (2020)
10. Fang, S., Li, K., Shao, J., Li, Z.: Snunet-cd: a densely connected siamese network for change detection of vhr images. IEEE Geosci. Remote Sens. Lett. **19**, 1–5 (2021)
11. Guo, E., et al.: Learning to measure change: fully convolutional siamese metric networks for scene change detection. arXiv:1810.09111 (2018)
12. He, K., Zhang, X., Ren, S., Sun, J.: Deep residual learning for image recognition. In: IEEE Conference on Computer Vision and Pattern Recognition, pp. 770–778 (2016)
13. He, X., Tan, E.L., Bi, H., Zhang, X., Zhao, S., Lei, B.: Fully transformer network for skin lesion analysis. Med. Image Anal. **77**, 102357 (2022)
14. Hou, B., Wang, Y., Liu, Q.: Change detection based on deep features and low rank. IEEE Geosci. Remote Sens. Lett. **14**(12), 2418–2422 (2017)
15. Huang, G., Liu, Z., Van Der Maaten, L., Weinberger, K.Q.: Densely connected convolutional networks. In: IEEE Conference on Computer Vision and Pattern Recognition, pp. 4700–4708 (2017)
16. Huo, C., Zhou, Z., Lu, H., Pan, C., Chen, K.: Fast object-level change detection for vhr images. IEEE Geosci. Remote Sens. Lett. **7**(1), 118–122 (2009)

17. Ioffe, S., Szegedy, C.: Batch normalization: accelerating deep network training by reducing internal covariate shift. In: International Conference on Machine Learning, pp. 448–456 (2015)
18. Ji, S., Wei, S., Lu, M.: Fully convolutional networks for multisource building extraction from an open aerial and satellite imagery data set. IEEE Trans. Geosci. Remote Sens. **57**(1), 574–586 (2018)
19. Jiang, H., Hu, X., Li, K., Zhang, J., Gong, J., Zhang, M.: Pga-siamnet: pyramid feature-based attention-guided siamese network for remote sensing orthoimagery building change detection. Remote Sens. **12**(3), 484 (2020)
20. Ke, Q., Zhang, P.: Hybrid-transcd: a hybrid transformer remote sensing image change detection network via token aggregation. ISPRS Int. J. Geo-Inf. **11**(4), 263 (2022)
21. Lei, Y., Peng, D., Zhang, P., Ke, Q., Li, H.: Hierarchical paired channel fusion network for street scene change detection. IEEE Trans. Image Process. **30**, 55–67 (2020)
22. Lei, Y., Liu, X., Shi, J., Lei, C., Wang, J.: Multiscale superpixel segmentation with deep features for change detection. IEEE Access **7**, 36600–36616 (2019)
23. Li, L., Li, X., Zhang, Y., Wang, L., Ying, G.: Change detection for high-resolution remote sensing imagery using object-oriented change vector analysis method. In: IEEE International Geoscience and Remote Sensing Symposium, pp. 2873–2876. IEEE (2016)
24. Li, Q., Zhong, R., Du, X., Du, Y.: Transunetcd: a hybrid transformer network for change detection in optical remote-sensing images. IEEE Trans. Geosci. Remote Sens. **60**, 1–19 (2022)
25. Lin, T.Y., Dollár, P., Girshick, R., He, K., Hariharan, B., Belongie, S.: Feature pyramid networks for object detection. In: IEEE Conference on Computer Vision and Pattern Recognition, pp. 2117–2125 (2017)
26. Liu, M., Shi, Q., Marinoni, A., He, D., Liu, X., Zhang, L.: Super-resolution-based change detection network with stacked attention module for images with different resolutions. IEEE Trans. Geosci. Remote Sens. **60**, 1–18 (2021)
27. Liu, X., Zhang, P., Yu, C., Lu, H., Qian, X., Yang, X.: A video is worth three views: trigeminal transformers for video-based person re-identification. arXiv:2104.01745 (2021)
28. Liu, Y., Pang, C., Zhan, Z., Zhang, X., Yang, X.: Building change detection for remote sensing images using a dual-task constrained deep siamese convolutional network model. IEEE Geosci. Remote Sens. Lett. **18**(5), 811–815 (2020)
29. Liu, Z., et al.: Swin transformer: hierarchical vision transformer using shifted windows. In: IEEE/CVF International Conference on Computer Vision, pp. 10012–10022 (2021)
30. Luo, Z., Mishra, A., Achkar, A., Eichel, J., Li, S., Jodoin, P.M.: Non-local deep features for salient object detection. In: IEEE Conference on Computer Vision and Pattern Recognition, pp. 6609–6617 (2017)
31. Lv, N., Chen, C., Qiu, T., Sangaiah, A.K.: Deep learning and superpixel feature extraction based on contractive autoencoder for change detection in sar images. IEEE Trans. Ind. Inf. **14**(12), 5530–5538 (2018)
32. Máttyus, G., Luo, W., Urtasun, R.: Deeproadmapper: extracting road topology from aerial images. In: IEEE International Conference on Computer Vision, pp. 3438–3446 (2017)
33. Peng, D., Guan, H.: Unsupervised change detection method based on saliency analysis and convolutional neural network. J. Appl. Remote Sens. **13**(2), 024512 (2019)

34. Peng, D., Zhang, Y., Guan, H.: End-to-end change detection for high resolution satellite images using improved unet++. Remote Sens. **11**(11), 1382 (2019)
35. Ronneberger, O., Fischer, P., Brox, T.: U-Net: convolutional networks for biomedical image segmentation. In: Navab, N., Hornegger, J., Wells, W.M., Frangi, A.F. (eds.) MICCAI 2015. LNCS, vol. 9351, pp. 234–241. Springer, Cham (2015). https://doi.org/10.1007/978-3-319-24574-4_28
36. Saha, S., Bovolo, F., Bruzzone, L.: Unsupervised deep change vector analysis for multiple-change detection in vhr images. IEEE Trans. Geosci. Remote Sens. **57**(6), 3677–3693 (2019)
37. Shi, Q., Liu, M., Li, S., Liu, X., Wang, F., Zhang, L.: A deeply supervised attention metric-based network and an open aerial image dataset for remote sensing change detection. IEEE Trans. Geosci. Remote Sens. **60**, 1–16 (2021)
38. Simonyan, K., Zisserman, A.: Very deep convolutional networks for large-scale image recognition. arXiv:1409.1556 (2014)
39. Song, F., Zhang, S., Lei, T., Song, Y., Peng, Z.: Mstdsnet-cd: multiscale swin transformer and deeply supervised network for change detection of the fast-growing urban regions. IEEE Geosci. Remote Sens. Lett. **19**, 1–5 (2022)
40. Tan, K., Zhang, Y., Wang, X., Chen, Y.: Object-based change detection using multiple classifiers and multi-scale uncertainty analysis. Remote Sens. **11**(3), 359 (2019)
41. Tang, Y., Zhang, L., Huang, X.: Object-oriented change detection based on the kolmogorov-smirnov test using high-resolution multispectral imagery. Int. J. Remote Sens. **32**(20), 5719–5740 (2011)
42. Vaswani, A., et al.: Attention is all you need. Adv. Neural Inf. Process. Syst. **30**, 1–11 (2017)
43. Wang, G., Li, B., Zhang, T., Zhang, S.: A network combining a transformer and a convolutional neural network for remote sensing image change detection. Remote Sens. **14**(9), 2228 (2022)
44. Wang, Y., et al.: End-to-end video instance segmentation with transformers. In: IEEE Conference on Computer Vision and Pattern Recognition, pp. 8741–8750 (2021)
45. Wang, Z., Zhang, Y., Luo, L., Wang, N.: Transcd: scene change detection via transformer-based architecture. Optics Exp. **29**(25), 41409–41427 (2021)
46. Wang, Z., Simoncelli, E.P., Bovik, A.C.: Multiscale structural similarity for image quality assessment. In: The Thrity-Seventh Asilomar Conference on Signals, Systems & Computers, vol. 2, pp. 1398–1402. IEEE (2003)
47. Wu, S., Wu, T., Lin, F., Tian, S., Guo, G.: Fully transformer networks for semantic image segmentation. arXiv:2106.04108 (2021)
48. Xiaolu, S., Bo, C.: Change detection using change vector analysis from landsat tm images in wuhan. Procedia Environ. Sci. **11**, 238–244 (2011)
49. Zhang, C., et al.: A deeply supervised image fusion network for change detection in high resolution bi-temporal remote sensing images. ISPRS J. Photogram. Remote Sens. **166**, 183–200 (2020)
50. Zhang, C., Wang, L., Cheng, S., Li, Y.: Swinsunet: pure transformer network for remote sensing image change detection. IEEE Trans. Geosci. Remote Sens. **60**, 1–13 (2022)
51. Zhang, G., Zhang, P., Qi, J., Lu, H.: Hat: hierarchical aggregation transformers for person re-identification. In: Proceedings of the 29th ACM International Conference on Multimedia, pp. 516–525 (2021)

52. Zhang, H., Gong, M., Zhang, P., Su, L., Shi, J.: Feature-level change detection using deep representation and feature change analysis for multispectral imagery. IEEE Geosci. Remote Sens. Lett. **13**(11), 1666–1670 (2016)
53. Zhang, P., Liu, W., Wang, D., Lei, Y., Wang, H., Lu, H.: Non-rigid object tracking via deep multi-scale spatial-temporal discriminative saliency maps. Pattern Recogn. **100**, 107130 (2020)
54. Zhang, P., Wang, D., Lu, H., Wang, H., Ruan, X.: Amulet: aggregating multi-level convolutional features for salient object detection. In: IEEE International Conference on Computer Vision, pp. 202–211 (2017)
55. Zhang, P., Wang, L., Wang, D., Lu, H., Shen, C.: Agile amulet: real-time salient object detection with contextual attention. arXiv:1802.06960 (2018)
56. Zheng, Z., Zhong, Y., Tian, S., Ma, A., Zhang, L.: Changemask: deep multi-task encoder-transformer-decoder architecture for semantic change detection. ISPRS J. Photogram. Remote Sens. **183**, 228–239 (2022)
57. Zhou, Z., Rahman Siddiquee, M.M., Tajbakhsh, N., Liang, J.: UNet++: a nested U-net architecture for medical image segmentation. In: Stoyanov, D., et al. (eds.) DLMIA/ML-CDS -2018. LNCS, vol. 11045, pp. 3–11. Springer, Cham (2018). https://doi.org/10.1007/978-3-030-00889-5_1

Self-supervised Augmented Patches Segmentation for Anomaly Detection

Jun Long, Yuxi Yang, Liujie Hua(✉), and Yiqi Ou

Big Data Research Institute of Central South University, Changsha, Hunan, China
{junlong,yuxiyang,liujiehua}@csu.edu.cn

Abstract. In this paper, our goal is to detect unknown defects in high-resolution images in the absence of anomalous data. Anomaly detection is usually performed at image-level or pixel-level. Considering that pixel-level anomaly classification achieves better representation learning in a finer-grained manner, we regard data augmentation transforms as a self-supervised segmentation task from which to extract the critical and representative information from images. Due to the unpredictability of anomalies in real scenarios, we propose a novel abnormal sample simulation strategy which augmented patches are randomly pasted to original image to create a generalized anomalous pattern. Following the framework of self-supervised, segmenting augmented patches is used as a proxy task in the training phase to extract representation separating normal and abnormal patterns, thus constructing a one-class classifier with a robust decision boundary. During the inference phase, the classifier is used to perform anomaly detection on the test data, while directly determining regions of unknown defects in an end-to-end manner. Our experimental results on MVTec AD dataset and BTAD dataset demonstrate the proposed SSAPS outperforms any other self-supervised based methods in anomaly detection. Code is available at https://github.com/BadSeedX/SSAPS.

Keywords: Anomaly detection · Self-supervised learning · Data augmentation

1 Introduction

Anomaly detection is a technique for outlier detection, commonly used for risk control [1] or workpiece surface defect detection [2]. It is a critical and long-standing problem for the financial and manufacturing industries [3]. Suffering from anomalous data scarcity and unexpected patterns, the modeling process

Supported by Network Resources Management and Trust Evaluation Key Laboratory of Hunan Province, Central South University.

Supplementary Information The online version contains supplementary material available at https://doi.org/10.1007/978-3-031-26284-5_6.

L. Wang et al. (Eds.): ACCV 2022, LNCS 13842, pp. 93–107, 2023.
https://doi.org/10.1007/978-3-031-26284-5_6

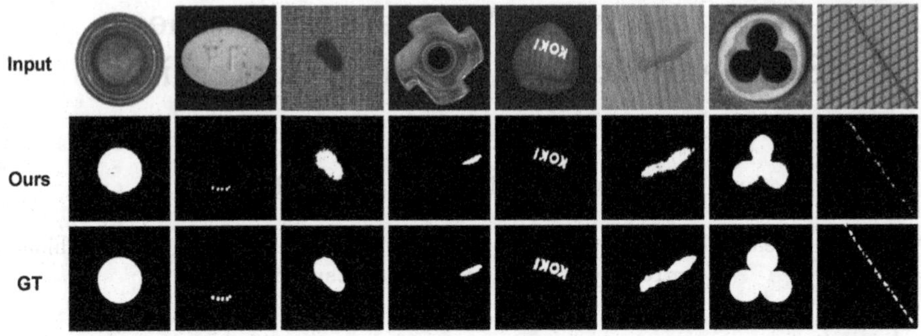

Fig. 1. Defect localization on some catagories of MVTec dataset. From top to bottom, input abnormal images, segmentation results generated by our method, and ground truth images.

of supervised learning is severely constrained. Therefore, unsupervised methods based on modeling of anomaly-free data are more general in industrial scenarios, and anomaly detection is usually expressed as a one-class classification problem [4–6], which attempts to construct the margin of normal samples. However, these approaches require complex feature matching during defect inference phase, resulting in excessive computational cost of the models. Furthermore, degradation occurs as a consequence of weak inter-class variation, also known as "hypersphere collapse" [5].

By designing a proxy task, self-supervised models [4,7,8] bypass the need for labeled data and serve as the effective proxy for supervised learning. These models preserve intra-class commonalities and extend inter-class variation by simulating anomalous patterns to learn visual representations. Nevertheless, the challenge raised by self-supervised methods is that relying on a pre-trained model risks feature extraction inconsistencies between the training source and target domains, which is insufficient to construct robust anomalous pattern margins. Due to the presence of tiny anomalous regions, models have a tendency to overfit and misclassify normal samples as abnormal samples [9]. Various data augmentation strategies [7,8,10] demonstrated a good ability to simulate anomalies in recent work. However, these approaches fail to overcome the limitation of the binary classification task for defect localization. Furthermore, these artificially created anomaly patterns are challenging to apply in real-world scenarios because of their sophisticated designs.

Given the shortcomings of above approaches, we propose an end-to-end self-supervised network based on segmentation of augmented patches. Our innovation lies in the design of a novel proxy task and the construction of a generative one-class classifier based on learned self-supervised representation to distinguish abnormal image from normal ones. Specifically, we use patches segmentation as a self-supervised proxy task to infer simulated anomalous regions by predicting augmented patches. Meanwhile, due to the diversity of real anomalies, the samples transformed by augmentation might only be considered as a rough approximation of the real defects. In fact, instead of introducing a method focused on simulating actual anomalies as is used in NSA [7], we design a generalized augmentation strategy that significantly increases irregularity of augmented patches in order to create more complex local irregular patterns with the aim of shaping a scalable normal pattern boundary by providing a common anomlous pattern that assists the network in learning outliers in a fine-grained manner.

In this work, we tackle the problem of one-class defect detection in high-resolution images using a proxy one-class classifier constructed during the inference phase based on visual representations obtained by segmenting augmented patches. We evaluate SSAPS on the MVTec AD [2] and BeanTech AD [11] dataset, both of which derive from real-life scenarios in industrial production. With 98.1% and 96.2% AUC on each dataset separately for image-level anomaly detection, our method outperforms existing self-supervised methods. SSAPS also exhibits strong anomaly segmentation abilities. We conduct an extensive study with various proxy tasks to prove the effectiveness of prediction augmented patches for unknown defect detection. The results also suggest that the introduction of a segmentation task to extract visual representations provides a well generalization ability for anomaly detection. The following are our contributions:

- We propose a novel data augmentation strategy which creates augmented samples for self-supervised learning by replacing irregular patches in images at random to simulate generalized anomalous patterns.
- We propose an augmented patches segmentation based proxy task to train a self-supervised residual deconvolution network to find the tiny variance between normal and abnormal patterns at pixel-level with the aim of practical needs of one-class anomaly detection.
- Through extensive experimental validation, SSAPS as a simulation of semantic segmentation in vision has high accuracy in surface defect detection while better meeting the requirements of industrial scenarios in a finer-grained manner.

2 Related Work

2.1 Embedding-Based Approach

Embeddding-based approaches [5,12–16] usually use a deep network pretrained on ImageNet [17] to extract visual representations in images, and define anomaly scores for test samples through various embedding-similarity metrics. Some works use one-class classification [5], multivariate Gaussian distribution [16] or improved k-nearest neighbour [15] to evaluate images that deviate from standard representation descriptions which are considered abnormal. Traditional embeddding based models [5] consider extracting features from the entire image and work with image-level embeddings, result in failing to explain the location of the defect. Recently some work extent embeddings to patch-level to obtain a more detailed visual representation. SPADE [15], based on KNN method, detects anomalies through the pixel-level correspondence between the normal images and the test images.

2.2 Reconstruction-Based Approach

Reconstruction-based approaches use auto-encoder [4,18,19] or generative adversarial network [10,20–23] to train a network to reconstruct the input images. They usually focus on the difference between the reconstruction output and the original input, where images containing defects tend to produce larger reconstruction errors. OCR-GAN [22], based on an omni-frequency reconstruction convolutions framework, hires a frequency decoupling module and a channel selection module to find differences in the frequency distribution of normal and abnormal images and complete the restruction process. InTra [23], based on a deep inpainting transformer, inpaints the masked patches in the large sequence of image patches, thereby integrating information across large regions of the input image and completing a image repair task. However, the downside of these methods is that it is hard to restore details of normal patterns, due to the strong learning representation ability and the huge capacity of neural network [24], reconstruction-based approaches are still affected by the uncertainty of the anomalous regions during inference which anomalous regions in the image are unexpectedly reconstructed well.

2.3 Self-supervised Learning-Based Approach

Self-supervised learning automatically generates supervisory signals under proxy tasks by mining the intrinsic properties of unlabelled data, thereby training the network to learn useful features that can be transfered to downstream tasks. Existing self-supervised anomaly detection methods, such as rotation prediction [25], geometric transformations prediction [26], block relative position prediction [4,27], and data augmentation classification prediction [7,8,28], share a common denominator that a classification task is preset based solely on the normal data, and the representation learned by network used for discovering unknown defects.

3 Methods

This section outlines the overall framework of our method. An overview of SSAPS is shown in Fig. 2. Following the general paradigm of self-supervised learning, SSAPS consists of a two-stage defect detection framework, aims at exploring local irregular patterns from the constructed augmented samples and attempts to segment the replaced patches, thus learning visual representations to construct a one-class anomaly classifier for inference of unknown defects in the test images. It contains several essential parts: AP module for augmenting samples by replacing irregular image patches in the original image (Sect. 3.1), self-supervised task based on patches segmentation and its loss function (Sect. 3.2), and inference phase (Sect. 3.3).

3.1 Add Augmented Patches to Image

This subsection will concentrate on the process of constructing an augmented sample. It seems hard to move irregular image patches directly, so we save different areas of the original input image I with the help of mask image M and

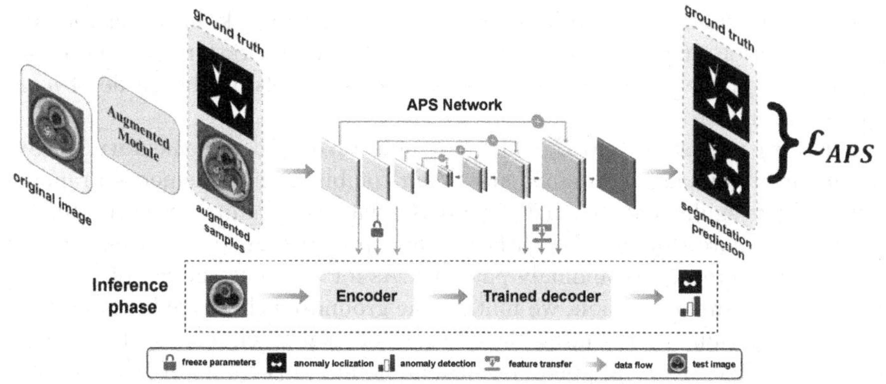

Fig. 2. An overview of our method. SSAPS is based on the framework for self-supervised learning. We get the augmented image and the ground truth image through Augmented module. The augmented image is then fed into the SSAPS network to obtain segmentation prediction. The process of the inference phase is similar to the training phase.

bottom image B, respectively, and then stack them up to achieve a partial irregular misalignment of the image. As shown in Fig. 3, the construction of augmented samples is mainly divided into the following steps.

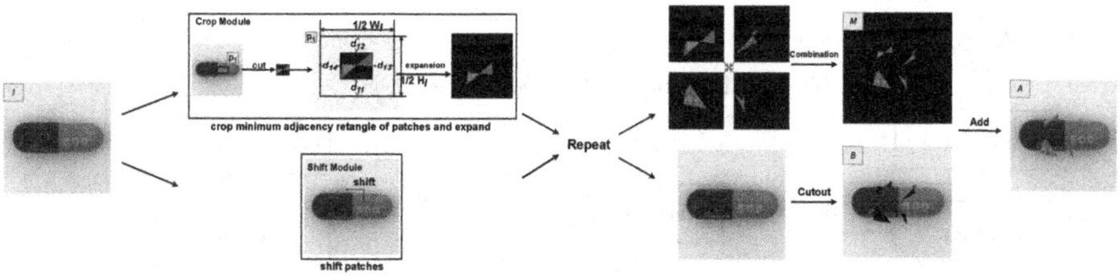

Fig. 3. Steps for the construction of simulation anomalous samples. In the first step, mask image M consists of multiple expansions of the minimum adjacency rectangle containing irregularly augmented patches. In the second step, bottom image B is made by masking in the corresponding position of the original image. Lastly, overlay mask image and bottom image.

For M, it comes from the expansion and combination of the augmented patches. We randomly select a series of coordinate points within a specified size region of I and connect them in turn to form an irregular polygon p. Then we cut out p by a minimum adjacent rectangle and randomly extend it around a distance d with RGB(0,0,0). Through a pair of "cut-expansion" operations, we get a quarter mask with half the length and width of the original image, which preserves a patch of I. The crop module is performed repeatedly and the quarter masks are combined into a mask image.

For B, we move p according to d and the quarter mask, ensuring that the relative position of p in B is the same as in M. In contrast to M, the image information on the outside of p is retained and all pixels on the inside of p are replaced using zero values.

For A, we overlay M and B, which resemble a pair of complementary images. The augmented image A obtained after their combination implements a random shift of the irregular patches in I. Compared with the traditional cutpaste, our method has more drastic edge dithering and provides greater convenience for the network to recognize anomaly patterns. As for the construction of the label identifying anomalous region, we imitate the ground-truth image in the test set with a binary value representing the category of each pixel, and create a matrix G of the same size as I, all points inside p are found in G to modify the category at the corresponding pixel location.

3.2 Augmented Patches Segmentation

In the traditional convolution architecture, the input image is converted into a multi-channel feature map after multiple downsampling. The pooling layer and the linear layer work collaboratively to convert it into a one-dimensional vector and provide it to the fully connected layer for downstream tasks. However, the flattened feature map fails to retain the anomaly regions information extracted by the convolution network, making defect localization difficult.

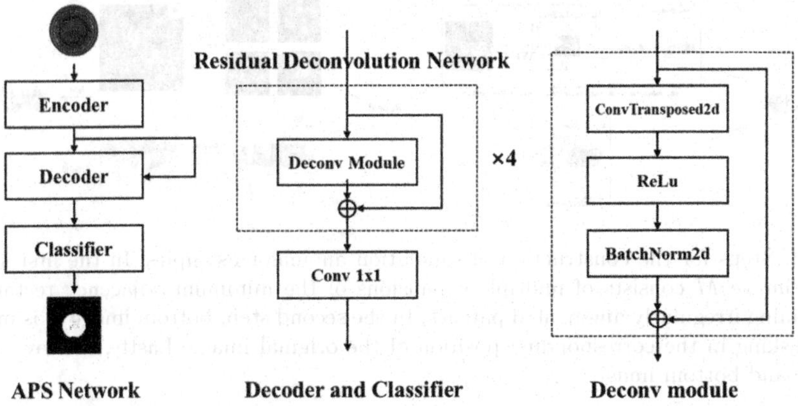

Fig. 4. The detailed structure of Augmented Patches Segmentation Network. Original image is extracted by the encoder(ConvNeXt) to obtain feature maps, and after 4 stacked inverse residual convolution modules to obtain a segmentation result of the same size as the input. Finally, the segmentation prediction is output by the classifier.

In our method, we make full use of the receptive field information from each stage of the convolution. Augmented Patches Segmentation network consists of a symmetrical encoder block and a decoder block. As shown in Fig. 4, we use ConvNeXt [29] as the encoder and save the feature map generated at each stage. Residual Deconvolution Network plays the role of decoder, which is stacked by several deconvolution modules, each module contains a BN (Batch Normalization) layer and a ReLU activation function. The deconvolution operation doubles the length and width of the feature map produced by the corresponding convolutional layers, while halving the channel. The output of each deconvolution module is then added with the upper stage's convolutional feature map as the input to the next layer. We treat the final output of the RDN as the extracted visual features, and perform anomaly classification on each pixel by a simple convolutional layer to obtain a binary image as the segmentation prediction of augmented patches.

$$\mathcal{L}_{APS} = E_{x \in X}\{BCE[seg(I), 0] + BCE[seg(A), (0,1)]\}. \tag{1}$$

The entire self-supervised network is optimized by SGD(Stochastic Gradient Descent), and the training data is obtained from Sect. 3.1, consisting of normal image and augmented image, with corresponding binary image labelled with the augmented regions. We use Eq. 1 as objective functions for self-supervised learning and compare the segmentation prediction with the corresponding ground truth, where X is the set of normal samples, $seg(\cdot)$ is the Augmented Patches Segmentation network, $E(\cdot)$ refers to the expectation and $BCE(\cdot,\cdot)$ refers to a binary cross-entropy loss. This loss enables model learn the representation of normal patterns whereas being sensitive to outliers.

This completes the "augmented or non-augmented" segmentation prediction for all pixels in the training image, which is a simulation of classifying "normal or abnormal" pixels in the test image. As demonstrated in Sect. 4.2, we believe that the completion of this siamese task is vital for anomaly detection and segmentation.

3.3 Inference

In this section, we will go over in detail how our proposed method detects and infers unknown defects. As demonstrated above, we use a holistic approach to generate artificially simulated abormal samples to provide supervised signals for the proxy task, and then learn visual representations by segmenting augmented regions to build a self-supervised feature extractor f. We introduce the Mahalanobis distance [30] to measure how many standard deviations away a given sample is from the mean of a distribution of normal samples [31], and use it as the anomaly score to detect real unknown defects in the test image.

We define $f(*)$ as the visual representation mapping obtained after the image has been processed by the self-supervised feature extractor. Equation 2 and Eq. 3 are used to compute the mean of distribution X_m and the covariance X_{conv} for the normal samples, where X is the set of normal samples with the length of N, $E(*)$ refers to the expectation.

$$X_m = E_{x \in X}(f(x)). \tag{2}$$

$$X_{conv} = \frac{1}{N-1}(f(x) - X_m) \cdot (f(x) - X_m)^{\top}. \quad (3)$$

At this point, we can obtain a feature hypersphere similar to the one obtained in the one-class classfication methods as a potential true pattern of train images. Then we calculate the anomaly score for the test image θ via Eq. 4, by comparing the mahalanobis distance between $f(\theta)$ and X_m.

$$score = \sqrt{(f(\theta) - X_m)^{\top} \cdot X_{conv}^{-1} \cdot (f(\theta) - X_m)}. \quad (4)$$

4 Experiment

In order to extensively evaluate the performance of SSAPS in anomaly detection, we conducted a series of ablation experiments on unsupervised datasets MVTec AD and BeanTech AD to evaluate the contribution of individual components of the proposed method and the effectiveness of segmentation-based self-supervised tasks. Additionally, performance of SSAPS on anomaly detection is compared with other recent models based on self-supervised learning, our method exhibit superior performance and outperform any of them.

4.1 Experiment Details

Our method follows the framework of the self-supervised learning [32], using ConvNeXt [29] as a feature extractor to encode the input image. We resize images in both the MVTec AD and BTAD datasets to 256×256, with each category trained for 300 iterations and the batch size set to 16, containing 8 normal samples and 8 abnormal samples. Their labels are binary images, with 0 for normal and 1 for abnormal. For the object categories in the dataset, we paste the irregular image patches closely to the centre of the image, while for the texture categories we pasted the patches evenly around because of the similarity of the entire image. More Experiment details are shown in Appendix.

4.2 Effectiveness of Segmentation Task

We utilize t-SNE [33] to evaluate the difference in distribution between the augmented samples and the real anomalous samples as well as the normal samples in order to better showcase the significance of the augmented patches in the segmentation task. t-SNE receives the output from the encoder and visualises representation by category. As shown in the Fig. 5, the three colors represent augmented samples, abnormal samples, and normal samples, respectively. The representation distributions of the augmented samples and the true abnormal samples overlap in space, while the augmented samples are separated from the normal data, which not only validates our proposed abnormal sample construction method, but also confirms the effectiveness of the segmentation task.

Notably, there are equally significant differences between the augmented samples and the true anomalous samples in some categories (Fig. 5.c), but this has no

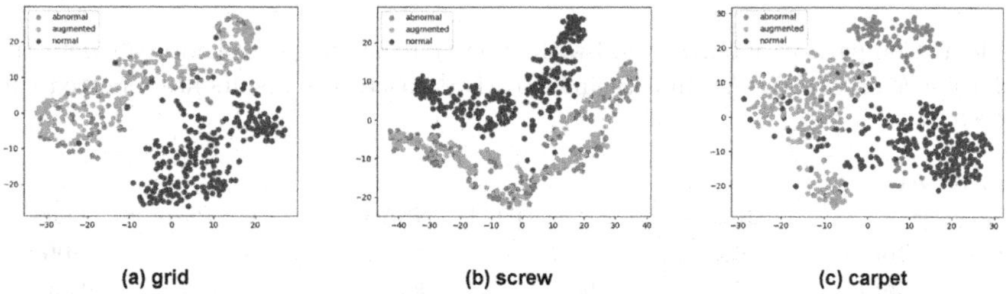

Fig. 5. t-SNE visualization of representations normal samples (green), real abnormal samples(blue), augmented samples constructed by our method (red) on some catagories in MVTEC dataset. (Color figure online)

effect on the anomaly detection results. Predicting the transformed pixels after augmentation is, in fact, a finer-grained binary classification task in which the model focuses on inter-class differences rather than requiring that the simulated anomalous samples have the same pattern distribution as the true anomalous samples. The inclusion of anomalous simulation samples in our proposed method allows the model to explicitly learn the difference between normal and abnormal patterns rather than constructing a supervised task that tends to overfit, thus ensuring excellent generalization of our method.

4.3 Comparison with Other SSL Models

We believe our proposed method based on segmentation of augmented patches outperforms other existing self-supervised proxy tasks in extracting difference between normal and abnormal patterns of images. To this end, we apply various proxy tasks for image-level anomaly detection on MVTec AD and BTAD dataset, and the results are shown in the Table 1. Our approach outperforms self-supervised methods that utilize other proxy training tasks. For the MVTec AD dataset, we achieve a near-perfect result in the texture category, as well as significant improvements in the majority of categories on objects comparing to other approaches. However, one of the poorly detected object categories is pill, where the analysis of the data suggests that some of the pill surface defects in the test set are small and barely different from the normal data, resulting in difficulties in being captured by the network and causing misclassification, and our model encounters some obstacles in dealing with these defects with weak anomalous features, as is also the case to some extent with the screw category, as is shown in Fig. 6. For the BeanTech AD dataset, the AUC detection score of SSAPS is still better than all the other methods, which demonstrates the strong generality of our method.

Table 1. Comparison of our method with others for the image-level anomaly detection performance on MVTec AD and BeanTech AD dataset. The results are reported with AUROC%

MVTec AD						
Category		R-Pred [26]	P-SVDD [4]	NSA [7]	CutPaste [8]	Ours
Object	**Bottle**	95.0	98.6	97.5	98.2	**100.0**
	Cable	85.3	90.3	90.2	81.2	96.9
	Capsule	71.8	76.7	92.8	**98.2**	97.8
	Hazelnut	83.6	92.0	89.3	98.3	**100.0**
	metal_nut	72.7	94.0	94.6	**99.9**	96.3
	Pill	79.2	86.1	94.3	**94.9**	90.6
	Screw	35.8	81.3	90.1	88.7	95.4
	Toothbrush	99.1	100.0	99.6	99.4	**100.0**
	Transistor	88.9	91.5	92.8	96.1	**97.9**
	Zipper	74.3	97.9	99.5	**99.9**	98.6
Object AVG		78.6	90.8	94.1	95.5	**97.4**
Texture	**Carpet**	29.7	92.9	90.9	93.9	**97.8**
	Grid	60.5	94.6	98.5	100.0	**100.0**
	Leather	55.2	90.9	100.0	100.0	**100.0**
	Tile	70.1	97.8	100.0	94.6	**100.0**
	Wood	95.8	96.5	97.8	99.1	**100.0**
Texture AVG		62.3	94.5	97.5	97.5	**99.6**
Overall AVG		73.1	92.1	95.2	96.1	**98.1**
BeanTech AD						
Category		**R-Pred** [26]	**P-SVDD** [4]	**MemSeg** [24]	**CutPaste** [8]	**Ours**
01		80.3	95.7	98.7	**96.7**	98.3
02		50.5	72.1	87	77.4	**90.2**
03		72.6	82.1	99.4	99.3	**100**
Overall AVG		67.8	83.3	95.0	91.1	**96.2**

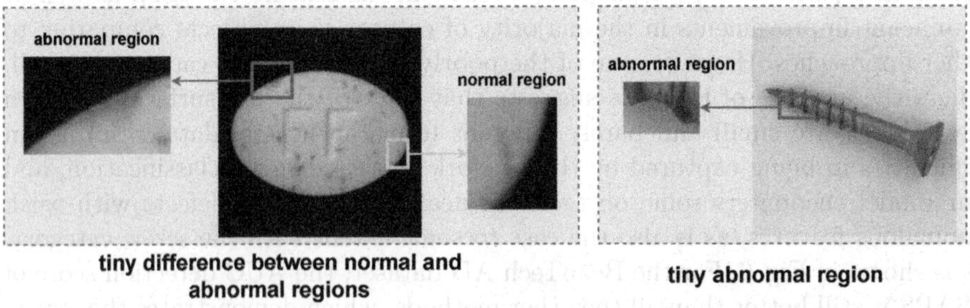

Fig. 6. Some failed detection examples of pill and screw. The red box on the left shows the small defective areas and the green box on the right shows a small part of the normal image. Surface defects are not only tiny but also differ minimally from normal regions. (Color figure online)

4.4 Ablation Study

We conduct various further studies to provide an explainable reason what contributes to discover anomalous regions in SSAPS. We begin with performing diverse patch replacement on the normal image to explore segmentation tasks of varying complexity. The impact of various data augmentation transforms based on cutX on anomaly detection and anomaly segmentation is then compared.

Hyperparameters. We evaluate the impact of the various augmented patch construction parameters, including the size, irregularity, and number of augmented patches, as well as the distribution of the pasted position, which determine the extent to the original image transformation. Table 2 indicates the detection results for various parameter on the object and texture categories, respectively. All experiments are performed under other optimal parameter conditions and the values in bold indicate the optimal results for each category of parameters. Figure 7 and show the examples of the various transformations on some categories in the MVTec dataset.

The results in Table 2 (top left) show that larger augmented patches tend to be captured. Moreover, sufficient information about the original image must be maintained in the constructed anomalous samples, due to the need to ensure a balance of positive and negative samples to avoid much more enhanced pixels than normal ones in order to allow the network to extract information about the anomalous differences from them to construct robust classification margins.

The results in Table 2 (bottom left) show that a moderate increase in complexity of the pasted patches improves anomaly detection. We analyse that high complexity augmented patches increase the local irregularity of the constructed anomalous samples and make them easier to be expressed in the receptive field; however, when the complexity is excessively high (shown in the Fig. 7), the model overfits and fails to learn the weak anomalous patterns. Furthermore, for the object class, we need to paste the augmented patches in the center of the image, these images consist of both object and background areas, the object usually locates in the center of the image and sometimes occupies a small proportion(e.g. capsule, screw, pill), while the background contains little anomalous valid information, when the patches are pasted into the background area (as shown in the Fig. 7), they interfere with the extraction of normal patterns in the form of noise, resulting in poor detection. For the texture category, due to the weak semantic differences in the entire image and the fact that defects can occur anywhere, the distribution of the augmented patches should be as dispersed and random as possible, and this conjecture is confirmed experimentally by the results of segmenting augmented patches randomly distributed over the entire image, which give significantly better results for defect detection than segmenting patches distributed in the center of the image.

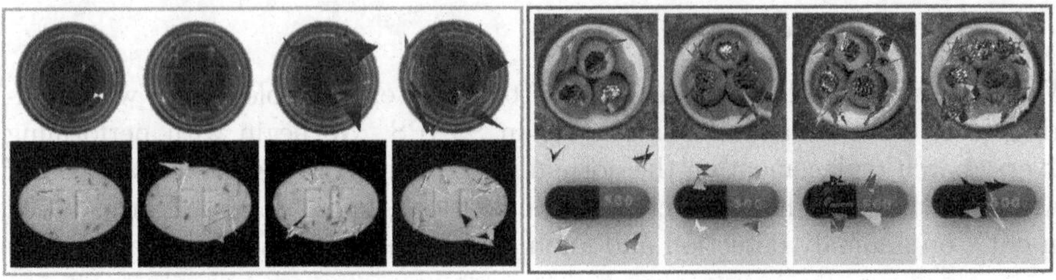

(a) Different size and number of the augmented patches

(b) Different irregularity and distribution of the augmented patches

Fig. 7. Visualization of different parameters of augmented patches on some categories in MVTEC dataset.

CutX Transformations. We explore other variants based on the CutX method, including cutout, cutmix and cutpaste. Cutout masks an area of the image, CutPaste masks and fills it with content from other areas of the same image while CutMix fills it with other image blocks, our method is described in Sect. 3.1. More details about cutX methods can be found in Appendix.

As the results shown in Table 3, our method demonstrates the best detection performance. On the one hand, pasting irregular patches of images increases the variability between the abnormal and normal patterns, making it easier for the network to extract features. On the other hand, capturing information from the same image allows the constructed anomalous patterns to be generalised, which prevents the network from learning the constructed anomalies as a new category and thus ignoring the real abnormal patterns.

4.5 Anomaly Segmentation

Our method demonstrates excellent anomaly detection performance at image-level through experimental validation. Furthermore, we experiment with defect localization in an end to end manner. Specifically, we put the test images into the APS network and compare the anomaly scores of the corresponding positions of the output to obtain the predicted segmentation anomaly maps. The segmentation results for some categories in MVTEC dataset are shown in Fig. 1(More detailed anomaly segmentation results are shown in Appendix). We achieved a result that is close to the ground truth, demonstrating the ability of our method to segment unknown defects in high-resolution images.

Table 2. Results of different irregularity and distribution of augmented patch. The values in the patch irregularity means the range of the number of vertices of irregular polygon, the patch distribution means the degree of proximity to the centre of the image. The results are reported with AUROCs.

Category	Size of patch				Number of patch			
	0.1	0.2	0.3	0.45	1	2	3	**4**
Object	96.3	96.7	**97.4**	95.9	96.5	97.1	96.8	**97.3**
Texture	98.1	99.2	**99.6**	98.9	97.9	98.5	98.6	**99.4**
Overall AVG	96.8	97.5	**98.1**	96.9	96.9	97.6	97.4	**98**
Category	Irregularity of patch				Distribution of patch			
	3–5	5–12	12–24	3–24	0.01–0.5	0.5–0.75	0.75–0.99	0.01–0.99
Object	95.9	**97.4**	97	96.8	94.3	95.9	**97.3**	95.6
Texture	98.3	**99.3**	98.9	99.1	98.7	98.6	98.3	**99.3**
Overall AVG	97.6	98	97.6	97.5	95.8	96.8	97.6	**96.8**

Table 3. Detection performance of variants of CutX on the MVTEC dataset. From left to right are results of cutout, cutmix, cutpaste and our proposed method. The results are reported with AUROCs.

Category	Transforms			
	Cutout	Cutpaste	Cutmix	Ours
Object	92.3	95.3	96.8	**97.4**
Texture	97.3	98.4	98.7	**99.6**
Overall AVG	93.9	96.3	97.4	**98.1**

5 Conclusion

We design a finer-grained self-supervised proxy task, where the key lies in segmenting the augmented patches so as to simulate segmentation task in vision to extract the differences between normal and abnormal patterns. Also our approach does not require the simulated anomalies to match those in the real scenario, ensuring that the model obtains more robust decision boundary. Extensive experimental results confirm that SSAPS achieves excellent results in detecting surface defects in high-resolution images of industrial scenes. Furthermore, SSAPS also demonstrates good defect localization capabilities due to the pixel-level differential perception of normal and abnormal patterns learned by the model during the self-supervised training phase.

Acknowledgements. The paper is supported by the Open Fund of Science and Technology on Parallel and Distributed Processing Laboratory under Grant WDZC20215250116.

References

1. Hilal, W., Andrew Gadsden, S., Yawney, J.: A review of anomaly detection techniques and recent advances, Financial fraud (2022)
2. Bergmann, P., Fauser, M., Sattlegger, D., Steger, C.: MVTec AD-a comprehensive real-world dataset for unsupervised anomaly detection. In: Proceedings of the IEEE/CVF Conference on Computer Vision and Pattern Recognition, pp. 9592–9600 (2019)
3. Tsai, C.-C., Wu, T.-H., Lai, S.-H.: Multi-scale patch-based representation learning for image anomaly detection and segmentation. In: Proceedings of the IEEE/CVF Winter Conference on Applications of Computer Vision, pp. 3992–4000 (2022)
4. Yi, J., Yoon, S.: Patch SVDD: patch-level SVDD for anomaly detection and segmentation. In: Proceedings of the Asian Conference on Computer Vision (2020)
5. Ruff, L.: Deep one-class classification. In International conference on machine learning, pp. 4393–4402. PMLR (2018)
6. Tax, D.M.J., Duin, R.P.W.: Support vector data description. Mach. Learn. **54**(1), 45–66 (2004)
7. Schlüter, H.M., Tan, J., Hou, B., Kainz, B.: Self-supervised out-of-distribution detection and localization with natural synthetic anomalies (NSA). arXiv preprint arXiv:2109.15222 (2021)
8. Li, C.-L., Sohn, K., Yoon, J., Pfister, T.: CutPaste: self-supervised learning for anomaly detection and localization. In: Proceedings of the IEEE/CVF Conference on Computer Vision and Pattern Recognition, pp. 9664–9674 (2021)
9. Cohen, N., Hoshen, Y.: Sub-image anomaly detection with deep pyramid correspondences. arXiv preprint arXiv:2005.02357 (2020)
10. Zavrtanik, V., Kristan, M., Skočaj, D.: DRAEM-a discriminatively trained reconstruction embedding for surface anomaly detection. In: Proceedings of the IEEE/CVF International Conference on Computer Vision, pp. 8330–8339 (2021)
11. Mishra, P., Verk, R., Fornasier, D., Piciarelli, C., Foresti, G.L.: VT-ADL: a vision transformer network for image anomaly detection and localization. In: 2021 IEEE 30th International Symposium on Industrial Electronics (ISIE), pp. 01–06. IEEE (2021)
12. Zheng, Y., Wang, X., Deng, R., Bao, T., Zhao, R., Wu, L.: Focus your distribution: coarse-to-fine non-contrastive learning for anomaly detection and localization. arXiv preprint arXiv:2110.04538 (2021)
13. Roth, K., Pemula, L., Zepeda, J., Schölkopf, B., Brox, T., Gehler, P.: Towards total recall in industrial anomaly detection. In: Proceedings of the IEEE/CVF Conference on Computer Vision and Pattern Recognition, pp. 14318–14328 (2022)
14. Defard, T., Setkov, A., Loesch, A., Audigier, R.: PaDiM: a patch distribution modeling framework for anomaly detection and localization. In: Del Bimbo, A., et al. (eds.) ICPR 2021. LNCS, vol. 12664, pp. 475–489. Springer, Cham (2021). https://doi.org/10.1007/978-3-030-68799-1_35
15. Cohen, N., Hoshen, Y.: Sub-image anomaly detection with deep pyramid correspondences. arxiv 2020. arXiv preprint arXiv:2005.02357
16. Rippel, O., Mertens, P., Merhof, D.: Modeling the distribution of normal data in pre-trained deep features for anomaly detection. In: 2020 25th International Conference on Pattern Recognition (ICPR), pp. 6726–6733. IEEE (2021)
17. Krizhevsky, A., Sutskever, I., Hinton, G.E.: ImageNet classification with deep convolutional neural networks. In: Advances in Neural Information Processing Systems, vol. 25 (2012)

18. Mei, S., Yang, H., Yin, Z.: An unsupervised-learning-based approach for automated defect inspection on textured surfaces. IEEE Trans. Instrum. Meas. **67**(6), 1266–1277 (2018)
19. Bergmann, P., Löwe, S., Fauser, M., Sattlegger, D., Steger, C.: Improving unsupervised defect segmentation by applying structural similarity to autoencoders. arXiv preprint arXiv:1807.02011 (2018)
20. Akcay, S., Atapour-Abarghouei, A., Breckon, T.P.: GANomaly: semi-supervised anomaly detection via adversarial training. In: Jawahar, C.V., Li, H., Mori, G., Schindler, K. (eds.) ACCV 2018. LNCS, vol. 11363, pp. 622–637. Springer, Cham (2019). https://doi.org/10.1007/978-3-030-20893-6_39
21. Sabokrou, M., Khalooei, M., Fathy, M., Adeli, E.: Adversarially learned one-class classifier for novelty detection. In: Proceedings of the IEEE Conference on Computer Vision and Pattern Recognition, pp. 3379–3388 (2018)
22. Liang, Y., Zhang, J., Zhao, S., Wu, R., Liu, Y., Pan, S.: Omni-frequency channel-selection representations for unsupervised anomaly detection. arXiv preprint arXiv:2203.00259 (2022)
23. Pirnay, J., Chai, K.: Inpainting transformer for anomaly detection. In: Sclaroff, S., Distante, C., Leo, M., Farinella, G.M., Tombari, F. (eds.) International Conference on Image Analysis and Processing, vol. 13232, pp. 394–406. Springer, Cham (2022). https://doi.org/10.1007/978-3-031-06430-2_33
24. Yang, M., Wu, P., Liu, J., Feng, H.: MemSeg: a semi-supervised method for image surface defect detection using differences and commonalities. arXiv preprint arXiv:2205.00908 (2022)
25. Komodakis, N., Gidaris, S.: Unsupervised representation learning by predicting image rotations. In: International Conference on Learning Representations (ICLR) (2018)
26. Golan, I., El-Yaniv, R.: Deep anomaly detection using geometric transformations. In: Advances in Neural Information Processing Systems, vol. 31 (2018)
27. Doersch, C., Gupta, A., Efros, A.A.: Unsupervised visual representation learning by context prediction. In: Proceedings of the IEEE International Conference on Computer Vision, pp. 1422–1430 (2015)
28. Song, J., Kong, K., Park, Y.-I., Kim, S.-G., Kang, S.-J.: AnoSeg: anomaly segmentation network using self-supervised learning. arXiv preprint arXiv:2110.03396 (2021)
29. Liu, Z., Mao, H., Wu, C.-Y., Feichtenhofer, C., Darrell, T., Xie, S.: A convnet for the 2020s. In: Proceedings of the IEEE/CVF Conference on Computer Vision and Pattern Recognition, pp. 11976–11986 (2022)
30. De Maesschalck, R., Jouan-Rimbaud, D., Massart, D.L.: The Mahalanobis distance. Chemom. Intell. Lab. Syst. **50**(1), 1–18 (2000)
31. Kim, J.-H., Kim, D.-H., Yi, S., Lee, T.: Semi-orthogonal embedding for efficient unsupervised anomaly segmentation. arXiv preprint arXiv:2105.14737 (2021)
32. Sohn, K., Li, C.-L., Yoon, J., Jin, M., Pfister, T.: Learning and evaluating representations for deep one-class classification. arXiv preprint arXiv:2011.02578 (2020)
33. Van der Maaten, L., Hinton, G.: Visualizing data using T-SNE. J. Mach. Learn. Res. **9**(11), 2579–2605 (2008)

Rethinking Low-Level Features for Interest Point Detection and Description

Changhao Wang, Guanwen Zhang(✉), Zhengyun Cheng, and Wei Zhou

Northwestern Polytechnical University, Xi'an, China
guanwen.zh@nwpu.edu.cn

Abstract. Although great efforts have been made for interest point detection and description, the current learning-based methods that use high-level features from the higher layers of Convolutional Neural Networks (CNN) do not completely outperform the conventional methods. On the one hand, interest points are semantically ill-defined and high-level features that emphasize semantic information are not adequate to describe interest points; On the other hand, the existing methods using low-level information usually perform detection on multi-level feature maps, which is time consuming for real time applications. To address these problems, we propose a Low-level descriptor-Aware Network (LANet) for interest point detection and description in self-supervised learning. Specifically, the proposed LANet exploits the low-level features for interest point description while using high-level features for interest point detection. Experimental results demonstrate that LANet achieves state-of-the-art performance on the homography estimation benchmark. Notably, the proposed LANet is a front-end feature learning framework that can be deployed in downstream tasks that require interest points with high-quality descriptors. (Code is available on https://github.com/wangch-g/lanet.).

1 Introduction

Interest point detection and description aims to propose reliable 2D points with representative descriptors for associating the 2D points projected from the same 3D point across images. It is an important task in many computer vision fileds, such as Camera Localization [28,38], Structure-from-Motion (SfM) [23,29], and Simultaneous Localization and Mapping (SLAM) [20,21].

The conventional methods mainly use hand-crafted features for interest point detection and description. These methods focus on low-level features such as edges, gradients, and corners of high-contrast regions of the images, and try to design the features that can be detectable even under changes in image viewpoint, scale, noise, and illumination [2,13,14,17,26]. In recent years, deep learning has been extensively applied in computer vision fields, and learning-based methods for interest point detection and description are becoming increasingly prevalent. These methods expect to leverage the feature learning ability of the deep neural network and to extract the high-level features to outperform hand-crafted methods. In earlier literature, Convolutional Neural Networks (CNN)

L. Wang et al. (Eds.): ACCV 2022, LNCS 13842, pp. 108–123, 2023.
https://doi.org/10.1007/978-3-031-26284-5_7

are used independently to learn the local descriptor based on the cropped image patches for detected points [9,18,22,32,40]. However, due to the weak capability of the existing detector and the patch-based descriptor, these methods easily produce inaccurate point locations and generate numerous wrong matches. Nowadays, researches perform the end-to-end manners to jointly learn detection and description based on the feature map extracted from the higher layers of CNN [6–8,19,24,37].

Although great efforts have been made for jointly learning detection and description, the learning-based methods fail to achieve significant improvements as expected as that in other computer vision tasks. The hand-crafted features still reveal high-quality description for interest points. Such as SIFT [17], its descriptors show distinctive and stable characteristics of interest points and are able to achieve remarkable performance [3]. Recently, some learning-based works proposed to leverage multi-level information for interest points detection and description [8,19]. These methods mainly focus on using the low-level and high-level features together as the multi-level features for detection and description. However, performing detection on multi-level feature maps is time consuming, and it is unfavourable for running interest points algorithm in real time for some downstream tasks. Besides, the interest points are semantically ill-defined and high-level features that emphasize semantic information are not adequate to describe interest points. Therefore, the low-level features tend to be appropriate for description, while the high-level features are effective for interest point detection. The two types of features exhibit complementary advantages relative to each other. From this perspective, one natural but less explored idea is to combine the advantages from both.

In this paper, we propose a Low-level descriptor-Aware Network (LANet) for interest point detection and description in self-supervised learning manner. The proposed LANet has a multi-task network architecture that consists of an interest point detection module, a low-level description-aware module, and a correspondence module. Similarly, the interest point detection module, following the previous works UnsuperPoint [6] and IO-Net [37], uses an encoder-decoder architecture to estimate locations and confidence scores of the interest points. Differently, we do not use a description decoder head after the encoder but a low-level description-aware module to directly exploit the low-level features from the lower layers of the encoder architecture for the learning of descriptors. Meanwhile, we introduce the learnable descriptor-aware scores to emphasize informative features of interest from different layers softly. Furthermore, to take full advantage of the pseudo-ground truth information, we introduce the correspondence module that takes the descriptor pairs to predict the correspondences between the detected points. The predicted correspondences is self-supervised by pseudo-ground truth labels for enhancing the learning of the interest point detection module and the low-level description-aware module. We evaluate the proposed LANet on the popular HPatches (Fig. 1), the experimental results demonstrate the proposed LANet is able to detect reliable interest points with high-quality descriptors

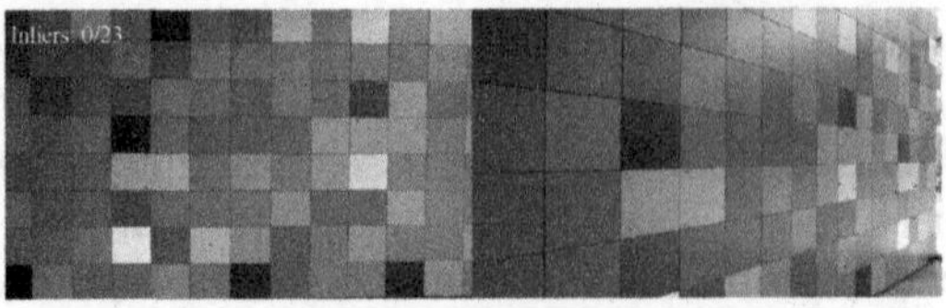

(a) SIFT + NN ($\epsilon = 1$)

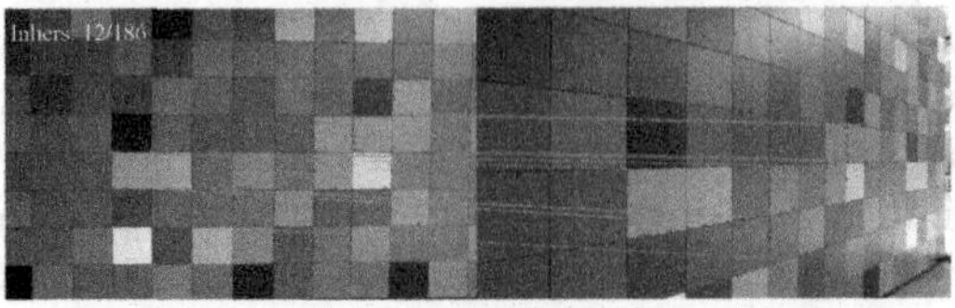

(b) IO-Net + NN ($\epsilon = 1$)

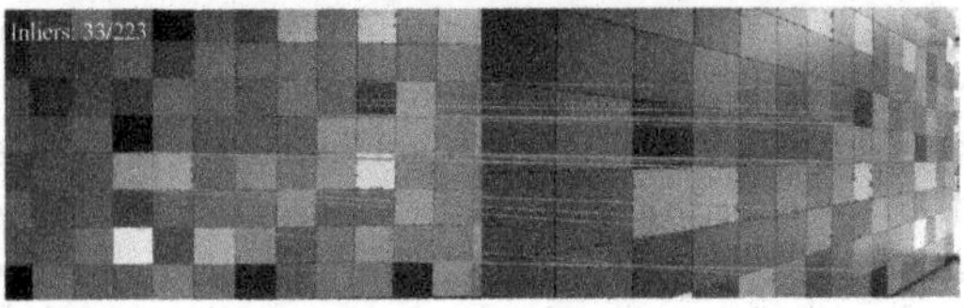

(c) LANet (ours) + NN ($\epsilon = 1$)

Fig. 1. A challenging example. Compared with SIFT and the most recent self-supervised method IO-Net, our proposed LANet can obtain stronger descriptors for interest point matching and achieve the best performance. The results are obtained by the nearest neighbor (NN) matcher with the correct distance threshold $\epsilon = 1$. The inlier matches are linked in green while the mismatched points are marked in red. (Color figure online)

that can be deployed in downstream tasks and outperforms the state-of-the-art methods on the homography estimation benchmark.

2 Related Work

Patch-Based Description. For interest point detection and description, both hand-crafted descriptors [2,14,17,26] and early learning-based descriptors [9,32,40] are computed from the local patches around the detected points. ORB [26] and SIFT [17] are considered to be the most representative conventional methods, and they are widely used in practical 3D computer vision applications [20,21]. With the progressive development of deep learning, the performance of CNN-based methods has been gradually improved. LF-Net [22] proposes to use the depth and relative camera pose cues as the supervisory signals to learn local descriptors from image patches. To leverage contextual information, ContextDesc [18] exploits the visual context from high-level image representation and the geometric context from the distribution of interest points for learning local descriptors.

Jointly Learned Detection and Description. Recent years, numerous works have paid increased attention to simultaneously learning detection and description within a single CNN architecture [6–8,19,24,37]. D2-Net [8] proposes to perform detection and description simultaneously via one CNN architecture, which tightly couples the learning of interest point locations and descriptors. To learn low-level details, ASLFeat [19] uses a multilevel learning mechanism based on the backbone of D2-Net [8] for interest point detection and description. In addition, instead of using expensive ground truth supervision, SuperPoint [7] employs an encoder-decoder architecture that contains two decoder branches, a detection decoder, and a description decoder, to learn the detector and descriptors jointly with pseudo-ground truth labels that are generated from MagicPoint [7]. On the basis of SuperPoint, UnsuperPoint [6], a self-supervised learning framework for interest point matching, is trained by the siamese scheme to learn the scores, locations, and descriptors of interest points automatically with unlabeled images. Most recently, IO-Net [37] is proposed based on UnsuperPoint, which learns a detector and descriptors with supervision from the proposed outlier rejection loss.

Finding Good Matches. With detected points and descriptors, using nearest neighbor search based on the similarity of descriptors can obtain a set of matched points. However, building correspondences on the basis of descriptors simply may cause a mass of outliers [3,27,33,41,42]. To alleviate the problem of false correspondences, some studies, such as RANSAC [10] and GMS [4,5], leverage geometric or statistical information to filter outliers. Recently, SuperGlue [27] proposes to address the matching problem with Graph Neural Network (GNN) in a learning-based manner. Inspired by SuperGlue, LoFTR [33], a detector-free approach, proposes to obtain high-quality matches with transformers [39]. Because our method is a front-end feature learning framework, it can be further enhanced by being embedded with a SuperGlue-like learnable matching algorithm.

3 Method

The proposed LANet consists of three substantial modules, *i.e.*, interest point detection module, low-level description-aware module, and correspondence module. An overview of the proposed LANet is depicted in Fig. 2.

3.1 Interest Point Detection Module

The interest point detection module is constructed based on the backbone of UnsuperPoint [6] and IO-Net [37] with several modifications. As shown in Fig. 2, the interest point detection module has a location decoder, a score decoder, and a shared encoder. It aims to estimate locations, confidence scores, and descriptor-aware scores of the interest points. During the training stage, in a self-supervised learning manner [6], the interest point detection module disposes a source image I_S and a target image I_T with a siamese framework to predict interest points.

Notably, the target image I_T is transformed from I_S by a random generated homography $\boldsymbol{H}_{S\to T}$, which allows us to train the proposed model without any human annotation.

Encoder. We use a convolutional structure as the encoder for extracting the features from the input images. The encoder consists of four convolutional blocks with 32, 64, 128, and 256 output channels. Each of the convolutional blocks contains two 3×3 convolutional layers followed by a batch normalization [11] and a ReLU activation function. The encoder uses three max-pooling layers with kernel size and stride of 2×2 among the four convolutional blocks to downsample the input image from $H \times W$ to $H/8 \times W/8$. We denote the pixels in the downsampled feature maps as cells as in [7]. Therefore, a 1×1 cell in the last feature map corresponds to 8×8 pixels in the raw image.

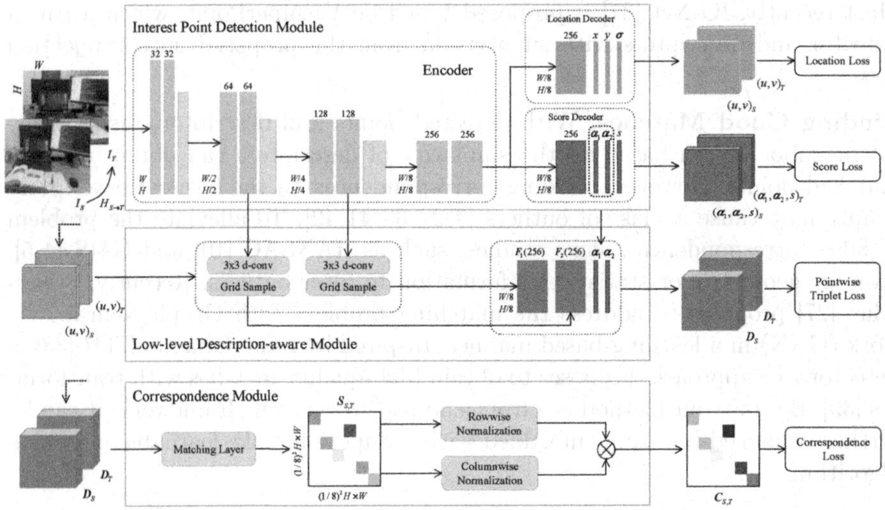

Fig. 2. Overview of LANet. The proposed LANet consists of an interest point detection module (Sect. 3.1), a low-level description-aware module (Sect. 3.2), and a correspondence module (Sect. 3.3). The interest point detection module is an encoder-decoder architecture that is used for estimating the point locations (u, v), the confidence scores s, and the descriptor-aware scores α. The low-level description-aware module exploits the low-level information from the lower layers of the encoder for producing the descriptors $\boldsymbol{D}$ of the detected points. The correspondence module is introduced to improve the supervision of the first two modules by estimating the correspondence matrix $\boldsymbol{C}$ of the detected points between the input image-pair.

Location Decoder. The location decoder contains two 3×3 convolutional layers with channels of 256 and 3 respectively. The location decoder takes the last feature map produced from the encoder as input and outputs a location map with 3 channels. The first two channels of a cell in the location map are denoted as (x, y), which indicates the normalized location of each interest point relative to the center of the corresponding 8×8 grid in the raw image. In addition, we denote the third channel of a cell in the location map by σ. σ is a learnable rectification factor for calculating the absolute pixel location of each interest point, which allows the predicted points across cell boundaries [37]. The pixel location of each interest point is defined as:

$$(u, v) = (u', v') + \frac{\sigma(r-1)}{2}(x, y), \forall (x, y) \in [-1, 1], \tag{1}$$

where (u, v) and (u', v') are the pixel location and the cell center location of the corresponding interest point, respectively. In Eq. 1, r is the down sample ratio of the interest point detection module ($r = 8$ in this paper).

Score Decoder. In the interest point detection module, the score decoder has the same structure as the location decoder. The score decoder outputs the score map with 3 channels. A cell in the first channel of the score map is the confidence score of the corresponding interest point. We normalize the confidence score to $[0, 1]$ and denote it by s. The score s is used for selecting the best K points that are most convinced for downstream tasks. Additionally, a cell in the last two channels of the score map is the descriptor-aware score α that is introduced to emphasize informative features of interest softly. We use Softmax to perform the channelwise normalization for the descriptor-aware scores.

3.2 Low-Level Description-Aware Module

From the core insights of this work, we do not learn descriptors with an extra description decoder but exploit the concrete low-level features directly from the lower layers of the backbone encoder. Specifically, as shown in Fig. 2, we take the output feature maps of the second and the third convolutional block of the encoder as the basic descriptor maps and employ two separate dilated 3×3 convolutional layers to increase the channel dimension of the basic descriptor maps to 256. The dilated 3×3 convolution has lager receptive filed with the dilation is set as 2, which is helpful for capturing more local information in the low-level feature maps. The descriptors, corresponding to the detected interest points from the basic descriptor maps, are warped by performing a differentiable grid sample operation [12,15]. Then, the warped descriptors are L2-normalized and denoted by $\boldsymbol{F}$. The warped descriptors $\boldsymbol{F}$ from different convolutional blocks are aggregated with the descriptor-aware scores α to generate the final descriptors $\boldsymbol{D}$ by taking a weighted sum.

3.3 Correspondence Module

The descriptors are used to correspond to the detected points based on matching algorithms for further applications. In recent years, several studies calculate the similarity matrix of descriptors between input pairs and use a differentiable matching layer to learn good correspondences [25,27,33].

Inspired by previous works, we introduce the correspondence module to predict the correspondence matrix by the descriptors of detected points between I_S and I_T, which is employed as an auxiliary task to supervise the training of the interest point detection module and the low-level description-aware module. First, we calculate the similarity matrix $\boldsymbol{S}_{S,T}$ between the source descriptors $\boldsymbol{D}_S$ and the target descriptors $\boldsymbol{D}_T$ as:

$$\boldsymbol{S}_{S,T}^{i,j} = <\boldsymbol{D}_S^i, \boldsymbol{D}_T^j>, \ \boldsymbol{A} \in \mathbb{R}^{(1/8)^2HW \times (1/8)^2HW}, \tag{2}$$

where $< \cdot \ , \ \cdot >$ is the inner product. With the similarity matrix $\boldsymbol{S}_{S,T}$ in size of $(1/8)^2HW \times (1/8)^2HW$, the correspondence module performs dual-softmax [25] to normalize the $\boldsymbol{S}_{S,T}$ in rowwise order and columnwise order respectively, and computes the correspondence matrix $\boldsymbol{C}_{S,T}$ as:

$$\boldsymbol{C}_{S,T}^{i,j} = \text{softmax}(\boldsymbol{S}_{S,T}^i)_i \cdot \text{softmax}(\boldsymbol{S}_{S,T}^j)_j, \ \boldsymbol{C} \in \mathbb{R}^{(1/8)^2HW \times (1/8)^2HW}. \tag{3}$$

The predicted correspondence matrix $\boldsymbol{C}_{S,T}$ reveals the interest point matching correlations between I_S and I_T. It is self-supervised by pseudo-ground truth matching labels as an auxiliary task to enhance the learning of the low-level description-aware module.

3.4 Optimization

The proposed LANet is optimized by four loss functions: the location loss L_{loc}, the score loss L_s, the pointwise triplet loss L_{tri}, and the correspondence loss L_{corr}. The final loss function L is balanced by trade-off parameters λ as:

$$L = \lambda_{loc}L_{loc} + \lambda_s L_s + \lambda_{tri}L_{tri} + \lambda_{corr}L_{corr}. \tag{4}$$

Location Supervision. During the training stage, we can obtain the location of detected points in raw input images through the interest point detection module. Following the self-supervised learning scheme [6], we warp the source points (detected points in the source image I_S) to the target image with the known homography $\boldsymbol{H}_{S\to T}$ and find the nearest neighbor of the warped points among the target points (detected points in the target image I_T) using an L2-distance. A source point and its nearest neighbor in the target image are associated as a nearest point pair if the distance between them is less than a threshold ϵ. We denote a source point and its associated target point in a nearest point pair as p_S and p_T, respectively. The distance between a nearest point pair is defined as:

$$d(p_S, p_T) = ||\boldsymbol{H}_{S\to T}(p_S) - p_T||_2, \tag{5}$$

and the location loss can then be formulated as:

$$L_{loc} = \frac{1}{N}\sum_i^N d(p_S^i, p_T^i), \tag{6}$$

where N is the number of nearest point pairs.

Score Supervision. Following [6,37], the score loss L_s should increase the confidence score s of the relatively convinced points. In addition, it should jointly keep the coherence of the confidence score s and the descriptor-aware score α between the nearest point pairs. Thus, we define the score loss as:

$$\begin{aligned} L_s =& \frac{1}{N}\sum_{i}^{N}[\frac{(s_S^i + s_T^i)}{2}(d(p_S^i, p_T^i) - \bar{d}) + (s_S^i - s_T^i)^2 \\ &+ (\alpha_S^{1,i} - \alpha_T^{1,i})^2 + (\alpha_S^{2,i} - \alpha_T^{2,i})^2], \end{aligned} \tag{7}$$

where $\bar{d}$ is the average distance of the nearest point pairs. In Eq. 7, the first term is the confidence constraint that allows the nearest point pairs with closer distances to have higher confidence scores, the last three terms are the coherence constraints that ensure the consistency between the confidence score and the descriptor-aware scores of the nearest point pairs.

Description Supervision. We use the pointwise triplet loss [30,34,36] for learning high-quality descriptors. In a triplet sample, the *anchor* and the *positive* are the descriptors of a source point and its matched target point, respectively. We denote the *anchor* and the *positive* as $\boldsymbol{D}_{p^i}$ and $\boldsymbol{D}_{p^i_+}$. The *negative* of the triplet is sampled among the descriptors of the unmatched target points with the hardest negative sample mining [36] and denoted by $\boldsymbol{D}_{p^i_-}$. The pointwise triplet loss is formulated as:

$$L_{tri} = \frac{1}{N}\sum_{i}^{N}[||\boldsymbol{D}_{p^i}, \boldsymbol{D}_{p^i_+}||_2 - ||\boldsymbol{D}_{p^i}, \boldsymbol{D}_{p^i_-}||_2 + \beta]_+, \tag{8}$$

where β is the distance margin for metric learning.

Correspondence Supervision. We introduce the correspondence loss to optimize the predicted correspondence matrix $\boldsymbol{C}_{S,T}$. The correspondence loss is a negative log-likelihood loss that contains a positive term and a negative term, which is formulated as:

$$\begin{aligned} L_{corr} =& -[\frac{1}{|\boldsymbol{M}_{pos}|}\sum_{(i,j)\in \boldsymbol{M}_{pos}} \log \boldsymbol{C}_{S,T}^{i,j} \\ &+ \gamma \cdot \frac{1}{|\boldsymbol{M}_{neg}|}\sum_{(i,j)\in \boldsymbol{M}_{neg}} \log(1 - \boldsymbol{C}_{S,T}^{i,j})], \end{aligned} \tag{9}$$

where the $\boldsymbol{M}_{pos}$ and the $\boldsymbol{M}_{neg}$ are the sets of positive matches and negative matches respectively, γ is a hyperparameter used for balance the two terms. Notably, the operations in the correspondence module are all differentiable. Therefore, the gradient of the correspondence loss can be propagated back to the low-level description-aware module for auxiliary supervision.

3.5 Deployment

During the testing stage, only the interest point detection module and the low-level description-aware module of the proposed LANet are employed to predict interest points with corresponding scores and descriptors. The scores are used to select the most convinced interest points for downstream applications. The proposed LANet is a novel feature extraction approach that provides interest points with high-quality descriptors, thus it can be introduced as a front-end solver embedded with existing matchers such as SuperGlue [27].

4 Experiments

4.1 Details

We use 118k images from COCO 2017 dataset [16] without any human annotation to train the proposed LANet. We perform spatial augmentation on the training dataset with scaling, rotation, and perspective transformation to generate the self-supervised signal for the siamese training scheme. The input images are resized to 240×320 and the augmentation settings are same as [37]. The proposed LANet is trained by Adam optimizer for 12 epochs with a batch size of 8. The learning rate is set to 3×10^{-4} and is reduced by a factor of 0.5 after 4 epochs and 8 epochs. The distance threshold ϵ between the point locations is set to 4.0 during the training process. The trade-off parameters of the loss function in Eq. 4 are set to $\lambda_{loc} = 1.0$, $\lambda_s = 1.0$, $\lambda_{\alpha-tri} = 4.0$, and $\lambda_{corr} = 0.5$, respectively. The pointwise triplet loss margin β in Eq. 8 is set to 1.0. The factor γ in Eq. 9 is set to 5×10^5.

4.2 Comparison

We evaluate the proposed LANet on HPatches dataset [1] which contains 116 scenes with dramatic changes in illumination or viewpoint. We use the metrics of Repeatability (Re), Localization Error (LE), Homography Estimation Accuracy with tolerance threshold $\epsilon = 1$ pixel (H-1), $\epsilon = 3$ pixels (H-3), $\epsilon = 5$ pixels (H-5), and Matching Score (MS) [7]. The evaluation metrics are measured with top P points selected according to the confidence scores. We report the results on testing images of 240×320 resolution ($P = 300$) and 480×640 resolution ($P = 1000$) in Table 1. Besides, we report the Mean Matching Accuracy (MMA) [8] with the error threshold of 3 pixels in Table 2.

Repeatability and Localization Error. The higher repeatability represents a higher probability that the same interest points can be detected in different images, and the lower localization error indicates that the pixel locations of detected points are more precise. The repeatability and the localization error are two basic metrics for evaluating the capability of an interest point detector. In the lower resolution settings, the proposed LANet achieves competitive performance with IO-Net and KP3D in repeatability. LANet has a lower localization error compared with that of IO-Net while approaching the performance of UnsuperPoint, SIFT, and KP3D. In the higher resolution settings, the proposed LANet achieves the second best performance both in repeatability and localization error.

Table 1. Comparisons on HPatches dataset with repeatability, localization error, homography estimation accuracy, and matching score.

Methods	240 × 320, 300 points						480 × 640, 1000 points					
	Re↑	LE↓	H-1↑	H-3↑	H-5↑	MS↑	Re↑	LE↓	H-1↑	H-3↑	H-5↑	MS↑
ORB [26]	0.532	1.429	0.131	0.422	0.540	0.218	0.525	1.430	0.286	0.607	0.710	0.204
SURF [2]	0.491	1.150	0.397	0.702	0.762	0.255	0.468	1.244	0.421	0.745	0.812	0.230
BRISK [14]	0.566	1.077	0.414	0.767	0.826	0.258	0.505	1.207	0.300	0.653	0.746	0.211
SIFT [17]	0.451	0.855	0.622	0.845	0.878	0.304	0.421	1.011	**0.602**	0.833	0.876	0.265
LF-Net (indoor) [22]	0.486	1.341	0.183	0.628	0.779	0.326	0.467	1.385	0.231	0.679	0.803	0.287
LF-Net (outdoor) [22]	0.538	1.084	0.347	0.728	0.831	0.296	0.523	1.183	0.400	0.745	0.834	0.241
SuperPoint [7]	0.631	1.109	0.491	0.833	0.893	0.318	0.593	1.212	0.509	0.834	0.900	0.281
UnsuperPoint [6]	0.645	0.832	0.579	0.855	0.903	0.424	0.612	0.991	0.493	0.843	0.905	0.383
IO-Net [37]	**0.686**	0.970	0.591	0.867	0.912	0.544	**0.684**	0.970	0.564	0.851	0.907	0.510
KP3D [35]	**0.686**	**0.799**	0.532	0.858	0.906	**0.578**	0.674	**0.886**	0.529	0.867	0.920	0.529
LANet (ours)	0.683	0.874	**0.662**	**0.910**	**0.941**	0.577	0.682	0.943	**0.602**	**0.874**	**0.924**	**0.543**

Table 2. Comparisons on HPatches dataset with mean matching accuracy (MMA).

MMA@3	SIFT [17]	D2Net(SS) [8]	D2Net(MS) [8]	SuperPoint [7]	ASLFeat [19]	**Ours**
Illumination	0.525	0.568	0.468	0.738	-	**0.822**
Viewpoint	0.540	0.354	0.385	**0.639**	-	0.620
Overall	0.533	0.457	0.425	0.686	**0.723**	0.717

Homography Estimation Accuracy. In practice, the correspondence between detected points is established based on the similarity of their descriptors. With the matched points, we can estimate the homography transformation between the input pairs with geometric constraint. In the experiments, we use the nearest neighbor matcher to associate the detected points across images and compute the homography matrix using OpenCV's *findHomography* method. As shown in Table 1, the proposed LANet surpasses the existing learning-based methods by a very large margin with different tolerance threshold in both lower and higher resolution settings.

Matching Score. Matching score measures the probability of the correct correspondences over the points detected in shared viewpoints, which evaluates the general performance of the whole detection and description pipeline. Our method achieves almost the same highest matching score as KP3D in lower resolution setting, while it outperforms the other methods by a large margin in higher resolution setting as shown in Table 1.

Mean Matching Accuracy. Mean matching accuracy is the ratio of correct matches for each image pair. As shown in Table 2, the proposed LANet shows surprisingly effectiveness on illumination sequences and achieves competitive result on viewpoint sequences. Compared with ASLFeat, the proposed LANet performs on par with it on overall performance.

Table 3. Ablation study.

Method	σ	LD-α	CL	240 × 320, 300 points					
				Re↑	LE↓	H-1↑	H-3↑	H-5↑	MS↑
Baseline	–	–	–	0.633	1.049	0.486	0.812	0.893	0.525
LANet	✓	–	–	0.685	0.874	0.545	0.862	0.914	0.580
	✓	✓	-	0.682	0.861	0.653	0.898	0.922	0.572
	✓	✓	✓	0.683	0.874	0.662	0.910	0.941	0.577

4.3 Ablation Study

In this section, we perform an ablation study on HPatches dataset to demonstrate the effectiveness of each module in our proposed method. The results are summarized in Table 3.

Baseline. We use the encoder-decoder architecture of the interest point detection module as the baseline. The baseline outputs the descriptors by a 3 × 3 convolutional layer that is connected after the 4th convolutional block of the backbone encoder. Besides, the rectification factor σ is fixed to 1 during the training process and the descriptor-aware score α is removed.

Ablation on the Learned σ. The learnable rectification factor σ enables the detection decoder to predict the locations across cell borders as described in [37]. Different from [37], we optimize the rectification factor σ with the predicted locations during the training stage rather than setting the σ as a default hyper-parameter. Compared with the baseline, using the learnable rectification factor σ improves the overall performance of the detection and description obviously.

Ablation on the Low-Level Descriptor-Aware Module. The low-level description-aware module (LD-α) learns descriptors from the lower convolution layers and the learned descriptors are weighted sum by the descriptor-aware score α. As shown in Table 3, compared with the results of second row, the low-level description-aware module boosts the homography estimation accuracy with +10.8% in H-1, +3.6% in H-3, and +0.8% in H-5 respectively, while only having slight influence on the repeatability, the localization error and the matching score. In Fig. 3, we visualize the interest points matching results that are obtained with and without the low-level descriptor-aware module. Compared with Fig. 3(b), Fig. 3(c) shows that using low-level features for description can learn stronger descriptors to acquire more precise and denser matches for interest point matching.

Ablation on the Correspondence Loss. The correspondence loss (CL) is introduced as an auxiliary supervision to enhance the learning of descriptors. With the correspondence loss, the proposed LANet achieves further improvement on homography estimation accuracy and achieves the best overall performance.

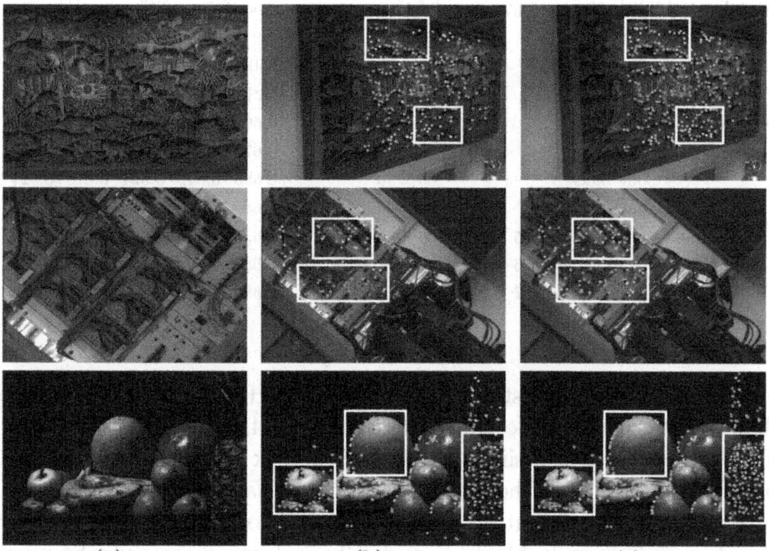

Fig. 3. Interest point matching with and without the low-level descriptor-aware module. (a) Source images. (b) Target images (processed using baseline + σ). (c) Target images (processed using baseline + σ + LD-α). The matched points are marked in different colors with the localization error. (Color figure online)

4.4 Validation on Low-Level Feature Learning

In this section, we conduct further study on learning descriptors from different layers individually to validate the effectiveness of low-level features. We use the interest point detection module of the proposed LANet as the basic network for the experiments. We extract the learned features from different convolutional blocks for evaluation. Meanwhile, we deepen the network by adding a descriptor decoder with upsample blocks for further comparisons. Each upsample block consists of an upsampling layer [31] and a basic convolutional block. Notably, the output layer is a 3×3 convolution to reshape the descriptors to the same size of 256 dimensions. The experimental results are summarized in Table 4.

Comparisons within the Backbone Encoder. From Table 4 we can see that the learned descriptors from the deeper layers have better performance on repeatability and matching score. However, the learned descriptors from different layers do not affect the localization error obviously. As for homography estimation accuracy, the learned descriptors from the lowest three convolutional blocks significantly boost the accuracy of H-1, H-3, and H-5 compared with that of the deeper layer. The learned descriptors from the convolutional block 2 achieve the best performance on H-1 (0.669), H-3 (0.895), and H-5 (0.934) at the same time.

Backbone Encoder vs. Description Decoder. With a deeper structure and higher-resolution feature map, the descriptors extracted from the upsample block

Table 4. Validation on low-level features extracted from different layers.

Output layer in description		Depth	Output resolution	240 × 320, 300 points					
				Re↑	LE↓	H-1↑	H-3↑	H-5↑	MS↑
Backbone encoder	Conv block 1	2	240 × 320	0.669	0.804	0.641	0.872	0.900	0.403
	Conv block 2	4	120 × 160	0.679	0.880	0.669	0.895	0.934	0.506
	Conv block 3	6	60 × 80	0.685	0.842	0.624	0.878	0.921	0.571
	Conv block 4	8	30 × 40	0.685	0.874	0.545	0.862	0.914	0.580
Description decoder	Upsample block 5	10	60 × 80	0.692	0.866	0.566	0.866	0.928	0.583
	Upsample block 6	12	120 × 160	0.685	0.857	0.562	0.872	0.922	0.571
	Upsample block 7	14	240 × 320	0.682	0.862	0.553	0.866	0.916	0.559

5 increase the homography estimation accuracy on H-1, H-3, and H-5 compared with that extracted from the convolutional block 4. However, compared with that of upsample block 5, using the deeper network could not further improve the performance. By using the feature map in the same resolution, the learned descriptors from the convolutional blocks 2 and 3 outperform those extracted from the deeper upsample blocks 5 and 6 by a large margin on homography estimation accuracy.

Observations. Based on the above comparisons, we can see that the low-level features can effectively improve the distinguishability of descriptors and significantly increase the homography estimation accuracy. Meanwhile, the low-level features have barely effects on the point repeatability and the localization error. It should be noted, although the descriptors extracted from the convolutional block 2 achieve the best performance on homography estimation accuracy, they do have an inferior matching score. Therefore, it is a promising solution to combine the low-level features extracted from the convolutional blocks 2 and 3 to balance the overall performance.

4.5 Further Analysis

Figure 4 shows a few qualitative matching results comparing the SIFT, IO-Net, and the proposed LANet on HPatches dataset. The results are obtained by using the nearest neighbor (NN) matcher with the correct distance threshold $\epsilon = 1$. The proposed LANet outperforms IO-Net and SIFT obviously in perspective scenes (Fig. 4 Row 1 to 4) and is on par with IO-Net in illumination scenes (Fig. 4 Row 5 and 6).

As shown in Row 7 and 8 of Fig. 4, when the images are rotated dramatically, the proposed LANet easily fails in the interest points detection and description. Whereas, the SIFT provides rotation invariant features and has the best performance compared with IO-Net and the proposed LANet. Using the largely rotated images during the training process could help to improve the performance of the proposed LANet in such cases. However, it will affect overall performance since most of the images in the dataset have moderate rotation. One promising solution is to introduce the smooth penalty function during the optimization.

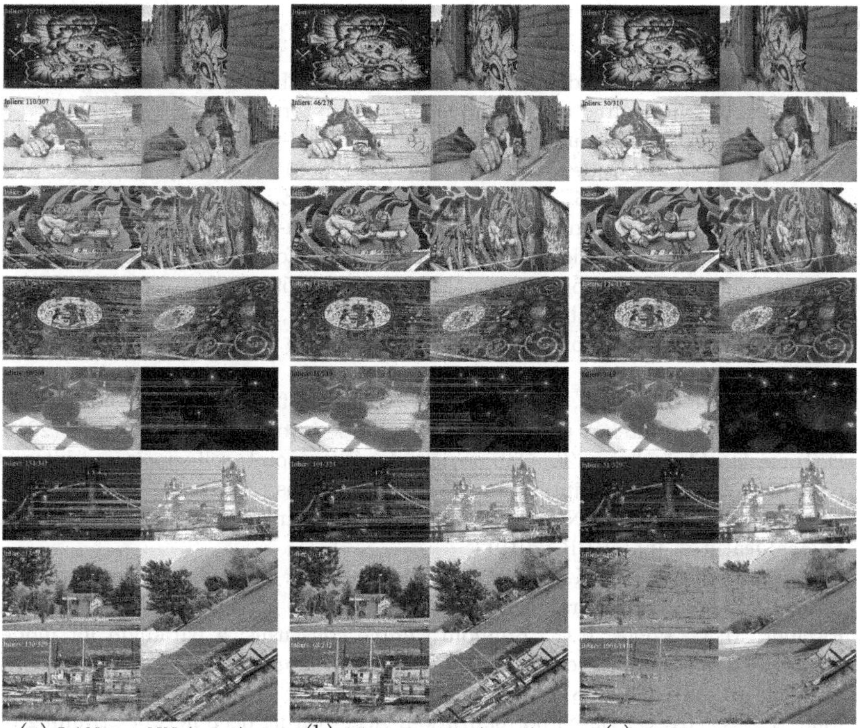

(a) LANet + NN ($\epsilon = 1$) (b) IO-Net + NN ($\epsilon = 1$) (c) SIFT + NN ($\epsilon = 1$)

Fig. 4. Matching results visualization. The examples of perspective scenes are shown in row 1–4 and the examples of illumination scenes are shown in row 5 and 6. We also show some inferior examples obtained in the largely rotated cases in row 7 and 8.

5 Conclusion

In this paper, we proposed a Low-level descriptor-Aware Network (LANet) for interest points detection and description in self-supervised scheme, which exploits the low-level features to learn adequate descriptors for interest points. Besides, we introduce a correspondence loss as an auxiliary supervision to enhance the learning of interest point descriptor. We show that using low-level features for description while using high-level features for detection is a feasible solution for interest point matching. Extensive experimental results demonstrate that the proposed LANet using low-level features can improve the distinguishability of interest point descriptor and outperform the popularly used methods.

Acknowledgements. This work was supported in part by the National Key R&D Program of China (2018AAA0102801 and 2018AAA0102803), and in part of the National Natural Science Foundation of China (61772424, 61702418, and 61602383).

References

1. Balntas, V., Lenc, K., Vedaldi, A., Mikolajczyk, K.: HPatches: a benchmark and evaluation of handcrafted and learned local descriptors. In: CVPR (2017)
2. Bay, H., Ess, A., Tuytelaars, T., Gool, L.V.: Speeded-up robust features (SURF). Comput. Vis. Image Underst. **110**(3), 346–359 (2008)
3. Bhowmik, A., Gumhold, S., Rother, C., Brachmann, E.: Reinforced feature points: optimizing feature detection and description for a high-level task. In: CVPR (2020)
4. Bian, J., et al.: GMS: grid-based motion statistics for fast, ultra-robust feature correspondence. Int. J. Comput. Vis. **128**(6), 1580–1593 (2020)
5. Bian, J., Lin, W., Matsushita, Y., Yeung, S., Nguyen, T., Cheng, M.: GMS: grid-based motion statistics for fast, ultra-robust feature correspondence. In: CVPR (2017)
6. Christiansen, P.H., Kragh, M.F., Brodskiy, Y., Karstoft, H.: Unsuperpoint: end-to-end unsupervised interest point detector and descriptor. arXiv: 1907.04011 (2019)
7. DeTone, D., Malisiewicz, T., Rabinovich, A.: Superpoint: self-supervised interest point detection and description. In: CVPR Workshops (2018)
8. Dusmanu, M., et al.: D2-Net: a trainable CNN for joint description and detection of local features. In: CVPR (2019)
9. Ebel, P., Trulls, E., Yi, K.M., Fua, P., Mishchuk, A.: Beyond cartesian representations for local descriptors. In: ICCV (2019)
10. Fischler, M.A., Bolles, R.C.: Random sample consensus: a paradigm for model fitting with applications to image analysis and automated cartography. Commun. ACM **24**(6), 381–395 (1981)
11. Ioffe, S., Szegedy, C.: Batch normalization: Accelerating deep network training by reducing internal covariate shift. In: ICML (2015)
12. Jaderberg, M., Simonyan, K., Zisserman, A., Kavukcuoglu, K.: Spatial transformer networks. In: NeurIPS (2015)
13. Ke, Y., Sukthankar, R.: PCA-SIFT: a more distinctive representation for local image descriptors. In: CVPR (2004)
14. Leutenegger, S., Chli, M., Siegwart, R.: BRISK: binary robust invariant scalable keypoints. In: ICCV (2011)
15. Li, X., et al.: Semantic flow for fast and accurate scene parsing. In: Vedaldi, A., Bischof, H., Brox, T., Frahm, J.-M. (eds.) ECCV 2020. LNCS, vol. 12346, pp. 775–793. Springer, Cham (2020). https://doi.org/10.1007/978-3-030-58452-8_45
16. Lin, T.-Y., et al.: Microsoft COCO: common objects in context. In: Fleet, D., Pajdla, T., Schiele, B., Tuytelaars, T. (eds.) ECCV 2014. LNCS, vol. 8693, pp. 740–755. Springer, Cham (2014). https://doi.org/10.1007/978-3-319-10602-1_48
17. Lowe, D.G.: Distinctive image features from scale-invariant keypoints. Int. J. Comput. Vis. **60**(2), 91–110 (2004)
18. Luo, Z., et al.: ContextDesc: local descriptor augmentation with cross-modality context. In: CVPR (2019)
19. Luo, Z., et al.: ASLFeat: learning local features of accurate shape and localization. In: CVPR (2020)
20. Mur-Artal, R., Montiel, J.M.M., Tardós, J.D.: ORB-SLAM: a versatile and accurate monocular SLAM system. IEEE Trans. Rob. **31**(5), 1147–1163 (2015)
21. Mur-Artal, R., Tardós, J.D.: ORB-SLAM2: an open-source SLAM system for monocular, stereo, and RGB-D cameras. IEEE Trans. Rob. **33**(5), 1255–1262 (2017)

22. Ono, Y., Trulls, E., Fua, P., Yi, K.M.: LF-Net: learning local features from images. In: NeurIPS (2018)
23. Pittaluga, F., Koppal, S.J., Kang, S.B., Sinha, S.N.: Revealing scenes by inverting structure from motion reconstructions. In: CVPR (2019)
24. Revaud, J., de Souza, C.R., Humenberger, M., Weinzaepfel, P.: R2D2: reliable and repeatable detector and descriptor. In: NeurIPS (2019)
25. Rocco, I., Cimpoi, M., Arandjelović, R., Torii, A., Pajdla, T., Sivic, J.: Neighbourhood consensus networks. In: NeurIPS (2018)
26. Rublee, E., Rabaud, V., Konolige, K., Bradski, G.R.: ORB: an efficient alternative to SIFT or SURF. In: ICCV (2011)
27. Sarlin, P., DeTone, D., Malisiewicz, T., Rabinovich, A.: Superglue: learning feature matching with graph neural networks. In: CVPR (2020)
28. Sarlin, P., et al.: Back to the feature: Learning robust camera localization from pixels to pose. In: CVPR (2021)
29. Schönberger, J.L., Frahm, J.: Structure-from-motion revisited. In: CVPR (2016)
30. Schroff, F., Kalenichenko, D., Philbin, J.: FaceNet: a unified embedding for face recognition and clustering. In: CVPR (2015)
31. Shi, W., et al.: Real-time single image and video super-resolution using an efficient sub-pixel convolutional neural network. In: CVPR (2016)
32. Simo-Serra, E., Trulls, E., Ferraz, L., Kokkinos, I., Fua, P., Moreno-Noguer, F.: Discriminative learning of deep convolutional feature point descriptors. In: ICCV (2015)
33. Sun, J., Shen, Z., Wang, Y., Bao, H., Zhou, X.: LoFTR: detector-free local feature matching with transformers. In: CVPR (2021)
34. Sun, Y., et al.: Perceive where to focus: Learning visibility-aware part-level features for partial person re-identification. In: CVPR (2019)
35. Tang, J., et al.: Self-supervised 3d keypoint learning for ego-motion estimation. In: CoRL (2020)
36. Tang, J., Folkesson, J., Jensfelt, P.: Geometric correspondence network for camera motion estimation. IEEE Rob. Autom. Lett. **3**(2), 1010–1017 (2018)
37. Tang, J., Kim, H., Guizilini, V., Pillai, S., Ambrus, R.: Neural outlier rejection for self-supervised keypoint learning. In: ICLR (2020)
38. Tang, S., Tang, C., Huang, R., Zhu, S., Tan, P.: Learning camera localization via dense scene matching. In: CVPR (2021)
39. Vaswani, A., et al.: Attention is all you need. In: NeurIPS (2017)
40. Yi, K.M., Trulls, E., Lepetit, V., Fua, P.: LIFT: Learned Invariant Feature Transform. In: Leibe, B., Matas, J., Sebe, N., Welling, M. (eds.) ECCV 2016. LNCS, vol. 9910, pp. 467–483. Springer, Cham (2016). https://doi.org/10.1007/978-3-319-46466-4_28
41. Yi, K.M., Trulls, E., Ono, Y., Lepetit, V., Salzmann, M., Fua, P.: Learning to find good correspondences. In: CVPR (2018)
42. Zhao, C., Cao, Z., Li, C., Li, X., Yang, J.: NM-Net: mining reliable neighbors for robust feature correspondences. In: CVPR (2019)

DILane: Dynamic Instance-Aware Network for Lane Detection

Zhengyun Cheng, Guanwen Zhang(✉), Changhao Wang, and Wei Zhou

Northwestern Polytechnical University, Xi'an, China

guanwen.zh@nwpu.edu.cn

Abstract. Lane detection is a challenging task in computer vision and a critical technology in autonomous driving. The task requires the prediction of the topology of lane lines in complex scenarios; moreover, different types and instances of lane lines need to be distinguished. Most existing studies are based only on a single-level feature map extracted by deep neural networks. However, both high-level and low-level features are important for lane detection, because lanes are easily affected by illumination and occlusion, *i.e.*, texture information is unavailable in non-visual evidence case; when the lanes are clearly visible, the curved and slender texture information plays a more important role in improving the detection accuracy. In this study, the proposed DILane utilizes both high-level and low-level features for accurate lane detection. First, in contrast to mainstream detection methods of predefined fixed-position anchors, we define learnable anchors to perform statistics of potential lane locations. Second, we propose a dynamic head aiming at leveraging low-level texture information to conditionally enhance high-level semantic features for each proposed instance. Finally, we present a self-attention module to gather global information in parallel, which remarkably improves detection accuracy. The experimental results on two mainstream public benchmarks demonstrate that our proposed method outperforms previous works with the F1 score of 79.43% for CULane and 97.80% for TuSimple dataset while achieving 148+ FPS.

Keywords: Lane detection · Dynamic head · Self-attention

1 Introduction

With the recent development of artificial intelligence, various autonomous driving technologies [1] have achieved satisfying results. Lane detection is a fundamental perception task in autonomous driving with a wide range of applications, such as adaptive cruise control, lane keeping assistance, and high-definition mapping. To ensure the safety of autonomous driving and ensure basic driving between lane lines, lane detection requires high accurate and real-time. However, in actual driving scenarios, lane lines are considerably affected by complex environments such as extreme lighting conditions, occlusion, and lane line damage;

Code is available at https://github.com/CZY-Code/DILane.

L. Wang et al. (Eds.): ACCV 2022, LNCS 13842, pp. 124–140, 2023.
https://doi.org/10.1007/978-3-031-26284-5_8

these are very common and make the task of lane detection challenging. The traditional lane detection methods [1,2] usually rely on texture information of the lane line for feature extraction; then they obtain the topology structure of the lane line through post-processing regression. Yet, traditional methods lack robustness in complex real-world scenarios.

Fig. 1. Illustration of different scenarios in lane detection. Blue lines are ground truth, while green and red lines are true positive and false positive cases. (a) Lane detection is accurate in normal scenes (despite the level of features we focus on). (b) Detailed localization of lanes is inaccurate when we only utilize high-level features in the curve scene. While facing the non-visual evidence case, *e.g.*, (c) occlusions and (d) no-line, only utilizing low-level features will increase false negative samples. (Color figure online)

Owing to the effective feature extraction capability of deep convolutional networks [3], most recent studies have focused on deep learning [4,5] to address the task of lane detection and achieved impressive performance on mainstream datasets [5,6]. However, there still lacks an efficient and effective lane representation. Given a front-view image captured by a camera mounted on the vehicle, segmentation-based method [7] outputs a segmentation map with per-pixel predictions and do not consider lanes as a whole unit. They predict a single feature map and overlook important semantic information because the lane itself has a slender topology and is easily occluded in the actual scene; moreover, it faces the problem of high label imbalance and time-consuming. Parameter-based [8,9] and keypoint-based [10,11] methods significantly improved inference speed, which model the lanes as holistic curves and treat lane line detection as a lane points localization and association problem respectively. However, those methods struggle to achieve higher performance because the polynomial coefficients are difficult to learn, and also the key points that are easily occluded will considerably affect the accuracy of lane line detection. Recently proposed anchor-based methods [12] regress the offsets between predefined fixed anchor points and lane points, then non-maximum suppression (NMS) [13] is applied to select lane lines with the highest confidence. However, the tremendous number of fixed anchors leads to inefficient calculations in NMS procedure.

To address the aforementioned issues, we propose a novel method named DILane. For efficiency, we perform learnable anchors to replace the fixed anchors. In our observations and analysis, the distribution of lane lines in the image is statistical, *i.e.*, most of lane lines start at the bottom or lower side of the image and end at the vanishing point [14]. Besides, the distance between lane lines is generally far from each other. Thus, the tremendous number of predefined fixed anchor is inefficient, the much less learnable anchors are expected to represent where lane lines are most likely to appear after the optimization.

For effectiveness, we perform dynamic head and self-attention in parallel to gather local and global information respectively. As shown in Fig. 1, the lane line has a high probability of being occluded and damaged in practical application scenarios which is different from other detection tasks [15,16]. Only leveraging low-level features that contain texture information will lead to an increase in false negative samples. Besides, lane lines are thin and have a long geometric shape at the same time, only using high-level features that mostly carry semantic information will cause localization inaccuracy. The proposed dynamic head conditionally enhanced high-level features with low-level features for each proposed lane instance. Inspired by iFormer [17], the self-attention is performed in parallel not only provide a possible implementation for instance interaction but also flexibly model discriminative information scattered within a wide frequency range.

Our contributions are as follows: 1) we propose learnable anchors that are the statistics of potential lane location in the image. 2) we develop a dynamic head to conditionally enhance high-level features with corresponding low-level features. 3) self-attention is utilized in parallel for global information aggregation. 4) the proposed method achieves state-of-the-art performance on two mainstream benchmarks of lane detection with high speed.

2 Related Works

According to the representation of lanes, current convolutional neural network (CNN)-based lane detection methods can be divided into four main categories: segmentation-based, anchor-based, keypoint-based, and parameter-based methods, as shown in Fig. 2.

2.1 Segmentation-Based Methods

Segmentation-based algorithms typically adopt a pixel-wise prediction formulation; *i.e.*, they treat lane detection as a semantic segmentation task, with each pixel classified as either a lane area or background. LaneNet [18] considered lane detection as an instance segmentation problem, a binary segmentation branch and embedding branch were included to disentangle the segmented results into

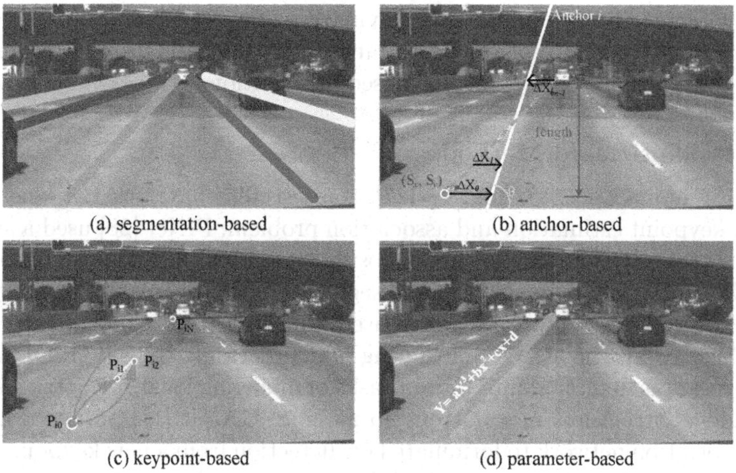

Fig. 2. Lane detection strategies. (a) Segmentation-based methods use per-pixel classification into several categories. (b) Anchor-based methods regress offsets between sampled points and predefined anchor points. (c) Keypoint-based methods use keypoint estimation and association. (d) Parameter-based methods regress parameters of lane curves.

lane instances. To distinguish different lane lines, SCNN [5] proposed a message-passing mechanism to address the no visual evidence problem, which captures the strong spatial relationship for lanes. Based on the SCNN, RESA [7] aggregates spatial information by shifting the sliced feature map recurrently in the vertical and horizontal directions. To achieve real-time requirements in practice, ENet-SAD [19] presented a knowledge distillation approach for transferring knowledge from large networks to small networks. CurveLane-NAS [20] proposed a neural architecture search (NAS) to find a better network for lane detection, which is extremely computationally expensive and requires 5,000 GPU hours per dataset.

2.2 Anchor-Based Methods

The methods of constructing an anchor can be divided into two types: line anchors and row anchors. In line anchor-based methods, a line-CNN [21] obtains a feature vector from each boundary position for regression and classification. LaneATT [12] predefined a set of fixed anchors for feature pooling, and proposed a attention mechanism for lane detection, which is potentially useful in other domains where the objects being detected are correlated. SGNet [22] proposed a vanishing point guided anchoring mechanism and multilevel structural constraints to improve performance. CLRNet [23] detected lanes with high-level semantic features, and then performed refinement based on low-level features. The row anchor-based approach selects the locations of lanes at predefined rows of the image, UFLD [24,25] proposed a simple formulation of lane detection aiming at extremely fast speeds and solving the non-visual cue problem. CondLaneNet [26] aimed to resolve lane instance-level discrimination based on conditional convolution and row anchor-based formulation.

2.3 Keypoint-Based Methods

Inspired by human pose estimation [27], some studies have treated lane detection as a keypoint estimation and association problem. PINet [28] used a stacked hourglass network to predict keypoint positions and cast a clustering problem of the predicted keypoints as an instance segmentation problem. In FastDraw [4], the author proposes a novel learning-based approach to decode the lane structures, which avoids the need for clustering post-processing steps. FOLOLane [10] decomposed lane detection into the subtasks of modeling local geometry and predicting global structures in a bottom-up manner. GANet [11] proposed a novel global association network to formulate lane detection from a new keypoint-based perspective that directly regresses each keypoint to its lane.

2.4 Parameter-Based Methods

The parameter-based methods treat lane lines as a parameter regression problem. PolyLaneNet [29] outputted polynomials that represent each lane marking in the image along with the domains for these polynomials and confidence scores for each lane. LSTR [9] developed a transformer-based network to capture long and thin structures for lanes and global context. BezierLaneNet [8] exploited the classic cubic Bezier curve because of its easy computation, stability, and high degree of freedom of transformation to model thin and long geometric shape properties of lane lines. Eigenlanes [30] propose a algorithm to detect structurally diverse road lanes in the eigenlane space which are data-driven lane descriptors, each lane is represented by a linear combination of eigenlanes.

3 Proposed Method

In this paper, we propose a novel learnable anchor-based method called DILane to detect structurally diverse road lanes. Figure 3 presents an overview of the proposed method.

The proposed DILane receives an input RGB image $I \in \mathbb{R}^{3\times H_i \times W_i}$ which is captured from a front-facing camera mounted on the vehicle, and predicts lanes $L = \{l_1, l_2, \ldots, l_N\}$. The $l_i = \{cls_i, S_{ix}, S_{iy}, \theta_i, len_i, (\Delta x_i^0, \Delta x_i^1, \ldots, \Delta x_i^{K-1})\}$, where the cls_i is classification outputs, the x- and y-coordinates of start point are defined as S_{ix} and S_{iy}, θ_i is the slope of the anchor, len_i is valid length of the lane, and K is the predefined maximum length of the lane with equally spaced coordinates in the y-axis, and $\Delta x_i \in \mathbb{R}^K$ is the horizontal offsets between the predictions and anchor lines. To generate these outputs, a backbone with FPN [31] is used to produce multi-level feature maps. The dynamic head and self-attention module receive multi-level features which are pooled by the learnable anchors from feature maps. The final feature for regression is concatenated by the outputs of dynamic head and self-attention module.

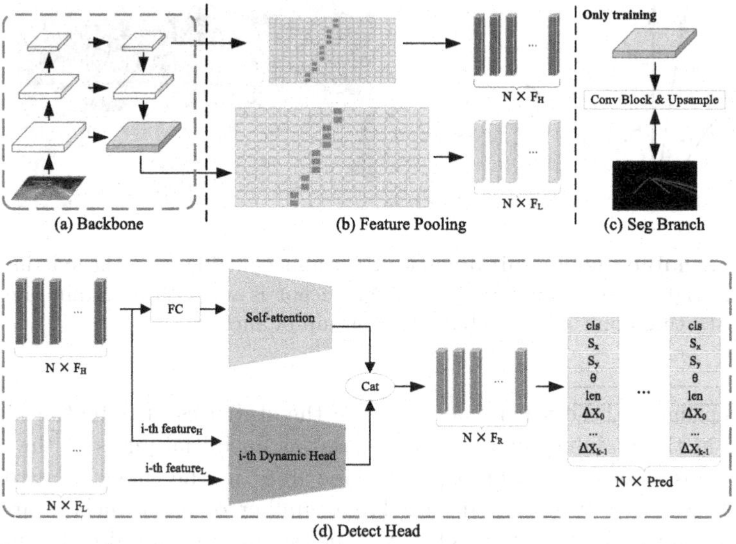

Fig. 3. Overview of DILane. (a) Backbone with FPN generates multi-level feature maps. (b) Each learnable anchor is used to pool high-level and low-level features from top and bottom layers, respectively. (c) Segmentation branch is only valid during training. (d) Detect head contains two components, the self-attention is used to aggregate global information, and dynamic head is used for local feature enhancement.

3.1 Learnable Anchor

Previous anchor-based methods predefined one anchor at every possible location. Obviously, the location of lanes is statistically distributed, thus we replace a large number of fixed anchors with a small set of learnable anchors. In common object detection, objects are represented by rectangular boxes, but a lane is thin and long with strong shape priors. Thus, we define each anchor by a 4-dimensional vector $reg_i = \{S_{ix}, S_{iy}, \theta_i, len_i\}$, which denotes the normalized x and y coordinates of the starting point, direction θ and length. For every fixed $y_i^k = k \cdot \frac{H_i}{K-1}$, each predicted x-coordinate $\hat{x}_i^k$ can be calculated as following:

$$\hat{x}_i^k = \frac{1}{\tan \theta_i} \cdot (y_i^k - S_{iy}) + S_{ix} + \Delta x_i^k. \tag{1}$$

The parameters of the learnable anchor reg_i were updated with the backpropagation of the algorithm during training. Following Line-CNN [21], we simply initialize the learnable anchor to make the algorithm converge faster, as shown in Fig. 4.

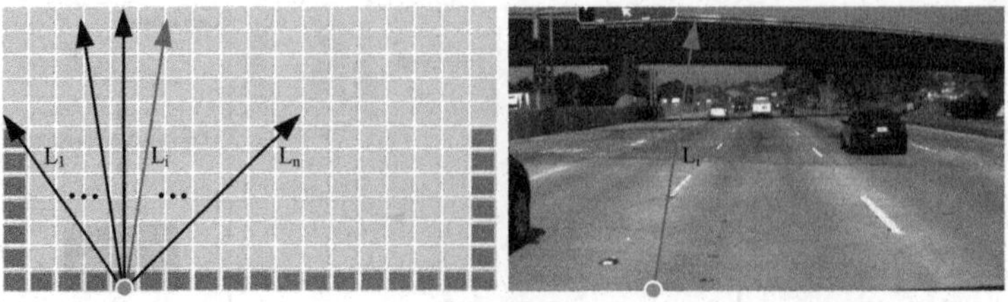

Fig. 4. Learnable anchor initialization. Anchors emitting from the starting points on the left/right/lower boundary. A straight anchor is set with a certain orientation and each starting point is associated with a group of anchors.

Conceptually, these learned anchors are the statistics of potential lane line locations in the training set and can be seen as an initial guess of the regions that are most likely to encompass the lane lines in the image, regardless of the input. Notably, proposals from a large number of fixed anchors are time-consuming and ineffective. Instead, a reasonable statistic can already qualify as a candidate. From this perspective, DILane can be categorized as a sparse lane detector.

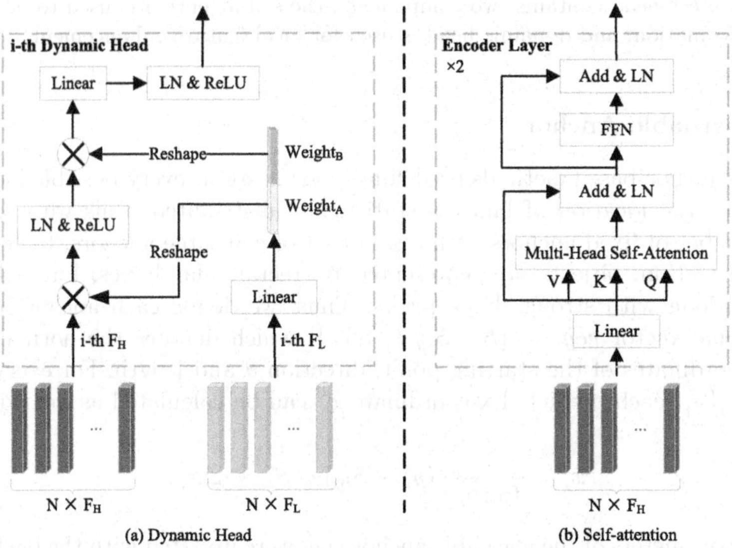

Fig. 5. Details of dynamic head and self-attention (LN: layer-norm; ⊗: 1D-convolution). (a) High-level feature and corresponding low-level feature are fed into its dynamic head to generate enhanced feature for each proposal. (b) Structure of self-attention which consists of two transformer encoder layers [32].

3.2 Dynamic Head

Motivated by dynamic algorithms [33,34], we propose the dynamic head based on conditional convolution - a convolution operation with dynamic kernel parameters [35,36]. Sparse R-CNN [37] introduced a concept termed proposal feature $F \in \mathbb{R}^d$, which is a high-dimensional latent vector that is expected to encode rich instance characteristics. For lane detection, texture information is critical to detection accuracy when the lane line is clearly visible or less affected by the background environment; still, there are more cases where lane lines are occluded or affected by extreme lighting conditions.

Considering that bottom layers contribute more in capturing details while top layers play a significant role in modeling semantic information. To obtain more useful information dynamically under a non-visual evidence situation, we propose a dynamic head to enhance instance features, which utilizes the extracted low-level texture features to conditionally enrich high-level features.

Every anchor i has its corresponding feature vector F_H pooled from the high-level feature map and F_L pooled from the low-level feature map, which carries different semantic information. Where the part of the anchor is outside the boundaries of the feature map, both F_H and F_L are zero-padded. For each proposal instance, the high-level feature is fed into its own head, which is conditioned on a specific low-level feature. Figure 5 illustrates the dynamic instance enhancement. in the k-th dynamic head, the k-th low-level feature F_L generates 1D-convolution kernel parameters instance-wisely for the corresponding k-th high-level feature F_H. The high-level feature $F_H \in \mathbb{R}^{H \times C}$ interact with the corresponding low-level proposal feature $F_L \in \mathbb{R}^{H \times C}$ to supplement texture information and output final enriched feature $F_R^{local} \in \mathbb{R}^C$ for next step.

3.3 Self-attention

Depending on the characteristics of the CNN to gather information from nearby pixels, each feature vector mostly carries local information. Thus, we utilize a global attention mechanism to gather the global context for each proposal feature to achieve better performance. Recent studies [38,39] have shown that the transformer has a strong capability to build long-range dependence, which achieves surprisingly high performance in many NLP tasks, *e.g.*, machine translation [40] and question answering [41]. Its success has led researchers to investigate its adaptation to the computer vision field, and Vision Transformer (ViT) [42] is a pioneer that is applied to image classification with raw image patches as input.

For each high-level feature $F_H \in \mathbb{R}^{H \times C}$, we carry out a fully connected layer to reduce the channels and generate $F'_H \in \mathbb{R}^C$. We regard each instance feature vector F'_H as a single token and put all feature vectors as a sequence into a self-attention module, which consists of two transformer encoders for gathering global information. The output of self-attention module $F_R^{global} \in \mathbb{R}^C$ concatenate with F_R^{local} which is generated by dynamic head in Sect. 3.2 and obtain $F_R = [F_R^{local}, F_R^{global}]$ for final regression.

3.4 Loss Function

Label Assignment. DILane infers a fixed-size set of N predictions in a single pass through the detector, where the N is set significantly larger than the typical number of lane lines in an image. The first difficulty is scoring the predicted lanes with respect to the ground truth. Our loss produces an optimal bipartite matching between the predicted and ground truth lanes, and then optimizes the lane-specific losses. We denote the ground truth as G and the set of N predictions as $P = \{p_1, p_2, \ldots, p_N\}$. Assuming N is larger than the maximum number of lanes in the single image, we consider $G = \{g_1, g_2, \ldots, g_N\}$ as a set of N padded with $\varnothing$ (no lane). To find a bipartite matching between these two sets, we search for a permutation of N elements with the lowest cost, which is expressed as:

$$\hat{\sigma} = \underset{\sigma}{arg\,min} \sum_{i}^{N} \mathcal{L}_{match}(p_i, g_{\sigma(i)}), \tag{2}$$

where $L_{match}(p_i, g_{\sigma(i)})$ is the pairwise matching cost between a prediction p_i and ground truth with index $\sigma(i)$. This optimal assignment is computed efficiently using the Hungarian algorithm described in [43].

The matching cost considers both the class prediction and similarity of the predicted and ground truth lanes. Each element i of the ground truth set can be seen as $g_i = \{cls_i, reg_i = \{S_{ix}, S_{iy}, \theta_i, len_i\}, \Delta X_i = \{\Delta x_i^0, \Delta x_i^1, \ldots, \Delta x_i^{K-1}\}\}$. For the prediction with index $\sigma(i)$, we define the probability of class cls_i as $\hat{p}_{\sigma(i)}(cls_i)$ and the predicted parameters as $\hat{reg}_{\sigma(i)}$. With these notations, we define $L_{match}(p_i, g_{\sigma(i)})$ as:

$$\mathcal{L}_{match}(p_i, g_{\sigma(i)}) = \left(\hat{p}_{\sigma(i)}(cls_i)\right)^{\alpha} \cdot \left(1 - L_1(reg_i, \hat{reg}_{\sigma(i)}) - \mathcal{L}_{\Delta X_i}\right)^{(1-\alpha)}, \tag{3}$$

where α is set to 0.2 by default, and the above equations can be efficiently solved using the Hungarian algorithm. In Eq. 3, $\mathcal{L}_{\Delta X_i}$ is used to measure the average L1-distance between the predicted x-direction offsets and ground truth, which is defined as:

$$\mathcal{L}_{\Delta X_i} = \frac{1}{K} \sum_{j=0}^{K} L_1(\Delta x_i^j, \hat{\Delta x}_i^j). \tag{4}$$

Training Loss. Our training loss includes three parts, *i.e.*, classification, matched sample, and segmentation loss, which is only used during training. The overall loss is the weighted sum of all losses:

$$\mathcal{L}_{Total} = \lambda_{cls} \cdot \mathcal{L}_{cls} + \lambda_{seg} \cdot \mathcal{L}_{seg} + \sum_{i=1}^{N} \mathbb{1}_{\{cls_i \neq \varnothing\}} \mathcal{L}_{match}(p_i, g_{\hat{\sigma}(i)}). \tag{5}$$

In Eq. 5, the $\mathcal{L}_{match}(p_i, g_{\hat{\sigma}(i)})$ is used to measure the distance between predicted lane p_i with corresponding ground truth with index $\hat{\sigma}(i)$, and is defined as:

$$\mathcal{L}_{match}(p_i, g_{\hat{\sigma}(i)}) = \lambda_{reg} \cdot \mathcal{L}_{reg}(p_i, g_{\hat{\sigma}(i)}) + \lambda_{IoU_i} \cdot \mathcal{L}_{IoU}(p_i, g_{\hat{\sigma}(i)}). \tag{6}$$

We used focal loss [44] $\mathcal{L}_{cls}$ to solve the imbalance between positive and negative examples. The regression loss $\mathcal{L}_{reg}$ is the smooth-$l1$ distance for positive lane parameters. For the x-direction offset loss $\mathcal{L}_{IoU}$, we followed the process in [23] to calculate the distance between the predicted positive samples and ground truth. Additionally, we added an auxiliary binary segmentation branch to segment the lane line or background during training, and expected the bottom-up detail perception to enforce the learning of spatial details.

4 Experiment

4.1 Datasets

To evaluate the performance of the proposed method, we conducted experiments on two well-known datasets: CULane [5] and TuSimple [6]. The CULane dataset contains 88,880 training images and $34,680$ test images, including 9 challenging scenarios collected on urban roads with 590×1640 pixels. The TuSimple dataset was collected only on highways with high-quality images, consisting of 3,626 images for training and $2,782$ images for testing, all of which had 720×1280 pixels.

4.2 Implementation Details

Except when explicitly indicated, all input images were resized to $H_i \times W_i = 320 \times 800$ pixels. For all training sessions, the AdamW optimizer was used for 20 and 70 epochs on CULane and TuSimple with an initial learning rate of 5e-3. The backbone parameters were initialized by the pretrained ResNet-18/34. Data augmentation was applied to the training phase, and random affine transformation was performed (with translation, rotation, and scaling) along with random horizontal flip. Moreover, we empirically and experimentally set the number of points $K = 72$, and the intersection over union (IoU) threshold of the NMS was set to 0.5. All experiments were performed on a machine with an Intel i7-10700K processor and a single RTX 2080Ti processor.

4.3 Metrics

The F1 score is the only metric for the CULane dataset, which is based on the IoU. Because the IoU relies on areas instead of points, a lane is represented as a thick line connecting the respective points. In particular, the official metric of the dataset considers the lanes as 30-pixels-thick lines. Only when the IoU between the prediction and ground truth is greater than 0.5, can the predicted lane lines be considered as positive. The F1 score is the harmonic mean of the precision and recall, which is defined as:

$$F1 = \frac{2 \times Precision \times Recall}{Precision + Recall}, \tag{7}$$

where $Precision = \frac{TP}{TP+FP}$ and $Recall = \frac{TP}{TP+FN}$.

For TuSimple dataset, the three standard official metrics are the false discovery rate $FDR = \frac{FP}{TP+FP}$, false negative rate $FNR = \frac{FN}{TP+FN}$, and the accuracy is defined as:

$$Acc = \frac{\sum_{clip} C_{clip}}{\sum_{clip} S_{clip}}, \tag{8}$$

where S_{clip} is the total number of points in the clip, and C_{clip} is the predicted lane points in the same clip. When the predicted lane point is within 20 pixels of the ground truth, it is regarded as a true positive. If a lane is to be considered as a true positive, 85% of the points must be correct.

4.4 Result

The results of our method on the CULane dataset compared with those of other popular methods are shown in Table 1. We can see that the lanes can be detected with accurate location and precise shape, even in complex scenarios. As demonstrated, our method achieves the state-of-the-art performance on the CULane benchmark with an F1 score of 79.43%. We can clearly see that the proposed method consistently outperforms other popular methods in most categories while maintaining 148+ FPS. After comparing our ResNet18-based method with other methods, the proposed method is faster than most other methods with 179 FPS and achieves an F1 score of 78.96%. These observations demonstrate the efficiency and robustness of our proposed method.

Table 1. Comparison with state-of-the-art methods on CULane dataset. F1 score ("%" is omitted) is used to evaluate the results of total and 9 sub-categories with IoU threshold equal to 0.5. For "Cross", only FP values are shown.

Method	Total	Normal	Crowd	Dazzle	Shadow	No-line	Arrow	Curve	Cross	Night	FPS
SCNN [5]	71.60	90.60	69.70	58.50	66.90	43.40	84.10	64.40	1990	66.10	7.5
RESA-R34 [7]	74.50	91.90	72.40	66.50	72.00	46.30	88.10	68.60	1896	69.80	45.5
UFLD-R18 [24]	68.40	87.70	66.00	58.40	62.80	40.20	81.00	57.90	1743	62.10	**282**
UFLD-R34 [24]	72.30	90.70	70.20	59.50	69.30	44.40	85.70	69.50	2037	66.70	170
PINet [28]	74.40	90.30	72.30	66.30	68.40	49.80	83.70	65.20	1427	67.70	25
LaneATT-R18 [12]	75.09	91.11	72.96	65.72	70.91	48.35	85.47	63.37	1170	68.95	250
LaneATT-R34 [12]	76.68	92.14	75.03	66.47	78.15	49.39	88.38	67.72	1330	70.72	171
SGNet-R18 [22]	76.12	91.42	74.05	66.89	72.17	50.16	87.13	67.02	1164	70.67	117
SGNet-R34 [22]	77.27	92.07	75.41	67.75	74.31	50.90	87.97	69.65	1373	72.69	92
Laneformer-R18 [45]	71.71	88.60	69.02	64.07	65.02	45.00	81.55	60.46	**25**	64.76	–
Laneformer-R34 [45]	74.70	90.74	72.31	69.12	71.57	47.37	85.07	65.90	26	67.77	–
[8]-R18	73.67	90.22	71.55	62.49	70.91	45.30	84.09	58.98	996	68.70	213
[8]-R34	75.57	91.59	73.20	69.20	76.74	48.05	87.16	62.45	888	69.90	150
Eigenlanes-R18 [30]	76.50	91.50	74.80	69.70	72.30	51.10	87.70	62.00	1507	71.40	–
Eigenlanes-R34 [30]	77.20	91.70	76.00	69.80	74.10	52.20	87.70	62.90	1507	71.80	–
Ours-R18	78.96	93.27	77.28	**71.98**	77.63	53.27	**89.96**	68.45	1372	74.56	179
Ours-R34	**79.43**	**93.81**	**77.70**	71.80	**78.99**	**54.12**	89.69	**70.11**	1230	**74.79**	148

Additionally, the comparison of our method on TuSimple dataset is shown in Table 2. Our method achieves the highest F1 score of 97.80% on TuSimple dataset. Comparing with the baseline LaneATT [12], our ResNet18 version surpasses 0.90% on F1 and 1.18% on accuracy respectively. Notably, our method performs best with 2.35% on FDR rate and achieves remarkable 2.05% on FNR rate.

Table 2. Comparison with state-of-the-art methods on TuSimple dataset. All measures were computed using the official source code [5].

Method	F1 (%)	Acc (%)	FDR (%)	FNR (%)
SCNN [5]	95.97	96.53	6.17	**1.80**
RESA-R34 [7]	96.93	**96.82**	3.63	2.48
UFLD-R18 [24]	87.87	95.82	19.05	3.92
UFLD-R34 [24]	88.02	95.86	18.91	3.75
LaneATT-R18 [12]	96.71	95.57	3.56	3.01
LaneATT-R34 [12]	96.77	95.63	3.53	2.92
Laneformer-R18 [45]	96.63	96.54	4.35	2.36
Laneformer-R34 [45]	95.61	96.56	5.39	3.37
BezierLaneNet-R18 [8]	95.05	95.41	5.30	4.60
BezierLaneNet-R34 [8]	95.50	95.65	5.10	3.90
Eigenlanes [30]	96.40	95.62	3.20	3.99
Ours-R18	<u>97.61</u>	96.75	<u>2.56</u>	2.22
Ours-R34	**97.80**	<u>96.82</u>	**2.35**	<u>2.05</u>

4.5 Ablation Analysis

We verify the impact of the major modules through experiments on CULane dataset to show the performance and analyze each part of the proposed method. The results of the overall ablation study is presented in Table 3.

Table 3. Ablation studies on each component.

Baseline	Learnable anchor	Dynamic head	Att-cascade	Att-parallel	F1 (%)
√	–	–	–	–	75.09
√	√	–	–	–	78.26
√	√	√	–	–	78.47
√	√	√	√	–	78.71
√	√	√	–	√	**78.96**

We take LaneATT [12] with Resnet-18 as our baseline, and gradually add the learnable anchor, dynamic head and self-attention. These three major components improve the F1 score by 3.17%, 0.21% and 0.24% respectively. The last row of the table indicates that constructing dynamic head and self-attention in parallel brings in a gain of 0.25% F1 comparing with cascade mode.

Learnable Anchor. The number of learnable anchors has a significant impact on the detection results. We explored the impact by controlling the number of learnable anchors, as shown in Table 4.

Table 4. Effectiveness of the number of learnable anchors.

Anchors	100	200	300	500	1000
F1 (%)	77.82	78.96	78.98	79.01	**79.02**
Recall (%)	73.11	73.67	73.78	73.79	**73.82**
FPS	**190**	179	160	122	96

As the proposal number increases, F1 and recall rate continue to improve, and the computational consumption also increases. In some cases, being efficient is crucial for lane detection, it might even be necessary to trade some accuracy to achieve the application's requirement. We set the number of anchors to 200 to balance efficiency and effectiveness in later experiments.

Feature Enhancement. We conducted comparative experiments using different level features. We first use features from the top and bottom layers to perform feature enhancement with learnable embedding, then combined the features of different layers for comparison. The experiment are summarized in Table 5.

Table 5. Different feature enhancement settings.

Setting	F_H	F_L	$F_L + F_H$	$F_H + F_L$
F1 (%)	77.68	78.26	78.64	**78.96**

In Table 5, the "F_H" and "F_L" mean enhancing high-level and low-level features with learnable embeddings. The "$F_H + F_L$" and "$F_L + F_H$" mean enhancing high-level and low-level features with another level feature. The experiment shows enhancing high-level feature with low-level feature performs best.

Visualization. The qualitative results for CULane and TuSimple datasets are shown in Fig. 6. Particularly, we select one image from each of the nine subcategories of CULane dataset for visualization. These visualizations demonstrate that our proposed DILane is able to provide high quality lane representation.

Fig. 6. Visualization on the Tusimple and the CULane dataset. Blue lines are ground-truth, while green and red lines are true-positive and false-positive. (Color figure online)

5 Conclusion

In this paper, we propose the Dynamic Instance-Aware Network (DILane) for lane detection. We introduce the learnable anchors which significantly improved the efficiency. Comparing with previous anchor-based methods [12,22] which predefined 1,000 fixed anchors, our method performs better (+2.14%) which only predefined 200 learnable anchors. Notably, our proposed dynamic head and self-attention in parallel, which focus on cross-level feature enhancement and instance interaction, improved the effectiveness remarkably (+0.7%). Our method achieves the state-of-the-art performance on both CULane and TuSimple datasets with F1 score of 79.43% and 97.80% while running at 148+ FPS.

Acknowledgements. This work was supported in part by the National Key R&D Program of China (2018AAA0102801 and 2018AAA0102803), and in part of the National Natural Science Foundation of China (61772424, 61702418, and 61602383).

References

1. Badue, C., et al.: Self-driving cars: a survey. Expert Syst. Appl. **165**, 113816 (2021)
2. Berriel, R.F., de Aguiar, E., De Souza, A.F., Oliveira-Santos, T.: Ego-lane analysis system (ELAS): dataset and algorithms. Image Vis. Comput. **68**, 64–75 (2017)
3. He, K., Zhang, X., Ren, S., Sun, J.: Deep residual learning for image recognition. In: Proceedings of the IEEE Conference on Computer Vision and Pattern Recognition, pp. 770–778 (2016)

4. Philion, J.: FastDraw: addressing the long tail of lane detection by adapting a sequential prediction network. In: Proceedings of the IEEE/CVF Conference on Computer Vision and Pattern Recognition, pp. 11582–11591 (2019)
5. Pan, X., Shi, J., Luo, P., Wang, X., Tang, X.: Spatial as deep: spatial CNN for traffic scene understanding. In: Proceedings of the AAAI Conference on Artificial Intelligence (2018)
6. TuSimple: Tusimple benchmark (2020). https://github.com/TuSimple/tusimple-benchmark/
7. Zheng, T., et al.: RESA: recurrent feature-shift aggregator for lane detection. In: Proceedings of the AAAI Conference on Artificial Intelligence, pp. 3547–3554 (2021)
8. Feng, Z., Guo, S., Tan, X., Xu, K., Wang, M., Ma, L.: Rethinking efficient lane detection via curve modeling. In: Proceedings of the IEEE/CVF Conference on Computer Vision and Pattern Recognition, pp. 17062–17070 (2022)
9. Liu, R., Yuan, Z., Liu, T., Xiong, Z.: End-to-end lane shape prediction with transformers. In: Proceedings of the IEEE/CVF Winter Conference on Applications of Computer Vision, pp. 3694–3702 (2021)
10. Qu, Z., Jin, H., Zhou, Y., Yang, Z., Zhang, W.: Focus on local: detecting lane marker from bottom up via key point. In: Proceedings of the IEEE/CVF Conference on Computer Vision and Pattern Recognition, pp. 14122–14130 (2021)
11. Wang, J., et al.: A keypoint-based global association network for lane detection. In: Proceedings of the IEEE/CVF Conference on Computer Vision and Pattern Recognition, pp. 1392–1401 (2022)
12. Tabelini, L., Berriel, R., Paixao, T.M., Badue, C., De Souza, A.F., Oliveira-Santos, T.: Keep your eyes on the lane: real-time attention-guided lane detection. In: Proceedings of the IEEE/CVF Conference on Computer Vision and Pattern Recognition, pp. 294–302 (2021)
13. Bodla, N., Singh, B., Chellappa, R., Davis, L.S.: Soft-NMS-improving object detection with one line of code. In: Proceedings of the IEEE International Conference on Computer Vision, pp. 5561–5569 (2017)
14. Lee, S., et al.: VPGNet: vanishing point guided network for lane and road marking detection and recognition. In: Proceedings of the IEEE International Conference on Computer Vision, pp. 1947–1955 (2017)
15. Zhou, X., Wang, D., Krähenbühl, P.: Objects as points. arXiv preprint arXiv:1904.07850 (2019)
16. Redmon, J., Farhadi, A.: Yolov3: an incremental improvement. arXiv preprint arXiv:1804.02767 (2018)
17. Si, C., Yu, W., Zhou, P., Zhou, Y., Wang, X., Yan, S.: Inception transformer. arXiv preprint arXiv:2205.12956 (2022)
18. Neven, D., De Brabandere, B., Georgoulis, S., Proesmans, M., Gool, V.L.: Towards end-to-end lane detection: an instance segmentation approach. In: IEEE Intelligent Vehicles Symposium (IV), IEEE 2018, pp. 286–291 (2018)
19. Hou, Y., Ma, Z., Liu, C., Loy, C.C.: Learning lightweight lane detection CNNs by self attention distillation. In: Proceedings of the IEEE/CVF International Conference on Computer Vision, pp. 1013–1021 (2019)
20. Xu, H., Wang, S., Cai, X., Zhang, W., Liang, X., Li, Z.: CurveLane-NAS: unifying lane-sensitive architecture search and adaptive point blending. In: Vedaldi, A., Bischof, H., Brox, T., Frahm, J.-M. (eds.) ECCV 2020. LNCS, vol. 12360, pp. 689–704. Springer, Cham (2020). https://doi.org/10.1007/978-3-030-58555-6_41
21. Li, X., Li, J., Hu, X., Yang, J.: Line-CNN: end-to-end traffic line detection with line proposal unit. IEEE Trans. Intell. Transp. Syst. **21**, 248–258 (2019)

22. Su, J., Chen, C., Zhang, K., Luo, J., Wei, X., Wei, X.: Structure guided lane detection. arXiv preprint arXiv:2105.05403 (2021)
23. Zheng, T., et al.: CLRNet: cross layer refinement network for lane detection. In: Proceedings of the IEEE/CVF Conference on Computer Vision and Pattern Recognition, pp. 898–907 (2022)
24. Qin, Z., Wang, H., Li, X.: Ultra fast structure-aware deep lane detection. In: Vedaldi, A., Bischof, H., Brox, T., Frahm, J.-M. (eds.) ECCV 2020. LNCS, vol. 12369, pp. 276–291. Springer, Cham (2020). https://doi.org/10.1007/978-3-030-58586-0_17
25. Qin, Z., Zhang, P., Li, X.: Ultra fast deep lane detection with hybrid anchor driven ordinal classification. IEEE Trans. Pattern Anal. Mach. Intell. (2022)
26. Liu, L., Chen, X., Zhu, S., Tan, P.: CondLaneNet: a top-to-down lane detection framework based on conditional convolution. In: Proceedings of the IEEE/CVF International Conference on Computer Vision, pp. 3773–3782 (2021)
27. Zhang, F., Zhu, X., Wang, C.: Single person pose estimation: a survey. arXiv preprint arXiv:2109.10056 (2021)
28. Ko, Y., Lee, Y., Azam, S., Munir, F., Jeon, M., Pedrycz, W.: Key points estimation and point instance segmentation approach for lane detection. IEEE Trans. Intell. Transp. Syst. **23**, 8949–8958 (2021)
29. Tabelini, L., Berriel, R., Paixao, T.M., Badue, C., De Souza, A.F., Oliveira-Santos, T.: PolyLaneNet: lane estimation via deep polynomial regression. In: 2020 25th International Conference on Pattern Recognition (ICPR), pp. 6150–6156. IEEE (2021)
30. Jin, D., Park, W., Jeong, S.G., Kwon, H., Kim, C.S.: Eigenlanes: Data-driven lane descriptors for structurally diverse lanes. In: Proceedings of the IEEE/CVF Conference on Computer Vision and Pattern Recognition, pp. 17163–17171 (2022)
31. Lin, T.Y., Dollár, P., Girshick, R., He, K., Hariharan, B., Belongie, S.: Feature pyramid networks for object detection. In: Proceedings of the IEEE Conference on Computer Vision and Pattern Recognition, pp. 2117–2125 (2017)
32. Vaswani, A., et al.: Attention is all you need. In: Advances in Neural Information Processing Systems, vol. 30 (2017)
33. Tian, Z., Shen, C., Chen, H.: Conditional convolutions for instance segmentation. In: Vedaldi, A., Bischof, H., Brox, T., Frahm, J.-M. (eds.) ECCV 2020. LNCS, vol. 12346, pp. 282–298. Springer, Cham (2020). https://doi.org/10.1007/978-3-030-58452-8_17
34. Wang, X., Zhang, R., Kong, T., Li, L., Shen, C.: SOLOv2: dynamic and fast instance segmentation. In: Advances in Neural Information Processing Systems (2020)
35. Jia, X., De Brabandere, B., Tuytelaars, T., Gool, L.V.: Dynamic filter networks. In: Advances in Neural Information Processing Systems (2016)
36. Yang, B., Bender, G., Le, Q.V., Ngiam, J.: CondConv: conditionally parameterized convolutions for efficient inference. In: Advances in Neural Information Processing Systems, vol. 32 (2019)
37. Sun, P., et al.: Sparse R-CNN: end-to-end object detection with learnable proposals. In: Proceedings of the IEEE/CVF Conference on Computer Vision and Pattern Recognition, pp. 14454–14463 (2021)
38. Wang, X., Girshick, R., Gupta, A., He, K.: Non-local neural networks. In: Proceedings of the IEEE Conference on Computer Vision and Pattern Recognition, pp. 7794–7803 (2018)

39. Carion, N., Massa, F., Synnaeve, G., Usunier, N., Kirillov, A., Zagoruyko, S.: End-to-end object detection with transformers. In: Vedaldi, A., Bischof, H., Brox, T., Frahm, J.-M. (eds.) ECCV 2020. LNCS, vol. 12346, pp. 213–229. Springer, Cham (2020). https://doi.org/10.1007/978-3-030-58452-8_13
40. Brown, T., et al.: Language models are few-shot learners. In: Advances in Neural Information Processing Systems, vol. 33, pp. 1877–1901 (2020)
41. Chowdhery, A., et al.: Palm: scaling language modeling with pathways. arXiv preprint arXiv:2204.02311 (2022)
42. Dosovitskiy, A., et al.: An image is worth 16 × 16 words: transformers for image recognition at scale. arXiv preprint arXiv:2010.11929 (2020)
43. Stewart, R., Andriluka, M., Ng, A.Y.: End-to-end people detection in crowded scenes. In: Proceedings of the IEEE Conference on Computer Vision and Pattern Recognition, pp. 2325–2333 (2016)
44. Lin, T.Y., Goyal, P., Girshick, R., He, K., Dollár, P.: Focal loss for dense object detection. In: Proceedings of the IEEE International Conference on Computer Vision, pp. 2980–2988 (2017)
45. Han, J., et al.: Laneformer: Object-aware row-column transformers for lane detection. arXiv preprint arXiv:2203.09830 (2022)

CIRL: A Category-Instance Representation Learning Framework for Tropical Cyclone Intensity Estimation

Dengke Wang[1], Yajing Xu[1(✉)], Yicheng Luo[1], Qifeng Qian[2], and Lv Yuan[1]

[1] Beijing University of Posts and Telecommunications, Beijing, China
{wangdk,xyj,luoyicheng,lvyuan}@bupt.edu.cn
[2] National Meteorological Centery, Beijing, China

Abstract. Tropical Cyclone (TC) intensity estimation is a continuous label classification problem, which aims to build a mapping relationship from TC images to intensities. Due to the similar visual appearance of TCs in adjacent intensities, the discriminative image representation plays an important role in TC intensity estimation. Existing works mainly revolve around the continuity of intensity which may result in a crowded feature distribution and perform poorly at distinguishing the boundaries of categories. In this paper, we focus on jointly learning category-level and instance-level representations from tropical cyclone images. Specially, we propose a general framework containing a CI-extractor and a classifier, inside which the CI-extractor is used to extract an instance-separable and category-discriminative representation between images. Meanwhile, an inter-class distance consistency (IDC) loss is applied on top of the framework which can lead to a more uniform feature distribution. In addition, a non-parameter smoothing algorithm is proposed to aggregate temporal information from the image sequence. Extensive experiments demonstrate that our method, with the result of 7.35 knots at RMSE, outperforms the state-of-the-art TC intensity estimation method on the TCIR dataset.

Keywords: Tropical cyclone · Intensity estimation · Representation learning

1 Introduction

Tropical Cyclone (TC) is one of the natural disasters that bring out severe threats to human society. The intensity of TC, which is defined as the largest continuous surface wind near the center of the TC, is an important indicator of its destructiveness. Estimating the intensity of TC can help mankind effectively reduce the damage caused by TC.

Supported by the National Natural Science Foundation of China (NSFC No. 62076031).

L. Wang et al. (Eds.): ACCV 2022, LNCS 13842, pp. 141–154, 2023.
https://doi.org/10.1007/978-3-031-26284-5_9

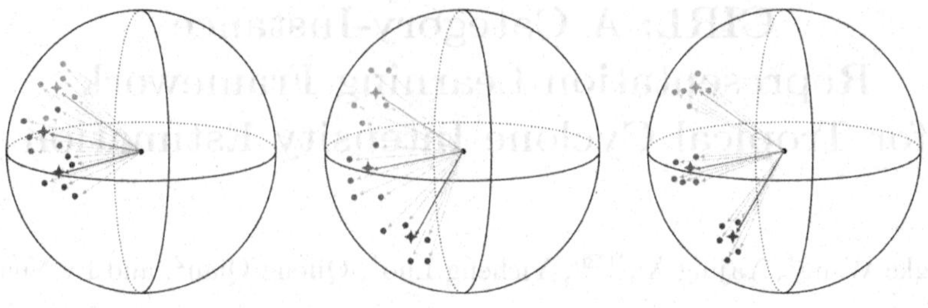

Fig. 1. Illustration of DC loss (a), DC loss + CE loss (b) and IDC + CE loss (c) based representation learning for TC intensity estimation. Stars represent the center of classes and circles represent samples. DC loss (a) learns the crowded feature distribution and has poor performance at boundaries. DC + CE loss (b) learns sparse embedding distribution, but it is scattered within the class. And IDC + CE loss (c) can lead to a uniform and discriminative feature distribution.

The essence of TC intensity estimation is establishing a mapping relationship from TC images to intensities. Since the change of TC is continuous during its life cycle, images of TC at adjacent moments may have similar visual appearances but different intensities. The appearance of TC with the same intensity may also vary greatly. These bring great challenges to TC intensity estimation.

The most classical method for TC intensity estimation is the Dvorak technique [7] which relies on TC cloud characteristics established on statistical experience. Over recent years, with the rapid development of deep learning, convolutional neural networks (CNNs) have achieved great success in TC intensity estimation. Among them, most methods [1–3,28] treat the intensity as continuous values and construct regression networks to estimate the exact intensity. [14] consider intensity estimation as a classification task. Different with most classification tasks, TC intensity estimation focuses on not only the right or wrong classifications but also the influence of different degrees of errors when classifying. Therefore, a distance consistency (DC) loss is proposed to keep the distance between representations in proportion to the distance between labels to reduce the errors. DC loss focuses on the relationship between instances to learn instance-level representations and achieved good performance. However, it neglects that the feature distribution of different classes should be separate, and the intra-class and inter-class samples should be treated differently when optimizing the embedding space.

On the other hand, considering the change of TC is a continuous process, the TC intensity at current time is related to those in the past. And generally, there is no violent shaking. Therefore, it is necessary to combine historical information to smooth the estimated intensity. [1] use fixed weight smooth methods to combine historical information. [14] adopts the transformer model to learn the change of intensity from a series of typhoon images. The application of transformer effectively utilizes the temporal information between typhoon samples but also introduces a large number of extra parameters to the estimation model.

In this paper, We focus on jointly learning category-level and instance-level representations from TC images. While instance-level learning aims to learn a uniform distribution between instances, the category-level representations attend to reduce the intra-class variance and distinguish boundaries between categories. Specifically, we describe our ideas in Fig. 1.

As shown in Fig. 1a, the feature distribution learned from DC loss can be crowded, since the ratio of distance between feature vectors to label distance is not supervised, which increases the difficulty of classification. In Fig. 1b, the distance between the feature of different classes still maintains a proportional relationship, but the inter-class feature distribution becomes separable with a CE loss. In Fig. 1c, an IDC loss with CE loss learns more intra-class compact and inter-class separable features, which further reduces the probability of the feature being classified incorrectly.

Motivated by above, we propose CIRL: a Category-Instance joint Representation Learning framework for TC intensity estimation with a CI-extractor and a classifier. As instance-level representation learning aims to obtain uniform distribution, category-level representation learning aims to distinguish the boundaries of categories and make the intra-class samples converge. Further, an IDC loss is proposed to optimize the backbone together with the CE loss. In IDC loss, the distance consistency is only maintained between categories. And the intra-class distance is optimized by the CE loss. Finally, we proposed a new smoothing algorithm, which can use historical information of any length to smooth the intensity estimate at the current moment. Without bells and whistles, our method achieved better performance than existing methods.

The contributions of our work can be summarized as follows:

- We propose a framework with a CI-extractor and a classifier aiming to learn a discriminative feature distribution which can take into account both instance-level and class-level representation learning. It is also general for continuous label classification problems.
- We propose a new inter-class distance consistency loss, which can learn a uniform feature distribution better.
- We explored a simple and fast smoothing post-processing algorithm without any parameters and find it is more accurate in real-time intensity estimation.
- Extensive experiments on the TCIR dataset demonstrate the effectiveness of our proposed approach, which outperforms the existing methods.

2 Related Works

Our work is closely related to both TC intensity estimation and metric learning.

2.1 Tropical Cyclone Intensity Estimation

Some results have been achieved by using CNN to estimate the intensity of the TC. [18] first propose to estimate the TC intensity by using a CNN-based classification network. However, only an approximate intensity range is obtained and the training data and test data are related in [18]. In [1,2], a regression network is further designed to to estimate the intensity accurately and more information beyond the image is taken into account, such as latitude, longitude, and date. [28] choose to divide the TC sample into three different categories and for each category, different regression networks are constructed to estimate the intensity. [27] proposed a context-aware cycleGAN to solve the problem of an unbalanced distribution of sample categories. [3] designed a Tensor Network to solve the asynchronous problem in remote sensing dataset to utilize more channels of data. In [14], a combined model of CNN and transformer is used to capture TC temporal information.

However, existing methods only notice the continuity of intensity and ignore that the sample of different intensities (labels) should be separable. [14] considers TC intensity estimation as a classification problem but still learns representations from instances. In contrast, our proposed framework jointly learns category-level and instance-level representations and aims to obtain a separable and discriminative feature distribution.

2.2 Metric Learning

Metric learning uses the distance to measure the similarity of samples and constraints to make similar samples close, and different samples far away. To our best knowledge, [4,8] introduced deep neural networks into metric learning for the first time. On this basis, the triple loss is proposed by [19,26] in which the relationship between inter-class samples and intra-class samples is further considered. [20] expands the number of positive and negative samples in a tuple and proposed an N-tuple loss. [21] integrated the above loss to cope with the situation of multiple positives. [13] combined the softmax and cross-entropy loss to softmax loss and proposed to increase the margin. [6,23–25] step further to optimize the margin in different opinions. The common idea of these methods is to minimize the intra-class distance and maximize the inter-class distance.

Recently, [12] noticed that the relationship between samples is not a simple positive and negative, and the distance between the features and the distance of their labels are connected to construct a triple log-ratio loss. [14] extend log-ratio loss to the case of N-tuples and proposed a DC loss. However, their method treats intra-class and inter-class samples equally, which will be harmful to the optimization.

3 The Proposed Approach

In this section, we firstly present our main idea about the Category-Instance fusion framework for TC intensity estimation. Then, we show the idea of IDC loss. Finally, a smoothing algorithm for eliminating fluctuations in intensity estimation is demonstrated.

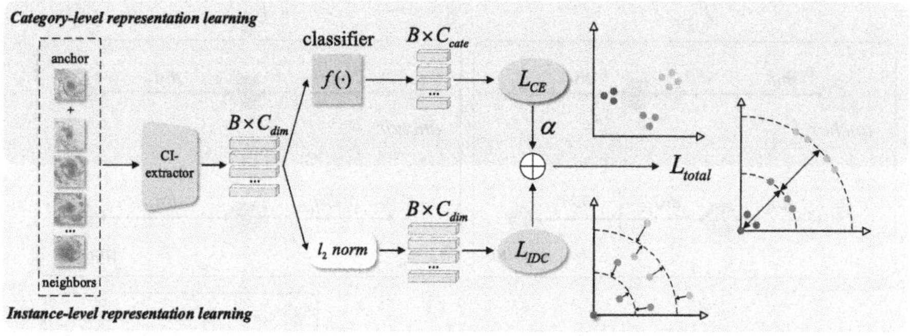

Fig. 2. Overview of the proposed Category-Instance representation learning framework. The framework consists of two loss functions: 1) CE loss for Category-level feature learning and 2) IDC loss for Instance-level feature learning. A CI-extractor is used to extract image representations, after which a multi-layer classifier $f(\cdot)$ is applied on top of the image representations to predict classification logits, and an l_2 normalization is adapted to translate the image representation for IDC loss. The total loss is obtained by the weighted summation of the two loss.

3.1 Category-Instance Representation Learning Framework

Figure 2 shows the overview of the proposed framework. There are two branches in our framework: one category-level representation learning in the upper part which aims to distinguish the boundaries of categories by CE loss and one instance-level representation learning in the lower part which aims to learn a uniform inter-class embedding distribution by IDC loss. The detail of IDC loss will be introduced in Sect. 3.2. The framework will learn a representation of the two levels at the same time to obtain the best feature distribution.

Formally, we adopt the method in [20] to constuct a mini-batch include M samples as anchors and N samples as neighbors. Each anchor is combined with the N neighbors to construct a set of N+1 tuple $\{x, y\} = \{(x_a, y_a), (x_1, y_1), \ldots, (x_N, y_N)\}$ with an anchor a and N neighbors randomly sampled from the remaining ones in which $x \in R^{2\times H\times W}$ is the images and y is the corresponding intensities. A CI-extractor is used to extract image repersentation $r = \{r_a, r_1, \ldots, r_k\} \in R^{D_E}$ from x. On the one hand, an ℓ_2 normalization is applied to r to get the normalized representaion $z = \{z_a, z_1, \ldots, z_k\} \in R^{D_E}$ to keep the vectors on the same sacle. After that, the IDC loss is applied on top of the normalized representations for instance-level representation learning. On the other hand, a classifier head $f_c(\cdot)$ is adopted to the image representation r to predict the class-wise logits $s = \{s_a, s_1, \ldots, s_k\} \in R^{D_C}$, which are used to compute the CE loss. Motivated by [13], the classifier, which is actually a full connection layers, is constructed without bias. Then, the logits can be writen as:

$$s = \|w\|\,\|r\|\, cos\theta. \tag{1}$$

w is the weight of the fully connected layer which can be regarded as the center of the class. Finally, a CE loss is applied to learn a separable feature which is

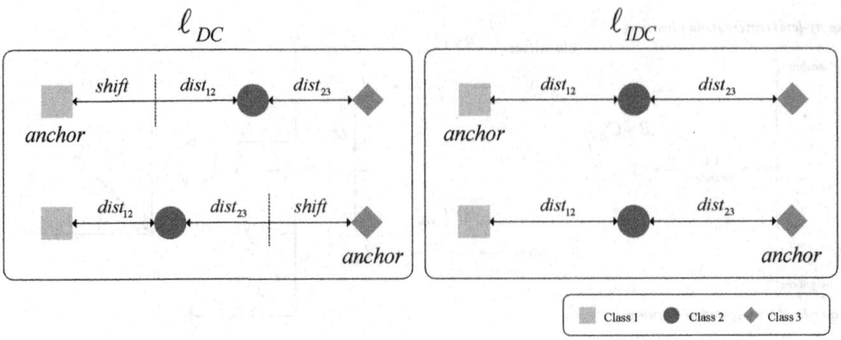

Fig. 3. Comparison of DC loss and IDC loss

important in discriminating class boundaries. As the CE loss aims to maximize the value of logits for the corresponding class, the cosine distance between r and w can also be increased and the intra-class variance is reduced. The final loss function for the framework is:

$$\ell_{total} = \ell_{IDC} + \alpha \ell_{CE}, \tag{2}$$

where ℓ_{IDC} is the IDC loss, ℓ_{CE} is the CE loss and α is a weighting parameter to balance the contribution of different losses.

3.2 Inter-class Distance Consistency Loss

The distance consistency loss [14] takes an anchor a and N neighbors as input. It is designed to penalize the sample for violating the rank constraint, namely, that the feature distance between samples in the embedding space should be consistency to the label distance. And the ratio, which is calculated by dividing the feature distance by the label distance, should be consistent across samples. However, when it comes to the case of intra-class samples, it is difficult to determine the label distance. Ideally, the feature distance between intra-class samples should be minimized and the label distance should be 0. But this would result in an infinite ratio of samples within the class. In [14], a constant is added to all label distances to avoid this which would cause a shift in label distances and be harmful to distance consistency.

Motivated by the above, we proposed an inter-class distance consistency loss to optimize the samples within and between classes respectively. Specially, we only maintain distance consistency across classes. The optimization of intra-class samples is left to CE loss. By doing so, the optimization of inter-class samples will no-longer be affected by the intra-class samples. For an anchor a and N neighbor, the IDC loss is formulated as:

$$\ell_{IDC} = -\sum_{\substack{i=1 \\ y_i \neq y_a}}^{N} log \frac{r_{ai}}{\sum_{j=1, y_j \neq y_a}^{N} r_{aj}}, \tag{3}$$

$$r_{ij} = \frac{D(f_i, f_j)}{D(y_i, y_j)}, \tag{4}$$

where f is the representation for the sample and $D(\cdot)$ denotes the Euclidean distance.

Compared with DC loss, IDC loss can obtain a more uniform distribution. As shown in Fig. 3, with a shift in label distance, in the same tuple, selecting different samples as anchors will generate different feature distributions in DC loss. This can lead to instability in the learning process and even oscillations in the feature distribution. In contrast, IDC loss can obtain a more uniform and stable distribution without being affected by anchor selection.

3.3 Inference Stage

After training the extractor, the inference stage is started. Considering the temporality of TCs, we sample the images in time order during the inference stage. The classifier is thrown away and only the CI-extractor is used. Following [1], each input image is rotated by four angles and fed into the extractor to obtain the feature embedding. An average operation is used to get the final embedding. We first utilize the backbone for extract features from the whole training dataset and averaged the features by class to construct a global class representation set $C = [c_1, c_2, \ldots]$. The class vectors were used for NN classification. Then, for a sequence of image $X = [x_{t-T}, x_{t-T+1}, \ldots, x_t]$, the process can be represented as

$$R = f(X, \theta) = [r_{t-T}, r_{t-T+1}, \ldots, r_t]. \tag{5}$$

Then, an ℓ_2 normalization was applied to R and get $Z = [z_{t-T}, z_{t-T+1}, \ldots, z_t]$. A nearest neighbor (NN) classifier is adopted to decide on the final intensity of Z, which is given by:

$$\hat{y}_t = \arg\min_i D(c_i, z_t), \tag{6}$$

where $D(\cdot)$ represents Euclidean distance, i represents the label and $\hat{y}_t$ is the estimated intensity at time t.

3.4 A Smooth Algorithm for Intensity Estimating

To reduce the shake of TC intensity estimation, we further adopted a weighted average method, named stage smooth, which uses the weighted average of the estimated TC intensity at the current time and the previous N moments as the final intensity estimation at the current time. Considering the TC intensity changes monotonously, we have adopted a basic algorithm, that is, the weight decays as the time increases.

Specifically, given a series of estimated intensites $\hat{Y} = [\hat{y}_{t-T}, \hat{y}_{t-T+1}, \ldots, \hat{y}_t]$, we use a sliding window to smooth the estimated intensity. The intensity estimation at time t was smoothed by the following Algorithm 1.

In the early stages of inference, the length of the sequence is less than the length of the window, we just repeat the first value of the sequence at the beginning.

Algorithm 1. Stage-smooth algorithm

Input: $\hat{y}_{t-T}, \hat{y}_{t-T+1}, \cdots, \hat{y}_t$
Output: $\tilde{y}_t$
Let $i = T$ and $\tilde{y}_{t-T} = \hat{y}_{t-T}$.
while $i > 0$ **do**
    $\tilde{y}_{t-i+1} = 0.5 * (\hat{y}_{t-i+1} + \tilde{y}_{t-i})$
    $i = i - 1$
end while
return $\tilde{y}_t$

4 Experiments

In this section, we firstly introduce the setting for our experiments which includes the dataset and implementation details. After that, we compare our proposed method with the state-of-the-art TC intensity estimating methods. Finally, some ablation studies are given to highlight some important properties of our framework.

4.1 Experimental Settings

Dataset. We conducted our experiment on the benchmark TCIR dataset [1] which contains 70501 TC images of 201 × 201 from 2003 to 2017. And There is 82 different intensity values. We used TC images from 2003 to 2016 for training and TC images from 2017 for testing to make sure the images in the training set and test set are from different typhoons. Followed [2], IR1 and PMW channels of the image in the dataset were used to train our model.

Implementation Details. Our method is implemented using Pytorch on an Nvidia RTX2080. A ResNet18 [10] is adopted as backbone to extract image representations, which has been pre-trained on the ImageNet ILSVRC 2012 dataset [5]. The classifier is consisted by one layer. And all images are resized to 224 × 224 before feeding into the network. Further, random rotation and random crops were adopted for data augmentation during training, and single-center crops were used for testing. We use SGD with a momentum of 0.9 and an initial learning rate of 5×10^{-4} as the optimizer to train the network. The network is trained for 50 epochs with the learning rate being decayed by a factor of 0.96 after each epoch. Following [2], We adopted a random sampling strategy to construct a mini-batch of a size of 12 from backbone training and the number of anchors is set to 4. The hyperparameter α is set to 1.3 and the T in Stage-smooth algorithm is set to 5.

Evaluation Metrics. We obtained the final results by searching for the nearest neighbors in the class vectors set. We adopted the root mean squared error (RMSE) and mean absolute error (MAE) as the evaluation metrics.

$$RMSE = \sqrt{\frac{1}{m}\sum_{i=1}^{m}(y_i - \tilde{y}_i)^2}, \tag{7}$$

Table 1. Tropical cyclone estimation results compared with other methods. Temporal means the way of handling temporal data. Bold numbers denote the best results.

	Approach	Temporal	RMSE (kt)
1	Cross-entropy	–	10.36
2	Npair [20]	–	10.75
3	log-ratio [12]	–	10.21
4	CNN-TC [1]	–	10.18
5	DR-extractor [14]	–	8.81
6	**Ours(CI-extractor)**	–	**8.49**
7	ADT [17]	linear	11.79
8	AMSU [11]	linear	14.10
9	SATCON [22]	linear	9.21
10	CNN-TC(S) [2]	five-point smooth	8.39
11	DR-transformer [14]	transformer	7.76
12	**Ours(CI-extractor)**	stage smooth	**7.35**

$$MAE = \frac{1}{m}\sum_{i=1}^{m} |y_i - \tilde{y_i}| . \tag{8}$$

4.2 Comparison to State-of-the-Art Methods

In this section, we compare the proposed CI-extractor to existing TC intensity estimation methods, such as traditional intensity estimation method [11,17,22], regreession-based method [1,2] and our main baseline [14] on TCIR dataset. The results are shown in Table 1. Note that kt is a unit commonly used in meteorology. 1kt ≈ 0.51 m/s.

The first three rows are some traditional methods that are reproduced on the TCIR dataset. The fourth and five rows are the typical regression-based method CNN-TC and our main baseline DR-extractor. Note that the DR-extractor here only includes the backbone and does not use any temporal strategy. The last six rows 7–12 compare experimental results using temporal information. Rows 7–9 are manual intensity estimation methods and linear interpolation is used to math the times of the dataset. CNN-TC(S) is the upgraded version of CNN-TC and a five-point smooth algorithm is applied. DR-transformer is the method that uses the DR-extractor as the backbone and a transformer model is further used to aggregate temporal information.

For rows 1–6, our CI-extractor outperforms other methods by a margin (8.49 vs 8.81) as none of the temporal information was used. Since our CI-extractor pays more attention to the reduction of intra-class variance in the feature space, fewer samples will be misclassified, which leads to a decrease in the RMSE metric. The last six rows (7–12) are results that make use of temporal information. From the table we can see, that our method is superior to the other approach by a

Table 2. Tropical cyclone estimation results with our main baseline. Intensites are devided into eight categories by SSHWS and the number of each is counted. The two methods on the left do not use temporal information which is used in the two on the right. Bold numbers denote the best results and the MAE and RMSE are reported as the final result.

Category	Intensity range	Numbers	Approach							
			DR-extractor		CI-extractor		DR-transformer		CI-extractor(S)	
			MAE	RMSE	MAE	RMSE	MAE	RMSE	MAE	RMSE
H5	$\geq 137kt$	41	6.04	7.53	7.03	8.48	10.39	11.93	**6.12**	**7.65**
H4	113–136 *kt*	93	6.65	8.95	7.24	9.16	6.45	8.17	**5.56**	**6.71**
H3	96–112 *kt*	130	9.97	12.44	9.61	11.95	7.68	9.31	**6.89**	**8.88**
H2	83–95 *kt*	243	10.10	12.68	9.84	12.25	8.13	10.33	**7.54**	**9.59**
H1	64–82 *kt*	468	10.26	12.47	9.18	11.47	8.77	10.96	**8.12**	**10.18**
TS	34–63 *kt*	1735	6.29	8.09	6.44	8.20	**5.47**	**7.08**	5.72	7.25
TD	20–33 *kt*	1501	5.33	6.92	4.70	6.28	5.04	6.38	**4.28**	**5.71**
NC	<20 *kt*	89	6.74	8.32	7.14	8.58	8.01	9.55	**5.93**	**7.19**
Total	–	4300	6.73	8.81	6.46	8.49	6.02	7.80	**5.62**	**7.35**

Table 3. Evaluation of the effects of different components in our framework. For row (a), the loss is used on top of the backbone. For row (b) and row (c), the category-level representation learning branch is further applied. And for row (d) and row (e), a stage smooth algorithm is adopted at inference stage.

	Methods	MAE	RMSE
(a)	DC	6.73	8.81
(b)	DC+CE	6.59	8.66
(c)	IDC+CE	6.46	8.49
(d)	DC+CE+stage smooth	5.80	7.54
(e)	IDC+CE+stage smooth	5.62	7.35

large margin, with the second-best model (DR-transformer) having circa 0.41 knots higher in the RMSE metric.

Performance in SSHWS. We further compare our method with our main baseline DR-transformer to show the performance in different intensity categories. As shown in Table 2, the left four columns are the results without temporal strategies which are used in the right four. The TC intensity is split by Saffir-Simpson Hurricane Wind Scale (SSHWS) along with intensity categorization for tropical storms and tropical depressions. The RMSE and MAE were reported. As we can see, in most categories, such as H5, H4, H3, H2, H1, TD, and NC, our method is superior to our baseline. In particular, the RMSE of our method for estimating the intensity of high-intensity (H5) typhoons and low-intensity(NC) typhoons is much lower which is greatly valuable in practical applications.

Table 4. Comparison of smooth methods. CI-extractor is used as backbone in all experiments. The RMSE and MAE is reported.

	Methods	MAE	RMSE
(a)	Transformer	5.89	7.69
(b)	Five-point	6.85	9.13
(c)	Stage-smooth	**5.62**	**7.35**

Table 5. Comparison of classifiers. For all experiments, CI-extractor is used as backbone.

	Classifier	MAE	RMSE
(a)	kNN (k = 1)	6.58	8.62
(b)	kNN (k = 3)	6.51	8.51
(c)	kNN (k = 5)	**6.51**	**8.50**
(d)	softmax	7.11	9.50

4.3 Ablation Studies and Discussions

In this section, we conduct some ablation studies to characterize our framework. Concretely, we study the effects of the IDC loss, CE loss, and the stage smooth algorithm to the result. We also discuss whether the KNN classifier is better than softmax and the advantage of our smooth algorithm to the transformer.

Effects of Components. For all experiments, the resnet-18 [10] is used as backbone to get the extractor. The estimation results are shown in Table 3. Results show that both category-level representation learning and inter-class calculation are necessary for CIRL. Especially, IDC loss reduces the RMSE (8.49 vs 8.66) by removing the intra-class instance distance which can cause the distance shifts and prevent the convergence of intra-class samples. The category-level representation learning also has a significant impact on the final result (8.66 vs 8.81) since it performs well in distinguishing class boundaries. And the smooth algorithm greatly improves the performance finally.

We further show the distribution of features which can be seen in Fig. 4. The features which are extracted by backbone are visualized by TSNE [15]. We sample ten categories in order to show them more clearly. In Fig. 4a, the feature distribution is crowded and it is difficult to distinguish different categories. In Fig. 4b, features of the same categories tend to cluster together with category-level representation learning. And in Fig. 4c, the intra-class features get closer which achieved the best performance.

KNN or Softmax. We further compare the kNN-based classifier and softmax-based classifier on the top of CI-extractor. For the second one, the classifier in our framework is reserved and the output of the classifier is regarded as the final result. The result is shown in Table 5. The kNN-based classifier performs better

Table 6. Effort of the sequence length in stage-smooth algorithm

Stage-smooth	T = 3	T = 5	T = 7	T = 9
MAE	5.76	5.66	5.65	5.65
RMSE	7.49	7.39	7.38	7.39

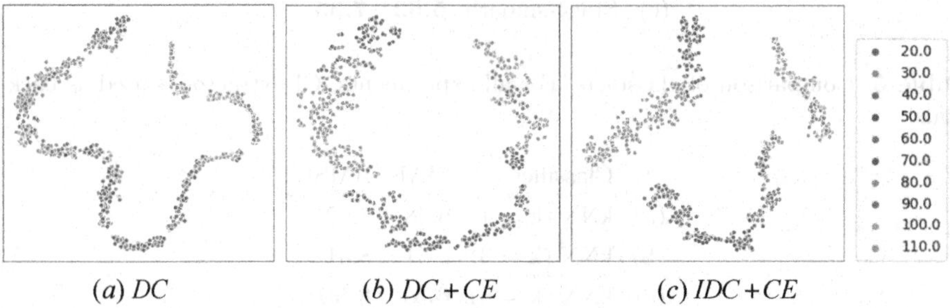

Fig. 4. The distribution of learned features under the different compose of loss, which lead to different distributions. The point with different colors denotes features of samples from different categories. The number in the legend represent labels and are also intensity values. Best viewed in color. (Color figure online)

than softmax-based classifier (8.50 vs 9.50). Since our framework aims to learn a uniform feature distribution, the softmax classifier is hard to find a dividing line for the embeddings. And a larger number of k has little effect on the results, which also owing to the uniform inter-class distribution.

Stage-Smooth or Transformer. In this work, we replace the stage-smooth algorithm with the transformer module. Specially, we use CI-extractor to obtain the feature embeddings from a sequence of images. Then, the feature embeddings are fed into the encoder of transformer and the one-hot vectors are fed to the decoder. The IDC loss is used to train the transformer. Following [14], the number of layers of encoder and decoder is set to 2 and the sequence length is set to 7. The result is shown in Table 4. Since transformer has been proven to effectively utilize temporal information to reduce errors, our method can also achieve the target with less computations. As can be seen, the transformer results in obviously inferior performance to our stage-smooth algorithm, since it only benefits from the continuity of the representations. Further, the effort of the sequence length in stage-smooth is explored in Table 6. A longer sequence has little effect on the results since the benefits of smoothing are limited.

5 Conclusion

In this paper, we proposed the CIRL framework which is general for continuous label classification problems. The framework focus on jointly learning a category-instance representation from images which can take full advantage of the separability and continuity of labels. Experiments on the TCIR dataset have convincingly demonstrated the effectiveness of our method. Additionally, we prove that in the continuous label classification problem, it is necessary to consider the separability between labels. We hope that this work will play a role in meteorological observations and in the future, our feature extraction framework can be extended to more continuous label classification problems.

References

1. Chen, B., Chen, B.F., Lin, H.T.: Rotation-blended CNNs on a new open dataset for tropical cyclone image-to-intensity regression. In: Proceedings of the 24th ACM SIGKDD International Conference on Knowledge Discovery & Data Mining, pp. 90–99 (2018)
2. Chen, B.F., Chen, B., Lin, H.T., Elsberry, R.L.: Estimating tropical cyclone intensity by satellite imagery utilizing convolutional neural networks. Weather Forecast. **34**(2), 447–465 (2019)
3. Chen, Z., Yu, X.: A novel tensor network for tropical cyclone intensity estimation. IEEE Trans. Geosci. Remote Sens. **59**(4), 3226–3243 (2021). https://doi.org/10.1109/TGRS.2020.3017709
4. Chopra, S., Hadsell, R., LeCun, Y.: Learning a similarity metric discriminatively, with application to face verification. In: 2005 IEEE Computer Society Conference on Computer Vision and Pattern Recognition (CVPR 2005), vol. 1, pp. 539–546. IEEE (2005)
5. Deng, J., Dong, W., Socher, R., Li, L.J., Li, K., Fei-Fei, L.: ImageNet: a large-scale hierarchical image database. In: 2009 IEEE Conference on Computer Vision and Pattern Recognition, pp. 248–255. IEEE (2009)
6. Deng, J., Guo, J., Xue, N., Zafeiriou, S.: ArcFace: additive angular margin loss for deep face recognition. In: Proceedings of the IEEE/CVF Conference on Computer Vision and Pattern Recognition, pp. 4690–4699 (2019)
7. Dvorak, V.F.: Tropical cyclone intensity analysis and forecasting from satellite imagery. Mon. Weather Rev. **103**(5), 420–430 (1975)
8. Hadsell, R., Chopra, S., LeCun, Y.: Dimensionality reduction by learning an invariant mapping. In: 2006 IEEE Computer Society Conference on Computer Vision and Pattern Recognition (CVPR 2006), vol. 2, pp. 1735–1742. IEEE (2006)
9. Harwood, B., Kumar BG, V., Carneiro, G., Reid, I., Drummond, T.: Smart mining for deep metric learning. In: Proceedings of the IEEE International Conference on Computer Vision, pp. 2821–2829 (2017)
10. He, K., Zhang, X., Ren, S., Sun, J.: Deep residual learning for image recognition. In: Proceedings of the IEEE Conference on Computer Vision and Pattern Recognition, pp. 770–778 (2016)
11. Kidder, S.Q., et al.: Satellite analysis of tropical cyclones using the advanced microwave sounding unit (AMSU). Bull. Am. Meteor. Soc. **81**(6), 1241–1260 (2000)
12. Kim, S., Seo, M., Laptev, I., Cho, M., Kwak, S.: Deep metric learning beyond binary supervision. In: Proceedings of the IEEE/CVF Conference on Computer Vision and Pattern Recognition, pp. 2288–2297 (2019)
13. Liu, W., Wen, Y., Yu, Z., Yang, M.: Large-margin softmax loss for convolutional neural networks. arXiv preprint arXiv:1612.02295 (2016)
14. Luo, Y., Xu, Y., Li, S., Qian, Q., Xiao, B.: DR-transformer: a multi-features fusion framework for tropical cyclones intensity estimation. Appl. Sci. **11**(13), 6208 (2021)

15. Van der Maaten, L., Hinton, G.: Visualizing data using t-SNE. J. Mach. Learn. Res. **9**(11), 2579–2605 (2008)
16. Mishchuk, A., Mishkin, D., Radenovic, F., Matas, J.: Working hard to know your neighbor's margins: local descriptor learning loss. arXiv preprint arXiv:1705.10872 (2017)
17. Olander, T.L., Velden, C.S.: The advanced Dvorak technique (ADT) for estimating tropical cyclone intensity: update and new capabilities. Weather Forecast. **34**(4), 905–922 (2019)
18. Pradhan, R., Aygun, R.S., Maskey, M., Ramachandran, R., Cecil, D.J.: Tropical cyclone intensity estimation using a deep convolutional neural network. IEEE Trans. Image Process. **27**(2), 692–702 (2017)
19. Schroff, F., Kalenichenko, D., Philbin, J.: FaceNet: a unified embedding for face recognition and clustering. In: Proceedings of the IEEE Conference on Computer Vision and Pattern Recognition, pp. 815–823 (2015)
20. Sohn, K.: Improved deep metric learning with multi-class N-pair loss objective. In: Advances in Neural Information Processing Systems, pp. 1857–1865 (2016)
21. Sun, Y., et al.: Circle loss: a unified perspective of pair similarity optimization. In: Proceedings of the IEEE/CVF Conference on Computer Vision and Pattern Recognition, pp. 6398–6407 (2020)
22. Velden, C.S., Herndon, D.: A consensus approach for estimating tropical cyclone intensity from meteorological satellites: Satcon. Weather Forecast. **35**(4), 1645–1662 (2020)
23. Wang, F., Cheng, J., Liu, W., Liu, H.: Additive margin softmax for face verification. IEEE Signal Process. Lett. **25**(7), 926–930 (2018)
24. Wang, F., Xiang, X., Cheng, J., Yuille, A.L.: NormFace: L2 hypersphere embedding for face verification. In: Proceedings of the 25th ACM International Conference on Multimedia, pp. 1041–1049 (2017)
25. Wang, H., et al.: CosFace: large margin cosine loss for deep face recognition. In: Proceedings of the IEEE Conference on Computer Vision and Pattern Recognition, pp. 5265–5274 (2018)
26. Weinberger, K.Q., Saul, L.K.: Distance metric learning for large margin nearest neighbor classification. J. Mach. Learn. Res. **10**(2), 207–244 (2009)
27. Xu, Y., Yang, H., Cheng, M., Li, S.: Cyclone intensity estimate with context-aware cyclegan. In: 2019 IEEE International Conference on Image Processing (ICIP), pp. 3417–3421. IEEE (2019)
28. Zhang, C.J., Wang, X.J., Ma, L.M., Lu, X.Q.: Tropical cyclone intensity classification and estimation using infrared satellite images with deep learning. IEEE J. Sel. Top. Appl. Earth Observations Remote Sens. **14**, 2070–2086 (2021)

Explaining Deep Neural Networks for Point Clouds Using Gradient-Based Visualisations

Jawad Tayyub[1], Muhammad Sarmad[2](✉), and Nicolas Schönborn[1]

[1] Endress+Hauser, Maulburg, Germany
{jawad.tayyub,nicolas.schoenborn}@endress.com
[2] Norwegian University of Science and Technology, Trondheim, Norway
muhammad.sarmad@ntnu.no

Abstract. Explaining decisions made by deep neural networks is a rapidly advancing research topic. In recent years, several approaches have attempted to provide visual explanations of decisions made by neural networks designed for structured 2D image input data. In this paper, we propose a novel approach to generate coarse visual explanations of networks designed to classify unstructured 3D data, namely point clouds. Our method uses gradients flowing back to the final feature map layers and maps these values as contributions of the corresponding points in the input point cloud. Due to dimensionality disagreement and lack of spatial consistency between input points and final feature maps, our approach combines gradients with points dropping to compute explanations of different parts of the point cloud iteratively. The generality of our approach is tested on various point cloud classification networks, including 'single object' networks PointNet, PointNet++, DGCNN, and a 'scene' network VoteNet. Our method generates symmetric explanation maps that highlight important regions and provide insight into the decision-making process of network architectures. We perform an exhaustive evaluation of trust and interpretability of our explanation method against comparative approaches using quantitative, quantitative and human studies. All our code is implemented in PyTorch and will be made publicly available.

Keywords: Point cloud · Explainability · Deep neural networks

1 Introduction

The black-box nature of deep neural networks is a major hindrance in their utilization and wide-acceptance in real-world and safety-critical scenarios. This trust deficit can be mitigated by interpreting the reasoning behind a network's

J. Tayyub, M. Sarmad and N. Schönborn—Contributed equally.

Supplementary Information The online version contains supplementary material available at https://doi.org/10.1007/978-3-031-26284-5_10.

L. Wang et al. (Eds.): ACCV 2022, LNCS 13842, pp. 155–170, 2023.
https://doi.org/10.1007/978-3-031-26284-5_10

behaviour. Researchers have made significant progress in proposing explanation methods [3,20,22,33] for demystifying CNN networks for image processing. However, interpretation of deep networks designed for 3D data, namely point clouds [15–17,26], remain an understudied area. Point clouds are an unstructured representation as opposed to regular grids such as images or voxel grids. Therefore, direct application of existing interpretation methods for image-based deep networks is not suitable for point cloud deep networks.

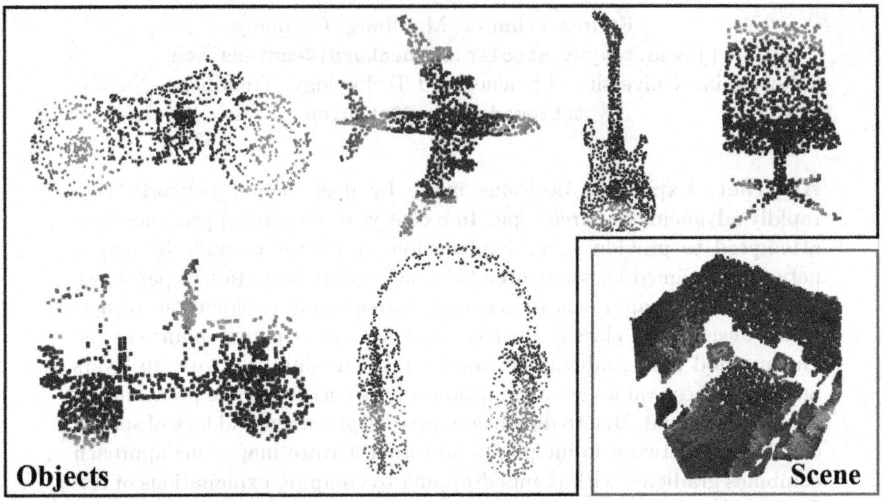

Fig. 1. Point Cloud Heatmaps. Our proposed approaches highlight the salient regions in single object and scene point clouds that are critical for the decision-making of a point cloud processing network.

Firstly, typical CNNs progressively apply convolution and pooling operations to images resulting in low-resolution final feature maps while preserving spatial consistency. Therefore, explanation values generated through logging gradients back to final feature layers of CNN's [20,33] can be mapped to the input image through bi-linear scaling. For unstructured point cloud data, such spatial consistency cannot be guaranteed, and therefore mapping logged gradients back to the input points is non-trivial. Second, the correspondence between each input point and the final feature layer neurons cannot be asserted in point cloud networks. Furthermore, image based explanation techniques such as gradients [22] or deconvolutions [30] operate directly at the input pixel-level resulting in fine and grainy explanations, making them difficult for humans to interpret.

In this work, we address these challenges and propose accumulated piecewise explanations (APE) which is a general explanation approach applicable to a wide variety of deep networks for point clouds. APE computes a point cloud heatmap which highlights each input point's contribution towards the network's decision. The generated heatmaps are interpretable and provide visual insight into the network's behavior. Our method logs gradients (computed for a preset target class) from each neuron in the final feature extraction layer of the network. These are then mapped to the input point cloud to generate the point cloud heatmap as seen in Fig. 1. Gradients are computed w.r.t. the final feature maps of a point cloud processing network architecture. These values can be used to demystify a network's inner workings. Since the feature extraction layers of a network reduce resolution through pooling operations, a direct mapping only reveals an explanation for a small segment of the point cloud. To resolve this, we propose to iteratively explain segments of the point clouds by dropping explained points from the previous iteration, allowing for explanation values to be computed for a different segment. Heatmaps gathered from each iteration are concatenated to generate a complete heatmap. Finally, this heatmap is refined through a second iterative process that recomputes heatmaps while dropping the lowest relevant points from the previous iteration. A weighted maximum over heatmaps from all iterations yields a high-fidelity point cloud heatmap. A key feature of our approach is that the generated heatmaps highlight semantic segments of a 3D shape regardless of the network architecture and therefore exhibit human friendly visual representation. We refer to this property as 'human interpretability'. The heatmaps generated by our method do not just look aesthetically pleasing, but also correctly highlight critical points. This is verified quantitatively in experiments. Our contributions are summarised below:

- We propose a general algorithm to explain 3D point cloud deep networks by generating human interpretable heatmaps.
- We extensively evaluate our explanation heatmaps on various point cloud classification architectures, namely PointNet, PointNet++, DGCNN, and VoteNet. Deteriorating performance of networks is shown by dropping high relevance points identified by our approach. We also evaluate our method against a existing work and demonstrate SOTA performance.
- We demonstrate that our approach outputs higher fidelity heatmaps than existing explanation methods for point clouds and images. Moreover, our strategy highlights biases and failure modes of point cloud networks and provides in-depth insights.

2 Related Work

Deep Learning on Point Cloud. Processing a 3D point cloud directly via deep networks has recently gained attention. Much work in this area has surfaced attempting to solve a wide array of vision problems for point clouds [1,16,17,19, 21,24,26]. Our work aims to analyse these opaque models and generate visual explanations for them. Seminal work on point clouds classification includes [15–17,26]. We demonstrate that these approaches benefit from our explainability method, providing clear insight into the network's decision-making process.

Explainability Methods for CNNs. The availability of large datasets [4,10,27] and discovery of CNNs have provided breakthroughs on various challenges in the vision community [6,8,9]. Efforts to visualize CNNs date back to their discovery, and many methods have been proposed in literature [3,11–13,20,22,33,34]. Pope et al. [14] have extended CNN explainability methods to graph CNNs. Selvaraju et al. [20] proposed Grad-CAM which generates gradient-based visual explanations. Grad-CAM proposes to log gradients at intermediate layers rather than input layers to generate a heatmap that reflects each activation's importance. This heatmap is overlaid onto the input image by simple bilinear scaling to represent contributions of pixels in the image for a given task, e.g., classification. Since later feature layers capture higher-level semantic features, observing these gradients allows for computing coarse explanation values that are human-interpretable. However, application of image-based explanation approaches for point clouds is non-trivial due to their unstructured nature.

Explainable Methods for Point Cloud Based Deep Networks. There is little work on visualizing deep networks that process 3D data such as point clouds. Zhang et al. [31] have visualized PointNet; however, they utilize class attentive features and modify PointNet's architecture. Our technique does not require any modification in the architecture. Qiu et al. [18] visualize the layers of their proposed point cloud processing pipeline, but they do not employ any gradient-based visualizations. Zheng et al. [32] propose a differentiable point shifting process that simulates point dropping and computes contribution scores to input points according to the loss value. This approach identifies highly accurate contributions of individual points; however, it compromises global interpretability. Computed heatmaps resemble fine details similar to gradients [22] or LRP [3] methods in images. In contrast, our method offers semantic visualisations which provide a global overview of the network's focal points. We argue that a heatmap that highlights point cloud's segments that correspond to semantic parts of a shape, e.g., the legs of a chair, builds higher trust than fine-grained explanations.

3 Preliminaries

A point cloud is defined as a set $\mathcal{P} = \{P_1, ..., P_n\}$ of n points where each point is denoted by its 3D coordinates (x, y, z). A deep neural network $f : \mathcal{X} \rightarrow C$ is any trained classifier which maps an input point cloud $\mathcal{P} \in \mathcal{X}$ to a class $c \in C$. Then, given a target class c, our goal is to find an explanation heatmap $\mathbb{L} = \{(P_1, \ell_1), ..., (P_n, \ell_n)\}$ where each $\ell_i \in [0, 1]$ represents the *contribution* of corresponding point P_i towards the network's decision. $\mathbb{L}$ denotes the point cloud heatmap of the input point cloud $\mathcal{P}$.

Point cloud classification networks are categorised into fixed and variable networks. As illustrated in Fig. 2, an input point cloud $\mathcal{P} \in \mathbb{R}^{n \times 3}$ of n points transforms into K feature maps $\mathcal{A} \in \mathbb{R}^{n' \times K}$ each of length n'. Fixed networks, such as PointNet [16] or DGCNN [26], preserve the dimensionality of $\mathcal{P}$ and subsequent feature maps. In these networks, convolutions are applied to each point resulting in a feature map per point hence satisfying $n = n'$. In variable networks, such as PointNet++ [17] and VoteNet [15], feature maps in subsequent

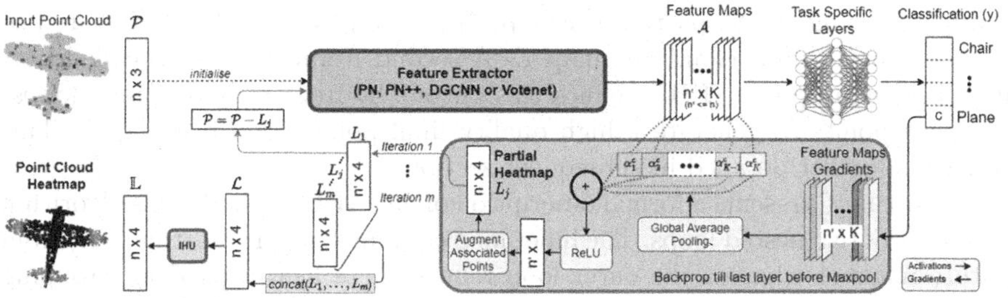

Fig. 2. Approach Overview. Our approach maps gradients logged at the feature maps layer to the point cloud for a target class e.g. 'Airplane'. Gradients are computed w.r.t to the feature maps $\mathcal{A}$. L_j denotes the computed partial heatmap with elements having the form (P_i, l_i) where each point coordinate is augmented with an explanation value. This is a weighted sum of the feature maps followed by ReLU. This partial heatmap only explains n' points of the point cloud since there are n' features per feature map. Explained points in L_j are removed from P iteratively to produce m partial heatmaps. A concatenation of all these partial heatmaps result in a complete *initial* heatmap $\mathcal{L}$. An IHU step, detailed in text, refines the initial heatmap to produce the final point cloud heatmap $\mathbb{L}$.

layers reduce or accumulate input points by various methods e.g. sampling and grouping, clustering, etc. In these networks, n' is often less than n. Most other models either fall into the first or the second category depending on their design.

We also define a *point dropping* operation as the removal of points from a point cloud $\mathcal{P}$. For consistency, our point dropping mechanism is similarly to [32] whereby point coordinates are shifted to the spherical core (centre) of the point cloud $\mathcal{P}$ neutralising their effect to a high degree.

4 Accumulated Piecewise Explanations (APE)

In this section, we present our approach called accumulated peiecewise explanations for generating highly interpretable point cloud heatmaps $\mathbb{L}$, Fig. 2. An input point cloud $\mathcal{P}$ is first classified using a network f, such as PointNet, PointNet++ etc. The final feature maps $\mathcal{A}$, before task-specific fully connected layers, capture high-level semantics which are used for generating coarse conceptual explanations. Therefore, we compute gradients with respect to the final feature maps $\mathcal{A}$. These feature map gradients can be used to gain insight into the network's decision. The gradients are then globally average pooled (GAP) per feature map to create a weighting α which reflects the contribution of neurons in each feature map. A *partial heatmap* L_j is then constructed by taking a weighted sum of feature maps $\mathcal{A}$ representing the aggregated contribution of corresponding input points n'. However, since feature extraction layers are of a lower resolution from the input point cloud, partial heatmap L_j at this stage only reveals contributions of a subset of the point cloud $\mathcal{P}$. To generate contributions of the other segments of $\mathcal{P}$, the subset of points explained by L_j are dropped from $\mathcal{P}$ iteratively where j denotes the iteration. This allows for explanation values to be computed for a different segment. All partial heatmaps $(L_1, ..., L_m)$ are then concatenated to form an initial point cloud heatmap $\mathcal{L}$. A second iterative process called iterative heatmap update (IHU) refines this output by iteratively dropping lowest relevance points to generate a high quality final point cloud heatmap $\mathbb{L}$. This process is further detailed in subsequent section.

Algorithm 1 presents a formal description of our APE method. This algorithm comprises of two nested loops. The inner loop computes partial heatmaps L_j from feature maps A whereas the outer loop refines these heatmaps $\mathcal{L}$ by dropping lowest relevance points n_L iteratively recomputing $\mathcal{L}$ in each iteration. This process iterates λ times and heatmaps gathered from different iterations are combined by a weighted maximum to produce the final point cloud heatmap $\mathbb{L}$. Next, we explain the partial heatmap L_j computation followed by the complete point cloud heatmap $\mathbb{L}$ in higher detail.

Algorithm 1 Accumulated Piecewise Explanations (APE)

Require: λ: number of IHU iterations. c: target class.
Input: Point cloud $\mathcal{P}$, Point Cloud classifier f.
Output: Point Cloud heatmap $\mathbb{L}$.

1: **for** $i = 1....\lambda$ **do**
2: $\quad j = 0$
3: $\quad$ **while** $\mathcal{P}$ is not empty **do**
4: $\qquad y^c = f(\mathcal{P})$
5: $\qquad \alpha_k^c = \frac{1}{n'} \sum_{n'} \frac{\delta y^c}{\delta A_{n'}^k}$
6: $\qquad L_j = ReLU(\sum_k \alpha_k^c A^k)$
7: $\qquad \mathcal{P} = \mathcal{P} - L_j$ (Drop explained points)
8: $\qquad$ Increment j
9: $\quad$ **end while**
10: $\quad \mathcal{L}_i = concatenate(L_1, L_2, ..., L_m)$
11: $\quad$ Drop n_L lowest contribution points from $\mathcal{P}$.
12: **end for**
13: $\mathbb{L} = \max_{i=1,..,\lambda} w_i l_j^i, \forall j = 1, .., m$ where $l^i \in L_i^A$

4.1 Partial Heatmap L_j

To compute the partial heatmap L_j, first gradient values corresponding to each of the neurons in the final feature map $\mathcal{A}$ are computed. Algorithm 1 details this step in lines 4–6. Input point cloud $\mathcal{P}$ is classified to produce y^c which denote the classification score of a given target class c (line 4). Gradients are then computed w.r.t to the final feature maps $\mathcal{A} = \{A^1, ..., A^K\}$. Each feature map $A^k \in \mathbb{R}^{n'}$ is of length n'. The calculated gradients are then globally average pooled (GAP) per A^k to generate a contribution weight α_k^c for the kth feature map (line 5). Point cloud $\mathcal{P}$ has n points whilst $\frac{1}{n'} \sum_{n'}$ is global average pooling and $\frac{\delta y^c}{\delta A_{n'}^k}$

are the gradients. This results in a weight vector where each weight α_k^c gives a relevance weighting of the neurons in the kth feature map. The *partial heatmap* L_j is then computed by taking the positively contributing gradients only through utilising *ReLU* (line 6) normalized in the range $[0, 1]$. L_j computed at this stage reflects the contributions of n' neurons in the final feature maps and not the input point cloud $\mathcal{P}$.

4.2 Point Cloud Heatmap

Depending on the network architecture (fixed or variable), L_j may be of lower dimensionality than $\mathcal{P}$, i.e. $n' \leq n$. We utilise an iterative mechanism whereby explained points are dropped in each iteration, and a new partial heatmap is computed for a different segment of the input point cloud. This is repeated until explanation values are computed for the complete point cloud. This method provides an exact contribution estimate for each point compared to trivial interpolation. Due to tracked associations from L_j to $\mathcal{P}$, the subset of points explained by L_j are identified and dropped from $\mathcal{P}$ (line 7). The resulting point cloud is again passed as input to the network in the next iteration to generate a new partial heatmap L_{j+1} explaining a different subset of points. This process is repeated $(j = 1, ..., m)$ until all points in the input point cloud have been explained. A concatenation of L_j from all iterations produces an initial point cloud heatmap $\mathcal{L}$ (line 10). Note that we concatenate the raw value of the partial heatmap and then normalise instead of normalising and concatenating. This procedure is valid since it preserves the relative importance of a particular point w.r.t others.

To create the final point cloud heatmap $\mathbb{L}$, the IHU step is introduced. This step computes initial point cloud heatmaps $\mathcal{L}_i$ iteratively, whereby in each ith iteration, lowest relevance value points n_L are dropped from $\mathcal{P}$ (line 11) where n_L is a empirically set hyperparameter. The resulting point cloud is reclassified to compute a new initial point cloud heatmap $\mathcal{L}_{i+1}$. After λ iterations, all points have been dropped from $\mathcal{P}$ and λ feature heatmaps $(L_1, ..., L_\lambda)$ have been computed. The final point cloud heatmap $\mathbb{L}$ is then computed by merging the different initial heatmaps (line 13), where l^i are components of $\mathcal{L}_i$ and weights w are hyperparameters which are empirically set. Figure 6 presents a visual illustration of this process. Removing the least significant points in each iteration allows for a better explanation of the remaining points. This enhances explanations by highlighting significant salient areas of the point cloud, which are suppressed by low-contributing points. Explanations over these suppressed areas are revealed by removing low-contributing points iteratively. Also, by taking the max explanation value of points over all iterations, the discovery of most liberal explanation is ensured.

Fixed and Variable Network Architectures. The proposed algorithm is applicable for both fixes and variable network architectures. Fixed networks preserve the dimensionality of input point cloud $\mathcal{P}$ to the feature maps $\mathcal{A}$, a simple bijective mapping of $\mathcal{A}$ to $\mathcal{P}$ is sufficient to create a final point cloud heatmap since $n' = n$. Therefore the inner loop only has one iteration in these cases, namely

PointNet and DGCNN. Note that the partial heatmap module can be applied to any layer before task-specific layers to generate the final point cloud heatmap. Variable networks produce lower-dimensional feature maps. This property is similar to image domain CNNs where, through convolution and pooling operations, computing an explanation heatmap at any intermediate layer results in a lower resolution heatmap than the input image. Similarly, partial heatmaps computation on intermediate feature maps of variable networks for point clouds produce a sparse heatmap since $n' \leq n$. In the case of images, since CNN preserves spatial consistency, a simple bi-linear scaling of partial heatmaps to the input image results in the final heatmap reflecting pixel-wise explanations. However, for point clouds, scaling is infeasible since points in a point cloud are unordered and precise association of all points to feature maps is not known. Our proposed algorithm effectively scales up sparse partial heatmaps to the input point cloud's size through the described iterative process.

Note that the inner loop in the APE algorithm vastly differs from the outer loop denoted as IHU. IHU outer loop requires full heatmaps to be computed at each iteration and is designed to refine the point cloud heatmap to gain additional explanatory power. The inner loop operates on variable networks and handles dimensionality disagreement between input point cloud $\mathcal{P}$ and final feature maps $\mathcal{A}$. Moreover, point dropping in the outer loop is guided by low-contribution values, whilst in the inner loop, points are dropped because they have acquired *some* explanation value. Finally, the generated point cloud heatmaps over all iterations are combined by weighted maximum selection in the outer loop and concatenating feature heatmaps in the inner loop.

5 Experiments

Datasets and Implementation Details. We evaluate our proposed approach APE on four different point cloud processing networks. Two fixed networks, namely PointNet and DGCNN, and two variable networks, name PointNet++ and VoteNet, are used to demonstrate our method's strength. We use three publically available datasets: ShapeNet-Part, Toy Flange, and SUN RGB-D dataset [23]. The first is the ShapeNet-Part dataset [29] used for point cloud object classification with 16 categories of everyday objects. The second is the Toy Flange dataset which comprises of 2 categories of mechanical flanges of either 4 holes (157 files) or 8 holes (280 files). This dataset will be made public. Finally, the SUN RGB-D dataset comprises of complete scene point clouds. The first two datasets are used to train PointNet, PointNet++, and DGCNN for classification. Our approach then generates heatmaps on the test set during inference. The third dataset is used to generate explanations on a pre-trained VoteNet [15]. We build on existing open-source implementations [5,7,25,28].

We compare our approach with existing techniques, namely Gradients [2,22] and Point cloud Saliency Maps (PcSN). Gradients is an early method for visualising networks in the images domain, whereby the gradient values of the loss function w.r.t to image pixels are computed and visualised. We adapt this approach to the point cloud processing networks. Point cloud Saliency Maps (PcSN)

[32] is the state-of-the-art approach for computing contributions of each point towards the network's decision. We demonstrate that our approach APE outperforms both of these by generating heatmaps which not only assign reasonable relevance scores to individual points but are also intuitive for an observer as it highlights semantic parts of shape.

5.1 Qualitative Experiments

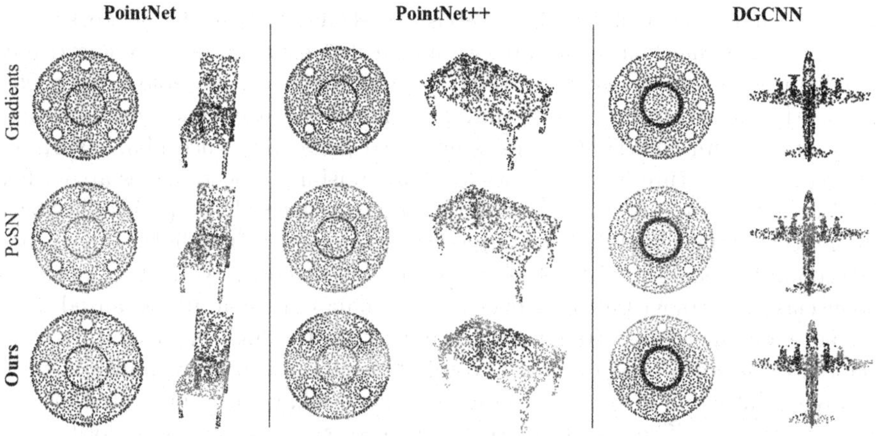

Fig. 3. Qualitative results of APE. The visualization show randomly selected objects which have been correctly classified for PcSN [32], Gradients [22] and **our** propose APE method. For each network, a Toy Flange dataset object is on the left, and a ShapeNet object is shown on the right. (Color figure online)

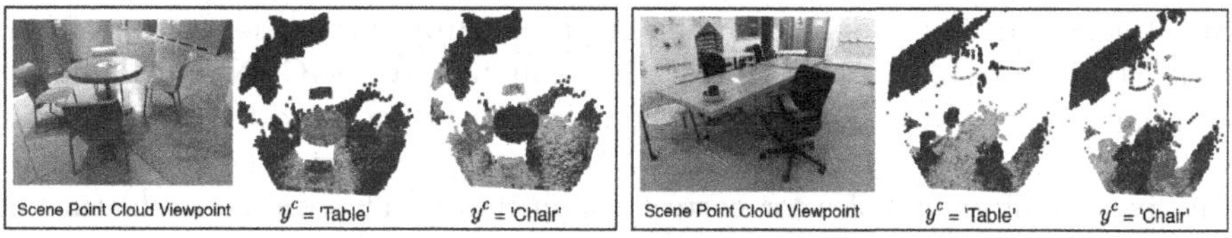

Fig. 4. Final Heatmaps for Scenes. For VoteNet scenes, the APE approach results in point cloud heatmaps highlighting points belonging to the queried target class y^c. Note that highlighted points correspond to semantic categories of the objects of the selected target class.

Point Cloud Heatmaps Evaluation. Our APE approach was applied to different point cloud networks for object classification and detection. A sample of point cloud heatmaps generated from our approach as compared to other methods is presented in Fig. 3. Point cloud heatmaps seen here assign a number to each point in the range [0,1], indicating low relevance (blue) to high relevance (red). Our method identifies significant segments of objects which are critical for decision-making for the network. It is apparent that our method generates highly interpretable heatmaps in comparison to Gradients and PcSN. Gradients approach produces a highly skewed explanation map where only a handful of points are identified as relevant whilst the majority of points remain insignificant. In contrast, PcSN creates an excessively high-resolution heatmap. Even though this method effectively identifies the most significant points, the resulting heatmaps are grainy and incomprehensible. Our method clearly generates intuitive point cloud heatmaps that can visually establish trust and faith in the networks.

From our point cloud heatmaps, we notice that extremities or geometrically varied features, such as wingtips, table corners, table corners, hole's edges, etc. are clearly highlighted as significant, and planar surfaces, such as tabletops, floors, etc. are unremarkable. This is expected behaviour since planar surfaces lack geometric texture, which allows for distinguishing object classes apart. For VoteNet, see Fig. 4, it is shown that our approach is general for applicability over a large scene point cloud. In such large scenes, the method highlights points that correspond to semantic objects as set by the target class. We further note the target class objects within the scenes (y^c ='Chair') are spatially localized. Furthermore, our method's strength is apparent from the high precision of spatial boundaries between objects seen in VoteNet results. For brevity, multiple other example point cloud heatmaps for different objects have been presented in supplementary work. The results presented here show consistent superiority over the basic approach of Gradients [22] and the state-of-the-art method PcSN [32].

Insight into Network's Decisions. Given the high veracity of point cloud heatmaps generated by our approach, it is possible to draw interesting insight into the network's decision-making process. Consider Fig. 3 DGCNN network architecture. Recall that the flange dataset poses a binary classification problem with 4-hole and 8-hole discrimination required. From visual inspection of our point cloud heatmaps in the figure, it is apparent that DGCNN and PointNet both have high focal points around the holes indicating correct network focus. A more interesting insight is that DGCNN heatmaps consistently show focus on only five out of eight holes for correct classification. This geometric feature is sensible for segregating between 4 and 8-hole binary class problem.

We provide further examples of insights in Fig. 5. Most prominently, the PointNet++ architecture classifies the 4-hole flange by focusing on the empty spaces between the holes rather than the holes themselves. In other words, the network has learned to detect the absence of holes rather than their presence. This is a clear contradiction to common human reasoning when identifying a 4-hole flange. We also note that important sections of the chair from PointNet and PointNet++ are the seats, whereas DGCNN (having the most superior

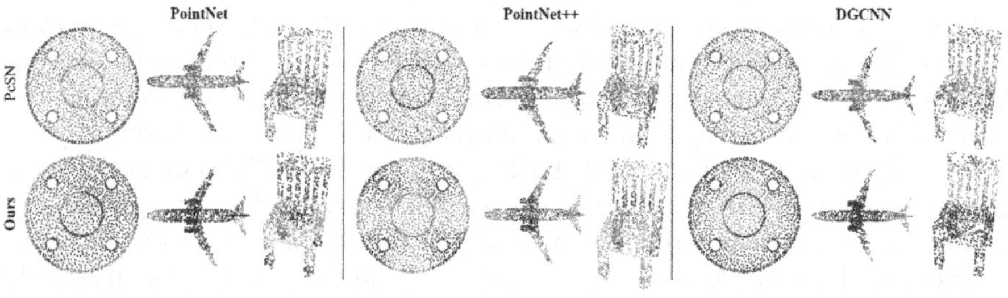

Fig. 5. In-depth insight into various network architectures are revealed using our APE approach for generating heatmaps. For example, a clear incorrect focal point is seen for PointNet++ when classifying the 4-hole flange. Corresponding heatmaps from PcSN are shown to lack human interpretability.

classification accuracy) has identified the unique pattern of the *seat back* as a discriminative feature. This focal point allows DGCNN to discriminate better amongst other similar furniture in the dataset, e.g., sofas, thereby achieving higher accuracy. Finally, the shape airplane shows consistent focal points, e.g., nose, wingtips, and tail, across all network architectures. Such in-depth insights cannot be drawn from the PcSN heatmaps as they lack human interpretability evident from Fig. 3 and 5.

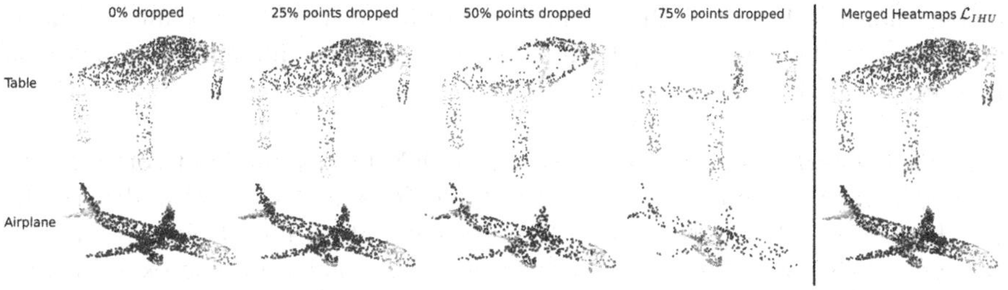

Fig. 6. Iterative Heatmap Updating (IHU). Visual example of the low-relevance dropping approach for PointNet at 0%, 25%, 50%, and 75% of points dropped. Note that the highest relevance points (red) differ as the heatmap is recalculated after each set of point drops. (Color figure online)

IHU Low-Relevance Point Dropping. Figure 6 demonstrates the effect of dropping low-relevance points on heatmaps which are recalculated on the remaining points at every iteration. This is the outer loop of Algorithm 1. Note that the highlighted regions change as points are dropped, indicating that a new set of points gain relevance for classification. It is further observed that more detailed explanations reveal on segments of objects in later iterations; for example, the table shows one leg being of high significance in the first column but all four legs are discovered as highly contributing segments with 75% points dropped. The last column in Fig. 6 shows the merged point cloud, which incorporates all discovered explanations across the different iterations. These results confirm our assertion of dropping low-relevance points to recompute heatmaps iteratively and generate a more representative merged final heatmap.

5.2 Quantitative Experiments

The point cloud heatmap obtained for various networks gives critical indications about the network's learned parameters. This point cloud heatmap can find critical points that can be used to assess our approaches quantitatively. In particular, for the classification task, the point cloud heatmap provides a way to determine the most and least relevant points in a given point cloud. To utilize the heatmap, we use the common evaluation measure point dropping curve (PDC) [12,32]. The PDCs show a drop in classification accuracy of the model as points are removed from the input point cloud. Points are dropped according to their computed relevance values, i.e., most relevant first (high-drop) and least relevant first (low-drop). The slope of these curves provides essential information about the quality of the heatmap generated. In particular, high-drop PDCs should fall steeper than the low-drop PDCs, and the accuracy should drop until its near-random guess. This metric can then be used to compare different approaches on the same network architecture.

Evaluation Results. Figure 7 presents the PDCs for our method compared to the PcSN approach on both fixed-size (PointNet and DGCNN) and variable-size (PointNet++) networks. Note that we compute a final heatmap once for both approaches and use these during the entire point dropping experimentation. Table 1 present the area under the curve (AUC) values of the corresponding PDCs in Fig. 7. It can be observed that when dropping high-contribution points (H.D.), the accuracy of the classification network significantly drops resulting in the AUC of our approach to be consistently lower than all comparative approaches on all networks. Furthermore, Fig. 7 illustrate this result whereby the high-drop curve of our approach sharply deteriorates compared to PcSN. This trend is indicative of the strength and correctness of our generated heatmaps as they have been assigned reasonable explanation values. In the case of PointNet++, our method's high drop line rises slightly above PcSN initially; however, the overall trend remains superior to PcSN.

We also report results obtained from dropping low-contribution points (L.D.). We observe a mixed trend and neither method has superiority over the other using this experimental scheme as seen from Fig. 7 and Table 1. An exception is seen when comparing against the Gradients approach only on PointNet++,

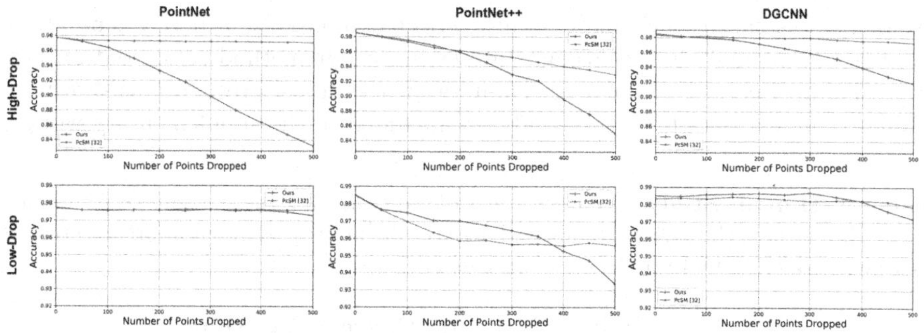

Fig. 7. Quantitative evaluation of APE. The point dropping curve (PDC) has been calculated for various networks with our proposed and existing SOTA methods. The top row presents a high-drop experiment, whereas the lower row presents low-drop experiments. A consistent overall superiority over the existing method is evident across all networks.

however, recall that qualitative results of Gradients (Fig. 3) displays no *meaningful* coloring of the point cloud heatmap. Human studies have been conducted to further validate the qualitative veracity of our approach, these are presented next.

Human Study: We train a multi-label PN++ classifier with 6 object classes namely 'chair', 'bike', 'table', 'plane', 'car' and 'skateboard'. The input data for training and testing this classifier are not single object point clouds but rather two objects concatenated with a certain distance ensuring no overlap. For any given point cloud pair, we generate explanation maps, using different approaches, by setting the target class to one of the two objects in the input. For e.g. consider an input containing an airplane and a car, then given a target class as 'airplane', generated heatmap is expected to highlight this class. We notice that our method consistently highlights the correct class in contrast to baselines. To confirm this we perform an extensive human study where each

Table 1. Area Under the Curve (AUC). AUC has been calculated from the point dropping curves for all point dropping experiments. H.D. stands for high-drop and L.D. for low-drop. A higher value for L.D. and a lower value for H.D. is better.

Method	Drop type	PointNet	PointNet++	DGCNN
Gradients [22]	H.D.↓	0.90	0.70	0.93
	L.D. ↑	0.91	**0.95**	0.96
PcSN [32]	H.D. ↓	0.89	0.80	0.93
	L.D ↑	**0.92**	0.83	0.96
Ours	H.D. ↓	**0.53**	**0.68**	**0.82**
	L.D. ↑	0.91	0.85	0.96

participant is asked to evaluate 'which object demonstrate a better separation of object segments?' in a heatmap. Interestingly, our method comes out on top with the most correct classes corresponding with the user response as shown in Table 2. Further details on this study are included in the supplementary work.

Table 2. Human Study. Accuracies in this table display the percentage of human users who selected the correct response. These results indicate that our approach creates heatmaps which are interpretable by human subjects.

	Saliency Maps [32]	Gradients [22]	**Ours**
Accuracy	0.44	0.51	0.72

6 Conclusions

In this work, we proposed a general approach to visually explain a wide variety of point cloud processing deep networks. We proposed the accumulated piece-wise explanation (APE) algorithm, which tracks gradients to the final feature maps to generate a partial heatmap. This heatmap indicates the contribution of each point towards the network decision. Often, networks reduce and aggregate the features in subsequent layers. Partial heatmaps at later layers are mapped to the input point cloud size by iteratively computing explanations for segments of the input point cloud. These partial heatmaps are then concatenated to create a initial point cloud heatmap. This heatmap is then refined iteratively by dropping low-relevance points at each iteration to discover deeper explanations. We evaluate this approach against existing approaches and demonstrate good performance qualitatively and quantitatively. In the future, we aim to generalise to a broader range of network architectures and tackle networks designed for other types of unstructured data such as meshes or graphs.

References

1. Achlioptas, P., Diamanti, O., Mitliagkas, I., Guibas, L.J.: Representation learning and adversarial generation of 3d point clouds. CoRR abs/1707.02392 (2017). https://arxiv.org/abs/1707.02392
2. Adebayo, J., Gilmer, J., Muelly, M., Goodfellow, I.J., Hardt, M., Kim, B.: Sanity checks for saliency maps. CoRR abs/1810.03292 (2018). https://arxiv.org/abs/1810.03292
3. Binder, A., Montavon, G., Bach, S., Müller, K., Samek, W.: Layer-wise relevance propagation for neural networks with local renormalization layers. CoRR abs/1604.00825 (2016). https://arxiv.org/abs/1604.00825
4. Deng, J., Dong, W., Socher, R., Li, L., Li, K., Fei-Fei, L.: ImageNet: a large-scale hierarchical image database. In: 2009 IEEE Conference on Computer Vision and Pattern Recognition, pp. 248–255 (June 2009). https://doi.org/10.1109/CVPR.2009.5206848

5. Facebook research: Deep Hough voting for 3d object detection in point clouds (2020). https://github.com/facebookresearch/votenet
6. He, K., Zhang, X., Ren, S., Sun, J.: Deep residual learning for image recognition. CoRR abs/1512.03385 (2015). https://arxiv.org/abs/1512.03385
7. Jiaxin, L.: Grad-CAM implementation in PyTorch (2020). https://github.com/jacobgil/pytorch-grad-cam
8. Krizhevsky, A., Sutskever, I., Hinton, G.E.: ImageNet classification with deep convolutional neural networks. In: Pereira, F., Burges, C.J.C., Bottou, L., Weinberger, K.Q. (eds.) Advances in Neural Information Processing Systems, vol. 25, pp. 1097–1105. Curran Associates, Inc. (2012). https://papers.nips.cc/paper/4824-imagenet-classification-with-deep-convolutional-neural-networks.pdf
9. LeCun, Y., et al.: Handwritten digit recognition with a back-propagation network. In: Touretzky, D.S. (ed.) Advances in Neural Information Processing Systems, vol. 2, pp. 396–404. Morgan-Kaufmann (1990). https://papers.nips.cc/paper/293-handwritten-digit-recognition-with-a-back-propagation-network.pdf
10. Lin, T., Maire, M., Belongie, S.J., Bourdev, L.D., Girshick, R.B., Hays, J., Perona, P., Ramanan, D., Dollár, P., Zitnick, C.L.: Microsoft COCO: common objects in context. CoRR abs/1405.0312 (2014), https://arxiv.org/abs/1405.0312
11. Oquab, M., Bottou, L., Laptev, I., Sivic, J.: Is object localization for free? - weakly-supervised learning with convolutional neural networks. In: 2015 IEEE Conference on Computer Vision and Pattern Recognition (CVPR), pp. 685–694 (June 2015). https://doi.org/10.1109/CVPR.2015.7298668
12. Petsiuk, V., Das, A., Saenko, K.: RISE: randomized input sampling for explanation of black-box models. CoRR abs/1806.07421 (2018). https://arxiv.org/abs/1806.07421
13. Pinheiro, P.H.O., Collobert, R.: Weakly supervised semantic segmentation with convolutional networks. CoRR abs/1411.6228 (2014). https://arxiv.org/abs/1411.6228
14. Pope, P.E., Kolouri, S., Rostami, M., Martin, C.E., Hoffmann, H.: Explainability methods for graph convolutional neural networks. In: The IEEE Conference on Computer Vision and Pattern Recognition (CVPR) (June 2019)
15. Qi, C.R., Litany, O., He, K., Guibas, L.J.: Deep Hough voting for 3d object detection in point clouds. CoRR abs/1904.09664 (2019). https://arxiv.org/abs/1904.09664
16. Qi, C.R., Su, H., Mo, K., Guibas, L.J.: PointNet: deep learning on point sets for 3d classification and segmentation. CoRR abs/1612.00593 (2016). https://arxiv.org/abs/1612.00593
17. Qi, C.R., Yi, L., Su, H., Guibas, L.J.: PointNet++: deep hierarchical feature learning on point sets in a metric space. CoRR abs/1706.02413 (2017). https://arxiv.org/abs/1706.02413
18. Qiu, S., Anwar, S., Barnes, N.: Geometric feedback network for point cloud classification (2019)
19. Sarmad, M., Lee, H.J., Kim, Y.M.: RL-GAN-Net: a reinforcement learning agent controlled GAN network for real-time point cloud shape completion. CoRR abs/1904.12304 (2019). https://arxiv.org/abs/1904.12304
20. Selvaraju, R.R., Das, A., Vedantam, R., Cogswell, M., Parikh, D., Batra, D.: Grad-CAM: why did you say that? Visual explanations from deep networks via gradient-based localization. CoRR abs/1610.02391 (2016). https://arxiv.org/abs/1610.02391

21. Simon, M., Milz, S., Amende, K., Gross, H.: Complex-YOLO: real-time 3d object detection on point clouds. CoRR abs/1803.06199 (2018). https://arxiv.org/abs/1803.06199
22. Simonyan, K., Vedaldi, A., Zisserman, A.: Deep inside convolutional networks: visualising image classification models and saliency maps (2013)
23. Song, S., Lichtenberg, S.P., Xiao, J.: SUN RGB-D: a RGB-D scene understanding benchmark suite. In: The IEEE Conference on Computer Vision and Pattern Recognition (CVPR) (June 2015)
24. Su, H., et al.: SPLATNet: sparse lattice networks for point cloud processing. CoRR abs/1802.08275 (2018). https://arxiv.org/abs/1802.08275
25. Wang, Y.: Dynamic graph CNN for learning on point clouds (2019). https://github.com/WangYueFt/dgcnn
26. Wang, Y., Sun, Y., Liu, Z., Sarma, S.E., Bronstein, M.M., Solomon, J.M.: Dynamic graph CNN for learning on point clouds. CoRR abs/1801.07829 (2018). https://arxiv.org/abs/1801.07829
27. Wu, Z., Song, S., Khosla, A., Tang, X., Xiao, J.: 3d ShapeNets for 2.5d object recognition and next-best-view prediction. CoRR abs/1406.5670 (2014). https://arxiv.org/abs/1406.5670
28. Xia, F.: PointNet.pytorch (2019). https://github.com/fxia22/pointnet.pytorch
29. Yi, L., et al.: A scalable active framework for region annotation in 3d shape collections. In: SIGGRAPH Asia (2016)
30. Zeiler, M.D., Fergus, R.: Visualizing and understanding convolutional networks. In: Fleet, D., Pajdla, T., Schiele, B., Tuytelaars, T. (eds.) ECCV 2014. LNCS, vol. 8689, pp. 818–833. Springer, Cham (2014). https://doi.org/10.1007/978-3-319-10590-1_53
31. Zhang, B., Huang, S., Shen, W., Wei, Z.: Explaining the PointNet: what has been learned inside the PointNet? In: The IEEE Conference on Computer Vision and Pattern Recognition (CVPR) Workshops (June 2019)
32. Zheng, T., Chen, C., Yuan, J., Li, B., Ren, K.: PointCloud saliency maps. In: Proceedings of the IEEE/CVF International Conference on Computer Vision, pp. 1598–1606 (2019)
33. Zhou, B., Khosla, A., Lapedriza, A., Oliva, A., Torralba, A.: Learning deep features for discriminative localization. In: 2016 IEEE Conference on Computer Vision and Pattern Recognition (CVPR), pp. 2921–2929 (June 2016). https://doi.org/10.1109/CVPR.2016.319
34. Ziwen, C., Wu, W., Qi, Z., Fuxin, L.: Visualizing point cloud classifiers by curvature smoothing (2019)

Multispectral-Based Imaging and Machine Learning for Noninvasive Blood Loss Estimation

Ara Abigail E. Ambita[1(✉)], Catherine S. Co[2], Laura T. David[3], Charissa M. Ferrera[3], and Prospero C. Naval Jr.[3]

[1] University of the Philippines Visayas, Iloilo, Philippines
aeambita@up.edu.ph
[2] University of the Philippines Manila, Manila, Philippines
csco2@up.edu.ph
[3] University of the Philippines Diliman, Quezon City, Philippines
{ltdavid,cmferrera}@msi.upd.edu.ph, pcnaval@up.edu.ph

Abstract. Blood loss estimation during surgical operations is crucial in determining the appropriate transfusion decisions. More practical emerging solutions, e.g. the Triton System, use image processing and artificial intelligence (AI) in quantifying blood loss from images of blood-soaked sponges. Triton utilizes an infrared or depth camera that's used to identify the region of color (RGB) image corresponding to a surgical textile. However, calculating depth is computationally expensive and can provide only the shape information. In this research, we propose a multispectral-based imaging and machine learning approach to directly quantify blood loss from images of surgical sponges. Near-infrared (NIR) and Visible (Vis) light sources in conjunction with an RGB imaging sensor without a NIR filter are used. With this, in addition to the improved focus and reduced background interference on the gauze image due to blood's IR absorption capacities, the color as well as the shape information may be utilized. Results show that the multispectral-based imaging approach rendered a +28.30%, +48%, +27.97%, and 25.72% improvement on the MAE, MSE, RMSE, and MAPE, compared to using a single Vis wavelength or RGB image.

Keywords: Multispectral imaging · Machine learning · Blood loss

1 Introduction

The estimation of blood loss in a patient during surgery is essential in determining the appropriate transfusion decision or it might lead to costly, invasive, and unnecessary treatments. Currently, there is a lack of a standardized approach to practically and accurately measure intraoperative blood loss [16,19]. The most frequently practiced method used by physicians in determining blood loss is a

Supported by DOST-ERDT.

L. Wang et al. (Eds.): ACCV 2022, LNCS 13842, pp. 171–186, 2023.
https://doi.org/10.1007/978-3-031-26284-5_11

visual estimation. However, aside from demanding expertise, the visual judgment may have been occluded or affected by other fluids, such as urine, amniotic fluid, or sterile water, that combine with blood [13].

Other alternatives to blood loss estimation is the gravimetric and the photometric/colorimetric method. While the gravimetric method is more accurate than visual estimation, it is time-consuming, laborious, and also heavily affected by the combined nonsanguineous fluids that cause overestimation [12]. On the other hand, the photometric/colorimetric analysis is the reference standard but the complexity of the clinical procedure and increased medical costs also limit the use of this method in clinical practice [15].

More recent blood loss estimation methods use computer vision and artificial intelligence (AI). An example is a US FDA-approved mobile application (Triton System, Gauss Surgical Inc, Los Altos, CA) that captures images of surgical sponges and then uses advanced algorithms to distinguish blood and non-sanguineous fluids [1]. Triton System utilizes an infrared or depth camera in conjunction with a color (RGB) image. From the depth image, an image mask may be generated using to identify a region of the color image that is corresponding to a surgical textile. This information from the color image may then be used to estimate a blood component characteristic [3]. However, calculating the depth as well as transforming the infrared image to the geometry perspective of the color image is computationally expensive. Moreover, the low light performance can be poor, and the accuracy of the depth is influenced by the baseline distance. True enough, in Triton, there are plenty of parameters in the depth image that must be passed in order to process the color image, i.e. perimeter classifier, planarity classifier, normality classifier, distance classifier, texture classifier, and color classifier. But, while there are IR cameras that can generate depth without calculation, they can only provide shape information. Moreover, since Triton estimates the hemoglobin (Hgb) loss per sponge using the patient's pre-procedure Hgb value, systemic biases from rinsing and the Hb analyzer could not be eliminated.

A direct measurement of blood loss on sponge using AI was also conducted (Li et al., 2020 [14]). Their research did not consider blood mixed with other non-sanguineous fluids, which is primarily the issue that impedes an accurate estimation. Moreover, the blood-soaked sponge was fully expanded in order to see all the details in the gauze, which generally costs 5x the time of capturing a folded gauze. As this is both time-consuming and inefficient, this may not be sustainable for the personnel in the long run.

Motivated by the discussed risks of inaccurate estimation and the shortcomings of each existing method to estimate blood loss, we aim to develop a direct, cost-effective approach for the accurate and real-time estimation of intraoperative well as the post-operative blood loss volume absorbed in surgical gauze and sponge. Triton uses optical systems that can provide depth information such as range-gated time-of-flight (ToF) cameras, RF-modulated ToF cameras, pulsed light ToF, and projected light stereo cameras [3]. Whereas, we propose that instead of using an infrared or a depth camera, a typical RGB imaging sensor without an infrared filter is used. With this, in addition to the improved focus and reduced background interference on the gauze image due to blood's IR absorption capacities [6], the color as well as the shape information may be utilized. We aim to improve (1) blood detection, (2) direct and non-invasive quantification of the blood volume absorbed on the gauze, and (3) distinguishing blood from non-blood samples or fluids.

More specifically, our proposed method is multispectral-based imaging that utilizes light sources with varying wavelengths. We employ an 850 nm near-infrared (NIR) in addition to the white visible (Vis) light, for which we term the *dual Vis-NIR method*. Additionally, we compare its estimation performance to models using only a Visible (white) light or NIR light. Also, because we aim to deploy the application in small devices, we utilize machine learning which requires less computational resources than traditional deep learning methods.

2 Methodology

The general flow of the method or research is presented in Fig. 1. The dataset is composed of surgical gauze RGB images captured using different lighting sources - Visible (Vis)/white light, 850 nm Near Infrared (NIR), and combined white light and NIR light (dual Vis-NIR). Before extracting features, the images are first converted to HSV. These images are the inputs for the model composed of three general steps: feature extraction, machine learning, and performance evaluation.

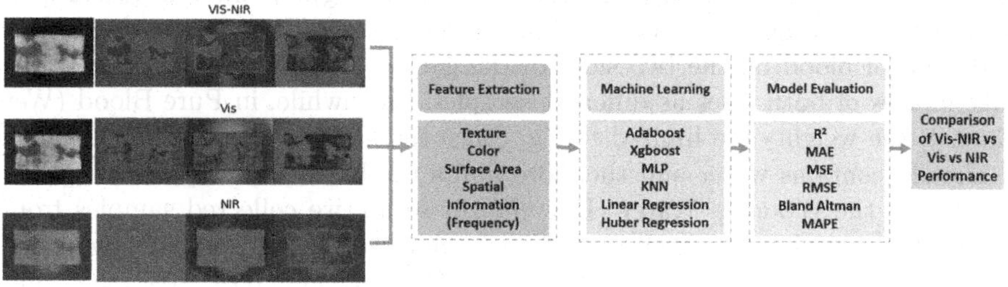

Fig. 1. Proposed Methodology

2.1 Hardware Setup

To lessen the varying lighting conditions, the device is enclosed in a box where no external light can seep in. This creates a controlled environment that has consistent lighting despite the different settings where the device can be used.

Table 1 shows the proposed dual Vis-NIR configuration inside the controlled environment. It contains an Infrared LED (850 nm), White LED (Visible Light), and (4) Imaging Sensor that can switch between an RGB mode and RGB-IR mode.

Table 1. Dual Vis-NIR imaging sensor and LED configurations

Sensor	Conditions		
	Vis	NIR	Vis-NIR
IR (LED 850 nm)	On	–	On
Vis (LED)	–	On	On
Vis (Cam)	–	On	–
IR (Cam)	On	–	On

Table 2. Dataset distribution

Setup	# Images
Pure Blood (Dry)	397
Pure Blood (Wet)	260
Blood + Water	355
Total	1012

2.2 Dataset

During surgical operations, the dry gauze may be soaked in water before use to avoid grazing the internal organs or may be rinsed with water and reused. We prepared pure blood solutions as well as different dilution ratios for the blood+water setup. These setups are summarized in Table 2. For the Pure Blood (Dry Gauze), we varied weights from 0.5–10.5 grams (g), 0.5 intervals, each with 10 repetitions. Since the gauze is folded, there are differences in the spread and color of blood on the two sides of the gauze. To augment our dataset, we take images of both sides as different samples. Meanwhile, in Pure Blood (Wet Gauze), the weights are limited to 7 g. Since the gauze was rinsed with water, it already contains water and therefore limits the blood capacity of the gauze. Lastly, for the Blood + Water (Dry Gauze) setup, we collected samples from 1–10 total g, with a 1 g interval. For each total weight, we took samples with varying blood + water ratio by 10%, 20%, 30%, 40%, 50%, 60%, 70%, 80%, and 90%.

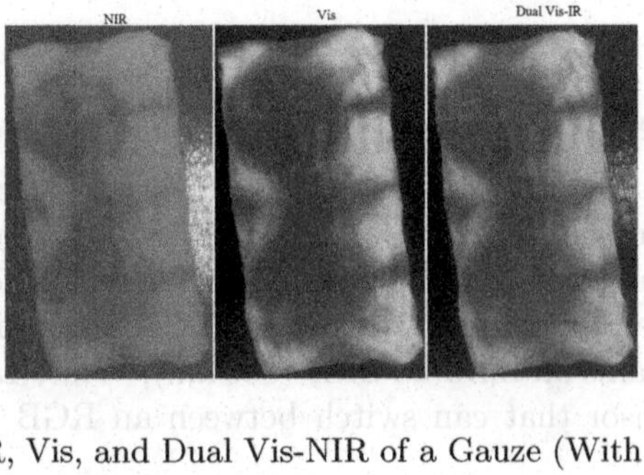

Fig. 2. NIR, Vis, and Dual Vis-NIR of a Gauze (With 2 g Blood)

Each gauze sample will have a set of three images of different configurations captured using: (1) a Vis light, (2) an NIR light, and (3) dual Vis-NIR light. An example of a 2 g blood is shown in Fig. 2. The resulting dataset is composed of gauze images with three channels (H, S, V), each two spatial (x, y) dimensions and one λ wavelength dimension, where the λ = VIS, VIS-NIR, NIR.

2.3 Feature Extraction Techniques (FET)

The color of blood appears desaturated as it gets diluted with more water and the volume is directly proportional to the surface area. In order to differentiate gauze with varying volumes and setups (e.g. dry vs wet gauze, pure blood vs blood + water), we often look at indicators such as color, texture, and as well as the surface area covered by the blood. We exploit color moments to represent colors, local binary patterns represent texture, Fourier transform descriptors for spatial information, and thresholding to compute for the surface area.

Color Moments. The color composition of an image can be viewed as a color distribution where the dominant features from this distribution can be extracted. These dominant features, the color moments, are measures that can be used to differentiate images based on their features of colors. The basis of this idea lies in the assumption that the color distribution in an image can be interpreted as a probability distribution, characterized by its moments, resp. central moments. Thus, it follows that if we interpret the color distribution of an image as a probability distribution, the color distribution can be characterized by its moments as well [21].

The feature descriptor will then be composed of the first four (4) moments of an image's color distribution - the mean, standard, deviation, skewness, and kurtosis. We define the mean as the average color of the image and the standard deviation as the square root of the variance. In the third and fourth moments, the skewness and kurtosis both provide an idea about the shape of the color distribution. The skewness measures how asymmetric the color distribution is while the kurtosis measures how extreme the tails are in comparison to the normal distribution. These moments are calculated independently for every color channel. Thus, if we compute for the moments of an HSV image, then will obtain a 12-dimensional feature vector - 4 features for each channel.

Local Binary Patterns. Local Binary Patterns (LBP) are based on the assumption that texture is based on a pattern and its strength, as proposed by Ojala et al. [17,18]. To compute for the LBP, a neighborhood of size r surrounding a center pixel is defined. Originally, LBP is defined in a 3×3 neighborhood with the gray value of the center pixel set as a threshold. Neighbors with intensities higher or equal to the value of the center pixel are given a value of 1, otherwise, they are set as 0. The thresholded values (0 or 1), are weighted, and by summing up the result, an LBP code that contains information about the local features of the texture of the image is obtained [20].

Surface Area. We define the surface area as the area of the gauze covered by the blood. In order to compute the area, we follow the process in Fig. 3 where we apply thresholding to segment the blood from the unused sections of the gauze and the background. We first compute the "C" (chroma) channel by finding the difference between the largest and smallest of the RGB values (for each pixel independently). A simple threshold of the chroma leads to finding the regions with blood.

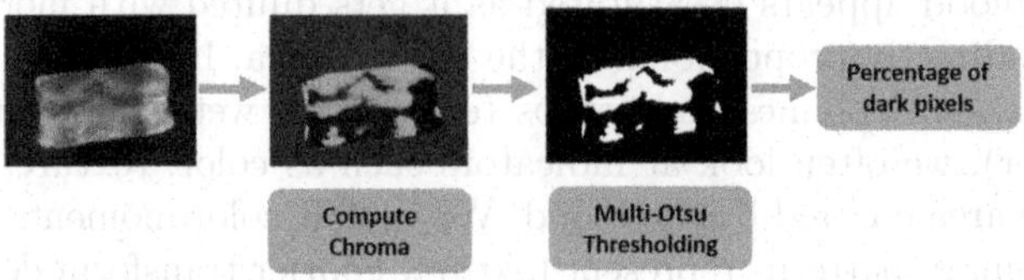

Fig. 3. Surface area

However, the issue with Otsu thresholding is it cannot filter the regions with only a minimal amount of blood and has a lighter shade of red. To overcome this problem, the threshold value returned by the Otsu binarization algorithm is multiplied by a certain factor or bias before using it in regular binary thresholding. After the thresholding, we have an image with black pixels that coincide with the blood and white pixels which we consider as the background. We simply compute for the number of the black pixels divided by the total number of pixels to compute for the surface area.

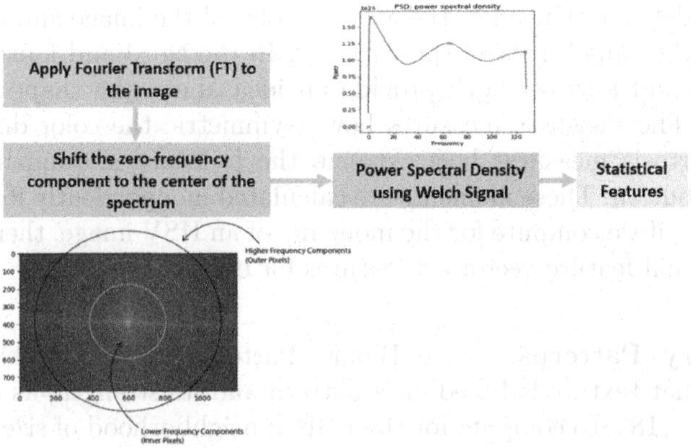

Fig. 4. Fourier transform

Fourier Transform Descriptors. We compute the two-dimensional fast Fourier transform of the input image. Since the result is a complex number array that is difficult to visualize directly, we take the absolute values of the output to be displayed as an image. We, therefore, transform it into a 2-dimensional spectrum. The output frequency domain image tells us how much each frequency component is included in the original image. The 2d may be converted to a summarized 1d by computing the Welch signal (Power Spectral Density). From this simplified representation, we can easily obtain statistical features such as the mean, standard deviation, skewness, kurtosis, minimum, and maximum value. The flowchart for computing these features is displayed in Fig. 4.

FET Keywords. As a standard, the keywords and acronyms used to represent features are listed in Table 3.

Table 3. Feature summary. n refers to the number of components extracted per channel

Feature	Keyword	Components	n
Local binary patterns	LBP	–	40
Color moments	CM	mean, std, skew, kurtosis	4
Power spectral density	PSD	mean, std, skew, kurtosis, min, max	6
Surface area	SA	–	1
LBP of Fast Fourier transform	LBP-FFT	mean, std, skew, kurtosis, min, max	6

2.4 Regression Modelling

Regression predictive modeling involves predicting a numeric variable given some input, often numerical input. Primarily, XGBoost was used as the model to predict the blood absorbed in surgical sponges using features discussed in Sect. 2.3. XGboost is an optimized scalable machine learning algorithm that uses a gradient boosting framework, like Adaboost [4]. Gradient boosting [9] refers to an ensemble of many decision trees, each of which is a weak learner because it only learns from several attributes from the dataset [8]. The boosted regression will obtain a strong predictor from this ensemble of multiple weak learners [23]. Each weak learner is based on random subsamples of the training set through several iterations, created one by one so that each subsequent learner is trained using the residuals of the previous learner. In other words, the new learner corrects the errors made previously by the previous learner and then predicts the outcome. In XGboost, each ensemble uses the sum of K functions to predict an output y_i using Eq. 1.

$$\hat{y}_i = \theta(x_i) = \sum_{k=1}^{K} f_k(x_i) \tag{1}$$

where f_k is the kth independent decision tree in the sample, $f_k(x_i)$ represents the prediction score generated by that tree for the ith sample. To train the weak learner, the set of f_ks is then used to minimize the objective function (loss function and regularization) at iteration t. Loss function, e.g. mean squared error (MSE) for regression measures the difference between the observed response and predicted response. Meanwhile, the tree pruning parameter that regulates the depth of the tree reduces the size and complexity of the decision tree, hence, preventing overfitting.

Its performance is compared to other state-of-the-art machine learning algorithms such as multilayer perceptron (MLP), K nearest neighbors (KNN), support vector machines (SVM), random forest (RF), linear regression, Huber regression, and AdaBoost.

2.5 Model Evaluation

To evaluate the performance of the regression model, the Mean Absolute Error (MAE), Root Mean Square Error (RMSE), R^2, and Mean Absolute Percentage Error (MAPE) are computed. The correlation only measures the linear association between the association of two sets of observations. Since this technique may be inadequate and misleading when assessing an agreement between two methods, the Bland-Altman method [10] is also included as an evaluation metric.

3 Results and Discussion

In this experiment, the performance of every single wavelength (visible (Vis), infrared (IR), and dual Vis-IR) is evaluated in order to determine the most effective wavelength for the detection and estimation of blood volume absorbed in gauze. Figure 1 displays the general overview of the setup involving a single wavelength training. Note that each dataset with a corresponding wavelength is trained independently. For each wavelength, different feature combinations (listed in Table 3) are extracted that serve as input to machine learning regressors. Five-fold cross-validation on the dataset was applied where we computed the R^2, MAE, and RMSE per split. Each train-test subset is comprised of 1012 and 203 images, respectively.

The result of running this experiment is displayed in Table 4 and Fig. 5. Note that we prefer a higher R^2 and a lower result for other metrics (MAE, MSE, RMSE, MAPE). Consistent on all the evaluation metrics and regardless of the feature set used, combining dual Vis-NIR wavelengths has dramatically reduced the errors in the prediction, as seen in the large difference between the dual Vis-NIR and just using the Vis or IR wavelength. For instance, simply using the CM (or the color moments extracted) as a feature, the improvements recorded a +28.30% improvement on the MAE, +48% on the MSE, +27.97% on RMSE, and 25.72% when using the dual-wavelength approach compared to just using visible (white) lighting.

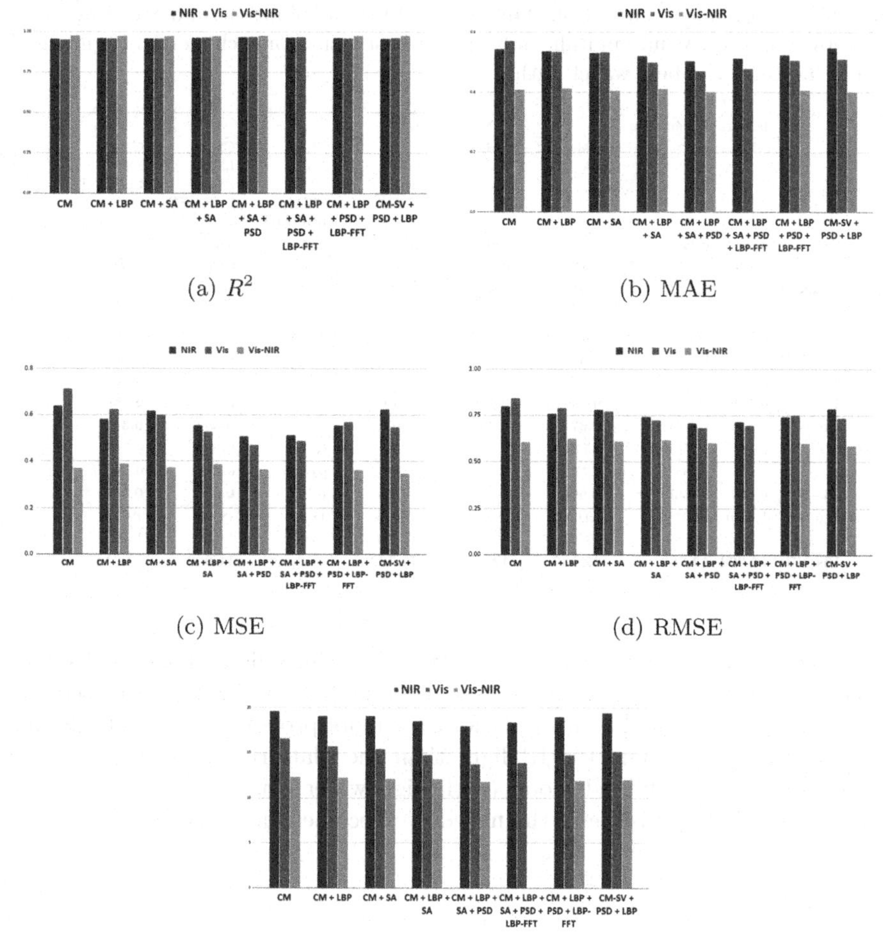

(a) R^2

(b) MAE

(c) MSE

(d) RMSE

(e) MAPE

Fig. 5. Performance comparison of light sources with varying wavelengths

From a spectroscopy perspective, specific molecular absorptions in a certain chemical or fluid cause NIR absorptions and therefore provide more information about a sample's chemical structure [7]. Since NIR wavelength then has been widely explored in blood-related medical or forensic applications, the result of this experiment further confirms the effectiveness of infrared lighting in medical imaging. Specifically, the proposed dual-wavelength technique that takes advantage of infrared lighting was proven to improve blood detection and estimation.

In addition, lower volumes and overly diluted blood are a source of spectral variations, where the absorbed light may be dominated by the background, hampering the quality of Vis images. The background interference may be reduced with the addition of infrared lighting [6,7]. Since blood absorbs infrared light, it

Table 4. Performance comparison of different feature combinations and wavelengths on the merged dataset. Values in italic indicate the best result for each metric. Highlighted in bold indicates the best wavelength.

Metric	W	Feature combination							
		CM	CM+ LBP	CM+ SA	CM+ LBP+ SA	CM+ LBP+ SA+ PSD	CM+ LBP+ SA+ PSD+ LBP-FFT	CM+ LBP+ PSD+ LBP-FFT	CM+ PSD+ LBP-FFT
R^2	NIR	0.954	0.959	0.956	0.961	0.964	0.964	0.961	0.956
	Vis	0.948	0.955	0.957	0.962	0.967	0.966	0.959	0.961
	Vis-NIR	**0.974**	**0.972**	**0.973**	**0.972**	**0.974**	*0.976*	**0.974**	**0.975**
MAE	NIR	0.543	0.536	0.530	0.520	0.504	0.512	0.524	0.547
	Vis	0.570	0.534	0.533	0.499	0.470	0.478	0.506	0.510
	Vis-NIR	**0.409**	**0.414**	**0.405**	**0.411**	**0.401**	*0.399*	**0.406**	**0.401**
MSE	NIR	0.638	0.580	0.617	0.552	0.506	0.510	0.554	0.623
	Vis	0.713	0.623	0.598	0.526	0.470	0.487	0.569	0.545
	Vis-NIR	**0.371**	**0.390**	**0.372**	**0.388**	**0.366**	*0.351*	**0.363**	**0.347**
RMSE	NIR	0.796	0.758	0.779	0.741	0.707	0.713	0.740	0.786
	Vis	0.842	0.788	0.770	0.723	0.682	0.696	0.752	0.736
	Vis-NIR	**0.607**	**0.623**	**0.607**	**0.619**	**0.601**	*0.590*	**0.600**	**0.586**
MAPE	NIR	19.560	19.044	19.052	18.472	17.935	18.342	18.939	19.379
	Vis	16.562	15.689	15.392	14.723	13.718	13.836	14.713	15.100
	Vis-NIR	**12.302**	**12.250**	**12.136**	**12.102**	**11.799**	*11.846*	**11.902**	**12.014**

then appears darker when compared to using Vis (white) light alone, as observed in the samples. The contrast between the blood and the background (unused sections of the gauze and the green platform) is more pronounced, essentially due to blood/water absorbing the IR light than the platform. Therefore, a small amount of blood or diluted blood (with high water content) absorbed in the gauze, that is hardly visible to the naked eye becomes more evident.

3.1 FET Results

We also evaluate the performance of local binary pattern (LBP), color moments (CM), statistical features (SF) from the power spectral density (PSD), and statistical features (SF) from the Fourier-transformed LBP (LBP-FFT) image used as the feature input. The features are extracted from at least one channel of an HSV image and concatenated to form one feature vector.

Both the experiments trained independently using the Vis (white) or the infrared wavelengths have achieved the best results with *CM + LBP + SA + PSD*, which are features that represent the color, texture, surface area, and spatial information in the image. The dual Vis-NIR configuration on the other hand has produced the best results with a similar feature set but with the addition of *LBP-FFT*.

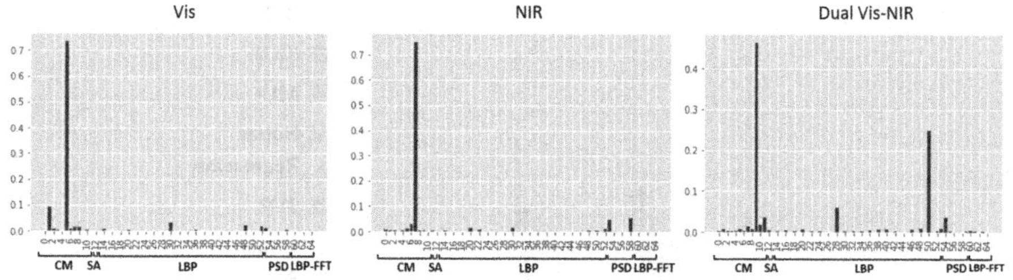

Fig. 6. Feature importance plots

While color representations are the primary indicators of blood detection and estimation, other features have also largely improved the estimations. Volume changes also result to change in the thickness of the blood which consequently influences the spectral variation [24]. However, based on our sample observations, volume changes also results to change in the pattern of blood spread. Spatial information then can also reveal useful information when estimating blood volume. This further highlights another contribution where we combine spectroscopy with imaging, thereby obtaining both spatial and spectral information in our gauze images.

In the Triton System approach, the infrared image is only used to generate depth that will be used as a mask to determine the surgical textile in an RGB image. But in our method, we utilize an RGB-IR image, taken under both white and IR lighting. Essentially, not only do we take advantage of the shape information, but also of the enhanced blood color due to the addition of the IR light. This claim is apparent in Fig. 6 where we visualize the importance scores for each feature. The score provides how useful or valuable each feature was in the construction of the boosted decision trees within the model. The more a feature is used to make key decisions with decision trees, the higher its relative importance. The importance score is calculated explicitly for each feature, allowing the features to be ranked and compared to each other. From Fig. 6, we plot the importance scores of each model independently trained on a specific wavelength. We cite a similar observation on all the wavelengths, i.e. the color moments (CM), which have highly contributed to the prediction. But, in the dual Vis-NIR setup, the model also takes advantage of texture and spatial features such as PSD - in contrast to Vis and NIR which focus on color as their primary indicator.

3.2 Machine Learning Algorithms Performance

We also evaluate the performance of other machine learning algorithms. As seen in Fig. 7, XGboost has outperformed other ML models on all the metrics. While we have shown only performances of the models trained on the dual Vis-NIR, we note that this result is consistent regardless of the wavelength and feature combination used. Similarly, apparent in Fig. 7, there's a substantial difference between XGboost and other models.

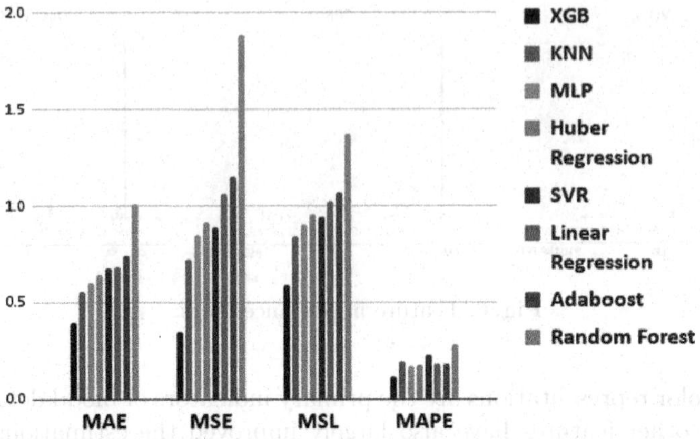

Fig. 7. Performance of machine learning models using LBP + CM + PSD SF + LBP-FFT as Feature and dual Vis-NIR as the Selected Wavelength

Table 5. Performance of machine learning models using LBP+ CM+ PSD + LBP-FFT as Feature and dual Vis-NIR as the Selected Wavelength

Model	R^2	MAE	MSE	RMSE	MAPE
XGB	**0.97526***	**0.39946***	**0.35082***	**0.59018***	**11.84578***
Adaboost	0.947	0.745	1.149	1.069	18.620
Random Forest	0.866	1.008	1.883	1.36	28.339
SVR	0.935	0.674	0.892	0.942	22.979
Linear regression	0.932	0.683	1.061	1.024	18.198
Huber regression	0.939	0.645	0.919	0.954	17.871
KNN	0.947	0.554	0.722	0.846	19.798
MLP	0.943	0.601	0.851	0.905	16.930

3.3 Sample Results

Lastly, we examine sample results using LBP + CM + PSD + LBP-FFT as features and XGBoost as predictor on the five (5) test sets generated using cross-validation. Each test set is comprised of at most 204 instances, with varying weights from 0.5 to 10.5 from varying blood/gauze setups. The average difference between the actual and predicted test sets is −6.6182 g which signifies that the majority of test sets resulted in an underestimation.

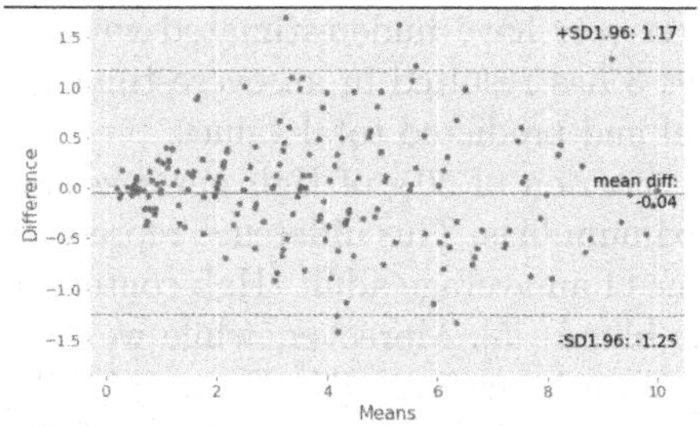

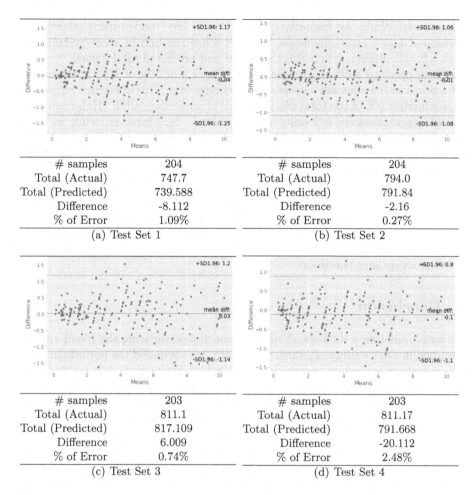

# samples	204
Total (Actual)	747.7
Total (Predicted)	739.588
Difference	-8.112
% of Error	1.09%

(a) Test Set 1

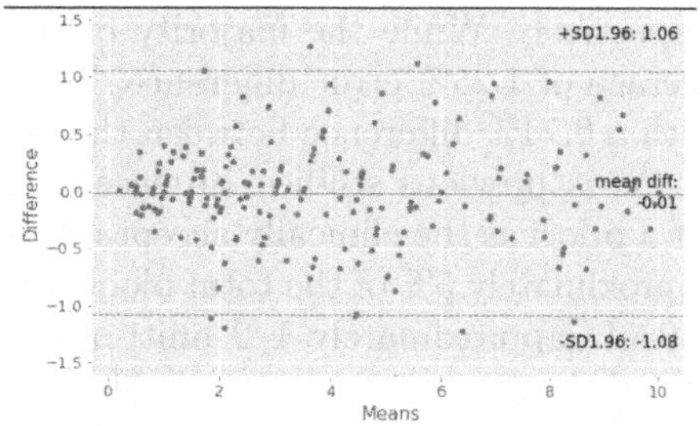

# samples	204
Total (Actual)	794.0
Total (Predicted)	791.84
Difference	-2.16
% of Error	0.27%

(b) Test Set 2

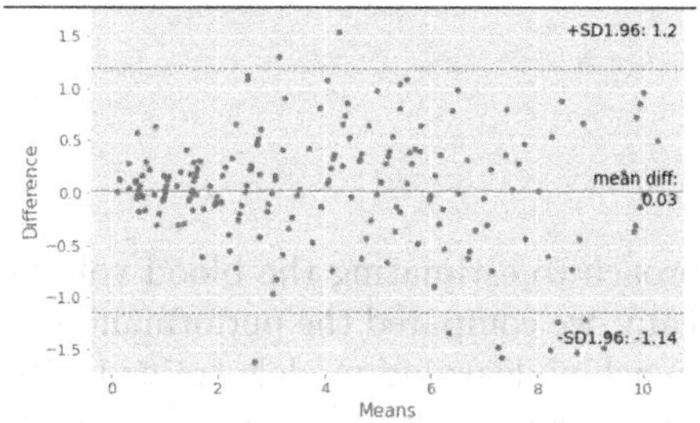

# samples	203
Total (Actual)	811.1
Total (Predicted)	817.109
Difference	6.009
% of Error	0.74%

(c) Test Set 3

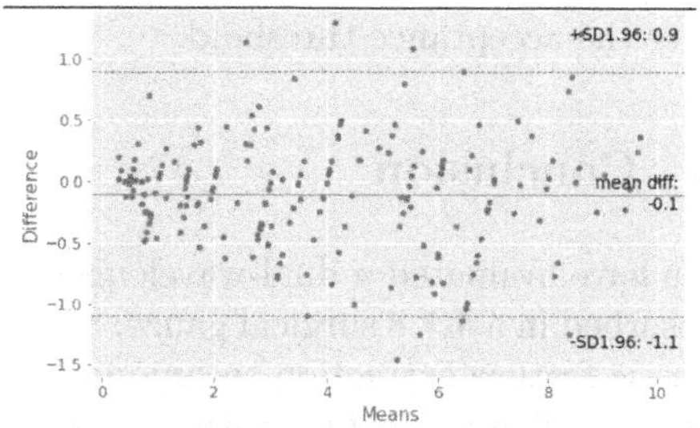

# samples	203
Total (Actual)	811.17
Total (Predicted)	791.668
Difference	-20.112
% of Error	2.48%

(d) Test Set 4

Fig. 8. Cross-validation bland alt-man plots Best Features (CM + LBP + SA + PSD + LBP-FFT) and Best ML Model (XGBoost)

The Bland-Altman plots of actual and the predicted volume (in g) per cross-validation subset are also shown in Fig. 8. This plot is a graphical method to analyze the quality of the predictions based on a bias between the mean differences and an agreement interval, within which 95% of the differences of the predicted, compared to the actual, fall [10]. In Fig. 8, *md* is the mean of the differences, and the limits of agreements are expressed in terms of the standard deviation (sd) of those differences, plotted as md ± sd_limit *sd. We would expect most of the differences to lie between md ± 2sd, or more precisely, 95% of differences will be between md ± 1.96sd [2].

As observed, the test sets have a mean difference close to 0 and the limits are very narrow. The highest upper and lower limit of agreement so far is generated by Test Set 5 with md $-0.05 + \text{sd} * 1.12$ and Test Set 1 with md $-0.05 - \text{sd} * 1.22$, respectively. While the majority of the test sets have underestimated with an average of 1.25% error difference, Test Set 3 has resulted in an overestimation with a 0.74% difference between the actual and predicted total values.

As in previous studies, an acceptance criterion of 30 g of Hgb per case was set a priori as the clinically acceptable maximum bias. This difference represents approximately 5% of the total blood volume of an average adult (Hgb content of 250 mL [approximately 1/2 unit] of whole blood) [5]. Moreover, while we have recorded an average of 11.48% Mean Absolute Percentage (%) Error (MAPE) for all the test sets, it is still lower than the recorded underestimations of 46–75%, 40–49%, and 32% (using visual estimation) in several literatures [11]. Also, our result may be acceptable given that the maximum allowable error is 20% in some studies [22]. However, as our objective is to have results as close as possible to the standard method, the spectrophotometric analysis having a 10% error, we still have to make re-adjustments with our model as well as data collection. Our results show that the blood loss estimations on the 4×8 gauze do not go over the acceptance threshold.

4 Conclusion

We have evaluated a dual-wavelength approach to estimating the blood volume absorbed in a 4×8 surgical gauze. Specifically, we compared the performances of classic feature extraction techniques and machine learning models trained with different lighting configurations. Results show that combining visible (white) light with infrared (IR) light outperforms the predictive capacities of machine learning models trained on images with just white or IR light.

While we have achieved a considerable performance, the proposed solution must be further validated with the addition of more datasets with varying setups. We want to also perform experiments further, e.g., combining features of images with different lighting configurations, fine-tuning the machine learning models, investigate possible overfitting and how to curb these, and optimizing the feature extraction process.

Acknowledgement. DOST-ERDT Philippines is acknowledged for the funding and support.

References

1. Obstetric Hemorrhage. https://www.gausssurgical.com/obstetric-hemorrhage/
2. statsmodels.graphics.agreement.mean_diff_plot - statsmodels. https://www.statsmodels.org/dev/generated/statsmodels.graphics.agreement.mean_diff_plot.html
3. US Patent for Method for estimating blood component quantities in surgical textiles Patent (Patent # 11,282,194 issued March 22, 2022) - Justia Patents Search. https://patents.justia.com/patent/11282194

4. Chen, T., Guestrin, C.: XGBoost: a scalable tree boosting system. In: Proceedings of the 22nd ACM SIGKDD International Conference on Knowledge Discovery and Data Mining, pp. 785–794. ACM, San Francisco (August 2016). https://doi.org/10.1145/2939672.2939785. https://dl.acm.org/doi/10.1145/2939672.2939785
5. Doctorvaladan, S., Jelks, A., Hsieh, E.W., Thurer, R.L., Zakowski, M.I., Lagrew, D.C.: Accuracy of blood loss measurement during cesarean delivery. AJP Rep. **7**, e93–e100 (2017)
6. Edelman, G.J., van Leeuwen, T.G., Aalders, M.C.G.: Hyperspectral imaging of the crime scene for detection and identification of blood stains, p. 87430A. Baltimore, Maryland, USA (May 2013). https://doi.org/10.1117/12.2021509. https://proceedings.spiedigitallibrary.org/proceeding.aspx?doi=10.1117/12.2021509
7. Edelman, G.J., Roos, M., Bolck, A., Aalders, M.C.: Practical implementation of blood stain age estimation using spectroscopy. IEEE J. Sel. Top. Quantum Electron. **22**(3), 415–421 (2016). https://doi.org/10.1109/JSTQE.2016.2536655, https://ieeexplore.ieee.org/document/7422679/
8. Freund, Y., Schapire, R.E.: A decision-theoretic generalization of on-line learning and an application to boosting. J. Comput. Syst. Sci. **55**(1), 119–139 (1997). https://doi.org/10.1006/jcss.1997.1504. https://linkinghub.elsevier.com/retrieve/pii/S002200009791504X
9. Friedman, J.H.: Stochastic gradient boosting. Comput. Stat. Data Anal. **38**(4), 367–378 (2002). https://doi.org/10.1016/S0167-9473(01)00065-2. https://linkinghub.elsevier.com/retrieve/pii/S0167947301000652
10. Giavarina, D.: Understanding Bland Altman analysis. Biochemia Medica **25**(2), 141–151 (2015). https://doi.org/10.11613/BM.2015.015. https://www.biochemia-medica.com/en/journal/25/2/10.11613/BM.2015.015
11. Hancock, A., Weeks, A.D., Lavender, D.T.: Is accurate and reliable blood loss estimation the 'crucial step' in early detection of postpartum haemorrhage: an integrative review of the literature. BMC Pregnancy Childbirth **15**(1), 230 (2015). https://doi.org/10.1186/s12884-015-0653-6. https://bmcpregnancychildbirth.biomedcentral.com/articles/10.1186/s12884-015-0653-6
12. Johar, R.S., Smith, R.P.: Assessing gravimetric estimation of intraoperative blood loss. J. Gynecol. Surg. **9**(3), 151–154 (1993). https://doi.org/10.1089/gyn.1993.9.151. https://www.liebertpub.com/doi/10.1089/gyn.1993.9.151
13. Kollberg, S.E., Häggström, A.C.E., Lingehall, H.C., Olofsson, B.: Accuracy of visually estimated blood loss in surgical sponges by members of the surgical team. Nurse Anesthesiol. **87**(4), 277–284 (2020)
14. Li, Y.J., et al.: A better method for the dynamic, precise estimating of blood/haemoglobin loss based on deep learning of artificial intelligence. Ann. Transl. Med. **8**(19), 1219 (2020). https://doi.org/10.21037/atm-20-1806. https://atm.amegroups.com/article/view/52195/html
15. Liumbruno, G.M., Bennardello, F., Lattanzio, A., Piccoli, P.L., Rossetti, G.: Recommendations for the transfusion of red blood cells. Blood Transfus. (2008). https://doi.org/10.2450/2008.0020-08. https://doi.org/10.2450/2008.0020-08
16. Nowicki, P.D., et al.: Measurement of intraoperative blood loss in pediatric orthopaedic patients: evaluation of a new method. JAAOS Glob. Res. Rev. **2**(5), e014 (2018). https://doi.org/10.5435/JAAOSGlobal-D-18-00014. https://journals.lww.com/01979360-201805000-00002
17. Ojala, T., Pietikäinen, M., Harwood, D.: A comparative study of texture measures with classification based on featured distributions. Pattern Recogn. **29**(1), 51–59 (1996). https://doi.org/10.1016/0031-3203(95)00067-4. https://linkinghub.elsevier.com/retrieve/pii/0031320395000674

18. Pietikäinen, M., Zhao, G.: Two decades of local binary patterns. In: Advances in Independent Component Analysis and Learning Machines, pp. 175–210. Elsevier (2015). https://doi.org/10.1016/B978-0-12-802806-3.00009-9. https://linkinghub.elsevier.com/retrieve/pii/B9780128028063000099
19. Sharareh, B., Woolwine, S., Satish, S., Abraham, P., Schwarzkopf, R.: Real time intraoperative monitoring of blood loss with a novel tablet application. Open Orthop. J. **9**, 422–6 (2015). https://doi.org/10.2174/1874325001509010422
20. Song, K.C., Yan, Y.H., Chen, W.H., Zhang, X.: Research and perspective on local binary pattern. Acta Automatica Sinica **39**(6), 730–744 (2013). https://doi.org/10.1016/S1874-1029(13)60051-8. https://linkinghub.elsevier.com/retrieve/pii/S1874102913600518
21. Stricker, M.A., Orengo, M.: Similarity of color images, San Jose, CA, p. 381 (March 1995). https://doi.org/10.1117/12.205308. https://proceedings.spiedigitallibrary.org/proceeding.aspx?doi=10.1117/12.205308
22. Sukprasert, M., Choktanasiri, W., Ayudhya, N.I.N., Promsonthi, P., O-Prasertsawat, P.: Increase accuracy of visual estimation of blood loss from education programme. J. Med. Assoc. Thai. (Chotmaihet Thangphaet) **89**(Suppl. 4), S54–S59 (2006)
23. Wu, S., Nagahashi, H.: Analysis of generalization ability for different AdaBoost variants based on classification and regression trees. J. Electr. Comput. Eng. **2015**, 1–17 (2015). https://doi.org/10.1155/2015/835357. https://www.hindawi.com/journals/jece/2015/835357/
24. Yang, J., Mathew, J.J., Dube, R.R., Messinger, D.W.: Spectral feature characterization methods for blood stain detection in crime scene backgrounds, Baltimore, Maryland, United States, p. 98400E (May 2016). https://doi.org/10.1117/12.2224099. https://proceedings.spiedigitallibrary.org/proceeding.aspx?doi=10.1117/12.2224099

Unreliability-Aware Disentangling for Cross-Domain Semi-supervised Pedestrian Detection

Wenhao Wu[1], Si Wu[2(✉)], and Hau-San Wong[1]

[1] Department of Computer Science, City University of Hong Kong, Kowloon, Hong Kong, China
wenhaowu5-c@my.cityu.edu.hk, cshswong@cityu.edu.hk

[2] School of Computer Science and Engineering, South China University of Technology, Guangzhou, China
cswusi@scut.edu.cn

Abstract. The rapid progress of pedestrian detection is supported by the ever-growing labeled training data and elaborate neural-network-based model. However, adequate labeled training data are not always accessible when it comes to a new scene. Semi-supervised learning is promising for the case where a small amount of manually annotated images and a large amount of unannotated images are handy. In the semi-supervised setting, data generation is a powerful technique as a type of data augmentation. Some methods conduct data generation by disentangling pedestrian instances into different codes in latent space and combining codes of different instances to reconstruct new instances. However, these methods either work in a single domain or cannot handle the case where some instances are partially represented in the images. In this work, we propose to solve code-level information transferring from reliable domains to unreliable domains by incorporating a domain classifier that competes with the disentangling module to generate domain-invariant codes. An external classifier is trained on appearance-enhanced instances and sends integrity signals to the generative module, which facilitates the generative module to recognize fully/partially represented pedestrian instances. The resulting classifier ultimately renders high-quality pseudo-annotations for the unannotated data. The pseudo-annotated data, combined with a small amount of manually annotated data, are used to achieve a detector with more generalization and accuracy. We perform extensive experiments on multiple challenging benchmarks to demonstrate the effectiveness of the proposed method.

Keywords: Pedestrian detection · Semi-supervised learning · Domain adaptation

1 Introduction

Pedestrian detection, which is a fundamental task in the computer vision community and well applied to a number of practical applications, like autonomous

L. Wang et al. (Eds.): ACCV 2022, LNCS 13842, pp. 187–203, 2023.
https://doi.org/10.1007/978-3-031-26284-5_12

driving and intelligent surveillance, has experienced significant progress in recent years, especially after the emergence of the deep network. However, pedestrian detection still experiences many challenges, and the existing methods mainly put their focus on learning robust features against occlusion, variant scales and illumination change. The remarkable performances of these works are at the cost of a huge number of labeled data, which is hugely time-consuming and human-resource-consuming. When it comes to limited labeled data, these performances face serious deterioration.

Semi-supervised method, which utilizes limited labeled data and a large number of unlabeled data to achieve an improved pedestrian detector, is a promising method to solve the challenge of annotated data deficiency. The key to semi-supervised pedestrian detection is to produce trustworthy annotations for unannotated images, which is used to re-training the detector combined with annotated images. Some efforts [27–29] have been proposed to improve the performance of pedestrian detectors through applying a pre-trained detector on unannotated images to extract high-confidence bounding boxes and re-training the detector. However, these methods have their limitations on the poor discrimination of the filtering mechanism over the hard positive and negative. Another attempt [3,33] is to synthesize pedestrian instances through generative adversarial networks (GANs) [10] to improve the generalization and differentiation of discriminative modules. However, these GAN-based methods usually need exhaustive fine-tuning and are inclined to produce unreliable pedestrian instances. Therefore, we need to propose a more controllable method to produce diverse and reliable pedestrian instances for improving the recognition ability of the discriminative module.

In this work, we focus on generating pedestrian instances with manageable latent codes. DG-Net [41] is a powerful framework for joint person disentangling and re-identification in single domains. However, the success of DG-Net is at the cost of the availability of identity information and high-quality person instance images. When given low-quality and partially-represented pedestrian instances without any identity knowledge in realistic detection scenes of different domains, DG-Net performs poorly in disentangling and reconstruction. To solve this problem, we propose to disentangle pedestrian instances into id-related appearance codes and id-unrelated structure codes through shared id-related and id-unrelated encoders, respectively. Different types of codes from different instances can be united to generate abundant unseen instances in the target domains. An additional code-level domain classifier is incorporated to compete with the id-related encoder, which accounts mainly for the diversity of generated pedestrian instances, of the generative module. An external classifier is appended to the generative module to absorb the knowledge of diverse pedestrian instances and return integrity signals of given instances to the generative module. The trained classifier is well discriminative on hard positives and negatives and is used to generate highly reliable pseudo annotations for unannotated images. Our goal is to re-train the base detector with a small amount of manually annotated and a large amount of pseudo-annotated data, and encourage the resulting detector to perform as closely as the model trained on fully-annotated data.

The main contributions of this work include: (1) We handle the case of transferring knowledge from calibrated pedestrian instances to uncalibrated pedestrian instances through adversarial training between the id-related encoder and domain classifier, and integrity signals from the external classifier; (2) The classifier trained on appearance-rich data can differentiate high-quality pseudo annotations from a bunch of pedestrian/background instances, which is used to achieve a powerful re-trained detector on several benchmark datasets.

2 Related Work

2.1 Scene-Specific Pedestrian Detection

The prevalence of deep neural networks drives a great advancement in both generic object detection and pedestrian detection tasks. Particularly, Faster R-CNN [25] is a revolutionary work and achieves an incredible performance on object detection. In Faster R-CNN, features generated by Region Proposal Network (RPN) are fed into the fully-connected-based classifier. Based on Faster R-CNN, a series of works have been developed on the pedestrian detection task. Zhang et al. [38] fed features generated through the RPN module into a boosted decision forest for pedestrian/background classification. Cai et al. [2] unified detections from different detection heads implemented on feature maps of different layers to achieve scale invariance. Mao et al. [23] incorporated extra segmentation information into features to enhance the semantic information. Brazil et al. [1] further enhance features with semantic information at the image-level and instance-level to improve detection performance.

Pedestrian detection has many challenges, such as scale variance and occlusion, which are also hot spots to explore. Wang et al. [30] developed a novel repulsion loss, which improves the model's regression ability and prevents predicted boxes from inaccurate shifting. Zhou and Yuan [42] instead explored the contribution of visible parts to occluded pedestrian detection and proposed a bi-box regression model. Liu et al. [20] developed an adaptive-NMS to dynamic suppress non-negative bounding boxes based on density scores generated from the density sub-network. Chi et al. [5,6] utilized the head information, which is hardly occluded, to detect occluded pedestrians. For scale variance, Li et al. [17] concatenated features from multiple parallel sub-networks, each of which accounts for pedestrian detection of different scale ranges. Wu et al. [31] proposed to force the features of small-scale pedestrians to mimic those of large-scale pedestrians, which therefore enhances the features of small-scale pedestrians. Kim et al. [13] proposed to memorize the features of large-scale pedestrians and recall similar features whenever meeting small-scale pedestrians.

Although plenty of works were developed to solve different challenges in the pedestrian detection task, there are limited existing methods that were originally designed to concentrate on the insufficient supervision problem. Rosenberg et al. [26] adopted the self-training method to utilize a pre-trained detector to obtain high-confidence detected boxes from unlabeled images, which are then used to re-train the detector. Wang et al. [28] transferred a detector to a new scene by applying the target samples with generated labels. Zeng et al. [37] simultaneously achieved classification and reconstruction tasks to analyze the data distribution of the target scene. Mhalla et al. [24] adopted the sequential Monte Carlo filter to estimate and approximate the data distribution in the target scene. Wu et al. [35] utilized a FCN-based verification module to improve the quality of pseudo annotations of unlabeled images. Wu et al. [36] proposed to reduce the domain gap between data from source scenes and target scenes by adversarial training in the feature space and developed a collaborative learning mechanism to make full use of the aligned source samples.

2.2 Pedestrian Synthesis

The emergence of generative adversarial networks (GANs) advances the development of image generation including pedestrian instance generation. Wu et al. [32] adopted a cascaded model on the multi-source information, including masked images, instance-level masks and edge maps, to generate scene-specific pedestrian patches. Cheung et al. [4] estimated camera parameters and the Spawn Probability Maps for given unannotated images to determine the location of generated pedestrians in scene images, constructing annotated images with generated annotations. Wu et al. [33] improved the Triple-GAN [16] to generate pedestrian instances from noise, enhancing the diversity of training samples. Lin et al. [18] translated unreliable instances generated from a pre-trained detector to reliable instances through a well-designed generator. Zheng et al. [41] proposed to disentangle person information into id-related appearance information and in-unrelated structure information, followed by combining appearance and structure information from different instances to construct brand new instances with known identity information. Zou et al. [44] further extended this method to combine information from instances of different domains to create id-known instances of a target domain. Our approach is different from these works: We focus on the case where transferring information from id-rich well-represented instances of the source domain to id-deficient partially-represented instances of the target domain.

3 Method

3.1 Overview

Our proposed model is trained on both the source dataset for the person re-identification task and the target dataset for the pedestrian detection task. To facilitate appearance and structure disentangling on target instances without identity information, we firstly apply the generative module to the source instances, followed by aligning the appearance code from different domains generated from the shared appearance encoder. This alignment facilitates to transfer appearance information from source instances to target instances. The pedestrian instances with appearance codes from the source dataset and structure codes from the target dataset can greatly augment the diversity of target instances. The generated pedestrian instances combined with reliable instances from the labeled and unlabeled images of the target domain are fed into the external classifier which is placed at the end of the generative module. The classifier returns integrity signals of given instances to the generative module to promote the discrimination of the generative module on partially-represented instances. The trained classifier is separated to generate pseudo annotations for unannotated images of the target dataset. Both annotated data and unannotated data with pseudo annotations are fed into the base detector to enhance its discrimination and localization. The whole framework of the generative module is shown in Fig. 1.

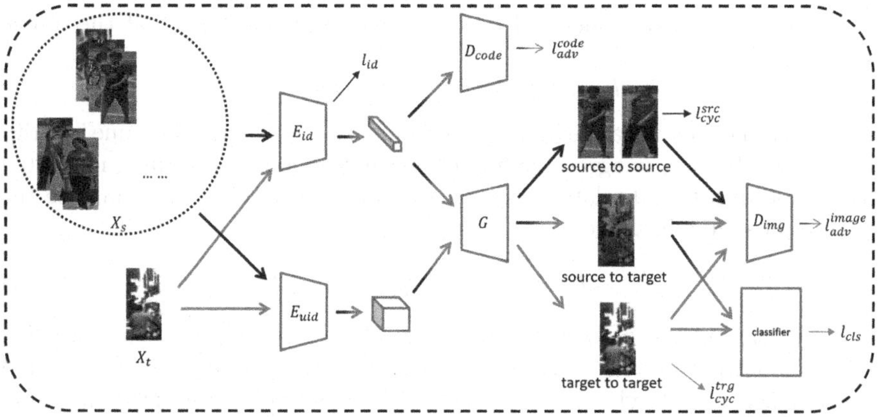

Fig. 1. An overview of the proposed framework. The generative module includes an id-related appearance encoder E_{id}, id-unrelated structure encoder E_{uid}, decoder G and image-level discriminator D_{img} shared across different domains. A code-level domain classifier D_{code} is introduced to compete with the id-related encoder for the alignment of id-related codes from different domains. An external classifier is introduced to train on appearance-rich instances and send integrity signals to the generative module to improve its discrimination on well/poorly represented instances. The different colored lines denote different dataflow to corresponding sub-modules and losses. Mixed colored lines denote the common dataflow of the source data and target data.

3.2 Information Transferring from ID-Rich Instances to ID-Absence Instances

In our setting, the training data includes a small amount of annotated images $\mathbb{I}_a$ and a large amount of unannotated images $\mathbb{I}_u$ from the target scene. In addition, source instances with identities $(x_i^s, y_i^s)_{i=1}^N \in X_s$ are necessary to facilitate information disentangling, where N indicates the number of images, $y_i^s \in [1, K]$ indicates the identities of corresponding instances, and K indicates the total amount of identities in the source dataset. In the beginning, trustworthy target instances $x_i^t \in X_t$ are generated from $\mathbb{I}_a$ and $\mathbb{I}_u$ through an initial detector pre-trained on $\mathbb{I}_a$. Our target is to transfer the appearance information from X_s to X_t while keeping the pedestrian shape in X_t. For this purpose, we should solve the problem of pedestrian disentangling and source-target domain discrepancy under the condition of lacking identity-level supervision and fully-represented instances in the target domain.

A shared id-related encoder $E_{id} : x_i \Rightarrow v_i$ and a shared id-unrelated encoder $E_{uid} : x_i \Rightarrow \omega_i$ are introduced to disentangle id-related appearance code and id-unrelated structure code from instances of different domains, and a shared decoder $G : (v_i, \omega_j) \Rightarrow x_{ij}$ is introduced to combine codes from different instances to generate unseen instances. To facilitate accurate disentangling, the generative module should be able to reconstruct an instance from the same structure code and the appearance codes of the same identities of itself at the source scene:

$$l_{recons} = \mathbb{E}[\|x_i^s - G(v_j^s, \omega_i^s)\|] . \tag{1}$$

where v_j^s is an id-related appearance code from instance x_j^s with the same identity as x_i^s. In Eq. 1, $i = j$ is agreeable for self-reconstruction. For keeping id-related information into the id-related code, E_{id} should also learn to discriminate the identities of source instances:

$$l_{id} = \mathbb{E}[-\log(p(y_i^s|x_i^s))] . \tag{2}$$

where $p(y_i^s|x_i^s)$ is the predicted probability of x_i^s from E_{id}. Encouraging E_{id} to learn about the identity information facilitate the disentangling of id-related code. Further, code-level cycling supervision is essential for preventing information loss from disentangling:

$$l_{cyc}^{src} = \mathbb{E}[\|v_i^s - E_{id}(G(v_i^s, \omega_j^s))\| + \|\omega_j^s - E_{uid}(G(v_i^s, \omega_j^s))\|] . \tag{3}$$

In this case, v_i^s and ω_i^s are well disentangled while together remaining all information of original instances.

Direct introduction of target instances can not help the information transferring from source instances to target instances because of the domain discrepancy. Although source instances and target instances share the id-related and id-unrelated encoder, the id-related encoder mainly accounts for the diversity of generative instances. In the experiments, we show that the domain adversarial training on the id-unrelated structure code overwhelms the information transferring on the id-related appearance code and further destroys the target style, like the background. A domain classifier D_{code} is introduced to match the distribution of source and target data at the id-related code-level based on the adversarial training with the id-related encoder. The domain-adversarial loss function l_{adv}^{code} is as follows:

$$l_{adv}^{code} = \mathbb{E}[\log D_{code}(E_{id}(x_i^s)) + \log(1 - D_{code}(E_{id}(x_j^t)))] . \tag{4}$$

To guarantee the information completion, code-level cycling supervision is also employed in the target instances, which is as follows:

$$l_{cyc}^{trg} = \mathbb{E}[\|v_i^t - E_{id}(G(v_i^t, \omega_j^t))\| + \|\omega_j^t - E_{uid}(G(v_i^t, \omega_j^t))\|] . \quad (5)$$

Further, to encourage the reality of produced images, an adversarial training is introduced to align the distribution of generated instances and real instances at the image-level of different domains, and the corresponding adversarial loss is as follows:

$$l_{adv}^{image} = \mathbb{E}[\log D_{img}(x_i) + \log(1 - D_{img}(G(v_i, \omega_j)))] . \quad (6)$$

where D_{img} denotes the image discriminator, and v_i and ω_j are extract from x_i and x_j, which are drawn from the union of X_s and X_t. The image discriminator shared across different domains focuses on flaws of images generated from the id-related code and the id-unrelated code of different domains at the same time. The generalization of the image discriminator leads to the generalization of the generative module on different domains, which can reduce domain discrepancy implicitly. The adversarial training at the code-level and image-level jointly promotes encoders to learn domain-invariant features.

Finally, we append a classifier C at the end of the generative module to learn about the appearance-rich instances directly. We experimentally explore that the anchor-free CSP-based [22] detector, which makes a prediction based on whether each location is the center of each pedestrian, is inclined to generate a large number of easy positive and negative instances, which are harmful to achieving a robust classifier. We adopt the focal loss [19] on the binary classification to fit this problem, which is shown as follows:

$$l_{cls} = \mathbb{E}[-\alpha(1 - C(x_i))^{\gamma} \log(C(x_i))] . \quad (7)$$

where α and γ are focusing factors, and x_i is drawn from the union of X_t and the set of generated instances with appearance codes from source instances and structure codes from target instances. The introduction of appearance-rich instances from the generative module and high-confidence instances from unannotated images encourages the classifier to become more robust to the appearance variance and specific scene variance in the target domain. In return, the classifier sends integrity signals for given pedestrian instances to the generative module, encouraging the generative module to discriminate the well/poorly represented instances.

We jointly train the id-related appearance encoder, id-unrelated structure encoder, decoder, image-level discriminator, domain classifier and external classifier using a weighted summation of the losses as follows:

$$l_{all} = \lambda_{recons} l_{recons} + \lambda_{id} l_{id} + \lambda_{cyc}(l_{cyc}^{src} + l_{cyc}^{trg}) + l_{adv}^{code} + l_{adv}^{image} + l_{cls} . \quad (8)$$

where λ_{recons}, λ_{id} and λ_{cyc} are the weighting factor to control the contributions of the reconstruction item, identity discrimination item and code-level cycling supervision item, respectively. As the common setting in [12,15,43], we set $\lambda_{recons} = 2$ and $\lambda_{cyc} = 1$ to prevent the information from losing when disentangling, and set $\lambda_{id} = 0.5$ to avoid the negative effect of low-quality generated images at the beginning. For the classifier, we set $\gamma = 2$ and $\alpha = 0.25$ as suggested in [19].

3.3 Re-training Detector

The introduction of appearance-rich information from X_s and diverse scene information from X_t promotes the classifier's ability to identify pedestrians from backgrounds. We adopt the trained classifier to generate pseudo annotations for the unannotated images, which are leveraged by the detector together with the limited annotated images. We adopt different strategies for learning different types of data.

In this work, We adopt a CSP-based detector D_{det} which consists of a ResNet-based [11] backbone and three heads for pedestrians' center recognition, height regression and height offset regression, respectively. Each mini-batch contains randomly sampled images from annotated images $\mathbb{I}_a$ and unannotated images $\mathbb{I}_u$ with pseudo annotations, which are generated from the initial detector followed by filtering through the trained classifier. The focal loss is used for binary center prediction, which is shown as follows:

$$l_{det} = -\sum_{u=1}^{W}\sum_{v=1}^{H} \beta_{uv}(1 - D_{det}(f(x_i)))^{\delta} \log(D_{det}(f(x_i))) \ . \tag{9}$$

where x_i denotes images from either $\mathbb{I}_a$ or $\mathbb{I}_u$, $f(*)$ denotes concatenated features from several layers of the backbone, W and H denote the width and height of the concatenated feature maps. In addition, β_{uv} and δ denote the focusing factors, in which δ is constantly set to 2 and β_{uv} is gained as follows:

$$\beta_{uv} = \begin{cases} 1_{\{y_{uv}=1\}} + 1_{\{y_{uv}=0\}}(1 - M_{uv})^{\tau} & x_i \in \mathbb{I}_a \ , \\ 1_{\{y_{uv}=1\}}\nu & x_i \in \mathbb{I}_u \ . \end{cases} \tag{10}$$

where y_{uv} is the label of each location in the feature map indicating whether the location is the center of each pedestrian, M_{uv} is a Gaussian-based mask centered at each object to weaken the ambiguity of background points around the center, τ is a penalty factor, and ν is a factor to control the contributions of the pseudo-annotated images. We set τ and ν to 4 and 1 in all experiments. For annotated images, we depress the contributions of locations around each object's center but keep other locations' contributions unchanged. For unannotated images, since there exists unavoidable missing on challenging pedestrians, the negatives are rejected while keeping only the contributions of positives for training.

Smoothing L1 loss is implemented for height regression and height offset regression, which is shown as follows:

$$l_{loc} = \sum_{\xi \in \{h,o\}} \phi(p^{\xi}, g^{\xi}) \ . \tag{11}$$

where

$$\phi(p^\xi, g^\xi) = \begin{cases} \frac{1}{2}\left|p^\xi - g^\xi\right|^2 & if \left|p^\xi - g^\xi\right| < 1 \ , \\ \left|p^\xi - g^\xi\right| - \frac{1}{2} & otherwise \ . \end{cases} \tag{12}$$

In Eq. 12, $p = (p^h, p^o)$ represents the predicted height and the offset to the height and $g = (g^h, g^o)$ represents the closest ground-truth's height and offset to the height, where $g^o = (\frac{g^x}{r} - \lfloor\frac{g^x}{r}\rfloor, \frac{g^y}{r} - \lfloor\frac{g^y}{r}\rfloor)$, (g^x, g^y) is the center location of the corresponding ground truth, and r is the downsampling factor for the feature map.

The final optimization is formulated as follows:

$$\min_\theta \eta_c \mathbb{E}[l_{det}] + \eta_r \mathbb{E}[l_{loc}] \ . \tag{13}$$

where θ denotes the parameters of the base detector and η_c and η_r are weighting factors to control the relative contributions of the classification item and regression item.

4 Experiments

In this section, we verify the effectiveness of the proposed method in transferring appearance information from integral instances in a source domain to incomplete instances in a target domain. We adopt Market-1501 [40] as the source dataset and conduct the validation through experiments on three wide-used benchmark datasets: Caltech1X [7], CityPersons [39], and KITTI [9]. The appearance-enhanced pedestrian instances further improve the performance of the classifier, which is used to produce reliable pseudo annotations for unannotated images. The pseudo-annotated images and manually annotated images are dependable for re-training an improved detector. We manage to surpass the previous state-of-the-art methods over all benchmark datasets in the semi-supervised setting.

4.1 Experimental Settings

In our semi-supervised setting, only a small portion of target images are annotated. Unless otherwise stated, 5% of training images are randomly sampled as fully annotated images and the remaining training images keep unannotated in the training stage. As annotations of KITTI's test set are not accessible, we randomly sample 2/3 of annotated images of the public training set as the training data and the remaining data are treated as validation data. For Caltech1X and CityPersons, we adopt annotated images of the public training set as the training data and verify the effectiveness of our proposed method on the public test set and validation set, respectively. Since different datasets adopt different protocols for evaluation, we follow the standard evaluation criterion of average-log Miss Rate (MR) [7], indicating the ratio of missing ground truths over all ground truths, for Caltech1X and CityPersons, and Average Precision (AP) [8], indicating the ratio of detected ground truths over the total detected boxes, for KITTI.

We follow the same network structure of DG-Net [41] with the proposed code-level domain classifier, composed of three fully-connected layers, for adversarial training and the VGGNet-based classifier, trained on the appearance-enhanced pedestrian instances, for delivering the signal of instances' completeness to the generative module. We implement the transferring task on datasets of different domains. The target datasets are composed of high-confidence but partially-represented pedestrian instances extracted from annotated images and unannotated images through the pre-training base detectors. The whole training stop at 100K iterations and each mini-batch includes 4 images from the source dataset and 4 images from the target dataset. Each image is scaled to 256 × 128. We adopt SGD to train E_{id} with a learning rate of 0.002 and momentum of 0.9, and Adam [14] to train E_{uid}, G, D_{code}, D_{img} with a learning rate of 0.0001 and set $\beta_1 = 0$, $\beta_2 = 0.999$. Another SGD optimizer is adopted to train the classifier with an initial learning rate of 0.001 and a momentum of 0.9. All learning rates are decreased by 10 times at 60K, 90K and 95K iterations. For the inference of the generative module, pedestrian instances from the source dataset and the target dataset with a uniform resolution of 256 × 128 are fed into our proposed framework under an inference speed of 87 ms per instance on the platform with one GTX 2080Ti.

We adopt CSP as our base detector with ResNet-50 [11] as the backbone and follow the official code with the original setting when pre-training the base detector on the annotated images. When re-training the base detector with annotated images and unannotated images with pseudo annotations, we set the $\eta_c = 0.01$ and $\eta_r = 0.1$ in Eq. 13 on three benchmark datasets. We use Adam to optimize the detector with initial learning rates of 10^{-4}, 2×10^{-4} and 2×10^{-4} and stop training at 120K, 120K and 160K iterations on Caltech1X, CityPersons and KITTI, respectively. We optimize the network on one GPU (GTX 2080Ti) with mini-batches of 8, 2 and 4 images on Caltech1X, CityPersons and KITTI, respectively. For the inference of the re-trained base detector, images in Caltech1X, CityPersons and KITTI, which are uniformly rescaled to resolutions of 480 × 640, 1024 × 2048, and 384 × 1280, respectively, are fed into the re-trained base detector under inference speeds of 41 ms per frame, 373 ms per frame, and 65 ms per frame on the platform with one GTX 2080Ti.

4.2 Evaluation of the Generative Ability

Our approach is built on the idea of disentangling and reconstructing instances from reliable to unreliable domains through adversarial training at image-level and code-level. To well reconstruct appearance information in the unreliable target domain in which instances are likely to be partial, we incorporate a classifier to backward integrity signals to the generative module, encouraging the generative module to recognize the body part in each target instance. We compare the proposed method with other representative methods, including DG-Net [41], DG-Net++ [44] and CycleGAN [43]. Since DG-Net can only work on single domains,

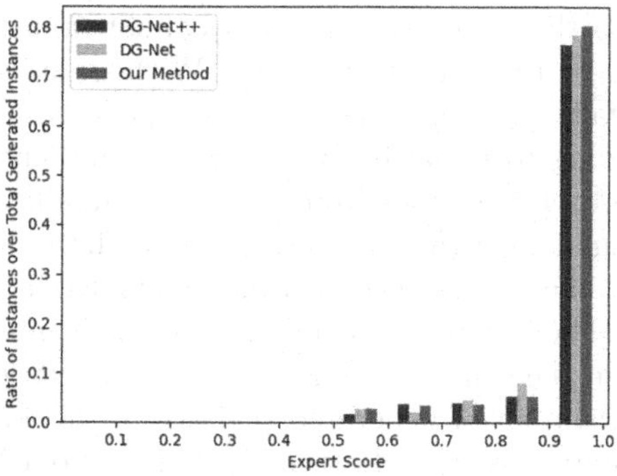

Fig. 2. The score distribution generated from the "expert" classifier over the generated pedestrian instances from DG-Net, DG-Net++, and our proposed method on KITTI.

we adapt DG-Net to be able to incorporate the instances from the target domain without identity information. As DG-Net++ is a multi-stage method for image transferring and person identification, we only adopt the results of the first stage which is the main stage for learning information transferring for comparison.

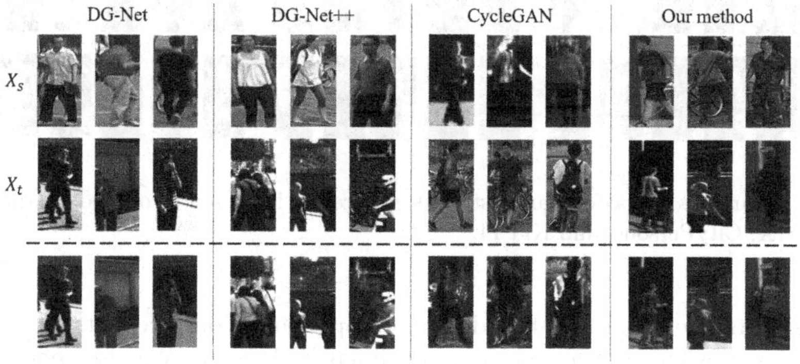

Fig. 3. Comparison of the generated instances from representative methods and our proposed method on KITTI.

We adopt a VGGNet-based expert model, which is trained on instances extracted from fully-annotated images, to evaluate the reality and reliability of the generated instances. Figure 2 shows the confidence score distribution of the generated instances by DG-Net, DG-Net++ and our proposed method on KITTI. Compared to DG-Net and DG-Net++, the ratio of high confidence instances is larger. It means that the expert model highly appraises the generated instances and verifies the reliability of the generated instances.

We also compare the generated instances with the representative methods on KITTI. As shown in Fig. 3, CycleGAN can only transfer the whole style from the source domain to the target domain and cannot transfer the appearance-encoded information to the target instances. Without adversarial training at code-level, DG-Net fails to accurately locate the pedestrian body parts in the target instances, leading to unnaturally appearance inpainting. Although DG-Net++ can encode id-related codes from different domains into the same code space, the model cannot identify the body parts in the target instances and is inclined to overpaint the appearance information to the target instances. Our method can locate the body part in each target instance and inpaint the appearance information from the source instance into the correct location.

Figure 4 shows examples of generated pedestrian instances of three benchmark datasets. Although some pedestrian instances from the target domains have incomplete body structures, the proposed model can still identify the body parts in the instances and inpaint the appearance information from the source instances into the body structures.

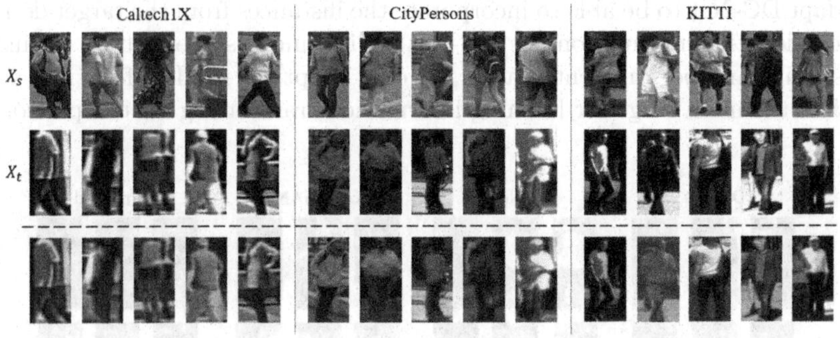

Fig. 4. Examples of pedestrian instances generated from our proposed method on Caltech1X, CityPersons, and KITTI.

Discussion. We claim that the id-related appearance code accounts mainly for the diversity when transferring between different domains. We also conduct an experiment to explore the case of deploying code-level domain classifiers to compete with both the id-related encoder and id-unrelated encoder. As shown in Fig. 5, the pedestrian instances generated from the generative module with separate code-level domain classifiers on both the id-related code and id-unrelated code are far from realistic. The competition between the id-unrelated encoder and the domain classifier is overwhelming, leading to the alignment of appearance codes of different domains and appearance information transferring more difficult. Therefore, an additional domain classifier for the id-unrelated encoder is helpless for generating appearance-enhanced pedestrian instances.

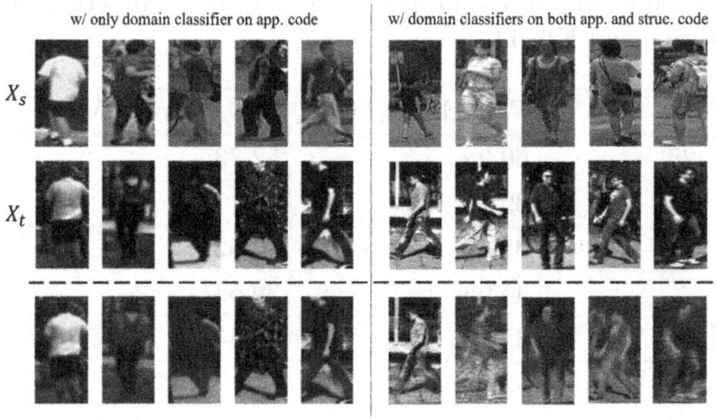

Fig. 5. Examples of pedestrian instances generated from the generative module with different types of domain classifiers on KITTI.

4.3 Evaluation of the Discriminative Ability

We adopt CSP model as the base detector with ResNet-50 as the backbone network, and follow the experimental setting as [22] except for the incorporation of unannotated images with pseudo annotations. We firstly achieve a baseline result, termed as 'Base Detector (Ann-only)', from the base detector trained on annotated images only, which is treated as the lower bound for evaluation. We also build a model, called 'Base Detector (Ful-sup)', which is trained on fully-annotated images as the upper bound for evaluation. The comparison is shown in Table 1. We can observe that 'Base Detector (Ann-only)' is far from satisfactory due to the heavy reduction of annotated images. However, the joining of pseudo-annotated images promotes the performance by a large margin. In particular, the performance is promoted from 29.61 to 18.41, and 32.67 to 18.03 in MR for Caltech1X and CityPersons, respectively. Similarly, the incorporation of pseudo-annotated images advances the performance from 60.76 to 73.57 in AP for KITTI. Finally, we can observe that the performance in the semi-supervised setting is close to the performance of the model under sufficient supervision on KITTI.

We perform a comprehensive comparison with the existing state-of-the-art pedestrian detection methods in Table 1. Most methods are devised to achieve pedestrian detection with full annotations. For a fair comparison, these methods with open-source codes are trained on the annotated images as strong baselines. We can note that performances of methods that are initially designed to train on fully-annotated images, like Faster R-CNN [25], RPN+BF [38], ALFNet [21] and PDOE+RPN [42], deteriorate when given limited annotations and underperform our proposed method. Semi-supervised methods like SPV-RPN [35], PISD-RPN [33] and SaPA [36] outperform the supervised methods when given annotated and unannotated images. Our proposed method still surpasses SaPA, the current state-of-the-art method that worked in the semi-supervised setting, among three datasets. Especially, the proposed method achieves performance improvement by

Table 1. Evaluation results of the proposed method and competing methods on Caltech1X, CityPersons, and KITTI.

Methods	Caltech1X (MR↓)	CityPersons (MR↓)	KITTI (AP↑)
Faster R-CNN [25]	60.98	52.73	50.77
RPN+BF [38]	39.16	–	–
SDS-RPN [1]	35.66	49.52	–
ALFNet [21]	45.95	39.54	–
PDOE+RPN [42]	35.68	42.44	–
Variant SemiBoost [34]	52.53	–	–
SPV-RPN [35]	32.05	–	–
PISD-RPN [33]	23.79	–	–
SaPA [36]	18.55	26.54	67.97
Base Detector (Ann-only)	29.61	32.67	60.76
Base Detector (Re-trained)	18.41	18.03	73.57
Base Detector (Ful-sup)	11.40	12.02	78.78

about 8% points and 6% points on CityPersons and KITTI, respectively, and performs comparably with SaPA on Caltech1X.

5 Conclusion

In this work, we propose a framework that disentangles pedestrian instances into id-related appearance codes and id-unrelated structure codes from different domains which can provide pedestrian instances at different body-integrity levels. To reduce the discrepancy of appearance codes from different domains, a domain classifier is introduced to compete with the id-related appearance encoder. The generative module can capture integrity signals sent from an external classifier trained on the appearance-enhanced pedestrian instances, and inpaint the appearance factor into the remaining body structure of each instance in the target domain. The trained classifier has improved generalization to discriminate hard positive/negative and leads to better pseudo-annotations. Both the annotated images and pseudo-annotated images further promote the discrimination and localization of the re-trained detector.

Acknowledgements. This work was supported in part by the Natural Science Foundation of Guangdong Province (Project No. 2020A1515010484, 2022A1515011160), in part by the National Natural Science Foundation of China (Project No. 62072189), and in part by the Research Grants Council of the Hong Kong Special Administration Region (Project No. CityU 11201220).

References

1. Brazil, G., Yin, X., Liu, X.: Illuminating pedestrians via simultaneous detection & segmentation. In: Proceedings of the IEEE International Conference on Computer Vision, pp. 4950–4959 (2017)
2. Cai, Z., Fan, Q., Feris, R.S., Vasconcelos, N.: A unified multi-scale deep convolutional neural network for fast object detection. In: Leibe, B., Matas, J., Sebe, N., Welling, M. (eds.) ECCV 2016. LNCS, vol. 9908, pp. 354–370. Springer, Cham (2016). https://doi.org/10.1007/978-3-319-46493-0_22
3. Chen, Z., Ouyang, W., Liu, T., Tao, D.: A shape transformation-based dataset augmentation framework for pedestrian detection. Int. J. Comput. Vis. **129**(4), 1121–1138 (2021)
4. Cheung, E., Wong, A., Bera, A., Manocha, D.: Mixedpeds: Pedestrian detection in unannotated videos using synthetically generated human-agents for training. Proc. AAAI Conf. Artif. Intell. **32**(1) (2018)
5. Chi, C., Zhang, S., Xing, J., Lei, Z., Li, S.Z., Zou, X.: PedHunter: occlusion robust pedestrian detector in crowded scenes. Proc. AAAI Conf. Artif. Intell. **34**(07), 10639–10646 (2020)
6. Chi, C., Zhang, S., Xing, J., Lei, Z., Li, S.Z., Zou, X.: Relational learning for joint head and human detection. Proc. AAAI Conf. Artif. Intell. **34**(07), 10647–10654 (2020)
7. Dollár, P., Wojek, C., Schiele, B., Perona, P.: Pedestrian detection: a benchmark. In: 2009 IEEE Conference on Computer Vision and Pattern Recognition, pp. 304–311. IEEE (2009)
8. Everingham, M., Van Gool, L., Williams, C.K., Winn, J., Zisserman, A.: The pascal visual object classes (voc) challenge. Int. J. Comput. Vis. **88**(2), 303–338 (2010)
9. Geiger, A., Lenz, P., Urtasun, R.: Are we ready for autonomous driving? The KITTI vision benchmark suite. In: 2012 IEEE Conference on Computer Vision and Pattern Recognition, pp. 3354–3361. IEEE (2012)
10. Goodfellow, I., et al.: Generative adversarial nets. In: Advances in Neural Information Processing Systems 27 (2014)
11. He, K., Zhang, X., Ren, S., Sun, J.: Deep residual learning for image recognition. In: Proceedings of the IEEE Conference on Computer Vision and Pattern Recognition, pp. 770–778 (2016)
12. Huang, X., Liu, M.-Y., Belongie, S., Kautz, J.: Multimodal unsupervised image-to-image translation. In: Ferrari, V., Hebert, M., Sminchisescu, C., Weiss, Y. (eds.) ECCV 2018. LNCS, vol. 11207, pp. 179–196. Springer, Cham (2018). https://doi.org/10.1007/978-3-030-01219-9_11
13. Kim, J.U., Park, S., Ro, Y.M.: Robust small-scale pedestrian detection with cued recall via memory learning. In: Proceedings of the IEEE/CVF International Conference on Computer Vision, pp. 3050–3059 (2021)
14. Kingma, D.P., Ba, J.: Adam: a method for stochastic optimization. arXiv preprint arXiv:1412.6980 (2014)
15. Lee, H.-Y., Tseng, H.-Y., Huang, J.-B., Singh, M., Yang, M.-H.: Diverse image-to-image translation via disentangled representations. In: Ferrari, V., Hebert, M., Sminchisescu, C., Weiss, Y. (eds.) ECCV 2018. LNCS, vol. 11205, pp. 36–52. Springer, Cham (2018). https://doi.org/10.1007/978-3-030-01246-5_3
16. Li, C., Xu, T., Zhu, J., Zhang, B.: Triple generative adversarial nets. In: Advances in Neural Information Processing Systems 30 (2017)

17. Li, J., Liang, X., Shen, S., Xu, T., Feng, J., Yan, S.: Scale-aware fast R-CNN for pedestrian detection. IEEE Trans. Multimedia **20**(4), 985–996 (2017)
18. Lin, S., Wu, W., Wu, S., Xu, Y., Wong, H.S.: Unreliable-to-reliable instance translation for semi-supervised pedestrian detection. IEEE Trans. Multimedia **24**, 728–739 (2021)
19. Lin, T.Y., Goyal, P., Girshick, R., He, K., Dollár, P.: Focal loss for dense object detection. In: Proceedings of the IEEE International Conference on Computer Vision, pp. 2980–2988 (2017)
20. Liu, S., Huang, D., Wang, Y.: Adaptive NMS: refining pedestrian detection in a crowd. In: Proceedings of the IEEE/CVF Conference on Computer Vision and Pattern Recognition, pp. 6459–6468 (2019)
21. Liu, W., Liao, S., Hu, W., Liang, X., Chen, X.: Learning efficient single-stage pedestrian detectors by asymptotic localization fitting. In: Ferrari, V., Hebert, M., Sminchisescu, C., Weiss, Y. (eds.) Computer Vision – ECCV 2018. LNCS, vol. 11218, pp. 643–659. Springer, Cham (2018). https://doi.org/10.1007/978-3-030-01264-9_38
22. Liu, W., Liao, S., Ren, W., Hu, W., Yu, Y.: High-level semantic feature detection: a new perspective for pedestrian detection. In: Proceedings of the IEEE/CVF Conference on Computer Vision and Pattern Recognition, pp. 5187–5196 (2019)
23. Mao, J., Xiao, T., Jiang, Y., Cao, Z.: What can help pedestrian detection? In: Proceedings of the IEEE Conference on Computer Vision and Pattern Recognition, pp. 3127–3136 (2017)
24. Mhalla, A., Maamatou, H., Chateau, T., Gazzah, S., Amara, N.E.B.: Faster R-CNN scene specialization with a sequential Monte-Carlo framework. In: 2016 International Conference on Digital Image Computing: Techniques and Applications (DICTA), pp. 1–7. IEEE (2016)
25. Ren, S., He, K., Girshick, R., Sun, J.: Faster R-CNN: towards real-time object detection with region proposal networks. In: Advances in Neural Information Processing Systems 28 (2015)
26. Rosenberg, C., Hebert, M., Schneiderman, H.: Semi-supervised self-training of object detection models. In: 2005 7th IEEE Workshops on Applications of Computer Vision (WACV/MOTION'05) - Volume 1, vol. 1, pp. 29–36 (2005)
27. Wang, M., Wang, X.: Automatic adaptation of a generic pedestrian detector to a specific traffic scene. In: CVPR 2011, pp. 3401–3408. IEEE (2011)
28. Wang, X., Wang, M., Li, W.: Scene-specific pedestrian detection for static video surveillance. IEEE Trans. Pattern Anal. Mach. Intell. **36**(2), 361–374 (2013)
29. Wang, X., Hua, G., Han, T.X.: Detection by detections: non-parametric detector adaptation for a video. In: 2012 IEEE Conference on Computer Vision and Pattern Recognition, pp. 350–357. IEEE (2012)
30. Wang, X., Xiao, T., Jiang, Y., Shao, S., Sun, J., Shen, C.: Repulsion loss: detecting pedestrians in a crowd. In: Proceedings of the IEEE Conference on Computer Vision and Pattern Recognition, pp. 7774–7783 (2018)
31. Wu, J., Zhou, C., Zhang, Q., Yang, M., Yuan, J.: Self-mimic learning for small-scale pedestrian detection. In: Proceedings of the 28th ACM International Conference on Multimedia, pp. 2012–2020 (2020)
32. Wu, J., Peng, Y., Zheng, C., Hao, Z., Zhang, J.: PMC-GANs: generating multi-scale high-quality pedestrian with multimodal cascaded GANs. arXiv preprint arXiv:1912.12799 (2019)
33. Wu, S., Lin, S., Wu, W., Azzam, M., Wong, H.S.: Semi-supervised pedestrian instance synthesis and detection with mutual reinforcement. In: Proceedings of the IEEE/CVF International Conference on Computer Vision, pp. 5057–5066 (2019)

34. Wu, S., Wong, H.S., Wang, S.: Variant semiboost for improving human detection in application scenes. IEEE Trans. Circ. Syst. Video Technol. **28**(7), 1595–1608 (2017)
35. Wu, S., Wu, W., Lei, S., Lin, S., Li, R., Yu, Z., Wong, H.S.: Semi-supervised human detection via region proposal networks aided by verification. IEEE Trans. Image Process. **29**, 1562–1574 (2019)
36. Wu, W., Jiao, Q., Wong, H.S., Li, G., Wu, S.: Learning scene-adaptive pseudo annotations for pedestrian detection in semi-supervised scenarios. Knowl. Based Syst. **243**, 108439 (2022)
37. Zeng, X., Ouyang, W., Wang, M., Wang, X.: Deep learning of scene-specific classifier for pedestrian detection. In: Fleet, D., Pajdla, T., Schiele, B., Tuytelaars, T. (eds.) ECCV 2014. LNCS, vol. 8691, pp. 472–487. Springer, Cham (2014). https://doi.org/10.1007/978-3-319-10578-9_31
38. Zhang, L., Lin, L., Liang, X., He, K.: Is faster R-CNN doing well for pedestrian detection? In: Leibe, B., Matas, J., Sebe, N., Welling, M. (eds.) ECCV 2016. LNCS, vol. 9906, pp. 443–457. Springer, Cham (2016). https://doi.org/10.1007/978-3-319-46475-6_28
39. Zhang, S., Benenson, R., Schiele, B.: CityPersons: a diverse dataset for pedestrian detection. In: Proceedings of the IEEE Conference on Computer Vision and Pattern Recognition, pp. 3213–3221 (2017)
40. Zheng, L., Shen, L., Tian, L., Wang, S., Wang, J., Tian, Q.: Scalable person re-identification: a benchmark. In: Proceedings of the IEEE International Conference on Computer Vision, pp. 1116–1124 (2015)
41. Zheng, Z., Yang, X., Yu, Z., Zheng, L., Yang, Y., Kautz, J.: Joint discriminative and generative learning for person re-identification. In: proceedings of the IEEE/CVF Conference on Computer Vision and Pattern Recognition, pp. 2138–2147 (2019)
42. Zhou, C., Yuan, J.: Bi-box regression for pedestrian detection and occlusion estimation. In: Ferrari, V., Hebert, M., Sminchisescu, C., Weiss, Y. (eds.) ECCV 2018. LNCS, vol. 11205, pp. 138–154. Springer, Cham (2018). https://doi.org/10.1007/978-3-030-01246-5_9
43. Zhu, J.Y., Park, T., Isola, P., Efros, A.A.: Unpaired image-to-image translation using cycle-consistent adversarial networks. In: Proceedings of the IEEE International Conference on Computer Vision, pp. 2223–2232 (2017)
44. Zou, Y., Yang, X., Yu, Z., Kumar, B.V.K.V., Kautz, J.: Joint disentangling and adaptation for cross-domain person re-identification. In: Vedaldi, A., Bischof, H., Brox, T., Frahm, J.-M. (eds.) ECCV 2020. LNCS, vol. 12347, pp. 87–104. Springer, Cham (2020). https://doi.org/10.1007/978-3-030-58536-5_6

Patch Embedding as Local Features: Unifying Deep Local and Global Features via Vision Transformer for Image Retrieval

Lam Phan, Hiep Thi Hong Nguyen(✉), Harikrishna Warrier, and Yogesh Gupta

HCL Technologies, Hanoi, Vietnam
{lam.phan,hiep.nguyen,harikrishna.w,yogeshg}@hcl.com

Abstract. Image retrieval is the task of finding all images in the database that are similar to a query image. Two types of image representations have been studied to address this task: global and local image features. Those features can be extracted separately or jointly in a single model. State-of-the-art methods usually learn them with Convolutional Neural Networks (CNNs) and perform retrieval with multi-scale image representation. This paper's main contribution is to unify global and local features with Vision Transformers (ViTs) and multi-atrous convolutions for high-performing retrieval. We refer to the new model as ViTGaL, standing for Vision Transformer based Global and Local features (ViT-GaL). Specifically, we add a multi-atrous convolution to the output of the transformer encoder layer of ViTs to simulate the image pyramid used in standard image retrieval algorithms. We use class attention to aggregate the token embeddings output from the multi-atrous layer to get both global and local features. The entire network can be learned end-to-end, requiring only image-level labels. Extensive experiments show the proposed method outperforms the state-of-the-art methods on the Revisited Oxford and Paris datasets. Our code is available at here

1 Introduction

Image retrieval is an important and long-standing task in computer vision that aims to effectively retrieve all images matching a query image over an (usually very large) image collection. This task is challenging due to various conditions, such as extreme viewpoint/pose, illumination change, occlusion, etc., especially on large-scale datasets. Therefore, image representations that are discriminative enough to deal with these challenges play a central role in this task. There are two types of image representations: global and local features.

Before deep learning revolutionized the field, various handcrafted features [7,24,28,33] have been proposed. With the introduction of deep learning to computer vision, both global feature [2,19,38,41,48] and local features [13,30,31,34,47] are extracted using deep neural network (DNN) in a data-driven paradigm. The global feature summarizes an image, usually as a high-dimensional vector. Due to its compact representation, the global feature can be

L. Wang et al. (Eds.): ACCV 2022, LNCS 13842, pp. 204–221, 2023.
https://doi.org/10.1007/978-3-031-26284-5_13

learned so that it is invariant to viewpoint and illumination with the risk of losing information about the spatial arrangement of visual elements. On the other hand, local features encode detailed spatial features and well preserve geometrical information about specific image regions. They are useful for patch-level matching between images and are shown to be essential for high retrieval precision [9,47]. Therefore, the best retrieval methods [9,44] typically use a global feature to first search for a list of candidate matching images, then re-rank them using local features matching. Recently, [55] proposed to integrate local features and a global feature into a compact descriptor, then perform retrieval in a single stage and showing promising results compared to two stages method.

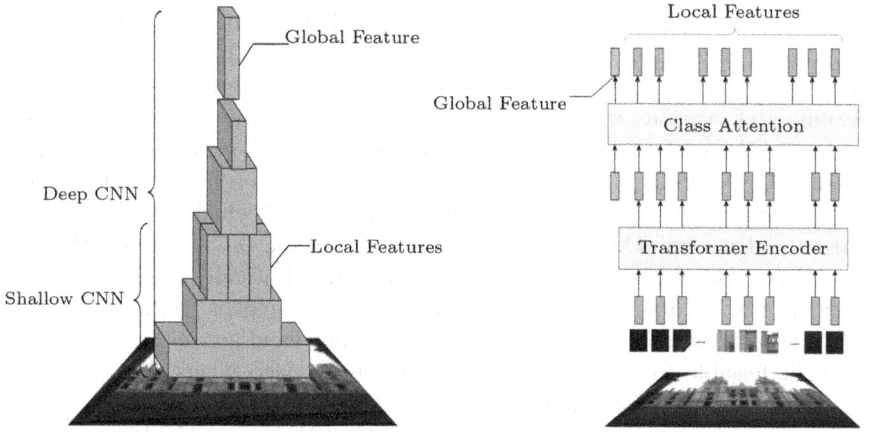

(a) CNN features extraction.   (b) ViT features extraction.

Fig. 1. Global and local features extraction pipeline of CNN and ViT

Today, state-of-the-art methods mainly utilize convolutional neural networks (CNNs) to extract local and global features [9,34,47]. Due to the hierarchical representations of CNNs, global features are associated with deep layers representing high-level cues, while local features are extracted at a shallow layer with high spatial dimensions to preserve spatial information. Recently, [15] demonstrated that Vision Transformers (ViT) are capable of equal or superior performance on image classification tasks at a large scale. Later works also show promising results of these models for other vision tasks, including object detection [26,27,57], semantic segmentation [52,53], video understanding [3,8], monocular depth estimation [25,40], to name a few. Unlike CNN, ViT first divides an image into non-overlapping patches and uses self-attention to aggregate information between these patches, making the spatial dimensions of ViT fixed across layers. Moreover, a recent study [39] found that spatial information from the input is preserved in ViT even as the final layer. In contrast, the representation from deep layers of CNNs is much less spatially discriminative. This property inspires us to utilize the patch embeddings of ViT and treat them as local features to perform geometric verification. Figure 1 shows local and global features extraction pipeline of CNN and ViT. With the ability to preserve spatial information across layers, patch embedding from arbitrary layers of the ViT model can be used as local features. Our experiments show that using patch embeddings from the final layer of ViT yields the best result. Compared to ViT, local features in CNN are extracted at shallow layers, thus may limit its capacity to extract discriminative and robust local features.

One major drawback of the vanilla ViT model is the time and memory complexity of the self-attention operation, which increases quadratically with the number of input patches. For fine-grained retrieval tasks such as landmark retrieval, the best practice is to train retrieval models with large input resolution [9,34,47,51,55], where the largest resolution can be up to 1024×1024 [9,47] during inference. These large resolutions make the direct usage of vanilla ViT models for fine-grained retrieval tasks infeasible. To alleviate the complexity of vanilla ViT, various strategies have been proposed. The dominant approach is reducing the spatial dimension of input resolutions at every block of layers, similar to CNN [26,27,50]. Recently, XCiT [1] replaced a self-attention between tokens with a "transposed" attention between channels which they call "cross-covariance attention" (XCA). Thanks to XCA operation, XCiT reduces the time and memory complexity of the vanilla ViT from quadratically to linear without sacrificing the accuracy of the classification task. Moreover, similar to vanilla ViT, the spatial dimension of input patches remains fixed across layers in XCiT, making it feasible to extract patch embeddings at deep layers. We, therefore, utilize XCiT as the backbone for our retrieval model.

To achieve high retrieval performance, state-of-the-art methods usually utilize an image pyramid at inference time to produce multi-scale representations, thus increasing time and memory complexity. To solve this problem, we proposed to simulate an image pyramid with multi-atrous convolutions [10]. Extensive experiments on Revisited Oxford and Pairs [37] show the effectiveness of the multi-atrous, as we observe significant performance improvements compared to the baseline. Moreover, since XCiT uses XCA operation to approximate the self-attention operation implicitly, local patch interaction implemented by small convolutional layers at each block is added. Therefore, the multi-atrous convolutions could help increase the receptive field of the model, leading to more robust local and global features. To summarize, our main contributions are as follows:

- We propose using the ViT model to perform image retrieval in a two-stage paradigm by using the patch embeddings and treating them as local features to perform geometric verification. Concretely, we use XCiT as our backbone to alleviate the complexity of vanilla ViT.
- We add multi-atrous convolutions to simulate the image pyramid used in standard retrieval algorithms, leading to state-of-the-art results using only single-scale representation.

- Extensive experiments are conducted, and comprehensive analyses are provided to validate the effectiveness of our solution. Our ViT-based model, referred to as ViTGaL (Vision Transformer based Global and Local features), significantly outperforms the previous state-of-the-art CNNs models.

2 Related Work

2.1 Local Features

Hand-crafted local features [7,28] based on low-level visual information were widely used in earlier retrieval works. To compare two images with local features, aggregation methods such as [22,58] are usually used. To improve precision and produce reliable and interpretable scores, a second reranking stage based on geometric verification via matching local features with RANSAC [18] is also widely adopted [4,35]. More recently, many methods have been proposed to learn local features [5,6,14,16,29,34,42,47,56] with deep neural networks. Those methods rely on CNN to perform local features extraction, where shallow layers from a CNN backbone are utilized. We go beyond the normal approach of using CNN and propose to use ViT patch embeddings as local features in traditional CNN-based methods. To the best of our knowledge, no previous works have studied whether patch embeddings of ViT can be utilized to perform geometric verification and how they perform compared to CNN.

2.2 Global Feature

Before deep learning revolutionalized the fields, conventional approaches to obtaining global features were via aggregating local features [22,24,46]. In deep learning era, most high-performing global features are extracted using neural networks, where the differentiable version of the traditional aggregating method [2,38,54] is used to enable end-to-end training which either ranking-based loss [11,20,41] or classification loss [12,49]. Unlike those widely used for the CNN-based model, our work uses multi-head self-attention operation at classification token to obtain global descriptor. For a fair comparison to previous state-of-the-art CNNs [9,55], we used ArcFace loss [12] to train our retrieval model.

2.3 Joint Local and Global CNN Features

Using both global and local features for retrieval is shown to be more efficient than using either global or local features alone [9,55]. Therefore, it is natural to consider learning both features jointly since using separate models may lead to high memory usage and increased latency. [43] distills pre-trained local and global features into a single model. DELG [9] takes a step further and proposes to train local and global features in an end-to-end manner jointly. We follow the work of DELG and also present a unifying model. The difference is our model is ViT-based, while DELG and other conventional models are CNN-based.

2.4 Transformers for High-Resolution Images

Since fine-grained retrieval tasks such as landmark retrieval require high-resolution input to achieve high retrieval performance, [9,55], using vanilla ViT for those tasks is infeasible. Recently, several works have adopted visual transformers for high-resolution images and proposed multiple strategies to alleviate the complexity of vanilla ViT. [50] designed a pyramidal architecture and addresses complexity by gradually reducing the spatial resolution of keys and values. Since then, several works have adopted the idea of lowering spatial resolution at each layer for efficient computations. [17] utilized pooling to reduce the resolution across the spatial and temporal dimensions, while [27] used local attention with shifted windows and patch merging. Recently, XCiT [1] proposed to replace the quadratic self-attention operation with a "transposed" attention operation between channels which they call "cross-covariance attention" (XCA). The advantage of XCiT is that it preserves the spatial dimension across layers, making it feasible to extract patch embeddings at deep layers. We, therefore, utilize XCiT as the backbone for our retrieval model.

3 Methodology

3.1 Model

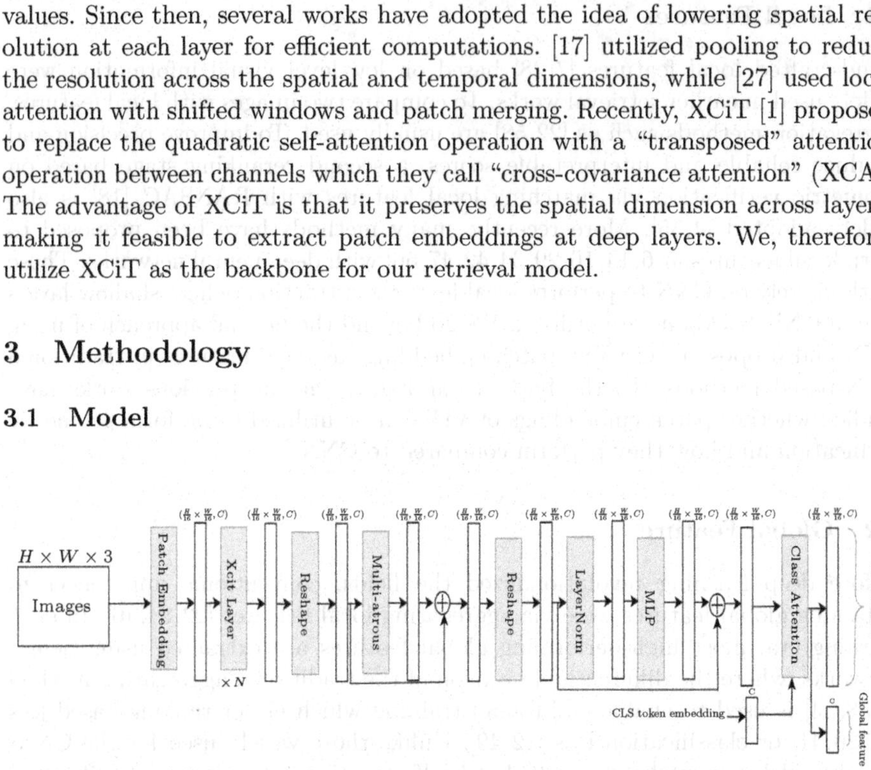

Fig. 2. The architecture of our ViTGal model. It consists of a ViT-based model XCiT [1] as a backbone; a multi-atrous convolution followed by a LayerNorm and an MLP layer to simulate an image pyramid, and a class attention layer to aggregate the token embeddings into a compact representation for global retrieval as well as new tokens embeddings for geometric verification.

Our VitGaL model is depicted in Fig. 2. We propose utilizing patch embeddings at the final layer of the ViTGaL model for geometric verification in the reranking stage. We also merge all the attention scores in different attention heads in the class attention layer and extract associated patch embeddings with the top scores.

Given an image $I \in R^{H \times W \times 3}$, we follow [1] and reshape the image into a sequence of flattened 2D patches $x_p \in R^{N \times (P^2.3)}$, where (H, W) is the resolution of the original image, (P, P) is the resolution of each image patch, and $N = H \times W/P^2$ is the resulting number of patches. These patches first undergo the patch embedding layer to transform into tokens of dimension D, which we follow [1] and implement by a small CNN. Those tokens are then added with the positional embeddings and fed into N XCiT layer, which we also implement following [1].

To simulate an image pyramid used in multiple state-of-the-art methods, we propose to use multi-atrous convolutions, as depicted in Fig. 2. We first reshape the 1d tokens sequence into a tensor of shape $H/P \times W/P \times C$. This tensor then goes to the multi-atrous layer and a skip connection to propagate information from the XCiT layer. The multi-atrous module contains five dilated convolution layers to obtain feature maps with different spatial receptive fields and a global average pooling branch. To save the computational and memory of our model, each dilated convolution have the output channel dimension of 1/6 of the input channel dimension. These features are concatenated and processed by a depth-wise convolution layer. We then reshape the 3d tensor back to the 1d sequence. Finally, we add layer norm, MLP, and skip connection layer following the design of the FFN module used in the self-attention layer in ViT.

To aggregate the token embeddings into a compact representation, we add a cls token and use a single class attention layer. This layer is identical to the transformer encoder block used in ViT, except the self-attention operation is only calculated between the cls token embedding (treated as a query) and the token embeddings of image patches (treated as keys and values). The tokens embeddings output from the class attention layer will be used for the reranking stage with geometric verification, while the cls token embedding is used as the global feature.

Furthermore, since hundreds to thousands of local features are used in the reranking stage, they must be represented compactly. Moreover, [23] shows that whitening down weights co-occurrences of local features, which is generally beneficial for retrieval applications. We, therefore, implement our local features dimensionality reduction using a small autoencoder (AE) module [21] following the state-of-the-art dimensionality reduction method used in [9], as depicted in Fig. 3. Both encoder and decoder are implemented by a simple multilayer perceptron (MLP), where we set the number of layers to 1. The new local features are obtained as $\mathcal{L} = \mathrm{E}(\mathcal{S})$, where $\mathcal{L} \in (H/16 \times W/16, C_E)$, $\mathcal{S} \in (H/16 \times W/16, C)$ is the local features from the trained ViTGaL model and E is the encoding part of the autoencoder. Note that the parameters of ViTGaL are kept fixed during the training of the autoencoder. The decoding part transforms L into $\mathcal{S}' = \mathrm{D(L)}$, where $\mathcal{S}' \in (H/16 \times W/16, C)$. We also use a single class attention layer to aggregate S' into a compact representation f_r and use cross-entropy on f_r as well as L2 loss between S' and S to train the autoencoder. We also use the attention scores from the autoencoder network as key point detection scores to extract top local descriptors, which we found are much cleaner than those from the ViTGaL model.

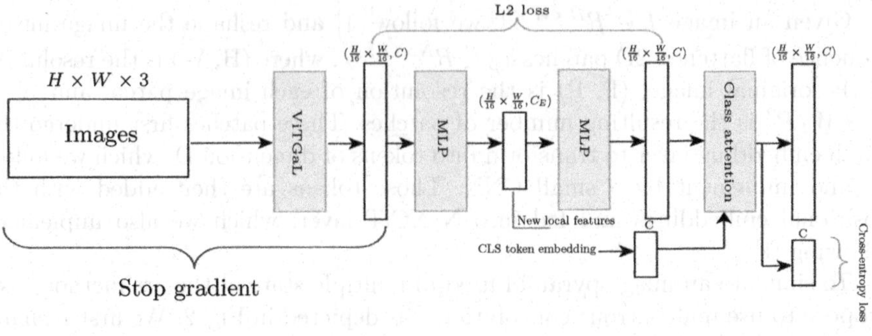

Fig. 3. The architecture of our dimensionality reduction model. It is an autoencoder model, where both encoder and decoder are a simple MLP. L2 loss and cross-entropy loss are used to train the network. The trained ViTGaL to extract high-dimensionality local features for reduction is fixed during training.

Each extracted local feature for the image patch at position h, w is represented with a token embedding $l_{h,w} \in \mathcal{L}$ and its corresponding keypoint detection score $a_{h,w}$ obtained by averaging the attention score from multiple heads at position h,w. These token embeddings are treated as local descriptors for the reranking stage, where their location in the input image is set to the center of their corresponding patch.

3.2 Training Objective

ViTGaL. Following state-of-the-art methods in [9,55], we propose to train our model using only image-level labels. To train both global and local features for the ViTGaL model, we use the ArcFace margin loss [12], where we add only one L2-normalized N class prediction head $W \in R^{C \times N}$

$$L = -log(\frac{exp(\gamma \times AF(\omega_t^T f_g, 1)}{\sum_i exp(\gamma \times AF(\omega_i^T f_g, y_i)))}) \quad (1)$$

where w_i is the ith row of W and f_g is the L2- normalized of the global feature output from the ViTGaL model, y is the one-hot label vector, and t is the index of the ground-truth class($y_t = 1$). γ is a learnable scalar, and AF denotes the ArcFace-adjusted cosine similarity, calculated as follows:

$$AF(s,c) = \begin{cases} cos(acos(s) + m), & \text{if c = 1} \\ s, & \text{if c = 0} \end{cases} \quad (2)$$

where s is the cosine similarity, m is the ArcFace margin, and c is a binary value which c = 1 means this is a ground-truth class.

Autoencoder Model. We follow [9] and use two losses: the mean-squared error regression loss and cross-entropy loss to train our autoencoder model. First, the mean-squared error regression loss measures how well the autoencoder reconstructs S:

$$L_r(\mathcal{S}, \mathcal{S}') = \frac{1}{H/16 \times W/16 \times C} \sum_{h,w} (\|\mathcal{S}_{h,w} - \mathcal{S'}_{h,w}\|)^2 \tag{3}$$

A cross-entropy loss is also used on top of the aggregation vector f_r of S' using a single class attention layer:

$$L_c(f_r, k) = -log(\frac{exp(v_t^T f_r + b_t)}{\sum_i exp(v_i^T f_r + b_i)}) \tag{4}$$

where v_i, b_i is the classifier weight and bias for class i and t is the ground-truth class for f_r. The total loss for autoencoder model is given by : $L_a = L_c + \lambda L_r$, where λ is a loss weight.

4 Experiments

4.1 Experimental Setup

Training Dataset. We use the cleaned version of Google landmarks dataset V2 (GLDv2-cleaned) [51] for training. It contains a total of 1,580,470 images and 81,313 classes. It is a subset of the bigger but more noisy dataset Google land-marks dataset V2 (the original dataset contains 5M images of 200K different landmarks). Google developed the original dataset to raise the challenges faced by the landmark identification system under real industrial scenarios. The cleaned version is built by the competitors from Google Landmark Retrieval Competition 2019, as they found the original dataset is too noisy.

Evaluation Datasets and Metrics. We use $\mathcal{R}$Oxf, $\mathcal{R}$Par, $\mathcal{R}$Oxf+$\mathcal{R}$1M, $\mathcal{R}$Par+$\mathcal{R}$1M to evaluate our method. $\mathcal{R}$Oxf, $\mathcal{R}$Par [37] are a special version of the original Oxford5k [35] and Paris6k [36] datasets with revisited annotations. Both datasets contain 70 query images and additionally include 4993 and 6322 database images, respectively. $\mathcal{R}$1M [37] refers to the dataset with additional 1M distractor images for evaluating large-scale retrieval. Mean Average Precision (mAP) is used to evaluate the performance of our method on the Medium and Hard splits of all datasets.

Implementation Details. We trained our model using the GLDv2-cleaned dataset. For a fair comparison with the state-of-the-art methods in [9,55], we follow them and randomly divide 80% of the dataset for training and the rest 20% for validation. We use XCiT-S12/16 and XCiT-S24/16 as our XCIT backbone models since they have a compatible number of parameters to the Restnet50 and Resnet101 backbones used in state-of-the-art CNN-based methods [9,55]. Our models are initialized from ImageNet pre-trained weights. The image first undergoes augmentation by randomly rotating, shifting, scaling, and cropping, then resizing to 512×512 resolution. We use a batch size of 64 and train our model using 2 Tesla V100 GPUs with 32 GB memory per card for 40 epochs. Adam optimizer and cosine learning rate decay strategy are adopted for training. We train our model with two warming-up epochs with the initial learning rate of $3e^{-5}$. We then train our model for additional 38 epochs, and the maximum learning rate is set as $5e^{-5}$. We set the ArcFace margin m = 0.1, the ArcFace scale $\gamma = 30$ and the loss weight for L_r to $\lambda = 10$.

As for feature extraction, while previous works [9,55] used multiple scales to extract global features, we use only a single scale. For local features extraction, since using a single scale image representation only can't produce enough local features for reranking, we used five scales, i.e., 0.3535, 0.5, 0.7071, 1.0, 1.4142 for the extraction, although the experiment using only single scale still gives a good result. Local features are selected based on their attention scores. We choose a maximum of 1k local features with the highest attention score. A minimum attention score threshold τ is also used, where we set τ to the median attention score in the last iteration of training following [9]. For local features matching, we use RANSAC [18] with an affine model. We follow [9] and tune the RANSAC parameters on $\mathcal{R}$Oxf, $\mathcal{R}$Par, then the best parameters are fixed for experiments on $\mathcal{R}$Oxf + $\mathcal{R}$1M, $\mathcal{R}$Par + $\mathcal{R}$1M. The top 100 ranked images from the first stage are considered for reranking, where the reranking is based on the number of inliers.

4.2 Comparison with the State-of-the-Art

mAP Comparison. We compare our model with the state-of-the-art methods in Table 1. All methods are tested on $\mathcal{R}$oxf and $\mathcal{R}$par datasets (and their large-scale versions $\mathcal{R}$oxf+1M, $\mathcal{R}$par+1M), with both Medium and Hard evaluation protocols. We follow previous works [9,55] and divide the previous state-of-the-art methods into three groups: (A) local feature aggregation and re-ranking; (B) global feature similarity search; (C) global feature search followed by re-ranking with local feature matching and spatial verification (SP). Note that although DOLG [55] proposed fusing global and local features into compact image representations, the search is conducted on the fused global features, so we grouped DOLG into the group (B).

Compared to methods in group B, our ViTGaL global feature variants are significantly better in all cases. In a standard evaluation setting, where query images are cropped [37], our model strongly outperforms state-of-the-art DELG trained on the GLDv2-clean dataset. For example, with the XCiT-S12/16 backbone (roughly the same parameters of ResNet50), the mAP is 79.64% v.s. 73.60% on $\mathcal{R}$oxf-Medium and 62.03% v.s. 51.00% on $\mathcal{R}$oxf-Hard. The gap is more significant in large-scale setting, with 13.48% absolute improvement on $\mathcal{R}$oxf-Hard+1M and 8.32% on $\mathcal{R}$oxf-Medium+1M. Our model with a global feature only also outperforms DELG with the second reranking stage. In evaluation settings where query images are not cropped, our ViTGaL global feature variants also outperform the state-of-the-art one-stage retrieval model DOLG. Note that unlike other methods, which use three [9] or five [55] image scales, our model use only a single image scale to perform retrieval.

Table 1. mAP comparison against the state-of-the-art retrieval methods on the $\mathcal{R}$oxf and $\mathcal{R}$par datasets (and their large-scale versions $\mathcal{R}$oxf+1M/$\mathcal{R}$par+1M), with both Medium and Hard evaluation protocols. ⋆ means feature quantization is used, and "†" means second-order loss is added. "GLDv1" and "GLDv2-clean" mark the difference in the training dataset. ˆdenotes evaluations where queries are not cropped. State-of-the-art performances and ours are marked bold.

Method	Medium				Hard			
	$\mathcal{R}$oxf	+1M	$\mathcal{R}$par	+1M	$\mathcal{R}$oxf	+1M	$\mathcal{R}$par	+1M
(A) Local feature aggregation + re-ranking								
HesAff-rSIFT-ASMK⋆ +SP [46]	60.60	46.80	61.40	42.30	36.70	26.90	35.00	16.80
HesAff-HardNet-ASMK⋆ +SP [31]	65.60	–	65.20	–	41.40	–	38.50	–
DELF-ASMK⋆ +SP [34,37]	67.80	53.80	76.90	57.30	43.10	31.20	55.40	26.40
DELF-R-ASMK⋆ +SP [45]	76.00	64.00	80.20	59.70	52.40	38.10	58.60	29.40
R50-How-ASMK, n = 2000 [47]	79.40	65.80	81.60	61.80	56.90	38.90	62.40	33.70
R50-MDA-ASMK [51]⋆	**81.80**	**68.70**	**83.30**	**64.70**	**62.20**	**45.30**	**66.20**	**38.90**
(B) Global features								
R101-R-MAC [19]	60.90	39.30	78.90	54.80	32.40	12.50	59.40	28.00
R101-GeM ↑ [44]	65.30	46.10	77.30	52.60	39.60	22.20	56.60	24.80
R101-GeM-AP [41]	67.50	47.50	80.10	52.50	42.80	23.20	60.50	25.10
R101-GeM-AP (GLDv1) [41]	66.30	–	80.20	–	42.50	–	60.80	–
R152-GeM [38]	68.70	–	79.70	–	44.20	–	60.30	–
ResNet101-GeM+SOLAR†[32]	69.90	53.50	81.60	59.20	47.90	29.90	64.50	33.40
R50-DELG [9]	69.70	55.00	81.60	59.70	45.10	27.80	63.40	34.10
R50-DELG (GLDv2-clean) [9]	**73.60**	**60.60**	**85.70**	**68.60**	**51.00**	**32.70**	**71.50**	**44.40**
R101-DELG [9]	73.20	54.80	82.40	61.80	51.20	30.30	64.70	35.50
R101-DELG(GLDv2-clean) [9]	**76.30**	**63.70**	**86.60**	**70.60**	**55.60**	**37.50**	**72.40**	**46.90**
XCiT-S12/16-ViTGaL (GLDv2-clean) (Ours)	**79.64**	**68.92**	**91.58**	**80.91**	**62.03**	**46.18**	**81.98**	**61.93**
XCiT-S24/16-ViTGaL (GLDv2-clean) (Ours)	**79.63**	**69.39**	**91.36**	**81.34**	**61.34**	**46.30**	**81.46**	**62.94**
R50-DOLG (GLDv2-clean)ˆ [55]	**80.50**	**76.58**	**89.81**	**80.79**	**58.82**	**52.21**	**77.70**	**62.83**
R101-DOLG (GLDv2-clean)ˆ [55]	**81.50**	**77.43**	**91.02**	**83.29**	**61.10**	**54.81**	**80.30**	**66.69**
XCiT-S12/16-ViTGaL (GLDv2-clean) ˆ (Ours)	**83.55**	**77.13**	**92.12**	**83.14**	**64.94**	**53.64**	**83.38**	**66.42**
XCiT-S24/16-ViTGaL (GLDv2-clean) ˆ (Ours)	**84.42**	**77.95**	**92.53**	**84.01**	**65.89**	**55.37**	**83.60**	**67.76**
(C) Global feature + Local features re-ranking								
R101-GeM ↑ + DSM [44]	65.30	47.60	77.40	52.80	39.20	23.20	56.20	25.00
R50-DELG [9]	75.10	61.10	82.30	60.50	54.20	36.80	64.90	34.80
R50-DELG(GLDv2-clean) [9]	**78.30**	**67.20**	**85.70**	**69.60**	**57.90**	**43.60**	**71.00**	**45.70**
R101-DELG [9]	78.50	62.70	82.90	62.60	59.30	39.30	65.50	37.00
R101-DELG (GLDv2-clean) [9]	**81.20**	**69.10**	**87.20**	**71.50**	**64.00**	**47.50**	**72.80**	**48.70**
XCiT-S12/16-ViTGaL + Autoencoder (GLDv2-clean) (Ours)	**83.17**	**74.04**	**91.51**	**81.42**	**66.72**	**51.62**	**80.87**	**62.14**
XCiT-S24/16-ViTGaL + Autoencoder (GLDv2-clean) (Ours)	**82.37**	**74.26**	**91.38**	**81.84**	**64.30**	**52.27**	**80.40**	**62.68**
XCiT-S12/16-ViTGaL + Autoencoder (GLDv2-clean) ˆ (Ours)	**85.62**	**78.79**	**92.34**	**83.60**	**69.37**	**58.09**	**83.25**	**66.97**
XCiT-S24/16-ViTGaL + Autoencoder (GLDv2-clean) ˆ (Ours)	**86.66**	**80.27**	**92.66**	**84.58**	**70.57**	**59.56**	**83.20**	**68.16**

For setup (C), we used both global and local features for retrieval. Local feature re-ranking boosts performance substantially for ViTGaL, especially in large-scale settings: gains of up to 6% (in $\mathcal{R}$oxf+Hard+1M). Our retrieval results also outperform the previous state-of-the-art DELG significantly, by more than 15% on $\mathcal{R}$paris+Hard+1M and 8% on $\mathcal{R}$oxf+Hard+1M. ViTGaL also outperforms local feature aggregation results from setup (A) in all cases, establishing a new state-of-the-art across the board.

Qualitative Results. We showcase the retrieval results of our model in Figs. 4a and 4b. Figure 4a illustrates the challenging cases where the gallery images show significant lighting changes and extreme viewpoint differences. These images are still capable of achieving relatively high ranks due to the effectiveness of our global feature, which captures the similarity well even in such challenging scenarios.

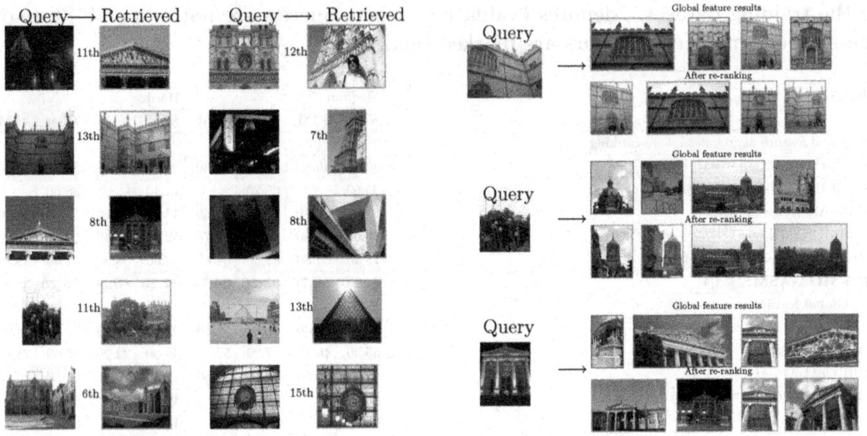

(a) Global feature retrieval. (b) Local feature re-ranking.

Fig. 4. Sample of ViTGaL results on $\mathcal{R}$oxf-Hard+1M and $\mathcal{R}$paris-Hard+1M. **(a)** Examples of challenging, high-ranked relevant retrieved images under the global feature retrieval. **(b)** Examples illustrating accuracy improvements using local features matching for re-ranking. For each query (left), two rows are presented on the right, the top one showing results based on global feature similarity and the bottom one showing results after re-ranking. All rows on the right show the top four retrieval results.

Figure 4b shows the effect of local feature re-ranking of our methods, where global features alone are not enough for high retrieval performance. Global features tend to retrieve images with a generally similar appearance but do not always depict the same object of interest. This can be significantly improved with local feature re-ranking, allowing stricter matching selectivity.

To further illustrate the power of the local feature of ViTGaL, we present qualitative results of local feature matching in Fig. 5 and Fig. 6. These visualizations depict the final obtained correspondences after RANSAC. We show the matching results between one query and two gallery images per row, where images with lines connecting the corresponding key points are on the figure's right. Figure 5 showcases the robustness of VitGaL in extreme cases such as strong viewpoint and illumination changes, where matches can be obtained across different scales, in occlusion cases, and in day-vs-night scenarios. Figure 6 presents the matching between images of different scenes/objects: matches are still found due to the similarity in patterns between query and index images (e.g., similar windows, arches, or roofs). Nevertheless, these do not affect retrieval results much because the number of inliers is low.

4.3 Ablation Experiments

We conducted experiments using the XCiT-S12/16 backbone to verify some of our design choices empirically.

Verification of the Multi-atrous Convolution. A multi-atrous convolution block is added to our model to simulate the image pyramid used in the standard retrieval algorithms. We provide experimental results to validate the contribution of the multi-atrous convolutions by removing them from the ViTGaL model. The results is shown in Table 2. It can be seen that adding the multi-atrous convolutions helps to improve the overall performance significantly. The mAP is improved from 57.80% to 62.03% and 79.18% to 81.98% on $\mathcal{R}$oxf-Hard and $\mathcal{R}$par-Hard, respectively. Moreover, our model can achieve state-of-the-art results using only single-scale representation at inference time. Table 3 reports the performance of the global retrieval of our model under different numbers of image scales. The scale rates are 0.7071, 1.0, 1.4142; 0.3535, 0.5, 0.7071, 1.0, 1.4142; 0.25, 0.3535, 0.5, 0.7071, 1.0, 1.4142, 2 for 3-scale, 5-scale and 7-scale setting respectively. To fuse these multi-scale features, we follow the previous works in [9,55] by firstly L2 normalizing them, then averaging the normalized features, and finally applying again an L2 normalization to produce the final descriptor. We can find from the empirical results that using single-scale only performs the best among four multi-scale settings. Such experimental results are also within our expectations. With multi-atrous convolution, we can simulate the image pyramid within the feature space directly. These results validate the effectiveness of the multi-atrous convolution in our model.

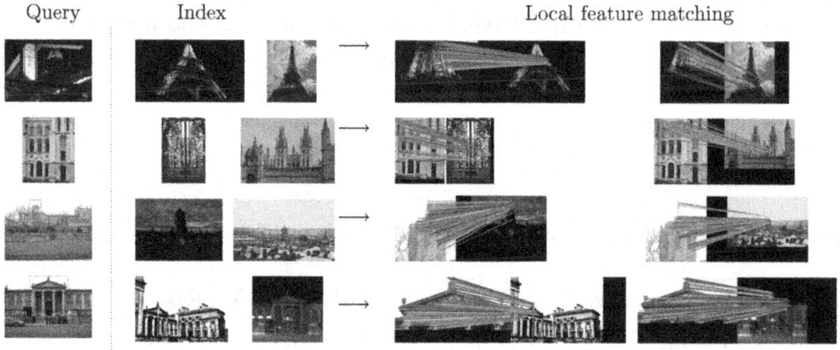

Fig. 5. Example of local feature matches for image pairs depicting the same objects/scenes.

Fig. 6. Example of local feature matches for image pairs depicting different objects/scenes.

Table 2. Ablation experiment on the multi-atrous convolution.

Config	$\mathcal{R}$oxf-M	$\mathcal{R}$oxf-H	$\mathcal{R}$par-M	$\mathcal{R}$par-H
w/o multi-atrous convolution	78.77	57.8	90.4	79.18
Full model	**79.64**	**62.03**	**91.58**	**81.98**

Re-ranking Experiment. Table 4 compares local features for re-ranking under different number of scales. The scaling rates are exactly the same as those used in the previous experiments presented in Table 3. All scales receive the identical retrieval short list of 100 images for re-ranking from the XCiT-S12/16-ViTGaL global retrieval result. For a fair comparison, all scales use 1k features and 2k RANSAC iterations. We also tune the matching hyperparameters separately for each scale. The tuning hyperparameters are the distance threshold for selecting correspondences, RANSAC reprojection error threshold, and RANSAC homography confidence. Unlike the global retrieval experiments where using a single-scale only shows superior performance, the best result in re-ranking is observed with the 3-scale, and 5-scale settings since using a single scale only can't produce enough local features (average number of local features in $\mathcal{R}$oxf dataset are 250.6 and 876.7 for 1-scale and 5-scales setting respectively). Nevertheless, using a single scale only gives competitive results compared to the best scale setting, with the largest gap with the best scale among the four evaluation benchmark being only 1.7% (at $\mathcal{R}$oxf-H).

Table 3. Global retrieval results of ViTGaL under different multi-scale settings.

Number of scales	$\mathcal{R}$oxf-M	$\mathcal{R}$oxf-H	$\mathcal{R}$par-M	$\mathcal{R}$par-H
7-scale	77.6	60.01	91.19	80.74
5-scale	77.73	59.90	91.18	80.75
3-scale	79.28	60.53	**91.68**	**82.18**
1-scale	**79.64**	**62.03**	91.58	81.98

Table 4. Re-ranking results of ViTGaL under different multi-scale settings.

Number of scales	$\mathcal{R}$oxf-M	$\mathcal{R}$oxf-H	$\mathcal{R}$paris-M	$\mathcal{R}$paris-H
7-scale	82.29	65.10	91.34	80.79
5-scale	**83.17**	**66.72**	91.51	80.87
3-scale	83.15	66.27	**91.61**	**81.20**
1-scale	82.53	65.02	91.38	80.73

Latency and Memory. In Table 5, we list the memory footprint and extraction latency required by different methods for $\mathcal{R}$1M; corresponding to the three settings from Table 1. Similar to DELG, using ViTGaL for joint extraction allows a significant speedup over using two separate local and global models. Moreover, our model performance is much better than DELG, especially for a single-scale setting, where our model is superior in both storage and speed (3x times faster than DELG and requires only 25% storage of DELG).

Table 5. Feature extraction latency and database memory requirements for different image retrieval models. Latency is measured on an NVIDIA Tesla P100 GPU.

Method	Extraction latency (ms)	Memory (GB)	
		$\mathcal{R}$Oxf+1M	$\mathcal{R}$Par+1M
(A) Local feature aggregation			
DELF-R-ASMK* [45]	2260	27.6	–
(B) Global features			
R50-GeM [38]	100	7.7	7.7
R101-GeM [38]	175	7.7	7.7
(C) Unified global + local features			
R50-DELG [9]	211	485.5	486.2
R101-DELG [9]	383	485.9	486.6
XCiT-S12/16-ViTGaL+Autoencoder [ours]	158	420.1	420.8
XCiT-S12/16-ViTGaL+Autoencoder (1 scale global & local) [ours]	63	120.1	120.2
XCiT-S24/16-ViTGaL+Autoencoder [ours]	302	420.5	421.2
XCiT-S12/16-ViTGaL+Autoencoder (1 scale global & local) [ours]	87	120.3	120.4

5 Conclusions

In this paper, we make the first attempt to learn a model that enables joint extraction of local and global image features with ViTs, referred to as ViTGaL. The model is based on an XCiT backbone, leveraging multi-atrous convolutions to simulate the spatial feature pyramid used in the standard image retrieval algorithms for high-performing retrieval using a single-scale image representation only. The entire network can be trained end-to-end using image-level labels. We also use an autoencoder to reduce the dimension of local features for an effective re-ranking step. Extensive experiments demonstrate the superior performance of our method on image retrieval, achieving state-of-the-art performance on the Revisited Oxford and Revisited Paris datasets.

References

1. Ali, A., et al.: XCiT: cross-covariance image transformers. In: Advances in Neural Information Processing Systems 34 (2021)
2. Arandjelović, R., Gronat, P., Torii, A., Pajdla, T., Sivic, J.: NetVLAD: CNN architecture for weakly supervised place recognition. In: IEEE Conference on Computer Vision and Pattern Recognition (2016)
3. Arnab, A., Dehghani, M., Heigold, G., Sun, C., Lučić, M., Schmid, C.: ViViT: a video vision transformer. In: Proceedings of the IEEE/CVF International Conference on Computer Vision, pp. 6836–6846 (2021)
4. Avrithis, Y., Tolias, G.: Hough pyramid matching: Speeded-up geometry re-ranking for large scale image retrieval. Int. J. Comput. Vision **107**, 1–19 (2014)
5. Balntas, V., Riba, E., Ponsa, D., Mikolajczyk, K.: Learning local feature descriptors with triplets and shallow convolutional neural networks. In: BMVC, p. 3 (2016)
6. Barroso-Laguna, A., Riba, E., Ponsa, D., Mikolajczyk, K.: Key.Net: keypoint detection by handcrafted and learned CNN filters. In: Proceedings of the IEEE/CVF International Conference on Computer Vision, pp. 5836–5844 (2019)
7. Bay, H., Ess, A., Tuytelaars, T., Gool, L.V.: Speeded-up robust features (surf). Comput. Vis. Image Underst. **110**(3), 346–359 (2008)
8. Bertasius, G., Wang, H., Torresani, L.: Is space-time attention all you need for video understanding. arXiv preprint arXiv:2102.05095 2(3), 4 (2021)
9. Cao, B., Araujo, A., Sim, J.: Unifying deep local and global features for image search. In: Vedaldi, A., Bischof, H., Brox, T., Frahm, J.-M. (eds.) ECCV 2020. LNCS, vol. 12365, pp. 726–743. Springer, Cham (2020). https://doi.org/10.1007/978-3-030-58565-5_43
10. Chen, L.C., Papandreou, G., Schroff, F., Adam, H.: Rethinking atrous convolution for semantic image segmentation. arXiv preprint arXiv:1706.05587 (2017)
11. Chopra, S., Hadsell, R., LeCun, Y.: Learning a similarity metric discriminatively, with application to face verification. In: 2005 IEEE Computer Society Conference on Computer Vision and Pattern Recognition, CVPR 2005, vol. 1, pp. 539–546. IEEE (2005)
12. Deng, J., Guo, J., Xue, N., Zafeiriou, S.: ArcFace: additive angular margin loss for deep face recognition. In: Proceedings of the IEEE/CVF Conference on Computer Vision and Pattern Recognition, pp. 4690–4699 (2019)
13. DeTone, D., Malisiewicz, T., Rabinovich, A.: SuperPoint: self-supervised interest point detection and description. In: 2018 IEEE/CVF Conference on Computer Vision and Pattern Recognition Workshops (CVPRW), pp. 337–33712 (2018). https://doi.org/10.1109/CVPRW.2018.00060
14. DeTone, D., Malisiewicz, T., Rabinovich, A.: SuperPoint: self-supervised interest point detection and description. In: Proceedings of the IEEE Conference on Computer Vision and Pattern Recognition Workshops, pp. 224–236 (2018)
15. Dosovitskiy, A., et al.: An image is worth 16 × 16 words: transformers for image recognition at scale. In: ICLR (2021)
16. Dusmanu, M., et al.: D2-Net: a trainable CNN for joint detection and description of local features. arXiv preprint arXiv:1905.03561 (2019)

17. Fan, H., et al.: Multiscale vision transformers. In: Proceedings of the IEEE/CVF International Conference on Computer Vision, pp. 6824–6835 (2021)
18. Fischler, M.A., Bolles, R.C.: Random sample consensus: a paradigm for model fitting with applications to image analysis and automated cartography. Commun. ACM **24**(6), 381–395 (1981)
19. Gordo, A., Almazán, J., Revaud, J., Larlus, D.: End-to-end learning of deep visual representations for image retrieval. CoRR abs/1610.07940 (2016)
20. He, K., Lu, Y., Sclaroff, S.: Local descriptors optimized for average precision. In: Proceedings of the IEEE Conference on Computer Vision and Pattern Recognition, pp. 596–605 (2018)
21. Hinton, G.E.: Connectionist learning procedures. In: Machine Learning, pp. 555–610. Elsevier (1990)
22. Jégou, H., Douze, M., Schmid, C., Pérez, P.: Aggregating local descriptors into a compact representation. In: IEEE Conference on Computer Vision and Pattern Recognition, pp. 3304–3311 (2010)
23. Jégou, H., Chum, O.: Negative evidences and co-occurrences in image retrieval: the benefit of PCA and whitening, pp. 774–787 (10 2012)
24. Jégou, H., Perronnin, F., Douze, M., Sánchez, J., Pérez, P., Schmid, C.: Aggregating local image descriptors into compact codes. IEEE Trans. Pattern Anal. Mach. Intell. **34**(9), 1704–1716 (2012)
25. Li, Z., Wang, X., Liu, X., Jiang, J.: BinsFormer: revisiting adaptive bins for monocular depth estimation. arXiv preprint arXiv:2204.00987 (2022)
26. Liu, Z., et al.: Swin Transformer V2: scaling up capacity and resolution. arXiv preprint arXiv:2111.09883 (2021)
27. Liu, Z., et al.: Swin transformer: hierarchical vision transformer using shifted windows. In: Proceedings of the IEEE/CVF International Conference on Computer Vision, pp. 10012–10022 (2021)
28. Lowe, D.G.: Distinctive image features from scale-invariant keypoints. Int. J. Comput. Vis. **60**(2), 91–110 (2004)
29. Luo, Z., et al.: ContextDesc: local descriptor augmentation with cross-modality context. In: Proceedings of the IEEE/CVF Conference on Computer Vision and Pattern Recognition, pp. 2527–2536 (2019)
30. Mishchuk, A., Mishkin, D., Radenović, F., Matas, J.: Working hard to know your neighbor's margins: local descriptor learning loss. In: Proceedings of the 31st International Conference on Neural Information Processing Systems, NIPS 2017, pp. 4829–4840. Curran Associates Inc., Red Hook, NY, USA (2017)
31. Mishkin, D., Radenović, F., Matas, J.: Repeatability is not enough: learning affine regions via discriminability. In: Ferrari, V., Hebert, M., Sminchisescu, C., Weiss, Y. (eds.) ECCV 2018. LNCS, vol. 11213, pp. 287–304. Springer, Cham (2018). https://doi.org/10.1007/978-3-030-01240-3_18
32. Ng, T., Balntas, V., Tian, Y., Mikolajczyk, K.: SOLAR: second-order loss and attention for image retrieval. In: Vedaldi, A., Bischof, H., Brox, T., Frahm, J.-M. (eds.) ECCV 2020. LNCS, vol. 12370, pp. 253–270. Springer, Cham (2020). https://doi.org/10.1007/978-3-030-58595-2_16
33. Nister, D., Stewenius, H.: Scalable recognition with a vocabulary tree. In: 2006 IEEE Computer Society Conference on Computer Vision and Pattern Recognition, CVPR 2006, vol. 2, pp. 2161–2168 (2006). https://doi.org/10.1109/CVPR.2006.264
34. Noh, H., Araujo, A., Sim, J., Weyand, T., Han, B.: Large-scale image retrieval with attentive deep local features. In: 2017 IEEE International Conference on Computer Vision (ICCV), pp. 3476–3485 (2017). https://doi.org/10.1109/ICCV.2017.374

35. Philbin, J., Chum, O., Isard, M., Sivic, J., Zisserman, A.: Object retrieval with large vocabularies and fast spatial matching. In: 2007 IEEE Conference on Computer Vision and Pattern Recognition, pp. 1–8. IEEE (2007)
36. Philbin, J., Chum, O., Isard, M., Sivic, J., Zisserman, A.: Lost in quantization: improving particular object retrieval in large scale image databases. In: 2008 IEEE Conference on Computer Vision and Pattern Recognition, pp. 1–8. IEEE (2008)
37. Radenović, F., Iscen, A., Tolias, G., Avrithis, Y., Chum, O.: Revisiting Oxford and Paris: large-scale image retrieval benchmarking. In: Proceedings of the IEEE Conference on Computer Vision and Pattern Recognition, pp. 5706–5715 (2018)
38. Radenovic, F., Tolias, G., Chum, O.: Fine-tuning CNN image retrieval with no human annotation. IEEE Trans. Pattern Anal. Mach. Intell. **41**(7), 1655–1668 (2019). https://doi.org/10.1109/TPAMI.2018.2846566
39. Raghu, M., Unterthiner, T., Kornblith, S., Zhang, C., Dosovitskiy, A.: Do vision transformers see like convolutional neural networks? In: Advances in Neural Information Processing Systems 34 (2021)
40. Ranftl, R., Bochkovskiy, A., Koltun, V.: Vision transformers for dense prediction. In: Proceedings of the IEEE/CVF International Conference on Computer Vision, pp. 12179–12188 (2021)
41. Revaud, J., Almazán, J., Rezende, R.S., Souza, C.R.: Learning with average precision: training image retrieval with a listwise loss. In: Proceedings of the IEEE International Conference on Computer Vision, pp. 5107–5116 (2019)
42. Revaud, J., et al.: R2d2: repeatable and reliable detector and descriptor. arXiv preprint arXiv:1906.06195 (2019)
43. Sarlin, P.E., Cadena, C., Siegwart, R., Dymczyk, M.: From coarse to fine: robust hierarchical localization at large scale. In: Proceedings of the IEEE/CVF Conference on Computer Vision and Pattern Recognition, pp. 12716–12725 (2019)
44. Simeoni, O., Avrithis, Y., Chum, O.: Local features and visual words emerge in activations. In: Proceedings of the IEEE/CVF Conference on Computer Vision and Pattern Recognition (CVPR) (June 2019)
45. Teichmann, M., Araujo, A., Zhu, M., Sim, J.: Detect-to-retrieve: efficient regional aggregation for image search. In: Proceedings of the IEEE/CVF Conference on Computer Vision and Pattern Recognition, pp. 5109–5118 (2019)
46. Tolias, G., Avrithis, Y., Jégou, H.: Image search with selective match kernels: aggregation across single and multiple images. Int. J. Comput. Vis. **116**(3), 247–261 (2016)
47. Tolias, G., Jenicek, T., Chum, O.: Learning and aggregating deep local descriptors for instance-level recognition. In: Vedaldi, A., Bischof, H., Brox, T., Frahm, J.-M. (eds.) ECCV 2020. LNCS, vol. 12346, pp. 460–477. Springer, Cham (2020). https://doi.org/10.1007/978-3-030-58452-8_27
48. Tolias, G., Sicre, R., Jégou, H.: Particular object retrieval with integral max-pooling of CNN activations (2016)
49. Wang, H., et al.: CosFace: large margin cosine loss for deep face recognition. In: Proceedings of the IEEE Conference on Computer Vision and Pattern Recognition, pp. 5265–5274 (2018)
50. Wang, W., et al.: Pyramid vision transformer: a versatile backbone for dense prediction without convolutions. In: Proceedings of the IEEE/CVF International Conference on Computer Vision, pp. 568–578 (2021)
51. Wu, H., Wang, M., Zhou, W., Li, H.: Learning deep local features with multiple dynamic attentions for large-scale image retrieval. In: 2021 IEEE/CVF International Conference on Computer Vision (ICCV), pp. 11396–11405 (2021). https://doi.org/10.1109/ICCV48922.2021.01122

52. Xie, E., Wang, W., Yu, Z., Anandkumar, A., Alvarez, J.M., Luo, P.: SegFormer: simple and efficient design for semantic segmentation with transformers. In: Advances in Neural Information Processing Systems 34 (2021)
53. Yan, H., Zhang, C., Wu, M.: Lawin transformer: improving semantic segmentation transformer with multi-scale representations via large window attention. arXiv preprint arXiv:2201.01615 (2022)
54. Yandex, A.B., Lempitsky, V.: Aggregating local deep features for image retrieval. In: 2015 IEEE International Conference on Computer Vision (ICCV), pp. 1269–1277 (2015)
55. Yang, M., et al.: DOLG: single-stage image retrieval with deep orthogonal fusion of local and global features. In: Proceedings of the IEEE/CVF International Conference on Computer Vision (ICCV), pp. 11772–11781 (October 2021)
56. Yi, K.M., Trulls, E., Lepetit, V., Fua, P.: LIFT: learned invariant feature transform. In: Leibe, B., Matas, J., Sebe, N., Welling, M. (eds.) ECCV 2016. LNCS, vol. 9910, pp. 467–483. Springer, Cham (2016). https://doi.org/10.1007/978-3-319-46466-4_28
57. Zhang, H., et al.: DINO: DETR with improved denoising anchor boxes for end-to-end object detection. arXiv preprint arXiv:2203.03605 (2022)
58. Zhang, Y., Jin, R., Zhou, Z.H.: Understanding bag-of-words model: a statistical framework. Int. J. Mach. Learn. Cybern. **1**(1), 43–52 (2010)

Improving Surveillance Object Detection with Adaptive Omni-Attention over Both Inter-frame and Intra-frame Context

Tingting Yu[1], Chen Chen[2], Yichao Zhou[1], and Xiyuan Hu[1](✉)

[1] School of Computer Science and Engineering, Nanjing University of Science and Technology, Nanjing 210094, China
{yutt,yczhou,huxy}@njust.edu.cn

[2] Institute of Automation, Chinese Academy of Sciences, Beijing 100190, China

Abstract. Surveillance object detection is a challenging and practical sub-branch of object detection. Factors such as lighting variations, smaller objects, and motion blur in video frames affect detection results, but on the other hand, the temporal information and stable background of a surveillance video are major advantages that does not exist in generic object detection. In this paper, we propose an adaptive omni-attention model for surveillance object detection, which effectively and efficiently integrates inter-frame contextual information to improve the detection of low-quality frames and intra-frame attention to suppress false positive detections in the background regions. In addition, the training of the proposed network can converge quickly with less epochs because during multi-frame fusion stage, the pre-trained weights of the single-frame network can be used to update simultaneously in reverse in both single-frame and multi-frame feature maps. The experimental results on the UA-DETRAC and the UAVDT datasets have demonstrated the promising performance of our proposed detector in both accuracy and speed. (Code is available at https://github.com/Yubzsz/Omni-Attention-VOD.)

1 Introduction

Surveillance object detection, as an important sub-branch of generic object detection, aims at localizing objects with tight bounding boxes in each frame of a surveillance video. It has been studied for many decades from traditional background modelling and subtraction algorithms to the recent deep learning-based models. Although deep learning-based object detection models [1–10] have made significant progress on both images and videos, detecting objects in the surveillance scenario still has its own set of challenges and difficulties. First, a more accurate object detection result is needed for subsequent tracking or re-identification tasks. For example, in many benchmarks of video object detection, such as [11,12], a threshold of 0.7 is set for calculating the average accuracy rather than a commonly used threshold of 0.5 for most image object detection. Then, the cases of occlusion, of smaller objects, and of motion blur will appear

L. Wang et al. (Eds.): ACCV 2022, LNCS 13842, pp. 222–237, 2023.
https://doi.org/10.1007/978-3-031-26284-5_14

more frequently in the surveillance videos than general images. Finally, considering the practical application of surveillance scenario, the inference speed of the detection model is another important issue, which puts high demands on the efficiency of surveillance object detection.

However, compared to the generic image object detection, surveillance object detection does have some additional prior information, such as relative stable background in the same video sequence and same object exists in consecutive frames, which can be used to improve the performance of object detection. For instance, for utilizing the intra-frame information, structural information or the layout of scenes have been used to refine the object detection results [13,14]; and for utilizing the inter-frame information, some background modeling and optical flow-based approaches have been proposed to improve the detection of small objects [15–17]. Considering these reasons, we propose an adaptive omni-attention model for improving surveillance object detection, which effectively and efficiently combine the temporal information of the consecutive frames as inter-frame attention and the spatial context of surveillance scenarios as intra-frame attention. The experimental results on the UA-DETRAC [11] dataset have demonstrated the efficacy of our proposed detector when compared with other state-of-the-art surveillance object detection models.

The main contributions of our approach are summarized as follows.

- An inter-frame attention module, with temporal adaptive convolutions, has been added into the backbone feature extraction phase to effectively incorporate the feature maps of consecutive frames to enhance the small object detection.
- An intra-frame attention module that weights the current frame feature maps in channel and spatial dimensions has been introduced to suppress the false positive detections in the background regions.
- An efficient feature fusion module has been proposed to integrate inter-frame and intra-frame feature maps at different scales which can achieve a high detection accuracy with less training cost.

2 Related Work

2.1 Generic Object Detection

Currently deep learning-based object detection networks can be classified into two-stage and one-stage detection methods according to the detection process. The classical two-stage methods, such as R-CNN [1], SPP-Net [2], Faster RCNN [3], etc., have independent region proposal module, which can support a relative high detection accuracy, but will affect the detection speed. The one-stage methods discard the process of region proposal and directly regress the detection boxes on the images, which gains advantages in speed and gradually approaches the two-stage methods in terms of accuracy. Typical representatives of one-stage methods include YOLO [4] and its variants, SSD [5], RetinaNet [6], etc. The recent rise of anchor-free object detection is an important improvement to one-stage detection models.

Compared with anchor-based method, anchor-free-based method refers to the detection by only dense prediction or keypoint estimation without manually designing anchors with different scale on the image. For example, CornerNet [7] and CenterNet [8], treat the object detection problem as key point detection, and respectively predict the corner and center points of the object to complete the detection. The selection of these points are determined by heatmap, and the loss function of heatmap is focal loss that proposed in [6]. What's more, FCOS [9] performs pixel-by-pixel prediction, RepPoints [10] learns a set of points to represent the object. Anchor-free-based methods solve the defects brought by the anchor, such as imbalance of positive and negative samples during training, high memory occupation, and difficulty in recognizing multi-scale objects.

Attention mechanism is another widely used technique for improving object detection, which has been developed rapidly as a training module that can significantly improve the accuracy of a model with only a small increase in model complexity or computational effort. For example, YOLOv4 [18] explores the impact of Squeeze-and-Excitation (SE) [19] and Spatial Attention Mechanisms (SAM) [20] methods for training the model, where SE is a channel attention mechanism and SAM is a spatial attention mechanism. EfficientDet [21] in backbone incorporates an SE attention module in each stage; TPH-YOLOv5 [22] integrates a convolutional attention module [23] (CBAM) to find attention regions in scenes with dense objects; in addition, attention mechanisms can also be used for feature aggregation [16].

2.2 Surveillance Object Detection

Compared with generic object detection, the biggest difference of surveillance object detection is the rich temporal and contextual information that can be utilized to improve the object detection. Existing surveillance object detection methods adopt this prior information in different ways, such as multi-frame feature aggregation, 3D convolution, dynamic foreground extraction, and so on.

Most multi-frame feature aggregation methods compute each frame detection intensively and then perform weighted average of features. FGFA [17] uses optical flow estimation to aggregate feature maps of neighboring frames. To reduce computational cost, THP [24] uses optical flow and sparse recursion to operate on sparse key frames. Because of the high computational complexity of estimating the optical flow, some non-optical flow-based feature aggregation methods have been proposed. For example, MEGA [25] integrates the precomputed features of the previous frame and stores them as global information in the remote memory module; FFAVOD [26] fuses feature matrices of the front and back frame of the current frame extracted by the backbone, and the output of the fusion module is used as the input of the current frame detection header.

3D-DETNet [27] uses 3D convolution to capture motion or temporal information encoded in multiple consecutive video frames, but the large number of 3D convolution parameters and low computational efficiency are rarely applied

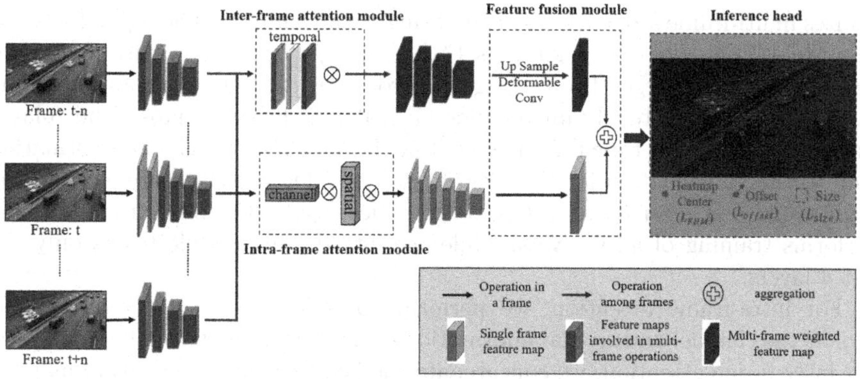

Fig. 1. Overview of our framework. The figure illustrates the detection process of 2n+1 frames video sequence. Figures 2 and 3 show the details of our inter-frame attention, intra-frame attention, and feature fusion modules, respectively.

directly in video detection. In the field of video understanding and object tracking, most of the temporal modeling methods usually decompose 3D convolution into the combination of 2D spatial convolution and 1D temporal convolution, such as P3D [28] and R(2+1)D [29]. In order to reduce the computation complexity of 3D convolution while giving 2D convolution the capability of temporal modeling, Huang et al. proposed Tada Conv [30], which performs temporal modeling by relaxing the temporal invariance of 2D convolution and superimposing adaptive temporal weights on it. TCTrack [31] is an example of using Tada Conv to improve the object tracking. Dynamic foreground extraction is another kind of approaches to utilize inter-frame information by updating a spatio-temporal background, which can effectively highlight moving objects, but has barely improvement for stationary objects or video jitter.

3 Proposed Method

Based on the consideration of effective and efficient use of attention in temporal, spatial and channel domain, we propose a multi-frame based omni-attention object detection network. Detailed illustration of the proposed network is provided in the following subsections.

3.1 Omni-Attention Based Surveillance Object Detection Architecture

The overall structure of the proposed omni-attention based surveillance object detection framework is illustrated in Fig. 1, which consists of three main submodules: the inter-frame attention module, the intra-frame attention module and the feature fusion module. Since detection on each frame independently usually ignores the connection between the contextual frames, the proposed framework utilizes multi-frame sequences instead of single-frame inputs. The input $L-$frame ($L = 2n + 1$) images are denoted as $\{I_e\}_{e=t-n,\ldots,t-1,t,t+1,\ldots t+n}$, with the target frame is I_e. The target frame I_e goes through 6 different convolution layers in the initial stage to obtain 6 different scales of feature maps. The 6-layer feature maps of the target frames contain both high-level global information and detailed information in the lower level. These feature maps are fed into the intra-frame and inter-frame attention modules, where the intra-frame module performs training of all 6 layers, while the inter-frame module trains only the last 4 layers.

For intra-frame attention, we perform operations at the target frame to exploit the channel and spatial information of the feature map at different scales. For inter-frame attention, we concatenate the last four layers of feature maps of the target frame and the involved contextual frames, assigning adaptive temporal weights to them. The reasons for selecting the last four layers of feature maps to apply temporal convolution here are: (1) the network structure of the first 2 layers of convolution is relatively simple (with only one convolution and normalization operation) and extracts low-level details; (2) the higher-level feature maps have a larger receptive field, which is more suitable for applying to multi-frame temporal contexts. Therefore, the temporal, channel and spatial information is aggregated at the same time, and then, the obtained fused feature maps are fed into 3 head branches: fusion heatmap, offset and size branches. The fusion heatmap (optimized by the loss L_{FHM}) represents the center point heatmap of each category object; offset (optimized by the loss L_{offset}) is the offset of the key point relative to the original image generated due to the image undergoing down-sampling, feature enhancement and other operations; and size (optimized by the loss L_{size}) is the distance between the object center point and the border. These three branches provide the final information to form the bounding box of the detection object.

3.2 Inter-frame Attention Module

Given the input image sequence, the multi-frame tandem feature map can be obtained at a certain stage of the backbone. Inspired by the temporal adaptive convolution in [30], for features of target frame enhance by the Tada Convolution with temporal modeling capability, we can obtain adaptive temporal weights specifically assigned to each frame. The input X represents the integration of all feature maps from multiple frames, and X_l represents the l-th of them. The output feature map $\tilde{X}_l$ corresponding to X_l obtained by passing through the Tada convolution is shown below:

$$\tilde{X}_l = W_l * X_l = (\alpha_l \cdot W_b) * X_l \tag{1}$$

where the $*$ indicates the convolution operation and $\cdot$ indicates the element-wise multiplication. For which the weight $W_l = \alpha_l \cdot W_b$, where W_b is the base weight shared by all frames in the network, and α_l is the calibration weight obtained from the temporal context different for each frame. Over the entire

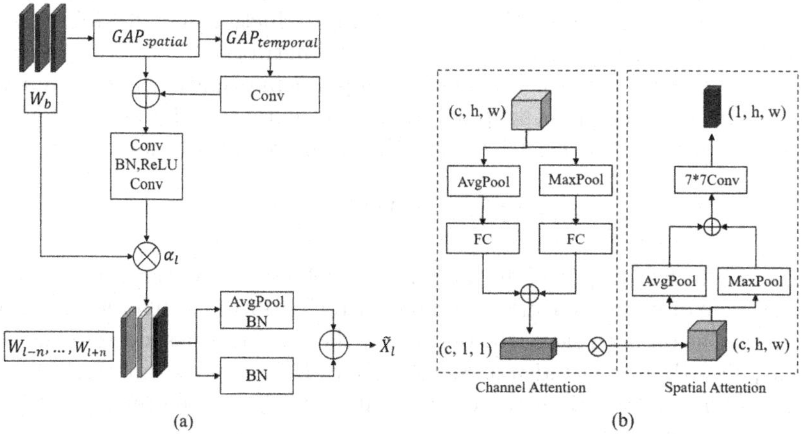

Fig. 2. (a) The inter-frame temporal attention architecture of the proposed method. GAP stands for global average pooling. (b) The channel and spatial attention architecture of the proposed method. FC stands for fully connected layer and the feature map dimensions are in the parentheses.

tandem feature map, feature map of each frame X_l is adaptively assigned with a specific calibration weight. The calibration weight α_l of the $l-$th frame not only considers the current frame, but also integrates the temporal attention among the related frame sequence. We show the process of generating the calibration weight α_l in Fig. 2(a).

Firstly, X_l is input to the Tada Convolution through a tandem operation. For obtaining the local temporal context information, a global average pooling (GAP) in the spatial dimension is first performed, such that $v_l = GAP_s(X_l)$, which is used as the frame descriptor. To obtain the global temporal context information, we perform a linear mapping of the global descriptors, superimposed on the frame descriptors, to further merge the global temporal information, i.e., $g = GAP_{st}(X)$. Here GAP_{st} represents the global average pooling in the spatial and temporal dimension. After aggregating global and local information, two convolutions to the frame descriptors as well as ReLU and normalization operations have been applied to obtain additional time-adaptive calibration weights. The whole process of spatial-temporal convolution can be described by the following equation:

$$\mathcal{F}_{\alpha}\left(\tilde{X}_l\right) = Conv\left(\rho\left(Conv\left(v_l + FC(g)\right)\right)\right) \tag{2}$$

where v_l and g stand for frame descriptor and global descriptor respectively, ρ stands for ReLU and batch normalization, FC stands for linear mapping.

The final weight W_l is the product of the calibration weight α_l and the base weight W_b, where the base weight can be replaced by the pre-trained weight of the single-frame network. The calibration weight is set to 0 during initialization, which has the advantage of the training time reduction and flexibility of the convolution module embedded into the existing network.

In order to better aggregate spatial-temporal information and compensate the problem of inadequate spatial information extracted by spatial-temporal convolution, we add two branches after the adaptive convolution module as:

$$y_l = \psi_\rho \tilde{X}_l + \psi_v AvgPool\left(\tilde{X}_l\right). \tag{3}$$

Specifically, the output of the adaptive convolution as $\tilde{X}_l$ is fed into these two branches, one of which is average pooling. After performing different normalization operations on both sides, the outputs of the two branches are aggregated. In this way, according to the idea of SmallBigNet [32], the pooling branch provides a larger receptive field than the other branch, which can better integrate core and contextual semantics.

3.3 Intra-frame Attention Module

Besides temporal attention, attention in the spatial and channel dimensions also provides possible enhancement for feature maps of single-frame images. For a particular layer of feature maps in convolutional neural networks, the attention mechanism learns an additional weight of corresponding pixels in a particular dimension, and these weights represent the importance of a certain information that strengthens the useful features and weakens the useless ones, thus facilitate the feature screening and enhancement. Since surveillance videos often have stable background, we can use spatial and channel attention to suppress the false positive detection in the background region.

CBAM [23] is a widely used hybrid attention mechanisms combining both spatial and channel dimensions. Inspired by CBAM [23], the proposed intra-frame channel and spatial attention mechanisms are shown in Fig. 2(b), where the attention information on channels and special allocation is weighted sequentially on the 6-layer feature map generated in backbone. For channel attention, both average-pooling and max-pooling are used to compress the information on the spatial dimension, and then the pooled features are fed into a multilayer perceptron network with shared weights. For spatial attention, the feature map weighted by channel attention performs average-pooling and max-pooling on each channel and concatenates them to generate valid feature descriptors. Then a 7×7 convolutional layer is applied to obtain the weighted feature map of spatial dimension. To sum up, the proposed intra-frame attention mechanism can be described as:

$$F_c = MLP\left(AvgPool\left(X_l\right) + MLP\left(MaxPool\left(X_l\right)\right)\right. \tag{4}$$

$$F_S = f^{7\times7}\left[AvgPool\left(F_c * X_l\right); MaxPool\left(F_c * X_l\right)\right] \tag{5}$$

where F_c and F_s denote attention in the channel and spatial direction and $f^{7\times7}$ denotes the convolution operation with the filter size of 7×7.

3.4 Feature Fusion Module

In order to aggregate multi-frame temporal information and single-frame channel and spatial attention while applying pre-trained weights from single-frame network training, we propose an additional feature fusion module.

As shown in Fig. 1, in the training phase, consecutive video sequential images are fed into the network, and each image is convolved through a 6-layer network to obtain 6 feature maps with different scales. In the intra-frame attention module, channel and spatial attention are applied to each feature map of the target frame. In the inter-frame attention module, the feature maps of the last four layers of target frame and its preceding and following frames are concatenated and then fed into the temporal adaptive convolution layer to obtain the temporal attention-weighted feature map of the target frame.

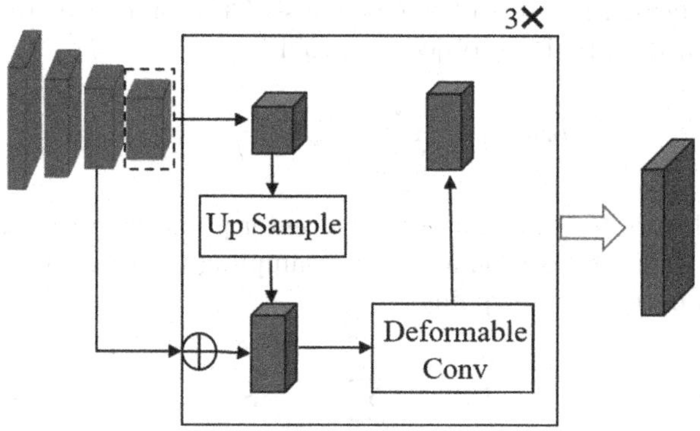

Fig. 3. Feature maps aggregation architecture after temporal adaptive convolution and intra-frame attention. 3× represents performing the fusion step for 3 times.

The four feature maps obtained after temporal weighting are shown as inputs of the feature fusion module in Fig. 3. In order to combine the semantic information of the higher-level feature map and the spatial information of the lower-level feature map, the proposed feature fusion module aggregates the four feature maps. First, the feature map of the highest dimension is selected, upsampled to the dimension of the second layer feature map, and then, the upsampled feature map and the second layer feature map are pixel-wise summed and sent to the deformable convolution for better adaption of different object shapes, sizes and other geometric deformations according to the feature maps. The feature map obtained completing the above operation is used as the updated feature map of the highest dimension, and then the same operation is performed. Since the input has a total of four layers of feature maps, the above-mentioned upsampling and convolution process is performed 3 times to get the final output.

Through the above process we obtain the channel and spatial attention feature map on a single frame and the temporal attention feature map among multiple frames, and then, we add them up to get the merged feature map. Specifically, we perform the computation of the fusion heatmap on the merged feature map, while size and offset of the bounding boxes are generated from the single-frame feature map, since the feature overlay of multi-frame has great influence on the width and height values of the bounding box.

The total loss function consists of a loss of fusion heatmap trained with focal loss, and two losses of offset and size regression trained with L1 loss. The loss function for each head block is as follows:

$$L_{FHM} = -\frac{1}{N}\sum_{xy}\begin{cases}(1-\hat{Y}_{xy})^{\alpha}\log(\hat{Y}_{xy}) & if\ Y_{xy}=1\\ (1-Y_{xy})^{\beta}(\hat{Y}_{xy})^{\alpha}\log(1-\hat{Y}_{xy}) & othrewise\end{cases} \tag{6}$$

L_{FHM} is the loss of fusion heatmap, where $\hat{Y}_{xy}$ represents the predict heatmap value for each pixel, and $\hat{Y}_{xy} = 1$ denotes the pixel is the center point of an object, α and β are hyper-parameters of the focal loss. Offset and size are calculated by the loss only on the predicted object centroid.

$$L_{\text{offset}} = \frac{1}{N}\sum_{p}\left|\hat{O}_{\tilde{p}} - \left(\frac{p}{R} - \tilde{p}\right)\right| \tag{7}$$

L_{offset} is the loss of heatmap offset to the center point, where $\hat{O}_{\tilde{p}}$ represents the predict offset, $\frac{p}{R}$ is the position after downsampling the original image, and $\tilde{p}$ is the true coordinate of center point.

$$L_{size} = \sum_{k=1}^{N}\left|\hat{S}_k - S_k\right| \tag{8}$$

L_{size} is the loss of the size of bounding box, where $\hat{S}_k$ represents the predicted size, while S_k is the size of ground truth bounding box. The overall training objective is:

$$L_{inter,intra} = L_{FHM} + \lambda_{\text{size}}\, L_{\text{size}} + \lambda_{off} L_{offset} \tag{9}$$

The hyper-parameters are set as $\lambda_{\text{size}} = 0.1$ and $\lambda_{off} = 1$ in experiments , which represent the weights of the loss for regression size and offset of the bounding box, respectively.

4 Experiments

4.1 Experimental Details

Datasets. As the proposed method utilizes the omni-attention including temporal information to improve surveillance object detection, the UA-DETRAC [11] dataset, a widely used real-world video object detection benchmark, is adopted to evaluate the performance. The dataset has over 140,000 frames and 1.21 million labeled object bounding boxes, with vehicle objects covering cars, buses, trucks, etc. We also perform experimental validation on the UAVDT [12] dataset.

For training, we sample one frame out of every 10 frames for the intra-frame context extraction, and the additional frames involved in inter-frame temporal information are selected from the intervening frames.

Models. In our experiments, the proposed network architecture is built based on the CenterNet [8] implementation, with the DLA-34 selected for the backbone network and all detectors optimized with Adam. The DLA-34 is selected as backbone because it achieves reasonable balance between efficiency and accuracy among the three pre-trained backbone networks including Hourglass, DLA-34 and ResNet-101. Table 1 shows the accuracy and efficiency of the three backbone networks in [8] for object detection on COCO validation.

Table 1. Object detection result comparison on COCO validation using different backbones, according to [8], flip and multi-scale represent the use of different data enhancements.

Backbone	AP/FPS	Flip AP/FPS	Multi-scale AP/FPS
Hourglass	40.3/14	42.2/7.8	45.1/1.4
DLA-34	37.4/52	39.2/28	41.7/4
ResNet-101	34.6/45	36.2/25	39.3/4

Training Details. The initialization of weights is set by the pre-training on COCO dataset, following the settings in [8]. The number of epochs is set to 50 and the learning rate is set to 2e−5, decreasing by a factor of 10 at the 30th and 40th epochs sequentially. The data augmentation operations including random scaling, random cropping, and flipping are used during training. Since the backbone network is first trained on UA-DETRAC dataset beforehand, the proposed method starts with decent feature representation capability and the losses can converge rapidly when training by the proposed method and the overall training time can be reduced.

4.2 Main Results

We compare the PR curve of our method on UA-DETRAC dataset under different settings in comparison with other state-of-the-art object detection methods, and show them in Figs. 4 and 5, respectively. The AP scores are calculated based on the literature [33], calculating the average precisions at the fixed 11 recall values from 0 to 1: [0, 0.1, 0.2, ..., 0.9, 1.0].

The different PR plots represent different settings in the dataset, which are shown in Fig. 4 and Fig. 5. The proposed method achieves a 3.33% improvement over the baseline in the overall setting. It shows that our method has provided significant average enhancement compared to the baseline in different difficulty and different lighting conditions, since the omni-attention strategy-based method

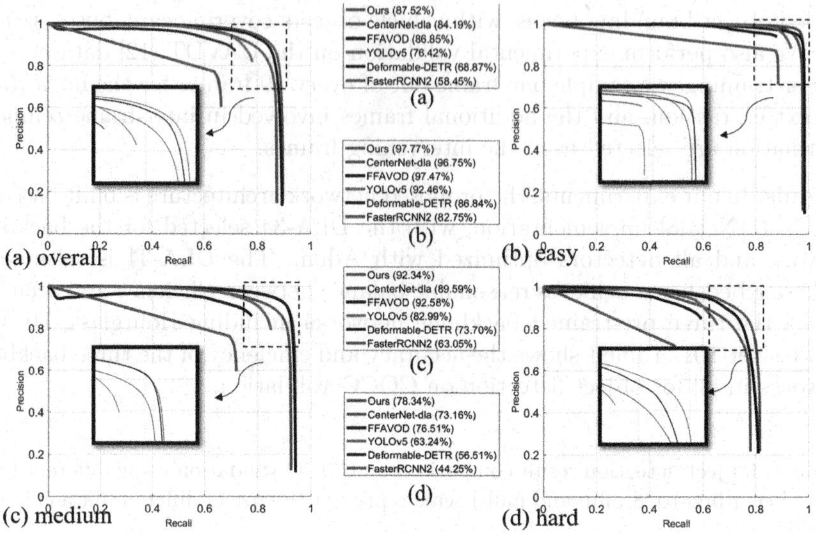

Fig. 4. PR curve comparison on UA-DETRAC dataset for object detection with the full test set and the breakdown by difficulty level: easy, medium, hard. Different colors represent different methods.

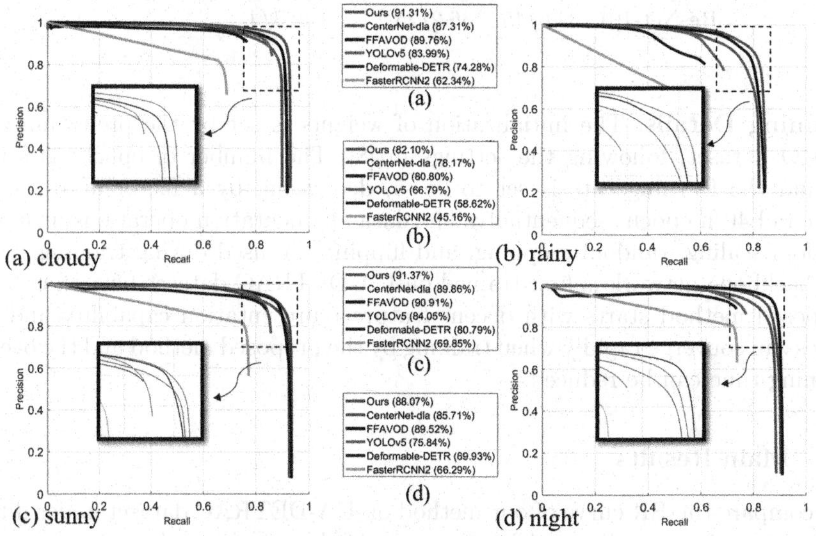

Fig. 5. PR curve comparison on UA-DETRAC dataset for object detection according to different light conditions: cloudy, rainy, sunny, night. Different colors represent different methods.

can accurately capture temporal and local contextual information to improve the effectiveness of vehicle detection per frame. The PR curve of the proposed method shows clear improvement in all settings compared with all methods.

We compare the proposed method with the state-of-the-art object detection methods and show the results in Table 2. The proposed method achieves quite competitive performance on all settings. It is worth noting that the proposed method outperforms 3D-DETnet, a method using temporal 3D convolution, with a large gap up to 34.22% in average precision. The proposed method surpasses all methods on all settings except for Medium and Night, and achieves the second-best performance on these two settings compared with FFAVOD with a slight decrease of 0.24% and 1.45% in precision. Considering the FFAVOD is a multi-frame feature fusion method using an hourglass backbone with deeper network layers, the proposed method with DLA-34 backbone verifies the effectiveness of itself. As the inference speed shown in Table 2, the proposed method achieves the optimal tradeoff between accuracy and efficiency.

The results on the UAVDT dataset are reported in Table 3. Our method achieves 2.87% gain compared with the sophisticated FFAVOD-SpotNet.

Table 2. Comparison with the state-of-the-art methods on UA-DETRAC dataset under different settings. The best result is shown in bold.

Method	Overall	Easy	Medium	Hard	Cloudy	Night	Rainy	Sunny	FPS	Environment
CenterNet(dla)	84.19	96.75	89.59	73.16	87.31	85.71	78.17	89.86	24	GPU@A30
FFAVOD(hourglass)	86.85	97.47	**92.58**	76.51	89.76	**89.52**	80.8	90.91	6	GPU@A30
FG-BR Net	79.96	93.49	83.6	70.78	87.36	78.42	70.5	89.8	10	GPU@M40
3D-DETnet	53.3	66.66	59.26	43.22	63.3	52.9	44.27	71.26	26	–
Illuminating	80.76	94.56	85.9	69.72	87.19	80.68	71.06	89.74	14	GPU@TitanX
FasterRCNN2	58.45	82.75	63.05	44.25	62.34	66.29	45.16	69.85	11.1	GPU@TitanX
Deformable DETR	68.87	86.84	73.7	56.51	74.28	69.93	58.62	80.79	2.5	GPU@A30
YOLOv5	76.42	92.46	82.99	63.24	83.99	75.84	66.79	84.05	–	–
Ours	**87.52**	**97.77**	92.34	**78.34**	**91.31**	88.07	**82.1**	**91.37**	11	GPU@A30

Table 3. Comparison of the mAP of our method on the UAVDT dataset with other state-of-the-art methods.

Method	Overall
FFAVOD	52.07%
FFAVOD-SpotNet	53.76%
CenterNet	51.18%
Ours	56.63%

The proposed method utilizes the temporal information over multiple frames in surveillance video for better detection of the objects in motion in the mean while combines the intra-frame channel and spatial attention information for accurate feature representation. Figure 6 shows the detection results of the proposed method compared with the baseline method. As we can see, the proposed method shows superior performance applied to surveillance video object detection with the enhancement in missed detection and false detection, especially on vehicle objects of small size or far distance and in environment with low light conditions.

Fig. 6. Comparison of detection performance of baseline and our method, yellow arrows indicate the missed detection, red arrows indicate the false detection. (Color figure online)

4.3 Ablation Study

Module Contribution. The proposed detector adaptively utilizes omni-attention within and between frames by the designed components of both inter-frame attention module, intra-frame attention module, and we perform the ablation studies to evaluate these components. We add the inter-frame temporal attention as well as intra-frame channel and spatial attention sequentially on top of the baseline, and show the effect of each component in Table 4. We find that the method with inter-frame temporal attention or the method with intra-frame attention both detects better than the baseline method, and the proposed method with both two modules performs the best. According to the proposed omni-attention mechanism, the intra-frame channel and spatial attention suppresses the false positive detection in the background region and improves the accuracy of the detection of the target objects within the single frame, while the temporal context features compensate for the unclear object caused by insufficient illumination.

Temporal Fusion Configurations. We also investigate the effect of temporal fusion configurations on the detection results with different number of frames and the frame sampling schemes. We evaluate the setting the number of fused frames

Table 4. Ablation study of the proposed method on UA-DETRAC dataset.

Baseline	✓	✓	✓	✓
+ Inter-frame Attention		✓		✓
+ Intra-frame Attention			✓	✓
AP (%)	84.19	87.05	86.58	87.52

Table 5. Temporal fusion configuration evaluation on the intra-frame module.

Frame sampling scheme	Numbers of frames			
	N = 3		N = 5	
Consecutive	✓		✓	
Interval		✓		✓
AP (%)	87.31	87.29	86.99	87.52

$n = 3$ or 5, and the fused frames are selected as consecutive frames or interval frames. The detection results for different number of frames and different interval frames are shown in Table 5. The best results are obtained when $n = 5$ and the fused frames are set as discontinuous, that is, the $l-1, l-3, l+1, l+3$ frames of the current frame l are selected as the temporal context. The table shows that when the number of fused frames is set as 3, there is little difference between consecutive or interval frames, it might because the information on the adjacent frames in surveillance video is similar and the weighting is valid for the current frame. And when the number of fused frames is set as 5, the effect of temporal weighting on consecutive frames becomes worse, which might be caused by the redundant information in multiple consecutive frames. While the performance achieves the best when the sampling scheme is changed into the interval frames, because the model integrates more temporally global information and has greater improvement in the object detection of the current frame.

5 Conclusion

In this work, we have proposed an adaptive omni-attention model for surveillance object detection based on anchor-free object detector. Using the proposed method, we train and evaluate our network on the dataset in the field of traffic surveillance. Our experiments demonstrate that our method compares favorably against the widely-used multi-frame and single frame methods. Our method makes efficient in the 3D temporal dimension, which has positive significance for subsequent research of video object detection. To improve the detection speed of our proposed model is another issue that warrants further study.

Acknowledgements. This work was supported by the National Natural Science Foundation of China (62172227) and National Key R&D Program of China (2021YFF0602101).

References

1. Girshick, R., Donahue, J., Darrell, T., Malik, J.: Rich feature hierarchies for accurate object detection and semantic segmentation. In: Proceedings of the IEEE Conference on Computer Vision and Pattern Recognition, pp. 580–587 (2014)
2. He, K., Zhang, X., Ren, S., Sun, J.: Spatial pyramid pooling in deep convolutional networks for visual recognition. IEEE Trans. Pattern Anal. Mach. Intell. **37**(9), 1904–1916 (2015)
3. Ren, S., He, K., Girshick, R., Sun, J.: Faster R-CNN: towards real-time object detection with region proposal networks. In: Advances in Neural Information Processing Systems 28 (2015)
4. Redmon, J., Divvala, S., Girshick, R., Farhadi, A.: You only look once: unified, real-time object detection. In: Proceedings of the IEEE Conference on Computer Vision and Pattern Recognition, pp. 779–788 (2016)
5. Liu, W., et al.: SSD: single shot multibox detector. In: Leibe, B., Matas, J., Sebe, N., Welling, M. (eds.) ECCV 2016. LNCS, vol. 9905, pp. 21–37. Springer, Cham (2016). https://doi.org/10.1007/978-3-319-46448-0_2
6. Lin, T.Y., Goyal, P., Girshick, R., He, K., Dollár, P.: Focal loss for dense object detection. In: Proceedings of the IEEE International Conference on Computer Vision, pp. 2980–2988 (2017)
7. Law, H., Deng, J.: CornerNet: detecting objects as paired keypoints. In: Ferrari, V., Hebert, M., Sminchisescu, C., Weiss, Y. (eds.) Computer Vision – ECCV 2018. LNCS, vol. 11218, pp. 765–781. Springer, Cham (2018). https://doi.org/10.1007/978-3-030-01264-9_45
8. Zhou, X., Wang, D., Krähenbühl, P.: Objects as points. arXiv preprint arXiv:1904.07850 (2019)
9. Tian, Z., Shen, C., Chen, H., He, T.: FCOS: fully convolutional one-stage object detection. In: Proceedings of the IEEE/CVF International Conference on Computer Vision, pp. 9627–9636 (2019)
10. Yang, Z., Liu, S., Hu, H., Wang, L., Lin, S.: RepPoints: point set representation for object detection. In: Proceedings of the IEEE/CVF International Conference on Computer Vision, pp. 9657–9666 (2019)
11. Wen, L., et al.: UA-DETRAC: a new benchmark and protocol for multi-object detection and tracking. Comput. Vis. Image Underst. **193**, 102907 (2020)
12. Du, D., et al.: The unmanned aerial vehicle benchmark: object detection and tracking. In: Ferrari, V., Hebert, M., Sminchisescu, C., Weiss, Y. (eds.) ECCV 2018. LNCS, vol. 11214, pp. 375–391. Springer, Cham (2018). https://doi.org/10.1007/978-3-030-01249-6_23
13. Loganathan, G.B., Fatah, T.H., Yasin, E.T., Hamadamen, N.I.: To develop multi-object detection and recognition using improved GP-FRCNN method. In: 2022 8th International Conference on Smart Structures and Systems (ICSSS), pp. 1–7. IEEE (2022)
14. Wang, T., He, X., Cai, Y., Xiao, G.: Learning a layout transfer network for context aware object detection. IEEE Trans. Intell. Transp. Syst. **21**(10), 4209–4224 (2019)
15. Fu, Z., Chen, Y., Yong, H., Jiang, R., Zhang, L., Hua, X.S.: Foreground gating and background refining network for surveillance object detection. IEEE Trans. Image Process. **28**(12), 6077–6090 (2019)
16. Wang, X., Hu, X., Chen, C., Fan, Z., Peng, S.: Illuminating vehicles with motion priors for surveillance vehicle detection. In: 2020 IEEE International Conference on Image Processing (ICIP), pp. 2021–2025. IEEE (2020)

17. Zhu, X., Wang, Y., Dai, J., Yuan, L., Wei, Y.: Flow-guided feature aggregation for video object detection. In: Proceedings of the IEEE International Conference on Computer Vision, pp. 408–417 (2017)
18. Bochkovskiy, A., Wang, C.Y., Liao, H.Y.M.: YOLOv4: optimal speed and accuracy of object detection. arXiv preprint arXiv:2004.10934 (2020)
19. Hu, J., Shen, L., Sun, G.: Squeeze-and-excitation networks. In: Proceedings of the IEEE Conference on Computer Vision and Pattern Recognition, pp. 7132–7141 (2018)
20. Zhu, X., Cheng, D., Zhang, Z., Lin, S., Dai, J.: An empirical study of spatial attention mechanisms in deep networks. In: Proceedings of the IEEE/CVF International Conference on Computer Vision, pp. 6688–6697 (2019)
21. Tan, M., Pang, R., Le, Q.V.: EfficientDet: scalable and efficient object detection. In: Proceedings of the IEEE/CVF Conference on Computer Vision and Pattern Recognition, pp. 10781–10790 (2020)
22. Zhu, X., Lyu, S., Wang, X., Zhao, Q.: TPH-YOLOv5: improved YOLOv5 based on transformer prediction head for object detection on drone-captured scenarios. In: Proceedings of the IEEE/CVF International Conference on Computer Vision, pp. 2778–2788 (2021)
23. Woo, S., Park, J., Lee, J.-Y., Kweon, I.S.: CBAM: convolutional block attention module. In: Ferrari, V., Hebert, M., Sminchisescu, C., Weiss, Y. (eds.) ECCV 2018. LNCS, vol. 11211, pp. 3–19. Springer, Cham (2018). https://doi.org/10.1007/978-3-030-01234-2_1
24. Zhu, X., Dai, J., Yuan, L., Wei, Y.: Towards high performance video object detection. In: Proceedings of the IEEE Conference on Computer Vision and Pattern Recognition, pp. 7210–7218 (2018)
25. Chen, Y., Cao, Y., Hu, H., Wang, L.: Memory enhanced global-local aggregation for video object detection. In: Proceedings of the IEEE/CVF Conference on Computer Vision and Pattern Recognition, pp. 10337–10346 (2020)
26. Perreault, H., Bilodeau, G.A., Saunier, N., Héritier, M.: FFAVOD: feature fusion architecture for video object detection. Pattern Recogn. Lett. **151**, 294–301 (2021)
27. Li, S., Chen, F.: 3D-DETNet: a single stage video-based vehicle detector. In: 3rd International Workshop on Pattern Recognition, vol. 10828, pp. 60–66. SPIE (2018)
28. Qiu, Z., Yao, T., Mei, T.: Learning spatio-temporal representation with pseudo-3d residual networks. In: Proceedings of the IEEE International Conference on Computer Vision, pp. 5533–5541 (2017)
29. Tran, D., Wang, H., Torresani, L., Ray, J., LeCun, Y., Paluri, M.: A closer look at spatiotemporal convolutions for action recognition. In: Proceedings of the IEEE Conference on Computer Vision and Pattern Recognition, pp. 6450–6459 (2018)
30. Huang, Z., et al.: TAda! temporally-adaptive convolutions for video understanding. arXiv preprint arXiv:2110.06178 (2021)
31. Cao, Z., Huang, Z., Pan, L., Zhang, S., Liu, Z., Fu, C.: TCTrack: temporal contexts for aerial tracking. In: Proceedings of the IEEE/CVF Conference on Computer Vision and Pattern Recognition, pp. 14798–14808 (2022)
32. Li, X., Wang, Y., Zhou, Z., Qiao, Y.: SmallBigNe: integrating core and contextual views for video classification. In: Proceedings of the IEEE/CVF Conference on Computer Vision and Pattern Recognition, pp. 1092–1101 (2020)
33. Everingham, M., Eslami, S., Van Gool, L., Williams, C.K., Winn, J., Zisserman, A.: The pascal visual object classes challenge: a retrospective. Int. J. Comput. Vision **111**(1), 98–136 (2015)

Lightweight Image Matting via Efficient Non-local Guidance

Zhaoxiang Kang, Zonglin Li(✉), Qinglin Liu, Yuhe Zhu, Hongfei Zhou, and Shengping Zhang

School of Computer Science and Technology, Harbin Institute of Technology, Weihai 264209, China
zonglin.li@hit.edu.cn

Abstract. Natural image matting aims to estimate the opacity of foreground objects. Most existing approaches involve prohibitive parameters, daunting computational complexity, and redundant dependency. In this paper, we propose a lightweight matting method termed LiteMatting, which learns the local smoothness of color space and affinities between neighboring pixels to estimate the alpha mattes. Specifically, a modified mobile block is adopted to construct an encoder-decoder framework, which reduces parameters while retaining sufficient spatial and channel information. In addition, a Long-Short Range Pyramid Pooling Module (LSRPPM) is introduced to extend the reception field by capturing long-range dependency between regions distributed discretely. Finally, an Efficient Non-Local Block (ENB) is presented for guiding high-level semantics propagation from low-level detail features to refine the alpha mattes. Extensive experiments demonstrate that our method achieves a favorable trade-off between accuracy and efficiency. Compared with most state-of-the-art approaches, our method attains an immense descent in parameters and FLOPs with 30% and 13%, respectively, while achieving an improvement of over 15% in SAD metrics. Code and model are available at https://github.com/kzx2018/LiteMatting.

Keywords: Image matting · Lightweight · Efficient non-local

1 Introduction

Natural image matting aims to estimate the opacity mask and has many applications, such as photo editing, compositing, and film production [1–5]. Mathematically, the observed image I is defined as a convex combination of the foreground image F and the background image B at each pixel i as

$$I_i = \alpha_i F_i + (1 - \alpha_i) B_i, \alpha_i \in [0, 1] \tag{1}$$

where α_i denotes the opacity of the foreground object at pixel i. This is a seriously ill-posed problem since I is known but F, B, and α are unknown. To address this problem, most existing methods take a trimap as additional input. However, there are still potential challenges in current approaches.

L. Wang et al. (Eds.): ACCV 2022, LNCS 13842, pp. 238–255, 2023.
https://doi.org/10.1007/978-3-031-26284-5_15

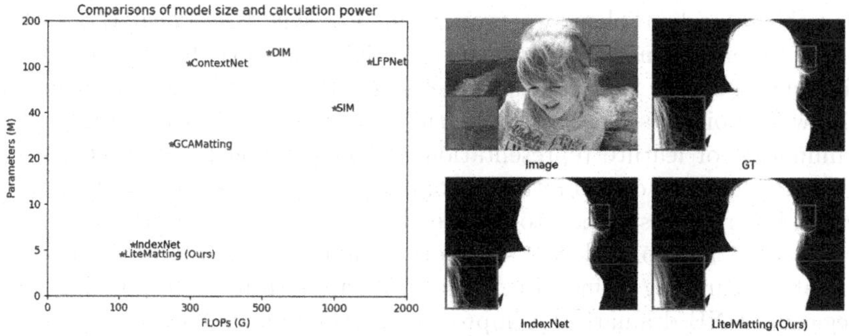

Fig. 1. Comparisons with SOTA methods. Fewer parameters and FLOPs mean that the method has better efficiency. IndexNet is very close to LiteMatting (Ours) in terms of efficiency, but our proposed method performs superiorly by viewing.

On the one hand, the most common way to improve matting performance is by scaling up their network depth and width, which is proved through subsequent work. For instance, Deep Image Matting [6] consists of an encoder-decoder stage and a refining stage. Context-Aware Matting [7] employs two encoders to predict both foreground and alpha mattes. GCA Matting [8] estimates the alpha mattes with a guided contextual attention mechanism. Networks mentioned above need more layers to increase the reception field and more channels to capture more fine-grained patterns, which do not avoid prohibitive parameters and computational complexity. Simultaneously, the vast GPU memory occupation and daunting computational cost hinder their usage in real application scenarios. Therefore, it is crucial to explore a lightweight image matting method when applied to devices with limited storage space and computing power.

On the other hand, constructing lightweight models require trading accuracy for efficiency. For example, IndexNet [9] using a learnable index pooling only captures local features due to the small reception field, which generates inaccurate alpha mattes. It is a challenge that achieves an accuracy gain on matting with better efficiency because of weights simplification and expensive tuning costs.

To address the first problem, we present a lightweight network LiteMatting where the opacity messages are transmitted efficiently across different semantic levels. We modify the original mobile block [10] via the group norm and ReLU/LeakyReLU. The modified mobile block employs depthwise filters with inverted residuals [11] and linear bottlenecks to substantially reduce the memory footprint needed and improve the computational efficiency. Also, our proposed network systematically explores the local smoothness assumptions [12] and extracts affinities between neighboring pixels to model the matte gradient intrinsically. Furthermore, we present a Long-Short Range Pyramid Pooling Module (LSRPPM) that utilizes an adaptive sampling strategy to gather the informative context with multi-scale kernels as a global prior. It improves the capability of modeling the long-range dependencies to extend the reception fields [13] and contributes to eliminating redundant perception information.

For the second problem, we propose an Efficient Non-Local Block (ENB) to refine the alpha mattes. To be specific, ENB models the spatial relevance of different pixels by guiding high-level semantics propagation from low-level details where pixels share similar texture features, which can strengthen the discrimination of feature representation and help refine blurring artifacts. To decrease the memory consumption of common non-local block [14], it introduces a pyramid sampling strategy to reduce the computational overhead of matrix multiplication. It also performs superiorly against the non-local block because the resulted sampling points are more informative from corresponding feature aggregation. ENB dramatically improves efficiency and achieves a considerable accuracy gain. As shown in Fig. 1, our proposed method is more lightweight and efficient than other trimap-based matting methods. Overall, our contributions can be summarized in the following aspects:

- We present a lightweight image matting architecture based on the modified mobile block and leverage LSRPPM to extend the reception field, which expands to more widespread application scenarios.
- We propose ENB that reduces the memory consumption of non-local blocks, which achieves a remarkable accuracy gain by guiding high-level semantics propagation from low-level detail features.
- We conduct extensive experiments on the Adobe Composition-1K dataset and AlphaMatting testing sets, which demonstrates that our method achieves a satisfactory trade-off between accuracy and efficiency.

2 Related Work

Natural image matting approaches are roughly categorized into sampling-based, propagation-based, and learning-based methods.

Sampling-Based Methods. Sampling-based methods [15–21] usually sample nearby the foreground and background colors for each unknown pixel, then design metrics to use the similarity of the pixels to estimate the alpha matte. Bayesian Matting [19] uses a well-defined Bayesian framework to predict the alpha value. Robust Matting [20] considers the spatial information and selects samples along the boundaries with confidence. Global Matting [21] samples all pixels in the image to prevent missing information.

Propagation-Based Methods. Propagation-based methods [22–28] are also known as affinity-based methods, which allow propagation of the alpha values from the known foreground and background regions to unknown regions. Close-form Matting [26] establishes a linear system to find the optimal solution by smoothness assumption on the foreground and background colors. KNN Matting [27] globally collects K nearest neighbors to increase the speed meanwhile keeping the accuracy of matting. Information-flow Matting [28] combines the local and non-local affinities of colors with spatial smoothness.

Learning-Based Methods. Learning-based methods [6–9, 29–38] utilize a deep network to directly estimate the alpha matte with the given image and trimap.

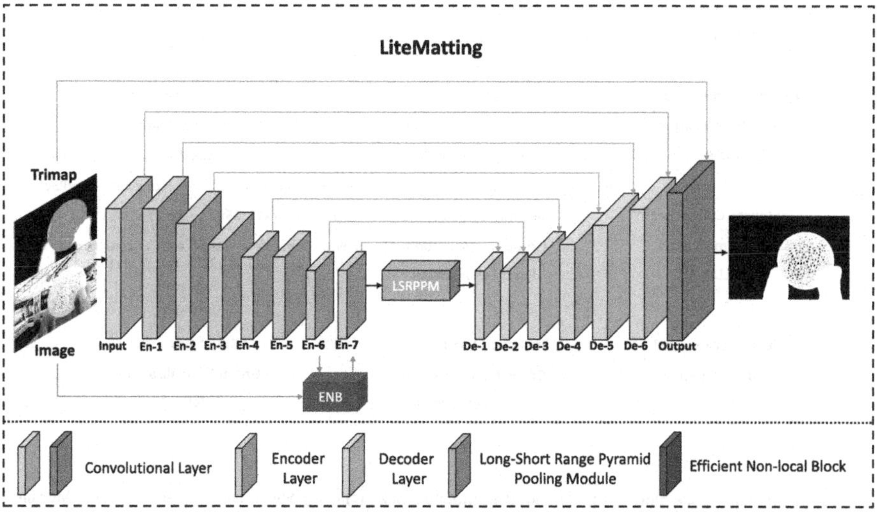

Fig. 2. The overview of the LiteMatting Network. Our proposed network consists of encoder, decoder, LSRPPM, and ENB four parts.

DIM [6] provides the first large-scale image dataset and presents the first end-to-end matting architecture with a refinement network. AlphaGan [29] introduces GAN into the image matting task. Context-Aware Matting [7] proposes a dual encoder-decoder structure to capture semantic information for foreground and alpha prediction. AdaMatting [30] uses trimap adaptation to refine the alpha matte. IndexNet [9] utilizes the pooling indices for unpooling operation. GCA Matting [8] presents a guided contextual attention mechanism to analyze image inpainting processing in matting. HDMatt [31] designs a cross-patch contextual module to improve accuracy in patch-based inference.

3 Proposed Method

In this section, we will first introduce the backbone of our network architecture and then illustrate the Long-Short Range Pyramid Pooling Module. Afterward, we will present the design of the Efficient Non-Local Block. Finally, we will describe the loss function.

3.1 Network Architecture

We construct an encoder-decoder architecture like U-net [39], which is illustrated in Fig. 2. Our proposed network is formed by stacking the modified mobile block. It utilizes the depthwise separable convolution to reduce parameters and computation, which splits convolution into two separate layers called Depthwise Conv and Pointwise Conv. The former performs light-weighted filtering by a single convolution per input channel, and the latter builds new features through

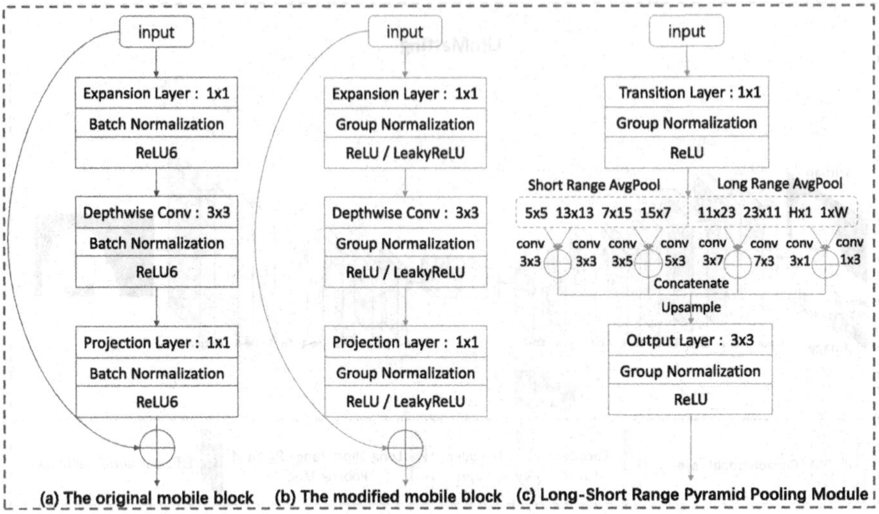

Fig. 3. Difference between the original mobile block and the modified mobile block. The detailed structure of the Long-Short Range Pyramid Pooling Module.

the linear combinations of the input channel. Specifically, the modified mobile block takes a low-dimensional compressed representation as an input, which is expanded to high-dimensional and filtered with a lightweight Depthwise Conv. These features are projected back to low-dimensional with a linear convolution at last. It reduces the parameters while retaining sufficient spatial and channel information. Overall, the network can efficiently explore the local smoothness of color space and learn affinities between neighboring pixels.

The Encoder. Firstly, the input layer is a conventional convolution layer that increases the number of input channels from 3 to 8 with a given trimap and a transformable map. The transformable map uses Gaussian blurs of the definite foreground and background masks at a prior scale to encode the given trimap [40]. Secondly, in contrast to the original mobile block [10], the modified mobile block replaces ReLU6 and the batch norm [41] with ReLU and the group norm, respectively, which is shown in Fig. 3(a & b). It benefits accelerating regression convergence and increasing accuracy. Moreover, ENB is embedded in the encoder to guide the information flow for refining the alpha mattes. Finally, the encoder has seven levels named En-1 to En-7, which help extract context features and propagate the semantic information to the decoder.

The Decoder. The decoder consists of the modified mobile block followed by up-sampling layers, which differs from the encoder because it employs the LeakyReLU instead of ReLU to avoid the dead ReLU issue. Specifically, the decoder first receives global priors from LSRPPM. Then it leverages six levels named De-1 to De-6 to upsample rich context features to the original size while fusing semantic information from each encoder. Finally, the output layer stacks three convolutional layers to estimate the alpha mattes.

Fig. 4. Schematic illustration of the Short-Range Block and the Long-Range Block.

3.2 Long-Short Range Pyramid Pooling Module

The bottleneck layer [42,43] between the encoder and the decoder simulates the receptive field of human vision to enhance the feature extraction capability of the network. We present the Long-Short Range Pyramid Pooling Module based on [43] to enlarge the receptive field of the network.

LSRPPM first takes the high-level feature from En-7 as input to feed into eight parallel pathways, each of which contains a different scale pooling layer followed by a convolution with a narrow kernel shape to obtain multiple representations. Afterward, it concatenates and upsamples them to the same size as the input. Finally, it generates a composite feature that combines multiple scales as output. Figure 3(c) depicts the detail of above steps.

LSRPPM is divided into the short-range block and the long-range block. The former extracts global information by pooling operations at four short-range scales (5×5, 13×13, 7×15, 15×7). However, there are limitations in capturing wide context scenes since the observed target may have a long-range structure (e.g., the cobweb in Fig. 4). Using short rectangle pooling windows cannot deal with this issue well, so we design the latter to capture long-range dependency by a longer pooling kernel. Inspired by [44], the long-range block expands long-range scales (11×23, 23×11, $H \times 1$, $1 \times W$) layers, where H and W are the spatial height and width, respectively. It improves the capability of capturing dependencies between regions distributed discretely and avoids contaminating information from irrelevant regions. Mathematically, given the two-dimensional tensor $\mathbf{x} \in \mathcal{R}^{H \times W}$, the output $\mathbf{y}^h \in \mathcal{R}^H$ from horizontal pooling ($H \times 1$) can be written as

$$y_i^h = \frac{1}{W} \sum_{0 \leq j < W} x_{i,j} \tag{2}$$

Similarly, the output $\mathbf{y}^v \in \mathcal{R}^W$ from vertical pooling ($1 \times W$) can be written as

$$y_j^v = \frac{1}{H} \sum_{0 \leq i < H} x_{i,j} \tag{3}$$

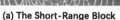

LSRPPM merges $\mathbf{y}^h$ and $\mathbf{y}^v$ together to obtain more useful global priors, which is shown in Fig. 5. Afterward, it repeats the same operation for other

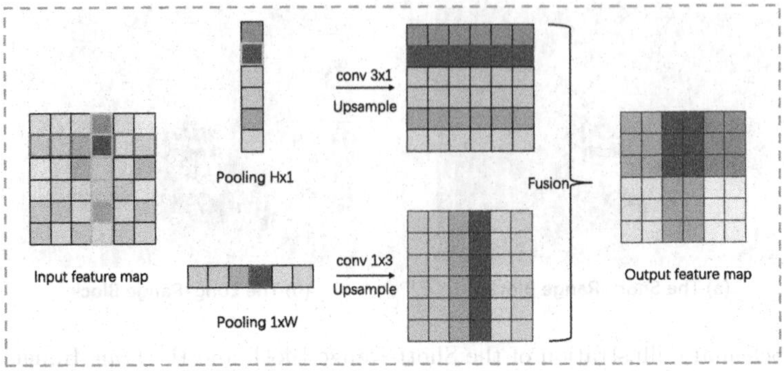

Fig. 5. The illustration of the Long-Range Block ($H \times 1$, $1 \times W$). It builds long-range dependencies between regions distributed discretely to extend the reception field.

scales as in horizontal and vertical pooling layers. Finally, it uses multi-scale feature aggregation to fuse contextual information. Overall, LSRPPM builds long-range dependencies to extend the reception field and eliminates redundant information, which is essential for improving performance.

3.3 Efficient Non-local Block

Most deep learning methods predict the alpha matte by learning the propagation of pixels in known regions to unknown regions according to their similarity of opacity. However, pixels in the unknown regions cannot be correlated with pixels in known regions because of the locality of the convolutional neural network. Non-Local [14] skillfully leverages the global dependencies to capture the relationship between pixels, which is beneficial to the matting task.

In fact, a common non-local operation is very time and memory-consuming, which is shown in Fig. 6(a). Firstly, it takes feature $X \in \mathcal{R}^{N \times H \times W \times C}$ as an input, where N, C, W, H indicate the batch, channel, width and height, respectively. Secondly, using convolutions W_θ, W_φ, W_g transforms X for obtaining the output of three embeddings θ, φ and g as

$$\theta = W_\theta(X), \varphi = W_\varphi(X), g = W_g(X) \tag{4}$$

Thirdly, the similarity matrix M is generated by the matrix multiplication and normalization, and then multiplied by g to obtain the attention layer A as

$$M = Softmax(\theta^{\mathrm{T}} \times \varphi) \tag{5}$$

$$A = M \times g^{\mathrm{T}} \tag{6}$$

Finally, it uses a weight parameter W_y to adjust the importance of the attention layer and merges the original input X, the final output Y is given by

$$Y = W_y\left(A^{\mathrm{T}}\right) + X \tag{7}$$

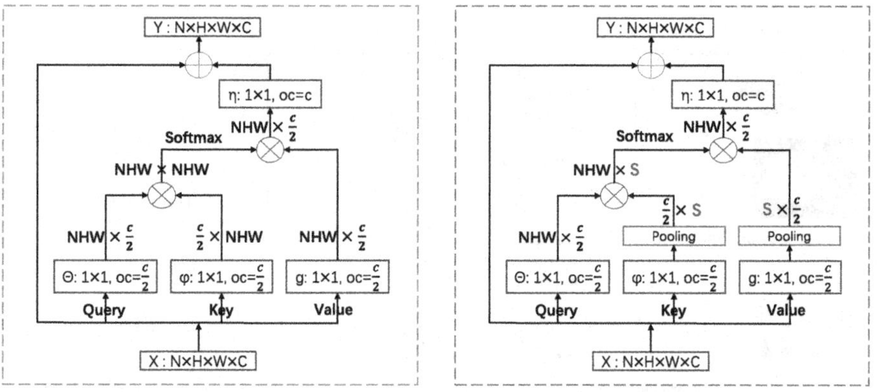

(a) Common Non-local Block **(b) Efficient Non-local Block**

Fig. 6. Difference between Common Non-Local Block and Efficient Non-Local Block.

We clearly find that the matrix multiplication of Eq. (5) and Eq. (6) dominate the heavy computation. The straightforward pipeline is

$$\underbrace{\mathcal{R}^{NHW\times\frac{C}{2}}\times\mathcal{R}^{\frac{C}{2}\times NHW}}_{Eq.\,(5)}\rightarrow\underbrace{\mathcal{R}^{NHW\times NHW}\times\mathcal{R}^{NHW\times\frac{C}{2}}}_{Eq.\,(6)}\rightarrow\mathcal{R}^{NHW\times\frac{C}{2}} \tag{8}$$

Inspired by [45], we choose the pyramid pooling operation (scales = 1, 3, 6, 8) to reduce the inefficiency of the non-local block. As shown in Fig. 6(b), it samples more important features after φ and g to filter out irrelevant information by changing NHW to number S ($S \ll NHW$), the pipeline after pooling is

$$\underbrace{\mathcal{R}^{NHW\times\frac{C}{2}}\times\mathcal{R}^{\frac{C}{2}\times S}}_{Eq.\,(5)}\rightarrow\underbrace{\mathcal{R}^{NHW\times S}\times\mathcal{R}^{S\times\frac{C}{2}}}_{Eq.\,(6)}\rightarrow\mathcal{R}^{NHW\times\frac{C}{2}} \tag{9}$$

According to the above method, ENB reduces the computational overhead of matrix multiplication to improve the efficiency of the non-local block. Pixels sharing similar texture information have similar opacity features [46]. However, blurs occur in the representation of each pixel as the network layers go deep. Therefore, ENB utilizes the low-level feature to guide high-level semantics propagation to increase the accuracy of the alpha mattes.

In our approach, ENB first receives a high-level feature X_h from the encoder En-6. Then it multiplies X_h with the values U involved in the pixels of the unknown area and known foreground object for adjusting weights to get the query feature, named as X_q. The key/value feature is extracted from the input image by the low-level feature guidance block, named as X_k and X_v.

The X_k and X_v are filtered to the sufficient feature statistics about global semantic cues by the pyramid pooling meanwhile decreasing the computational cost of non-local. ENB utilizes the similarity of the X_k and X_q by the related matrix and takes the normalizing function applied to them. Afterward, it multiplies X_v with the result from the previous step and merges the origin value X_h

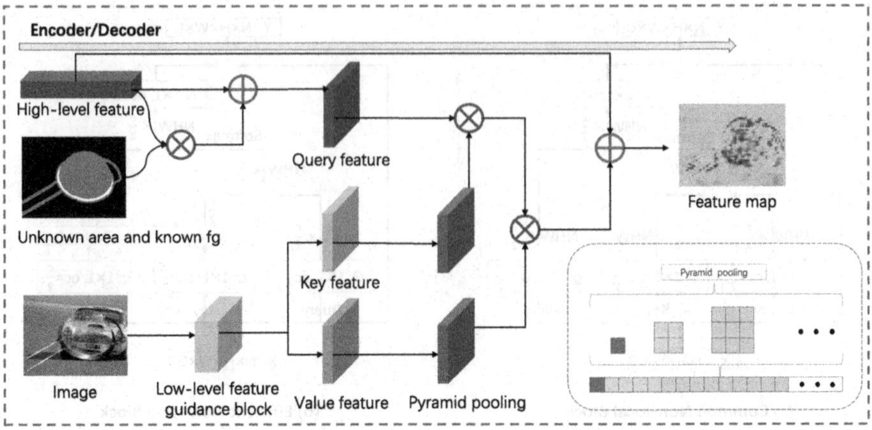

Fig. 7. The architecture of the Efficient Non-Local Block. ENB guides high-level semantics propagation from low-level features and introduces pyramid pooling to reduce the computational overhead of the common non-local block.

to obtain the output feature maps. Finally, it feeds the output into the encoder En-7 to strengthen the discrimination of feature representation. The architecture of our proposed ENB is depicted in Fig. 7. Overall, ENB successfully achieves a considerable improvement of alpha mattes with high efficiency.

3.4 Loss Function

A combination of loss functions is used in the network training, including the reconstruction $\mathcal{L}_1$ loss named $\mathcal{L}_u$, the composition loss $\mathcal{L}_c$ and the Laplacian pyramid loss $\mathcal{L}_{\text{lap}}$. Given the original image C, the estimated alpha matte α, and the ground truth α^{gt}, F^{gt}, B^{gt}. The loss $\mathcal{L}_u$ is defined as

$$\mathcal{L}_u = \frac{1}{|\mathcal{T}_u|} \sum_{i \in \mathcal{T}_u} \left\| \alpha_i - \alpha_i^{gt} \right\|_1 \tag{10}$$

where $\mathcal{T}_u$ is the set of unknown pixels in the trimap. The loss $\mathcal{L}_c$ is defined as

$$\mathcal{L}_c = \sum_i \left\| C_i - \alpha_i F_i^{gt} - (1 - \alpha_i) B_i^{gt} \right\|_1 \tag{11}$$

The Laplacian pyramid loss is calculated by the Laplacian pyramid L_{pyr}^s with multiple scales s [47], which is defined as

$$\mathcal{L}_{\text{lap}} = \sum_{s=1}^{5} 2^{s-1} \left\| L_{\text{pyr}}^s(\alpha) - L_{\text{pyr}}^s(\alpha^{gt}) \right\|_1 \tag{12}$$

Finally, our total loss function is computed as

$$\mathcal{L}_\alpha = \lambda_1 \mathcal{L}_u + \lambda_2 \mathcal{L}_c + \lambda_3 \mathcal{L}_{\text{lap}} \tag{13}$$

where λ_1, λ_2, λ_3 are proportion factor to balance the loss function weights.

4 Experiments

4.1 Datasets

Adobe Composition-1k. The Composition-1k testing set contains 1,000 composed images with a unique trimap. These images are synthesized by 50 foreground images and 1,000 background images from the PASCAL VOC dataset.

AlphaMatting. The AlphaMatting dataset is a matting dataset that consists of real-world images for the online benchmark. There are eight testing images, which of each has three different trimaps (i.e. 'small', 'large', and 'user').

4.2 Implementation Details

We train our network on the Adobe Composition-1K training dataset with end-to-end mode. We use several common data augmentation ways [48], including affine transformation, flip transformation, contrast transformation, saturation transformation, and random foreground composition. Images are cropped into patch of dimensions $1{,}024 \times 1{,}024$. In addition, their trimaps are generated by the alpha matte ground truth with random erosion and dilation of 3 to 35 pixels. We use the RAdam optimizer with $\beta_1 = 0.5$ and $\beta_2 = 0.999$. The learning rate is initialized as 5×10^{-5} and proportion factor of loss function λ_1, λ_2, λ_3 are set to 1. We train our network for 150 epochs with a batch size of 4. And our model is trained from scratch without any pretrained models.

4.3 Comparison with the SOTA Methods

There are four metrics used in the evaluation: the sum of absolute difference (SAD), the mean square error (MSE), the gradient error (Grad), and the connectivity error (Conn). Furthermore, we count the number of parameters and computational cost at $1{,}024 \times 1{,}024$ resolution as shown in Table 1.

Table 1. The quantitative results on the Adobe Composition-1k testing set. The best results are highlighted in bold.

Methods	SAD	MSE	Grad	Conn	Params	FLOPs
KNN Matting	175.4	103.0	124.1	176.4	–	–
Closed-Form	168.1	91.0	126.9	167.9	–	–
DIM	50.4	14.0	31.0	50.8	130.6M	511.0G
IndexNet	45.8	13.0	25.6	43.7	6.0M	116.6G
ContextNet	35.8	8.2	17.3	33.2	107.5M	292.5G
GCAMatting	35.3	9.1	16.9	32.5	25.3M	257.3G
LiteMatting (ours)	30.1	6.1	13.1	26.2	**4.3M**	**101.5G**
SIM	27.7	5.6	10.7	24.4	44.5M	1001.9G
LFPNet	**23.6**	**4.1**	**8.4**	**18.5**	112.2M	1539.4G

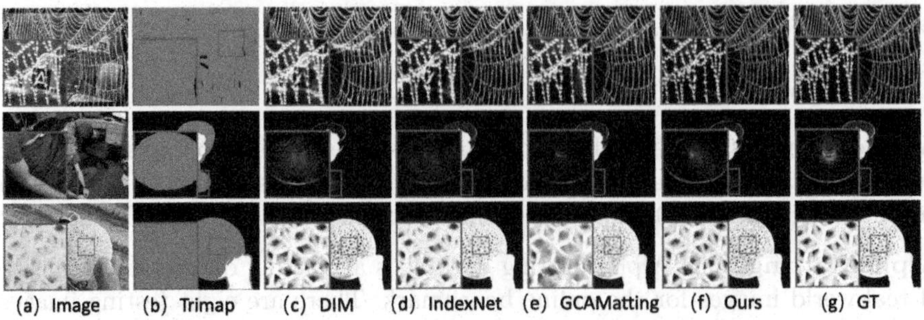

Fig. 8. The qualitative comparison results on the Adobe Composition-1k testing set.

We compare our model with traditional matting methods: KNN Matting [27], Closed-Form [26], and learning-based methods: DIM [6], IndexNet [9], ContextNet [7], GCAMatting [8], SIM [32] and LFPNet [48] on Composition-1k testing set. In addition, we also compare with AdaMatting [30], A2U Matting [49] and SampleNet [50] on the AlphaMatting benchmark. To be specific, we utilize the patch-based crop-and-stitch method for inference [31], where the images are cropped into patches and then fed into the network.

As shown in Table 1 and Table 2, our LiteMatting attains an immense descent in parameters and FLOPs with 30% and 13%, respectively, while achieving an improvement of over 15% in SAD metrics of most methods (e.g., GCAMatting). Although our method performs inferior to SIM and LFPNet in terms of SAD, we are more lightweight (at 3.8%–9.7% in Params and 6.6%–10.1% in FLOPs) than them. It shows that our method has effectiveness in practical application scenarios, especially in resource-limited environments. As shown in Fig. 8 and Fig. 9, we qualitatively compare with SOTA approaches. Our method generates more meticulous alpha mattes in background interference and shows robustness in performance where it estimates the opacity of foreground objects.

Table 2. Our average ranking results on the AlphaMatting testing set. S, L, and U denote the small trimap, large trimap, and user trimap respectively. The best results are highlighted in bold.

Methods	SAD				MSE				Grad			
	Overall	S	L	U	Overall	S	L	U	Overall	S	L	U
Ours	**13.8**	**12.5**	**11.0**	**17.8**	**14.6**	13.1	**11.6**	**19.0**	**12.2**	**10.4**	**8.8**	**17.5**
AdaMatting	15.2	13.6	14.1	**17.8**	16.0	13.1	14.9	19.9	16.0	11.5	13.8	22.6
A2U Matting	15.4	14.0	12.8	19.4	18.2	15.8	15.3	23.6	14.9	13.9	11.9	19.0
SampleNet	15.8	12.8	16.0	18.8	16.7	**12.6**	17.4	20.0	18.2	13.8	16.3	24.6
GCAMatting	17.3	18.0	15.3	18.5	18.3	18.1	17.3	19.4	17.0	17.1	15.9	18.1
DIM	19.2	20.1	18.8	18.8	22.2	20.4	21.4	24.8	27.0	24.0	23.9	33.0
IndexNet	22.5	24.4	21.5	21.5	26.5	29.0	25.1	25.3	22.1	20.6	21.1	24.6

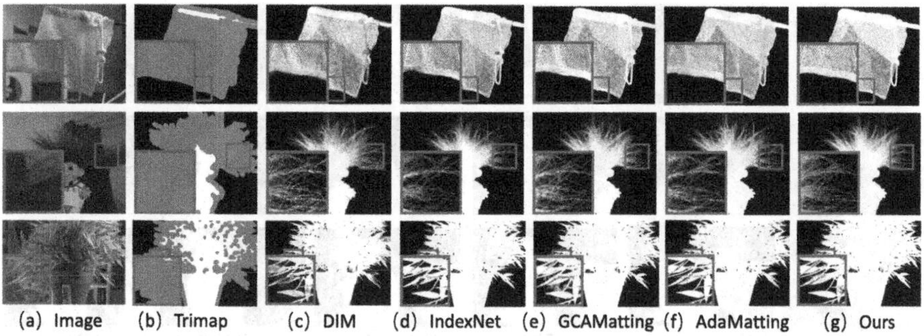

Fig. 9. The qualitative comparison results on the AlphaMatting testing set.

Finally, we pay attention to the real-world high-resolution images. In Fig. 10, we test DIM [6], IndexNet [9], GCAMatting [8] and our proposed LiteMatting. These images are too large to be fed into a single GPU, so we implement inference on the CPU to avoid insufficient memory. It has to spend a long inference time for each image where the usage scenarios for high-resolution images are limited. The results demonstrate that our method extracts finer details and outperforms other state-of-the-art matting methods with a far faster inference speed. In conclusion, we achieve a promising trade-off between accuracy and efficiency through the above experiments.

4.4 Ablation Study

To validate the efficacy of our proposed backbone based on the modified mobile block, we first compare it to other light-weighted backbones, including ShuffleNet [51], EfficientNet [52] and ConvNetXt [53]. These backbones are known for being lightweight and efficient. As shown in Table 3, we notice that our backbone achieves better performance with fewer parameters and FLOPs than others while training for the same epochs.

Afterward, we confirm that the group norm is more suitable for matting than the batch norm. The matting task demands pixel-level relationships, but

Table 3. Ablation study of existing lightweight backbones, normalization method, and loss function on the Adobe Composition-1k testing set.

Backbone	Normalization	Loss Function	SAD	MSE	Params	FLOPs
ShuffleNet	GroupNorm	Alpha Loss	37.3	8.3	10.6M	45.2G
EfficientNet			38.7	9.1	4.7M	55.0G
ConvNetXt			39.9	9.5	5.2M	52.0G
Ours	GroupNorm	Alpha Loss	36.8	7.9	3.6M	52.8G
	BatchNorm	Alpha Loss	37.9	8.6	3.6M	51.7G
	GroupNorm	F, B, A Loss	40.6	10.3	3.6M	52.8G

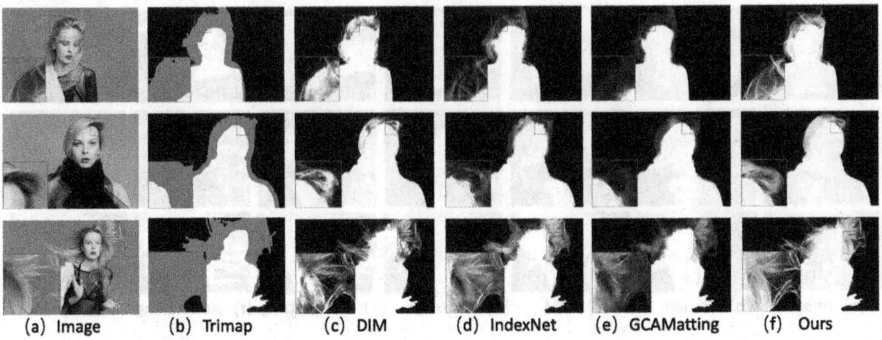

Fig. 10. The qualitative comparison results on the real-world high resolution images. Image sizes from top to bottom: 5863 × 3909, 5636 × 3757, 6240 × 4160.

the high-resolution training samples lead to a mini-batch size of 1–16 in GPU. The group norm contributes to increasing the accuracy of alpha mattes since it is an expert in dealing with the mini-batch size questions. We also attempt to use F, B, A loss [54] which means training with the loss of foreground and background in addition to the alpha, but the result gets worse than training with the Alpha Loss. We think it may be easy to learn redundant information for a lightweight model because of the irrelevant feature interference. Therefore, we choose the modified mobile block and the group norm to build up our backbone, training our network with the alpha loss.

Moreover, we further verify the reasonableness of our efficient non-local block. We compare our model with the common non-local block and then explore different sampling methods (max pooling, average pooling, and pyramid pooling) on ENB. These non-local blocks help gain a significant performance improvement through our ablation experiments. It is effective for image matting to build a long-range contextual dependency by modeling the relevance between pixels. As shown in Table 4, compared to the common non-local block, ENB based on the pyramid pooling not only regresses more accurate alpha values but also performs more efficiently by reducing the FLOPs. The reason is that the sampling points are more informative by receiving the provided features from the pooling kernel. Then we conduct several experiments to study the effect of sampling strategies

Table 4. Ablation study of Efficient Non-Local Block in terms of different sampling methods. 's' represents the scale of a pooling layer.

Configuration	Sampling Method	SAD	MSE	FLOPs
Common non-local block	–	31.3	6.6	144.2G
Efficient non-local block (ours)	Max pooling (s = 15)	32.0	6.9	101.4G
	Average pooling (s = 15)	31.6	6.8	101.4G
	Pyramid pooling (s = 1, 3, 6, 8)	30.1	6.1	101.5G

Table 5. Ablation study of LSRPPM and ENB on the Adobe Composition-1k testing set. The best results are highlighted in bold.

Backbone	LSRPPM	ENB	SAD	MSE	Grad	Conn	Params	FLOPs
Ours			36.8	7.9	19.4	34.0	**3.6M**	**52.8G**
	✓		34.2	7.2	16.1	31.0	3.8M	53.0G
	✓	✓	**30.1**	**6.1**	**13.1**	**26.2**	4.3M	101.5G

by altering the scales of pooling layers. ENB based on the pyramid pooling (scales = 1, 3, 6, 8) performs better than others based on the max pooling and the average pooling. The second ablation experiments show that ENB improves accuracy and decreases the computational cost of the model.

In the final ablation study, we reveal the effectiveness of each component in LiteMatting on the Adobe Composition-1k testing set. As shown in Table 5, we make a remarkable gain in the accuracy with the combination of LSRPPM and ENB. They are not only efficient but also help refine the alpha mattes. LSRPPM captures the long-range context features and utilizes multi-scale information to enlarge the reception field. Furthermore, ENB guides rich high-level semantic information from low-level details by modeling the relevance between different pixels for regressing the high-precision alpha mattes. And we think ENB that improves performance is positive for light-weighted matting despite the increase in computation. These two architectures are verified to be beneficial to the prediction of the alpha mattes. In conclusion, the ablation study suggests that our method is effective for the image matting task.

5 Conclusion

In this paper, we propose a lightweight method termed LiteMatting for image matting. Our method leverages the modified mobile block to extract sufficient spatial and channel representations with fewer parameters. LSRPPM captures the long-range dependencies to extend the reception field. ENB guides high-level semantics propagation from low-level detail features. With the proposed LSRPPM, ENB can efficiently estimate more accurate alpha mattes with less computational cost. Extensive experiments on the Adobe Composition-1k and AlphaMatting testing set demonstrate that our LiteMatting is more lightweight and performs superiorly against most SOTA approaches, which attains an immense descent in parameters and FLOPs with 30% and 13%, respectively, while achieving an improvement of over 15% in SAD metrics. Overall, we successfully achieve a trade-off between accuracy and efficiency.

Acknowledgement. This work was supported by the National Natural Science Foundation of China (Nos. 61872112 and 6207071409) and the Taishan Scholars Program of Shandong Province (No. tsqn201812106).

References

1. Sengupta, S., Jayaram, V., Curless, B., Seitz, S.M., Kemelmacher-Shlizerman, I.: Background matting: the world is your green screen. In: Proceedings of the IEEE/CVF Conference on Computer Vision and Pattern Recognition, pp. 2291–2300 (2020)
2. Lin, S., Ryabtsev, A., Sengupta, S., Curless, B.L., Seitz, S.M., Kemelmacher-Shlizerman, I.: Real-time high-resolution background matting. In: Proceedings of the IEEE/CVF Conference on Computer Vision and Pattern Recognition, pp. 8762–8771 (2021)
3. Lin, S., Yang, L., Saleemi, I., Sengupta, S.: Robust high-resolution video matting with temporal guidance. In: Proceedings of the IEEE/CVF Winter Conference on Applications of Computer Vision, pp. 238–247 (2022)
4. Chen, Y., Guan, J., Cham, W.K.: Robust multi-focus image fusion using edge model and multi-matting. IEEE Trans. Image Process. **27**, 1526–1541 (2017)
5. Ke, Z., et al.: Is a green screen really necessary for real-time portrait matting? arXiv preprint arXiv:2011.11961 (2020)
6. Xu, N., Price, B., Cohen, S., Huang, T.: Deep image matting. In: Proceedings of the IEEE Conference on Computer Vision and Pattern Recognition, pp. 2970–2979 (2017)
7. Hou, Q., Liu, F.: Context-aware image matting for simultaneous foreground and alpha estimation. In: Proceedings of the IEEE/CVF International Conference on Computer Vision, pp. 4130–4139 (2019)
8. Li, Y., Lu, H.: Natural image matting via guided contextual attention. Proc. AAAI Conf. Artif. Intell. **34**, 11450–11457 (2020)
9. Lu, H., Dai, Y., Shen, C., Xu, S.: Indices matter: learning to index for deep image matting. In: Proceedings of the IEEE/CVF International Conference on Computer Vision, pp. 3266–3275 (2019)
10. Sandler, M., Howard, A., Zhu, M., Zhmoginov, A., Chen, L.C.: MobileNetV2: inverted residuals and linear bottlenecks. In: Proceedings of the IEEE Conference on Computer Vision and Pattern Recognition, pp. 4510–4520 (2018)
11. He, K., Zhang, X., Ren, S., Sun, J.: Deep residual learning for image recognition. In: Proceedings of the IEEE Conference on Computer Vision and Pattern Recognition, pp. 770–778 (2016)
12. Wang, R., Xie, J., Han, J., Qi, D.: Improving deep image matting via local smoothness assumption. arXiv preprint arXiv:2112.13809 (2021)
13. Liu, Y., Yu, J., Han, Y.: Understanding the effective receptive field in semantic image segmentation. Multimedia Tools Appl. **77**(17), 22159–22171 (2018). https://doi.org/10.1007/s11042-018-5704-3
14. Wang, X., Girshick, R., Gupta, A., He, K.: Non-local neural networks. In: Proceedings of the IEEE Conference on Computer Vision and Pattern Recognition, pp. 7794–7803 (2018)
15. Ruzon, M.A., Tomasi, C.: Alpha estimation in natural images. In: Proceedings of the IEEE Conference on Computer Vision and Pattern Recognition, CVPR 2000 (Cat. No. PR00662). Volume 1, pp. 18–25. IEEE (2000)
16. Gastal, E.S., Oliveira, M.M.: Shared sampling for real-time alpha matting. In: Computer Graphics Forum, vol. 29, pp. 575–584. Wiley Online Library (2010)

17. Guan, Y., Chen, W., Liang, X., Ding, Z., Peng, Q.: Easy matting-a stroke based approach for continuous image matting. In: Computer Graphics Forum, vol. 25, pp. 567–576. Wiley Online Library (2006)
18. Feng, X., Liang, X., Zhang, Z.: A cluster sampling method for image matting via sparse coding. In: Leibe, B., Matas, J., Sebe, N., Welling, M. (eds.) ECCV 2016. LNCS, vol. 9906, pp. 204–219. Springer, Cham (2016). https://doi.org/10.1007/978-3-319-46475-6_13
19. Chuang, Y.Y., Curless, B., Salesin, D.H., Szeliski, R.: A Bayesian approach to digital matting. In: Proceedings of the 2001 IEEE Computer Society Conference on Computer Vision and Pattern Recognition, CVPR 2001, vol. 2, pp. 264–271. IEEE (2001)
20. Wang, J., Cohen, M.F.: Optimized color sampling for robust matting. In: 2007 IEEE Conference on Computer Vision and Pattern Recognition, pp. 1–8. IEEE (2007)
21. He, K., Rhemann, C., Rother, C., Tang, X., Sun, J.: A global sampling method for alpha matting. In: CVPR 2011, pp. 2049–2056. IEEE (2011)
22. Sun, J., Jia, J., Tang, C.K., Shum, H.Y.: Poisson matting. In: ACM SIGGRAPH 2004 Papers, pp. 315–321 (2004)
23. Bai, X., Sapiro, G.: A geodesic framework for fast interactive image and video segmentation and matting. In: 2007 IEEE 11th International Conference on Computer Vision, pp. 1–8. IEEE (2007)
24. Levin, A., Rav-Acha, A., Lischinski, D.: Spectral matting. IEEE Trans. Pattern Anal. Mach. Intell. **30**, 1699–1712 (2008)
25. Lee, P., Wu, Y.: Nonlocal matting. In: CVPR 2011, pp. 2193–2200. IEEE (2011)
26. Levin, A., Lischinski, D., Weiss, Y.: A closed-form solution to natural image matting. IEEE Trans. Pattern Anal. Mach. Intell. **30**, 228–242 (2007)
27. Chen, Q., Li, D., Tang, C.K.: KNN matting. IEEE Trans. Pattern Anal. Mach. Intell. **35**, 2175–2188 (2013)
28. Aksoy, Y., Ozan Aydin, T., Pollefeys, M.: Designing effective inter-pixel information flow for natural image matting. In: Proceedings of the IEEE Conference on Computer Vision and Pattern Recognition, pp. 29–37 (2017)
29. Lutz, S., Amplianitis, K., Smolic, A.: AlphaGAN: generative adversarial networks for natural image matting. arXiv preprint arXiv:1807.10088 (2018)
30. Cai, S., et al.: Disentangled image matting. In: Proceedings of the IEEE/CVF International Conference on Computer Vision, pp. 8819–8828 (2019)
31. Yu, H., Xu, N., Huang, Z., Zhou, Y., Shi, H.: High-resolution deep image matting. Proc. AAAI Conf. Artif. Intell. **35**, 3217–3224 (2021)
32. Sun, Y., Tang, C.K., Tai, Y.W.: Semantic image matting. In: Proceedings of the IEEE/CVF Conference on Computer Vision and Pattern Recognition, pp. 11120–11129 (2021)
33. Liu, Y., et al.: Tripartite information mining and integration for image matting. In: Proceedings of the IEEE/CVF International Conference on Computer Vision, pp. 7555–7564 (2021)
34. Jiang, W., Yu, D., Xie, Z., Li, Y., Yuan, Z., Lu, H.: Trimap-guided feature mining and fusion network for natural image matting. arXiv preprint arXiv:2112.00510 (2021)
35. Cheng, H., Xu, S., Jiang, X., Wang, R.: Deep image matting with flexible guidance input. arXiv preprint arXiv:2110.10898 (2021)
36. Goel, A., Kumar, M., Sudheendra, P., Team, V., et al.: IamAlpha: instant and adaptive mobile network for alpha matting (2021)

37. Liu, Y., Xie, J., Qiao, Y., Tang, Y., Yang, X.: Prior-induced information alignment for image matting. IEEE Trans. Multimedia **24**, 2727–2738 (2021)
38. Dai, Y., Price, B., Zhang, H., Shen, C.: Boosting robustness of image matting with context assembling and strong data augmentation. In: Proceedings of the IEEE/CVF Conference on Computer Vision and Pattern Recognition, pp. 11707–11716 (2022)
39. Ronneberger, O., Fischer, P., Brox, T.: U-Net: convolutional networks for biomedical image segmentation. In: Navab, N., Hornegger, J., Wells, W.M., Frangi, A.F. (eds.) MICCAI 2015. LNCS, vol. 9351, pp. 234–241. Springer, Cham (2015). https://doi.org/10.1007/978-3-319-24574-4_28
40. Le, H., Mai, L., Price, B., Cohen, S., Jin, H., Liu, F.: Interactive boundary prediction for object selection. In: Ferrari, V., Hebert, M., Sminchisescu, C., Weiss, Y. (eds.) Computer Vision – ECCV 2018. LNCS, vol. 11218, pp. 20–36. Springer, Cham (2018). https://doi.org/10.1007/978-3-030-01264-9_2
41. Wu, Y., He, K.: Group normalization. In: Ferrari, V., Hebert, M., Sminchisescu, C., Weiss, Y. (eds.) ECCV 2018. LNCS, vol. 11217, pp. 3–19. Springer, Cham (2018). https://doi.org/10.1007/978-3-030-01261-8_1
42. Chen, L.-C., Zhu, Y., Papandreou, G., Schroff, F., Adam, H.: Encoder-decoder with Atrous separable convolution for semantic image segmentation. In: Ferrari, V., Hebert, M., Sminchisescu, C., Weiss, Y. (eds.) ECCV 2018. LNCS, vol. 11211, pp. 833–851. Springer, Cham (2018). https://doi.org/10.1007/978-3-030-01234-2_49
43. Zhao, H., Shi, J., Qi, X., Wang, X., Jia, J.: Pyramid scene parsing network. In: Proceedings of the IEEE Conference on Computer Vision and Pattern Recognition, pp. 2881–2890 (2017)
44. Hou, Q., Zhang, L., Cheng, M.M., Feng, J.: Strip pooling: rethinking spatial pooling for scene parsing. In: Proceedings of the IEEE/CVF Conference on Computer Vision and Pattern Recognition, pp. 4003–4012 (2020)
45. Zhu, Z., Xu, M., Bai, S., Huang, T., Bai, X.: Asymmetric non-local neural networks for semantic segmentation. In: Proceedings of the IEEE/CVF International Conference on Computer Vision, pp. 593–602 (2019)
46. Zhong, Y., Li, B., Tang, L., Tang, H., Ding, S.: Highly efficient natural image matting. arXiv preprint arXiv:2110.12748 (2021)
47. He, K., Sun, J., Tang, X.: Fast matting using large kernel matting Laplacian matrices. In: 2010 IEEE Computer Society Conference on Computer Vision and Pattern Recognition, pp. 2165–2172. IEEE (2010)
48. Liu, Q., Xie, H., Zhang, S., Zhong, B., Ji, R.: Long-range feature propagating for natural image matting. In: Proceedings of the 29th ACM International Conference on Multimedia, pp. 526–534 (2021)
49. Dai, Y., Lu, H., Shen, C.: Learning affinity-aware upsampling for deep image matting. In: Proceedings of the IEEE/CVF Conference on Computer Vision and Pattern Recognition, pp. 6841–6850 (2021)
50. Tang, J., Aksoy, Y., Oztireli, C., Gross, M., Aydin, T.O.: Learning-based sampling for natural image matting. In: Proceedings of the IEEE/CVF Conference on Computer Vision and Pattern Recognition, pp. 3055–3063 (2019)
51. Ma, N., Zhang, X., Zheng, H.-T., Sun, J.: ShuffleNet V2: practical guidelines for efficient CNN architecture design. In: Ferrari, V., Hebert, M., Sminchisescu, C., Weiss, Y. (eds.) Computer Vision – ECCV 2018. LNCS, vol. 11218, pp. 122–138. Springer, Cham (2018). https://doi.org/10.1007/978-3-030-01264-9_8
52. Tan, M., Le, Q.: EfficientNet: rethinking model scaling for convolutional neural networks. In: International Conference on Machine Learning, pp. 6105–6114. PMLR (2019)

53. Liu, Z., Mao, H., Wu, C.Y., Feichtenhofer, C., Darrell, T., Xie, S.: A ConvNet for the 2020s. arXiv preprint arXiv:2201.03545 (2022)
54. Forte, M., Pitié, F.: f, b, alpha matting. arXiv preprint arXiv:2003.07711 (2020)

RGB Road Scene Material Segmentation

Sudong Cai, Ryosuke Wakaki, Shohei Nobuhara, and Ko Nishino(✉)

Graduate School of Informatics, Kyoto University, Kyoto, Japan
kon@i.kyoto-u.ac.jp
https://www.vision.ist.i.kyoto-u.ac.jp/

Fig. 1. Materials vs. Semantics. Top: One semantic object may consist of multiple materials and different semantic objects may contain the same material. Middle: The same "Road" can be made of "asphalt," "concrete," or "brick." Bottom: A metal obstacle which is unclear in the semantic annotations, can cause hazard for driving.

Abstract. We address RGB road scene material segmentation, *i.e.*, per-pixel segmentation of materials in real-world driving views with pure RGB images, by building a new tailored benchmark dataset and model for it. Our new dataset, KITTI-Materials, based on the well-established KITTI dataset, consists of 1000 frames covering 24 different road scenes of urban/suburban landscapes, annotated with one of 20 material categories for every pixel in high quality. It is the first dataset tailored to RGB material segmentation in realistic driving scenes which allows us to train and test any RGB material segmentation model. Based on an analysis on KITTI-Materials, we identify the extraction and fusion of texture and context as the key to robust road scene material appearance. We introduce Road scene Material Segmentation Network (**RMSNet**), a new Transformer-based framework which will serve as a baseline for this challenging task. RMSNet encodes multi-scale hierarchical features with self-attention. We construct the decoder of RMSNet based on a novel lightweight self-attention model, which we refer to as **SAMixer**. SAMixer achieves adaptive fusion of informative texture and context cues across multiple feature levels. It also significantly accelerates self-attention for feature fusion with a balanced query-key similarity measure.

Supplementary Information The online version contains supplementary material available at https://doi.org/10.1007/978-3-031-26284-5_16.

L. Wang et al. (Eds.): ACCV 2022, LNCS 13842, pp. 256–272, 2023.
https://doi.org/10.1007/978-3-031-26284-5_16

We also introduce a built-in bottleneck of local statistics to achieve further efficiency and accuracy. Extensive experiments on KITTI-Materials validate the effectiveness of our RMSNet. We believe our work lays a solid foundation for further studies on RGB road scene material segmentation.

1 Introduction

Recognition of materials, what things are made of, in an image is critical for many computer vision applications. Materials inform the physical properties of the objects and regions in a scene which are otherwise inaccessible from just knowing the object categories. The way an action is planned for a paper cup as opposed to a ceramic cup would be different and gauging from sight would be advantageous. The importance of material recognition becomes even more significant for road scenes, particularly for self-driving vehicles to successfully navigate in daily environments. Despite its potential contributions to safety, and past work on object-level material recognition, little has been studied on regular color-image-level visual material understanding in road scenes.

Per-pixel material recognition (*i.e.*, material segmentation, in contrast to semantic segmentation) in a regular color image would be particularly informative for self-driving and driving assistance. Knowing that the asphalt-made road turns into gravel or brick would help an autonomous system to plan its speed, discerning a twig from a metal bar would help decide whether to avoid it, and telling a bronze statue from a live pedestrian would help anticipate its movements. Material segmentation, however, is not yet another semantic segmentation problem with a different set of labels. The challenge lies in the fact that the same object category can have different material categories, *e.g.*, a road can be made of asphalt, concrete, or even dirt and brick, yet they have the same shapes. The difficulty is exacerbated by the fact a single object can have multiple regions of different materials, *e.g.*, a bicycle made of metal, rubber, plastic, and leather. In contrast, objects can mostly be discerned with shape cues for category-level recognition, *i.e.*, semantic segmentation and object recognition.

In this paper, we address RGB road scene material segmentation by introducing a new benchmark dataset and a novel network that exploits the unique properties of material appearance and serve as a baseline model for this challenging task. We build a new dataset tailored to RGB road scene material segmentation by annotating images from the KITTI dataset [12]. We refer to this new dataset as the *KITTI-Materials* dataset. By building on a widely adopted road scene dataset, we are able to establish a dataset guaranteed to be relevant for autonomous driving research. KITTI-Materials consists of 1000 frames densely annotated with one of 20 material categories covering 24 different road scenes of common urban/suburban landscapes. The KITTI-Materials is the first tailored benchmark dataset for pure RGB road scene material segmentation which enables us to train and evaluate our ideas for the task and others to follow.

Figure 1 illustrates the key differences of RGB road scene material and semantic segmentation. A careful examination of the new dataset reveals that effective texture and context information extraction and fusion is essential for robust

RGB road scene material recognition. The characteristic textures of materials provide vital visual cues for their identification. The appearance of material texture, however, changes dramatically with scale (*i.e.*, distance from viewpoint) and occlusion. We may incorporate structural dependencies of local texture features to arrive at representations robust to these scale and occlusion variations. Such structural context, however, is unreliable for material appearance in contrast to global shape cues often exploited for object recognition. For this, effective fusion of texture and context cues to produce discriminative joint representations becomes vital for discerning road scene materials.

We introduce Road scene Material Segmentation Network (**RMSNet**), a new Transformer-based material segmentation network which generates discriminative material appearance representations through joint texture-context learning with low computational cost. RMSNet adopts the efficient hierarchical encoder introduced by SegFormer [40] to extract features of local textures and long-range context from multi-level hierarchies. It then merges multi-level multi-scale features with a novel self-attention-based feature fusion model which we refer to as **SAMixer**. SAMixer introduces a new balanced query-key similarity (Q-K-Sim) measure with a container feature generated by aggregating all input feature maps. This results in a highly efficient self-attention mechanism with only $O(N+1)$ complexity, where N denotes the number of input feature maps. SAMixer also uses a bottleneck local statistics encoding-decoding (BLSED) strategy for additional efficiency and accuracy.

We evaluate the effectiveness of RMSNet through extensive quantitative analysis and ablation studies on KITTI-Materials RGB RMS, and compare its accuracy with existing RGB material segmentation and road scene semantic segmentation methods. The results clearly demonstrate the effectiveness of RMSNet. We believe our work can contribute to richer visual understanding, particularly of road scenes, for safer driving. RMSNet will serve as a sound baseline model for this important task. We disseminate our project[1] to catapult this emerging avenue of research.

2 Related Work

Bell *et al.* [1] demonstrated material segmentation with a fully convolutional network cascaded with a fully-connected CRF [19,20,34], which is essentially semantic segmentation with material categories applied to mainly architectural photographs. Schwartz and Nishino introduced the use of material attributes as an intermediate representation for per-pixel material recognition without regard to shape features [29,30,32]. Later they introduced the integration of global contextual information in the form of semantic segmentation and place recognition and demonstrated its application to material segmentation on a material dataset consisting of local image patches sourced from COCO dataset and ImageNet [33].

Xue *et al.* [42] introduced the GTOS dataset, consisting of over 30k images of 40 ground surface material taken as top-down fronto-parallel images. They investigated the advantage of differential angular imaging for material

[1] https://github.com/kyotovision-public/RGB-Road-Scene-Material-Segmentation.

recognition. Zhang *et al.* [48] proposed the Deep-TEN model by using "orderless" texture encoding [25]. More recently, Xue *et al.* [43] incorporated texture encoding showing superior results to DAIN [42]. These methods, however, focus on image-wise material recognition.

Recently, Demir *et al.* [9] introduced the DeepGlobe dataset which consists of satellite images mainly for road and building extraction. Purri *et al.* [26] also proposed building material segmentation datasets from satellite images [2] and proposed reflectance residual encoding. Xue *et al.* [41] derived AngLNet which uses per-pixel angular luminance from multiple views. Material segmentation on road scenes is distinct from these bird-eye-view material segmentation as scale variation due to a dynamic perspective is inevitable.

Road scene semantic segmentation is a popular research field which provides us with inspirations for model design. Cordts *et al.* [8] introduced the Cityscapes dataset for scene understanding of urban driving environments. Many works have tackled road scene semantic segmentation using this dataset [4,6,7,24,27,28,36,40,44,50,52]. In contrast, road scene material segmentation with regular RGB images has not been intensively explored.

More related to our work, Liang *et al.* [22] introduced the MCubeS dataset, a multimodal material segmentation dataset consisting of RGB, NIR, and polarization images of city scenes, where the material categories are identical to our KITTI-Materials. Based on the dataset, they proposed the MCubeSNet modified on DeepLabv3+ [6] equipped with the RGFS layer to jointly apply various imaging modalities for improving material segmentation with the help of semantic segmentation annotations. In contrast, our KITTI-Materials dataset is tailored to pure RGB road scene material segmentation and comprises more images and scenes covering both city and suburban landscapes.

Vision transformers leverage multi-head self-attention (MSA) [35] to model long-range visual cues [18,38,45]. Full MSA to $2D$ spatial features, however, incurs excessive computational cost. Ramachandran *et al.* [27] modified the transformer to work on a fixed region and added positional biases. Wang *et al.* [36] introduced Stand-Alone Axial-MSA which processes feature maps along the height- and the width-axis separately to balance computational cost and accuracy for semantic segmentation. Zhang *et al.* [49] showed that co-occurrence of semantics including object categories exhibit long-range dependencies.

ViT [11] computes MSA within each non-overlapping image patch (*i.e.*, window) to achieve a speed-accuracy tradeoff for object recognition. PVT [37] introduced the first pyramid transformer architecture and demonstrated its potential for dense prediction tasks. Liu *et al.* [23] suggested applying MSA within fine-grained shifted windows and model cross-window connections to enhance local cues. Related models LeViT [13] and TNT [14] also improved window-MSA by infusing extra local details. Pure window-MSA, however, is computationally expensive for high-resolution features.

SegFormer [40] built a hierarchical transformer encoder with an efficient MSA in which the keys and values with reduced resolution were computed from condensed features with convolutions. It also introduced a lightweight All-MLP decoder and demonstrated its advantage over existing heavy decoders. Relevant idea was suggested in CvT [39], where Q-K-V projections were realized by

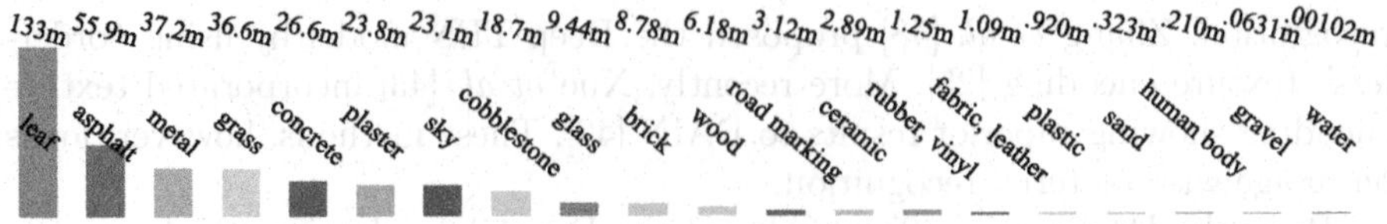

Fig. 2. Per-class pixel statistics (in "millions") of KITTI-Materials. Pixel labels show a clear long-tail distribution of material categories.

convolutions. In contrast, RMSNet introduces a novel SAMixer model that fuses multi-level features of local textures and long-range contextual cues to generate robust representations for road scene material segmentation.

Past works have explored multi-scale feature learning in object recognition and semantic segmentation. Chen *et al.* [5] plugged a spatial attention layer into the bottom of a two-branch network to learn weights for features at different scales. SKNet [21] expanded SE-Net [17] to aggregate multi-scale features. The Deeplab family [3,4,6,44] used atrous spatial pyramid pooling to learn scale-invariance through global statistics and a set of convolutions with different dilations. To enhance multi-level scale-aware feature learning, our RMSNet selectively activates meaningful features of local textures and non-local contextual interactions to form discriminative representations with SAMixer.

3 KITTI-Materials Dataset

We introduce the KITTI-Materials dataset, the first comprehensive RGB road scene material segmentation dataset. The images used in KITTI-Materials are sourced from the KITTI raw data [12]. It consists of 1000 images covering 24 different driving scenes including downtown, campus, residential area, highway, and other city/suburban landscapes captured from a car. In the 24 road scenes, there are 19 scenes consisting of 50 images sampled for every 5 consecutive frames, and other 5 scenes that contain 1, 5, 20, 15, and 9 images.

We annotated per-pixel material labels of 20 categories by professional paid annotators. All annotations are 1216×320 in resolution, and the raw images are center-cropped to this size beforehand. Figure 1 includes an example of road scene color images with their corresponding material annotations. Note that more visual examples can be found in the supplementary material.

Naturally reflecting the real world, our dataset has a very strong imbalance in the material categories, which is a significant challenge for accurate segmentation. Note that there is no manual selection applied to adjust the class-wise distribution of pixels in our dataset. Figure 2 shows pixel statistics with respect to each of the material categories. It shows that 16 material categories span $0.9 \times 10^6 - 1.4 \times 10^8$ pixels, *i.e.*, 99.84% of the total number of pixels. In contrast, 4 categories including "sand," "gravel," "water," and "human body," accounts for 0.083%, 0.016%, 0.00026%, and 0.054% of the overall pixels, respectively.

For evaluation on KITTI-materials, we define two training-test data splits (*i.e.*, Split-1 and -2), where the test set of Split-1 contains more scenes with highways and rural areas while Split-2 is biased to city scenes. Both splits contain 800

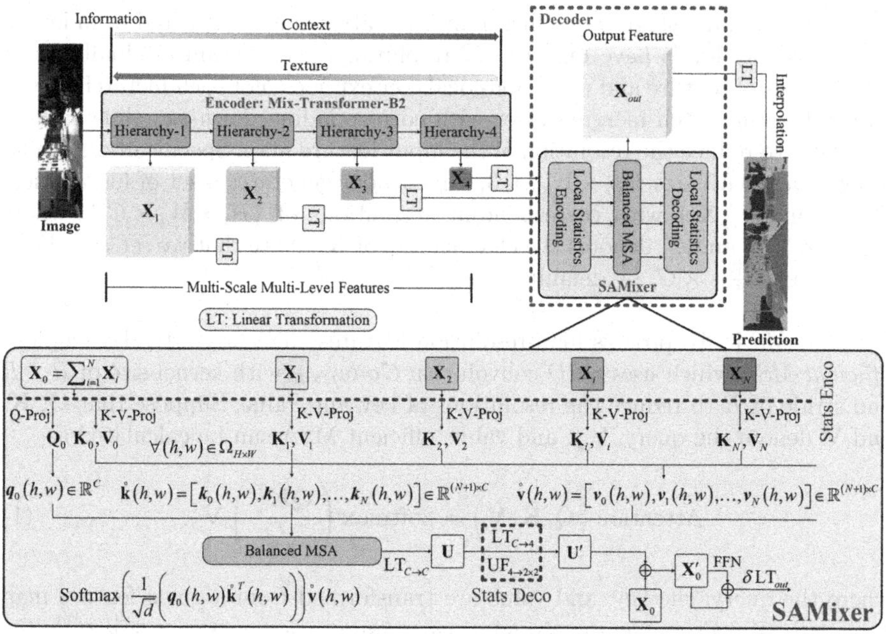

Fig. 3. Overview of RMSNet. "LT" denotes "Linear Transformation" layer with corresponding input and output channel-sizes. "Q-Proj" and "K-V-Proj" are "Query-Projection" and "Key- and Value-projection," respectively. "UF" means "Unfold" operation. After obtaining the output feature $\mathbf{X}_{out}$ of SAMixer, we employ a linear layer to generate the segmentation mask from $\mathbf{X}_{out}$ to achieve per-pixel material recognition.

images for training and 200 images for testing, but with different combinations of scenes. Both training and test sets of these two splits show very strong imbalance in the material categories. Further discussions and details including specific components of scenes, visual examples exhibiting their different characteristics of test sets, and per-class statistics are shown in the supplementary material.

4 RMSNet

We introduce RMSNet as a new baseline model for road scene material segmentation. RMSNet effectively fuses texture and contextual cues of material appearance with SAMixer. To the best of our knowledge, it is the first model to realize multi-level feature fusion with the MSA mechanism. Figure 3 depicts the overall architecture of RMSNet.

4.1 Mix-Transformer as Encoder

Hierarchical Feature Encoding. Our RMSNet adopts the middle-size hierarchical transformer encoder introduced by SegFormer [40], namely Mix-Transformer-B2 (MiT-B2), to extract a set of multi-level multi-scale feature maps, from 4 sequential learning stages (*i.e.*, hierarchies). Feature maps extracted from low to high hierarchy levels have high to low resolutions and contain gradually fewer local details of texture and more non-local context cues. For each hierarchy level, an overlapping patch merging layer with corresponding down-sampling ratio is employed to reduce the resolution of the input feature map. Specifically, given an input image with size $H_{in} \times W_{in} \times 3$, the encoder generates a set of hierarchical feature maps $\{\mathbf{X}_i\}$ with corresponding resolutions of $\{H_i \times W_i \times C_i\}$, where $i \in \{1, 2, 3, 4\}$ and C_i denotes the channel-size of $\mathbf{X}_i$. Note that we set $H_i \times W_i \times C_i = \frac{H_{in}}{2^{i+1}} \times \frac{W_{in}}{2^{i+1}} \times C_i$ by default.

Efficient MSA. To process high-resolution features efficiently, MiT-B2 employs *efficient MSA* which uses a $2D$ convolution $Conv_{R\times R}$ with kernel-size of $R \times R$ and stride of R to reduce the resolutions of key and value. Suppose that $\mathbf{Q}$, $\hat{\mathbf{K}}$, and $\hat{\mathbf{V}}$ denote the query, key, and value, efficient MSA can be calculated as

$$\text{Attention}\left(\mathbf{Q}, \hat{\mathbf{K}}, \hat{\mathbf{V}}\right) = \text{Softmax}\left(\frac{\mathbf{Q}\hat{\mathbf{K}}^T}{\sqrt{d}}\right)\hat{\mathbf{V}}, \quad (1)$$

where the query, the key, and value are transformed from a given feature map $\mathbf{X}_I \in \mathbb{R}^{H_I \times W_I \times C_I}$ and its condensed feature map $Conv_{R\times R}(\mathbf{X}_I) \in \mathbb{R}^{\frac{H_I}{R} \times \frac{W_I}{R} \times C_I}$. Note that feature maps $\mathbf{X}_I$ and $Conv_{R\times R}(\mathbf{X}_I)$ are reshaped to sizes $M_I \times C_I$ and $\frac{M_I}{R^2} \times C_I$ (*i.e.*, $M_I = H_I W_I$), respectively. Here, $d = \frac{C_I}{g}$, where g denotes the number of heads for an MSA computation. In this way, the computational complexity of an MSA can be controlled with the resolution reduction ratio R. For hierarchy-1 to = 4, MiT-B2 assigns $R = 8, 4, 2, 1$, respectively.

Position-Aware FFN. MiT-B2 inserts a 3×3 depth-wise convolution $\text{DWConv}_{3\times3}$ in each feed-forward network (FFN), at the top of the first linear layer, to enforce position awareness without additional positional encodings. With this modification, local details can be preserved without the sacrificing accuracy due to interpolation for matching resolutions. The position-aware FFN is defined as

$$\mathbf{X}'' = \text{LT}_2\left(\delta\left(\text{DWConv}_{3\times3}\left(\text{LT}_1\left(\mathbf{X}'\right)\right)\right)\right) + \mathbf{X}', \quad (2)$$

where $\mathbf{X}'$ denotes the attended feature by the MSA layer and $\mathbf{X}''$ is the output feature of the FFN; δ denotes the assigned nonlinear activation (GELU [16] by default); LT_1 and LT_2 are the first and second linear layers, respectively.

4.2 SAMixer-Based Decoder

Through careful examination of KITTI-Materials images and also in agreement to past works on material recognition (*e.g.*, [29–31,33,43,46]), we find that efficient and faithful encoding of local texture patterns of different materials is critical for per-pixel material recognition. The appearance of textures, however, vary significantly with scale and occlusion. Structural dependencies and co-occurrences of local texture features may help extract a representation robust to this variability. Unlike semantic objects, however, materials often show more complicated spatial distributions (*i.e.*, more fragmented) and lack prominent shape cues. This makes fusion of local textures and long-range context cues even challenging. To realize effective fusion for Transformer-induced features, we propose a novel multi-level multi-scale feature fusion model based on MSA, which we refer to as **SAMixer**. Figure 3 depicts the diagram of the SAMixer-based decoder. SAMixer can efficiently fuse local and non-local features to generate robust representations for road scene materials.

Discussion on the Challenge of MSA-Based Feature Fusion. MSA introduces informative context dependencies to deep learning representations. For multi-scale feature fusion, however, it inevitably causes excessive computational overhead. Suppose that $\chi = \{\mathbf{X}_i \in \mathbb{R}^{H_i \times W_i \times C_i} \mid i = 1, 2, \ldots, N\}$ is a set of feature maps for fusion ($N = 4$ in our experiments). The fused feature map $\mathbf{X} \in \mathbb{R}^{H \times W \times C}$ is generated by mixing all element feature maps $\mathbf{X}_i \in \chi$ at each aligned position $(h, w) \in \Omega_{H \times W}$, where $\Omega_{H \times W}$ denotes the spatial lattice of $\mathbf{X}$. Note that before fusion, each of the feature maps $\mathbf{X}_i$ of different sizes should be transformed and interpolated to the same size $H \times W \times C$ which we refer to as the *anchor size*.

With MSA feature fusion, each transformed and interpolated $\mathbf{X}_i$ with the anchor size is projected to $\mathbf{Q}_i$, $\mathbf{K}_i$, $\mathbf{V}_i$, where $\boldsymbol{q}_i(h, w)$, $\boldsymbol{k}_i(h, w)$, $\boldsymbol{v}_i(h, w)$ with the unified length C are corresponding feature vectors of $\mathbf{Q}_i$, $\mathbf{K}_i$, $\mathbf{V}_i$ at the given spatial position (w, h), respectively. To fuse each feature vector at an aligned position (w, h), MSA can be defined as

$$\text{Attention}(\mathbf{q}(h, w), \mathbf{k}(h, w), \mathbf{v}(h, w)) = \text{Softmax}\left(\frac{\mathbf{q}(h, w)\,\mathbf{k}^T(h, w)}{\sqrt{d}}\right)\mathbf{v}(h, w), \quad (3)$$

where $\mathbf{q}(h, w), \mathbf{k}(h, w), \mathbf{v}(h, w) = [\boldsymbol{q}_i(h, w)], [\boldsymbol{k}_i(h, w)], [\boldsymbol{v}_i(h, w)] \in \mathbb{R}^{N \times C}$ are the query, key, and value of the position (h, w), respectively, formed by arranging the corresponding feature vectors along the row-axis. Let $\mathbf{F}(h, w) \in \mathbb{R}^{N \times C}$ denote the attended feature descriptor at position (h, w) computed by the MSA layer, each fused feature vector $\mathbf{X}(h, w) \in \mathbb{R}^C$ can be obtained by applying a simple average aggregation along the row-axis or a linear projection on $\mathbf{F}(h, w)$. In this way, the computational complexity is $O(N^2)$. General MSA can be excessively expensive in computational cost for feature fusion since coarse features of local texture patterns are usually of large sizes.

Proposed SAMixer. We construct SAMixer with MSA consuming only $O(N+1)$ computational complexity by deriving a new balanced query-key similarity (Q-K-Sim) measure in which a container feature is introduced by simply aggregating (*i.e.*, summing) all input features to trigger the MSA computation. SAMixer also introduces a new built-in bottleneck local encoding-decoding (BLSED) strategy to realize further efficiency and accuracy. Figure 3 depicts the SAMixer. In the following paragraphs, we present the two core components of SAMixer, *i.e.*, the balanced Q-K-Sim measure and BLSED strategy.

Balanced Query-Key Similarity Measure. Vanilla query-key similarity measure $\mathbf{q}(h,w)\,\mathbf{k}^T(h,w)$ defines a balanced (*i.e.*, symmetric) computation on the feature set χ for fusion. In contrast, for each query vector $\boldsymbol{q}_i(h,w)$ (where $i=1,2,\ldots,N$), its corresponding decomposed group of query-key similarity measure $\boldsymbol{q}_i(h,w)\,\mathbf{k}^T(h,w)$ is imbalanced for different key vectors. That is, for $\forall i,j\in\{1,2,\ldots,N\}, i\neq j$, the query vector $\boldsymbol{q}_i(h,w)$ is likely closer to the corresponding key vector $\boldsymbol{k}_i(h,w)$ than a key vector $\boldsymbol{k}_j(h,w)$, because both $\boldsymbol{q}_i(h,w)$ and $\boldsymbol{k}_i(h,w)$ are generated from the same feature vector $\boldsymbol{x}_i(h,w)\in\mathbb{R}^C$. As a result, employing a decomposed group of query-key similarity measures independently, although efficient, leads to imbalanced feature fusion and the representational ability of the fused feature is limited.

Our goal is to calculate an efficient balanced MSA on the feature set χ by applying only one group of query-key similarity measures. As depicted in Fig. 3, We achieve this by introducing a novel query-key similarity measure which we refer to as the *balanced query-key similarity measure.*

The core idea of this balanced Q-K-Sim measure is the new tailored element feature referred to as the *container feature* that enables balanced computation on a single group of query-key similarity measures. We generate this container feature $\mathbf{X}_0\in\mathbb{R}^{H\times W\times C}$ by aggregating each of the features in χ with a simple summation (*i.e.*, $\mathbf{X}_0=\sum_{i=1}^{N}\mathbf{X}_i$). Then, the feature set χ can be expanded into a new set $\mathring{\chi}$ comprising of $N+1$ feature elements by introducing $\mathbf{X}_0$.

Similarly, for $\forall(h,w)\in\Omega_{H\times W}$, we generate the key and value descriptors $\mathring{\mathbf{k}}(h,w), \mathring{\mathbf{v}}(h,w)=[\mathring{\boldsymbol{k}}_i(h,w)], [\mathring{\boldsymbol{v}}_i(h,w)]\in\mathbb{R}^{(N+1)\times C}$ from features in $\mathring{\chi}$ and the single query vector $\boldsymbol{q}_0(h,w)\in\mathbb{R}^C$ from the container feature vector $\boldsymbol{x}_0(h,w)\in\mathbb{R}^C$, respectively. With this, we can compute an efficient balanced MSA on $\mathring{\chi}$

$$\mathrm{Attention}\left(\boldsymbol{q}_0(h,w), \mathring{\mathbf{k}}(h,w), \mathring{\mathbf{v}}(h,w)\right)=\mathrm{Softmax}\left(\frac{\boldsymbol{q}_0(h,w)\,\mathring{\mathbf{k}}^T(h,w)}{\sqrt{d}}\right)\mathring{\mathbf{v}}(h,w). \quad (4)$$

With the proposed balanced Q-K-Sim measure, we can preserve the balance of MSA while reducing the quadratic complexity of $O(N^2)$ to only $O(N+1)$. As a result, our model can effectively fuse high-resolution features to produce discriminative representations for road scene materials.

Bottleneck Local Statistics Encoding-Decoding Strategy. We achieve further efficiency and accuracy by introducing a lightweight embedded encoder-decoder strategy in the SAMixer. Figure 3 depicts the process of the proposed BLSED strategy. We first assign an anchor size $H\times W\times C$, where $H=\frac{H_1}{2^l}$ and $W=\frac{W_1}{2^l}$. Here, $l\in\mathbb{Z}^+$; H_1 and W_1 are the largest height and width of all the input features extracted by the hierarchical encoder. Note that we apply $l=1$ such that $S=2^l=2$ by default in our experiments.

Before the MSA computation, we encode local statistics $\mathbf{U}_i\in\mathbb{R}^{H_i\times W_i\times C}$ from each of the input feature maps $\mathbf{X}_i$ in $\mathring{\chi}$ whose spatial resolutions $H_i\times W_i$ are higher than $H\times W$, respectively, by employing corresponding $2D$ convolutions $Conv_{S_i\times S_i}$ with kernel-size of $S_i\times S_i$ and stride of S_i ($S_i=\frac{H_i}{H}$ is divisible by 2). To reduce computational cost, each $Conv_{S_i\times S_i}$ is replaced by splicing a depth-wise convolution with a linear layer. We interpolate all the feature maps to the anchor size whose spatial resolution is smaller than $H \times W$. We preserve the size of any feature naturally possessing spatial resolution of $H \times W$. Note that $\mathring{\chi}$ comprises the container feature $\mathbf{X}_0$, so index $i = 0, 1, \ldots, N$. Particularly, since the anchor size is smaller than the highest resolution of features, we produce the container feature by $\mathbf{X}_0 = \sum_{i=1}^{N} \mathrm{UP}_i(\mathbf{X}_i)$, where UP_i denotes up-sampling via an bilinear interpolation with a scale factor of $\frac{H_1}{H_i}$ (equivalent to $\frac{W_1}{W_i}$). For $i = 1$, UP degrades to an identity mapping.

After the MSA computation, we decode the attended fused feature map $\mathbf{U} \in \mathbb{R}^{H \times W \times C}$ to a high-resolution feature map $\mathbf{U}' \in \mathbb{R}^{H_1 \times W_1 \times C}$ with a channel-spatial decoupled combination scheme which significantly reduces the computational cost. We first generate the spatial mask $\mathbf{P} \in \mathbb{R}^{H_1 \times W_1 \times 1}$ from $\mathbf{U}$ by applying a linear transformation LT_{deco} with input channels of C and output channels of S^2. That is, for each feature vector $\boldsymbol{u}(h, w) \in \mathbb{R}^C$ of $\mathbf{U}$, LT_{deco} generates corresponding S^2 spatial feature units and unfolds these feature units into a spatial feature patch $\mathbf{p}(h, w) \in \mathbb{R}^{S \times S \times 1}$. Then, each feature vector $\boldsymbol{u}(h, w)$ and its corresponding spatial feature patch $\mathbf{p}(h, w)$ are combined by element-wise summation $\oplus$ to produce the feature patch $\mathbf{u}'(h, w) \in \mathbb{R}^{S \times S \times C}$ of $\mathbf{U}'$:

$$\mathbf{u}'(h, w) = (\mathbf{1}_{S \times S} \otimes \boldsymbol{u}(h, w)) \oplus (\mathbf{1}_C \otimes \mathbf{p}(h, w)) \;, \tag{5}$$

where $\otimes$ denotes Kronecker product; $\mathbf{1}_{S \times S}$ and $\mathbf{1}_C$ each denotes a ones matrix of the corresponding size. Then, we obtain $\mathbf{U}'$ by arranging each of the feature patches $\mathbf{u}'(h, w)$ according to their spatial position order.

With the proposed BLSED strategy, SAMixer achieves higher efficiency by operating on condensed feature maps.

Segmentation Mask Generation. The output feature $\mathbf{X}_{out} \in \mathbb{R}^{H_1 \times W_1 \times C}$ of SAMixer is generated by applying a linear layer LT_{out} (with a GELU activation δ) over the output of the FFN layer:

$$\mathbf{X}_{out} = \delta\left(\mathrm{LT}_{out}\left(\mathrm{LT}_2\left(\delta\left(\mathrm{LT}_1\left(\mathrm{DWConv}_{3 \times 3}\left(\mathbf{X}'_0\right)\right)\right)\right) + \mathbf{X}'_0\right)\right) \;, \tag{6}$$

where $\mathbf{X}'_0 = \mathbf{X}_0 + \mathbf{U}'$, and LT_1 and LT_2 denote the first and second linear layers of the FFN, respectively. Unlike the FFN in MiT-B2, we employ a depth-wise convolution before LT_1, which increases the speed of the FFN. The segmentation mask is obtained by employing a linear layer with an output channel-size of the number of material classes (*i.e.*, 20) on $\mathbf{X}_{out}$.

5 Experiments and Discussions

We evaluate the effectiveness of our method on the KITTI-Materials dataset with detailed ablation studies and also thorough comparison with past material segmentation methods [1,32], road scene semantic segmentation methods with CNN encoders [6,21,44,50], related state-of-the-art transformers [11,39,40] and a gating-based dynamic network [51] that have been applied to semantic segmentation. Note that DeepLabv3+ [6] also represents Liang *et al.* [22] without semantic segmentation masks and using RGB only.

Table 1. Material segmentation results on KITTI-Materials dataset for our methods and other methods. "Trs" denotes "Transformer"; "DLv3+" denotes DeepLabv3+; Symbols "◇" and "⋆" denote "modified" and "our implementation," respectively; Symbol "‡" denotes methods whose original code cannot support multi-GPU training/inference settings. ViT [11] and CvT [39] are applied with the All-mlp decoder [40].

Method	Backbone	Params	Fps↑	Split-1	Split-2
				mIoU(%)↑	
MINC‡ [1]	VGG16 [34]	134.34M	15.88	29.73	32.12
Matcontext‡ [32]	VGG16◇ [32]	25.42M	7.40	30.87	33.16
DLv3+ [6]	ResNet101 [15]	59.34M	14.60	41.35	46.09
	SK-ResNet101 [21]	60.47M	14.08	41.96	46.04
DeeperLab [44]	ResNet101 [15]	240.58M	11.29	42.56	47.12
PSPNet [50]	ResNet101 [15]	43.38M	14.22	31.92	37.11
DDF-DL [51]	DDFNet101 [51]	42.94M	12.66	41.55	46.41
ViT [11]	ViT-B/16 [11]	89.03M	13.69	40.02	46.06
CvT [39]	CvT-13 [39]	21.89M	18.02	41.72	47.54
SegFormer [40]	Mix-Trs-B2 [40]	27.36M	18.87	44.47	48.32
RMSNet **(Ours)**	Mix-Trs-B2 [40]	31.53M	16.81	**46.82**	**50.34**

5.1 Implementation Details

Two different training-test data splits (denoted by *Split*-1 and -2, respectively) of KITTI-Materials with different characteristics are used for evaluation. The test set of "split-1" contains more scenes with highways and rural areas while "split-2" is biased to city scenes (see the supplemental material for details). Both splits consist of all 1000 images of KITTI-Materials where 800 images for training and 200 images for testing with different split rules. For all models, we apply the AdamW optimizer with a weight decay of 0.01 for 300 epochs including 10 epochs of linear warm-up. Following [40], we start the learning rate from 6×10^{-5} and 6×10^{-4} for encoders and decoders, respectively, with a cosine decay scheduler and a mini-batch of 16. We adopt standard image augmentation settings [6]. In the training phase, images are randomly center-cropped and then resized to 512×512 pixels, while in the testing, images are fixed to the original size (*i.e.*, 1216×320 pixels). To reduce the negative effect of extreme data imbalance, we calculated balancing weights based on class frequencies of materials and applied them to CE-losses of all models. Experiments are conducted on a computer with 4 × RTX A5000 GPUs. For fair comparisons, all encoders of our method and compared methods use ImageNet [10] pre-trained weights obtained from corresponding open-sourced projects or websites. All methods are evaluated in the raw image size without multi-scale averaging augmentation [47]. We use mean intersection of union (mIoU) to evaluate the performance of each model.

5.2 Experimental Results on KITTI-Materials

Based on the proposed KITTI-Materials dataset, we verify the effectiveness of our network designs by comparing with (1) existing general material segmentation

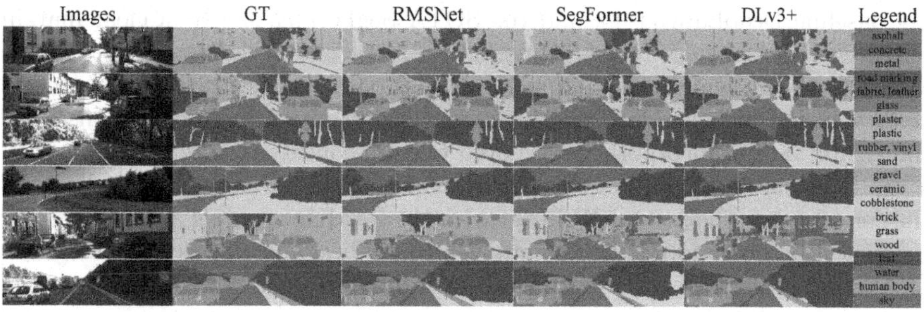

Fig. 4. Visual examples on KITTI-Materials. Compared with DeepLabv3+ [6] (denoted by "DLv3+") and SegFormer [40]. "GT" denotes "ground truth".

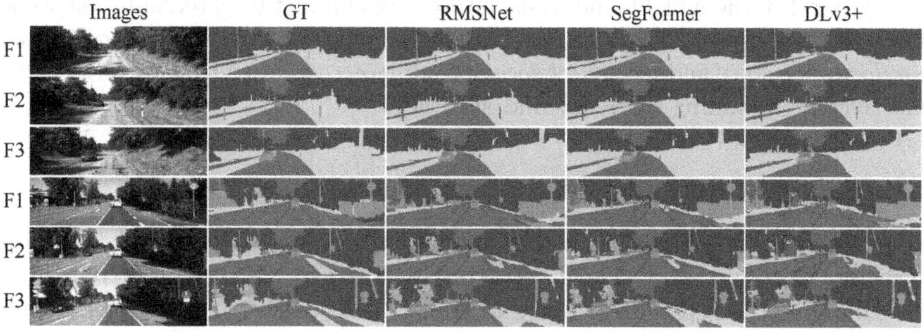

Fig. 5. Visual examples of moving cars of different scales. "F" denotes "frame."

methods for RGB images [1,32]; (2) popular road scene semantic segmentation methods with CNN encoders [6,15,44,50]; (3) enhanced DeepLabv3+ [6] with a multi-scale fusion method (*i.e.*, SKNet [21]) and a state-of-the-art (SOTA) gating-induced dynamic networks [51]; (4) related SOTA transformers [11,39,40] that have been validated on semantic segmentation, where SegFormer [40] is the closest model to our RMSNet.

As shown in Table 1, our RMSNet enjoys clear improvements over all compared methods for general material segmentation and road scene semantic segmentation in accuracy. Note that the major difference between RMSNet and SegFormer-B2 is the replacement of the All-MLP decoder with our SAMixer-based decoder. The results demonstrate the effectiveness of our network for road scene material segmentation. Compared with popular semantic segmentation frameworks [6,44,50] with CNN encoders [15,21] and the SOTA gating-induced dynamic network [51], RMSNet yields significant gains in accuracy. Our RMSNet also shows further accuracy improvements over other compared SOTA Transformers [11,39,40] with competitive efficiency.

To demonstrate detailed performance differences on each material, we report the per-class comparative results (see the supplemental material) with visual examples shown in Fig. 4, where we find that our network outperforms competing baseline DeepLabv3+ [6] and the SOTA SegFormer [40] by a clear margin on categories "fabric," "glass," "metal," "rubber," and "human body." These materials span a wide range of appearances as part of different semantic objects (*e.g.*, vehicles, bicycles, road markings, and pedestrians). Figure 5 demonstrates the significance of incorporating tailored texture-context feature fusion for road scene material segmentation through visual comparison between RMSNet and the compared methods. Our RMSNet with the SAMixer module achieves cleaner segmentation on windows, headlights, vehicle bodies, and wheels of moving cars of different scales (*i.e.*, different distances from viewpoint).

6 Ablation Study

Using split-1 of the KITTI-Materials dataset, we conduct targeted ablation studies on the proposed SAMixer and its core ingredients, *i.e.*, the *balanced query-key similarity measure* and *bottleneck local encoding-decoding strategy*, to verify their effectiveness in adding efficiency and accuracy.

6.1 Balanced Query-Key Similarity Measure

Here we independently discuss and validate the proposed balanced Q-K-Sim measure. We build an abridged SAMixer (denoted by "SAM-a") by removing the BLSED strategy, and conduct a comparative experiment by introducing three targeted control groups of feature fusion models built on (1) the raw MSA mechanism (denoted by "MSA-raw"); (2) a series of imbalanced partial MSA mechanisms where the queries are only transformed from one of the four feature maps for fusion (denoted by "MSA-ib1" to "-ib4"); (3) the All-MLP module of SegFormer [40] (denoted by "SegF"). We use SAM-a and these control groups to replace the original SAMixer to build abridged RMSNets and compare them on our KITTI-Materials dataset. To prevent excessive computational overhead, all methods SAM-a, MSA-raw, and "MSA-ib1" to "-ib4" employ the resolution reduction strategy suggested in the Mix-Transformer [40] with reduction ratio $R = 2$. We also add DeepLabv3+ [6] (denoted by "DLv3+") as a baseline method. Table 2(a) and (b) show that SAM-a and MSA-raw introduce similar accuracy gains while SAM-a consumes far fewer computational costs; MSA-ib1 to -ib4 lead to close/less accuracy gains to the SegFormer, demonstrating the effectiveness of the balanced Q-K-Sim measure. This confirms the effectiveness of our proposed multi-scale feature fusion model.

6.2 Bottleneck Local Statistics Encoding-Decoding Strategy

Effectiveness. We propose this tailored built-in strategy to achieve further efficiency and accuracy. To evaluate its effectiveness, we compare the original SAMixer (denoted by "SAM") with two targeted control groups (1) the abridged SAMixer without the BLSED strategy (*i.e.*, "SAM-a" introduced in Sect. 6.1); (2) the SegFormer All-MLP module (denoted by "SegF"). The comparative results

Table 2. Ablation studies on (a) and (b) the balanced Q-K-Sim measure, (c) effectiveness and (d) resolution reduction ratio setting of the BLSED strategy. "DLv3+", "SegF," and "SAM" denote "DeepLabv3+ [6]," "SegFormer [40]," and our "SAMixer," respectively. "SAM-a" denotes the abridged SAMixer without BLSED strategy.

(a)

Method	Fps↑	mIoU↑
DLv3+	14.60	41.35
SegF	18.87	44.47
MSA-raw	10.98	**45.51**
SAM-a	15.39	45.33

(b)

Method	Fps↑	mIoU↑
MSA-ib1	17.47	**44.17**
MSA-ib2	17.24	43.08
MSA-ib3	17.55	43.94
MSA-ib4	17.71	43.29

(c)

Method	Fps↑	mIoU↑
DLv3+	14.60	41.35
SegF	18.87	44.47
SAM-a	15.39	45.33
SAM	16.81	**46.82**

(d)

Ratio	Fps↑	mIoU↑
DLv3+	14.60	41.35
W/o	15.39	45.33
2	16.81	**46.82**
4	15.75	45.74

are reported in Table 2(c). Our RMSNet with SAMixer outperforms the abridged SAMixer SAM-a in both accuracy and efficiency. RMSNet also improves the SOTA SegFormer by a clear margin on accuracy. The results verify the effectiveness of our network design.

Reduction Ratio Setting. We assign an unified resolution reduction ratio (denoted by "Ratio") as the stride and kernel-size for corresponding depth-wise convolutions to control the encoding process of the local statistics of feature maps for fusion. We conduct this ablation study to evaluate the effectiveness of the BLSED strategy with different Ratio settings. Table 2(d) reports the comparative results with reduction ratios of "W/o", 2, and 4, where "W/o" denotes removing BLSED strategy (*i.e.*, "SAM-a" introduced in Sect. 6.1). Based on the results, we set "ratio = 2" by default, since it reaches high accuracy with competitive efficiency, compared with other counterparts.

7 Conclusion

We address RGB road scene material segmentation by constructing a new benchmark dataset, KITTI-Materials, and by deriving a new network that effectively fuses texture and contextual cues for accurate per-pixel material recognition. The network, *i.e.*, RMSNet, achieves this with a newly derive SAMixer module built on a balanced Q-K-Sim measure and a BLSED strategy. Experimental validations and ablation studies on KITTI-Materials dataset confirm the effectiveness of our proposed designs. We believe our data and model can contribute to further studies on leveraging rich visual material information for road scene understanding and will serve as a sound baseline to tackle this challenging task. As a limitation, RMSNet falls short in extracting very high-resolution features of abundant local textures with adequate computational efficiency, which we plan to address in future work. We hope dissemination of our data and code will catalyze further studies on this important and challenging visual task.

Acknowledgements. This work was in part supported by JSPS 20H05951, 21H04893, JST JPMJCR20G7, and SenseTime Japan.

References

1. Bell, S., Upchurch, P., Snavely, N., Bala, K.: Material recognition in the wild with the materials in context database. In: Proceedings of the CVPR, pp. 3479–3487 (2015)
2. Brown, M., et al.: Large-scale public LiDAR and satellite image data set for urban semantic labeling. In: Laser Radar Technology and Applications XXIII, vol. 10636 (2018)
3. Chen, L.C., Papandreou, G., Kokkinos, I., Murphy, K., Yuille, A.L.: DeepLab: Semantic Image Segmentation with Deep Convolutional Nets, Atrous Convolution, and Fully Connected CRFs. TPAMI **40**(4), 834–848 (2018)
4. Chen, L.C., Papandreou, G., Schroff, F., Adam, H.: Rethinking Atrous Convolution for Semantic Image Segmentation. arXiv preprint arXiv:1706.05587 (2017)
5. Chen, L.C., Yang, Y., Wang, J., Xu, W., Yuille, A.L.: Attention to scale: scale-aware semantic image segmentation. In: Proceedings of the CVPR, pp. 3640–3649 (2016)
6. Chen, L.-C., Zhu, Y., Papandreou, G., Schroff, F., Adam, H.: Encoder-decoder with atrous separable convolution for semantic image segmentation. In: Ferrari, V., Hebert, M., Sminchisescu, C., Weiss, Y. (eds.) ECCV 2018. LNCS, vol. 11211, pp. 833–851. Springer, Cham (2018). https://doi.org/10.1007/978-3-030-01234-2_49
7. Choi, S., Kim, J.T., Choo, J.: Cars can't fly up in the sky: improving urban-scene segmentation via height-driven attention networks. In: Proceedings of the CVPR, pp. 9373–9383 (2020)
8. Cordts, M., et al.: The cityscapes dataset for semantic urban scene understanding. In: Proceedings of the CVPR, pp. 3213–3223 (2016)
9. Demir, I., et al.: DeepGlobe 2018: a challenge to parse the earth through satellite images. In: Proceedings of the CVPR Workshop, pp. 172–17209 (2018)
10. Deng, J., Dong, W., Socher, R., Li, L.J., Li, K., Fei-Fei, L.: ImageNet: a large-scale hierarchical image database. In: Proceedings of the CVPR, pp. 248–255 (2009)
11. Dosovitskiy, A., et al.: An image is worth 16×16 words: transformers for image recognition at scale. In: Proceedings of the ICLR (2021)
12. Geiger, A., Lenz, P., Urtasun, R.: Are we ready for autonomous driving? The KITTI vision benchmark suite. In: Proceedings of the CVPR (2012)
13. Graham, B., et al.: LeViT: a vision transformer in ConvNet's clothing for faster inference. In: Proceedings of the ICCV (2021)
14. Han, K., Xiao, A., Wu, E., Guo, J., Xu, C., Wang, Y.: Transformer in transformer. In: Proceedings of the NeurIPS (2021)
15. He, K., Zhang, X., Ren, S., Sun, J.: Deep residual learning for image recognition. In: Proceedings of the CVPR, pp. 770–778 (2016)
16. Hendrycks, D., Gimpel, K.: Gaussian error linear units (GELUs). arXiv preprint arXiv:1606.08415 (2016)
17. Hu, J., Shen, L., Sun, G.: Squeeze-and-excitation networks. In: Proceedings of the CVPR (2018)
18. Huang, Z., Wang, X., Huang, L., Huang, C., Wei, Y., Liu, W.: CCNet: criss-cross attention for semantic segmentation. In: Proceedings of the ICCV, pp. 603–612 (2019)
19. Kraehenbuehl, P., Koltun, V.: Parameter learning and convergent inference for dense random fields. In: Proceedings of the ICML, pp. 513–521 (2013)
20. Krizhevsky, A., Sutskever, I., Hinton, G.E.: ImageNet classification with deep convolutional neural networks. In: Proceedings of the NeurIPS, pp. 1097–1105 (2012)

21. Li, X., Wang, W., Hu, X., Yang, J.: Selective kernel networks. In: Proceedings of the CVPR, pp. 510–519 (2019)
22. Liang, Y., Wakaki, R., Nobuhara, S., Nishino, K.: Multimodal material segmentation. In: Proceedings of the CVPR (2022)
23. Liu, Z., et al.: Swin transformer: hierarchical vision transformer using shifted windows. In: Proceedings of the ICCV (2021)
24. Neuhold, G., Ollmann, T., Bulo, S.R., Kontschieder, P.: The mapillary vistas dataset for semantic understanding of street scenes. In: Proceedings of the ICCV, pp. 5000–5009 (2017)
25. Perronnin, F., Sánchez, J., Mensink, T.: Improving the fisher kernel for large-scale image classification. In: Daniilidis, K., Maragos, P., Paragios, N. (eds.) ECCV 2010. LNCS, vol. 6314, pp. 143–156. Springer, Heidelberg (2010). https://doi.org/10.1007/978-3-642-15561-1_11
26. Purri, M., et al.: Material segmentation of multi-view satellite imagery. arXiv preprint arXiv:1904.08537 (2019)
27. Ramachandran, P., Parmar, N., Vaswani, A., Bello, I., Levskaya, A., Shlens, J.: Stand-alone self-attention in vision models. In: Proceedings of the NeurIPS, vol. 32, pp. 68–80 (2019)
28. Reda, F.A., et al.: SDC-Net: video prediction using spatially-displaced convolution. In: Ferrari, V., Hebert, M., Sminchisescu, C., Weiss, Y. (eds.) ECCV 2018. LNCS, vol. 11211, pp. 747–763. Springer, Cham (2018). https://doi.org/10.1007/978-3-030-01234-2_44
29. Schwartz, G., Nishino, K.: Visual material traits: recognizing per-pixel material context. In: IEEE Color and Photometry in Computer Vision Workshop (2013)
30. Schwartz, G., Nishino, K.: Automatically discovering local visual material attributes. In: Proceedings of the CVPR (2015)
31. Schwartz, G., Nishino, K.: Integrating local material recognition with large-scale perceptual attribute discovery. arXiv preprint arXiv:1604.01345 (2016)
32. Schwartz, G., Nishino, K.: Material recognition from local appearance in global context. arXiv preprint arXiv:1611.09394 (2016)
33. Schwartz, G., Nishino, K.: Recognizing material properties from images. TPAMI **42**(8), 1981–1995 (2020)
34. Szegedy, C., et al.: Going deeper with convolutions. In: Proceedings of the CVPR, pp. 1–9 (2015)
35. Vaswani, A., et al.: Attention is all you need. In: Proceedings of the NeurIPS, vol. 30, pp. 5998–6008 (2017)
36. Wang, H., Zhu, Y., Green, B., Adam, H., Yuille, A., Chen, L.-C.: Axial-DeepLab: stand-alone axial-attention for panoptic segmentation. In: Vedaldi, A., Bischof, H., Brox, T., Frahm, J.-M. (eds.) ECCV 2020. LNCS, vol. 12349, pp. 108–126. Springer, Cham (2020). https://doi.org/10.1007/978-3-030-58548-8_7
37. Wang, W., et al.: Pyramid vision transformer: a versatile backbone for dense prediction without convolutions. In: Proceedings of the ICCV (2021)
38. Wang, X., Girshick, R., Gupta, A., He, K.: Non-local neural networks. In: Proceedings of the CVPR, pp. 7794–7803 (2018)
39. Wu, H., et al.: CvT: introducing convolutions to vision transformers. In: Proceedings of the ICCV, pp. 22–31 (2021)
40. Xie, E., Wang, W., Yu, Z., Anandkumar, A., Alvarez, J.M., Luo, P.: SegFormer: simple and efficient design for semantic segmentation with transformers. In: Proceedings of the NeurIPS (2021)

41. Xue, J., Purri, M., Dana, K.: Angular luminance for material segmentation. In: IGARSS 2020–2020 IEEE International Geoscience and Remote Sensing Symposium (2020)
42. Xue, J., Zhang, H., Dana, K., Nishino, K.: Differential angular imaging for material recognition. In: Proceedings of the CVPR, pp. 6940–6949 (2017)
43. Xue, J., Zhang, H., Nishino, K., Dana, K.: Differential viewpoints for ground terrain material recognition. TPAMI **44**, 1205–1218 (2020)
44. Yang, T.J., et al.: DeeperLab: single-shot image parser. arXiv preprint arXiv:1902.05093 (2019)
45. Zhang, H., Goodfellow, I.J., Metaxas, D.N., Odena, A.: Self-attention generative adversarial networks. In: Proceedings of the ICML, pp. 7354–7363 (2018)
46. Zhang, H., Dana, K., Nishino, K.: Reflectance hashing for material recognition. In: Proceedings of the CVPR, pp. 3071–3080 (2015)
47. Zhang, H., et al.: Context encoding for semantic segmentation. In: Proceedings of the CVPR, pp. 7151–7160 (2018)
48. Zhang, H., Xue, J., Dana, K.: Deep TEN: texture encoding network. In: Proceedings of the CVPR, pp. 2896–2905 (2017)
49. Zhang, H., Zhang, H., Wang, C., Xie, J.: Co-occurrent features in semantic segmentation. In: Proceedings of the CVPR, pp. 548–557 (2019)
50. Zhao, H., Shi, J., Qi, X., Wang, X., Jia, J.: Pyramid scene parsing network. In: Proceedings of the CVPR (2017)
51. Zhou, J., Jampani, V., Pi, Z., Liu, Q., Yang, M.H.: Decoupled dynamic filter networks. In: Proceedings of the CVPR (2021)
52. Zhu, Y., et al.: Improving semantic segmentation via video propagation and label relaxation. In: Proceedings of the CVPR, pp. 8856–8865 (2019)

Self-distilled Vision Transformer for Domain Generalization

Maryam Sultana[1,2(✉)], Muzammal Naseer[1,3], Muhammad Haris Khan[1], Salman Khan[1,3], and Fahad Shahbaz Khan[1,4]

[1] Mohamed Bin Zayed University of AI, Abu Dhabi, UAE
{maryam.sultana,muzammal.naseer,muhammad.haris, salman.khan,fahad.khan}@mbzuai.ac.ae

[2] VAIL, Oxford Brookes University, Oxford, UK

[3] Australian National University, Canberra, Australia

[4] Linköping University, Linköping, Sweden

Abstract. In the recent past, several domain generalization (DG) methods have been proposed, showing encouraging performance, however, almost all of them build on convolutional neural networks (CNNs). There is little to no progress on studying the DG performance of vision transformers (ViTs), which are challenging the supremacy of CNNs on standard benchmarks, often built on i.i.d assumption. This renders the real-world deployment of ViTs doubtful. In this paper, we attempt to explore ViTs towards addressing the DG problem. Similar to CNNs, ViTs also struggle in out-of-distribution scenarios and the main culprit is overfitting to source domains. Inspired by the modular architecture of ViTs, we propose a simple DG approach for ViTs, coined as *self-distillation for ViTs*. It reduces the overfitting of source domains by easing the learning of input-output mapping problem through curating non-zero entropy supervisory signals for intermediate transformer blocks. Further, it does not introduce any new parameters and can be seamlessly plugged into the modular composition of different ViTs. We empirically demonstrate notable performance gains with different DG baselines and various ViT backbones in five challenging datasets. Moreover, we report favorable performance against recent state-of-the-art DG methods. Our code along with pre-trained models are publicly available at: https://github.com/maryam089/SDViT.

Keywords: Domain generalization · Vision transformers · Self distillation

1 Introduction

Since their inception, transformers have displayed remarkable performance in various natural language processing (NLP) tasks [1–3]. Owing to their success

Supplementary Information The online version contains supplementary material available at https://doi.org/10.1007/978-3-031-26284-5_17.

L. Wang et al. (Eds.): ACCV 2022, LNCS 13842, pp. 273–290, 2023.
https://doi.org/10.1007/978-3-031-26284-5_17

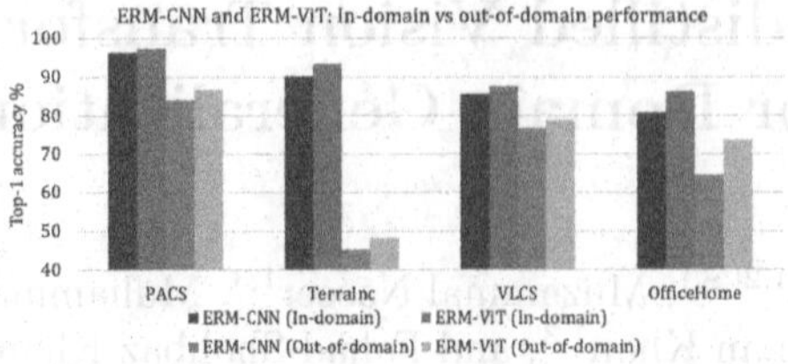

Fig. 1. In-domain (validation) and out-of-domain (target) classification accuracy of ERM-CNN and ERM-ViT in four DG datasets. Similar to ERM-CNN, ERM-ViT also shows performance degradation in out-of-domain scenarios.

in NLP, recently, transformer design has been adopted for vision tasks [4]. Since then, we have been witnessing several vision transformer (ViT) models for image recognition [4], object detection [5,6] and semantic segmentation [7,8]. ViTs are intrinsically different in design compared to convolution neural networks (CNNs), since they lack explicit inductive biases such as spatial connectivity and translation equivariance. They process (input) image as a sequence of patches that is enhanced via successive transformer blocks (comprised of self-attention mechanisms), thereby allowing the network to model the relationship between any parts of the image. A useful consequence of such processing is a wide receptive field that facilitates capturing the global context in contrast to a limited receptive field modeled in CNNs.

Many deep learning models are usually deployed in real-world scenarios where the test data is unknown in advance. When their predictions are used for decision making in safety-critical applications, such as medical diagnoses or self-driving cars, an erroneous prediction can lead to dangerous consequences. This typically occurs because there is a distributional gap between the training and testing data. Hence, it is critical for deep learning models to provide reliable predictions that generalize across different domains. Domain generalization (DG) is a problem setting in which data from multiple source domains is leveraged for training to generalize to a new (unseen) domain [9–18]. Existing DG methods aim to explicitly reduce domain gap in the feature space [9,19,20], learn well-transferable model parameters through meta-learning [21–24], propose different data augmentation techniques [15,25–27], or leverage auxiliary tasks [12,28]. Lately, Gulrajani and Lopez-Paz [13] show that a simple Empirical Risk Minimization (ERM) method obtains favorable performance against previous methods under a fair evaluation protocol termed as "Domainbed". To our knowledge, almost all aforementioned DG approaches are based on CNNs, and there is little to no work on studying the DG performance of ViTs. So, in effect, despite ViTs demonstrating state-of-the-art performance on some standard benchmarks, often rooted in i.i.d assumption, their real-world deployment remains doubtful. To this end, we attempt to explore ViTs towards addressing the DG problem.

We note that ViT-based ERM (ERM-ViT), similar to its CNN-based counterpart (ERM-CNN), also suffers from performance degradation when facing out-of-distribution (OOD) target domain data (see Fig. 1). In the absence of any explicit overfitting prevention mechanism coupled with the one-hot encoded

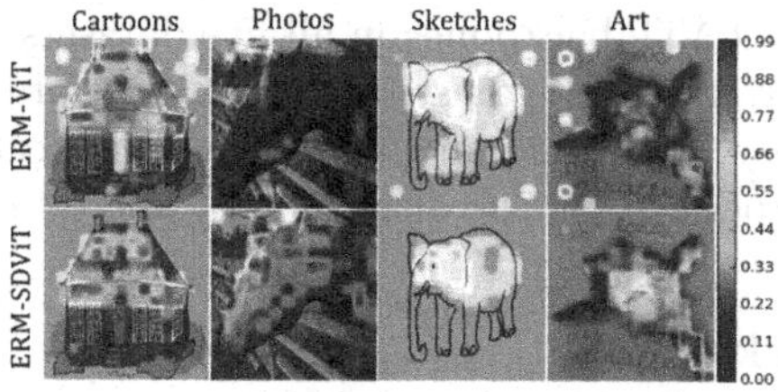

Fig. 2. Attention maps from ERM-ViT (top) and ERM-SDViT (bottom) corresponding to images of the target domains in `PACS`. ERM-ViT is prone to exploit non-generalizable features e.g., background. Whereas ERM-SDViT is capable of learning cross-domain generalizable features e.g., object shape and its semantics.

ground-truth supervision, which is essentially zero-entropy signals, as such it is challenging for a simple ERM-ViT model to obtain favorable OOD generalization. Under the hood, since the mapping problem is difficult, the model is prone to inadvertently exploiting non-generalizable, brittle features for making predictions. Figure 2 visualizes attention maps from ERM-ViT on arbitrary images of four target domains in `PACS` dataset. ERM-ViT has the tendency to rely on non-object related features such as the background features, which are potentially non-transferable between the source and the target domains.

Inspired by the modular architecture of ViTs, we propose a light-weight plug-and-play DG approach for ViTs, namely *self-distillation for ViT (SDViT)*. It explicitly encourages the model towards learning generalizable, comprehensive features. ViTs process a sequence of input image patches repeatedly by multiple multi-headed self-attention layers, a.k.a transformer blocks [1]. These image patches are also known as *patch tokens*. A randomly initialized class token is usually appended to the set of image patches (tokens). This group is then passed through a sequence of transformer blocks followed by the passing of class token through a linear classifier to get final predictions. The class token can learn information that is useful while making a final prediction. So, it can be extracted from the output of each transformer block and can be leveraged to get class-specific logits using the final classifier of the pretrained model [29]. Armed with this insight, we propose to transfer the so-called dark knowledge from the final classifier output to the intermediate blocks by developing a self-distillation strategy for ViT (Sect. 3.2). It alleviates the overfitting of source domains by moderating the learning of input-output mapping problem via non-zero entropy supervision of intermediate blocks. We show that improving the intermediate blocks, which are essentially multiple feature pathways, through soft supervision from the final classifier facilitates the model toward learning cross-domain generalizable features (see Fig. 2). Our approach naturally fits into the modular and compositional architecture of different ViTs, and does not introduce any new parameters. As such it adds a minimal training overhead over the baseline. Extensive experiments have been conducted on five diverse datasets from DomainBed suite [13], including `PACS`, `VLCS`, `OfficeHome`, `TerraIncognita`, and `DomainNet`. We empirically show better performance across different DG baselines as well as different ViT backbones in all five datasets. Further, we demonstrate competitive performance against the recent state-of-the-art DG methods. With CvT-21 backbone, we obtain an (overall) average accuracy (five datasets) of 68.6%, thereby outperforming the existing best [30] by 1.8%.

2 Related Work

Domain Generalization: A prevalent motivation of several existing DG methods is to learn the underlying domain-invariant representations from the available source data. The seminal work of Vapnik *et al.* [31] introduced Empirical Risk Minimization (ERM), which minimizes the sum of squared errors across domains. Following this route, we observe several variants aimed at uncovering the domain-invariant features via matching distributions across domains. For instance, Muandet *et al.* [9] employed maximum mean discrepancy (MMD) constraint, Ghifary *et al.* [10] proposed a multi-task autoencoder, and Yang *et al.* [32] used canonical correlation analysis (CCA). Arjovsky *et al.* [33] proposed the learning of invariant predictors across various source domains. A few methods used low-rank regularization to extract invariant features e.g., [11,34]. Meta-learning based methods have also been used as regularizers. Li *et al.* [24] switched domain-specific feature extractors and classifiers across domains via episodic training. Balaji *et al.* [23] learned a regularization function in an episodic training paradigm. Furthermore, some DG methods masked features via ranking gradients [14], utilized auxiliary tasks [12,28], employed domain-specific masks [35], and exploited domain-specific normalizations [36]. A few DG approaches proposed contrastive semantic alignment and self-supervised contrastive formulations [17,22,37]. Another class of DG methods employs various data augmentation techniques to improve the diversity of source domains. Shankar *et al.* [25] proposed Crossgrad training, Volpi *et al.* [26] imposed wasserstein constraint in semantic space, Zhou *et al.* [27] learned a generator to generate new examples, and Khan *et al.* [15] estimated class-conditional covariance matrices for generating novel source features. Recently, Gulrajani *et al.* [13] demonstrated that, under a fair evaluation protocol, a simple empirical risk minimization (ERM) method can achieve state-of-the-art DG performance. Cha *et al.* [30] proposed stochastic weight averaging in a dense manner to achieve flatter minima for DG. We note that, all aforementioned DG methods are based on CNN architecture, however, little to no attention has been paid to investigating the DG performance of ViTs. To this end, we choose to study the performance of ViTs under domain generalization with ERM as a simple, but strong DG baseline.

Vision Transformers: operate in a hierarchical manner by processing input images as a sequence of non-overlapping patches via the self-attention mechanism. Recently, we have seen some ViT-based methods for image classification [4,38,39], object detection [5,40], and semantic segmentation [41,42]. Dosovitskiy *et al.* [4] proposed the first fully functional ViT model for image classification. Despite its promising performance, its adoption remained limited because it requires large-scale datasets for model training and huge computation resources.

Towards improving data efficiency in ViTs, Touvron *et al.* [39] developed Data-efficient image Transformer (DeiT); it attains competitive results against the CNN by training only on ImageNet and without leveraging external data. Similarly, Yuan *et al.* [43] proposed Tokens-To-Token Vision Transformer (T2T-ViT) strategy. It progressively structurizes the patch tokens in a way that the local structure represented by surrounding tokens can be modeled while reducing the tokens length. Furthermore, Wu *et al.* [38] proposed a hybrid approach, namely Convolutional Vision Transformer (CvT), by combining the strengths of CNNs and ViTs aimed at improving the performance and robustness of ViTs, while maintaining computational and memory efficiency. Recently, Zhang *et al.* [44] studied the performance of ViTs under distribution shifts and proposed a generalization-enhanced vision transformer from the outlook of self-supervised learning and information theory. They concluded that by scaling the capacity of ViTs the out-of-distribution (OOD) generalization performance can be enhanced, mostly under the domain adaptation settings. On the other hand, we show that it is possible to improve the OOD generalization performance of ViTs without introducing any new parameters under the established DG protocols [13].

Knowledge Distillation: was initially designed for model compression and aims at matching the output of a teacher model to a student model whose capacity is smaller than the teacher model [45]. Zhang *et al.* [46] partitioned a CNN model into several blocks, and the knowledge from the full (deeper part) of the model is squeezed into the shallow parts. Yun *et al.* [47] proposed a self-distillation approach based on penalizing the predictive distributions between similar data samples. In particular, it distills the predictive distribution between different samples of the same label during training. Towards addressing DG problem, Wang *et al.* [48] proposed a teacher-student distillation strategy, based on CNNs, and a gradient filter as an efficient regularization term. In contrast, we propose a new self-distillation strategy to enhance the DG capabilities of ViTs. It prevents introducing any new parameters via seamlessly exploiting the modular architecture of ViTs.

3 Proposed Approach

In this paper, we aim to explore ViTs towards tackling the domain generalization problem. We observe that a simple, but competitive DG baseline (ERM) built on ViT displays notable performance decay in a typical DG setting (Fig. 1). Towards this end, we propose a simple plug-and-play DG approach for ERM-ViT, termed as self-distillation for ViTs, that explicitly facilitates the model towards exploiting cross-domain transferable features (Fig. 2).

3.1 Preliminaries

Problem Settings: In traditional domain generalization (DG) setting [13], we assume the availability of data from a set of training (source) domains $\mathcal{D} = \{\mathcal{D}\}_{k=1}^{K}$. Where $\mathcal{D}_k$ denotes a distribution over the input space $\mathcal{X}$ and K is the total number of training domains. From a domain k, we sample J training datapoints that comprise of input x and label y as pairs $(x_j^k \in \mathcal{X}, y_j^k \in \mathcal{Y})_{j=1}^J$. Besides a set of training (source) domains, we also assume a set of target domains $\{\mathcal{T}\}_{t=1}^T$, where T is the total number of target domains and is typically set to 1. The goal in DG is to learn a mapping $\mathcal{F}_\theta : \mathcal{X} \rightarrow \mathcal{Y}$ that provides accurate predictions on data from an unseen target domain $\mathcal{T}_t$.

Empirical Risk Minimization (ERM) for DG: Assume a loss function $\mathcal{L} : \mathcal{Y} \times \mathcal{Y}$ which can quantify the prediction error, such as standard Cross Entropy (CE) for the image recognition task. A simple DG baseline accumulates the data from multiple source domains $\mathcal{D}$ and searches for a predictor minimizing the following empirical risk [31]: $\frac{1}{N}\sum_{i=1}^{N} \mathcal{L}(\mathcal{F}_\theta(x_k^j, y_k^j))$. Where $N = K \times J$ is the total number of data points from all source domains. Recently, Gulrajani and Lopez-Paz [13] demonstrated that this simple ERM based DG baseline shows competitive results or even performs better than many previous state-of-the-art DG methods under a fair evaluation protocol.

ViT Based ERM: While exploring ViTs for DG, we observe that ViT-based ERM (ERM-ViT) shows notable performance drop, similar to their CNN-based ERM counterpart (Fig. 1). This is likely due to the lack of any explicit overfitting mechanism and the supervision from one-hot encoded ground truth labels. It renders the overall learning of the mapping problem, from input space to label space, rather difficult. As a result, the model is more prone to exploiting non-generalizable, brittle features, such as the specific background of a domain (Fig. 2 and 6). To tackle this problem, in the next section, we propose a new self-distillation technique for improving the DG capabilities of ViTs. The core idea is to ease the mapping problem by generating non-zero entropy supervision for multiple feature pathways in ViTs. This enables the model towards utilizing more generalizable features (Fig. 2 & 6), that are mostly shared across source and target domains.

3.2 Self-distilled Vision Transformer for Domain Generalization

ViTs have Modular Architectures: Assume the model $\mathcal{F}$ is composed of n intermediate layers and the final classifier h such as $\mathcal{F} = (f_1 \circ f_2 \circ f_3 \circ \ldots f_n) \circ h$, where f_i represents an intermediate block or layer. In the case of ViT (*e.g* Deit-Small [39]), f_i is based on a self-attention transformer block, and such network design is monolithic as any transformer block produces equi-dimensional features that are $\mathbb{R}^{m \times d}$, where m represent the number of input features or tokens and each token has d dimensions. The monolithic design approach of ViT allows a self-ensemble behavior [29], where the output of each block can be processed by the final classifier h to create an intermediate classifier[1].

$$\mathcal{F}_i = f_i \circ h \tag{1}$$

[1] For non-monolithic ViT designs and CNNs, where feature dimension changes across the layers, the intermediate classifier can be obtained via $\mathcal{F}_i = (f_i \circ g_i) \circ h$, where g projects the output of f_i to the same dimension as h.

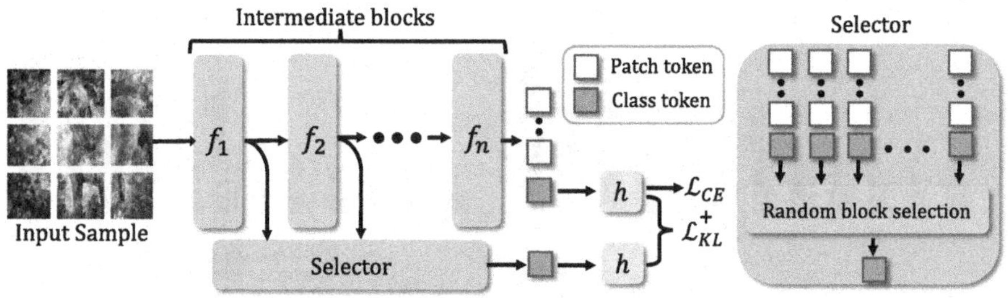

Fig. 3. Proposed self-distillation in ViTs for domain generalization (ERM-SDViT). ViTs build upon a modular and hierarchical architecture, where a model is comprised of n intermediate blocks/layers (f_i) and a final classifier h. The 'Selector' chooses a random block from the range of intermediate blocks and makes a prediction after passing its classification token through the final classifier. This way the dark knowledge, as non-zero entropy signals, is distilled from the final classification token to the intermediate class tokens during training.

Our goal is to induce the so-called dark knowledge, non-zero entropy supervisory signals, from the final classifier to these sub-models, manifesting multiple feature pathways.

Self-distillation in ViTs: As discussed earlier, ViTs can be easily dissected into a number of sub-models due to their monolithic architectural design as each transformer block produces a classification token that can be processed by the final classification head (Eq. 1) to produce a class-specific score. Each sub-model represents a discriminative feature pathway through the network. We believe that inducing dark knowledge from the final classifier output to these sub-models via soft supervision during training can enhance the overall model's capability towards learning object semantics.

Random Sub-model Distillation: The number of sub-models within a given ViT depends on the number of Transformer blocks (see Fig. 3) and distilling the knowledge to all of the sub-models at once poses optimization difficulties during online training. Therefore, we introduce a simple technique that randomly samples one sub-model based on Eq. 1 from all the possible set of sub-models (see Fig. 3). In this manner, our approach trains all sub-models but knowledge is transferred to only a single sub-model at any step of the training. This strategy eases the optimization and leads to better domain generalization.

Impact on Internal Representations: In Fig. 4, we plot the block-wise accuracy of baseline (ERM-ViT) and our method (ERM-SDViT). Random sub-model distillation improves the accuracy of all blocks, in particular, the improvement is more pronounced for the earlier blocks. Besides later blocks, it also encourages earlier blocks to bank on transferable representations, yet discriminative representations. Since these earlier blocks manifest multiple discriminative fea-

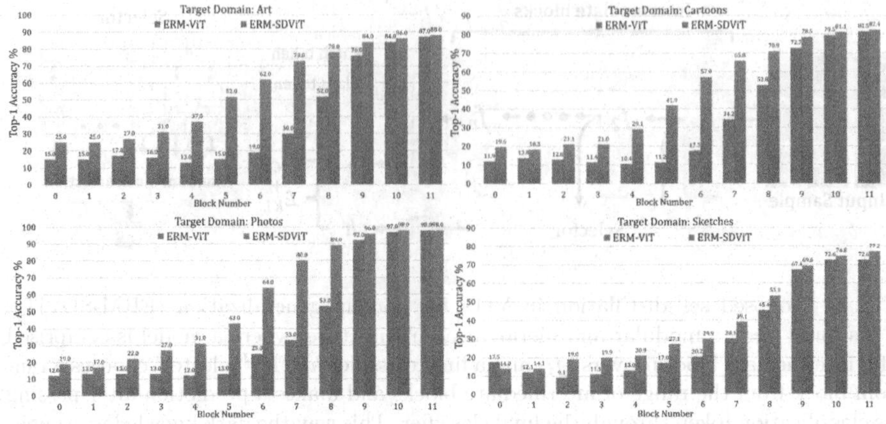

Fig. 4. Block-wise accuracy of baseline (ERM-ViT [39]) and our method (ERM-SDViT), featuring random sub-model distillation for DG. Results are reported on four challenging target domains from PACS dataset.

ture pathways, we believe that they better facilitate the overall model towards capturing the semantics of the object class.

Training Objective: For a given input x, the prediction error from the final classification token of the ViT is computed using cross-entropy loss in comparison with one-hot encoded ground-truths as follows.

$$\mathcal{L}_{\mathrm{CE}}(\mathcal{F}(x), y) = -\sum_{j=1}^{n} y_j log(\mathcal{F}(x)_j), \tag{2}$$

where n is the output dimension of the final classifier. We randomly sample an intermediate block as shown in Fig. 3 and produce logits from the classification token of a sub-model by applying Eq. 1.

We then compute the difference between the final and randomly sampled intermediate classification token by comparing the KL divergence between their logit distributions as follows:

$$\mathcal{L}_{\mathrm{KL}}(\mathcal{F}(x) \| \mathcal{F}_i(x)) = \sum_{j=1}^{n} \sigma\left(\mathcal{F}(x)/\tau\right)_j \log \frac{\sigma\left(\mathcal{F}(x)/\tau\right)_j}{\sigma\left(\mathcal{F}_i(x)/\tau\right)_j}, \tag{3}$$

where σ denotes the softmax operation and τ represent temperature used to rescale the logits [45]. The model is optimized by minimizing the overall loss based on Eqs. 2 and 3 and given as follows:

$$\mathcal{L} = \mathcal{L}_{\mathrm{CE}} + \lambda \mathcal{L}_{\mathrm{KL}}, \tag{4}$$

where λ balances the contribution of $\mathcal{L}_{\mathrm{KL}}$ towards the overall loss $\mathcal{L}$.

4 Experiments

Datasets: Following the work of Gulrajani and Lopez-Paz [13], we rigorously evaluate the effectiveness of our proposed method and draw comparisons with the existing state-of-the-art on five benchmark datasets including `PACS` [11], `VLCS` [49], `OfficeHome` [50], `TerraIncognita` [51] and `DomainNet` [52]. `PACS` [11] contains four domains $d \in$ {Art, Cartoons, Photos, Sketches},7 classes and a total of 9,991 images. `VLCS` [49] comprises of four domains as well $d \in$ {Caltech101, LabelMe, SUN09, VOC2007}, 5 classes and offers 10,729 images. `OfficeHome` [50] also contains four domains $d \in$ {Art, Clipart, Product, Real}, 65 classes and a total of 15,588 images. `TerraIncognita` [51]: has four camera-trap domains $d \in$ {L100, L38, L43, L46}, 10 classes and offers 24,778 wild photographs. `DomainNet` [52] contains six domains $d \in$ {Clipart, Infograph, Painting, Quickdraw, Real, Sketch}, 345 classes and 586,575 images.

Implementation and Training/Testing Details: To allow fair comparisons, we follow the training and evaluation protocol of Gulrajani and Lopez-Paz [13]. We use the training domain validation protocol for model selection. After partitioning each training domain data into the training and validation subsets (80%/20%), the validation data from each training domain are pooled to obtain an overall validation set. The model that maximizes the accuracy on this overall validation set is considered the best model which is then evaluated on the target domain to report classification (top-1) accuracy. For all our ViT-based methods, including the proposed approach, we use AdamW [53] optimizer and use the default hyperparameters (HPs) of ERM from [13][2], including the batch size of 32, the learning rate of $5e$–05, and the weight decay of 0.0. Note that, only the values of our method-specific HPs, λ and τ, are sought via grid search in the ranges $\{0.1, 0.2, 0.5\}$ and $\{3.0, 5.0\}$, respectively, using the validation set. We report accuracy for each target domain and their average where a model is trained/validated on training domains and evaluated on an (unseen) target domain. Each accuracy on the target domain is an average over three different trials with different train-validation splits.

Evaluation with Different ViT Backbones: We establish the generalizability of our method by experimenting with three different ViT backbones, namely **DeiT** [39], **CvT** [38], and **T2T-ViT** [43]. **DeiT** is a data-efficient image transformer and was trained on 1.2 million ImageNet examples. We use the DeiT-Small model having 22M parameters, which can be regarded as a ResNet-50 counterpart containing 23.5M parameters. Note that, we utilize the DeiT-Small model without the distillation token and a student-teacher formulation.

CvT introduces convolutions into ViT to improve accuracy and efficiency. We use CvT-21 which contains 32M parameters in our baselines and proposed self-distilled ViTs. **T2T-ViT** relies on a progressive tokenization to aggregate neighboring Tokens to one Token; it can encode the local structure information

[2] They are default parameters in the pre-defined ranges [13] for random HP search.

Table 1. Comparison with the several (17) existing SOTA DG methods. The best results are in bold and the second best is underlined.

Algorithm	Backbone	# Params	VLCS	PACS	OfficeHome	TerraInc	DomainNet	Average
ERM [13]	ResNet-50	23.5M	77.5 ± 0.4	85.5 ± 0.2	66.5 ± 0.3	46.1 ± 1.8	40.9 ± 0.1	63.3
IRM [33]	ResNet-50	23.5M	78.5 ± 0.5	83.5 ± 0.8	64.3 ± 2.2	47.6 ± 0.8	33.9 ± 2.8	61.5
GroupDRO [54]	ResNet-50	23.5M	76.7 ± 0.6	84.4 ± 0.8	66.0 ± 0.7	43.2 ± 1.1	33.3 ± 0.2	60.7
Mixup [55]	ResNet-50	23.5M	77.4 ± 0.6	84.6 ± 0.6	68.1 ± 0.3	47.9 ± 0.8	39.2 ± 0.1	63.4
MLDG [21]	ResNet-50	23.5M	77.2 ± 0.4	84.9 ± 1.0	66.8 ± 0.6	47.7 ± 0.9	41.2 ± 0.1	63.5
CORAL [56]	ResNet-50	23.5M	78.8 ± 0.6	86.2 ± 0.3	68.7 ± 0.3	47.6 ± 1.0	41.5 ± 0.1	64.5
MMD [20]	ResNet-50	23.5M	77.5 ± 0.9	84.6 ± 0.5	66.3 ± 0.1	42.2 ± 1.6	23.4 ± 9.5	58.8
DANN [19]	ResNet-50	23.5M	78.6 ± 0.4	83.6 ± 0.4	65.9 ± 0.6	46.7 ± 0.5	38.3 ± 0.1	62.6
CDANN [57]	ResNet-50	23.5M	77.5 ± 0.1	82.6 ± 0.9	65.8 ± 1.3	45.8 ± 1.6	38.3 ± 0.3	62.0
MTL [58]	ResNet-50	23.5M	77.2 ± 0.4	84.6 ± 0.5	66.4 ± 0.5	45.6 ± 1.2	40.6 ± 0.1	62.8
SagNet [16]	ResNet-50	23.5M	77.8 ± 0.5	86.3 ± 0.2	68.1 ± 0.1	48.6 ± 1.0	40.3 ± 0.1	64.2
ARM [59]	ResNet-50	23.5M	77.6 ± 0.3	85.1 ± 0.4	64.8 ± 0.3	45.5 ± 0.3	35.5 ± 0.2	61.7
VREx [60]	ResNet-50	23.5M	78.3 ± 0.2	84.9 ± 0.6	66.4 ± 0.6	46.4 ± 0.6	33.6 ± 2.9	61.9
RSC [14]	ResNet-50	23.5M	77.1 ± 0.5	85.2 ± 0.9	65.5 ± 0.9	46.6 ± 1.0	38.9 ± 0.5	62.6
SelfReg [17]	ResNet-50	23.5M	77.5 ± 0.0	86.5 ± 0.3	69.4 ± 0.2	<u>51.0 ± 0.4</u>	44.6 ± 0.1	65.8
mDSDI [18]	ResNet-50	23.5M	79.0 ± 0.3	86.2 ± 0.2	69.2 ± 0.4	48.1 ± 1.4	42.8 ± 0.1	65.0
SWAD [30]	ResNet-50	23.5M	79.1 ± 0.1	88.1 ± 0.1	70.6 ± 0.2	50.0 ± 0.3	46.5 ± 0.1	66.8
ERM-ViT [39]	DeiT-Small	22M	78.8 ± 0.5	84.9 ± 0.9	71.4 ± 0.1	43.4 ± 0.5	45.5 ± 0.0	64.8
ERM-ViT + T3A	DeiT-Small	22M	81.6 ± 0.2	85.5 ± 0.7	72.6 ± 0.2	43.6 ± 0.4	46.8 ± 0.1	66.0
ERM-SDViT	DeiT-Small	22M	78.9 ± 0.4	86.3 ± 0.2	71.5 ± 0.2	44.3 ± 1.0	45.8 ± 0.0	65.3
ERM-SDViT + T3A	DeiT-Small	22M	<u>81.6 ± 0.1</u>	86.7 ± 0.2	72.5 ± 0.3	44.9 ± 0.4	47.4 ± 0.1	66.6
ERM-ViT [38]	CvT-21	32M	79.0 ± 0.3	86.9 ± 0.3	75.5 ± 0.0	48.7 ± 0.4	50.4 ± 0.1	68.1
ERM-ViT + T3A	CvT-21	32M	80.6 ± 0.3	<u>88.5 ± 0.1</u>	<u>76.2 ± 0.0</u>	49.7 ± 0.5	52.0 ± 0.1	<u>69.4</u>
ERM-SDViT	CvT-21	32M	79.2 ± 0.4	88.3 ± 0.2	75.6 ± 0.2	49.7 ± 1.4	<u>50.4 ± 0.0</u>	68.6
ERM-SDViT + T3A	CvT-21	22M	**81.9 ± 0.4**	**88.9 ± 0.5**	**77.0 ± 0.2**	**51.4 ± 0.7**	**52.0 ± 0.0**	**70.2**
ERM-ViT [43]	T2T-ViT-14	21.5M	78.9 ± 0.3	86.8 ± 0.4	73.7 ± 0.2	48.1 ± 0.2	48.1 ± 0.1	67.1
ERM-ViT + T3A	T2T-ViT-14	21.5M	81.0 ± 0.6	87.7 ± 0.4	75.3 ± 0.1	47.8 ± 0.2	50.0 ± 0.1	68.3
ERM-SDViT	T2T-ViT-14	21.5M	79.5 ± 0.8	88.0 ± 0.7	74.2 ± 0.3	<u>50.6 ± 0.8</u>	48.2 ± 0.2	68.1
ERM-SDViT + T3A	T2T-ViT-14	21.5M	81.2 ± 0.3	87.8 ± 0.6	75.5 ± 0.2	50.5 ± 0.6	50.2 ± 0.1	69.0

of surrounding tokens and reduce the length of tokens iteratively. We use T2T-ViT-14 model, containing 21.5M parameters, which is approx. equivalent to the capacity of the ResNet-50 model.

4.1 Comparison with the State-of-the-Art

We compare our approach to several (in particular, 17) existing state-of-the-art algorithms for DG (see Table 1) listed in Domainbed suite [13]. Specifically, we include the following DG algorithms: Empirical Risk Minimization (ERM) [13], Invariant Risk Minimization (IRM) [33], Group Distributionally Robust Optimization (GroupDRO) [54], Inter-domain Mixup (Mixup) [55], Meta-Learning for Domain Generalization (MLDG) [21], Deep CORrelation ALignment (CORAL) [56], Maximum Mean Discrepancy (MMD) [20], Domain Adversarial Neural Networks (DANN) [19], Class-conditional DANN (CDANN) [57], Marginal Transfer Learning (MTL) [58], Style-Agnostic Networks (SagNet) [16], Adaptive Risk Minimization (ARM) [59], Variance Risk Extrapolation (VREx) [60], Representation Self Challenging (RSC) [14], Self-supervised con-

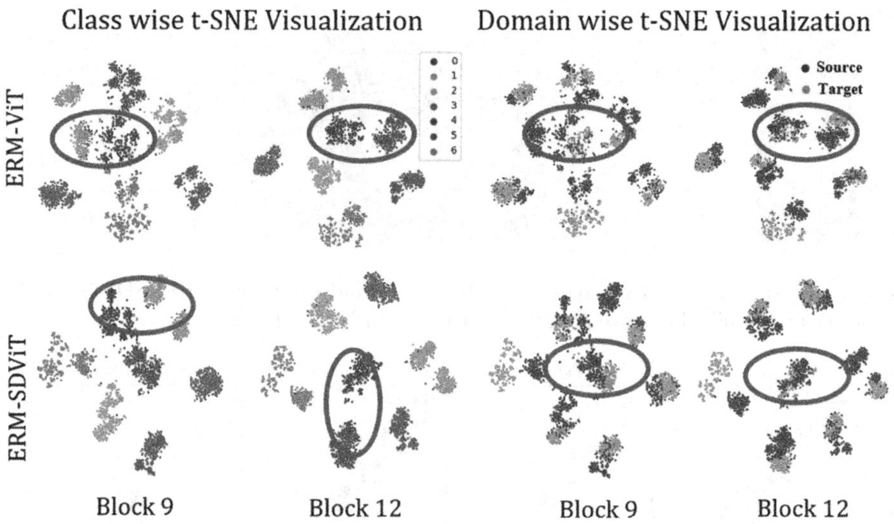

Fig. 5. t-SNE visualization of features from different blocks (9 & 12) in baseline and our approach. Left: Features are colored corresponding to their class labels (classes: 7,`PACS` dataset). Right: Features are colored corresponding to their domain labels. Our approach has performed well for instance, in class-wise t-SNE in block 9, the features of class 0 and 1 (highlighted in red circle) are well separated as compared to ERM-ViT baseline. Similarly in class 0 and 4 in the final 12th block features of our ERM-SDViT approach are also separated clearly. While in domain-wise t-SNE, a similar pattern is observed, as source and target domain features are more overlapped with each other and clearly separated as well. See Appendix for more t-SNE results. (Color Figure Online)

trastive regularization (SelfReg) [17], meta-Domain Specific-Domain Invariant (mDSDI) [18], and Stochastic Weight Averaging Densely (SWAD) [30].

`VLCS` and `OfficeHome`: In `VLCS`, our approach (ERM-SDViT) records the best classification accuracy of 79.5% with T2T-ViT-14 backbone outperforming the baseline (ERM-ViT) and DG SOTA algorithms. Similarly, in `OfficeHome`, our method outperforms all other methods under all three ViT backbones. In particular, it displays the best accuracy of 75.6% with CvT-21 backbone.

`PACS`, `DomainNet` and `TerraInc`: In `PACS` our approach delivers the top accuracy of 88.3% and in `DomainNet` it achieves an accuracy of 50.4% with CvT-21 backbone. In `TerraIncognita`, our method achieves a competitive accuracy of 50.6% with T2T-ViT-14 backbone against the top performing method of SelfReg [17]. In the overall average accuracy over five datasets, our method outperforms the existing state-of-the-art in DG with CvT-21 and T2T-ViT-14 backbones. Moreover, it provides notable gains over the baseline (ERM-ViT) under the three recent ViT backbones architectures.

Fig. 6. Attention maps from baseline (ERM-ViT) and proposed (ERM-SDViT, backbone: DeiT-Small). They are computed at the final block of ViT.

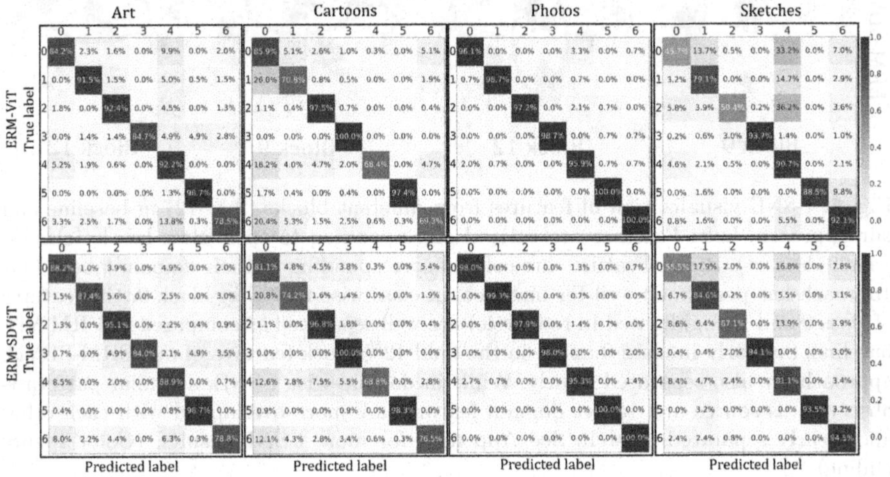

Fig. 7. Confusion matrices for baseline and our method. The classes in the Figure are '0':Dog, '1': Elephant, '2':Giraffe, '3':Guitar, '4':Horse, '5':House, and '6':Person.

4.2 Ablation Study and Analysis

In all experiments, unless stated otherwise, the baseline method is ERM-ViT with DeiT-Small backbone [39].

With a Recent DG Baseline: We show that our proposed approach is also effective in further improving the performance of a strong DG baseline namely T3A [61] (see Table 1 and 2). T3A computes a pseudo-prototype representation for each class using unlabeled data augmented by the base classifier trained in the source domains in an online manner. A test image is classified based on its distance to the pseudo-prototype representation. Our proposed approach with T3A (ERM-SDViT+T3A) consistently improves the performance of the baseline (ERM-ViT+T3A) with all three ViT backbones.

Feature Visualizations: Figure 5 (left) visualizes the class-wise feature representations of different blocks using t-SNE in baseline and our approach. Our

Table 2. Our proposed approach is also effective in further improving the performance of a strong DG baseline namely T3A [61]. Results are reported with different ERM-ViT backbone architectures: DeiT-Small, CvT-21, and T2T-ViT-14.

Model	Backbone	# of Params	Art	Cartoons	Photos	Sketches	Average
ERM	ResNet-50	23.5M	81.3 ± 0.6	80.9 ± 0.3	96.3 ± 0.6	78.0 ± 1.6	84.1 ± 0.4
ERM-ViT	DeiT-Small	22M	87.4 ± 1.2	81.5 ± 0.8	98.1 ± 0.1	72.6 ± 3.3	84.9 ± 0.9
ERM-ViT + T3A	DeiT-Small	22M	88.1 ± 1.5	81.8 ± 0.8	98.3 ± 0.1	73.8 ± 2.7	85.5 ± 0.7
ERM-SDViT	DeiT-Small	22M	87.6 ± 0.3	82.4 ± 0.4	98.0 ± 0.3	77.2 ± 1.0	86.3 ± 0.2
ERM-SDViT + T3A	DeiT-Small	22M	88.2 ± 0.4	82.9 ± 0.5	98.3 ± 0.1	77.2 ± 0.9	86.7 ± 0.2
ERM-ViT	CvT-21	32M	89.0 ± 1.0	84.8 ± 0.6	98.8 ± 0.2	78.6 ± 0.3	87.8 ± 0.1
ERM-ViT + T3A	CvT-21	32M	90.1 ± 0.7	85.3 ± 0.6	99.0 ± 0.1	79.6 ± 0.4	88.5 ± 0.1
ERM-SDViT	CvT-21	32M	90.8 ± 0.1	84.1 ± 0.5	98.3 ± 0.2	80.0 ± 1.3	88.3 ± 0.2
ERM-SDViT + T3A	CvT-21	32M	91.2 ± 0.8	83.5 ± 0.2	98.3 ± 0.1	82.5 ± 1.5	**88.9 ± 0.5**
ERM-ViT	T2T-ViT-14	21.5M	89.6 ± 0.9	81.0 ± 0.9	98.9 ± 0.2	77.6 ± 2.6	86.8 ± 0.4
ERM-ViT + T3A	T2T-ViT-14	21.5M	90.7 ± 1.0	82.4 ± 0.6	99.1 ± 0.1	78.5 ± 2.2	87.7 ± 0.4
ERM-SDViT	T2T-ViT-14	21.5M	90.2 ± 1.2	82.7 ± 0.7	98.6 ± 0.2	80.5 ± 2.2	88.0 ± 0.7
ERM-SDViT + T3A	T2T-ViT-14	21.5M	89.2 ± 1.8	84.0 ± 0.1	98.7 ± 0.1	79.3 ± 1.1	87.8 ± 0.6

Table 3. Performance with different block selection strategies in our self-distillation.

Model	Art	Cartoons	Photos	Sketches	Average
ERM-ViT	87.4 ± 1.2	81.5 ± 0.8	98.1 ± 0.1	72.6 ± 3.3	84.9 ± 0.9
ERM-SDViT[0-5]	87.3 ± 0.2	82.1 ± 0.4	98.3 ± 0.1	76.6 ± 2.1	86.1 ± 0.5
ERM-SDViT[6-11]	86.8 ± 0.8	81.4 ± 0.3	98.6 ± 0.1	74.3 ± 1.7	85.3 ± 0.3
ERM-ViT[self.dist all blocks]	87.8 ± 1.9	82.2 ± 0.7	97.9 ± 0.1	75.0 ± 1.1	85.7 ± 0.3
ERM-SDViT[0-11] Ours	87.6 ± 0.3	82.4 ± 0.4	98.0 ± 0.3	77.2 ± 1.0	**86.3 ± 0.2**

approach facilitates learning more discriminative features. As a result, the intra-class features are compactly clustered while the inter-class features are far apart. Likewise, Fig. 5 (right) visualizes the same features, however, they are colored based on their domain labels. Our method promotes a greater overlap between the features of source and target domains. Moreover, we quantify the domain overlap between the source/target features using cosine similarity (see Table 4).

SDViT with ERM Based on ResNet: Table 5 shows that our SD can also improve the performance of a competitive DG baseline based on ResNet-50.

Table 4. Domain overlap quantified based on cosine similarity between source and target domain class tokens.

Model	Art	Cartoons	Photos	Sketches
ERM-ViT	0.908	0.882	0.921	0.748
ERM-SDViT	**0.948**	**0.904**	**0.950**	**0.805**

Table 5. Our self distillation (SD) improves the performance of ResNet-50 over DG baseline (ERM).

ResNet50		DeiT-Small	
ERM	Ours (ERM-SD)	ERM	Ours (ERM-SD)
84.1 ± 0.4	**85.7 ± 0.4**	84.9 ± 0.9	**86.3 ± 0.2**

On Different Block Selection Techniques: We show performance with different ways of selecting blocks in our self-distillation method (see Table 3). First, we restrict the random sampling of blocks to earlier blocks i.e. from block 0 to block 5. Second, we limit the random sampling of blocks to later blocks i.e. from block 6 to block 11. Finally, we do not perform any sampling in any range and rather include all the blocks [46]. Compared to all these block selection techniques, our proposal of random sampling from the full range of blocks shows the best (overall) average accuracy of 86.3%. Sampling from earlier blocks seems beneficial as compared to the later blocks. When earlier blocks, with relatively longer feature pathways to the final block, are provided with soft target labels, there is potentially greater room for exploring cross-domain generalizable features.

Confusion Matrices: Figure 7 visualizes the confusion matrices for the baseline and our method on PACS dataset. Compared to baseline, our method produces less number of false positives, particularly in 'Sketches' as the target domain.

What Kind of Features our DG Approach Facilitates?: We visualize the features used by the baseline and our method to make predictions through visualizing attention maps (see Fig. 6). In all target domains, our method mostly attends to features that mainly capture the semantics and the shape of the object class. Whereas the baseline has a greater tendency to attend background features and thus focus less on the object class features.

Training Overhead: Table 6 reports training overhead, computed as a relative % increase in training time (hrs.), introduced by our method on top of the baseline. Our method adds very little training overhead over the baseline.

Performance Under Different Domain Shifts: We benchmark the performance under various domain shifts, including background shifts, corruption shifts, texture shifts, and style shifts. For instance, background shifts do not affect pixel, shape, texture, and structures in the foreground objects. Whereas style shifts typically depict variance at different stages of concepts, including texture, shape, and object part [44]. To this end, we classify five DG datasets, including `PACS`, `VLCS`, `OfficeHome`, `TerraIncognita`, and `DomainNet`, into these four different domain shift categories based on the type of shift(s) exhibited by them.

Table 6. Training overhead, computed as relative % increase in training time (hrs.), introduced by our method on top of the baseline.

Model	Art	Cartoons	Photos	Sketches
ERM-ViT	0.266	0.270	0.278	0.267
ERM-SDViT	0.279	0.276	0.279	0.278
Rel.overhead	4.88	2.22	0.35	4.11

Table 7. Performance under various domain shifts. Each entry is the accuracy (%) averaged over the datasets belonging to a certain domain shift category.

Methods	Shift type			
	Background shifts	Corruption shifts	Texture shifts	Style shifts
	(VLCS,Terra)	(Terra)	(PACS,DomainNet)	(OH,PACS,DomainNet)
ERM-ViT	60.0	43.4	65.2	67.2
ERM-SDViT	**61.6**	**44.3**	**66.0**	**67.8**

Table 7 reports the results by ERM-ViT (baseline) and ERM-SDViT (ours) under four different kinds of domain shifts. We observe that ERM-SDViT outperforms ERM-ViT across the whole spectrum of domain shifts. See Appendix for more results.

5 Conclusion

We propose a simple, plug-and-play approach namely self-distillation in ViTs for tackling DG. It provides soft supervision to the intermediate blocks of ViTs to strengthen their internal representations, thereby moderating the learning of input-output mapping problem. Extensive experiments on five datasets with different DG baselines and ViT backbones, including comparisons with the recent SOTA, validate the effectiveness of our approach for ViTs tackling DG problem.

References

1. Vaswani, A., et al.: Attention is all you need. Adv. Neural Inf. Process. Syst. **30**, 1–11 (2017)
2. Devlin, J., Chang, M.W., Lee, K., Toutanova, K.: Bert: pre-training of deep bidirectional transformers for language understanding. arXiv preprint arXiv:1810.04805 (2018)
3. Brown, T., et al.: Language models are few-shot learners. Adv. Neural Inf. Process. Syst. **33**, 1877–1901 (2020)
4. Dosovitskiy, A., et al.: An image is worth 16×16 words: transformers for image recognition at scale. arXiv preprint arXiv:2010.11929 (2020)
5. Carion, N., Massa, F., Synnaeve, G., Usunier, N., Kirillov, A., Zagoruyko, S.: End-to-end object detection with transformers. In: Vedaldi, A., Bischof, H., Brox, T., Frahm, J.-M. (eds.) ECCV 2020. LNCS, vol. 12346, pp. 213–229. Springer, Cham (2020). https://doi.org/10.1007/978-3-030-58452-8_13

6. Zhu, X., Su, W., Lu, L., Li, B., Wang, X., Dai, J.: Deformable detr: deformable transformers for end-to-end object detection. arXiv preprint arXiv:2010.04159 (2020)
7. Zheng, S., et al.: Rethinking semantic segmentation from a sequence-to-sequence perspective with transformers. In: Proceedings of the IEEE/CVF Conference on Computer Vision and Pattern Recognition, pp. 6881–6890 (2021)
8. Wang, W., et al.: Pyramid vision transformer: a versatile backbone for dense prediction without convolutions. In: Proceedings of the IEEE/CVF International Conference on Computer Vision, pp. 568–578 (2021)
9. Muandet, K., Balduzzi, D., Schölkopf, B.: Domain generalization via invariant feature representation. In: International Conference on Machine Learning, pp. 10–18 (2013)
10. Ghifary, M., Bastiaan Kleijn, W., Zhang, M., Balduzzi, D.: Domain generalization for object recognition with multi-task autoencoders. In: Proceedings of the IEEE International Conference on Computer Vision, pp. 2551–2559 (2015)
11. Li, D., Yang, Y., Song, Y.Z., Hospedales, T.M.: Deeper, broader and artier domain generalization. In: Proceedings of the IEEE International Conference on Computer Vision, pp. 5542–5550 (2017)
12. Carlucci, F.M., D'Innocente, A., Bucci, S., Caputo, B., Tommasi, T.: Domain generalization by solving jigsaw puzzles. In: Proceedings of the IEEE Conference on Computer Vision and Pattern Recognition, pp. 2229–2238 (2019)
13. Gulrajani, I., Lopez-Paz, D.: In search of lost domain generalization. ArXiv abs/2007.01434 (2021)
14. Huang, Z., Wang, H., Xing, E.P., Huang, D.: Self-challenging improves cross-domain generalization (2020)
15. Khan, M.H., Zaidi, T., Khan, S., Khan, F.S.: Mode-guided feature augmentation for domain generalization (2021)
16. Nam, H., Lee, H., Park, J., Yoon, W., Yoo, D.: Reducing domain gap by reducing style bias. In: Proceedings of the IEEE/CVF Conference on Computer Vision and Pattern Recognition (CVPR), pp. 8690–8699 (2021)
17. Kim, D., Yoo, Y., Park, S., Kim, J., Lee, J.: Selfreg: self-supervised contrastive regularization for domain generalization. In: Proceedings of the IEEE/CVF International Conference on Computer Vision (ICCV), pp. 9619–9628 (2021)
18. Bui, M.H., Tran, T., Tran, A., Phung, D.: Exploiting domain-specific features to enhance domain generalization. Adv. Neural Inf. Process. Syst. **34**, 21189–21201 (2021)
19. Ganin, Y., et al.: Domain-adversarial training of neural networks. J. Mach. Learn. Res. **17**, 2030–2096 (2016)
20. Li, H., Pan, S.J., Wang, S., Kot, A.C.: Domain generalization with adversarial feature learning. In: Proceedings of the IEEE Conference on Computer Vision and Pattern Recognition, pp. 5400–5409 (2018)
21. Li, D., Yang, Y., Song, Y.Z., Hospedales, T.M.: Learning to generalize: meta-learning for domain generalization. In: Thirty-Second AAAI Conference on Artificial Intelligence (2018)
22. Dou, Q., de Castro, D.C., Kamnitsas, K., Glocker, B.: Domain generalization via model-agnostic learning of semantic features. Adv. Neural Inf. Process. Syst. **32**, 6450–6461 (2019)
23. Balaji, Y., Sankaranarayanan, S., Chellappa, R.: Metareg: towards domain generalization using meta-regularization. Adv. Neural Inf. Process. Syst. **31**, 998–1008 (2018)

24. Li, D., Zhang, J., Yang, Y., Liu, C., Song, Y.Z., Hospedales, T.M.: Episodic training for domain generalization. In: ICCV (2019)
25. Shankar, S., Piratla, V., Chakrabarti, S., Chaudhuri, S., Jyothi, P., Sarawagi, S.: Generalizing across domains via cross-gradient training. arXiv preprint arXiv:1804.10745 (2018)
26. Volpi, R., Namkoong, H., Sener, O., Duchi, J.C., Murino, V., Savarese, S.: Generalizing to unseen domains via adversarial data augmentation. Adv. Neural Inf. Process. Syst. **31**, 5334–5344 (2018)
27. Zhou, K., Yang, Y., Hospedales, T., Xiang, T.: Learning to generate novel domains for domain generalization (2020)
28. Wang, S., Yu, L., Li, C., Fu, C.W., Heng, P.A.: Learning from extrinsic and intrinsic supervisions for domain generalization (2020)
29. Naseer, M., Ranasinghe, K., Khan, S., Khan, F.S., Porikli, F.: On improving adversarial transferability of vision transformers. arXiv preprint arXiv:2106.04169 (2021)
30. Cha, J., et al.: Swad: domain generalization by seeking flat minima. Adv. Neural Inf. Process. Syst. **34**, 22405–22418 (2021)
31. Vapnik, V.: The Nature of Statistical Learning Theory. Springer, Heidelberg (1999). https://doi.org/10.1007/978-1-4757-3264-1
32. Yang, P.Y., Gao, W.: Multi-view discriminant transfer learning (2013)
33. Arjovsky, M., Bottou, L., Gulrajani, I., Lopez-Paz, D.: Invariant risk minimization. arXiv preprint arXiv:1907.02893 (2019)
34. Xu, Z., Li, W., Niu, L., Xu, D.: Exploiting low-rank structure from latent domains for domain generalization. In: Fleet, D., Pajdla, T., Schiele, B., Tuytelaars, T. (eds.) ECCV 2014. LNCS, vol. 8691, pp. 628–643. Springer, Cham (2014). https://doi.org/10.1007/978-3-319-10578-9_41
35. Chattopadhyay, P., Balaji, Y., Hoffman, J.: Learning to balance specificity and invariance for in and out of domain generalization (2020)
36. Seo, S., Suh, Y., Kim, D., Kim, G., Han, J., Han, B.: Learning to optimize domain specific normalization for domain generalization (2020)
37. Motiian, S., Piccirilli, M., Adjeroh, D.A., Doretto, G.: Unified deep supervised domain adaptation and generalization. In: Proceedings of the IEEE International Conference on Computer Vision, pp. 5715–5725 (2017)
38. Wu, H., et al.: Cvt: introducing convolutions to vision transformers. In: Proceedings of the IEEE/CVF International Conference on Computer Vision, pp. 22–31 (2021)
39. Touvron, H., Cord, M., Douze, M., Massa, F., Sablayrolles, A., Jégou, H.: Training data-efficient image transformers & distillation through attention. In: International Conference on Machine Learning, pp. 10347–10357. PMLR (2021)
40. Dai, X., Chen, Y., Yang, J., Zhang, P., Yuan, L., Zhang, L.: Dynamic detr: end-to-end object detection with dynamic attention. In: Proceedings of the IEEE/CVF International Conference on Computer Vision (ICCV), pp. 2988–2997 (2021)
41. Strudel, R., Garcia, R., Laptev, I., Schmid, C.: Segmenter: transformer for semantic segmentation. In: Proceedings of the IEEE/CVF International Conference on Computer Vision (2021) 7262–7272
42. Lu, Z., He, S., Zhu, X., Zhang, L., Song, Y.Z., Xiang, T.: Simpler is better: few-shot semantic segmentation with classifier weight transformer. In: Proceedings of the IEEE/CVF International Conference on Computer Vision, pp. 8741–8750 (2021)
43. Yuan, L., et al.: Tokens-to-token vit: training vision transformers from scratch on imagenet. In: Proceedings of the IEEE/CVF International Conference on Computer Vision, pp. 558–567 (2021)
44. Zhang, C., et al.: Delving deep into the generalization of vision transformers under distribution shifts. arXiv preprint arXiv:2106.07617 (2021)

45. Hinton, G., Vinyals, O., Dean, J.: Distilling the knowledge in a neural network (2015). arXiv preprint arXiv:1503.02531
46. Zhang, L., Song, J., Gao, A., Chen, J., Bao, C., Ma, K.: Be your own teacher: improve the performance of convolutional neural networks via self distillation. In: Proceedings of the IEEE/CVF International Conference on Computer Vision, pp. 3713–3722 (2019)
47. Yun, S., Park, J., Lee, K., Shin, J.: Regularizing class-wise predictions via self-knowledge distillation. In: Proceedings of the IEEE/CVF Conference on Computer Vision and Pattern Recognition, pp. 13876–13885 (2020)
48. Wang, Y., Li, H., Chau, L.P., Kot, A.C.: Embracing the dark knowledge: domain generalization using regularized knowledge distillation. In: Proceedings of the 29th ACM International Conference on Multimedia, pp. 2595–2604 (2021)
49. Fang, C., Xu, Y., Rockmore, D.N.: Unbiased metric learning: on the utilization of multiple datasets and web images for softening bias. In: Proceedings of the IEEE International Conference on Computer Vision, pp. 1657–1664 (2013)
50. Venkateswara, H., Eusebio, J., Chakraborty, S., Panchanathan, S.: Deep hashing network for unsupervised domain adaptation. In: Proceedings of the IEEE Conference on Computer Vision and Pattern Recognition, pp. 5018–5027 (2017)
51. Beery, S., Van Horn, G., Perona, P.: Recognition in terra incognita. In: Proceedings of the European conference on computer vision (ECCV), pp. 456–473 (2018)
52. Peng, X., Bai, Q., Xia, X., Huang, Z., Saenko, K., Wang, B.: Moment matching for multi-source domain adaptation. In: Proceedings of the IEEE International Conference on Computer Vision, pp. 1406–1415 (2019)
53. Loshchilov, I., Hutter, F.: Fixing weight decay regularization in adam (2018)
54. Sagawa, S., Koh, P.W., Hashimoto, T.B., Liang, P.: Distributionally robust neural networks for group shifts: on the importance of regularization for worst-case generalization. arXiv preprint arXiv:1911.08731 (2019)
55. Yan, S., Song, H., Li, N., Zou, L., Ren, L.: Improve unsupervised domain adaptation with mixup training. arXiv preprint arXiv:2001.00677 (2020)
56. Sun, B., Saenko, K.: Deep CORAL: correlation alignment for deep domain adaptation. In: Hua, G., Jégou, H. (eds.) ECCV 2016. LNCS, vol. 9915, pp. 443–450. Springer, Cham (2016). https://doi.org/10.1007/978-3-319-49409-8_35
57. Li, Y., et al.: Deep domain generalization via conditional invariant adversarial networks. In: Proceedings of the European Conference on Computer Vision (ECCV), pp. 624–639 (2018)
58. Blanchard, G., Deshmukh, A.A., Dogan, U., Lee, G., Scott, C.: Domain generalization by marginal transfer learning. arXiv preprint arXiv:1711.07910 (2017)
59. Zhang, M., Marklund, H., Dhawan, N., Gupta, A., Levine, S., Finn, C.: Adaptive risk minimization: learning to adapt to domain shift. Adv. Neural Inf. Process. Syst. **34**, 23664–23678 (2021)
60. Krueger, D., et al.: Out-of-distribution generalization via risk extrapolation (rex). In: International Conference on Machine Learning, pp. 5815–5826. PMLR (2021)
61. Iwasawa, Y., Matsuo, Y.: Test-time classifier adjustment module for model-agnostic domain generalization. Adv. Neural Inf. Process. Syst. **34**, 2427–2440 (2021)

CV4Code: Sourcecode Understanding via Visual Code Representations

Ruibo Shi(✉), Lili Tao, Rohan Saphal, Fran Silavong, and Sean Moran

JP Morgan Chase, New York, USA
ruibo.shi@jpmchase.com

Abstract. We present CV4Code[1], a compact and effective *computer vision* method for sourcecode understanding. Our method leverages the contextual and the structural information available from the code snippet by treating each snippet as a two-dimensional image, which naturally encodes the context and retains the underlying structural information through an explicit spatial representation. To codify snippets as images, we propose an ASCII codepoint-based image representation that facilitates fast generation of sourcecode images and eliminates redundancy in the encoding that would arise from an RGB pixel representation. Furthermore, as sourcecode is treated as images, neither lexical analysis (tokenisation) nor syntax tree parsing is required, which makes the proposed method agnostic to any particular programming language and lightweight from the application pipeline point of view. CV4Code can even featurise syntactically incorrect code which is not possible from methods that depend on the Abstract Syntax Tree (AST). We demonstrate the effectiveness of CV4Code by learning Convolutional and Transformer networks to predict the functional task, *i.e.* the problem it solves, of the source code directly from its two-dimensional representation, and using an embedding from its latent space to derive a similarity score of two code snippets in a retrieval setup. Experimental results show that our approach achieves state-of-the-art performance in comparison to other methods with the same task and data configurations. For the first time we show the benefits of treating sourcecode understanding as a form of image processing task. ([1] https://github.com/jpmorganchase/cv4code)

Keywords: Sourcecode understanding · ResNet · Transformer · ViT

1 Introduction

Machine Learning on Sourcecode (MLOnCode) promises to redefine how software is delivered through intelligent augmentation of the software development lifecycle (SDLC). Automation of routine tasks with software makes our lives more comfortable and efficient. For example, software drives the global economy and transfer of value worldwide making the purchase of goods and the management of finances a seamless experience. Furthermore, at the touch of a few buttons on our smartphone we can communicate to friends and family worldwide. Reliable software can also change healthcare outcomes by making it easier

L. Wang et al. (Eds.): ACCV 2022, LNCS 13842, pp. 291–306, 2023.
https://doi.org/10.1007/978-3-031-26284-5_18

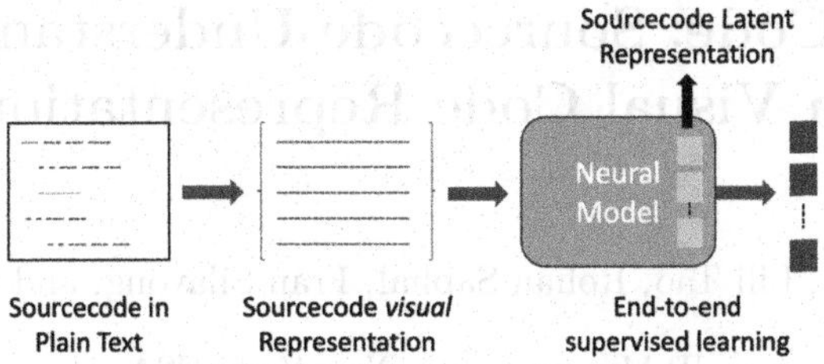

Fig. 1. The proposed CV4Code code understanding pipeline.

for doctors to diagnose disease. A key to accomplishing these feats of automation and being prepared for future complex use-cases is accelerating the software development process without sacrificing software quality, robustness and time-to-market. Augmenting the SDLC with machine learning holds this promise, in which the developer's capabilities are magnified through predictive analytics driven by the vast quantities of exhaust data naturally produced by the SDLC. Machine learning can potentially enhance every stage of the SDLC including requirements gathering, build and test and deployment. For example, AI-driven code auto-completion and enhanced code search are near-term possibilities for enhancing developer productivity with startups and established companies alike productionising such capabilities for mass consumption.

The academic field of MLOnCode explores the application of machine learning techniques for mining the massive amount of sourcecode and associated metadata available in public and private repositories [3]. Indicative tasks in this field include code search using natural language keywords [28] and sourcecode [17] as queries, automated bug finding [21], vulnerability detection [24], design pattern detection [29], program repair [7] and code auto-completion [27]. Core to the field of MLOnCode is the learning of expressive sourcecode latent feature representations ("code vectors") that capture semantics of programs and can be flexibly used in generic machine learning classifiers to support a myriad of downstream tasks, such as code search and repository annotation with semantic keywords. Code is unique from natural language in many respects, for example more distantly spaced tokens may be highly related (*e.g.* opening and closing brackets) and code that looks very similar can have very different behaviour (and vice-versa). Capturing these subtleties and intricacies of sourcecode to learn program semantics requires methods that can understand the underlying context (sequence of tokens) and structure (as presented by syntax parse tree) of the language. Prior methods for sourcecode feature extraction can be differentiated to the extent on which they capture context, structure or both when learning code vectors. For example, early methods treat sourcecode either as a set of independent tokens [2], a sequence of tokens (processed sequentially by an RNN or CNN) or generate an Abstract Syntax Tree (AST) from the code snippet before linearising the tree into vector form, thereby capturing local structure in the snippet [17]. Drawing a parallel between the word2vec model in natural language processing (NLP), Alon *et al.* [5] propose a code2vec alternative that uses the proxy task of method name prediction based on paths in the AST tree to learn expressive code vectors from a shallow neural network architecture. More recent research has explored transformer architectures to learn effective code vectors [31] that capture structure and context. In contrast to prior research, we represent sourcecode in a visual way as a set of *images* that explicitly, through the 2-dimensional spatial representation, present both the code structure and context directly to the learning algorithm. We adapt well-known image processing techniques to process these sourcecode images. We argue that, with no assumption of naturalness in programming language [3], treating sourcecode understanding as a computer vision problem can not only produce more effective code vectors, but can also address key limitations of existing methods such as their inability to featurise partial code snippets and syntactically incorrect code.

In more detail we introduce a series of vision models, including Residual Convolutional Neural Networks (CNNs) [13] and Vision Transformers (ViT) variants [11,14,15], adapted for sourcecode understanding (Fig. 1). We contribute to the sparse amount of prior research that draws a parallel between successful image understanding models in the field of Computer Vision and their application to representation learning for sourcecode [6,9,23]. Different to this closely related prior research, CV4Code does not require any language-specific pre-processing (*e.g.* extraction of syntax parse trees). To represent sourcecode in a visual form, we propose a novel compact encoding of sourcecode as a two dimensional spatial grid of numeric values that represent the characters in the code by their ASCII codepoints. This representation is advantageous over a standard RGB pixel representation of the code for two key reasons: 1) *elimination of redundancy;* for a pixel representation many pixels would be dedicated to encoding a single character; and 2) *fast feature generation:* we find it multiple order of magnitude slower and less scalable to render a pixel representation of code (sub-second) compared to the proposed code representation (sub-millisecond). The computational speed-up enables real-time applications over large codebases, such as code search directly within the Integrated Development Environment (IDE). The proposed image encoding for sourcecode is therefore practical for real-world machine learning pipelines. The CV4Code architecture is designed for learning representations effectively from the ASCII-based codepoint encoded sourcecode images. CV4Code ingests the image representation of code and encodes each pixel as either a one-hot or learnable embedding. The encoded input is subsequently processed by a neural network that learns features expressive for the predictive task *e.g.* the language of code, the task being solved by the code *etc.*. And the learned latent embedding from CV4Code can be used as code embeddings for other MLOnCode tasks, similar to VGG features [26] that have been shown to be a powerful and flexible embedding of images for many computer vision tasks.

Our contributions in this paper can be summarised as:

- **CV4Code Deep Neural Models:** We introduce and compare modern deep vision models adapted for language-agnostic sourcecode understanding that learns from a novel ASCII codepoint representation of sourcecode. We report state-of-the-art performance on a public benchmark dataset compared to competitive baselines. Compared to a NLP-inspired Transformer strong base-

line, we achieve 4.06% absolute gain in top-1 accuracy on a language-agnostic problem classification task, and 0.011 gain in mAP@R on a similarity based sourcecode retrieval task.
- **ASCII Sourcecode Encoding:** We introduce an ASCII codepoint image representation for sourcecode that efficiently (low redundancy, fast generation) encodes snippets capturing both code structure and context in a single representation.

2 Related Work

Machine learning for sourcecode analysis aims at learning semantically meaningful representation of the code and then apply the embeddings on downstream tasks, such as code quality detection, code summarisation, defect prediction and code duplication detection [25]. We include the research that are most relevant to our contribution.

AST Based Representation. Machine learning based intelligent code analysis relies on extracting representative features from sourcecode. The majority of studies leverage structured graphical models for sourcecode, through parse trees. AST carries rich semantic and syntactic information and provides a unique representation of a sourcecode snippet in a given language and grammar [4,30,32]. The paper [32] learned jointly from the AST and the sourcecode of programs while relying on language-agnostic features, and performed on code summarisation task on five programming languages. Although the model does not rely on language-specific features, parsing sourcecode to a tree structure is language dependent.

NLP Based Techniques. A sourcecode sample can be treated as a piece of text. A code snippet can be represented by a vector of frequencies of token occurrences, similar to the bag of word model. The frequently used tokens include regex, keywords and operators [20,22]. Considering the lack of sequential information retained in the bag of tokens method, a sequence of token method uses the same set of tokens but keeps the order information to form a sequence [22]. Such a token embedding layer is then input into a CNN based model.

Computer Vision for Code. Despite the fact that more effort has been made on automated sourcecode analysis using machine learning, representing and processing sourcecode in the form of image data is still an under-explored area in the field. Dey *et al.* [9] automatically convert program sourcecode to visual images via an intermediate representation of code created by the LLVM compiler. The ASCII value of the remaining characters are treated as a pixel value in a predefined empty image canvas. Bilgin *et al.* [6] use a coloured image of syntax trees for representing code, where the tokens are plotted in a rectangular shape and are completed with specific colours to indicate the type and content of the token.

The most recent work by Rabin *et al.* [23] on Java code analysis transforms the original code by removing comments and empty lines using a JavaParser tool, and then redacts snapshots of input by replacing any alphanumeric characters in the reformatted input programs with a single letter 'x' to emphasise the structure of code snippets, rather than their content (*e.g.* the specific naming of variables). While the aforementioned research exploited image-based representation for sourcecode, they all require a parser for specific programming language, which cannot process the sentences with incorrect syntax.

3 CV4Code

We propose CV4Code, an end-to-end learning framework for sourcecode understanding by treating sourcecode snippets as *images i.e.*. 2-dimensional matrix.

3.1 Sourcecode Representation

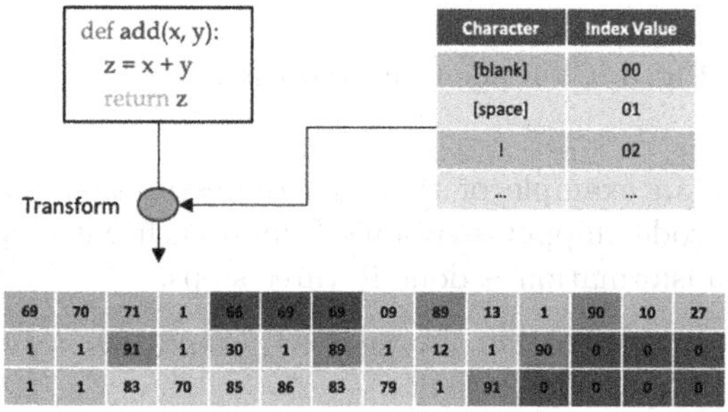

Fig. 2. Example of 2D code representation generation

While the sourcecode of most modern programming languages can be written in plain text from an extensive character set, only a small set of tokens and their composing characters have syntactic and semantic roles. In CV4Code, code snippets are transformed into 2-dimensional (matrix) representation by mapping each printable ASCII character to their unique index values and padding the special *[blank]* token wherever necessary to retain the rectangular shape of the output. The set of valid printable ASCII characters together with the special padding token $\mathbb{V}_c$, $|\mathbb{V}_c| = 96$, consists of the following:

```
abcdefghijklmnopqrstuvwxyz
ABCDEFGHIJKLMNOPQRSTUVWXYZ
0123456789
!"#$%&'()*+,-./:;<=>?@[\]^_`{}|~
[space][blank]
```

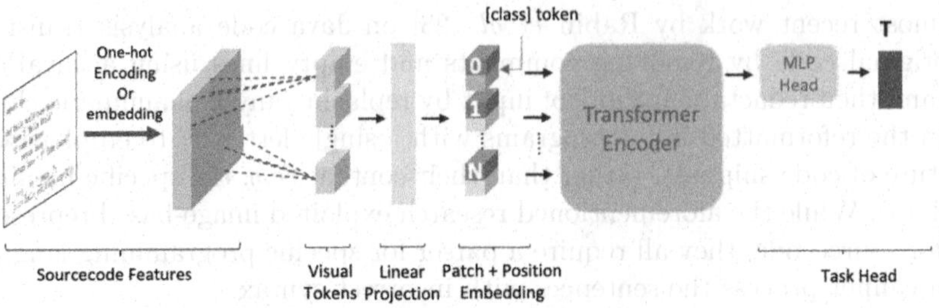

(a) CV4Code ViT model overview.

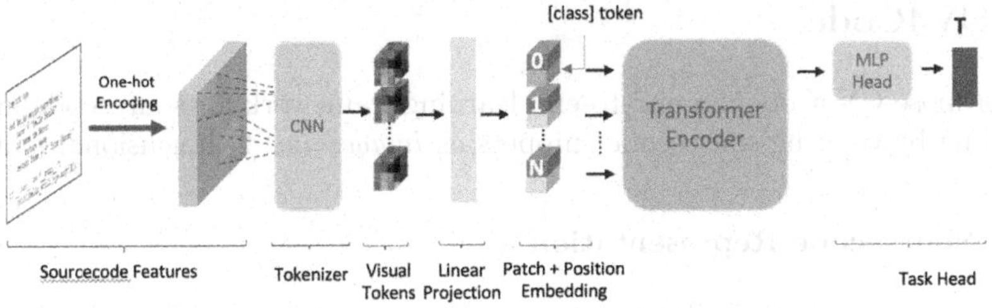

(b) CV4Code Conv-ViT model overview.

Fig. 3. CV4Code transformer model variants.

Figure 2 shows an example of the code representation generation process. Specifically, for a code snippet spanning L lines each with C_l, $l \in 0, ..., L-1$ characters, the transformation is done in three steps:

1. Remove characters not within the valid set, output has $\hat{L}$ lines each with $\hat{C}_l$, $l \in 0, ..., \hat{L}-1$ characters;
2. Map each input character $v_k \in \mathbb{V}_c$ to its index value k;
3. Pad each line to $M = \max_{l=0}^{\hat{L}-1} \hat{C}_l$ long with the index value of *[blank]*, generate the output 2-dimensional code matrix $X \in \mathbb{R}^{L \times M}$.

Compared to human-readable images of sourcecode, *e.g.* screenshots of code snippets, which usually are sparse in semantics and require multiple pixels to represent a single character, the proposed sourcecode representation is compact and do not introduce any unnecessary information other than the blank padding that is required to keep the spatial relations.

As the code image, *i.e.* the compact 2-dimensional sourcecode representation, encodes the character index values which do not form a numerically continuous space, unless otherwise specified, one-hot encoding is used in this work to transform each *pixel* in the code image to a vector of fixed dimension equal to the size of the set of valid characters, i.e. $\mathbf{X} \in \mathbb{R}^{L \times M} \rightarrow \hat{\mathbf{X}} \in \mathbb{R}^{L \times M \times |\mathbb{V}_c|}$.

3.2 Model

We apply and adapt state-of-the-art vision models, including ResNet [13], ViT [14], ViT for small-size datasets (ViT-fsd) [15] and hybrid Convolutional Transformer

(Conv-ViT) [11], on the proposed sourcecode representation. And we show the effectiveness of the proposed method through experiments on a supervised multi-class classification task. Figure 3 shows an overview of the CV4Code transformer model variants. While Fig. 3a shows a general architecture of CV4Code-ViT model, differences exist in ViT, ViT-fsd which we briefly describe below.

ViT. We follow [14] and split images into non-overlapping fixed-size patches and prepend a learnable *[class]* embedding whose state at the ViT output serves as the sourcecode representation. This sourcecode representation is then passed to a single-layer MLP head for the classification task.

ViT-fsd. While the same setup as ViT is used, we apply shifted patch tokenization and Locality Self-Attention proposed in [15]. In addition, as the tokenization process creates 4 extra shifted images leading to a largely increased input dimensionality after concatenation, to control the number of parameters in the linear projection layer, instead of one-hot encoding, *i.e.* $\hat{\mathbf{X}} \in \mathbb{R}^{L \times M \times |\mathbb{V}_c|}$, an 32-dimensional learnable embedding is used such that $\hat{\mathbf{X}} \in \mathbb{R}^{L \times M \times 32}$.

Conv-ViT. Shown in Fig. 3b. To leverage CNN's inductive bias, *e.g.* locality, similar to [11] and hybrid in [14] we use convolutional layers to create soft visual tokens but keep the use of *[class]* embedding to generate sourcecode representation at the output. Furthermore, as the soft tokenization process does not require fixed-size input, similar to NLP applications of transformers where the input sequence is of variable length, for smaller sourcecode image we append learnable *[pad]* embeddings to the generated visual token sequence, *i.e.* at the output of the CNN, to form equal-length input to the transformer encoder.

3.3 Implementation Details

Variable Code Snippet Size. It is expected that the sourcecode 2-dimensional representation will vary in size. To address this, using the *[blank]* token, we batch up sourcecode snippets of different sizes with *interleaved* padding vertically and constant padding horizontally. While constant padding appends constant values from the end of an array, *interleaved* padding avoids leaving large continuous blank region by inserting *[blank]* tokens between original input code lines. In contrast, if an image exceeds the maximum size limit, we crop and keep the top left corner of the code image, following the raster order to retain most information. For instance, given input of size $L \times M$, $\mathbf{X}_i = [\mathbf{x}_0, \mathbf{x}_1, ..., \mathbf{x}_{L-1}]$ where $\mathbf{x}_l$ for $l = 0, ..., L-1$ each is a row vector of length M, then the cropped output of size $\hat{L} \times \hat{M}$ is $\hat{\mathbf{X}}_\mathbf{i} = [\hat{\mathbf{x}}_0, \hat{\mathbf{x}}_1, ..., \hat{\mathbf{x}}_{\hat{L}}]$ where $\hat{\mathbf{x}}_\mathbf{l} = \{x_{l,m}\}_{m=0}^{\hat{M}}$ for $l = 0, ..., \hat{L}-1$. In addition, we pad or limit the input images to be the same size of 96×96 for ResNet, ViT and ViT-fsd, while in Conv-ViT, along with an global minimum of 12×12 and maximum of 96×96 on all batches, we dynamically limit the maximum image size per minibatch at training time, *i.e.* instead of setting a constant maximum limit in all minibatches, it is configured to the 95th-percentile, if smaller than the global maximum, of the width and height

Table 1. CodeNetBench data summary. Balanced distribution among C++, Python and Java.

Dataset	Summary		
	#problems	#samples	#languages
CodeNetBench-Train	237	171300	3
CodeNetBench-Validation	237	21000	3
CodeNetBench-Test	237	21000	3

Table 2. Similarity evaluation datasets.

Dataset	Summary		
	#problems	#samples	#languages
CodeNetBench-Sim	100	2000	2

independently of the images within the minibatch. Note, while the input code image size in this work is limited to fit for the data distribution, the proposed CV4Code models are not limited by its architecture to process larger snippets.

Training. We use AdamW [16] optimiser with an learning rate of 10^{-3} and weight decay is set to 0.0001. We also use a 5-epoch warm-up along with a Cosine Learning Rate Annealing. Unless otherwise specified, all models are trained for 100 epochs and the model with the highest validation accuracy is selected to report results on the test set. Our model is implemented in PyTorch and trained on 1x NVIDIA V100 GPU with a batch size of 256.

4 Experimental Evaluation

In order to benchmark the performance and capabilities of our proposed framework, we conduct experiments on a real-world sourcecode dataset and compare against a set of competitive baselines, including character-, token- and AST-based representations respectively. Furthermore, test results are reported on Code Classification and Code Similarity tasks. The goal of these experiments is to answer the following research questions :

- RQ1: How well does CV4Code perform on the tasks in comparison to baselines using alternative forms of source code representation?
- RQ2: How scalable and flexible is CV4Code to baselines using alternative forms of source code representation?
- RQ3: How useful are the latent features learnt by CV4Code for alternate downstream tasks?

4.1 Setup

Dataset. We use CodeNet [22], a high-quality real-world dataset with code samples scraped from online coding platforms. The dataset provides code samples submitted by students and developers from across the world, resulting in a sundry pool of source code. The dataset is categorised based on the problems presented in the platform and for each such problem it provides the submissions in multiple programming languages. Our choice of CodeNet stems from the fact that we aim to benchmark the sourcecode understanding capability of CV4Code against baseline models in both language-agnostic and language-specific setups, and high-quality CodeNet benchmark set supporting 3 popular languages, including C++, Java and Python, makes it the ideal choice. We use the curated benchmark set of CodeNet [22] as it is made to be challenging with duplicated and dead code samples filtered. First, we extracted a multilingual set composed of code solutions to 237 overlapping *problem_id*s from C++1400, Python800 and Java250 and it is split into train, validation and test sets following 80%, 10% and 10% sample distributions for each *problem_id*. For convenience, we name them CodeNetBench-Train/Validation/Test. Furthermore, for the code similarity retrieval task, we test on *CodeNetBench-Sim* set, in which we randomly sample 100 problems and each problem with 10 code snippets in C++ and Python respectively, *i.e.* 20 code snippets per problem, from CodeNetBench-Test. Finally, we create One-versus-All test pairs, *i.e.* each test sample is paired with all other samples and positive pairs are those of the same *problem_id*.

Tasks. To evaluate the efficacy of our proposed framework, we evaluate on two tasks as follows:

- Code Classification: the goal is to classify source code samples based on their respective programming problem i.e. *problem_id*. Code samples belonging to the same programming problem would have high structural and semantic information overlap. As a result, it provides a solid ground for comparing the effectiveness of different source code representations.
- Code Similarity: the goal is to compare the efficacy of the various latent sourcecode representations for retrieving *similar* code samples.

Table 1 and 2 summarise the datasets classification model training, evaluation and similarity task evaluation.

Loss Function. As for the loss function for *problem_id*s classification, we adopt Additive Angular Margin (AAM) Softmax loss [8], as shown in Eq. 1, which has been shown to perform well by explicitly optimising similarity for intra-class samples and diversity for inter-class samples.

$$L = -\frac{1}{N}\sum_{i=1}^{N} \ln \frac{\exp\{s \cdot \cos(\theta_{y_i,i} + m)\}}{\exp\{s \cdot \cos(\theta_{y_i,i} + m)\} + \sum_{j \neq y_i} \exp\{s \cdot \cos(\theta_{j,i})\}}) \tag{1}$$

where $\theta_{y_i,j}$ is the angle between i-th sample to the y_i-th class and N is the batch size. We use angular margin penalty $m = 0.2$ and feature scale $s = 30$. For consistency and fairness, the same loss function is used in all models trained on *problem_id* classification.

Table 3. Bag of Character (BoC) and SPTR-Java MLP model configurations. ReLU activation and BatchNorm are used in fc layers. $N = 95$, 256 and 512 respectively in BoC, SPTR-Java-S and SPTR-Java-L.

Input Size	fc0	fc1	fc2	Output
N	128	256	512	237

Table 4. Token transformer model configurations. Learnable position embedding is used.

Model	Vocab	Depth	Hidden size D	MLP size	Heads	Params
a-Transformer	30K	12	128	512	4	7.1M
k-Transformer	120	12	128	512	4	3.3M

4.2 Baseline Methods

In this section we describe baseline methods to which we compare the proposed method against, including language agnostic and language specific ones. Specifically, we categorise a method as language-agnostic or language-specific 1) if any language-specific pre- or post-processing, including feature extraction, technique is required; 2) if any language-specific assumption is imposed on the model.

Language Agnostic Models

Bag of Characters. (BoC) A code sample is represented by the relative frequencies of character occurrence. Specifically, the feature vector of a code snippet is formed by counting the number of occurrences of each valid character introduced in Sect. 3.1 (excluding *[blank]*). Table 3 summarises the configuration of the MLP network that was selected after experiments with an array of setups. The training strategy described in Sect. 3.3 is used (Table 4).

Token Transformer. Similar to CodeBERT [10], relying on the assumption of naturalness in programming language, we build state-of-the-art NLP Transformer with text-based tokens input as baseline. Specifically, we learn two Token Transformer models via supervised training on the *problem_id* task, one with input of all tokens, including *all* operators and words and another with only 120 combined *keywords* and key operators from Python, Java and C++ provided in [22], which we call a-Transformer and k-Transformer respectively. To avoid noisy and sparse input embedding for a-Transformer, we extract a vocabulary in which each token occurs at least twice in the training dataset. In addition, both models have maximum input token length of 512. Table 6 summarises the configurations of the Transformers that was empirically selected. While the same training strategy described in Sect. 3.3 is followed, due to the higher model memory footprint compared to other models, we use a smaller batch size of 32.

Table 5. CV4Code-ResNet model configuration. Conv layer weights are annotated as number of filters and stride step size. 7×7 kernel is used in conv0 and 3×3 in others. 3×3 with stride 2 and 6×6 (global) max_pool is used after conv0 and conv3. Total #params=3.25M.

conv0	conv1	conv2	conv3	fc	Output
$16, 2$	$\begin{bmatrix}64\\64\end{bmatrix} \times 2, 2$	$\begin{bmatrix}128\\128\end{bmatrix} \times 2, 2$	$\begin{bmatrix}256\\256\end{bmatrix} \times 2, 1$	128	237

Table 6. CV4Code-ViT and CV4Code-ViT-fsd configurations. All configurations use depth of 8, hidden size of 128, MLP size of 512 with 4 heads. Learnable position embedding is used.

Model	Patch size	Params
CV4Code-ViT-S	16×16	5.32M
CV4Code-ViT-L	8×8	2.98M
CV4Code-ViT-fsd-S	16×16	13.97M
CV4Code-ViT-fsd-L	8×8	4.58M

Table 7. Conv-ViT configurations. All configurations use depth of 8, hidden size of 128, MLP size of 512 with 4 heads. Each convolutional layer is followed by a 2×2 max_pool layer with stride of 2. Fixed Sinusoidal position embedding is used.

Model	Convolutional tokenizer	Visual tokens	Params
Conv-ViT-S	$\begin{bmatrix}7 \times 7, 64\end{bmatrix} \times 2$, stride 2	49	2.35M
Conv-ViT-L	$\begin{bmatrix}3 \times 3, 64\end{bmatrix} \times 3$, stride 1	169	1.83M

Language Specific Models

Simplified Parse Tree Relation. (SPTR) Leveraging Simplified Parse Tree (SPT) features originally presented in [17], SPTR directly exploits the underlying structure from code snippets. We extract SPT and build a vocabulary from all training Java code samples. Then a sparse binary count vector is extracted from each SPT that defines the existence of a particular vocabulary from a SPT. Finally dense feature vectors are obtained through Truncated Single Value Decomposition (tSVD). We experiment with the two MLP models, as summarised in Table 3, with tSVD output dimensions set to 256 and 512 respectively (explained variance ratios of 0.768 and 0.834). We train and report results on Java samples in CodeNetBench-Train and CodeNetBench-Test, following the same training strategy described in Sect. 3.3.

4.3 CV4Code Setup

Table 5 reports the model configuration for CV4Code-ResNet, we experiment with various configurations and report result from the model with the best validation accuracy. For ViT and ViT-fsd, we experiment with two different patch

Table 8. *problem_id* classification results on CodeNetBench-Test.

Model	Multilingual		Java-only	
	Top-1	Top-5	Top-1	Top-5
BoC	80.97	90.16	80.56	89.74
k-Transformer	90.30	95.42	89.54	94.97
a-Transformer	93.58	96.63	94.04	97.40
SPTR-Java-S	–	–	91.09	96.98
SPTR-Java-L	–	–	92.95	96.78
ResNet	92.93	96.50	91.17	95.50
ViT-S	85.45	93.64	80.50	90.95
ViT-L	92.85	96.86	90.27	95.46
ViT-fsd-S	86.04	93.80	80.60	90.80
ViT-fsd-L	92.27	96.47	88.99	94.49
Conv-ViT-S	96.08	98.45	94.63	98.01
Conv-ViT-L	**97.64**	**98.99**	**97.13**	**98.79**

Table 9. Code similarity evaluation result on CodeNetBench-Sim.

Model	mAP@R
a-Transformer	0.980
ResNet	0.983
Conv-ViT-L	**0.991**

sizes, *i.e.* 16×16 and 8×8, which result in 36 and 144 visual tokens respectively, given input sourcecode image of size 96×96. For Conv-ViT, we compare two convolutional soft tokenization setups which result in similar length of visual tokens. Table 7 summarises the configurations for all vision transformer variants.

4.4 Results

Evaluation Metrics. For classification tasks, top-1 and top-5 accuracy are reported as the evaluation metrics on CodeNetBench-Test. For code similarity, we consider a retrieval task where a code snippet is used as query to search for similar snippets and mAP@R [19] is reported as the main evaluation metric.

Language-agnostic Classification. As summarised in Table 8, Conv-ViT-L outperforms all other models in multilingual test, including the strong NLP-based a-Transformer. We attribute the gain of CV4Code over k-Transformer to exploitation of the contextual and structural information in the spatial relationships, which is not directly available through the sequential token input in k-Transformer. Comparing CV4Code ResNet and transformer variants, ViT and ViT-fsd variants all perform worse than ResNet. This potentially implies that the inductive bias, *e.g.* locality, associated with convolutional networks are critical for sourcecode understanding from our proposed sourcecode image representation. This is further supported by the gain obtained in Conv-ViT-S/L which inherit the inductive bias through its soft convolutional tokenizer and leverage the expressiveness of the Transformer network. In addition, it is noticed that the

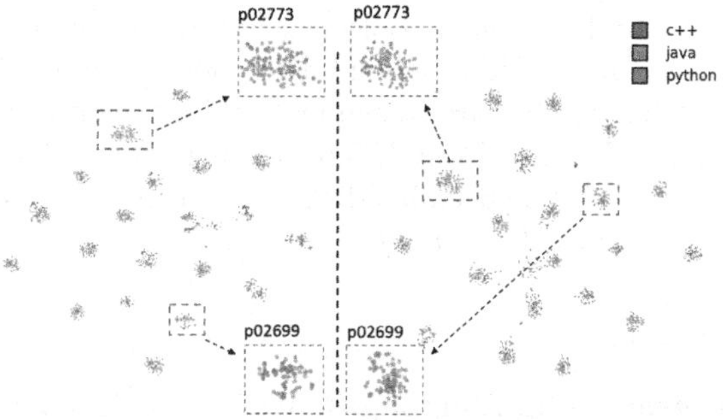

Fig. 4. t-SNE 2D projection of **left**: Conv-ViT-L and **right**: a-Transformer embeddings. Colour-labeled by unique programming *languages*.

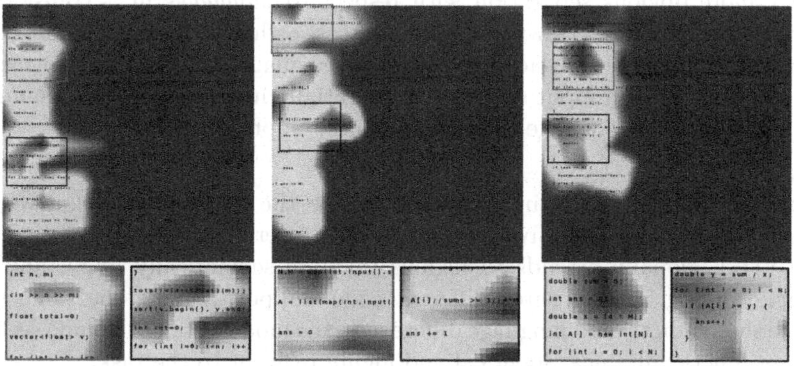

Fig. 5. Attention maps (rollout) of Conv-ViT-L. 1) C++, 2) Python, 3) Java

size of patch and the subsequently generated visual tokens have a strong influence on the performance of all CV4Code transformer variants, which is exhibited through the gap between all -S/-L model pairs. In addition, for Conv-ViT-L, we also experimented with 2- and 8-head attentions achieving 97.65% and 96.98% top-1 accuracy scores respectively, showing minor variations.

Language-Specific (Java) Classification. The Java-only column in Table 8 summaries model test results on Java samples in CodeNetBench-Test. Although SPTR-Java-L/S both achieve strong performance, which demonstrates the effectiveness of the input features that exploit the underlying code structures via SPTs, it is outperformed by a-Transformer by 1.09% in terms of Top-1 accuracy. And Conv-ViT-L, as a language-agnostic model trained on C++, Python and Java, still achieves the strongest result overall with a 97.13% Top-1 accuracy.

Code Similarity. We test the learned embeddings from the models that achieve strong results on the classification task, with respect to code similarity and report similarity mAP@R scores on datasets summarised in Table 2. We extract *[class]* embedding at the transformer output from a-Transformer, pre-ReLU bottleneck output from ResNet and sequence pooled embedding from transformer output from Conv-ViT-L. Cosine similarity is used to compute the pairwise similarity of the embeddings. Test results are shown in Table 9. We observe that CV4Code models, including ResNet and Conv-ViT-L, outperform a-Transformer on this task, implying that the learned embeddings from proposed models are highly discriminative and encodes the *semantics* of sourcecode.

4.5 Ablation Studies

Influence of Programming Languages. We look at the influence of programming languages on the latent representations from Conv-ViT-L, compared with a-Transformer. t-SNE [18] is used for projection. As shown in Fig. 4, while both embedding spaces show clusters formed with respect to *problem_ids*, we observe slightly more obvious separation with respect to *languages* in a-Transformer. Given that distinctive syntax, use of operators, coding styles and naming conventions are often used in each programming language, this implies that models with text-based token input sequence is potentially more sensitive to the underlying programming language than the proposed approach.

Attention Maps. Following [1], we compute the average attention weights across all heads and recursively multiply them over all layers. We notice that Conv-ViT-L globally attends to regions that are semantically important. For example, in Fig. 5, with three different code snippets solving the same task, despite the syntactic and language differences, the model attends to code sections that are functionally near identical and highly relevant to the task.

Future Studies. Following recent advancement of self-supervised learning for Vision models [12], we would like to scale up CV4Code with the abundance of unlabelled sourcecode snippets in the public domain. With such a setup, we look to establish a study of self-supervised approaches with the proposed CV4Code method and compare to NLP approaches, *e.g.* BERT, for additional MLOnCode problems. Furthermore, to capture richer semantics in the latent space and test on more generic downstream applications, *e.g.* code quality assessment, we plan to extend the training framework to include multi-task learning configurations.

5 Conclusion

In this work, we propose an idea for sourcecode understanding via a novel visual representation. Compared to traditional syntax parse tree or token based methods, the proposed approach is programming language agnostic and does not depend on syntax correctness in the preprocessing stage. Finally, along with a thorough study of vision models applied on the proposed representations and a comparison with syntax tree and NLP-based approaches, using a Compact Convolution Transformer CV4Code model, we show state-of-the-art performance in terms of both classification and code similarity retrieval.

References

1. Abnar, S., Zuidema, W.: Quantifying attention flow in transformers. In: Proceedings of the 58th Annual Meeting of the Association for Computational Linguistics, pp. 4190–4197. Association for Computational Linguistics (2020). https://doi.org/10.18653/v1/2020.acl-main.385, https://aclanthology.org/2020.acl-main.385/
2. Allamanis, M., Sutton, C.: Mining source code repositories at massive scale using language modeling. In: 2013 10th Working Conference on Mining Software Repositories (MSR), pp. 207–216 (2013). https://doi.org/10.1109/MSR.2013.6624029
3. Allamanis, M., Barr, E.T., Devanbu, P., Sutton, C.: A survey of machine learning for big code and naturalness. ACM Comput. Surv. (CSUR) **51**(4), 81 (2018)
4. Alon, U., Brody, S., Levy, O., Yahav, E.: code2seq: generating sequences from structured representations of code. In: International Conference on Learning Representations (2019). https://openreview.net/forum?id=H1gKYo09tX
5. Alon, U., Zilberstein, M., Levy, O., Yahav, E.: Code2vec: learning distributed representations of code. Proc. ACM Program. Lang. **3**(POPL) (2019). https://doi.org/10.1145/3290353
6. Bilgin, Z.: Code2image: intelligent code analysis by computer vision techniques and application to vulnerability prediction. arXiv preprint arXiv:2105.03131 (2021)
7. Chen, Z., Kommrusch, S., Tufano, M., Pouchet, L.N., Poshyvanyk, D., Monperrus, M.: Sequencer: Sequence-to-sequence learning for end-to-end program repair. IEEE Trans. Softw. Eng. **47**(9), 1943–1959 (2021). https://doi.org/10.1109/TSE.2019.2940179
8. Deng, J., Guo, J., Xue, N., Zafeiriou, S.: Arcface: additive angular margin loss for deep face recognition. In: Proceedings of the IEEE/CVF Conference on Computer Vision and Pattern Recognition, pp. 4690–4699 (2019)
9. Dey, S., Singh, A.K., Prasad, D.K., Mcdonald-Maier, K.D.: Socodecnn: program source code for visual CNN classification using computer vision methodology. IEEE Access **7**, 157158–157172 (2019)
10. Feng, Z., et al.: CodeBERT: a pre-trained model for programming and natural languages. In: Findings of the Association for Computational Linguistics: EMNLP 2020, pp. 1536–1547. Association for Computational Linguistics (2020). https://doi.org/10.18653/v1/2020.findings-emnlp.139, https://aclanthology.org/2020.findings-emnlp.139/
11. Hassani, A., Walton, S., Shah, N., Abuduweili, A., Li, J., Shi, H.: Escaping the big data paradigm with compact transformers. ArXiv abs/2104.05704 (2021)
12. He, K., Chen, X., Xie, S., Li, Y., Dollár, P., Girshick, R.: Masked autoencoders are scalable vision learners. arXiv preprint arXiv:2111.06377 (2021)
13. He, K., Zhang, X., Ren, S., Sun, J.: Deep residual learning for image recognition. In: 2016 IEEE Conference on Computer Vision and Pattern Recognition (CVPR), pp. 770–778 (2016). https://doi.org/10.1109/CVPR.2016.90
14. Kolesnikov, A., et al.: An image is worth 16×16 words: transformers for image recognition at scale (2021)
15. Lee, S.H., Lee, S., Song, B.C.: Vision transformer for small-size datasets. arXiv preprint abs/2112.13492 (2021)
16. Loshchilov, I., Hutter, F.: Decoupled weight decay regularization. In: ICLR (2019)
17. Luan, S., Yang, D., Barnaby, C., Sen, K., Chandra, S.: Aroma: code recommendation via structural code search. Proc. ACM Program. Lang. **3**(OOPSLA) (2019)
18. van der Maaten, L., Hinton, G.: Visualizing data using t-sne. J. Mach. Learn. Res. **9**(86), 2579–2605 (2008). https://jmlr.org/papers/v9/vandermaaten08a.html

19. Musgrave, K., Belongie, S., Lim, S.-N.: A metric learning reality check. In: Vedaldi, A., Bischof, H., Brox, T., Frahm, J.-M. (eds.) ECCV 2020. LNCS, vol. 12370, pp. 681–699. Springer, Cham (2020). https://doi.org/10.1007/978-3-030-58595-2_41
20. Ochodek, M., Hebig, R., Meding, W., Frost, G., Staron, M.: Recognizing lines of code violating company-specific coding guidelines using machine learning. Empir. Softw. Eng. **25**(1), 220–265 (2020)
21. Pradel, M., Sen, K.: Deepbugs: a learning approach to name-based bug detection. Proc. ACM Program. Lang. **2**(OOPSLA) (2018). https://doi.org/10.1145/3276517
22. Puri, R., et al.: Project codenet: a large-scale AI for code dataset for learning a diversity of coding tasks (2021)
23. Rabin, M.R.I., Alipour, M.A.: Encoding program as image: evaluating visual representation of source code. arXiv preprint arXiv:2111.01097 (2021)
24. Russell, R.L., et al.: Automated vulnerability detection in source code using deep representation learning. CoRR abs/1807.04320 (2018). https://arxiv.org/abs/1807.04320
25. Sharma, T., Kechagia, M., Georgiou, S., Tiwari, R., Sarro, F.: A survey on machine learning techniques for source code analysis. arXiv preprint arXiv:2110.09610 (2021)
26. Simonyan, K., Zisserman, A.: Very deep convolutional networks for large-scale image recognition. In: International Conference on Learning Representations (2015)
27. Svyatkovskiy, A., Lee, S., Hadjitofi, A., Riechert, M., Franco, J.V., Allamanis, M.: Fast and memory-efficient neural code completion. In: 18th IEEE/ACM International Conference on Mining Software Repositories, MSR 2021, Madrid, Spain, 17–19 May 2021, pp. 329–340. IEEE (2021). https://doi.org/10.1109/MSR52588.2021.00045
28. Yan, S., Yu, H., Chen, Y., Shen, B., Jiang, L.: Are the code snippets what we are searching for? a benchmark and an empirical study on code search with natural-language queries. In: 2020 IEEE 27th International Conference on Software Analysis, Evolution and Reengineering (SANER), pp. 344–354 (2020). https://doi.org/10.1109/SANER48275.2020.9054840
29. Zanoni, M., Arcelli Fontana, F., Stella, F.: On applying machine learning techniques for design pattern detection. J. Syst. Softw. **103**(C), 102–117 (2015). https://doi.org/10.1016/j.jss.2015.01.037
30. Zhang, J., Wang, X., Zhang, H., Sun, H., Wang, K., Liu, X.: A novel neural source code representation based on abstract syntax tree. In: 2019 IEEE/ACM 41st International Conference on Software Engineering (ICSE), pp. 783–794. IEEE (2019)
31. Zügner, D., Kirschstein, T., Catasta, M., Leskovec, J., Günnemann, S.: Language-agnostic representation learning of source code from structure and context. In: International Conference on Learning Representations (ICLR) (2021)
32. Zügner, D., Kirschstein, T., Catasta, M., Leskovec, J., Günnemann, S.: Language-agnostic representation learning of source code from structure and context. arXiv preprint arXiv:2103.11318 (2021)

PPR-Net: Patch-Based Multi-scale Pyramid Registration Network for Defect Detection of Printed Label

Dongming Li, Yingjian Li, Jinxing Li, and Guangming Lu(✉)

Harbin Institute of Technology Shenzhen, Shenzhen, China
{lijinxing158,luguangm}@hit.edu.cn

Abstract. Detecting defects in printed labels is essential to ensure product quality. Reference-based comparison is a potential method to challenge this task, which is widely used for defect detection. However, this method gets poor performance under large deformation, due to the lack of ability of registering the testing image with the reference image. Therefore, accurate image registration is an urgent case for defect detection of printed labels. In this paper, a patch-based multi-scale pyramid registration network (PPR-Net) is proposed. First, an image patch splitting and stitching strategy is proposed, which is scalable in image resolution. Second, a multi-scale pyramid registration module is designed to fuse multiple convolutional features to enhance the registration capability for large deformation, which gradually refines multi-scale deformation fields in a coarse-to-fine manner. Third, a distortion loss function is introduced to improve text distortions of registered images. Finally, a synthetic database is generated based on real printed labels, to simulate defective printed labels with large deformation for performance comparison. Extensive experimental results show that our method dramatically outperforms other comparable approaches.

Keywords: Printed labels · Multi-features fusion · Artifact · Deformable registration · Defect detection

1 Introduction

Nowadays, manufacturing enterprises of electronic products are facing fierce competition, and improving product quality is crucial to enhance competitiveness. As an essential part of electronic products, printed labels are widely used. Therefore, it is of great significance to perform defect detection of printed labels during manufacturing. Up to now, one of the most widely used techniques is the reference-based comparison method. This method stores a reference image in advance and then compares it with the testing image at pixel-by-pixel or feature-by-feature level. If any pixels or features do not match, defects may exist [14,20]. However, most printed labels are manufactured from non-rigid materials. Due to the mechanical vibration and distortion of printed labels, large deformations

L. Wang et al. (Eds.): ACCV 2022, LNCS 13842, pp. 307–323, 2023.
https://doi.org/10.1007/978-3-031-26284-5_19

are inevitably introduced. At present, there is no uniform specification to define large deformation. Here, we use the average mean square error (MSE) to evaluate large deformation. Subsequently, these deformations may result in obvious artifacts after image subtraction, increasing the difficulty of defect detection.

Based on this aforementioned analysis, accurate image registration becomes an urgent case for defect detection of printed labels. At present, image registration algorithms can be categorized into two groups: traditional and deep learning methods. Traditional methods can be roughly divided into two categories: pixel grayscale-based and feature-based registration methods. Pixel grayscale-based method is stable and usually measures the similarity of two images based on full grayscale information, such as the normalized cross-correlation registration algorithm [19]. However, the pixel grayscale-based method is often computationally expensive and sensitive to illumination. The feature-based method usually extracts image features, such as corner, edge, shape, and texture. The existing feature extraction operators mainly include SIFT operator [11], SURF operator [3], AKAZE operator [1], and so on. These feature-based methods have a low computational cost and strong robustness. However, such methods seem to be ineffective [7] for images with inconspicuous features.

In recent decades, many deep learning-based methods have been innovated to achieve image registration. There are several recent works [9,15,17,18] to learn a function for image registration based on neural networks. However, most of them rely on ground truth deformation fields for training models. By contrast, our PPR-Net is an unsupervised method. Since the success of the spatial transformer network [6], unsupervised deep learning techniques have achieved state-of-the-art performance in many registration tasks. For example, Balakrishnan et al. [2] proposed an unsupervised registration network (VoxelMorph), which makes a straightforward prediction of the deformation field based on the U-net [16] structure. However, VoxelMorph is usually limited to small deformations. To solve this issue, recent works, such as DDN [13], PRDFE [21], and Dual-PRNet [5], introduced new improvements to handle large deformations. Nevertheless, few of them pay attention to the text distortion of registered images, which has a great impact on the accuracy of defect detection.

To address the above challenges, we focus on the enhancement of large deformable image registration. Our main contributions are:

1) We introduce a patch-based image splitting and stitching strategy, which is scalable in image resolution and overcomes the limited memory of GPUs.
2) We propose a multi-scale pyramid registration module with a multi-feature fusion strategy. Multi-scale deformation fields are refined gradually in a coarse-to-fine manner, to boost the registration performance for large deformations.
3) We design a novel distortion loss function, which is incorporated with shift-invariant loss, to improve text distortion of registered images.

4) We generate a synthetic database based on real printed labels, to simulate defective printed labels with large deformation. Furthermore, we evaluate the accuracy of registration from two aspects: pixel level and defect detection. Experimental results show that our method performs better than other compared methods.

The remainder of this paper is organized as follows. Section 2 presents the details of our proposed method. Section 3 presents our experimental results, followed by a conclusion in Sect. 4.

2 Method

2.1 The Proposed Framework

Figure 1 shows the overall framework of our method. It consists of an image registration network with a patch-based image splitting and stitching strategy, followed by reference-based comparison defect detection (RCDD) to evaluate the accuracy of image registration. Fixed image, moving image and warped image represent reference image (defect-free), testing image and registered image, respectively. Patches are first extracted from fixed and moving images based on the image splitting strategy, then fed into the registration network. After image registration and stitching, a final registered image is obtained. In this paper, we focus on image registration, which is used for defect detection of printed labels.

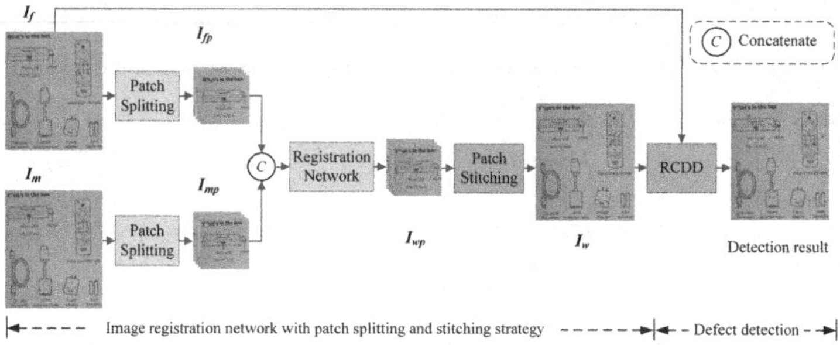

Fig. 1. The proposed framework for image registration. I_f and I_m are the fixed and moving image, which are split into patches I_{fp} and I_{mp}. I_{wp} are patches warped by registration. I_w is the image stitched from I_{wp}.

2.2 Patch Splitting and Stitching Strategy

Table 1 shows that resolutions of testing images are different, and a common approach is to resize them to the same size before image registration. However, resizing the image is not recommended, as it may result in some inconspicuous defects being missed. In addition, GPU memory is limited in the real-world industrial scene. To deal with these problems, we introduce a patch-based image splitting and stitching strategy, as shown in Fig. 2. For image splitting, slide patches are generated by sliding a window on fixed and moving images, respectively. A sliding window starts from the red frame and slides $w-l$ pixels per step along the X axis. After reaching the right edge of the testing image, the sliding window returns to the location of the yellow frame and slides along the X axis again. q is the number of slide patches, and $q = \lceil (W-l)/(w-l) \rceil * \lceil (H-l)/(w-l) \rceil$. $W \times H$ and $w \times w$ are the size of the testing image and sliding window, respectively. l is the overlap pixels. The location of image stitching is at half of the overlap of two adjacent slide patches. Algorithm 1 and 2 gives the details.

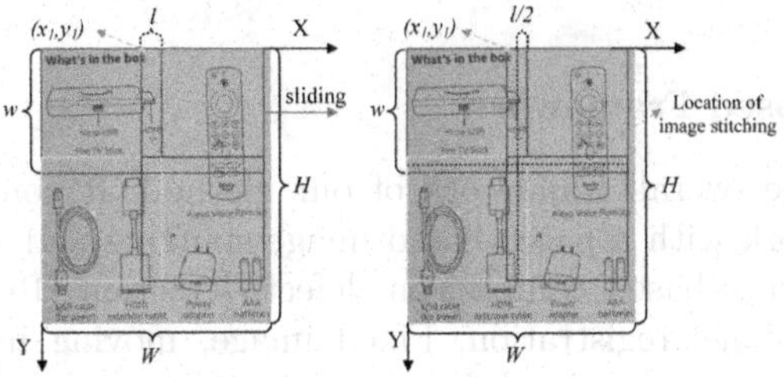

(a) Image splitting (b) Image stitching

Fig. 2. Patch splitting and stitching strategy. (x_1, y_1) is the upper-left coordinate of the sliding window.

Algorithm 1: Image splitting

Input: I_m, I_f, l, w, W, H, q
Output: Sets $\mathbb{P}^m$, $\mathbb{P}^f$, $\mathbb{C}$, contain moving, fixed patches and corresponding coordinate.

for $i \in [0, \lceil (H-l)/(w-l) \rceil - 1]$ **do**
    if $((w-l)*i+w) > H$ **then**
        $y_1 = H - w$
    else
        $y_1 = (w-l)*i$
    end
    for $j \in [0, \lceil (W-l)/(w-l) \rceil - 1]$ **do**
        if $((w-l)*j+w) > W$ **then**
            $x_1 = W - w$
        else
            $x_1 = (w-l)*j$
        end
        Put a $w \times w$ sliding window at (x_1, y_1) of I_m and I_f, put the areas of the two sliding windows into $\mathbb{P}^m$ and $\mathbb{P}^f$. Put (y_1, x_1) into $\mathbb{C}$.
    end
end
Output $\mathbb{P}^m = \{P_1^m, P_2^m, ..., P_q^m\}$, $\mathbb{P}^f = \{P_1^f, P_2^f, ..., P_q^f\}$, $\mathbb{C} = \{C_1, C_2, ..., C_q\}$.

Algorithm 2: Image stitching

Input: I_{wp}^k, $\mathbb{C}$, l, w, q. I_{wp}^k means the k-th warped image patch, $k = 1, ..., q$.
Output: I_w
for $C_k \in \mathbb{C}$ **do**
  $y = C_k[0]$, $x = C_k[1]$ // $C_k[0]$,$C_k[1]$ meaning the coordinates of Y and X axes.
  if ($y == 0$ *and* $x == 0$) **then**
    $I_w[0 : w - l/2, 0 : w - l/2] = I_{wp}^k[0 : w - l/2, 0 : w - l/2]$;
  else if ($y == 0$ *and* $x \neq 0$) **then**
    $I_w[y : y + w - l/2, x + l/2 : x + w] = I_{wp}^k[0 : w - l/2, l/2 : w]$;
  else if ($y \neq 0$ *and* $x == 0$) **then**
    $I_w[y + l/2 : y + w, x : x + w - l/2] = I_{wp}^k[l/2 : w, 0 : w - l/2]$;
  else
    $I_w[y + l/2 : y + w, x + l/2 : x + w] = I_{wp}^k[l/2 : w, l/2 : w]$;
  end
end
Output the final warped image I_w.

2.3 Multi-scale Pyramid Registration Network

Given a pair of fixed image and moving image which are defined over a $2D$ spatial domain $\Omega \subset \mathcal{R}^2$, image registration seeks to find a coordinate transform between the image pair. A convolutional neural network (CNN) is adopt to model a function f to estimate the optimal deformation field $\Phi = f_\theta(I_{fp}, I_{mp})$. θ represents the learnable parameters of function f. I_{fp} and I_{mp} are corresponding patches extracted from the fixed and moving image. Figure 3 illustrates the architecture of our proposed registration network, which consists of an encoder module, a decoder module, and a multi-scale pyramid module.

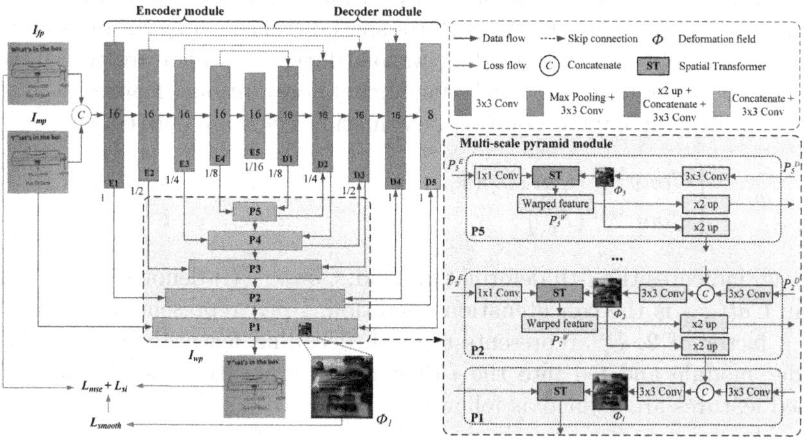

Fig. 3. The architecture of our proposed registration network. P_i means the i-th pyramid layer. Details of the multi-scale pyramid module are shown on the right.

Encoder Module. We adopt an encoder with the same architecture as VoxelMorph [2], which consists of five encoding blocks, named E_1, ..., E_5. Fixed and moving patches are concatenated and fed into the first encoding block, with a 3×3 convolution. The rest encoding blocks consist of a 3×3 down-sampling convolution with a stride of 2, which reduces the resolution of input image pairs by a factor of 16 in total.

Decoder Module. There are also five decoding blocks in the decoder module, named D_1, ..., D_5. We apply skip connections to the corresponding encoding and decoding blocks. The first four decoding blocks consist of a 3×3 up-sampling convolution with a stride of 2, followed by a concatenation and 3×3 convolution. Unlike VoxelMorph [2], to allow the network to learn stronger and more discriminative convolutional features, we fuse the multi-scale warped features into the corresponding decoding blocks. The warped features are generated by spatial transformation on the i-th pyramid layer, as shown on the right of Fig. 3.

Multi-scale Pyramid Module. VoxelMorph [2] only generates a single deformation field, which limits its ability to deal with large deformations. To overcome this, we introduce a multi-scale pyramid module with multi-features fusion, as follows: 1) We apply 1×1 convolution to the corresponding encoding blocks to extract multi-scale convolutional features, as the first input to spatial transformer (ST). 2) We estimate multi-resolution deformation fields Φ_i on the i-th pyramid layer, as the second input to ST. Especially, we adopt a multi-resolution deformation field fusion (MDFF) mechanism: a low-resolution deformation field is gradually fused into the high-resolution one by a series of operations, such as up-sampling, concatenation, and convolution. 3) Multi-scale warped features fusion (MWFF) mechanism. We perform spatial transform by ST on each pyramid layer and obtain corresponding warped features, which are concatenated into corresponding decoding blocks for further convolutional feature fusion.

Specifically, the multi-resolution deformation fields are defined as:

$$\Phi_i = \begin{cases} Conv^{3\times 3}\left(Cat\left(Up(\Phi_{i+1}), Conv^{3\times 3}\left(P_i^D\right)\right)\right), & i = 1,2,3,4 \\ Conv^{3\times 3}\left(P_i^D\right), & i = 5 \end{cases} \tag{1}$$

where Φ_i indicates the i-th deformation field, $Conv^{3\times 3}(.)$ denotes a 3×3 convolution, $Cat(.,.)$ is the concatenation operation, $Up(.)$ represents a up-sampling with a factor of 2, P_i^D represents the i-th decoding feature generated by the decoder module and fed into the i-th pyramid layer. Besides, the multi-scale warped features are defined as follows:

$$P_i^W = ST\left(Conv^{1\times 1}\left(P_i^E\right), \Phi_i\right), i = 2, ..., 5 \tag{2}$$

where P_i^W represents the i-th warped feature on the i-th pyramid layer, $ST(.,.)$ is a spatial transformer function implemented by VoxelMorph [2], $Conv^{1\times 1}(.)$ indicates a 1×1 convolution, P_i^E represents the i-th encoding feature generated by the encoder module.

The above multi-scale features P_i^E, P_i^D and P_i^W are implemented sequentially to estimate the final deformation field, which fuses multiple convolutional features with multi-scale deformation fields gradually in a coarse-to-fine manner. Finally, the warped image patch is obtained by:

$$I_{wp} = ST\left(I_{mp}, \Phi_1\right) \tag{3}$$

where Φ_1 is the final deformation field.

2.4 Loss Function

Based on our observations, text distortions may occur after image registration, which directly affect the accuracy of subsequent defect detection. To this end, we design a distortion loss L_{dist}, which consists of the MSE loss L_{mse}, the local spatial variation loss L_{smooth}, and the shift-invariant loss L_{si}. The distortion loss is calculated as follows:

$$L_{dist}(I_{fp}, I_{wp}, \Phi_1) = L_{mse}(I_{fp}, I_{wp}) + \lambda_1(L_{smooth}(\Phi_1) + L_{si}(I_{fp}, I_{wp})) \tag{4}$$

whered λ_1 is a regularization trade-off parameter. The L_{mse} and L_{smooth} are defined as follows:

$$L_{mse}(I_{fp}, I_{wp}) = \frac{1}{n}\sum_{i}^{n}(I_{fp}(i) - I_{wp}(i))^2 \tag{5}$$

$$L_{smooth}(\Phi_1) = \sum \| \bigtriangledown (\Phi_1)\|^2 \tag{6}$$

where $L_{mse}(.,.)$ measures image similarity between its two inputs, and n is the number of elements in I_{wp}. I_{fp} is the corresponding patch extracted from the fixed image, and I_{wp} is the image patch warped by registration. $L_{smooth}(.)$ is a regularization constraint on the final deformation field Φ_1 to enforce spatial smoothness, $\bigtriangledown$ represents the spatial gradients. Motivated by [12], we also use a shift-invariant loss L_{si} to handle the shift problem, which is defined as follows:

$$L_{si}(I_{fp}, I_{wp}) = \frac{1}{n}\sum_{i}|d_i| - \frac{\lambda_2}{n}|\sum_{i} d_i| \tag{7}$$

where $d_i = I_{fp}(i) - I_{wp}(i)$, $I_{fp}(i)$ is the corresponding ground truth value and $I_{wp}(i)$ is the predicted value at index i, and λ_2 controls the strength of the second term. Notably, the L_{si} does not care about the absolute value of $I_{wp}(i)$. It enforces that the difference between $I_{wp}(i)$ and $I_{wp}(j)$ should be close to that between $I_{fp}(i)$ and $I_{fp}(j)$. Therefore, L_{si} is helpful to alleviate the influence of shift.

2.5 Reference-Based Comparison Defect Detection (RCDD)

As illustrated in Sect. 1, RCDD is one of the widely used method to detect defects of printed labels. In this paper, we first utilize PPR-Net to perform image registration, then conduct RCDD. Figure 4 shows the process of RCDD. The input is a pair of fixed and warped images. First, a median filter is applied for denoising. Subsequently, the warped image is subtracted from the fixed image to get a difference image. Second, a morphological opening operation with a 3×3 structure element is performed on the difference image for further denoising. Finally, a low-threshold filter with threshold T_{filter} is applied to binarize the image, where a pixel value larger than T_{filter} is set to 255, otherwise 0. In the defect discrimination stage, if the area of a white pixel region in the binary image is larger than the threshold T_{area}, this region will be judged as a defect. Based on our observation, a pixel value lower than 20 is difficult to be distinguished by human beings, and defects with a pixel area less than 5 can be ignored. Therefore, we set T_{filter} and T_{area} to 20 and 5, respectively.

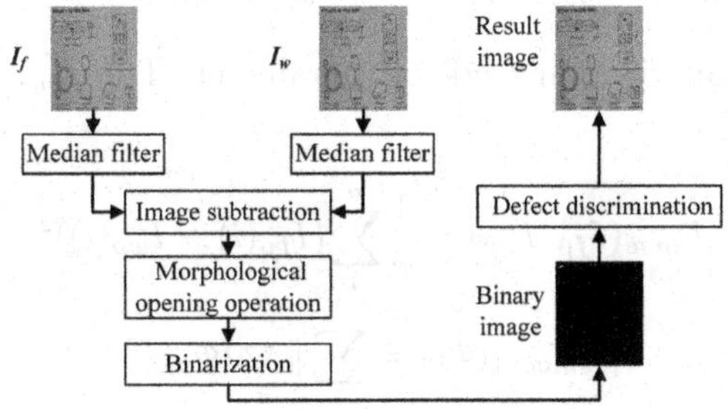

Fig. 4. The process of reference-based comparison defect detection.

2.6 Discussion

Herein, we discuss the differences between our PPR-Net and other compared deep learning methods, such as VoxelMorph [2], DDN [13], and PRDFE [21].

1) Difference with VoxelMorph [2] and DDN [13]: Both VoxelMorph and DDN predict only a single deformation field, which limits their ability to deal with large deformations. Differently, PPR-Net can predict multi-scale deformation fields based on the multi-level pyramid, which refines the deformation fields gradually and boosts the registration performance of large deformations. In addition, VoxelMorph takes the entire image as input rather than patches, which can not handle the images with different sizes. By contrast, PPR-Net uses a patch-based image splitting and stitching strategy, with better scalability of image resolution.
2) Difference with PRDFE [21]: The methods of generating deformation fields are different. PRDFE uses the warped moving and fixed features to estimate their residual deformation field at each pyramid scale. Differently, PPR-Net

adopts the multi-scale warped features fusion (MWFF) mechanism to fuse warped features into corresponding decoding blocks, which further enhances the registration robustness of large deformation. Furthermore, PPR-Net is more accurate than PRDFE with less registration time, which will be reported in the experimental section.

3) All above compared methods ignore the text distortion in registered images, which may result in a high false positive rate. Differently, we introduce a novel distortion loss function incorporated with a shift-invariant loss to penalize the text distortion, which helps to reduce the false positive rate.

3 Experiments

3.1 Dataset Generation

As illustrated in Sect. 1, printed label images captured from the production line may exhibit large deformation compared with the reference image. Thus, to evaluate the performance of image registration, we need an industrial printed labels database with large deformations and defects. However, there is no such public database. To this end, we create a synthetic database in which deformable images are generated via the perturbed mesh generation algorithm [12]. We adjust the parameters d and α in [12] empirically to get proper images. Table 1 shows the summary of 10 training sets in our database. Each training set consists of one reference image and 200 synthetic images with large deformations. During the training stage, we randomly extract image patches with a size of 512×512 from synthetic images and corresponding reference images.

For testing sets, we collect four kinds of defect-free printed labels from a factory. To evaluate the registration performance with defect samples, we randomly generate 10 artificial defects for each image based on [10]. Table 1 gives a summary of our database and Fig. 5 shows some representative samples in testing sets.

Table 1. Summary of our database.

Dataset	Dataset name	Image size ($W \times H \times C$)	Number of images
Training set	Consr	1210 × 1396 × 1	201
	Devr	2482 × 1457 × 1	201
	Donr	3300 × 1316 × 1	201
	Jackr	2694 × 1568 × 1	201
	Nomr	1644 × 1414 × 1	201
	Tvboxr	1267 × 1214 × 1	201
	Ubiqr	900 × 873 × 1	201
	Warr	833 × 833 × 1	201
	Xg2r	1338 × 1338 × 1	201
Testing set	Label-A	1092 × 714 × 1	1,475
	Label-B	969 × 1133 × 1	1,325
	Label-C	2774 × 1509 × 1	1,400
	Label-D	1389 × 680 × 1	1,472

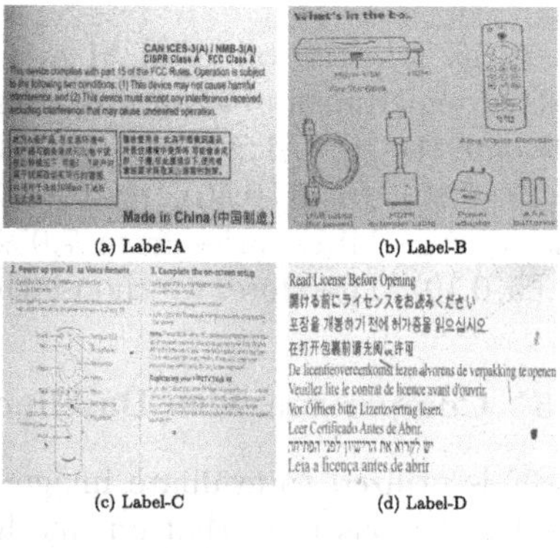

(a) Label-A (b) Label-B (c) Label-C (d) Label-D

Fig. 5. Some representative samples in testing sets.

3.2 Evaluation Metrics

In this paper, we evaluate the registration performance based on three metrics: MSE, F_1 score, and registration time. Among them, MSE is introduced to evaluate the registration accuracy at the pixel level, and the F_1 score is used for defect detection. For RCDD, better registration means better defect detection. The MSE and F_1 score are defined as follows:

$$MSE(I_f, I_w) = \frac{1}{N}\sum_{i}^{N}(I_f(i) - I_w(i))^2 \tag{8}$$

$$F_1 = 2PR/(R+P) \tag{9}$$

where N is the number of elements in I_w, $R = TP/(TP+FN)$, R means the recall, $P = TP/(TP+FP)$, P means the precision, true positive (TP) indicates real defects that are correctly predicted as defects, false positive (FP) represents the non-defects that are incorrectly predicted as defects, and false negative (FN) means real defects that are incorrectly identified as non-defects. A defect is determined if the intersection over union (IoU) with the ground truth box is larger than 0.001.

3.3 Implementation Details

The network is implemented using Keras [4], and trained with the Adam optimizer [8] with a learning rate of 0.001 on an Nvidia RTX2080 Ti GPU. The min-batch size is set to 4. During the training and testing stages, the image patch size $w \times w$ is set to 512×512. The overlap size l is set to 16. The generation of the synthetic dataset and defect detection are conducted in Python 3.6 with OpenCV. The channel numbers of encoding and decoding blocks are $[16, 16, 16, 16, 16]$ and $[16, 16, 16, 16, 8]$.

3.4 The Parameter Choice of λ1 and λ2

To verify how the values of $\lambda 1$ and $\lambda 2$ affect registration performance, we conducted experiments by varying these two parameters. As shown in Table 2, we evaluate the set $(\lambda 1, \lambda 2)$ in {(0.05, 0.05), (0.05, 0.10), (0.05, 0.50), (0.05, 1.00), (0.10, 0.10), (0.10, 0.50), (0.10, 1.00), (0.50, 0.10)} on all testing sets, where the average MSE varies from 7.059 to 11.471. The best result is achieved by (0.05, 0.10). Therefore, we choose the best-optimized set ($\lambda 1 = 0.05, \lambda 2 = 0.10$).

3.5 Comparison with Baselines and Time Complexity

MSE is utilized to evaluate image registration performance at the pixel level. Table 3 demonstrates that our method has the lowest average MSE (7.059) and gets the lowest value for each testing set. PRDFE (8.492) and VoxelMorph (9.377) are close to our method. By contrast, the average MSE of traditional

Table 2. Effect of varying parameters λ1 and λ2 on MSE ($\times 10^{-4}$) (Bold: best).

λ1	λ2	Label-A	Label-B	Label-C	Label-D	Average MSE
0.05	0.05	17.170	5.460	3.961	11.928	9.630
0.05	**0.10**	**13.812**	**3.252**	**2.625**	**8.548**	**7.059**
0.05	0.50	15.245	4.339	3.095	10.287	8.242
0.05	1.00	15.212	3.816	2.833	10.311	8.043
0.10	0.10	16.494	5.393	3.910	11.713	9.378
0.10	0.50	15.655	4.424	3.203	10.880	8.541
0.10	1.00	19.367	6.698	5.011	14.806	11.471
0.50	0.10	17.804	6.793	4.823	13.691	10.778

methods, such as SIFT (23.728), SURF (23.134), and AKAZE (30.806), are relatively high.

In terms of F_1 score, Table 4 shows several existing methods to illustrate the effectiveness of our proposed method. Our method yields the highest average F_1 score of 0.7916, which is a new state-of-the-art performance. Further, our method clearly outperforms these existing methods, for PRDFE (0.6520), VoxelMorph (0.6378), SIFT (0.4227), SURF (0.4210), AKAZE (0.3682) and DDN (0.2629) with improvements of 13.96%, 15.38%, 36.89%, 37.06%, 42.34%, and 52.87%, respectively. In addition, our method also achieves the highest F_1 score for each testing set.

Table 5 shows the comparison of registration time. Traditional methods take a longer time than deep learning methods. Although VoxelMorph achieves the best result (0.3087 s), our registration time (0.3913 s) is close to VoxelMorph which is still in real-time.

Table 3. Comparison on MSE ($\times 10^{-4}$) (Bold: best).

Method	Label-A	Label-B	Label-C	Label-D	Average MSE
SIFT [11]	34.682	16.133	14.234	29.864	23.728
SURF [3]	34.027	15.575	13.678	29.256	23.134
AKAZE [1]	38.823	17.550	34.414	32.436	30.806
VoxelMorph [2]	17.396	4.877	3.832	11.402	9.377
DDN [13]	30.030	11.100	11.139	16.546	17.204
PRDFE [21]	15.530	4.724	3.539	10.173	8.492
PPR-Net	**13.812**	**3.252**	**2.625**	**8.548**	**7.059**

Table 4. Comparison in terms of average F_1 score (Bold: best).

Method	Label-A	Label-B	Label-C	Label-D	Average F_1
SIFT [11]	0.6275	0.3830	0.1545	0.5257	0.4227
SURF [3]	0.6262	0.3747	0.1520	0.5312	0.4210
AKAZE [1]	0.5615	0.3626	0.1425	0.4061	0.3682
VoxelMorph [2]	0.6477	0.6890	0.4145	0.8000	0.6378
DDN [13]	0.2722	0.3896	0.0677	0.3221	0.2629
PRDFE [21]	0.7203	0.6867	0.4258	0.7753	0.6520
PPR-Net	**0.7462**	**0.8571**	**0.7004**	**0.8625**	**0.7916**

Table 5. Comparison in terms of registration time (s) (Bold: best).

Method	Label-A	Label-B	Label-C	Label-D	Average time (s)
SIFT [11]	0.9877	0.6125	3.0090	0.6807	1.3225
SURF [3]	0.6032	0.4320	2.4815	0.5781	1.0237
AKAZE [1]	0.4636	0.5166	1.7136	0.5161	0.8025
VoxelMorph [2]	**0.1665**	**0.1679**	**0.7037**	**0.1965**	**0.3087**
DDN [13]	0.1731	0.1752	0.7854	0.2190	0.3382
PRDFE [21]	0.2359	0.2367	0.9437	0.2168	0.4083
PPR-Net	0.1996	0.2236	0.8997	0.2421	0.3913

3.6 Ablation Studies for PPR-Net

For PPR-Net, we improve the distortion loss function L_{dist} and the multi-scale pyramid module with multi-features fusion. To figure out the performance of each part, we discard some parts of PPR-Net and get some models. Table 6 shows the settings for different models. Model "PPR-Net-w/o-L_{si}" means removing the shift-invariant loss L_{si} from the distortion loss L_{dist}. Model "PPR-Net-w/o-MDFF" represents PPR-Net without the gradual fusion from the low-resolution deformation field to the high-resolution one. Model "PPR-Net-w/o-MWFF" means removing warped features at pyramid layers from P_2 to P_5. Model "PPR-Net-Single-scale" represents that the multi-scale pyramid module is changed to single-scale by removing pyramid layers from P_2 to P_5.

Table 7 and Table 8 show the comparison results in terms of MSE and F_1 score. After the pyramid module is changed from multi-scale to single-scale, the average MSE increases significantly from 7.059 to 8.973 and the F_1 score decreases from 0.7916 to 0.6440. After removing MWFF, the average MSE increases from 7.059 to 8.951, and the F_1 score drops from 0.7916 to 0.6609. Therefore, the most significant improvement is the multi-scale pyramid, and the second one is the MWFF. Meanwhile, compared with other models, our PPR-Net has the best MSE and F_1 score on all testing sets.

Table 6. The settings of different models.

Model	L_{si}	Multi-scale pyramid		Single-scale pyramid
		Multi-resolution deformation fields fusion (MDFF)	Multi-scale warped features fusion (MWFF)	
PPR-Net-w/o-L_{si}	No	Yes	Yes	No
PPR-Net-w/o-MDFF	Yes	No	Yes	No
PPR-Net-w/o-MWFF	Yes	Yes	No	No
PPR-Net-Single-scale	Yes	No	No	Yes
PPR-Net	Yes	Yes	Yes	No

3.7 Qualitative Results

Figure 6 shows that our PPR-Net achieves more accurate alignment than the compared methods. For SIFT, SURF, and AKAZE, a black region is introduced on the left of the registered images, resulting in false detection. Further, DDN and PRDFE get significant text distortion compared with PPR-Net. Although the performance of VoxelMorph is close to ours, a certain text distortion can

Table 7. Ablation study in terms of MSE ($\times 10^{-4}$) (Bold: best).

Model	Label-A	Label-B	Label-C	Label-D	Average MSE
PPR-Net-w/o-L_{si}	15.311	3.641	2.890	10.103	7.986
PPR-Net-w/o-MDFF	15.398	3.401	2.813	10.038	7.913
PPR-Net-w/o-MWFF	16.362	4.814	3.733	10.893	8.951
PPR-Net-Single-scale	16.344	4.704	3.887	10.955	8.973
PPR-Net	**13.812**	**3.252**	**2.625**	**8.548**	**7.059**

Table 8. Ablation study in terms of F_1 score (Bold: best).

Model	Label-A	Label-B	Label-C	Label-D	Average F_1
PPR-Net-w/o-L_{si}	0.7265	0.8567	0.6976	0.8329	0.7784
PPR-Net-w/o-MDFF	0.7432	0.8563	0.6884	0.8114	0.7748
PPR-Net-w/o-MWFF	0.7071	0.7126	0.4123	0.8117	0.6609
PPR-Net-Single-scale	0.7012	0.7208	0.3543	0.7996	0.6440
PPR-Net	**0.7462**	**0.8571**	**0.7004**	**0.8625**	**0.7916**

be observed, such as the letter "T" of "This device" in Fig. 6e. Figure 7 gives a comparison of subtraction images, and our method is the best with the fewest artifacts. By contrast, DDN yields significant artifacts, and the traditional methods, such as SIFT, SURF, and AKAZE, all have artifacts generated by the above black regions. Figure 8 visualizes a comparison among compared methods and PPR-Net (Label-B), and the compared methods do exist with more FP or FN.

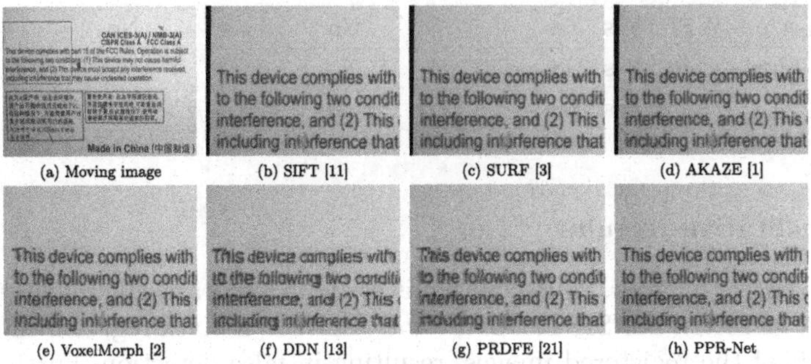

Fig. 6. Registration performance visualization (Label-A). (a) is the moving image with an ROI indicated by the red frame, (b)–(h) are the corresponding warped images of the red frame of (a). (Color figure online)

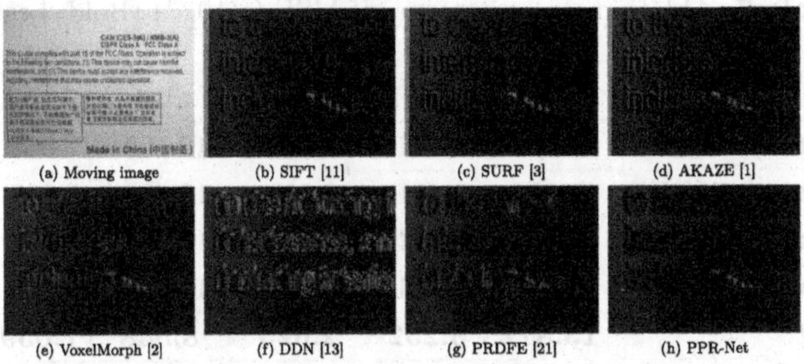

Fig. 7. The visualization of difference images. After image registration, warped images are subtracted from the fixed image. (a) is a moving image from Label-A. (b)–(h) are corresponding difference images of the red frame of (a). (Color figure online)

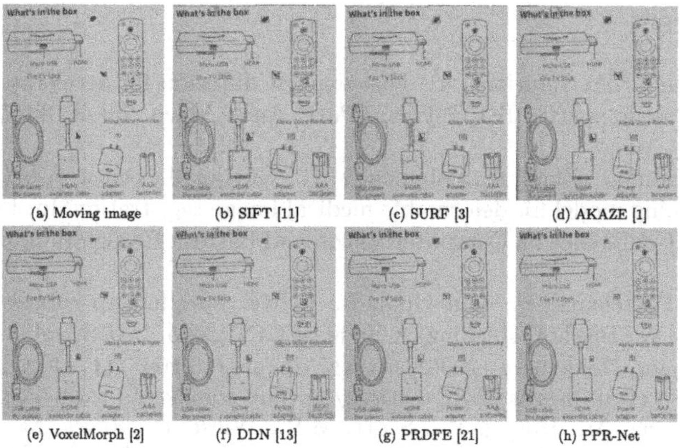

Fig. 8. Visualization of defect detection results (Label-B). The results of TP, FP, and FN are marked with red, green, and yellow frames, respectively. (Color figure online)

4 Conclusions

In this paper, we proposed a patch-based multi-scale pyramid registration network for printed labels with large deformation. We adopt a patch-based image splitting and stitching strategy, which can scale to larger image resolution. PPR-Net allows recurrently refining the deformation fields with multi-scale convolutional features in a coarse-to-fine manner, which helps handle large deformations. Further, we introduced a novel distortion loss function incorporated with shift-invariant loss, which can improve text distortion after image registration. Finally, we generated a synthetic database based on real printed labels and evaluated registration performance. The experimental result shows that our method outperforms other comparable approaches. The precision of reference-based comparison defect detection is significantly improved by using our registration network.

Acknowledgements. This work was supported in part by NSFC fund (62176077, 62272133, 61906162), in part by the Shenzhen Colleges and Universities Stable Support Program No. GXWD20220811170100001, in part by the Guangdong Basic and Applied Basic Research Foundation under Grant 2019B1515120055, in part by the Shenzhen Key Technical Project under Grant 2020N046, in part by the Shenzhen Fundamental Research Fund under Grant JCYJ20210324132210025, in part by Shenzhen Science and Technology Program (RCBS20200714114910193).

References

1. Alcantarilla, P.F., Solutions, T.: Fast explicit diffusion for accelerated features in nonlinear scale spaces. IEEE Trans. Patt. Anal. Mach. Intell. **34**(7), 1281–1298 (2011)
2. Balakrishnan, G., Zhao, A., Sabuncu, M.R., Guttag, J., Dalca, A.V.: An unsupervised learning model for deformable medical image registration. In: Proceedings of the IEEE Conference on Computer Vision and Pattern Recognition, pp. 9252–9260 (2018)
3. Bay, H., Tuytelaars, T., Van Gool, L.: SURF: speeded up robust features. In: Leonardis, A., Bischof, H., Pinz, A. (eds.) ECCV 2006. LNCS, vol. 3951, pp. 404–417. Springer, Heidelberg (2006). https://doi.org/10.1007/11744023_32
4. Chollet, F., et al.: Keras (2015). https://github.com/keras-team/keras
5. Hu, X., Kang, M., Huang, W., Scott, M.R., Wiest, R., Reyes, M.: Dual-stream pyramid registration network. In: Shen, D., et al. (eds.) MICCAI 2019. LNCS, vol. 11765, pp. 382–390. Springer, Cham (2019). https://doi.org/10.1007/978-3-030-32245-8_43
6. Jaderberg, M., Simonyan, K., Zisserman, A., et al.: Spatial transformer networks. Adv. Neural. Inf. Process. Syst. **28**, 2017–2025 (2015)
7. Ke, Y., Sukthankar, R.: PCA-SIFT: a more distinctive representation for local image descriptors. In: Proceedings of the 2004 IEEE Computer Society Conference on Computer Vision and Pattern Recognition, CVPR 2004, vol. 2, pp. II-II. IEEE (2004)
8. Kingma, D.P., Ba, J.: Adam: a method for stochastic optimization. arXiv preprint arXiv:1412.6980 (2014)
9. Krebs, J., et al.: Robust non-rigid registration through agent-based action learning. In: Descoteaux, M., Maier-Hein, L., Franz, A., Jannin, P., Collins, D.L., Duchesne, S. (eds.) MICCAI 2017. LNCS, vol. 10433, pp. 344–352. Springer, Cham (2017). https://doi.org/10.1007/978-3-319-66182-7_40
10. Li, D., Li, J., Fan, Y., Lu, G., Ge, J., Liu, X.: Printed label defect detection using twice gradient matching based on improved cosine similarity measure. Expert Syst. Appl. **204**, 117372 (2022)
11. Lowe, D.G.: Distinctive image features from scale-invariant keypoints. Int. J. Comput. Vision **60**(2), 91–110 (2004)
12. Ma, K., Shu, Z., Bai, X., Wang, J., Samaras, D.: Docunet: document image unwarping via a stacked U-Net. In: Proceedings of the IEEE Conference on Computer Vision and Pattern Recognition, pp. 4700–4709 (2018)
13. Nazib, A., Fookes, C., Perrin, D.: Dense deformation network for high resolution tissue cleared image registration. arXiv preprint arXiv:1906.06180 (2019)
14. Peng, X., Chen, Y., Xie, J., Liu, H., Gu, C.: An intelligent online presswork defect detection method and system. In: 2010 Second International Conference on Information Technology and Computer Science, pp. 158–161. IEEE (2010)
15. Rohé, M.-M., Datar, M., Heimann, T., Sermesant, M., Pennec, X.: SVF-Net: learning deformable image registration using shape matching. In: Descoteaux, M., Maier-Hein, L., Franz, A., Jannin, P., Collins, D.L., Duchesne, S. (eds.) MICCAI 2017. LNCS, vol. 10433, pp. 266–274. Springer, Cham (2017). https://doi.org/10.1007/978-3-319-66182-7_31
16. Ronneberger, O., Fischer, P., Brox, T.: U-Net: convolutional networks for biomedical image segmentation. In: Navab, N., Hornegger, J., Wells, W.M., Frangi, A.F. (eds.) MICCAI 2015. LNCS, vol. 9351, pp. 234–241. Springer, Cham (2015). https://doi.org/10.1007/978-3-319-24574-4_28

17. Sokooti, H., de Vos, B., Berendsen, F., Lelieveldt, B.P.F., Išgum, I., Staring, M.: Nonrigid image registration using multi-scale 3D convolutional neural networks. In: Descoteaux, M., Maier-Hein, L., Franz, A., Jannin, P., Collins, D.L., Duchesne, S. (eds.) MICCAI 2017. LNCS, vol. 10433, pp. 232–239. Springer, Cham (2017). https://doi.org/10.1007/978-3-319-66182-7_27
18. Yang, X., Kwitt, R., Styner, M., Niethammer, M.: Quicksilver: fast predictive image registration-a deep learning approach. Neuroimage **158**, 378–396 (2017)
19. Yoo, J.C., Han, T.H.: Fast normalized cross-correlation. Circ. Syst. Sig. Process. **28**(6), 819–843 (2009)
20. Zhang, E., Chen, Y., Gao, M., Duan, J., Jing, C.: Automatic defect detection for web offset printing based on machine vision. Appl. Sci. **9**(17), 3598 (2019)
21. Zhou, Y., et al.: Unsupervised deformable medical image registration via pyramidal residual deformation fields estimation. arXiv preprint arXiv:2004.07624 (2020)

A Prototype-Oriented Contrastive Adaption Network for Cross-Domain Facial Expression Recognition

Chao Wang, Jundi Ding(✉), Hui Yan, and Si Shen

Nan Jing University of Science and Technology, Nanjing, China
dingjundi2010@njust.edu.cn

Abstract. Numerous well-performing facial expression recognition algorithms suffer from severe slippage when trained on one dataset and tested on another, due to inconsistencies in facial expression datasets caused by different acquisition conditions and subjective biases of annotators. In order to improve the generalization ability of the model, in this paper we propose a simple but effective Prototype-Oriented Contrastive Adaptation Network (POCAN) unified contrastive learning and prototype networks for cross-domain facial expression recognition. We employ a two-stage training pipeline. Specifically, in the first stage, we pre-train on the source domain to obtain semantically meaningful features and obtain good initial conditions for the target domain. In the second stage, we perform intra-domain feature learning and inter-domain feature fusion by narrowing the distance between samples and their corresponding prototypes and widening the distance with other prototypes, and we also use an adversarial loss function for domain-level alignment. In addition, we also consider the problem of data category imbalance, and category weights are introduced into our method so that the categories of the two domains are in a uniform distribution. Extensive experiments show that our method can yield competitive performance on both lab-controlled and in-the-wild datasets.

Keywords: Cross-domain facial expression recognition · Prototype network · Contrastive learning

1 Introduction

Facial expressions(referring to the emotions shown on the face) are one of the most direct and natural ways to convey human emotions. How understanding human beings' emotion is a key problem that needs to be solved in the fields of human-computer interaction, perceptual psychology, intelligent control, medical care, security, etc. Therefore, the research of facial expression recognition(FER) is getting growing attention. Facial expressions are a complex signal, humans have many expressions, and the types of expressions in different regions and cultures are different. However, some facial expressions are universal across cultures. The researchers summarize these expressions as basic expressions. In the

L. Wang et al. (Eds.): ACCV 2022, LNCS 13842, pp. 324–340, 2023.
https://doi.org/10.1007/978-3-031-26284-5_20

1990s,s, Ekman et al. [6] proposed six basic expressions through cross-cultural research: fear, sadness, anger, disgust, surprise, and happiness. Later researchers added neutral to the basic expression.

Early studies of facial expression recognition were conducted on lab-controlled datasets using hand-crafted features (*e.g.*, local binary patterns (LBP [22])) and shallow learning (*e.g.*, sparse learning [34]). Lab-controlled datasets with acted facial expressions collect in artificial environments and do not correspond to the spontaneous, unconstrained situation in practice. Moreover, Shallow learning can only extract shallow features and cannot extract deep features, resulting in poor algorithm performance. As deep learning has achieved remarkable results in many fields, deep learning technology, especially CNN, has been applied to facial expression recognition, and has achieved remarkable progress. At the same time, many facial expression datasets from unconstrained scenarios were collected. Both contributed to the development of facial expression recognition.

Although today's excellent facial expression algorithms can achieve satisfactory performance, these algorithms are trained and tested on a single dataset. When using these algorithms to test on other datasets, they suffer from dramatic performance deterioration [15]. The reason for this phenomenon is due to the existence of domain shift: different facial expression datasets are collected in different contexts, and different annotators perceive expressions differently. Therefore, cross-domain facial expression recognition is a more meaningful and challenging task, which is closely related to the field of domain adaptation in the field of transfer learning. Recently, many studies have begun to focus on cross-domain facial expression recognition and have proposed many effective methods. They either use MMD [13,15] to minimize domain distribution differences, or borrow the idea of adversarial learning [2] to learn cross-domain invariant features, but there are no methods using contrastive learning.

In this paper, motivated by the success of contrastive learning in representation learning, we use the idea of contrastive learning to solve the problem of cross-domain facial expression recognition and propose POCAN[1] , a two-stage algorithm.

- In the first stage, we use supervised contrastive learning to perform pre-training on the source domain. By making the samples of the same category close and the samples of different categories separate, we can learn the category information of the source domain and obtain a better initial conditions for the target domain. Inspired by [23], we parameterize the category prototype and learn the category prototypes by bringing the prototype close to the samples belonging to the class it represents. Compared with computing prototypes by clustering, this learnable prototype acquisition method can reduce computational consumption and improve training speed.
- In the second stage, we learn the features of the source and target domains separately. For the target domain features, we use self-supervised contrastive

[1] The source code and the trained model are publicly available at https://github.com/Winter-is-coming-wow/ACCV2022_POCAN.git..

learning and prototype contrastive learning to learn, which can compensate for the lack of self-supervised contrastive learning ability to obtain high-level semantic knowledge, while maintaining the local smoothness of the learned representations. We use supervised contrastive learning and prototype contrastive learning to learn the features of the source domain to keep the model's low error rate in the source domain. At the same time, we carry out a two-level domain fusion strategy, one is to make the samples of one domain close to the prototypes of the same category in the other domain to perform domain alignment at the category level, and the other cleverly uses classifiers to form a domain discriminator and then uses the adversarial method for domain-level alignment.

2 Related Works

In this section, we briefly review two related topics: cross-domain facial expression recognition and contrastive learning.

2.1 Cross-Domain Facial Expression Recognition

The serious decline of traditional facial expression recognition algorithms in cross-dataset experiments has made researchers realize the importance of improving the generalization ability of algorithms. Many cross-domain facial expression recognition algorithms have been proposed in the last few years. In [29], Transfer Subspace Learning is introduced into cross-domain facial expression recognition, this method learns a subspace in which the knowledge learned from the source domain is transferred to the target domain. In [30], the author proposed unsupervised domain adaptive dictionary learning (UDADL) to learn a reliable synthesis dictionary for both source and target samples such that the distribution mismatch between two different domains could be alleviated. In [35], a discriminative feature adaptation method is proposed to minimize the mismatch between source and target distribution by learning a new feature space. In [13], the author embeds maximum mean inconsistency (MMD) into deep networks to reduce inconsistency between datasets, while introducing a learnable parameter to reduce the impact of the source domain and target and class distribution differences. In [26], a generative adversarial network (GAN) is introduced to generate more target domain data, and a distributed pseudo-label method is applied to achieve domain adaptation with limited target data without ground truth labels. In [15], the author found that the conditional probabilities between different datasets are different by measuring the deviation between facial expression datasets. Therefore, ECAN is proposed to learn domain-invariant features, align the marginal probability distribution and conditional probability distribution of the source and target domains at the same time, and consider the class imbalance problem between domains. Recently, In [2] joint Adversarial Learning and Graph Representation Learning for simultaneous domain adaptation of global and local Features.

2.2 Contrastive Learning

Unsupervised/self-supervised contrastive learning has great success in representation learning. The basic idea of contrastive learning is to make pairs of positive samples more similar and pairs of negative samples less similar via a contrastive loss in the latent space. The common contrastive loss function is InfoNEC loss [20], the loss for a sample is defined as:

$$L_{InfoNEC}(z_i) = -log(\frac{exp(sim(z_i, z_i')/\tau)}{\sum_{j=1}^{n} exp(sim(z_i, z_j)/\tau)}), \tag{1}$$

where sim() is the similarity function (cosine similarity is commonly used), n is number of sample pairs, τ denotes a temperature parameter, it is an important parameter that controls the strength of the penalty for hard negative samples [25].

In instance-based self-supervised contrastive learning, samples from the same instance are regarded as pairs of positive samples, and samples from different instances are regarded as pairs of negative samples. Positive samples always come from another views of a instance, negative samples come from a memory bank [27], or a queue [9], or a minibatch [3]. Although instance-based self-supervised contrastive learning has good performance, this method has an inherent disadvantage: it treats two samples as a negative sample pair as long as they come from different instances, which unavoidably determines some samples with the same semantics as the anchor as negative samples. This negatively affects the performance of the algorithm. To solve this problem, [4]develop a debiased contrastive objective that corrects for the sampling of same-label datapoints without knowledge of the true labels. [12]propose prototypical contrastive learning (PCL) which encodes the semantic structure of data into the embedding space through make samples similar with its prototype.

3 Method

In the context of unsupervised cross-domain facial expression recognition, we have a labeled source dataset $D_s = \{x_i^s, y_i^s\}_{i=1}^{N_s}$ with marginal probability distribution $P^s(X)$ and an unlabeled target dataset $D_t = \{x_i^t\}_{i=1}^{N_t}$ with marginal probability distribution $P^t(X)$, where N_s and N_t are the number of images in source and target domains respectively. Both x_i^s and x_i^t belong to one of the 7 predefined expressions: anger, disgust, fear, happiness, sadness, surprise, and neutral. The two datasets are drawn from different distributions (i.e., $P^s(X)! = P^t(X)$). Our task is to train a model using the labeled source dataset and the unlabeled target dataset, reducing the domain offset between the source and target domains, and minimizing the error rate of the model on the target domain.

As show in Fig. 1, our model consists of a encoder $f(\cdot)$, a projection head $g(\cdot)$, a learnable source domain prototype matrix M_s and a learnable target domain prototype matrix M_t.

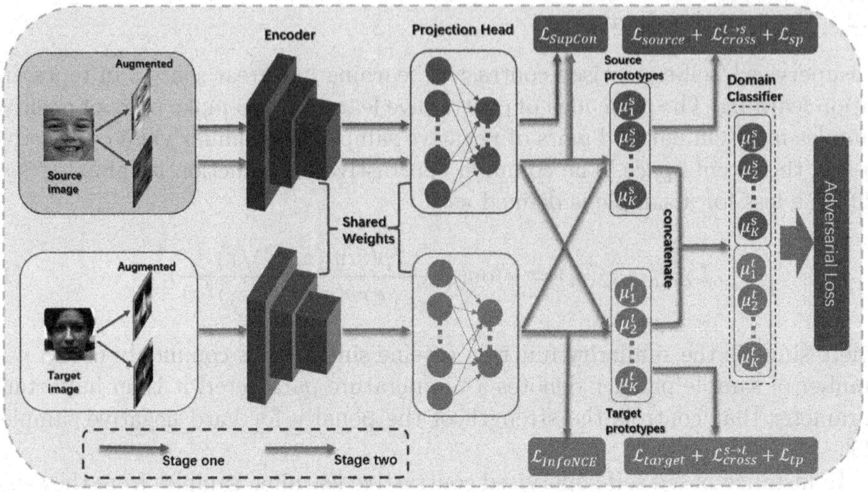

Fig. 1. Overview of the proposed Prototype-Oriented Contrastive Adaptation Network (POCAN) structure. POCAN contains two stages: **(1)pre-train on source domain**: using supervised contrastive learning, an encoder and a projection head are trained to learn semantically meaningful features and obtain good initial conditions for the target domain. **(2)transfer to target domain**: learning domain specific feature via L_{in}, while performing category-level alignment via L_{cross} and domain-level alignment via L_{ad}.

3.1 Learnable Category Prototypes

Most of the methods [12,24,31] using prototypes obtain the prototypes of the categories through the k-means clustering algorithm. However, this method consumes huge computing resources, and when the model parameters change, the prototypes of the categories cannot be updated in time, which affects the convergence speed and accuracy of the algorithm. Inspired by [23], we parameterize the prototypes so that they can be learned using the backpropagation algorithm. Specifically, we define the prototype as a vector $\mu \in \mathbb{R}^m$, where m is the dimension of each prototype that is the same as the embedding dimension of the sample output vector through the encoder and projection head. All prototypes form a prototype matrix $M = \{\mu_1, \mu_2, ..., \mu_k\}$, where k is the number of categories. The prototype matrix is closely connected to the linear classification layer because the neural network weights in the classification layer can be interpreted as class prototypes. Given a sample x_i with label y_i, we compute its feature vector $z_i = g(f(x_i))$, then we compute similarity matirx vector between z_i and M as $P_i = [p_{i,1}, p_{i,2}, ..., p_{i,k}]$, with $P_{i,j} = \frac{exp(\mu_j \cdot z_i)}{\sum_{r=1}^{k} exp(\mu_r \cdot z_i)}$. We learn source and target prototypes by minimize cross entropy between P and true label using:

$$\mathcal{L}_{sp} = \frac{1}{N_s} \sum_{i=1}^{N_s} w_{y_i^s}^s \cdot \mathcal{L}_{CE}(P_i^s, y_i^s), \tag{2}$$

$$\mathcal{L}_{tp} = \frac{1}{N_t^{trust}} \sum_{i=1}^{N_t^{trust}} w_{y_i^t}^t \cdot \mathcal{L}_{CE}(P_i^t, y_i^t), \tag{3}$$

where w is class weight, its calculation method is in Sect. 3.5. Because the target domain data has no label information, we first generate pseudo-labels(we use hard labels) for the target domain. It is worth noting that we only generate pseudo-labels for samples whose confidence is higher than a certain threshold, denoted as $D_t^{trust} = \{x_i\}_{i=1}^{N_t^{trust}}$ $(x_i \in D_t, confidence(x_i) > threshold_i)$. And the initial prototypes of the target domain are initialized to the prototypes of the source domain. We train the extractor and prototypes in turns, in the same fashion as by performing iterative clustering and representation learning. We freeze the parameters of the prototypes when training the extractor, and freeze the parameters of the extractor when training the prototypes.

3.2 Pre-train on Source Domain

In the traditional field of fully supervised learning, the cross-entropy loss function is usually used to train the model. Recently [11] introduced contrastive learning into the field of supervised learning and proposed supervised contrastive learning, and obtained better results. Compared with traditional supervised learning based on cross-entropy, supervised contrastive learning is more stable to hyper-parameters such as optimizers and data augmentation, and has better robustness to data corruption and training data reduction, and is inherently capable of hard positive and negative mining attributes [11]. Based on these advantages, we choose supervised contrastive learning for pre-training, and the experimental results also show the correctness of our choice.

The core idea of supervised contrastive learning is to shorten the distance between samples belonging to the same category and to widen the distance between samples of different categories. Its loss takes the following form:

$$\mathcal{L}_{SupCon} = \sum_{i=1}^{n} \frac{-1}{|P(i)|} \sum_{p \in P(i)} log \frac{exp(sim(v_i \cdot v_p)/\tau')}{\sum_{a \in A(i)} exp(sim(v_i \cdot v_a)/\tau')}, \tag{4}$$

where A(i) denotes the set of samples in a minibatch, $P(i) \equiv \{p \in A(i) : \widetilde{y}_p = \widetilde{y}_i\}$ indicates the set of samples share the same labels with the anchor in a minibatch. After pre-training, we can use the method in Sect. 3.1 to get the prototype matrix of the source domain M_s.

3.3 Domain-Specific Feature Learning

The goal of this part is to learn the discriminative features of each domain. Inspired by self-supervised contrastive learning using instance discrimination in [3] to learn knowledge from unlabeled data, we use it to learn the features of the target domain. However, instance discrimination-based self-supervised contrastive learning treats samples from different instances as negative sample pairs without considering the semantic similarity of the samples, resulting in the learned features lacking high-level semantic structure. Inspired by some algorithms that learn semantic structure through clustering, we learn the semantic structure of a category by keeping a sample close to its corresponding prototype while staying away from other prototypes. This is exactly what prototype contrastive learning does. So the domain-specific loss in target domain can be written as:

$$\mathcal{L}_{target} = \frac{1}{N_t}\sum_{i=1}^{N_t} L_{InfoNCE}(z_i^t) + \lambda_t \cdot \frac{1}{N_t^{trust}}\sum_{j=1}^{N_t^{trust}} w_{y_j^t}^t \cdot \mathcal{L}_{CE}(P_j^t, y_j^t), \quad (5)$$

where w is class weight, Its calculation method is in Sect. 3.5. λ_t is a weight from 0 to 1 to suppress the noisy signal from the pseudo-label of target data. Specifically, we set $\lambda_t = \frac{2}{1+exp(-10p)} - 1$, where p is the training progress linearly changing from 0 to 1.

Following the suggestion of work [21], we use both supervised contrastive learning and prototype contrastive learning to learn source domain features to ensure that the model maintains a low error rate in the source domain. So the domain-specific loss in source domain can be written as:

$$\mathcal{L}_{source} = \frac{1}{N_s}\sum_{i=1}^{N_s}(\mathcal{L}_{SupCon}(z_i^s) + w_{y_i^s}^s \cdot \mathcal{L}_{CE}(P_i^s, y_i^s)). \quad (6)$$

And the loss for domain-specific feature learning is:

$$\mathcal{L}_{in} = \mathcal{L}_{target} + \mathcal{L}_{source}. \quad (7)$$

3.4 Cross-Domain Feature Fusion

Category-Level Alignment. Many existing unsupervised domain adaptation methods use MMD to minimize domain distribution differences or use adversarial methods to maximize domain fusion to perform domain alignment. Both methods are performed at the domain level, however, domain distribution alignment does not guarantee alignment between categories. Performing category-level alignment can better align the features of the two domains. We consider that if the source domain and target domain are already aligned, the prototypes of the source domain should also have a good classification effect on the target domain, and vice versa, the prototypes of the target domain should also classify the source domain samples well. Based on the above assumptions, we perform category-level domain alignment by forcing samples to be close to the corresponding prototype of another domain and away from the prototypes of other categories.

In particular, given a source domain sample x_i^s and target domain prototype matrix $M_t = \{\mu_j^t\}_{j=1}^k$, we first compute the features of x_i^s Vector $z_i^s = g(f(x_i^s))$, and then calculate the similarity distribution between z_i^s and M_t, with $P_{ij}^{s \to t} = softmax(z_i^s \cdot M_t)$. Then we use the cross-entropy loss function to maximize the similarity of x_i^s with the corresponding prototype and minimize the similarity with other prototypes, which is:

$$\mathcal{L}_{cross}^{s\to t} = \sum_{i=1}^{N_s} w_{y_i^s}^s \cdot \mathcal{L}_{CE}(P_{ij}^{s\to t}, y_i^s). \tag{8}$$

Similarly,we can compute $\mathcal{L}_{cross}^{t\to s}$, and the final loss for category-level alignment is:

$$\mathcal{L}_{cross} = \mathcal{L}_{cross}^{s\to t} + \mathcal{L}_{cross}^{t\to s}. \tag{9}$$

Domain-Level Alignment. Adversarial domain adaptation methods usually include a discriminator to identify which domain the sample comes from. But in our method, we do not set an additional domain discriminator. Inspired by [28], we utilize the trainable prototypes of the source and target domains to form a domain discriminator. Specifically, given a sample x_i, we first obtain its feature vector $z_i = g(f(x_i))$ and then calculate the similarity between z_i and the prototype matrices of the two domains respectively: $sim(z_i, M_s) = z_i \cdot M_s, sim(z_i, M_t) = z_i \cdot M_t$, Then we concatenate $sim(z_i, M_s)$ and $sim(z_i, M_t)$ and apply softmax to form $P^{st} = softmax([sim(z_i, M_s), sim(z_i, M_t)])$. Define $\alpha_i^s = \sum_{i=1}^{k} P^{st}$ as the score of sample belonging to the source domain and $\alpha_i^t = \sum_{i=k+1}^{2k} P^{st}$ as the score of sample belonging to target domain. For a sample of the source domain, α_i^s should be greater than α_i^t. Similarly, For the target domain sample, α_i^t should be greater than α_i^s. The domain discriminator should try to distinguish whether the samples are from the source domain or the target domain. Therefore, we train the domain discriminator with

$$\mathcal{L}_{dis} = -\frac{1}{N_s}\sum_{j=1}^{N_s} log(\sum_{i=1}^{k} P_j^{st}(x_j^s)) - \frac{1}{N_t}\sum_{j=1}^{N_t} log(\sum_{i=k+1}^{2k} P_j^{st}(x_j^t)). \tag{10}$$

According to the idea of an adversarial network, the feature extractor is designed to learn domain-invariant features, which does the opposite to the domain discriminator, so we optimize the feature extractor by the following formula to make it have an adversarial relationship with the domain discriminator.

$$\mathcal{L}_{st} = -\frac{1}{N_t}\sum_{j=1}^{N_t} log(\sum_{i=1}^{k} P_j^{st}(x_j^t)) - \frac{1}{N_s}\sum_{j=1}^{N_s} log(\sum_{i=k+1}^{2k} P_j^{st}(x_j^s)). \tag{11}$$

So the adversarial loss for domain-level alignment is:

$$\mathcal{L}_{ad} = \mathcal{L}_{dis} + \mathcal{L}_{st}. \tag{12}$$

3.5 Obtain Class Weights and a Adaptative Threshold

Class imbalance is a common problem in facial expression datasets, and understandably: it is much easier to collect a happiness or sadness expression than an anger or disgust expression. Some previous works [13,15] set a learnable class-wise weighting parameter to explore the class distribution of the target domain to match the distribution of the source domain. We believe that it is inappropriate to simply make the distribution of the target domain match the distribution of the source domain, because the distribution of the source domain may itself be unbalanced, the impact of imbalance is that the classifier tends to discriminate the samples as those expressions that are easy to recognize. If the distribution of the target domain matches the distribution of the source domain, this classifier bias will be transmitted to the target domain. So we set the category weights of the two domains separately so that they are both on a relatively balanced distribution. Specifically, we calculate the class weights in the following way:

$$w_i = 1 - \left(\frac{n_i}{\sum_{j=1}^{k} n_j}\right)^2, \tag{13}$$

where n_i is the number of category i.

When generating pseudo-labels for the target domain, we take samples with label confidence greater than a certain threshold as trust samples, and use these samples to train prototypes. How to set an appropriate threshold is a relatively difficult matter, because different datasets have different degrees of difficulty in identification. For the more difficult datasets, we hope to set a lower threshold to ensure that the trust samples can occupy a certain proportion. Moreover, for different categories in a dataset, the samples that are easier to identify will always get a higher confidence level, and the difficult categories have a low confidence level, which leads to a small proportion of difficult samples in the trust sample set, which is not conducive to training prototypes. Considering these we set up an adaptive threshold generation method, specifically, for the category i of target domain D_t, its threshold is set as

$$threshold_i = avgc_i + (maxc_i - avgc_i) * (1 - avgc_i^{\beta}), \tag{14}$$

where $avgc_i$ is average confidence of samples judged to be category i, $maxc_i$ is the maximum confidence of the sample that is judged to be category i. β a is a hyperparameter that trades off pseudo-label noise and the number of trusted samples, a larger value indicates a stricter threshold policy.

3.6 Overall Objective

In the second stage, we update the four modules shown in Fig. 1: encoder $f(\cdot)$, profection head $g(\cdot)$, source prototypes matrix M_s and target prototypes matrix M_t. Based on the above, we have the following training objective for POCAN :

$$\min_{f,g} \mathcal{L}_{in} + \mathcal{L}_{cross} + \mathcal{L}_{st}. \tag{15}$$

$$\min_{M_s,M_t} \mathcal{L}_{sp} + \mathcal{L}_{tp} + \mathcal{L}_{dis}. \tag{16}$$

We iteratively take turns training the feature extractor and prototype matrixes. First, we train the encoder and the projection head via Eq. (15), and second, we update the source and target domain prototype matrix via Eq. (16).

4 Experiments

4.1 Databases

CK+ [16]: The Extended CohnKanade (CK+) dataset is a lab-controlled dataset that consists of 593 video sequences from 123 subjects. Only 309 sequences are labeled with six basic expression labels based on the Facial Action Coding System (FACS). Following previous work [2], we select the three frames with peak formation and the first frame(neutral face) from 309 sequences, resulting in 1,236 images. The dataset is divided into a training set of 1,125 and a test set of 129 images.
JAFFE [17]: The Japanese Female Facial Expression (JAFFE) database is a laboratory-controlled dataset that contains 213 samples of posed expressions from 10 Japanese females. Each person has 3–4 images annotated with one of the six basic expressions and one image annotated with a neutral expression. This database is challenging because it is a highly biased dataset in terms of gender and ethnicity. All images were used in our experiments.
Oulu-CASIA [33]: The Oulu-CASIA is a lab-controlled facial expression dataset consisting of six expressions (except neutral) from 80 people between 23 and 58 years old. Similar to CK+, we select the last three frames with peak information and the first frame (neutral face) from the 480 videos with the VIS System under normal indoor illumination, resulting in 1,920 images.
FER2013 [7]: It is a large-scale and unconstrained dataset collected automatically by the Google image search API. It contains 35,887 gray images of size 48×48 pixels, and each image is annotated with seven basic expressions. The dataset is further divided into a training set of 28,709 images, a validation set of 3,589 images, and a test set of 3,589 images.
SFEW2.0 [5]: The Static Facial Expressions in the Wild(SFEW) 2.0 is a real-world facial expression dataset that was created by selecting static frames from different films with spontaneous expressions, various head pose, age range, occlusions, and illuminations. This challenging dataset is divided into three groups: 958 training sets, 436 validation sets, and 372 test sets. In our experiments, only the training set and validation set provided with labels is used.
RAF-DB [14]: The Real-World Affective Face Database(RAF-DB) is a real-world dataset collected from the Internet. With manually crowd-sourced annotation and reliable estimation, seven basic and eleven compound emotion labels are provided for the samples. The dataset contains 15,339 images from thousands of individuals and is divided into two groups (12,271 training samples and 3,068 testing samples) for evaluation.

4.2 Experimental Details

We use ResNet-50 pre-trained on the VGG-FACE2 [1] dataset without fully connected layers as the encoder. A two-layer MLP is used as the projection head to map the output of the encoder into a 256-dimensional vector. Both the encoder and projection head outputs are normalized using L2-normalizing. All images are resized to 112 × 112. In the first stage, we train for 300 epochs using stochastic gradient descent (SGD) with an initial learning rate of 0.001, a momentum of 0.9, and weight decay of 0.00001. We use cosine descent to adjust the learning rate. In the second stage, we alternately train the encoder, projection head, and prototypes for about 20 epochs. We use SGD with the same momentum and weight decay as the first stage. The learning rates for the feature extractor and projection head are set to 0.00001, and the learning rates for the source and target prototypes are initialized to 0.0001, divided by 10 after about 10 epochs.

4.3 Comparison Results

It is difficult to compare the results of cross-domain facial expression recognition across literature because the feature extractors used in different literature and the selection of source domains are different. Fortunately, the work [2] constructs a unified CD-FER evaluation benchmark by re-implementing some excellent cross-domain recognition algorithms in the literature, ensuring that these algorithms use the same source/target datasets and feature extractors to achieve fair CD-FER evaluation. Therefore, for the fairness of the comparison, according to the previous work [2], we use ResNet-50 as the encoder and RAF-DB as the source domain.

Table 1 report the results of cross-domain recognition on three lab-controlled datasets. Table 2 report the results of cross-domain recognition on two in-the-wild datasets. Note that DT refers to directly transferring the model to the target domain after pre-training on the source domain (that is, after the first stage). As shown in the tables, when using the same encoder and source domain datasets, our method achieves the best accuracy on JAFFE and the second-best results on CK+, FER2013 and SFEW2.0 datasets. There is only one dataset on which our method does not show better results: Oulu-CASIA. Only ECAN [15] conducts cross-domain recognition experiments on Oulu-CASIA, and it uses a different pre-trained model and source domain than ours. It has a DT accuracy of 59.39%, which is much higher than our DT accuracy 53.90%, which may be responsible for the difference in performance. It should be pointed out that AGRA [2] used landmark information as local features, and we do not use any additional feature information. Nonetheless, our method achieves competitive performance on both lab-controlled and in-the-wild datasets.

4.4 Empirical Analysis

Ablation Study. Our POCAN have three important loss: domain-specific feature learning $\mathcal{L}_{in}$, category-level alignment loss $\mathcal{L}_{cross}$ and adversarial loss for domain-level alinment $\mathcal{L}_{ad}$. Therefore, We examine the effectiveness of each component of our method by remove each loss while keeping other loss. We also detect the effect of class imbalance. The experimental results are shown in Table 3. From the results of ablation experiments we can see that both in-domain feature

Table 1. Cross-domain accuracy(%) comparison on CK+, JAFFE and Oulu-CASIA dataset.

Method	Source	Backbone	CK+	JAFFE	Oulu-CASIA
STCNN [8]	MMI+FERA	Inception-ResNet	73.91	–	–
GDFER [19]	*Sixdatasets*†	Inception	64.20	–	–
ICID [10]	RAF-DB+MMI+SFEW	DarkNet-19	88.7	–	–
ICID [10]	RAF-DB+MMI	DarkNet-19	84.5	–	–
DFA [35]	CK+	Customized	–	63.38	–
FTDNN [32]	*Sixdatasets*⋆	VGGNet	88.58	44.32	–
DETN [13]	RAF-DB	Customized	78.83	57.75	–
ECAN [15]	RAF-DB 2.0	VGGNet	86.49	61.94	**63.97**
AGRA [2]	RAF-DB	ResNet-50	**85.27**	61.50	–
- ICID [10]	RAF-DB	ResNet-50	74.42	50.70	–
DFA [35]	RAF-DB	ResNet-50	64.26	44.44	–
FTDNN [32]	RAF-DB	ResNet-50	79.07	52.11	–
DETN [13]	RAF-DB	ResNet-50	78.22	55.89	–
ECAN [15]	RAF-DB	ResNet-50	79.77	57.28	–
DT	RAF-DB	ResNet-50	76.74	52.11	53.90
POCAN(Ours)	RAF-DB	ResNet-50	<u>84.50</u>	**64.32**	<u>55.83</u>

The upper part is taken from the corresponding papers, the middle part is taken from [2], the bottom part are generated by our implementation.
† MultiPIE, CK+, DISFA, FERA, SFEW, and FER2013.
⋆ CK+, JAFFE, MMI, RaFD, KDEF, BU3DFE, ARFace.

Table 2. Cross-domain accuracy(%) comparison on FER2013 and SFEW2.0 dataset.

Method	Source	Backbone	FER2013	SFEW2.0
GDFER [19]	*Sixdatasets*†	Inception	34.00	39.80
DETN [13]	RAF-DB	Customized	52.37	47.55
ECAN [15]	RAF-DB 2.0	VGGNet	58.21	54.34
AGRA [2]	RAF-DB	ResNet-50	**58.95**	**56.43**
ICID [10]	RAF-DB	ResNet-50	53.70	48.85
DFA [35]	RAF-DB	ResNet-50	45.79	43.07
FTDNN [32]	RAF-DB	ResNet-50	55.98	47.48
DETN [13]	RAF-DB	ResNet-50	52.29	49.40
ECAN [15]	RAF-DB	ResNet-50	56.46	52.29
DT	RAF-DB	ResNet-50	54.73	47.49
POCAN(Ours)	RAF-DB	ResNet-50	<u>58.87</u>	<u>53.44</u>

The upper part is taken from the corresponding papers,the middle part is taken from [2],the bottom part are generated by our implementation.
† MultiPIE, CK+, DISFA, FERA, MMI, and FER2013 or SFEW.

Table 3. Ablation Study of the role of each loss function and category weights in our method.

Method	CK+	JAFFE	Oulu	FER2013	SFEW2.0	AVG
POCAN(w/o $\mathcal{L}_{in}$)	80.62	57.75	55.26	57.70	51.65	60.60
POCAN(w/o $\mathcal{L}_{cross}$)	76.74	58.21	52.18	58.34	53.30	59.75
POCAN(w/o $\mathcal{L}_{ad}$)	83.72	61.97	55.78	58.76	53.37	62.72
POCAN(w/o weights)	83.72	60.56	55.68	**58.87**	52.37	62.24
POCAN	**84.50**	**64.32**	**55.83**	58.74	**53.44**	**63.37**

learning and domain alignment are important in our method. Just doing domain alignment or in-domain feature learning yields mediocre results. Category-level alignment is more important than domain-level alignment, which contributes little to the improvement of our experimental results. We can also find that category-wise weights are beneficial to our approach.

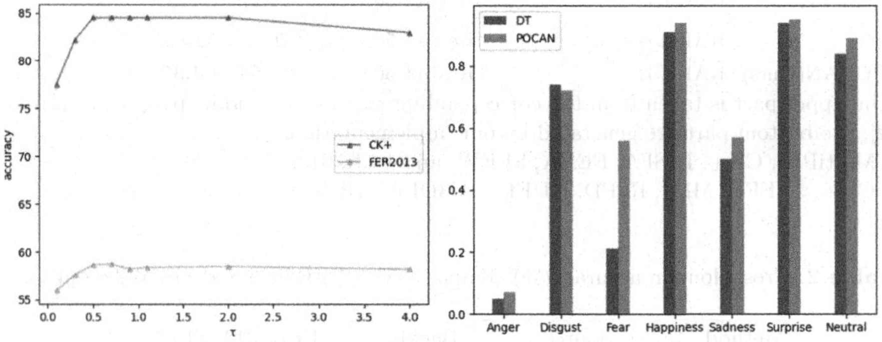

(a) (b)

Fig. 2. (a) Performances w.r.t τ on CK+ and FER2013. (b) CK+ F1 scores after the first stage and the second stage

Parameter Analysis. The hyperparameter τ has an important impact on feature learning in the target domain. To examine the effect of this parameter in our experiments, we show the accuracy of our method on the lab-controlled dataset CK+ and in-the-wild dataset FER2013 under different τ. As shown in Fig. 2(a), our method benefits from a relatively large temperature parameter, and 0.5–0.7 is a suitable range. A extremely low temperature will seriously degrade the performance of the model, and too high one will also degrade the performance of the model. The smaller the temperature is, the more the loss function pays attention to the difficult negative samples. The larger the temperature, the more

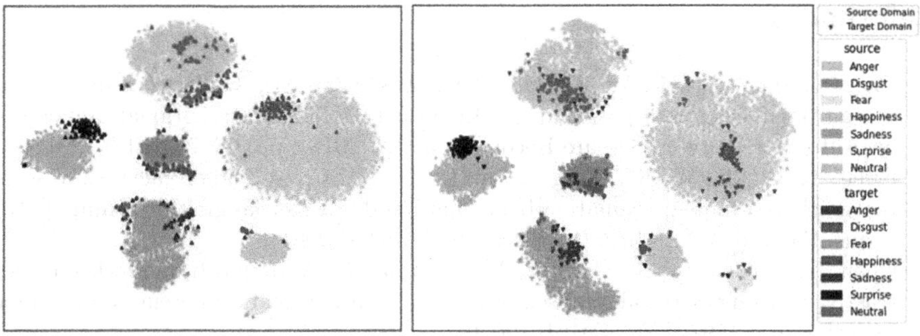

Fig. 3. The t-SNE visualization of the feature representations learned by our proposed method after the first stage (left) and the second stage (right) on CK+.

evenly the loss function pays attention to negative samples. Difficult negative samples are likely to be samples of the same category as the anchor and too much attention to these samples will destroy the high-level class semantics learned by the model. Therefore the temperature is a parameter of a trade-off between learning new features and retaining already learned features.

Visualization. We utilize t-SNE [18] to visualize the feature representations of different domains at different training stages. As shown in Fig. 3, after the first stage, there is a difference in the distributions of the source and target domains, and after domain adaptation, the distributions of the same classes in different domains become closer, and the distribution of different categories in the same domain is more separated. And Fig. 2(b) shows the specific changes in the accuracy of each category. The effectiveness of our method is demonstrated intuitively.

5 Conclusion

In this paper, we propose POCAN for unsupervised cross-domain facial expression recognition research. We parameterize the prototypes to make them learnable variables, reducing the computational effort of traditional prototype computing methods. We carry out a two-stage training approach, wherein in the first stage we pre-train on the source domain to acquire semantically meaningful features, and in the second stage, we simultaneously perform domain-specific feature learning and two levels of domain alignment. Moreover, the problem of category imbalance in facial expression datasets is considered and we propose an adaptive threshold method to select trusted target samples. We conduct extensive experiments on both lab-controlled and in-the-wild datasets, and our method exhibits competitive results.

Acknowledgements. This work was supported in part by the NSF of china (Grant No. 61773215).

References

1. Cao, Q., Shen, L., Xie, W., Parkhi, O.M., Zisserman, A.: Vggface2: a dataset for recognising faces across pose and age. In: 2018 13th IEEE International Conference on Automatic Face & Gesture Recognition (FG 2018), pp. 67–74. IEEE (2018)
2. Chen, T., Pu, T., Wu, H., Xie, Y., Liu, L., Lin, L.: Cross-domain facial expression recognition: a unified evaluation benchmark and adversarial graph learning. IEEE Trans. Pattern Anal. Mach. Intell. **44**, 9887–9903 (2021)
3. Chen, T., Kornblith, S., Norouzi, M., Hinton, G.: A simple framework for contrastive learning of visual representations. In: International Conference on Machine Learning, pp. 1597–1607. PMLR (2020)
4. Chuang, C.Y., Robinson, J., Lin, Y.C., Torralba, A., Jegelka, S.: Debiased contrastive learning. Adv. Neural Inf. Process. Syst. **33**, 8765–8775 (2020)
5. Dhall, A., Goecke, R., Lucey, S., Gedeon, T.: Static facial expression analysis in tough conditions: data, evaluation protocol and benchmark. In: 2011 IEEE International Conference on Computer Vision Workshops (ICCV Workshops), pp. 2106–2112. IEEE (2011)
6. Ekman, P., Friesen, W.V.: Constants across cultures in the face and emotion. J. Pers. Soc. Psychol. **17**(2), 124 (1971)
7. Goodfellow, I.J., et al.: Challenges in representation learning: a report on three machine learning contests. In: Lee, M., Hirose, A., Hou, Z.-G., Kil, R.M. (eds.) ICONIP 2013. LNCS, vol. 8228, pp. 117–124. Springer, Heidelberg (2013). https://doi.org/10.1007/978-3-642-42051-1_16
8. Hasani, B., Mahoor, M.H.: Spatio-temporal facial expression recognition using convolutional neural networks and conditional random fields. In: 2017 12th IEEE International Conference on Automatic Face & Gesture Recognition (FG 2017), pp. 790–795. IEEE (2017)
9. He, K., Fan, H., Wu, Y., Xie, S., Girshick, R.: Momentum contrast for unsupervised visual representation learning. In: Proceedings of the IEEE/CVF Conference on Computer Vision and Pattern Recognition, pp. 9729–9738 (2020)
10. Ji, Y., Hu, Y., Yang, Y., Shen, F., Shen, H.T.: Cross-domain facial expression recognition via an intra-category common feature and inter-category distinction feature fusion network. Neurocomputing **333**, 231–239 (2019)
11. Khosla, P.: Supervised contrastive learning. Adv. Neural Inf. Process. Syst. **33**, 18661–18673 (2020)
12. Li, J., Zhou, P., Xiong, C., Hoi, S.C.: Prototypical contrastive learning of unsupervised representations. arXiv preprint arXiv:2005.04966 (2020)
13. Li, S., Deng, W.: Deep emotion transfer network for cross-database facial expression recognition. In: 2018 24th International Conference on Pattern Recognition (ICPR), pp. 3092–3099. IEEE (2018)
14. Li, S., Deng, W.: Reliable crowdsourcing and deep locality-preserving learning for unconstrained facial expression recognition. IEEE Trans. Image Process. **28**(1), 356–370 (2018)
15. Li, S., Deng, W.: A deeper look at facial expression dataset bias. IEEE Trans. Affect. Comput. **13**, 881–893 (2020)
16. Lucey, P., Cohn, J.F., Kanade, T., Saragih, J., Ambadar, Z., Matthews, I.: The extended cohn-kanade dataset (ck+): a complete dataset for action unit and emotion-specified expression. In: 2010 IEEE Computer Society Conference on Computer Vision and Pattern Recognition-Workshops, pp. 94–101. IEEE (2010)

17. Lyons, M.J., Akamatsu, S., Kamachi, M., Gyoba, J., Budynek, J.: The Japanese female facial expression (jaffe) database. In: Proceedings of Third International Conference on Automatic Face and Gesture Recognition, pp. 14–16 (1998)
18. Van der Maaten, L., Hinton, G.: Visualizing data using t-sne. J. Mach. Learn. Res. **9**(11) (2008)
19. Mollahosseini, A., Chan, D., Mahoor, M.H.: Going deeper in facial expression recognition using deep neural networks. In: 2016 IEEE Winter Conference on Applications of Computer Vision (WACV), pp. 1–10. IEEE (2016)
20. Van den Oord, A., Li, Y., Vinyals, O.: Representation learning with contrastive predictive coding. arXiv e-prints pp. arXiv-1807 (2018)
21. Qiu, Z., et al.: Source-free domain adaptation via avatar prototype generation and adaptation. arXiv preprint arXiv:2106.15326 (2021)
22. Shan, C., Gong, S., McOwan, P.W.: Facial expression recognition based on local binary patterns: a comprehensive study. Image Vision Comput. **27**(6), 803–816 (2009)
23. Tanwisuth, K., et al.: A prototype-oriented framework for unsupervised domain adaptation. Adv. Neural Inf. Process. Syst. **34**, 17194–17208 (2021)
24. Tian, L., Tang, Y., Hu, L., Ren, Z., Zhang, W.: Domain adaptation by class centroid matching and local manifold self-learning. IEEE Trans. Image Process. **29**, 9703–9718 (2020)
25. Wang, F., Liu, H.: Understanding the behaviour of contrastive loss. In: Proceedings of the IEEE/CVF Conference on Computer Vision and Pattern Recognition, pp. 2495–2504 (2021)
26. Wang, X., Wang, X., Ni, Y.: Unsupervised domain adaptation for facial expression recognition using generative adversarial networks. Comput. Intell. Neurosci. **2018** (2018)
27. Wu, Z., Xiong, Y., Yu, S.X., Lin, D.: Unsupervised feature learning via non-parametric instance discrimination. In: Proceedings of the IEEE Conference on Computer Vision and Pattern Recognition, pp. 3733–3742 (2018)
28. Xia, H., Zhao, H., Ding, Z.: Adaptive adversarial network for source-free domain adaptation. In: Proceedings of the IEEE/CVF International Conference on Computer Vision, pp. 9010–9019 (2021)
29. Yan, H.: Transfer subspace learning for cross-dataset facial expression recognition. Neurocomputing **208**, 165–173 (2016)
30. Yan, K., Zheng, W., Cui, Z., Zong, Y.: Cross-database facial expression recognition via unsupervised domain adaptive dictionary learning. In: Hirose, A., Ozawa, S., Doya, K., Ikeda, K., Lee, M., Liu, D. (eds.) ICONIP 2016. LNCS, vol. 9948, pp. 427–434. Springer, Cham (2016). https://doi.org/10.1007/978-3-319-46672-9_48
31. Yue, X., et al.: Prototypical cross-domain self-supervised learning for few-shot unsupervised domain adaptation. In: Proceedings of the IEEE/CVF Conference on Computer Vision and Pattern Recognition, pp. 13834–13844 (2021)
32. Zavarez, M.V., Berriel, R.F., Oliveira-Santos, T.: Cross-database facial expression recognition based on fine-tuned deep convolutional network. In: 2017 30th SIBGRAPI Conference on Graphics, Patterns and Images (SIBGRAPI), pp. 405–412. IEEE (2017)

33. Zhao, G., Huang, X., Taini, M., Li, S.Z., PietikäInen, M.: Facial expression recognition from near-infrared videos. Image Vision Comput. **29**(9), 607–619 (2011)
34. Zhong, L., Liu, Q., Yang, P., Liu, B., Huang, J., Metaxas, D.N.: Learning active facial patches for expression analysis. In: 2012 IEEE Conference on Computer Vision and Pattern Recognition, pp. 2562–2569. IEEE (2012)
35. Zhu, R., Sang, G., Zhao, Q.: Discriminative feature adaptation for cross-domain facial expression recognition. In: 2016 International Conference on Biometrics (ICB), pp. 1–7. IEEE (2016)

Multi-stream Fusion for Class Incremental Learning in Pill Image Classification

Trong-Tung Nguyen[1,2], Hieu H. Pham[1,3(✉)], Phi Le Nguyen[4], Thanh Hung Nguyen[4], and Minh Do[1,3,5]

[1] VinUni-Illinois Smart Health Center, VinUniversity, Hanoi, Vietnam
{tung.nt,hieu.ph,minh.do}@vinuni.edu.vn
[2] John von Neumann Institute, University of Science, VNU-HCM, Ho Chi Minh City, Vietnam
[3] College of Engineering and Computer Science, VinUniversity, Hanoi, Vietnam
[4] School of Information and Communication Technology, Hanoi University of Science and Technology, Hanoi, Vietnam
{lenp,hungnt}@soict.hust.edu.vn
[5] University of Illinois at Urbana-Champaign, Champaign, USA
minhdo@illinois.edu

Abstract. Classifying pill categories from real-world images is crucial for various smart healthcare applications. Although existing approaches in image classification might achieve a good performance on fixed pill categories, they fail to handle novel instances of pill categories that are frequently presented to the learning algorithm. To this end, a trivial solution is to train the model with novel classes. However, this may result in a phenomenon known as catastrophic forgetting, in which the system forgets what it learned in previous classes. In this paper, we address this challenge by introducing the class incremental learning (CIL) ability to traditional pill image classification systems. Specifically, we propose a novel incremental multi-stream intermediate fusion framework enabling incorporation of an additional guidance information stream that best matches the domain of the problem into various state-of-the-art CIL methods. From this framework, we consider color-specific information of pill images as a guidance stream and devise an approach, namely "*Color Guidance with Multi-stream intermediate fusion*" (CG-IMIF) for solving CIL pill image classification task. We conduct comprehensive experiments on real-world incremental pill image classification dataset, namely VAIPE-PCIL, and find that the CG-IMIF consistently outperforms several state-of-the-art methods by a large margin in different task settings. Our code, data, and trained model are available at https://github.com/vinuni-vishc/CG-IMIF.

1 Introduction

Pill image recognition task has attracted various studies recently with the aim to design high-quality algorithm for visual-based assistance system on pill images.

L. Wang et al. (Eds.): ACCV 2022, LNCS 13842, pp. 341–356, 2023.
https://doi.org/10.1007/978-3-031-26284-5_21

This can help the healthcare community automatically identify unknown pill categories by taking several real-world pictures with mobile devices. It is noteworthy that real-world scenarios of pill images are often challenging due to the changing background as well as variances of pill instances in terms of shape, color, and texture. There have been several works that are developed to mitigate such challenges, most of them are based on hand-crafted features [3,5,6,10]. These works are then utilized by Ling et al. [16] and combined with a two-stage training strategy to create a novel framework for the pill recognition model in few-shot learning. Another approach is to explore external knowledge from medical text data (e.g. prescription) to improve the detection performance of visual-based models [18,19]. However, existing models are often limited by novel instances of pill categories which frequently arrive at a pill recognition system. This often happens when a novel class of pill instance is introduced by images uploaded from the end-user using mobile devices or from the healthcare community. A report in [1] shows that there are roughly 40–50 novel drugs being approved each year. In such a scenario, the core learning model of the system, which is often deployed in a lightweight device (*e.g*, mobile phones), might need to rewind the training process on the whole training data (in which novel categories participate). This is not an effective strategy for many reasons. Memory allocated for such extensively training data is often limited. Acquiring novel knowledge while maintaining what the model has learned so far requires the system to store a huge amount of samples for both old and new classes, which is infeasible. Another solution for this is to provide an initial training dataset for the model. The model is then fine-tuned on novel categories to update the model's knowledge about new pill instances. However, this fine-tuning scheme suffers from a serious behavior of the learning system which is widely known as catastrophic forgetting [8,9] (degrading performance on old tasks while accessing data of novel tasks). This system, therefore, is in need of a flexible and effective strategy to handle the novel real-world object categorization of pill image instances. In this way, it would be able to incrementally learn from new classes without exhaustively storing old category samples. This scenario is called class-incremental learning (CIL).

The progress of studies on class incremental learning (CIL) for visual tasks has been developed significantly for many years. The general setting of CIL is that the disjoint sets of different classes arrive at the learning algorithm gradually. Many works such as [4,13,21–23] have proposed several methods which employed available techniques to tackle the mutual challenge: catastrophic forgetting. Knowledge distillation [12] is the most common technique which is widely adopted to tackle catastrophic forgetting and was first applied to the CIL setting by Li et al. [15]. After that, a derived version [21] with additional usage of representation learning was proposed, in which valuable herding exemplars are replayed frequently to keep track of the old knowledge. The strategy of herding is to pick those neighbors which are nearest to the mean sample of the class. Using this herding strategy, Castro et al. [4] managed to build an end-to-end framework with an additionally balanced fine-tuning strategy. On the other hand, Wu et al. [22] introduced a bias correction approach by adding a bias correction layer. This is conducted at the last layer of each incremental learning task to refine the overall scores for the final prediction. Meanwhile, Hou et al. [13] identified the imbalance between previous and new data as the main issue leading to catastrophic forgetting. They tackled this imbalanced scenario by incorporating three main components: cosine normalization, less-forget constraint, and inter-class separation.

In this research, we aim to investigate the application of CIL methods in a pill classification system. Figure 1 illustrates the effect of such a system with and without class incremental learning capability. To the best of our knowledge, we are the first to explore incremental learning on the pill classification system. Existing single stream incremental learning methods [4,13,21–23], when being applied to a domain of application for practical usage, can be improved with the help of some domain-specific knowledge. This serves as additional information which might collaborate well with the original RGB image to alleviate catastrophic forgetting. The introduction of a supplementary information stream requires a prudent strategy to incorporate such information. Based on this motivation, we propose a novel integration framework that serves as a plug-in technique for any available class incremental learning algorithms. Our fusion framework enables the incremental learning methods to receive additional information streams as cues. This will then help to flexibly update corresponding feature representations in an optimal way for each learning task through the intermediate stage. To demonstrate the usage of such an integration framework, we consider color information as additional stream and devise an approach, named "*Color Guidance with Multi-stream intermediate fusion*" (CG-IMIF). Experimental results on a real-world incremental pill image classification dataset called VAIPE-PCIL show that the proposed learning framework consistently surpasses most metric scores of various state-of-the-art methods in different task settings.

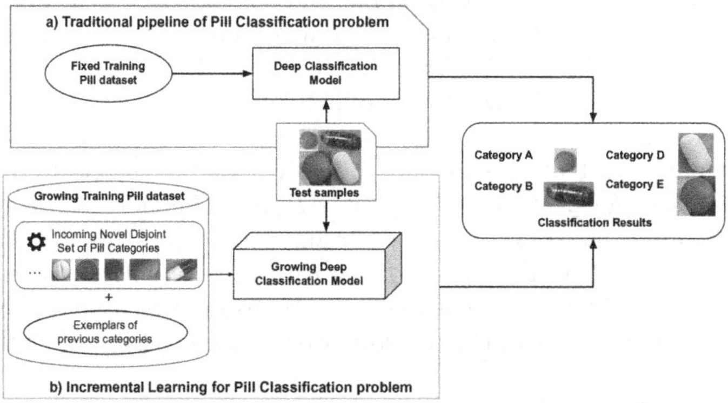

Fig. 1. The pipeline for a learning algorithm to acquire knowledge of pill categories could be divided into two options: (a) feeding a fixed pill images database to an off-the-shelf deep learning algorithm; (b) maintaining a few samples of old categories as exemplars, combining with novel categories to form a growing pill image dataset, and finally feeding into a growing deep classification model.

Our contributions can be summarized in the following three aspects:

1. We introduce CG-IMIF, a novel incremental learning framework based on multiple streams for the task of pill classification from images. To the best of our knowledge, we are the first to introduce the incremental learning capability to this task and provide a new approach to tackle challenges in learning novel pill classes.
2. We conduct thorough experiments and in-depth ablation studies to demonstrate the effectiveness of the proposed approach on a real-world incremental pill image classification dataset. Experimental results show that the CG-IMIF consistently outperforms previous state-of-the-art methods by a large margin.

The rest of this paper is organized as follows. We briefly formulate the problem setting of pill CIL, which we aim to solve in Sect. 2. Details of our proposed CG-IMIF framework are described in Sect. 3. Experimental results and further analysis are presented in Sect. 4 and 5. Finally, we conclude the paper with our discussion on strengths and limitations in Sect. 6, and 7.

2 Preliminaries

2.1 Problem Definition and Notation

Generally, the class incremental learning (CIL) problem represented by τ consists of a sequence of n image classification learning tasks

$$\tau = [(C^1, P^1_{train}, P^1_{test}), (C^2, P^2_{train}, P^2_{test}), ..., (C^n, P^n_{train}, P^n_{test})], \quad (1)$$

where each tuple $(C^t, P^t_{train}, P^t_{test})$ depicts a task t. C^t is a set of m^t categories, *i.e.*, $C^t = \{c^t_1, c^t_2, ..., c^t_{m^t}\}$, P^t_{train} and P^t_{test} denote the training and testing data, respectively. To represent the total number of classes up to the current task, we define $M^t = \sum_{i=1}^{t} |C^i|$. The training, and testing data is defined as $P^t = \{(X^t, Y^t)\}$ where X^t and Y^t denote the training images and their corresponding labels, respectively. During the training phase, the learning model at stage t is presented with categories set C^t, training samples P^t_{train}, and an exemplar set K_t. In practice, K_t is a fixed-size set acting as a support set which helps to retain a partial set of images and the corresponding labels from previous training data, *i.e.*, $K_t \subseteq P^1_{train} \cup P^2_{train} \cup ... \cup P^{t-1}_{train}$. Therefore, a revised version of training samples at stage t can be obtained by combining K_t and P^t_{train}, $K_t \cup P^t_{train} = V^t_{train}$. It is also assumed that categories of different learning tasks do not overlap ($i.e$ $C^i \cap C^j = \varnothing$ where $i \neq j$). At testing time, the performance of learner t is evaluated on all of the previous seen categories $\bigcup_{i=1}^{t} C^t$ with samples from $\bigcup_{i=1}^{t} P^t_{test}$.

2.2 Conventional CIL Methods

Several CIL methods have been proposed which consider various properties of CIL problem to tackle mutual challenge: catastrophic forgetting. Most CIL works are divided into two branches: exemplar-based, and non-exemplar-based approaches. While the latter is much more challenging, the former is more practical since it is reasonable to maintain a few samples of old classes to avoid performance degrading. Within the scope of this research, we aim to exploit the capability of exemplar-based CIL methods by attaching these to our proposed framework. Therefore, we first describe a few core components of exemplar-based CIL approaches as follows.

Representative Memory is a set of samples from categories of old tasks and is represented by K_{t-1}. It serves as an exemplar set to support model in revisiting knowledge acquired from old tasks. In an exemplar-based approach, the learning model can only access the previous category set C^{t-1} through K_{t-1}. The size of the support set is often limited and mainly divided into two memory settings: 1) a constant number of exemplars per class, and 2) and a limited capacity of S samples. In the first setting, the size of the support set K_t grows with the number of classes. In addition, the size of K_t in the second memory setting is constant over the time t. Samples across categories are manipulated frequently with two main operations: new sample selection and old sample removal. For each class, a sorted list of its samples is maintained based on their distances to the class' mean feature vector. Hence, the most representative samples for each class are selected as members in the next support set K_{t+1}. Meanwhile, the remaining samples are ignored to reserve slots for novel samples from new classes.

Growing Deep Neural Networks in an exemplar-based method is constructed by two main factors: the common feature extractor backbone, and a growing classification layer module. At a specific learning stage t, new classification head CL_t is initiated to allocate corresponding parameters W_t. Feature vectors, after being extracted by the feature extractor, are fed into CL_t to produce prediction logits for the current category set C^t. The size of the logits after being input to CL_t is equal to the size of the category set C^t. The prediction vectors are then utilized to compute the traditional cross-entropy loss which represents the training loss on the set of pill images P^t_{train} for the current task. On the other hand, the old classification head CL_{t-1} can be used to represent the old knowledge of the model. Samples from support set K_t can be passed through a list of classification head module CL_i from the first task to the latest old task $t-1$ ($i.e$ $i \in [0, t-1]$) in order to obtain prediction logits.

Cross-Distillation Loss Function is common in most of exemplar-based methods. This is constructed by combining cross-entropy, and distillation loss function. The cross-entropy loss function helps minimize the overall empirical errors when learning on new category set C_t at task t. Meanwhile, the distillation loss function plays a role in distilling the old model M_{t-1} from previous tasks into the current model M_t to avoid catastrophic forgetting. Let's consider the incremental learning model at a specific learning stage t where it has obtained $t-1$ numbers of classification heads. New classification head CL_t, which is now added to learn on new task t, produce the prediction logits as $o(x) = [o_1(x), o_2(x), ..., o_t(x)]$ for any input x. Similarly, output logits which are produced by old classification head can be represented as

$\hat{o}(x) = [\hat{o}_1(x), \hat{o}_2(x), ..., \hat{o}_{t-1}(x)]$. With these representation, the distillation loss can be computed for all samples from exemplar sets K_{t-1} and from new classes P^t_{train} (*i.e.* $K_{t-1} \cup P^t_{train} = V^t_{train}$) as follows:

$$L_d = \sum_{x \in V^t_{train}} \sum_{k=1}^{t-1} -\hat{\pi}_k(x) \log \left[\pi_k(x)\right], \tag{2}$$
$$\hat{\pi}_k(x) = \frac{e^{\hat{o}_k(x)/T}}{\sum_{j=1}^{t-1} e^{\hat{o}_j(x)/T}}, \quad \pi_k(x) = \frac{e^{o_k(x)/T}}{\sum_{j=1}^{t-1} e^{o_j(x)/T}},$$

where T plays as the temperature scaling factor. Meanwhile, the groundtruth label for each sample x (*i.e.* y(x)) for new category sets along with softmax of logits of the k-th category set (i.e. $p_k = softmax(o_k(x))$) can be used to compute the cross-entropy loss function as follows

$$L_c = \sum_{x \in V^t_{train}} \sum_{k=1}^{t} -y(x) \log \left[p_k(x)\right], \tag{3}$$

The final cross-distilled loss function; therefore, can be obtained by combining distillation loss function in Eq. 2 and cross-entropy loss function in Eq. 3

$$L = \alpha L_d + (1 - \alpha) L_c, \tag{4}$$

where the scalar value α controls the balance between the two functions.

3 Methodology

The majority of current pill identification methods rely on RGB images. Therefore, to the best of our knowledge, most existing systems fail to address hard examples (*e.g*, pills with very similar shapes and colors) [16]. This problem becomes more challenging in the context of the class incremental learning. In this problem, we have to cope with two issues at the same time: 1) recognizing pill instances that belong to the novel classes, and 2) not forgetting the previously learned knowledge of the old ones. We seek in this study robust domain-specific knowledge, which could be in good companion to traditional RGB image stream. However, the introduction of an additional stream issue a different challenge; the significant need for a stream integration method. To tackle such a challenge, we propose an Incremental Multi-stream Intermediate Fusion framework (IMIF). The IMIF allows additional information streams to be effectively propagated during the incremental learning phase. In the following subsections, we briefly define the multi-stream class incremental learning method and describe how it can be decomposed into different components.

3.1 Multi-stream Class Incremental Learning Model

We define a multi-stream class incremental learning model **M** as a combination of three key components: 1) a single stream base method **X**, 2) an additional stream of information **Y**, and 3) a method of fusing stream **Z**.

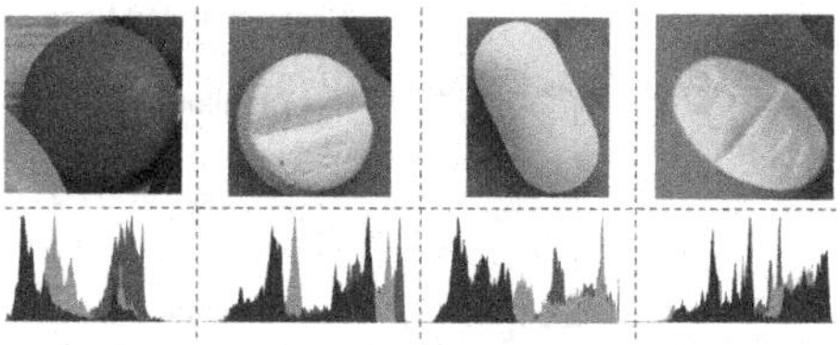

Fig. 2. Samples of pill images from VAIPE-PCIL dataset are shown on the first row. The second row exhibits the corresponding color histogram information for images on the upper row respectively.

$$\mathbf{M} = \text{Base method } \mathbf{X} + \text{Feature stream } \mathbf{Y} + \text{Fusion mechanism } \mathbf{Z}$$

At this point, the base method represents any method that follows the general setting described in Sect. 2.2. **Y** serves as a piece of additional domain information that gives cues to the learning model apart from RGB images. Normally, **Y** is specific to the domain of the task. Lastly, **Z** presents a fusion mechanism that enables method **M** to incorporate additional information stream **Y** into the incremental learning process. From this decomposition, our CG-IMIF replaces: 1) the representative stream **Y** with color-specific information, and 2) the fusion technique **Z** with the proposed IMIF. In the following, we describe our proposed Color histogram guidance stream and Incremental Multi-stream Intermediate Fusion technique in Sect. 3.2 and Sect. 3.3, respectively.

3.2 Color Histogram Guidance Stream

Pill images compose of various features which can be used as discriminative factors in classification problems. Among those is color distribution information which can be approximated by the color histogram of pill images. The color histogram represents the color distribution of the input image in terms of three different channels: red, green, and blue. In detail, it is often encoded into a single vector where the indexes of entries are mapped to a set of all possible colors. In grey-scale images, values of vector entries store the frequency that counts the total number of pixels having the intensity (*i.e.*, color value), which is hashed by the corresponding index of the vector. However, in a three-dimensional image, color ranges for each channel are associated across color channels to formulate a unique combination. This combination accounts for those pixels of which color ranges lie in three discrete ranges of value: $[r_i, r_j]; [g_k, g_l]; [b_m, b_n]$. The color feature vectors are useful to make a distinction among pill instances that have similar shapes but different colors. The color ranges for each channel of the RGB images are divided into eight segments in our problem, where each segment represents 32 different consecutive color values. After that, the color histogram vector can be obtained by accumulating the quantities of pixels assigned to a specific color range for each channel to result in a vector with $8 \times 8 \times 8 = 512$ elements. Figure 2 illustrates a few examples of extracting the color histogram stream for each corresponding cropped pill image.

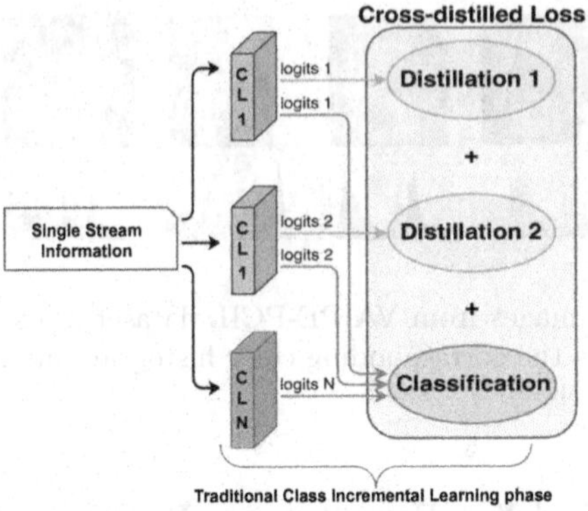

Fig. 3. The traditional training paradigm with only single-stream information used by almost existing exemplar-based CIL methods.

3.3 Incremental Multi-stream Intermediate Fusion Technique

Traditional Early Fusion. A naive fusion technique $\mathbf{Z}$ concatenates different streams of information right after the feature extraction phase. Specifically, feature vectors $f_r \in \mathbb{R}^{d_r}$, and $f_Y \in \mathbb{R}^{d_Y}$ are extracted from raw RGB images, and additional information stream $\mathbf{Y}$, respectively. Both of these features are then fed into separate projection layers to project into the same latent space. In practice, the projection layers are implemented by a single hidden layer controlled by parameters $\Theta^p = [W^p, b^p]$ as follows

$$s_r = \sigma(W^p_r.f_r + b^p_r), s_Y = \sigma(W^p_Y.f_Y + b^p_Y), s_g = [s_r, s_Y]. \tag{5}$$

The projected vectors are then concatenated to obtain the global feature vector $s_g \in \mathbb{R}^{d_g}$. This global feature s_g can be considered as a single input into any traditional single stream CIL methods as shown in Fig. 3.

Intermediate Fusion. We observed that fusing information stream in an early manner for a class incremental learning problem is not optimal. The global feature s_g is problematic since it is regularly updated at each incremental task. As a result, the projection layer in the early phase can not find good parameters that balance the performance of old and novel categories in different tasks. Therefore, we propose to relocate the projection layer to the intermediate phase instead (*i.e*, incremental learning phase) by initiating an entirely novel projection layer in an incremental manner. When C^t from new task t arrives at the system, a new classification layer CL_t is created. This layer accompanies by an attached projection layer specific to the information stream of a specific task. Therefore, the parameters controlling the projection layer for each information stream are different from those defined in the early fusion (Fig. 4).

$$\Theta^{p^t}_r = [W^{p^t}_r, b^{p^t}_r], \Theta^{p^t}_Y = [W^{p^t}_Y, b^{p^t}_Y]. \tag{6}$$

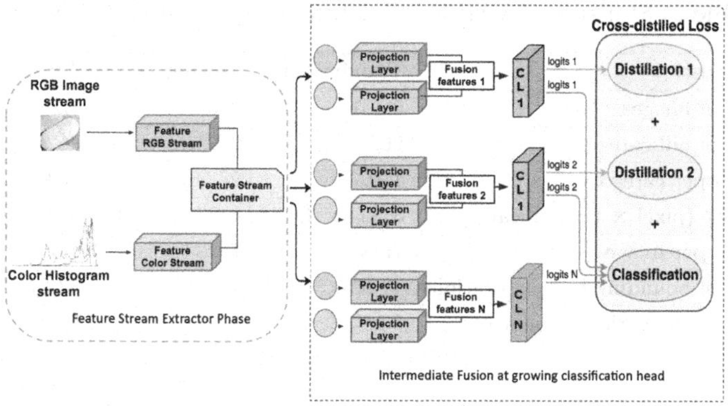

Fig. 4. Our proposed CG-IMIF architecture composes of: 1) color histogram feature extraction (orange block), and 2) intermediate fusion framework (purple block) to incorporate additional information stream. (Color figure online)

4 Experiments

4.1 Dataset

We employ a real-world image dataset, namely VAIPE-Pill (VAIPE Pill Identification) [2] to exploit CIL capability on the pill image classification problem. This dataset is created to promote the research on recognizing distinct types of medicines from mobile devices. The dataset contains 7,294 pill images of 262 categories taken in real-world scenarios. The characteristics of VAIPE-Pill dataset are illustrated in Table 1.

To facilitate research of CIL in pill image classification tasks, we derive a dataset version, namely VAIPE-PCIL (VAIPE Pill Class Incremental Learning) dataset from the original VAIPE-Pill data. VAIPE-PCIL is obtained by cropping pill instances from the original data. We only select those categories which satisfy either of the following conditions: 1) the number of samples should not be too small (*i.e.*, and 2) larger than 10 samples), image size of samples should be at least 64×64. Samples of pill image from VAIPE-PCIL can be found in Fig. 1. All of our experiments are conducted on the VAIPE-PCIL dataset to study the performance of CG-IMIF.

4.2 Experimental Protocol

Settings. We follow the standard benchmark protocol proposed in [13]. We fix class arrangements in random order. After each training stage t, the resulting learner is evaluated on the testing data $\bigcup_{i=1}^{t} P_{test}^{t}$ which represents for all of the testing data up to the current task t. Since no test data from the previous learning stage are hidden from the learner, it is guaranteed that no overfitting can occur.

Table 1. Statistics of VAIPE-Pill dataset on different characteristics.

Characteristic	Training set	Testing set	Total
Number of images	6,461	833	7,294
Number of pill categories	262	262	262
Instances per category	179.75	23.56	203.2
Image size (pixel × pixel, mean)	$3,311 \times 3,276$	$3,276 \times 3,469$	$3,300 \times 3,400$
Instances per image	7.28	7.4	7.3
Number of bounding box annotations	47,097	6,174	53,271
Number of categories per image	5.18	5.76	5.32

There are two commonly different task evaluation settings in class incremental learning: task-awareness and task-agnostic. The first setting is much easier for the algorithm since it has access to the task ID (*i.e.*, ID or set of categories) about the incoming test data. Therefore, it is reasonable to only use the corresponding classification head in the incremental learning phase, which is trained on that task-ID to evaluate the performance. This task setting, however, is not practical in many real-world circumstances since task-ID is not always available. We evaluate our performance in terms of task-agnostic instead. In task-agnostic, the model is not given the task identities of the test data. Hence, the evaluation results are achieved by taking the results of all prediction logits, which are predicted by all of the classification heads. In this way, the model has to learn to resolve the confusion among classes from a different set of classes.

Evaluation Metrics. We adopt two commonly used benchmark metrics from [13] for CIL problems: average accuracy and average forgetting rate. The average accuracy and forgetting rate records of performance for each incremental learning phase are often if a single number is preferable. Meanwhile, the average phase accuracy and forgetting rate would be used to observe learning behaviors during incremental tasks for each method.

4.3 Implementation Details

All of our experiments and methods are implemented with Pytorch [20] and trained on a single NVIDIA GeForce RTX 3090. We inherit the codebase from FACIL [17]. They have already implemented various state-of-the-art methods for CIL problems in a well-structured manner. Details of base models as well as the implementation of our IMIF framework are discussed below. In all experiments, we attach our IMIF framework to several state-of-the-art methods in CIL: BiC [22], EEIL [4], and LUCIR [13]. Since these methods followed the common prototype of exemplar-based methods, we discussed some of the general settings of exemplar-based methods in our experiments before diving into details about the setting of each base one. There are two common strategies to reserve samples for old classes: 1) exemplar-management stores a constant number of samples for each old class, or 2) it maintains a fixed capacity (*e.g*, $R_{total} = 2,000$ for

CIFAR-100 [14] and $R_{total} = 20,000$ for ImageNet [7]). In our experiments, we follow the first setting since it is usually more challenged than the second one. In addition, the exemplars are randomly selected among different categories. For the training network, we made use of 50-layer ResNet [11] with no pre-trained weights as the backbone for feature extraction module, which is applicable to the VAIPE-PCIL dataset. To fairly compare the performance, we fixed the number of training epochs (200 epochs) across different methods. The learning rate is initialized with 0.1 and is divided by 1.5 if the loss function suffers from non-decreasing circumstances for a specific number of attempts (*e.g*, $lr_{patience} = 5$). The networks are trained using stochastic gradient descent with mini-batches of 32 samples. The training images are resized to the same shape of $256 \times 256 \times 3$ with only one transformation (*e.g*, flipping). The class orders across different methods are randomly fixed for a fair comparison.

In terms of configuration for base methods, we follow the same settings for the original version BiC [22], and our improved version BiC-CG-IMIF. BiC [22] proposed to integrate a bias correction layer attached to the end of each classification head to adjust the classification score. The number of training epochs for the bias correction layer is 200 epochs in our setting. Moreover, we set 0.1 as the ratio of the number of exemplars that are used for the validation. EEIL [4] performs an additional fine-tuning phase after each official training phase to balance the performance between old and novel categories. In our experiments, we fix 40 as the number of epochs for fine-tuning and the learning rate fine-tuning factor as 0.01 across different methods. We also adhere to the base setting of LUCIR [13] method where they removed ReLU in the penultimate layer to take both positive and negative values. For the IMIF framework, the projection layer implemented is represented by a single hidden layer. Therefore, the output size for different projection layers should be the same so that the transformed feature vectors, then can be fused in the shared space. In terms of the color-guided information, color ranges for each channel of the RGB images are divided into 8 segments where each segment represents 32 different consecutive pixel values.

4.4 Experimental Results

We evaluate our proposed CG-IMIF approach and report the overall performance in comparison with several state-of-the-art approaches in Table 2 Experimental results show that most of the state-of-the-art approaches attached with our proposed IMIF tool and color-specific information as additional stream help to achieve consistent improvements over task settings. The setting consists of three tasks in total where the number of categories is uniformly distributed for 5, 10, and 15 tasks. It is noticeable that the lower score of forgetting rate indicates that the model is more unlikely to forget about old knowledge. In addition, average phase accuracy and forgetting rate is also illustrated in Fig. 5, 6 to inspect the learning behaviour of each method through incremental phases. Dashed and solid lines with different colors are utilized to differentiate the base ones (X) and our CG-IMIF, respectively. In terms of the average phase accuracy, LUCIR-CG-IMIF obtains the highest performance where it can consistently and

Table 2. Average accuracy $\bar{\mathcal{A}}$ (%) and forgetting rate $\bar{\mathcal{F}}$ (%) of CG-IMIF compared to other state-of-the-art results in different task settings. Best scores are marked in bold for both evaluation metrics.

Metric	Method	Task settings		
		$N=5$	$N=10$	$N=15$
Average acc. (%) ↑ $\bar{\mathcal{A}} = \frac{1}{n}\sum_{i=1}^{n} \mathcal{A}_i$	EEIL [4]	63.83	62.40	57.41
	EEIL-CG-IMIF	**70.80**	**64.85**	**60.93**
	BiC [22]	53.83	55.75	53.77
	BiC-CG-IMIF	**65.53**	**63.59**	**54.83**
	LUCIR [13]	69.63	62.90	55.49
	LUCIR-CG-IMIF	**76.85**	**69.94**	**64.97**
Forgetting rate. (%) ↓ $\bar{\mathcal{F}} = \frac{1}{n}\sum_{i=1}^{n} \mathcal{F}_i$	EEIL [4]	49.82	45.46	48.27
	EEIL-CG-IMIF	**46.68**	**44.64**	**46.23**
	BiC [22]	20.05	30.50	26.93
	BiC-CG-IMIF	**7.75**	**22.01**	**27.35**
	LUCIR [13]	44.13	44.32	47.11
	LUCIR-CG-IMIF	**33.15**	**37.88**	**39.79**

◇ Using the similar exemplar settings and selection for fair comparison.

significantly surpass other methods (also the base one-LUCIR). On the other hand, BiC-CG-IMIF is better at mitigating the forgetting constraint. However, it is not consistent over tasks and the curve fluctuates. One possible explanation is that the bias layer inside the traditional BiC method and BiC-CG-IMIF might cause the model to sacrifice the performance of the current task to maintain the memory of the old ones.

5 Ablation Studies

To examine the effect of additional information stream usage and fusion framework, we perform extensive ablation studies. This is aim to observe the effect of different components in our proposed framework where LUCIR is chosen as the base method. LUCIR is preferable because of the consistency and high performance of LUCIR across task settings in the experimental results which have been discussed in Sect. 4.4. In addition to color information, edge signals might be a good candidate to discriminate different pill categories based on their shape. To understand the importance of different stream usage in our method, we compare 4 different settings: 1) RGB image only, 2) RGB and edge images, 3) RGB and color histogram, and 4) a combination of all three streams. Each separated row in Table 3 refers to each scenario of information stream usage with two different fusion techniques. Concretely, the setting that combines RGB and color histogram streams achieves the highest score. One possible explanation for this result is that the edge signal might not be sufficiently strong to push the performance.

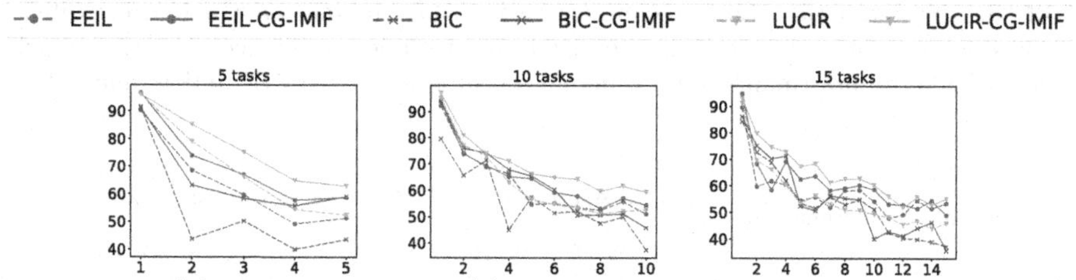

(a) 5 tasks (b) 10 tasks (c) 15 tasks

Fig. 5. Incremental accuracy for different task settings among the original version and our method CG-IMIF.

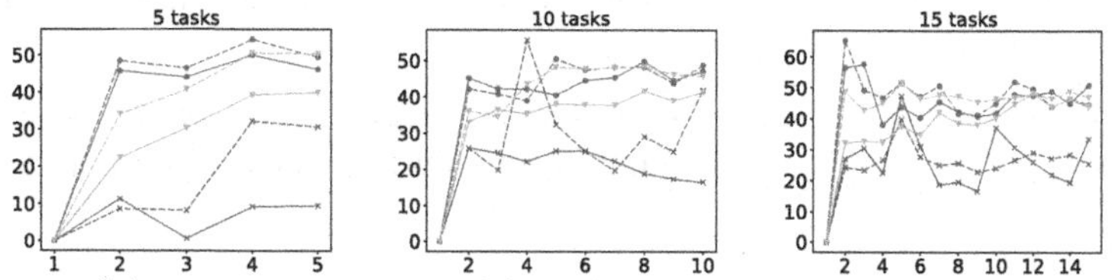

(a) 5 tasks (b) 10 tasks (c) 15 tasks

Fig. 6. Incremental forgetting rate for different task settings among the original version and our method CG-IMIF.

In addition, we implement the basic fusion technique where additional information streams are fused in an early manner. Each separated row in Table 3 illustrates the results of two different fusion mechanisms. Our fusion technique (on the second line of each row) outperforms the traditional one in various metrics and task settings. The best result is LUCIR-CG-IMIF which integrates color histogram information into the traditional LUCIR method with IMIF.

6 Discussions

Key Findings. To the best of our knowledge, this work is the first to tackle the class incremental learning problem for the pill image domain, which is crucial and applicable for real-world pill recognition systems. Also, we empirically showed that the technique of intermediate fusion with the additional stream is superior to the early fusion technique. One plausible explanation for this effect is the flexibility of the fusion layer after it has been relocated to the intermediate stage. This allows the additional information to maintain its optimal performance for old tasks while learning to adapt to new tasks.

Limitations. Though the proposed framework has superior performance over several state-of-the-art methods in CIL, it contains some limitations in different aspects. The new fusion layer at the intermediate phase might enlarge the model's size in terms of the number of parameters. Considering the scenario when a massive amount of tasks are encountered in the learning progress, the learning

Table 3. Ablation performance to compare variants of combination which utilize different information streams as well as different fusion techniques. The combination which achieves highest performance over different tasks is our CG-IMIF and is marked in green.

Variant of Combination	*Average acc.* (%) ↑			*Forgetting rate.* (%) ↓		
	N=5	*N*=10	*N*=15	*N*=5	*N*=10	*N*=15
RGB only	69.93	62.90	55.49	44.13	44.32	47.1
RGB-Edge + Early	70.94	63.90	55.28	42.4	42.13	45.80
RGB-Edge + Intermediate	**72.58**	**68.38**	**62.90**	**38.78**	**38.19**	**41.02**
RGB-Color + Early	73.58	64.57	53.56	37.825	42.86	46.15
RGB-Color+ Intermediate	76.85	69.94	64.97	33.15	37.88	39.79
RGB-Edge-Color + Early	69.99	63.33	56.34	42.35	44.17	46.24
RGB-Edge-Color+ Intermediate	**73.65**	**68.32**	**62.15**	**36.30**	**38.48**	**40.58**

model could create sequences of abundant layers. This might create a side effect when too much memory is reserved for storing the model's parameters. Such reservation is unreasonable in a real-world deployment. Another restriction with the proposed framework is related to additional stream utilization. Apart from the traditional RGB stream, another information channel that is specific to the domain of usage might impose a disagreement with the original RGB channel. This requires a careful study of a different combination of streams accompanying the traditional stream to observe its effect.

7 Conclusion

This paper introduces the incremental learning capability to the traditional pill image classification systems. To this end, we propose a novel framework, namely Incremental Multi-stream Intermediate Fusion (IMIF) which integrates an additional stream of information to improve the performance of the single stream CIL method. We then devise CG-IMIF which utilizes IMIF along with a color histogram as guidance information. Our CG-IMIF is flexible and can be attached to any exemplar-based approach to improve the performance of the base ones. We experimentally show that CG-IMIF outperforms many existing state-of-the-art methods on the VAIPE-PCIL dataset. We hope our work would lay the foundation and could benefit several types of future research into the continual learning ability of intelligent machines in smart health applications.

Acknowledgements. This work was funded by Vingroup Joint Stock Company (Vingroup JSC), Vingroup, and supported by Vingroup Innovation Foundation (VINIF) under project code VINIF.2021.DA00128. Trong-Tung Nguyen was funded and supported by the Master, PhD Scholarship Programme/Post-Doctoral Scholarship Programme of Vingroup Innovation Foundation (VINIF), Vingroup Big Data Institute (VinBigdata), code VINIF.2021.ThS.JVN.07.

References

1. CDER's New Molecular Entities and New Therapeutic Biological Products – fda.gov. https://www.fda.gov/drugs/development-approval-process-drugs/new-drugs-fda-cders-new-molecular-entities-and-new-therapeutic-biological-products. Accessed 04 July 2022
2. Anh Duy, N., Dung Thuy, N., Thanh Hung, N., Phi Le, N., Hieu H., P., Minh N., D.: VAIPE: A Large-scale and Real-World Open Pill Image Dataset for Visual-based Medicine Inspection. https://vaipe.org/
3. Caban, J.J., Rosebrock, A., Yoo, T.S.: Automatic identification of prescription drugs using shape distribution models. In: 2012 19th IEEE International Conference on Image Processing, pp. 1005–1008 (2012). https://doi.org/10.1109/ICIP.2012.6467032
4. Castro, F.M., Marín-Jiménez, M.J., Guil, N., Schmid, C., Alahari, K.: End-to-end incremental learning. CoRR abs/1807.09536 (2018). https://arxiv.org/abs/1807.09536
5. Chen, Z., Kamata, S.I.: A new accurate pill recognition system using imprint information 9067 (2013). https://doi.org/10.1117/12.2051168
6. Chen, Z., Yu, J., Kamata, S.I., Yang, J.: Accurate system for automatic pill recognition using imprint information. IET Image Process. **9**, 1039–1047 (2015). https://doi.org/10.1049/iet-ipr.2014.1007
7. Deng, J., Dong, W., Socher, R., Li, L.J., Li, K., Fei-Fei, L.: Imagenet: a large-scale hierarchical image database. In: 2009 IEEE Conference on Computer Vision and Pattern Recognition, pp. 248–255. IEEE (2009)
8. French, R.: Catastrophic forgetting in connectionist networks. Trends Cogn. Sci. **3**, 128–135 (1999). https://doi.org/10.1016/S1364-6613(99)01294-2
9. Goodfellow, I.J., Mirza, M., Xiao, D., Courville, A., Bengio, Y.: An empirical investigation of catastrophic forgetting in gradient-based neural networks (2013). https://doi.org/10.48550/ARXIV.1312.6211. https://arxiv.org/abs/1312.6211
10. Hartl, A.: Computer-vision based pharmaceutical pill recognition on mobile phones (2012)
11. He, K., Zhang, X., Ren, S., Sun, J.: Deep residual learning for image recognition. CoRR abs/1512.03385 (2015). https://arxiv.org/abs/1512.03385
12. Hinton, G.E., Vinyals, O., Dean, J.: Distilling the knowledge in a neural network. CoRR abs/1503.02531 (2015). https://arxiv.org/abs/1503.02531
13. Hou, S., Pan, X., Loy, C.C., Wang, Z., Lin, D.: Learning a unified classifier incrementally via rebalancing. In: Proceedings of the IEEE/CVF Conference on Computer Vision and Pattern Recognition (CVPR) (2019)
14. Krizhevsky, A., Nair, V., Hinton, G.: CIFAR-100 (Canadian institute for advanced research). https://www.cs.toronto.edu/~kriz/cifar.html
15. Li, Z., Hoiem, D.: Learning without forgetting. CoRR abs/1606.09282 (2016). https://arxiv.org/abs/1606.09282
16. Ling, S., et al.: Few-shot pill recognition. In: 2020 IEEE/CVF Conference on Computer Vision and Pattern Recognition (CVPR), pp. 9786–9795 (2020). https://doi.org/10.1109/CVPR42600.2020.00981
17. Masana, M., Liu, X., Twardowski, B., Menta, M., Bagdanov, A.D., van de Weijer, J.: Class-incremental learning: survey and performance evaluation. arXiv preprint arXiv:2010.15277 (2020)
18. Nguyen, A.D., Nguyen, T.D., Pham, H.H., Nguyen, T.H., Nguyen, P.L.: Image-based contextual pill recognition with medical knowledge graph assistance. arXiv preprint arXiv:2208.02432 (2022)

19. Nguyen, T.T., Nguyen, H.D., Nguyen, T.H., Pham, H.H., Ide, I., Nguyen, P.L.: A novel approach for pill-prescription matching with GNN assistance and contrastive learning. arXiv preprint arXiv:2209.01152 (2022)
20. Paszke, A., et al.: Pytorch: an imperative style, high-performance deep learning library. In: Wallach, H., Larochelle, H., Beygelzimer, A., d'Alché-Buc, F., Fox, E., Garnett, R. (eds.) Advances in Neural Information Processing Systems, vol. 32, pp. 8024–8035. Curran Associates, Inc. (2019). https://papers.neurips.cc/paper/9015-pytorch-an-imperative-style-high-performance-deep-learning-library.pdf
21. Rebuffi, S., Kolesnikov, A., Lampert, C.H.: ICARL: incremental classifier and representation learning. CoRR abs/1611.07725 (2016). https://arxiv.org/abs/1611.07725
22. Wu, Y., et al.: Large scale incremental learning. CoRR abs/1905.13260 (2019). https://arxiv.org/abs/1905.13260
23. Zhang, J., et al.: Class-incremental learning via deep model consolidation. CoRR abs/1903.07864 (2019). https://arxiv.org/abs/1903.07864

Decision-Based Black-Box Attack Specific to Large-Size Images

Dan Wang and Yuan-Gen Wang(✉)

School of Computer Science and Cyber Engineering, Guangzhou University, Guangzhou, China
wangyg@gzhu.edu.cn

Abstract. Decision-based black-box attacks can craft adversarial examples by only querying the target model for hard-label predictions. However, most existing methods are not efficient when attacking large-size images due to optimization difficulty in high-dimensional space, thus consuming lots of queries or obtaining relatively large perturbations. In this paper, we propose a novel decision-based black-box attack to generate adversarial examples, which is Specific to Large-size Image Attack (SLIA). We only perturb on the low-frequency component of discrete wavelet transform (DWT) of an image, reducing the dimension of the gradient to be estimated. Besides, when initializing the adversarial example of the untargeted attack, we remain the high-frequency components of the original image unchanged, and only update the low-frequency component with the randomly sampled uniform noise, thereby reducing the distortion at the beginning of the attack. Extensive experimental results demonstrate that the proposed SLIA outperforms state-of-the-art algorithms when attacking a variety of different threat models. The source code is publicly available at https://github.com/GZHU-DVL/SLIA.

1 Introduction

At present, deep neural networks (DNNs) have been widely applied in various fields due to their ability to efficiently solve complex tasks. However, DNN is highly uninterpretable, making it difficult to control [1]. The safety of its applications in specific fields deserves attention, such as military, autonomous driving, and medical treatment. The concept of adversarial example was first proposed by Szegedy et al. [1] in 2014. That is, adding a small perturbation to an original image can generate an adversarial example that makes the DNN model misclassified with high confidence. According to the accessible knowledge of the structure and parameters of the target model, adversarial attack can be divided into white-box attack and black-box attack. Since the black-box attack is more practicable, thus attracting more attentions than the former [2]. The black-box attack includes the transfer-based [3] and query-based attack [4].

In the transfer-based black-box attack, Dong et al. [5] make the generated adversarial examples more transferable by increasing the momentum in the gradient direction. However, this approach has low attack success rate. In [11], Papernot et al. propose a dataset expansion method based on the Jacobian matrix to

L. Wang et al. (Eds.): ACCV 2022, LNCS 13842, pp. 357–372, 2023.
https://doi.org/10.1007/978-3-031-26284-5_22

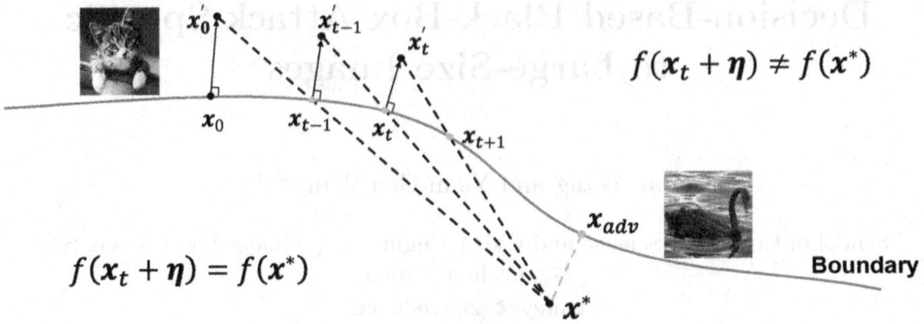

Fig. 1. Pipeline of SLIA. The attack is initialized with an example that has already been adversarial, and then generating adversarial examples iteratively move along the decision boundary. Taking targeted attack as an example, we aim to obtain an adversarial image that visually looks like a cat but be misclassified as a black swan.

iteratively expand and improve surrogate model. However, when the dimension of the sampled image is large, the calculation of the Jacobian matrix will consume huge resources. Besides, it is difficult to completely imitate the decision boundary of the attacked model, which causes a low attack success rate.

Since the surrogate model cannot fully imitate the target model, many researchers tend to directly estimate the structure and parameter information of the target model. The focus of black-box attacks is gradient estimation by querying model. Chen et al. [4] utilize the finite difference based Zero-Order Optimization (ZOO) algorithm to estimate the gradient of the loss function by accessing predicted probabilities of the target model. This method needs to estimate each pixel one by one, which requires numerous queries to generate accurate gradient estimation in each iteration, causing the low attack efficiency. Bhagoji et al. [13] use the finite difference method and the random grouping method to reduce the amount of calculation. However, the reduced calculation causes the low attack success rate on the large-size image dataset.

When the model's prediction probabilities are accessible, attackers will typically prefer score-based attack. While in more realistic scenarios where only top-1 class predictions are available, attackers will have to resort to decision-based attack. The concept of boundary-based black-box attack was first proposed by Brendel et al. [19]. It only needs to utilize the final classification output of the model to craft adversarial example. The method works by randomly walking in the direction of the original example along the decision boundary until it is closest to the original example, while remaining adversarial. This attack requires less model knowledge but can achieve comparable attack effects to white-box attack. However, the perturbation sampling strategy in [19] has great randomness, and the convergence of perturbation cannot be guaranteed. To address this problem, [6,20] were proposed to carry out decision-based black-box attack. However, these attacks often require numerous queries to converge or have large perturbations under a given number of query budget, which makes the attack process consume heavy computation, especially when attacking large-size images.

To improve the query efficiency, we propose a decision-based boundary adversarial attack, which is specific to large-size images, termed SLIA. SLIA optimizes both l_2-norm and l_∞-norm distortion. The main contributions of this paper are as follows: (1) We propose a decision-based black-box attack for large-size images (named SLIA), wherein adversarial images can be crafted by sending a few queries to the model; (2) When performing untargeted attack, SLIA replaces the low-frequency component of the original image with random uniform noise, and reconstructs it back to the original image space with high-frequency components. This can fool the model while retaining as much key information of the original image as possible; (3) SLIA performs discrete wavelet decomposition on adversarial example at the boundary, only estimates and updates the gradient of low-frequency component, greatly reduces the number of dimensions to be estimated with fewer model queries. Experiments show that our algorithm can be successfully used to attack different ImageNet models with less distortion than state-of-the-art algorithms under the same number of queries.

2 Related Work

According to the available knowledge of the network model, adversarial attack is classified into white-box attack and black-box attack. In a white-box setting, the attacker has all knowledge about the network. Since Szegedy et al. [1] discovered vulnerability of DNNs, various white-box attacks [8–10,12] have been developed. In practice, the attacker may not be able to access the structure and parameters of the model, which is more in line with the actual attack situation. Hence black-box attacks have received more attention recently. It is often divided into three families: transfer-based, score-based, and decision-based attacks.

2.1 Transfer-Based Black-Box Attacks

Transfer-based black-box attack algorithms are mainly based on the phenomenon of transferability: adversarial example against a certain model is often misclassified by other models. Papernot et al. [10,11] trained a local substitute model by querying the target model and used backpropagation gradient from the substitute network to craft adversarial examples. These examples can also successfully fool the target model with high probability. The follow-up work [3] showed that adversarial example generated on substitute network tends not to have better transferability for targeted attack, but can be developed on an ensemble of models. However, query-based algorithms that directly estimate the gradient of the target network outperform these methods. In addition, it is difficult to find a suitable surrogate model to learn the decision boundary of the target model.

2.2 Score-Based Black-Box Attacks

In the score-based black-box setting, the attacker utilizes the corresponding predicted probabilities to make adversarial examples by querying the target model.

Chen et al. [21] applied zeroth order optimization and coordinate descent to estimate the gradient, but requred a large number of queries on the target model. The method in [6] performs gradient estimation via Natural Evolutionary Strategy (NES) and then uses Projected Gradient Descent (PGD) [7], further reduces the query complexity.

2.3 Decision-Based Black-Box Attacks

As an important category of adversarial attacks, an initial attempt named Boundary Attack [19] is highly relevant to real-world applications. It starts from an adversarial point and tries to reduce the distortion by walking towards the original image along the decision boundary while keeping adversarial. The main issue is the trade-off between the number of queries and the quality of adversarial example. HopSkipJumpAttack [21] significantly improves the former [19] in terms of query efficiency. This method can balance both the accuracy of gradient estimation and query complexity well. However, when attacking large-size images, the number of queries required to produce adversarial examples still is in the tens of thousands.

3 Problem Definition

We consider an image classifier $f : \boldsymbol{x} \rightarrow c$, where $\boldsymbol{x} \in \mathbb{R}^n$ is a normalized RGB image and c is its corresponding true label such as the top-1 classification label. $F(\boldsymbol{x})$ is a k-dimensional vector, referring to the probability distribution over classes. $c := \arg\max_{c\in[k]} F_c(\boldsymbol{x})$ represents the label of $\boldsymbol{x}$. Given an original image $\boldsymbol{x}^*$, c^* represents its label. Denote the adversarial perturbation as $\boldsymbol{\mu} \in \mathbb{R}^n$, the goal of untargeted attack is to make the model misclassified wherein $c(\boldsymbol{x}^* + \boldsymbol{\mu}) \neq c^*$, and targeted attack aims to change the original classifier decision c^* into a pre-specified class c^+.

The process of generating adversarial examples can be formulated as an optimization problem by defining the function $\mathcal{L}$:

$$\mathcal{L}_{\boldsymbol{x}^*}(\boldsymbol{x}) := \begin{cases} \max\limits_{c\neq c^*} F_c(\boldsymbol{x}) - F_{c^*}(\boldsymbol{x}) \text{ (Untargeted)} \\ F_{c^+}(\boldsymbol{x}) - \max\limits_{c\neq c^+} F_c(\boldsymbol{x}) \text{ (Targeted)} \end{cases} \tag{1}$$

Gradient-based methods can be used to efficiently optimize this problem under the white-box setting. However, in the decision-based black-box attack, models only provide attackers with a hard label, even without any output probabilities. In other words, only the value of $\mathrm{sign}(\mathcal{L})$ is available, while the value of $\mathcal{L}$ is unknown. We denote the indicator function $\mathcal{I}$ as:

$$\mathcal{I}_{\boldsymbol{x}^*}(\boldsymbol{x}) = \mathrm{sign}(\mathcal{L}_{\boldsymbol{x}^*}(\boldsymbol{x})) = \begin{cases} 1 & \text{if } \mathcal{L}_{\boldsymbol{x}^*}(\boldsymbol{x}) > 0 \\ -1 & \text{otherwise} \end{cases} \tag{2}$$

In our decision-based attack, the goal of the adversary is to find an adversarial perturbation $\boldsymbol{\mu}$ which satisfies $\mathcal{I}(\boldsymbol{x}^* + \boldsymbol{\mu}) = 1$ by sending queries to model. That

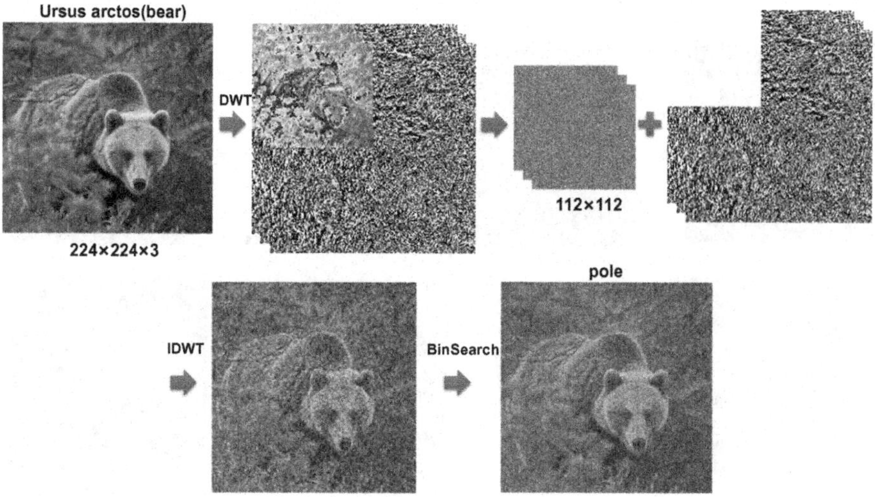

Fig. 2. Initialization for untargeted attacks.

is, only when $\mathcal{I}(\boldsymbol{x}^* + \boldsymbol{\mu}) = 1$ can it be considered as a successful attack. Generating adversarial examples under decision-based black-box setting can be defined as the following optimization problem:

$$\min \mathcal{D}(\boldsymbol{x}^*, \boldsymbol{x}^* + \boldsymbol{\mu}) \quad s.t. \quad \mathcal{I}(\boldsymbol{x}^* + \boldsymbol{\mu}) = 1 \tag{3}$$

where $\mathcal{D}(\cdot, \cdot)$ is l_2-norm or l_∞-norm distance metric. We strive to find an example with as little distortion as possible from the original example under the condition of guaranteed adversarial.

4 Decision-Based Black-Box Attack Specific to Large-Size Images (SLIA)

In this section, we propose to utilize discrete wavelet transform (DWT) to decompose the low-frequency component of the attacked image, and only adds perturbation to this part, while maintaining a 100% attack success rate. The pipeline of SLIA is shown in Fig. 1, which includes three steps: gradient estimation by querying the model, moving along the estimated gradient direction, and projecting new example to the decision boundary by binary search towards the original example. Details of each step are given below.

4.1 Initialization

Our SLIA starts from an adversarial image outside the boundary, and gradually reduces the distortion by moving towards the original image along the decision boundary while remaining adversarial.

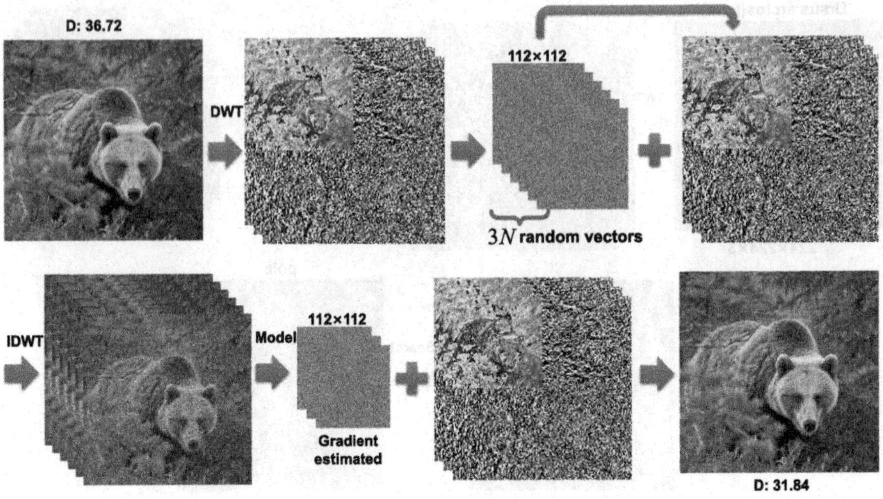

Fig. 3. Overview of estimating gradient at decision boundary.

Given a correctly classified original image $\boldsymbol{x}^*$, the first step is to generate an initial adversarial example: (1) As shown in Fig. 2, for untargeted attacks, we perform 1-level discrete wavelet decomposition on the original image. Then the low-frequency component $\boldsymbol{LL}^*$ is reset to a random uniform noise $\boldsymbol{u} \sim \mathcal{U}(\min(\boldsymbol{LL}^*), \max(\boldsymbol{LL}^*))$. Next, we combine the low-frequency noise with the original high-frequency components to reconstruct the image through inverse DWT. We make queries to the target model, until the new image is misclassified. Different from the previous attack methods that use a uniform random noise as the initialization image, the advantage of SLIA is that the new image can retain more original image information without causing large distortion. Finally, we project it to the boundary through the binary search algorithm and identify it as the initial adversarial example $\boldsymbol{x}_0$; (2) For targeted attacks, the image is randomly selected from a pre-specified class which is different from the class of the original image. Similarly, we leverage the binary search algorithm to search for the decision boundary, and take the image as initial adversarial example $\boldsymbol{x}_0$.

4.2 Gradient Direction Estimate at the Decision Boundary

In this subsection, we will elaborate the gradient estimation part in the proposed method in detail. Suppose that at the t-th iteration, the adversarial example on the boundary is $\boldsymbol{x}_t$. As shown in Fig. 3, $\boldsymbol{x}_t$ is decomposed into low-frequency and high-frequency components by DWT. Therefore, the gradient direction of loss function $\mathcal{L}$ at this point is estimated by sending queries to the target model,

$$\nabla \mathcal{L}(\boldsymbol{x}_t) := \frac{1}{N} \sum_{i=1}^{N} \mathcal{I}_{\boldsymbol{x}^*}[\text{IDWT}(\boldsymbol{LL}_t + \delta \boldsymbol{\eta}_i, \boldsymbol{HL}_t, \boldsymbol{LH}_t, \boldsymbol{HH}_t)]\boldsymbol{\eta}_i, \tag{4}$$

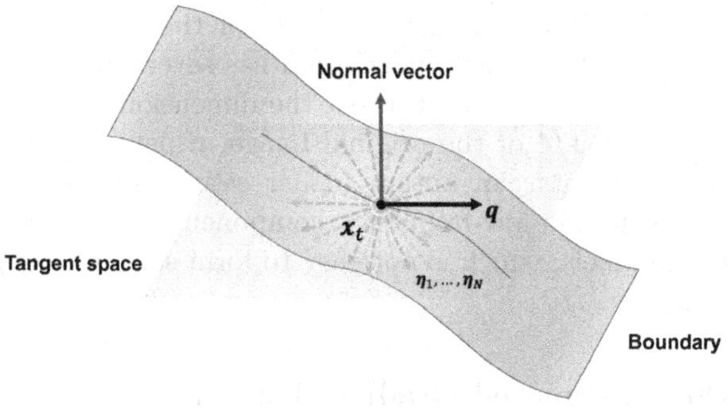

Fig. 4. Illustration of estimating gradient at $\boldsymbol{x}_t$ by sampling N Gaussian noises. $\boldsymbol{q}$ is an arbitrary in the tangent space.

where δ and t are probe step size which are a small positive parameter and t is the current number of iteration. $\boldsymbol{\eta}_{i=1}^{N}$ are normalized random noise vectors drawn from the Gaussian distribution over the 1/4-dimensional sphere as shown in Fig. 4 ($\boldsymbol{x}^* \in \mathbb{R}^n$). $\delta\boldsymbol{\eta}_{i=1}^{N}$ is added to the low-frequency component. By combining with high-frequency components of $\boldsymbol{x}_t$, inverse DWT is utilized to reconstruct N samples with unknown labels.

We determine the directions of the noise vectors by accessing the model to observe whether these samples have the same labels as the original example: (1) If $\mathcal{I}_{\boldsymbol{x}^*} = -1$, the noise vector will be updated to its opposite direction; (2) If $\mathcal{I}_{\boldsymbol{x}^*} = 1$, the noise vector will remain unchanged. Finally, we average the above noises and use the mean as the normal vector of tangent hyperplane, *i.e.*, the gradient direction $\nabla\mathcal{L}(\boldsymbol{x}_t)$ at the decision boundary.

Due to the flatness of the boundary, it is theoretically likely that the noise vectors are symmetrically distributed on both sides of the decision boundary. Therefore, the updated and unchanged noise vectors can be clustered around the true gradient as much as possible, the mean vector is also closer to the true gradient.

The gradient estimation in SLIA is essentially a Monte Carlo estimation method. When the dimension of the gradient to be estimated is large, using the Monte Carlo method requires more sampling points to make the estimated gradient closer to the true gradient. In an RGB color image, each pixel is represented by three channels. Moreover, as the size of the image becomes larger, the dimensionality of the image increases dramatically (e.g., the data dimension on ImageNet is over 150k), resulting in a low accuracy of estimating gradient. To reduce the dimension of the gradient to be estimated and further minimize the visual effect of adversarial perturbation, SLIA applies DWT to decompose the sample into low-frequency components and high-frequency components. Note that most of the key content-defining information in natural images exists at the low-frequency end of the spectrum, while high-frequency signals are often associated with noise. That is, adversarial examples are more likely to be generated by adding noise to low-frequency component. Therefore, we keep the high-frequency components unchanged, and only perturb the low-frequency component, which reduces the dimension to be perturbed to 1/4 of the original image. Adding perturbation to the low-frequency information has several advantages: (1) Only the low-frequency component is perturbed, the dimension of the gradient to be estimated is reduced to 1/4 of the original image, which means that the same number of sampling points can obtain higher estimation accuracy; (2) Only adding perturbation to the low-frequency component, the perturbation is distributed in multiple pixels, which is not easy to form salt and pepper noise and has less visual impact.

4.3 Move Along Estimated Gradient Direction

In this part, we will move one step along the gradient direction estimated in Eq. (4) to obtain an example located in the adversarial area,

$$5\boldsymbol{x}'_t = \boldsymbol{x}_t + \epsilon_t \cdot \frac{\nabla \mathcal{L}(\boldsymbol{x}_t)}{\|\nabla \mathcal{L}(\boldsymbol{x}_t)\|_2}, \tag{5}$$

where ϵ_t is perturbation magnitude at t-th iteration. It is computed from the distortion result of the last iteration and the geometric progression related to current iteration number t. We multiply the normalized estimated gradient by ϵ_t, and add it to $\boldsymbol{x}_t$ to obtain an adversarial example $\boldsymbol{x}'_t$, which is slightly away from the boundary, shown in Fig. 1. Note that $\boldsymbol{x}'_t$ is at the opposite side of the boundary to $\boldsymbol{x}^*$.

4.4 Project to Decision Boundary

Since the proposed gradient direction estimation works only at the boundary, we adopt binary search algorithm to quickly find the decision boundary and project $\boldsymbol{x}'_t$ to it. We use the following formula to adjust the value of the parameter γ to control the relative position of the adversarial example from the original example, until the stopping condition is satisfied. Hence, we move the adversarial image $\boldsymbol{x}'_t$ towards the direction of the original image $\boldsymbol{x}^*$ via

$$\boldsymbol{x}^{t+1} = \gamma_t \cdot \boldsymbol{x}^* + (1 - \gamma_t) \cdot \boldsymbol{x}'_t, \tag{6}$$

where γ_t is a changing positive parameter between 0 and 1 so that $\boldsymbol{x}'_t$ projected back to the decision boundary. We denote the example projected back on the boundary as $\boldsymbol{x}^{t+1}$, and let it enter to the next iteration as a new boundary adversarial example. The pseudo code of the complete process in generating adversarial images is outlined in Algorithm 1.

5 Experiments

5.1 Experimental Settings

Dataset and Target Models. We experiment on ImageNet [18], a public large-scale labeled image dataset, to demonstrate the efficiency of our proposed

Algorithm 1 Boundary attack specific to large-size images (SLIA)

Input: Indicator function $\mathcal{I}$, the original example $\boldsymbol{x}^*$, the number of normalized random noises N, iteration number T, constraint l_p (p=0 or p=∞), attack objective (untargeted or targeted), stopping threshold of binary search.
Output: Adversarial example.

if *objective is untargeted* **then**
  $\boldsymbol{LL}^*, \boldsymbol{LH}^*, \boldsymbol{HL}^*, \boldsymbol{HH}^* \leftarrow \text{DWT}(\boldsymbol{x}^*)$.
  Sample noise $\boldsymbol{u} \sim \mathcal{U}(\min(\boldsymbol{LL}^*), \max(\boldsymbol{LL}^*))$.
  while $\mathcal{I}(IDWT(\boldsymbol{u}, \boldsymbol{LH}^*, \boldsymbol{HL}^*, \boldsymbol{HH}^*)) = -1$ **do**
    Sample noise $\boldsymbol{u} \sim \mathcal{U}(\min(\boldsymbol{LL}^*), \max(\boldsymbol{LL}^*))$.
  end
  $\boldsymbol{x}_{initail} = \text{IDWT}(\boldsymbol{u}, \boldsymbol{LH}^*, \boldsymbol{HL}^*, \boldsymbol{HH}^*)$.
else
  A randomly sampled image $\boldsymbol{x}_{initail}$ belonging to the target class.
end
Search starting point $\boldsymbol{x}_0 = \text{BinarySearch}(\boldsymbol{x}_{initail}, \boldsymbol{x}^*, \mathcal{I})$ which lies on the boundary.
for $t = 0$ *to* $T-1$ **do**
  $\boldsymbol{LL}_t, \boldsymbol{LH}_t, \boldsymbol{HL}_t, \boldsymbol{HH}_t \leftarrow \text{DWT}(\boldsymbol{x}_t)$.
  Sample N noise vectors: $\boldsymbol{\eta}_{i=1}^{N} \sim \mathcal{N}(0,1)$.
  Estimate gradient direction of $\boldsymbol{LL}_t$: $\nabla\mathcal{L}(\boldsymbol{x}_t)$ with the rule defined in Eq.(??).
  if *constraint is* l_∞ **then**
    $\nabla\mathcal{L}(\boldsymbol{x}_t) = \text{sign}(\nabla\mathcal{L}(\boldsymbol{x}_t))$.
  end
  Initialize $\epsilon_t = \|\boldsymbol{x}_t - \boldsymbol{x}^*\|_p/\sqrt{t} \times 4$ for obtaining attack step size.
  while $\mathcal{I}(\boldsymbol{x}_t + \epsilon_t \cdot \frac{\nabla\mathcal{L}}{\|\nabla\mathcal{L}\|_p}) = -1$ **do**
    $\epsilon_t = \epsilon_t \,/\, 2$.
  end
  Compute $\boldsymbol{x}_t' = \boldsymbol{x}_t + \epsilon_t \cdot \frac{\nabla\mathcal{L}}{\|\nabla\mathcal{L}\|_p}$.
  Update adversarial image $\boldsymbol{x}_{t+1} = \text{BinarySearch}(\boldsymbol{x}_t', \boldsymbol{x}^*, \mathcal{I})$ on the boundary.
end
Return an adversarial example $\boldsymbol{x}_{T-1}$;

method. For ImageNet, we randomly sample 100 correctly classified test images, evenly distributed among 10 randomly selected classes. The whole images are clipped into [0,1] by default for all experiments. We perform both untargeted attacks and targeted attacks to a random class against three prevailing models: ResNet-50 [22], VGG16 [23] and DenseNet-201 [24]. All models are pretrained on ImageNet and provided by Keras online[1].

Compared Baseline Methods. To demonstrate the effectiveness of our method, we compare SLIA with several state-of-the-art decision-based attacks including Boundary Attack method [19], HopSkipJumpAttack (HSJA [21] and

[1] https://keras.io/applications/#resnet50.
https://keras.io/applications/#vgg16.
https://keras.io/applications/#densenet201.

Table 1. Mean l_2-norm distortions for performing untargeted and targeted attacks with different query budgets.

Objective	Victim Model	Method	1 K	5 K	10 K	20 K
Untargeted	ResNet-50	Boundary Attack [19]	54.67	27.03	14.89	10.34
		HopSkipJumpAttack [21]	28.69	9.12	5.46	3.31
		LHS-BA [25]	23.84	6.39	4.92	3.20
		Ours	**14.73**	**4.85**	**3.59**	**2.98**
	VGG16	Boundary Attack [19]	60.06	24.76	18.63	14.83
		HopSkipJumpAttack [21]	26.35	12.22	9.78	7.97
		LHS-BA [25]	22.84	10.20	7.51	7.32
		Ours	**13.41**	**5.11**	**3.68**	**2.86**
	DenseNet-201	Boundary Attack [19]	78.83	33.29	15.90	10.64
		HopSkipJumpAttack [21]	35.20	7.74	4.52	2.92
		LHS-BA [25]	27.09	7.36	3.74	2.28
		Ours	**17.64**	**6.83**	**2.84**	**0.80**
Targeted	ResNet-50	Boundary Attack [19]	83.10	49.24	31.85	22.59
		HopSkipJumpAttack [21]	54.85	27.54	17.04	9.34
		LHS-BA [25]	50.29	26.81	16.70	9.25
		Ours	**49.10**	**26.21**	**16.16**	**9.06**
	VGG16	Boundary Attack [19]	97.23	58.94	39.27	28.25
		HopSkipJumpAttack [21]	67.36	40.49	27.47	18.17
		LHS-BA [25]	60.64	36.72	25.70	16.38
		Ours	**56.25**	**26.27**	**15.08**	**10.18**
	DenseNet-201	Boundary Attack [19]	92.78	54.86	26.41	17.03
		HopSkipJumpAttack [21]	67.92	30.63	15.79	8.62
		LHS-BA [25]	61.85	27.49	15.66	8.40
		Ours	**54.32**	**19.56**	**13.13**	**7.70**

Latin Hypercube Sampling based Boundary Attack (LHS-BA) [25]. We mainly focus on attack method LHS-BA, which outperforms all of other Boundary Attack [19], Limited Attack [6], and HSJA [21]. We use the implementation of the three algorithms with the suggested hyperparameters from the publicly available source code online. We fixed the number of queries at 1K, 5K, 10K and 20K and magnitude of the average distortion is what we mainly observe when performing untargeted and targeted attacks respectively.

Evaluation Metrics. Effective querying is the most important indicator to evaluate the decision-based adversarial attack, which requires the method to craft adversarial example with smaller model queries at the same distortion. SLIA's attack success rate is 100%, so we quantify the performance in terms of two dimensions: average l_p-norm distortion and specified query numbers. It can be formulated as:

$$\|\boldsymbol{x}\|_p = \left(\sum_{i=1}^{n} |\boldsymbol{x}_i|^p \right)^{\frac{1}{p}}, \tag{7}$$

Table 2. Mean l_∞-norm distortions for performing untargeted and targeted attacks with different query budgets.

Objective	Victim Model	Method	1 K	5 K	10 K	20 K
Untargeted	ResNet-50	Boundary Attack [19]	0.553	0.411	0.247	0.193
		HopSkipJumpAttack [21]	0.231	0.129	0.103	0.098
		LHS-BA [25]	0.164	0.082	0.070	0.047
		Ours	**0.089**	**0.039**	**0.032**	**0.023**
	VGG16	Boundary Attack [19]	0.475	0.349	0.257	0.124
		HopSkipJumpAttack [21]	0.291	0.185	0.121	0.087
		LHS-BA [25]	0.166	0.095	0.073	0.038
		Ours	**0.067**	**0.032**	**0.024**	**0.018**
	DenseNet-201	Boundary Attack [19]	0.431	0.318	0.234	0.109
		HopSkipJumpAttack [21]	0.267	0.132	0.107	0.076
		LHS-BA [25]	0.204	0.116	0.085	0.061
		Ours	**0.152**	**0.074**	**0.058**	**0.035**
Targeted	ResNet-50	Boundary Attack [19]	0.780	0.618	0.372	0.244
		HopSkipJumpAttack [21]	0.370	0.267	0.199	0.137
		LHS-BA [25]	0.310	0.229	0.163	0.125
		Ours	**0.253**	**0.146**	**0.120**	**0.091**
	VGG16	Boundary Attack [19]	0.739	0.584	0.301	0.236
		HopSkipJumpAttack [21]	0.441	0.238	0.186	0.133
		LHS-BA [25]	0.405	0.210	0.169	0.117
		Ours	**0.361**	**0.182**	**0.128**	**0.090**
	DenseNet-201	Boundary Attack [19]	0.683	0.553	0.291	0.255
		HopSkipJumpAttack [21]	0.410	0.216	0.175	0.117
		LHS-BA [25]	0.381	0.188	0.146	0.099
		Ours	**0.315**	**0.133**	**0.098**	**0.078**

where l_2-norm and l_∞-norm are are two most commonly used metrics in the adversarial attack field. l_2-norm means Euclidean distance between the original example and the adversarial one, and l_∞-norm represents perturbation's maximum changeable degree.

Hyperparameters. In our proposed attack, the number of iteration and the maximum queries are set to 76 and 20,000, respectively. At the t-th iteration, we compute probe step size in each gradient direction estimation by $\delta_t = \|\boldsymbol{x}_{t-1} - \boldsymbol{x}^*\|_2/n\times 4$ and $\epsilon_t = \|\boldsymbol{x}_{t-1} - \boldsymbol{x}^*\|_2/\sqrt{t}\times 4$ as perturbation step size in moving along estimated gradient direction, where $n = 224\times 224\times 3$ is the input dimension. Random vectors N is set to 100 first, and we gradually increase it by $N = N \times (t+1)^{\frac{1}{4}}$. Stopping threshold θ when performing binary search is set to $n^{-\frac{3}{2}}$.

5.2 Experimental Results

To evaluate SLIA's performance, we report mean l_2-norm and l_∞-norm distortion results in Tables 1 and 2 when performing untargeted and targeted attacks. The distortion descending curves of various algorithms under different query budgets are given in Fig. 5. Two qualitative example processes of attacking the ResNet-50 by different attack methods are shown in Figs. 6 and 7, respectively.

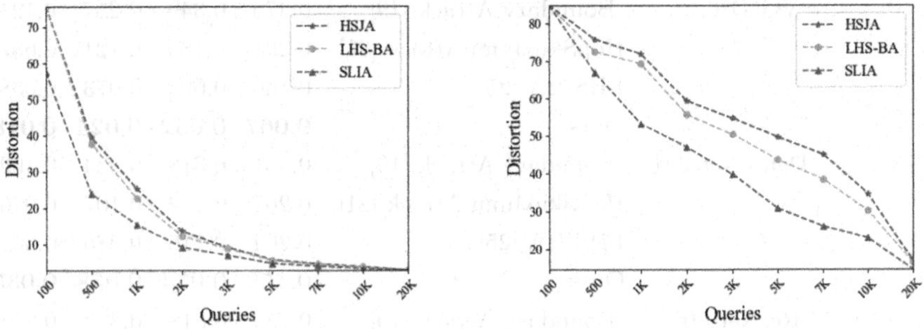

Fig. 5. l_2-norm distortions across various model queries on ImageNet with ResNet-50. 1st column: untargeted attacks. 2nd column: targeted attacks.

Untargeted Attacks. As shown in the untargeted attack section of Tables 1 and 2, it is obvious that our method outperforms existing decision-based attacks by a large margin under all fixed number of model queries. SLIA also converges in a fewer number of queries, as shown in Fig. 5.

Especially in the early stages of the attack, the advantages of SLIA are more obvious. When the number of fixed model queries does not exceed 10K: (1) Under the l_2-norm distance metric, SLIA can reduce the distortion to 56% of HSJA and about 67% of LHS-BA; (2) Under the l_∞-norm distance metric, the distortion of adversarial examples constructed via SLIA is about 64% lower than that of HSJA and about 45% lower than that of LHS-BA. Experimental data demonstrates that the adversarial examples can be crafted by our method rather quickly without using too many queries.

This is due to two reasons: (1) In the initialization part, we replace the low-frequency component of the original example with a uniform noise, and do not update other high-frequency components. In this way, more details of the original example can be preserved in the case of making the model misclassify; (2) When estimating the gradient, we consider DWT to decompose the low-frequency component of the example, and estimate the gradient of it. This greatly reduces the dimension of the gradient to be estimated to 1/4 of the original space. When sampling the same amount of Gaussian noises, the gradient can be estimated with higher accuracy than that of the original full space.

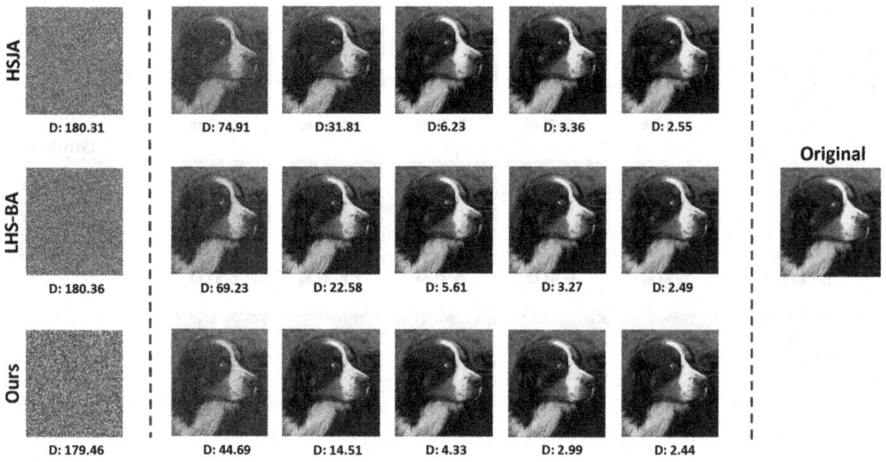

Fig. 6. Visualized trajectories of HSJA [21], LHS-BA [25] and SLIA for performing untargeted attacks on ResNet-50. 1st column: initialization. 2nd-9th columns: images after blended with original images and at 1 K, 5 K, 10 K, 20 K model queries. 10th column: original image. D is l_2-norm metric to compute the distortion between adversarial image and original image.

Targeted Attacks. We randomly select a target label and pick one image belonging to the target label. Then we use it as initialization image for all targeted attacks. The results for targeted attacks are presented in the lower parts of Tables 1 and 2. We can see that SLIA not only outperforms HSJA [21], but also surpasses the latest gradient estimation-based boundary attack LHS-BA [25]. From a qualitative example comparison using different methods shown in Fig. 7, when model queries is fixed at 5,000 (4-th column), the adversarial example crafted by SLIA is visually closer to the original example than the other two attacks. It can be seen that under a limited number of queries, SLIA is able to make adversarial examples with significantly smaller distortions from the corresponding original example. In other words, under the same distortion condition, SLIA requires fewer number of queries than the state-of-the-art methods. We can also find that SLIA requires a larger number of model queries to achieve a comparable distortion when performing targeted attacks than untargeted attacks. This phenomenon is evident on the ImageNet dataset which has many categories. There is often an order-of-magnitude difference in the average l_p-norm distortion between untargeted and targeted attacks for the same number of queries.

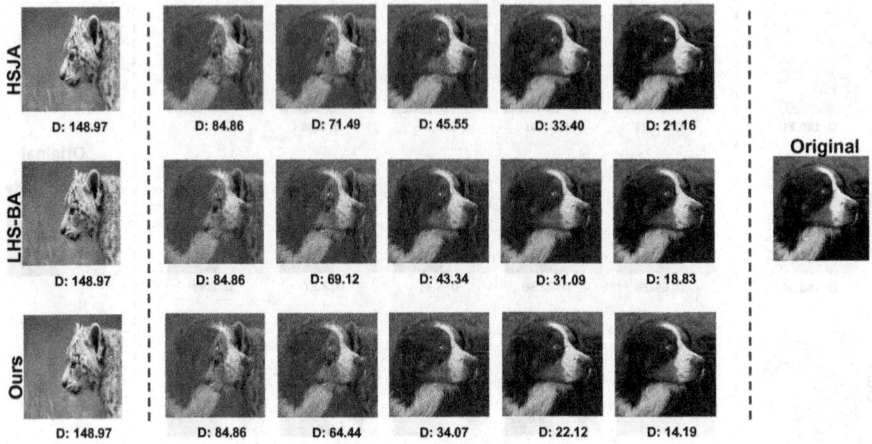

Fig. 7. Visualized trajectories of HSJA [21], LHS-BA [25] and SLIA for performing targeted attacks on ResNet-50. 1st column: initialization. 2nd-9th columns: images after blended with original images and at 1K, 5K, 10K, 20K model queries. 10th column: original image. D is l_2-norm metric to compute the distortion between adversarial image and original image.

6 Conclusion

In this work, we present a query-efficient adversarial example generation algorithm (SLIA), which is specific to ImageNet with a large image size. SLIA can be performed to ensure 100% attack success rate for settings where the attacker only has access to the final decisions of a model. We generate adversarial examples by estimating the gradient of the low-frequency component, which greatly reduces the dimension of the gradient to be estimated. When attacking a variety of different ImageNet models, the distortion can be reduced faster with our method compared to state-of-the-art attacks with different query budgets.

Acknowledgement. This work was supported in part by the NSFC (61872099, 62272116) and in part by the Scientific Research Project of Guangzhou University (YJ2021004). The authors acknowledge the Network Center of Guangzhou University for providing HPC computing resources.

References

1. Szegedy, C., et al.: Intriguing properties of neural networks. In: Proceedings of International Conference on Learning Representations (2014)
2. Goodfellow, I.J., Shlens, J., Szegedy, C.: Explaining and harnessing adversarial examples. In: Proceedings of International Conference on Learning Representations (2015)
3. Liu, Y., Chen, X., Liu, C., Song, D.: Delving into transferable adversarial examples and black-box attacks. In: Proceedings of International Conference on Learning Representations (2016)

4. Chen, P.Y., Zhang, H., Sharma, Y., Yi, J., Hsieh, C.J.: ZOO: zeroth order optimization based black-box attacks to deep neural networks without training substitute models. In: Proceedings of the 10th ACM Workshop on Artificial Intelligence and Security, pp. 15–26. ACM (2017)
5. Dong, Y., et al.: Boosting adversarial attacks with momentum. In: Proceedings of the IEEE Conference on Computer Vision and Pattern Recognition, pp. 9185–9193. IEEE (2018)
6. Ilyas, A., Engstrom, L., Athalye, A., Lin, J.: Black-box adversarial attacks with limited queries and information. In: Proceedings of International Conference on Machine Learning, pp. 2142–2151. ACM (2018)
7. Madry, A., Makelov, A., Schmidt, L., Tsipras, D., Vladu, A.: Towards deep learning models resistant to adversarial attacks. In: Proceedings of International Conference on Learning Representations (2017)
8. Fan, Y., et al.: Sparse adversarial attack via perturbation factorization. In: Vedaldi, A., Bischof, H., Brox, T., Frahm, J.-M. (eds.) ECCV 2020. LNCS, vol. 12367, pp. 35–50. Springer, Cham (2020). https://doi.org/10.1007/978-3-030-58542-6_3
9. Moosavi-Dezfooli, S.M., Fawzi, A., Frossard, P.: DeepFool: a simple and accurate method to fool deep neural networks. In: Proceedings of the IEEE Conference on Computer Vision and Pattern Recognition, pp. 2574–2582. IEEE (2016)
10. Papernot, N., McDaniel, P., Jha, S., Fredrikson, M., Celik, Z.B., Swami, A.: The limitations of deep learning in adversarial settings. In: Proceedings of the IEEE European Symposium on Security and Privacy (Euro S&P), pp. 372–387. IEEE (2016)
11. Papernot, N., McDaniel, P., Goodfellow, I., Jha, S., Celik, Z.B., Swami, A.: Practical black-box attacks against machine learning. In: Proceedings of the 2017 ACM on Asia Conference on Computer and Communications Security, pp. 506–519. ACM (2017)
12. Carlini, N., Wagner, D.: Towards evaluating the robustness of neural networks. In: Proceedings of the IEEE Symposium on Security and Privacy (SP), pp. 39–57. IEEE (2017)
13. Bhagoji, A.N., He, W., Li, B., Song, D.: Practical black-box attacks on deep neural networks using efficient query mechanisms. In: Ferrari, V., Hebert, M., Sminchisescu, C., Weiss, Y. (eds.) ECCV 2018. LNCS, vol. 11216, pp. 158–174. Springer, Cham (2018). https://doi.org/10.1007/978-3-030-01258-8_10
14. Tu, C.-C., et al.: AutoZOOM: autoencoder-based zeroth order optimization method for attacking black-box neural networks. In AAAI Conference on Artificial Intelligence (2018)
15. Al-Dujaili, A., O'Reilly, U.M.: Sign bits are all you need for black-box attacks. In: Proceedings of International Conference on Learning Representations (2020)
16. Guo, C., Gardner, J.R., You, Y., Wilson, A.G., Weinberger, K.Q.: Simple black-box adversarial attacks. arXiv preprint arXiv:1905.07121 (2019)
17. Moon, S., An, G., Song, H.O.: Parsimonious black-box adversarial attacks via efficient combinatorial optimization. In: Proceedings of International Conference on Machine Learning (2019)
18. Deng, J., Dong, W., Socher, R., Li, L.J., Li, K., Fei-Fei, L.: ImageNet: a large-scale hierarchical image database. In: Proceedings of the IEEE Conference on Computer Vision and Pattern Recognition, pp. 248–255 (2009)
19. Brendel, W., Rauber, J., Bethge, M.: Decision-based adversarial attacks: reliable attacks against black-box machine learning models. In: Proceedings of International Conference on Learning Representations (2018)

20. Cheng, M., Le, T., Chen, P.Y., Yi, J., Zhang, H., Hsieh, C.J.: Query-efficient hard-label black-box attack: an optimization-based approach. In: Proceedings of International Conference on Learning Representations (2019)
21. Chen, J., Jordan, M.I., Wainwright, M.: HopSkipJumpAttack: a query-efficient decision-based attack. In: Proceedings of the IEEE Symposium on Security and Privacy (SP), pp. 1277–1294. IEEE (2020)
22. He, K., Zhang, X., Ren, S., Sun, J.: Identity mappings in deep residual networks. In: Leibe, B., Matas, J., Sebe, N., Welling, M. (eds.) ECCV 2016. LNCS, vol. 9908, pp. 630–645. Springer, Cham (2016). https://doi.org/10.1007/978-3-319-46493-0_38
23. Simonyan, K., Zisserman, A.: Very deep convolutional networks for large-scale image recognition. arXiv preprint arXiv:1409.1556. (2014)
24. Huang, G., Liu, Z., van der Maaten, L., Weinberger, K.Q.: Densely connected convolutional networks. In: Proceedings of the IEEE Conference on Computer Vision and Pattern Recognition, pp. 2261–2269. IEEE (2017)
25. Wang, D., Lin, J., Wang, Y.-G.: Query-efficient adversarial attack based on Latin hypercube sampling. arXiv preprint arXiv: 2207.02391. (Accept for presentation in IEEE International Conference on Image Processing 2022)
26. Mallat, S.: The theory for multiresolution signal decomposition: the wavelet representation. In: Proceedings of the IEEE Transactions on Pattern Analysis and Machine Intelligence, pp. 654–693. IEEE (1989)
27. Guo, C., Frank, J.S., Weinberger, K.Q.: Low frequency adversarial perturbation. In: International Conference on Uncertainty in Artificial Intelligence, pp. 1127–1137. AUAI (2019)

RA Loss: Relation-Aware Loss for Robust Person Re-identification

Kan Wang[1,2], Shuping Hu[2], Jun Cheng[1], Jun Cheng[3], Jianxin Pang[2(✉)], and Huan Tan[2]

[1] Guangdong Provincial Key Laboratory of Robotics and Intelligent System, Shenzhen Institute of Advanced Technology, Chinese Academy of Sciences, Shenzhen, China
jun.cheng@siat.ac.cn

[2] Guangdong Provincial Key Laboratory of Robot Localization and Navigation Technology, UBTech Robotics Corp Ltd., Shenzhen, China
{shuping.hu,walton,huan.tan}@ubtrobot.com

[3] Institute for Infocomm Research, A*STAR, Singapore, Singapore
cheng_jun@i2r.a-star.edu.sg

Abstract. Previous relation-based losses in person re-identification (ReID) typically comprise two sequential steps: they firstly sample both positive pair and negative pair and then deploy constraints to simultaneously improve intra-identity compactness and inter-identity separability. However, existing relation-based losses usually place emphasis on exploring the relation between images and therefore consider only several pairs during each optimization. This inevitably leads to different convergence status for pairs of the same kind and brings about the intra-pair variance problem. Accordingly, we propose a novel Relation-Aware (RA) loss to address the intra-pair variance via exploring the informative relation across pairs. In brief, we introduce a macro-constraint and a micro-constraint. The macro-constraint encourages the separation of positive pair and negative pair via pushing far apart the two "centers" of the positive pair and the negative pair. The "center" of each kind of pair are obtained via averaging all the pairs of the same kind. The micro-constraint further enhances the compactness by minimizing the discrepancies among pairs of the same kind. The two constraints work cooperatively to relieve the intra-pair variance and improve the quality of pedestriansąŕ representation. Results of extensive experiments on three widely used ReID benchmarks, *i.e.*, Market-1501, DukeMTMC-ReID and CUHK03, demonstrate that the RA loss brings improvements over existing relation-based losses.

Keywords: Deep learning · Person re-identification · Metric learning

1 Introduction

Person re-identification (ReID) intends to retrieve pedestrian images belonging to the same identity from viewpoints across multiple cameras. In recent years,

L. Wang et al. (Eds.): ACCV 2022, LNCS 13842, pp. 373–390, 2023.
https://doi.org/10.1007/978-3-031-26284-5_23

due to the widespread range of potential applications in video surveillance, *e.g.*, multi-camera tracking [60] and forensic search [54,64], ReID has drawn a lot of attention from both academia and industry [8,15,43–45,47,57,70].

The key to robust ReID lies in high-quality representation for pedestrian image. Recently, many loss functions have been proposed to improve the quality of pedestrians representation and achieve superior performance [4,6,13,16,33, 35,41,51,63]. Among them, a kind of loss functions that explore the relation between pedestrian images emerge and become popular [4,6,13,16,27,33,41]. In common, existing relation-based loss functions typically comprise two steps. First, they sample structures from a batch of pedestrian images. The structure commonly contains two categorizes of pairs, *i.e.*, the positive pair and negative pair that comprise two pedestrian images with identical identity and different identities, respectively. Second, the losses improve both intra-identity compactness and the inter-identity separability via deploying constraints on the two kinds of pairs. For example, as one of the most widely-used relation-based loss in ReID, the triplet loss [16,31] samples triplet from a batch of pedestrian images. Each triplet comprises one positive pair and one negative pair. Afterwards, the triplet loss enlarges intra-identity similarity while shrinking inter-identity similarity via encouraging the distance between positive pair and negative pair[1] to be greater than a predefined threshold. Besides, the contrastive loss [6,13] directly samples positive pairs and negative pairs from a batch of pedestrian images, and then minimizes the embedding distance of positive pairs but demands the distance of negative pairs to be consistently larger than a predefined threshold.

However, although brings performance improvements, during optimization, existing relation-based losses [16,31] typically increase the distance between individual pairs. As a result, the consideration of only several pairs usually leads to the intra-pair variance problem: pairs of the same kind show significant variation. More specifically, as illustrated in Fig. 1(a,b,c), the positive pairs have significant variance in appearance similarity, due to viewpoint variations and inaccurate detection. Besides, appearance similarity of the negative pairs presented in Fig. 1(d,e,f) also varies, since the two pedestrians in different pairs may wear similar or dissimilar clothes. Accordingly, we propose a novel Relation-Aware (RA) loss which relieves the intra-pair variance via exploring the relation across pairs from both macro- and micro-perspectives.

Though several subsequent methods proposed to utilize information from more pairs [4,33], the global cues from all pairs are not sufficiently explored. Accordingly, we propose the macro-constraint to improve the separation between positive pairs and negative pairs from a global perspective. In brief, the macro-constraint pushes far apart the two "centers" of the two kinds of pairs. More specifically, at first, we average the distance of all pairs of the same kind as the corresponding "center". Afterwards, distance between the two "centers" is encouraged to be larger than a per-defined threshold. By using this method, the

[1] In this paper, the distance **of** one pair denotes the distance between the two pedestrian images contained in this pair. In comparison, the distance **between** two pairs indicates the difference value in the two distances of the two pairs.

Fig. 1. Illustration of the challenges of intra-pair variance in ReID. (a, b, c) Three positive pairs show high, moderate and low similarity in pedestrian appearance, respectively. (d, e, f) Three negative pairs show high, moderate and low similarity in pedestrian appearance, respectively.

macro-constraint is able to explore the useful information from all pairs sampled from a batch; therefore it is more effective in improving the separation between positive pairs and negative pairs and therefore stabilizes the training procedure. Since the introduced constraint optimizes pairs from a global perspective; we therefore name it macro-constraint.

We further propose the micro-constraint to minimize the discrepancies among the positive pairs and negative pairs. As presented in Fig. 1, there exist significant intra-pair variance for both kinds of pairs. Accordingly, in order to enhance the compactness of each kind of pairs, we impose constraint on these "unqualified" pairs, which denotes these pairs that away from the corresponding "center". More specifically, we drives these "unqualifie" pairs within both kinds to be close to the corresponding "center" to a certain extent. By this way, the micro-constraint is able to explicitly alleviate the intra-pair variance; and therefore improves the generalization ability of the pedestrian representation. Besides, different with the macro-constraint which optimizes pairs from a global perspective, this constraint is imposed on individual pairs; therefore we name it micro-constraint.

In conclusion, the macro- and micro-constraints are complementary to each other and work cooperatively to relieve the intra-pair variance and therefore improve the quality of pedestrian representation. From the methodological point of view, the contributions of this work can be summarized as follows:

- First, to the best of our knowledge, this is the first attempt to study the intra-pair variance from a comprehensive perspective for robust ReID.
- Second, we propose a novel and simple-yet-effect loss named RA loss, which releives the intra-pair variance via exploring the informative relation across pairs from both macro- and micro-perspectives.

- Third, we demonstrate the effectiveness of the proposed RA loss on three popular large-scale ReID benchmarks, *i.e.*, Market-1501 [60], DukeMTMC-ReID [64] and CUHK03 [22], and the results show that RA loss brings significant improvements over existing relation-based losses.

2 Related Works

2.1 Person Re-identification

Over the past few years, deep learning-based methods [5,8,11,18,38,42,46,49,50] have come to dominate in the ReID community. We here categorize existing deep learning-based ReID methods into two groups: the methods based on feature learning and the methods using metric learning, according to the manner adopted to improve the quality of pedestrian representation.

Feature Learning-Based ReID. Previous methods belonging to this category typically target on learning discriminative holistic representation for pedestrian image [2,3,5,17,23,28,53,62,67,68]. However, the holistic representation-based approaches usually suffer from the overfitting problem [26,38,47]. Accordingly, part-level representations [8,34,37,38,57] have been adopted for robust ReID, as their features contain fine-grained information to distinguish identity. In common, existing part-level representations-based approaches usually use some auxiliary tools, *e.g.*, pose estimators [21,26,59], human parsing algorithms [12,19,70] or attention modules [45,58,68] to infer the body parts' regions, from which the part-level representations are subsequently extracted.

Metric Learning-Based ReID. According to the manner to employ supervision during the training stage, methods in this category can be further subdivided into two categories, *i.e.*, approaches that optimized with identity-level labels [36,51,52,63] and methods using pair-wise labels [4,16,33,41]. The former methods typically see ReID as an image classification task [36,63] and employ classification losses to optimize the similarity between representations of pedestrian images and the identity-related weight vectors. Unfortunately, the classification losses may harm the generalization ability of pedestrian representations, as classification losses in ReID usually encourage features to overfit to the identities. However, ReID is actually a zero-shot task: the identities encountered during testing has no overlap with that in the training stage.

In order to enhance the generalization ability of pedestrian representation, subsequent works [16,33,41] attempted to explore the relation across pedestrian images and formulate pair-wise labels as supervision during the training stage. In common, existing relation-based losses typically comprise two steps. First, they adopt various strategies to construct loss-specific structures [16,31,33,41]. Second, constraints are designed in order to improve both intra-identity compactness and inter-identity discrepancy. For example, the triplet loss [16] first samples triplet which contains a positive pair and a negative pair, then pushes away the distance of the negative pair to be larger than that of the positive pair. Moreover, quadruplet loss [4] proposed to generalizes the triplet loss via

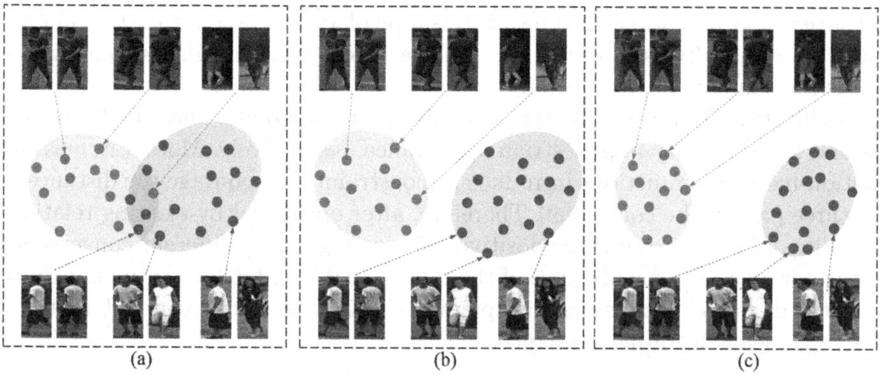

Fig. 2. Distribution of the positive pairs (green dots) and negative pairs (red dots). (a) Significant intra-pair variance for the pairs of the same kind. (b) The macro-constraint improves the separation between positive pairs and negative pairs. (c) The micro-constraint enhances the compactness of each of the two kinds of pairs. Pairs in this and the next figure are demonstrated in the two-dimensional space for better illustration. The vertical coordinates has no indication. (Color figure online)

introducing another negative pair for better approximation of the inter-identity distance. However, although brings performance improvements, existing relation-based loss functions consider only several pairs; therefore the relation across pairs are not sufficiently explored.

In comparison, we deploy constraints on all of the positive and negative pairs from both macro- and micro-perspectives; therefore, we make use of the informative relation across pairs in a more comprehensive manner.

2.2 Intra-pair Variance

A few previous methods [20,35,55] have attempted to study the intra-pair variance problem. For example, [55] introduced an intra-pair loss which learns a class-independent distance metric by minimizing the intra-pair variance in both positive and negative pairs. Besides, [35] addressed the intra-pair variance via setting up a definite optimization target for all pairs of the same kind. Afterwards, it utilized the re-weighting strategy to highlight the less-optimized pairs in order to benefit the deep feature learning with flexible optimization.

We propose RA loss, which directly pushes apart the positive pair and the negative pair from a comperhensive perspective, and promotes the compactness of each kind of pairs via deploying constraint on each individual pair.

3 Method

3.1 Intra-pair Variance in ReID

Existing loss functions in ReID commonly lay emphasis on exploring the relation between pedestrian images, and only several pairs are taken into consideration during each optimization. As a result, this type of local constraint for the relation across pairs usually brings the intra-pair variance, *i.e.*, ambiguous convergence status for pairs of the same kind.

As illustrated in Fig. 2(a), three positive pairs inherently show high, moderate and low similarity in pedestrian appearance, respectively. Moreover, existing relation-based loss functions introduce no constraint to encourage the distance of the three pairs to be consistent. Therefore, after optimized by existing relation-based loss functions, the three positive pairs may get into different convergence status: for example, the distance of them are 0.9, 0.7 and 0.3, respectively. Similarly, the distance of three negative pairs in Fig. 2(a) also obviously different and show significant intra-pair variance.

3.2 The Macro-constraint

The macro-constraint targets on improving the separation between positive pairs and negative pairs from a global perspective. To this end, the macro-constraint comprises two sequential steps to explore the informative cues from all the positive and negative pairs that sampled from a batch.

First, the macro-constraint respectively computes the "center" for both the category of positive pairs and negative pairs via averaging the distances of all the pairs within each of the two categories, as follows:

$$C_{pos} = \frac{1}{|\mathcal{P}|}\sum_{i=1}^{|\mathcal{P}|} \mathcal{D}(\mathbf{f}_i^{p_1}, \mathbf{f}_i^{p_2}), \tag{1}$$

$$C_{neg} = \frac{1}{|\mathcal{N}|}\sum_{i=1}^{|\mathcal{N}|} \mathcal{D}(\mathbf{f}_i^{n_1}, \mathbf{f}_i^{n_2}). \tag{2}$$

Here $\mathcal{P}$ and $\mathcal{N}$ represent the set of positive pairs and negative pairs that are sampled from the same batch, respectively. Besides, $|\mathcal{P}|$ and $|\mathcal{N}|$ denotes the number of elements in $\mathcal{P}$ and $\mathcal{N}$, respectively. At last, $\mathcal{D}(\mathbf{f}_i^{p_1}, \mathbf{f}_i^{p_2})$ and $\mathcal{D}(\mathbf{f}_i^{n_1}, \mathbf{f}_i^{n_2})$ indicates the cosine distance between two pedestrian representations for the i-th positive pair $(\mathbf{f}_i^{p_1}, \mathbf{f}_i^{p_2})$ in $\mathcal{P}$ and the i-th negative pair $(\mathbf{f}_i^{n_1}, \mathbf{f}_i^{n_2})$ in $\mathcal{N}$, respectively.

Second, in order to improve the separation between positive pairs and negative pairs, the macro-constraint encourages the distance between the two "centers" to be larger than a per-defined margin, as follows:

$$\mathcal{L}_{macro} = [C_{pos} - C_{neg} + \alpha]_+, \tag{3}$$

where α represents the margin of the proposed macro-constraint. Besides, $[\cdot]_+ = \max(\cdot, 0)$ denotes the hinge loss.

During each optimization, previous methods [16,33,41] commonly utilize the information from a part of pairs; therefore, they improve the separation between only several positive pairs and negative pairs. Compared with them, the proposed macro-constraint shows its superiority in taking full advantage of the comprehensive cues from all the pairs and improves the separation between positive pairs and negative pairs from a global perspective.

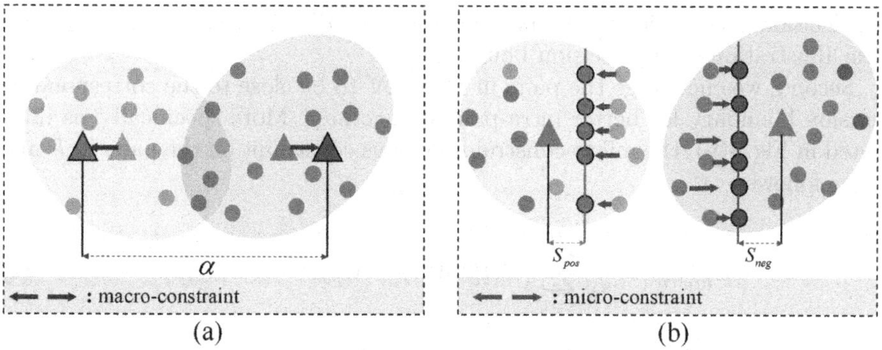

Fig. 3. (a) The macro-constraint pushes away the two "centers" of positive pairs and negative pairs. Rectangles without and with black border represent the "centers" before and after macro-constraint. (b) The micro-constraint encourages pairs within each category to be close to the corresponding decision boundary, which is denoted by the green/red vertical lines. Dots without and with black border represent the pairs before and after micro-constraint. (Color figure online)

3.3 The Micro-constraint

We further propose the micro-constraint, which is complementary to the macro-constraint, to explicitly address the intra-pair variance. We can observe that, as presented in Fig. 1, due to the significant variance in appearance similarity, the distance of pairs within the same category varies. Accordingly, we propose the micro-constraint to explicitly minimize the discrepancies among pairs for both the category of positive pair and negative pair. As illustrated in Fig. 3(b), the micro-constraint is imposed on a part of individual pairs. More specifically, the micro-constraint consists two sequential steps.

First, we identify the pairs on which the micro-constraint should be deployed. We name these pairs as "unqualified" pairs. Intuitively, in order to address the intra-pair variance, it seems reasonable to encourage all pairs of the same kind to be close to corresponding "center". However, the above constraint is too rigid and unreasonable. This is because the distance of different pairs are inherently inconsistent, as illustrated in Fig. 1. Besides, it is also counterintuitive to further increase distance of these positive pairs whose distance is already smaller than its corresponding "center" (C_{pos}), and decrease distance of these negative pairs whose distance is already larger than its corresponding "center" (C_{neg}).

Based on the above analysis, as presented in Fig. 3(b), the sets of "unqualified" pairs for the positive pairs and negative pairs are defined respectively as:

$$\hat{\mathcal{P}} = \left\{ (\mathbf{f}_i^{p_1}, \mathbf{f}_i^{p_2}) \in \mathcal{P} : \mathcal{D}(\mathbf{f}_i^{p_1}, \mathbf{f}_i^{p_2}) > (C_{pos} + \beta \cdot S_{pos}) \right\}, \tag{4}$$

$$\hat{\mathcal{N}} = \left\{ (\mathbf{f}_i^{n_1}, \mathbf{f}_i^{n_2}) \in \mathcal{N} : \mathcal{D}(\mathbf{f}_i^{n_1}, \mathbf{f}_i^{n_2}) < (C_{neg} + \beta \cdot S_{neg}) \right\}. \tag{5}$$

Here S_{pos} and S_{neg} denotes the standard deviation for the category of positive pair and negative pair, respectively. β is a hyper-parameter which determines the decision boundaries of the micro-constraint. Please refer to Fig. 3(b) for a clear illustration of the decision boundaries.

Second, we encourage the pairs in $\hat{\mathcal{P}}$ and $\hat{\mathcal{N}}$ to be close to the corresponding decision boundary for better intra-pair compactness. More specifically, as illustrated in Fig. 3(b), the micro-constraint deploys constraint on the pairs in $\hat{\mathcal{P}}$ and $\hat{\mathcal{N}}$, as follows:

$$\mathcal{L}_{micro}^{pos} = \frac{1}{|\hat{\mathcal{P}}|} \sum_{i=1}^{|\hat{\mathcal{P}}|} \{\mathcal{D}(\mathbf{f}_i^{p_1}, \mathbf{f}_i^{p_2}) - (C_{pos} + \beta \cdot S_{pos})\}, \tag{6}$$

$$\mathcal{L}_{micro}^{neg} = \frac{1}{|\hat{\mathcal{N}}|} \sum_{i=1}^{|\hat{\mathcal{N}}|} \{(C_{neg} + \beta \cdot S_{neg}) - \mathcal{D}(\mathbf{f}_i^{n_1}, \mathbf{f}_i^{n_2})\}. \tag{7}$$

Based the above, the micro-constraint is summarized as follows:

$$\mathcal{L}_{micro} = \mathcal{L}_{micro}^{pos} + \mathcal{L}_{micro}^{neg}. \tag{8}$$

Finally, the proposed RA loss which comprises both macro-constraint and micro-constraint is formulated as follows:

$$\mathcal{L}_{RA} = \mathcal{L}_{macro} + \lambda_1 \mathcal{L}_{micro}. \tag{9}$$

Here λ_1, whose value is set to 1 for the sake of simplicity, indicates the hyper-parameter which balances the macro-constraint and micro-constraint.

3.4 Person ReID via RA Loss

During the training stage, the proposed RA loss is deployed on the ReID feature extractor (*i.e.*, the PCB model [38]) and optimized simultaneously with the popular Cross-Entropy (CE) loss as well as the triplet loss. Therefore, the overall objective function can be written as:

$$\mathcal{L} = \mathcal{L}_{ID} + \lambda_2 \mathcal{L}_{TP} + \lambda_3 \mathcal{L}_{RA}. \tag{10}$$

Here $\mathcal{L}_{ID}$ and $\mathcal{L}_{TP}$ represents the cross-entropy loss and triplet loss, respectively. λ_2 and λ_3 are represents the weights of loss functions. For the sake of simplicity, they are consistently set to 1.

During the testing stage, the cosine metric is adopted in order to measure the similarity between two pedestrian representations $\mathbf{f}_1$ and $\mathbf{f}_2$:

$$\rho = \frac{\mathbf{f}_1{}^{\mathsf{T}} \mathbf{f}_2}{||\mathbf{f}_1||\ ||\mathbf{f}_2||}. \tag{11}$$

Here $||*||$ indicates the L2 norm of $*$.

4 Experiments

4.1 Datasets and Evaluation Protocols

In order to verify the effectiveness of RA loss, we perform extensive experiments on three large-scale ReID benchmarks, *i.e.*, Market-1501 [60], DukeMTMC-ReID [64] and CUHK03 [22]. The official protocol for each database is respectively followed. The widely utilized Rank-1 accuracy and mean Average Precision (mAP) are consistently adopted as metrics for performance evaluation.

Market-1501 includes images of 1,501 identities. A total of 12,936 images of 751 identities are utilized as the training set, while images of the other 750 identities are used for testing. The testing set is further split into a gallery set containing 19,732 images and a query set including the other 3,368 images.

DukeMTMC-reID comprises pedestrian images belonging to 1,404 identities. This dataset is divided into a training set containing 16,522 images belonging to 702 identities, and a testing set comprising images of the remaining 702 identities. Similar to that of Market-1501, the testing set is further subdivided into a gallery set of 17,661 images and a query set of 2,268 images.

CUHK03 contains 14,097 images of 1,467 identities in total. This dataset provides both hand-labeled and DPM-detected bounding boxes. We follow the train/test protocol proposed in [65] to split this dataset into a training set of 767 identities and a testing set of the remaining 700 identities.

4.2 Implementation Details

We use Pytorch to implement the proposed RA loss and adopt PCB [38], one of the most popular part-level representation-based approach in ReID, to extract the pedestrian representations. Besides, in this paper, we denote the PCB model which optimized with the CE loss and triplet loss as the baseline model. We implement the PCB model and the triplet loss following [47]. For the proposed RA loss, we empirically set the hyper-parameters α (in Eq. 3) and β (in Eq. 6 and Eq. 7) to 0.5 and 1, respectively.

During training, all images are resized to 384×128 pixels. The training sets of all three datasets are augmented by means of offline translation [22], online horizontal flipping, and random erasing [66] with ratio 0.5. We construct a batch as follows: each batch comprises 6 identities and each identity has 8 random sampled images; therefore the size of a batch is 48. The stochastic gradient descent optimizer with a weight decay of 5×10^{-4} and a momentum [39] value of 0.9 is utilized for model optimization. The PCB model is fine-tuned from the IDE model [61] and trained in an end-to-end manner for 70 epochs, with the learning rate is initially set to 0.01 and then multiplied by 0.1 every 20 epochs.

4.3 Ablation Study

In this subsection, we first validate the effectiveness of the macro- and micro-constraints. Afterwards, we evaluate the values of two important hyper-parameters and conclude this subsection by evaluating the universality of RA Loss.

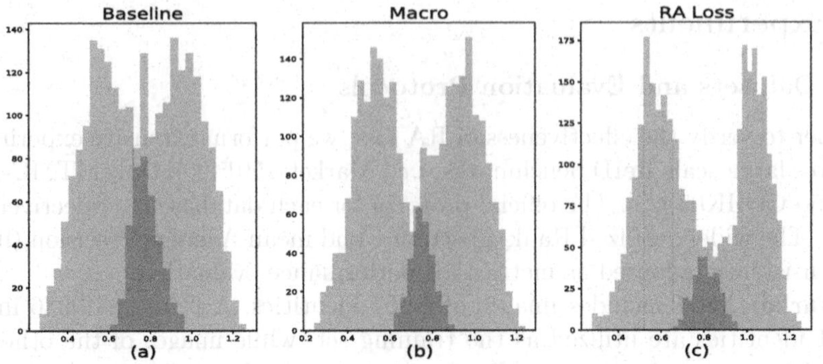

Fig. 4. Distribution of the positive pairs (green) and negative pairs (red) for: (a) baseline, (b) macro-constraint (c) RA loss. Values of the horizontal and vertical coordinates denotes the distance and number of pairs within each histogram, respectively. (Color figure online)

Table 1. Ablation study on each key component of RA loss

Dataset	Components		Market-1501		DukeMTMC		CUHK03-D		CUHK03-L	
Metric	$\mathcal{L}_{macro}$	$\mathcal{L}_{micro}$	Rank-1	mAP	Rank-1	mAP	Rank-1	mAP	Rank-1	mAP
Baseline	–	–	94.2	84.4	88.2	77.4	70.9	66.7	75.6	71.3
Macro	✓	-	94.9	85.7	89.0	78.8	74.1	69.3	77.2	73.8
Micro	–	✓	95.2	86.2	89.8	79.2	74.9	69.8	78.6	74.0
RA Loss	✓	✓	95.7	86.6	90.6	79.8	76.6	70.6	79.6	74.5

Effectiveness of the Macro-constraint. We equip the baseline model with the macro-constraint only to demonstrate its effectiveness. The experimental results presented in Table 1 show that the macro-constraint consistently promote the performance of baseline model on all benchmarks. For example, performance improvements of 0.7% and 1.3% can be observed on Market-1501 in terms of Rank-1 accuracy and mAP, respectively. These experimental results firmly justify the effectiveness of macro-constraint.

Effectiveness of the Micro-constraint. We then verify the effectiveness of micro-constraint in this experiment via equipping the baseline model with the micro-constraint only. The experimental results tabulated in Table 1 clearly demonstrate that micro-constraint achieves significantly superior results than those of the baseline model. For example, it improves the Rank-1 accuracy and mAP on DukeMTMC-ReID from 88.2% and 77.4% to 89.8% and 79.2%, respectively. These experimental results validate the effectiveness of micro-constraint.

Effectiveness of the RA Loss. We finally equip the baseline model with both the macro- and micro-constraints. The results can be found in the row "RA Loss"

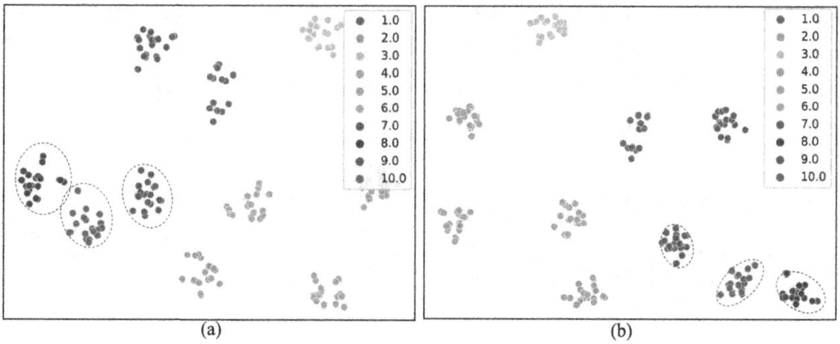

Fig. 5. Visualization using t-SNE [25] for the feature embeddings produced by (a) the baseline model and (b) the baseline model with the proposed RA loss. We sample 20 pedestrian images for each of 10 random identities, which are denoted using different colors, from the testing set of Market-1501.

of Table 1. After assessing the results, we can observe that the combination of the two components creates a considerable performance boost relative to the use of one component. Finally, the combination of the two constraints improves the performance of baseline by 1.5%, 2.4%, 5.7%, and 4.0% in terms of Rank-1 accuracy, as well as by 2.2%, 2.4%, 3.9%, and 3.2% in terms of mAP on each dataset, respectively, These results show that the macro- and micro-constraints are complementary and convincingly demonstrate the effectiveness of RA loss.

Visualization of the Distribution of Pairs. To support the above quantitative experimental results, we compare the distributions of the positive pairs and negative pairs produced by three representative models: (a) baseline, (b) baseline with macro-constraint, and (c) baseline with RA loss, in Fig. 4 .

As can be seen in Fig. 4(a), the distribution of positive pairs significantly overlapps to that of the negative pairs. Moreover, we note that the compactness of both kinds of pairs is unsatisfactory. By contrast, Fig. 4(b) illustrates that the macro-constraint enlarges the separation between positive pairs and negative pairs. Moreover, as shown in Fig. 4(c), the micro-constraint further improves the compactness of both kinds of pairs. These above qualitative visualization firmly justify the effectiveness of the two components in RA loss.

Visualization of the Pedestrian Representations. In addition, in order to further justify the effectiveness of the RA loss, we visualize the pedestrain representations produced by (a) baseline and (b) baseline with RA loss.

After assessing the visualizations presented in Fig. 5, we can draw the conclusion that RA loss effectively improve the intra-identity compactness of feature embeddings as well as enhance the separability between different identities. These visualization results qualitatively justify the capability of RA loss in improving the robustness of pedestrian representations.

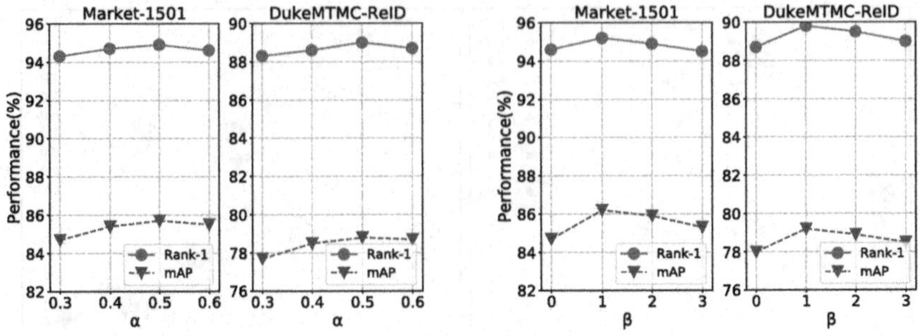

Fig. 6. Evaluation on the value of hyper-parameters α and β.

Table 2. Evaluation on the universality of RA loss

Dataset	Market-1501		DukeMTMC	
Metric	Rank-1	mAP	Rank-1	mAP
CE + Contrastive	93.1	82.9	86.8	75.4
CE + Contrastive + RA	95.2	84.5	88.7	77.4
CE + Quadruplet	94.4	84.8	88.5	77.5
CE + Quadruplet + RA	95.6	86.7	90.6	79.7

Evaluation on the Value of Hyper-parameters α and β. In this experiment, we evaluate the performance of RA loss at different values of α and β. To facilitate clean comparison, we equip the baseline model with only the proposed macro-constraint and micro-constraint for two hyper-parameters, respectively.

We can make three observations from Fig. 6. First, the performance becomes better when α increases to a certain extent while it drops when α further increases. This is because a moderately large α is helpful for the macro-constraint to separate the positive pairs and negative pairs. Second, the performance tends better when β increases from 0 to 1. This is because the micro-constraint demands all pairs of the same kind to be consistent when β is set to 0; therefore the micro-constraint becomes too strict. Third, the performance drops when β further increases; this is because a large β lowers the effect of micro-constraint.

Evaluation on the Universality of RA Loss. We finally validate the universality of RA loss. For a clean comparison, we simply replace the triplet loss in baseline with two other popular relation-based losses, *i.e.*, contrastive loss [6, 13] and quadruplet loss [4], and then add the RA loss.

It can be seen from the experimental results in Table 2 that the RA loss consistently improve the performance of both losses. For example, it promotes the performance of two losses by 2.1% and 1.2% in terms of Rank-1 accuracy, as well as by 1.6% and 1.9% in terms of mAP on Market-1501. These results firmly verify the universality of RA loss in improving existing relation-based losses.

4.4 Comparisons with State-of-the-Art Methods

Table 3. Performance comparisons on Market-1501, DukeMTMC-ReID and CUHK03. blue and Red indicates the best results obtained by HF- and PF-based methods, respectively. Results of our methods are marked in **bold**. "-" represents these results are not available, "RR" denotes the re-ranking operation in [65].

	Methods	Market-1501		DukeMTMC		CUHK03-D		CUHK03-L	
		Rank-1	mAP	Rank-1	mAP	Rank-1	mAP	Rank-1	mAP
HF-based	PSE [30]	87.7	69.0	79.8	62.0	87.7	69.0	–	–
	DuATM [32]	91.4	76.6	81.8	64.6	–	–	–	–
	SFT [24]	93.4	82.7	86.9	73.2	–	–	71.7	60.8
	BDB [7]	94.2	84.3	86.8	72.1	72.8	69.3	73.6	71.7
	Circle loss [35]	94.2	84.9	–	–	–	–	–	–
	IANet [17]	94.4	83.1	87.1	73.4	–	–	–	–
	Res50+NFormer[44]	94.7	87.7	87.4	74.9	–	–	–	–
	ViT + DCAL [69]	94.7	87.5	89.0	80.1	–	–	–	–
	CFPR [56]	94.8	87.7	87.4	74.9	–	–	–	–
	OSNet [67]	94.8	84.9	88.6	73.5	72.3	67.8	–	–
	TransReID [15]	94.9	88.1	90.2	81.3	–	–	–	–
	BDB-Cut [7]	95.3	86.7	89.0	76.0	76.4	73.5	79.4	76.7
	AAFormer [71]	95.6	87.7	90.1	80.0	77.6	74.8	79.9	77.8
PF-based	HA-CNN [23]	91.2	75.7	80.5	63.8	41.7	38.6	44.4	41.0
	PCB [38]	92.3	77.4	81.7	66.1	61.3	54.2	–	–
	PCB+RPP [38]	93.8	81.6	83.3	69.2	63.7	57.5	–	–
	HPM [10]	94.2	82.7	86.6	74.3	63.9	57.5	–	–
	Auto-ReID [29]	94.5	85.1	88.7	78.4	73.3	69.3	77.9	73.0
	BIN [40]	94.8	87.2	89.4	79.6	72.6	69.8	74.3	72.4
	BAT-net [9]	95.1	81.4	87.7	77.3	76.2	73.2	78.6	76.1
	MHN-6[2]	95.1	85.0	–	–	71.7	65.4	77.2	72.4
	DSLNet [48]	95.1	87.3	90.4	78.5	76.3	72.4	–	–
	AdaMine [1]	95.2	85.9	89.9	79.0	–	–	–	–
	CDPM [45]	95.2	86.0	88.2	77.5	71.9	67.0	75.8	71.1
	MuDeep [28]	95.3	84.7	88.2	75.6	71.9	67.2	75.6	70.5
	FPR [14]	95.4	86.6	88.6	78.4	76.1	72.3	–	–
	MGN [43]	95.7	86.9	88.7	78.4	66.8	66.0	68.0	67.4
	DSA-reID [57]	95.7	87.6	86.2	74.3	78.2	73.1	78.9	75.2
	RA Loss	**95.7**	**86.6**	**90.6**	**79.8**	**76.6**	**70.6**	**79.6**	**74.5**
	RA Loss + RR	**96.3**	**94.1**	**93.1**	**90.9**	**85.1**	**85.2**	**87.9**	**88.2**

We here compare the performance of RA loss with that of state-of-the-art methods on three ReID benchmarks: namely, Market-1501 [60], DukeMTMC-ReID [64] and CUHK03 [22]. Moreover, to facilitate fair comparison, we divide existing approaches into two categories: holistic feature-based (HF) methods and part feature-based (PF) methods, according to the properties of the pedestrian representation. The comparisons are tabulated in Table 3.

First, the performance of RA loss is comparable with that of the state-of-the-art methods on **Market-1501**. For example, when compared TransReID [15] and AAFormer [71], RA loss outperforms them by 0.8% (95.7% - 94.9%) and 0.1% (95.7% - 95.6%) for Rank-1 accuracy, respectively. Moreover, compared with the two methods, RA loss shows superiority in efficiency by using a much simpler feature extractor (*i.e.,* PCB). Besides, RA loss also surpasses Circle Loss [35], one of the most recent loss for ReID, by a significant margin in terms of both Rank-1 accuracy (95.7% vs 94.2%) and mAP (86.6% vs 84.9%). At last, we note that the Re-ranking [65] further promotes the performance of RA loss to 96.3% and 94.1% for Rank-1 accuracy and mAP, respectively. These results firmly validate the effectiveness of RA loss. Second, the RA loss consistently beats all PF-based methods on both Rank-1 accuracy and mAP on **DukeMTMC-ReID**. For example, RA loss suppresses DSLNet [48], one of the most recent methods, by 0.2% (90.6% - 90.4%) for Rank-1 accuracy and 1.3% (79.8% - 78.5%) for mAP. Moreover, the Rank-1 accuracy obtained by the RA loss is also superior to that scored by all HF-based methods, particularly, including two recent transformer-based methods, *i.e.,* TransReID [15] and AAFormer [71]. These comparisons clearly justify the overall effectiveness of RA loss. Third, the performance scored by RA loss on **CUHK03** are comparable to that obtained by other PF-based methods: in particular, RA loss achieves the best Rank-1 accuracy of 79.6% on the CUHK03-Labeled dataset, beating DSA-reID [57], the second best PF-based method, by 0.7% (79.6% - 78.9%). These above comparisons convincingly demonstrate the effectiveness of RA loss.

5 Conclusion

In this work, we propose a novel and simple-yet-effective loss function, named RA loss, which addresses the intra-pair variance for robust ReID. The proposed RA loss constructs a pair of constraints to explore the relation across pairs. First, the macro-constraint improves the discrepancy between positive pairs and negative pairs. Second, the micro-constraint promotes the compactness for both the positive pairs and negative pairs. The macro- and micro-constraints are complementary to each other and work collaboratively to address the intra-pair variance; it therefore improves the quality of pedestrian representation. We conduct extensive experiments on three popular large-scale ReID benchmarks, thereby demonstrating the effectiveness of the proposed RA loss.

Acknowledgements. This research is supported by the National Natural Science Foundation of China (U2013601), and Key-Area Research and Development Program of Guangdong Province, China (2019B010154003), and the Program of Guangdong Provincial Key Laboratory of Robot Localization and Navigation Technology (2020B121202011), and the Natural Science Foundation of China (U21A20487), and Shenzhen Technology Project (JCYJ20180507182610734, KCXFZ20201221173411032, Y795001001), and CAS Key Technology Talent Program, and Guangdong Technology Project (No. 2016B010125003).

References

1. Carvalho, M., Cadene, R., Picard, D., Soulier, L., Thome, N., Cord, M.: Cross-modal retrieval in the cooking context: learning semantic text-image embeddings. In: The 41st International ACM SIGIR Conference on Research & Development in Information Retrieval, pp. 35–44 (2018)
2. Chen, B., Deng, W., Hu, J.: Mixed high-order attention network for person re-identification. In: ICCV, pp. 371–381 (2019)
3. Chen, T., et al.: Abd-net: attentive but diverse person re-identification. In: ICCV, pp. 8350–8360 (2019)
4. Chen, W., Chen, X., Zhang, J., Huang, K.: Beyond triplet loss: a deep quadruplet network for person re-identification. In: CVPR, pp. 403–412 (2017)
5. Chen, X., et al.: Salience-guided cascaded suppression network for person re-identification. In: CVPR, pp. 3300–3310 (2020)
6. Chopra, S., Hadsell, R., LeCun, Y.: Learning a similarity metric discriminatively, with application to face verification. In: CVPR, vol. 1, pp. 539–546 (2005)
7. Dai, Z., Chen, M., Gu, X., Zhu, S., Tan, P.: Batch dropblock network for person re-identification and beyond. In: ICCV, pp. 3690–3700 (2019)
8. Ding, C., Wang, K., Wang, P., Tao, D.: Multi-task learning with coarse priors for robust part-aware person re-identification. IEEE Trans. Pattern Anal. Mach. Intell. **44**, 1474–1488 (2020)
9. Fang, P., Zhou, J., Roy, S.K., Petersson, L., Harandi, M.: Bilinear attention networks for person retrieval. In: ICCV, pp. 8029–8038 (2019)
10. Fu, Y., et al.: Horizontal pyramid matching for person re-identification. In: AAAI, pp. 8295–8302 (2019)
11. Gu, X., Ma, B., Chang, H., Shan, S., Chen, X.: Temporal knowledge propagation for image-to-video person re-identification. In: ICCV, pp. 9647–9656 (2019)
12. Guo, J., Yuan, Y., Huang, L., Zhang, C., Yao, J.G., Han, K.: Beyond human parts: dual part-aligned representations for person re-identification. In: ICCV, pp. 3641–3650 (2019)
13. Hadsell, R., Chopra, S., LeCun, Y.: Dimensionality reduction by learning an invariant mapping. In: CVPR, vol. 2, pp. 1735–1742 (2006)
14. He, L., Wang, Y., Liu, W., Zhao, H., Sun, Z., Feng, J.: Foreground-aware pyramid reconstruction for alignment-free occluded person re-identification. In: ICCV, pp. 8449–8458 (2019)
15. He, S., Luo, H., Wang, P., Wang, F., Li, H., Jiang, W.: Transreid: transformer-based object re-identification. In: ICCV, pp. 15013–15022 (2021)
16. Hermans, A., Beyer, L., Leibe, B.: In defense of the triplet loss for person re-identification. arXiv preprint arXiv:1703.07737 (2017)
17. Hou, R., Ma, B., Chang, H., Gu, X., Shan, S., Chen, X.: Interaction-and-aggregation network for person re-identification. In: CVPR, pp. 9317–9326 (2019)
18. Hou, R., Ma, B., Chang, H., Gu, X., Shan, S., Chen, X.: Vrstc: occlusion-free video person re-identification. In: CVPR, pp. 7183–7192 (2019)
19. Kalayeh, M.M., Basaran, E., Gökmen, M., Kamasak, M.E., Shah, M.: Human semantic parsing for person re-identification. In: CVPR, pp. 1062–1071 (2018)
20. Kim, S., Kim, D., Cho, M., Kwak, S.: Proxy anchor loss for deep metric learning. In: CVPR, pp. 3238–3247 (2020)
21. Li, J., Zhang, S., Tian, Q., Wang, M., Gao, W.: Pose-guided representation learning for person re-identification. IEEE Trans. Pattern Anal. Mach. Intell. **44**, 622–635 (2019)

22. Li, W., Zhao, R., Xiao, T., Wang, X.: Deepreid: deep filter pairing neural network for person re-identification. In: CVPR, pp. 152–159 (2014)
23. Li, W., Zhu, X., Gong, S.: Harmonious attention network for person re-identification. In: CVPR, pp. 2285–2294 (2018)
24. Luo, C., Chen, Y., Wang, N., Zhang, Z.: Spectral feature transformation for person re-identification. In: ICCV, pp. 4975–4984 (2019)
25. Maaten, L.V.D., Hinton, G.: Visualizing data using t-sne. JMLR **9**, 2579–2605 (2008)
26. Miao, J., Wu, Y., Liu, P., Ding, Y., Yang, Y.: Pose-guided feature alignment for occluded person re-identification. In: ICCV, pp. 542–551 (2019)
27. Nguyen, B., De Baets, B.: Kernel distance metric learning using pairwise constraints for person re-identification. IEEE Trans. Image Process. **28**(2), 589–600 (2019)
28. Qian, X., Fu, Y., Xiang, T., Jiang, Y.G., Xue, X.: Leader-based multi-scale attention deep architecture for person re-identification. IEEE Trans. Pattern Anal. Mach. Intell. **42**(2), 371–385 (2020)
29. Quan, R., Dong, X., Wu, Y., Zhu, L., Yang, Y.: Auto-reid: searching for a part-aware convnet for person re-identification. In: ICCV, pp. 3749–3758 (2019)
30. Saquib Sarfraz, M., Schumann, A., Eberle, A., Stiefelhagen, R.: A pose-sensitive embedding for person re-identification with expanded cross neighborhood re-ranking. In: CVPR, pp. 420–429 (2018)
31. Schroff, F., Kalenichenko, D., Philbin, J.: Facenet: a unified embedding for face recognition and clustering. In: CVPR, pp. 815–823 (2015)
32. Si, J., et al.: Dual attention matching network for context-aware feature sequence based person re-identification. In: CVPR, pp. 5363–5372 (2018)
33. Sohn, K.: Improved deep metric learning with multi-class n-pair loss objective. In: NIPS, vol. 29 (2016)
34. Suh, Y., Wang, J., Tang, S., Mei, T., Lee, K.M.: Part-aligned bilinear representations for person re-identification. In: ECCV, pp. 402–419 (2018)
35. Sun, Y., et al.: Circle loss: a unified perspective of pair similarity optimization. In: CVPR, pp. 6398–6407 (2020)
36. Sun, Y., Zheng, L., Deng, W., Wang, S.: Svdnet for pedestrian retrieval. In: ICCV, pp. 3800–3808 (2017)
37. Sun, Y., Zheng, L., Li, Y., Yang, Y., Tian, Q., Wang, S.: Learning part-based convolutional features for person re-identification. IEEE Trans. Pattern Anal. Mach. Intell. **43**, 902–917 (2019)
38. Sun, Y., Zheng, L., Yang, Y., Tian, Q., Wang, S.: Beyond part models: person retrieval with refined part pooling (and a strong convolutional baseline). In: ECCV, pp. 480–496 (2018)
39. Sutskever, I., Martens, J., Dahl, G.E., Hinton, G.E.: On the importance of initialization and momentum in deep learning. In: ICML, pp. 1139–1147 (2013)
40. Tang, Z., Huang, J.: Branch interaction network for person re-identification. In: ACCV (2020)
41. Tang, Z., Huang, J.: Harmonious multi-branch network for person re-identification with harder triplet loss. ACM Trans. Multimedia Comput. Commun. Appl. **18**(4), 1–21 (2022)
42. Tao, D., Guo, Y., Yu, B., Pang, J., Yu, Z.: Deep multi-view feature learning for person re-identification. IEEE Trans. Circ. Syst. Video Technol. **28**(10), 2657–2666 (2017)

43. Wang, G., Yuan, Y., Chen, X., Li, J., Zhou, X.: Learning discriminative features with multiple granularities for person re-identification. In: ACM MM, pp. 274–282 (2018)
44. Wang, H., Shen, J., Yongtuo, L., Gao, Y., Gavves, E.: Nformer: robust person re-identification with neighbor transformer. In: CVPR, pp. 7297–7307 (2022)
45. Wang, K., Ding, C., Maybank, S.J., Tao, D.: CDPM: convolutional deformable part models for semantically aligned person re-identification. IEEE Trans. Image Process. **29**, 3416–3428 (2020)
46. Wang, K., Ding, C., Pang, J., Xu, X.: Context sensing attention network for video-based person re-identification. arXiv preprint arXiv:2207.02631 (2022)
47. Wang, K., Wang, P., Ding, C., Tao, D.: Batch coherence-driven network for part-aware person re-identification. IEEE Trans. Image Process. **30**, 3405–3418 (2021)
48. Wang, L., Fan, B., Guo, Z., Zhao, Y., Zhang, R., Li, R., Gong, W.: Dense-scale feature learning in person re-identification. In: ACCV (2020)
49. Wang, P., Ding, C., Shao, Z., Hong, Z., Zhang, S., Tao, D.: Quality-aware part models for occluded person re-identification. IEEE Trans. Multimedia (2022)
50. Wang, P., Ding, C., Tan, W., Gong, M., Jia, K., Tao, D.: Uncertainty-aware clustering for unsupervised domain adaptive object re-identification. IEEE Trans. Multimedia (2022)
51. Wen, Y., Zhang, K., Li, Z., Qiao, Yu.: A discriminative feature learning approach for deep face recognition. In: Leibe, B., Matas, J., Sebe, N., Welling, M. (eds.) ECCV 2016. LNCS, vol. 9911, pp. 499–515. Springer, Cham (2016). https://doi.org/10.1007/978-3-319-46478-7_31
52. Wen, Y., Zhang, K., Li, Z., Qiao, Y.: A comprehensive study on center loss for deep face recognition. IJCV **127**(6), 668–683 (2019)
53. Xia, B.N., Gong, Y., Zhang, Y., Poellabauer, C.: Second-order non-local attention networks for person re-identification. In: ICCV, pp. 3759–3768 (2019)
54. Yao, H., Zhang, S., Hong, R., Zhang, Y., Xu, C., Tian, Q.: Deep representation learning with part loss for person re-identification. IEEE Trans. Image Process. **28**(6), 2860–2871 (2019)
55. Yu, B., Tao, D.: Deep metric learning with tuplet margin loss. In: ICCV, pp. 6490–6499 (2019)
56. Zhang, A., Gao, Y., Niu, Y., Liu, W., Zhou, Y.: Coarse-to-fine person re-identification with auxiliary-domain classification and second-order information bottleneck. In: CVPR, pp. 598–607 (2021)
57. Zhang, Z., Lan, C., Zeng, W., Chen, Z.: Densely semantically aligned person re-identification. In: CVPR, pp. 667–676 (2019)
58. Zhao, L., Li, X., Zhuang, Y., Wang, J.: Deeply-learned part-aligned representations for person re-identification. In: ICCV, pp. 3219–3228 (2017)
59. Zheng, L., Huang, Y., Lu, H., Yang, Y.: Pose invariant embedding for deep person re-identification. IEEE Trans. Image Process. **28**(9), 4500–4509 (2019)
60. Zheng, L., Shen, L., Tian, L., Wang, S., Wang, J., Tian, Q.: Scalable person re-identification: a benchmark. In: ICCV, pp. 1116–1124 (2015)
61. Zheng, L., Zhang, H., Sun, S., Chandraker, M., Yang, Y., Tian, Q.: Person re-identification in the wild. In: ICCV, pp. 1367–1376 (2017)
62. Zheng, M., Karanam, S., Wu, Z., Radke, R.J.: Re-identification with consistent attentive siamese networks. In: CVPR, pp. 5735–5744 (2019)
63. Zheng, Z., Zheng, L., Yang, Y.: A discriminatively learned CNN embedding for person reidentification. ACM Trans. Multimedia Comput. Commun. Appl. **14**(1), 1–20 (2017)

64. Zheng, Z., Zheng, L., Yang, Y.: Unlabeled samples generated by GAN improve the person re-identification baseline in vitro. In: ICCV, pp. 3754–3762 (2017)
65. Zhong, Z., Zheng, L., Cao, D., Li, S.: Re-ranking person re-identification with k-reciprocal encoding. In: CVPR, pp. 1318–1327 (2017)
66. Zhong, Z., Zheng, L., Kang, G., Li, S., Yang, Y.: Random erasing data augmentation, pp. 13001–13008 (2020)
67. Zhou, K., Yang, Y., Cavallaro, A., Xiang, T.: Omni-scale feature learning for person re-identification. In: ICCV, pp. 3701–3711 (2019)
68. Zhou, S., Wang, F., Huang, Z., Wang, J.: Discriminative feature learning with consistent attention regularization for person re-identification. In: ICCV, pp. 8039–8048 (2019)
69. Zhu, H., Ke, W., Li, D., Liu, J., Tian, L., Shan, Y.: Dual cross-attention learning for fine-grained visual categorization and object re-identification. In: CVPR, pp. 4692–4702 (2022)
70. Zhu, K., Guo, H., Liu, Z., Tang, M., Wang, J.: Identity-guided human semantic parsing for person re-identification. In: Vedaldi, A., Bischof, H., Brox, T., Frahm, J.-M. (eds.) ECCV 2020. LNCS, vol. 12348, pp. 346–363. Springer, Cham (2020). https://doi.org/10.1007/978-3-030-58580-8_21
71. Zhu, K., et al.: Aaformer: auto-aligned transformer for person re-identification. arXiv preprint arXiv:2104.00921 (2021)

Class Specialized Knowledge Distillation

Li-Yun Wang[1(✉)], Anthony Rhodes[2], and Wu-chi Feng[1]

[1] Portland State University, Portland, OR 97201, USA
{liyuwang,wuchi}@pdx.edu
[2] Intel Labs, Santa Clara, CA 95054, USA
anthony.rhodes@intel.com

Abstract. Knowledge Distillation (KD) is a compression framework that transfers distilled knowledge from a teacher to a smaller student model. KD approaches conventionally address problem domains where the teacher and student network have equal numbers of classes for classification. We provide a knowledge distillation solution tailored for class specialization, where the user requires a compact and performant network specializing in a subset of classes from the class set used to train the teacher model. To this end, we introduce a novel knowledge distillation framework, Class Specialized Knowledge Distillation (CSKD), that combines two loss functions: Renormalized Knowledge Distillation (RKD) and Intra-Class Variance (ICV) to render a computationally-efficient, specialized student network. We report results on several popular architectural benchmarks and tasks. In particular, CSKD consistently demonstrates significant performance improvements over teacher models for highly restrictive specialization tasks (e.g., instances where the number of subclasses or datasets is relatively small), in addition to outperforming other state-of-the-art knowledge distillation approaches for class specialization tasks.

Keywords: Neural network compression · Class specialization · Knowledge distillation

1 Introduction

Researchers have demonstrated the success of Deep Convolutional Neural Networks (DCNNs) on a wide range of computer vision applications including image recognition [5,35,63], instance-based and pixel-level image segmentation [33,50,53], and object localization in images and videos [7,25,47,49]. Oftentimes, state-of-the-art DCNN models are unwieldy and require a significant amount of computation time and memory space for training and inference, which can limit their real-world usability, particularly for mobile and edge applications. Neural network compression techniques have been dedicated to alleviating these issues by removing less activated parameters in complex models [11,15,19,27,38,55,64] or leveraging knowledge distillation [1,17,18,34,40,51,57] to train a smaller student network.

L. Wang et al. (Eds.): ACCV 2022, LNCS 13842, pp. 391–408, 2023.
https://doi.org/10.1007/978-3-031-26284-5_24

Knowledge distillation is a teacher-student learning methodology that aims to train a compact student neural network by replicating the implicit knowledge encoded in a larger teacher model. In general, KD generates a pre-trained neural network (i.e., a teacher network) and transfers the teacher's knowledge to a student network by minimizing the difference between the outputs of the two networks. Yun et al. [56] combine a self-knowledge distillation technique with class-wise prediction regularization to tackle the issue of overfitting for neural network training. By penalizing the predictive distribution between similar samples, their approach achieves competitive classification accuracy. Muller et al. [32] show that the student network can experience sub-optimal knowledge transfer from the teacher network when using coarsely-defined class labels. Their work improved the knowledge transfer between the teacher and student by enabling the teacher to partition class labels into multiple subclasses.

There exist a wide range of practical applications and use cases for class specialized neural networks [20,22,45]. Many general-purpose ensembling methods in machine learning [9,36] leverage specialized *experts*, including [43,59]. In addition, most AI-assisted real-world manufacturing processes require fine-grain model specialization [3,4,8,13,39,48,60], as do a variety of deployed models in Medicine [13,60], Biology [4,39], Agriculture [3,8], and vital supply chain operations [48]. These specialization domain challenges are also frequently exacerbated due to inherent data scarcity and annotation costs.

Several recent works have called attention to class or task specification problems in relation to knowledge distillation. Shen et al. [43] and Zaras et al. [59] aggregate the knowledge from multiple teacher networks and transfer it to the student. Morgado et al. [31] use a teacher network fine-tuned on a specific task as guidance to train task-specific proxy layers in a student network. This method also focuses on specialized tasks, but it requires a fine-tuning of the remaining parameters in the proxy layers. Kao et al. [21] present a KD technique to improve the overall accuracy on weak classes by transferring distilled outputs from multiple teacher networks to a single student network. Notably, this approach does not produce a compact, specialized network for specific subclasses.

In this paper, we focus on the problem of training a compact student network for explicit specialized classes applicable in real-world specialization tasks to simultaneously reduce compute overhead and improve data efficiency costs. We present a novel KD framework to generate a compact student network for class specialization by restricting knowledge transfer from the teacher model to a subset of *classes of interest.* This specialized knowledge transfer is effected primarily using Renormalized Knowledge Distillation (RKD) loss. We furthermore regularize this knowledge distillation process by simultaneously minimizing the intra-class variance for latent representations among all subclasses in the student network with the introduction of Intra-Class Variation (ICV) loss. We show that these two loss functions work in tandem to bolster class specialization performance for compact student networks through empirical experiments and qualitative analyses. Our proposed technique is generalizable across a variety of different model architectures and vision tasks, including image classification and transfer learning applications.

The contributions of our work are as follows:

1. We introduce a novel KD technique using the proposed RKD and the ICV loss functions for class specialization problems.
2. We empirically evaluate our proposed technique on standard benchmarks models and image datasets. Our experiments show that the proposed technique is competitive with, and frequently outperforms, the state-of-the-art KD techniques for specialized student networks.
3. We further demonstrate the generalizability of our proposed technique by generating specialized neural networks on both image classification and transfer learning tasks.

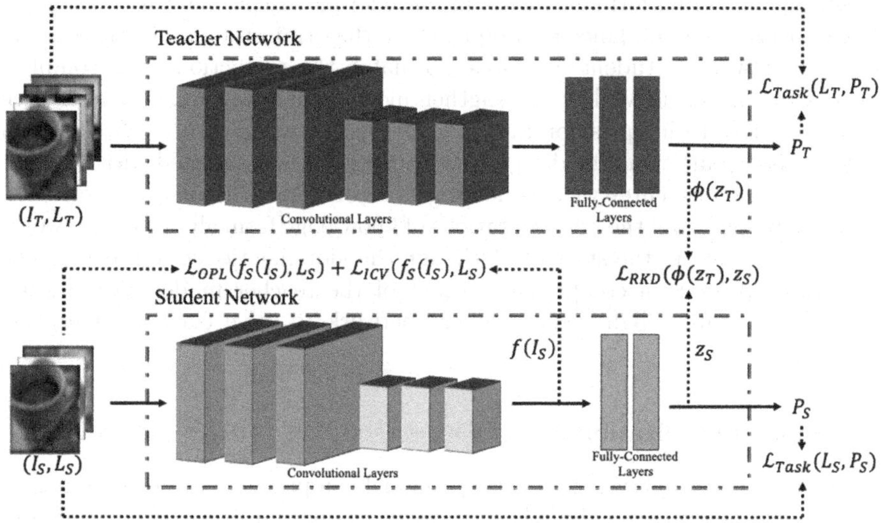

Fig. 1. Overview of our CSKD approach. During the knowledge distillation stage, the student network calculates the Renormalized Knowledge Distillation Loss ($\mathcal{L}_{RKD}$), Orthogonal Projection Loss ($\mathcal{L}_{OPL}$), and the Intra-Class Variance Loss ($\mathcal{L}_{ICV}$) given the teacher network trained on training data with all classes. $f_S(I_S)$ represents a feature extractor that outputs feature embeddings for the student network given an image batch. ϕ is a mapping function that chooses a subset of logits from the teacher and transfers it to the student. P_T and P_S are a prediction for the teacher and the student. z_T and z_S represent a logit for the teacher and the student, respectively.

2 Related Work

2.1 Knowledge Distillation

Many knowledge distillation approaches have been applied to network model compression problems by training a student network with fewer parameters that nevertheless achieve competitive performance with large-scale models [1, 6, 17, 34, 40, 51, 57, 65]. These approaches mainly focus on the transfer of probability-based knowledge [6, 17], latent representation knowledge [1, 34, 40, 51, 57], or combinations of both types of knowledge [65] to the student. Another common research tactic for knowledge distillation centers around self-knowledge transfer approaches. Yun et al. [56] leverage the concept of self-distillation to propose a class-wise prediction regularization for reducing overfitting and improving model generalization. Zhang et al. [61] propose a self-distillation framework that extracts representations of knowledge from different depths of attention modules to enhance model performance without considering teacher networks. Zheng et al. [62] utilize a self-guidance technique where they train the predictions of multiple sub-networks (student networks) to match the predictions of a complete network (a teacher network) to strengthen model generalization. Other research focuses on KD techniques for multiple teacher networks [29, 44, 59]. Each of these methods partitions the data into multiple subsets associated with different classes and then executes various heterogenous or homogeneeous KD processes. Lastly, they aggregate the representations of knowledge from all teacher networks and transfer them to the student. Although the aforementioned approaches can successfully produce a compressed variant of the teacher model, they do not explicitly generate a lightweight student network to solve class specialization problems.

2.2 Task Specialization

Some relevant research work on network model compression for class specialization applications propose the KD framework among multiple teacher networks and one or more student networks [21, 43, 59] or prune a neural network via a non-KD technique [12]. Shen et al. [43] propose a knowledge amalgamation framework to combine with teacher outputs from multiple pre-trained teacher models and leverage the combined teacher outputs to learn a lightweight student network for comprehensive classification. Like Shen's work [43], Zaras et al. [59] utilize a similar idea to ensemble multiple teacher networks trained on non-overlapping subclasses given the entire class and efficiently distill the knowledge from all the teachers to a compact student network for the whole class. Kao et al. [21] also propose an ensemble-based KD framework for making a student network by leveraging multiple expert networks (teacher networks) to better guide student knowledge acquisition. As a result of aggregating the predictions of multiple teacher-student frameworks that specialize in specific tasks, the trained student network achieves better classification performance. Gabbay et al. [12] define a value-locality-based network compression algorithm to search for the specific neurons of the activations associated with high degree values given specialized tasks and replace these neurons with the associated average arithmetic values. Since this method does not use fine-tuning, their approach can only match the performance of the uncompressed neural network. Each of these prior approaches require specialization of the teacher network while distilling knowledge injectively (that is, for an equal number of classes) to the student. By contrast, our proposed KD framework focuses on explicitly tailoring the distillation process itself to the required task specialization problem.

3 Our Approach

In this section we provide an overview of our CSKD framework and give details of each loss function used to train our student network. Our method is designed to generate a lightweight student network for specialized classes (e.g., 5 or 10 specific classes) given the entire set of class (e.g., 100 or 200 classes) using the RKD and ICV loss functions. Figure 1 shows the overview of the proposed knowledge distillation framework for neural network compression tasks on specialized classes. RKD loss effectively transfers the teacher logit information associated with the target classes of interest to the student, so we can integrate the latent data representations learned by the teacher into the student training process. In addition, we introduce ICV loss to regularize student training by minimizing the intra-class variance of feature embeddings output by the penultimate layer of the student network.

3.1 Knowledge Distillation

In this section, we revisit the orthodox knowledge distillation methodology [2,17]. Given a teacher network T and a student network S, we define f_T and f_S as a function approximated by a deep neural network for the teacher and the student. We also define z_T and z_S as a logit for the teacher and the student network. Then, we consider X^2, the set of tuples with two distinct elements, as a set of data points. Specifically, we denote $X^2 = \{(x_i, x_j) \mid i \neq j\}$ henceforth.

In the teacher-student framework setting, knowledge distillation [17] aims to minimize the following objective function, given the logits of the teacher and the student network:

$$\mathcal{L}_{KD} = \alpha \sum_{x_i \in \mathcal{X}} L(z_T,\ z_S) \tag{1}$$

where L is a loss function minimizing the difference between the output of the teacher and the student. Additionally, α is a hyperparameter used to control the severity of penalty for the knowledge distillation loss.

Researchers have proposed a variety of different loss functions to calculate the difference between two logits or feature embeddings. Hinton et al. [17] normalize logits of the teacher and the student via softmax and leverage Kullback-Leibler (KL) divergence to calculate the difference between them for the loss function L:

$$\sum_{x_i \in \mathcal{X}} \mathcal{KL}(softmax(\frac{z_T}{\tau}), softmax(\frac{z_S}{\tau})) \quad (2)$$

where τ is a hyperparameter, temperature, that controls the smoothness of the probability distribution. As τ increases, the $\mathcal{KL}$ loss is more sensitive to differences between the teacher and student logits.

Unlike the Hinton et al. work, Romero et al. [40] transfer knowledge from the teacher to the student by minimizing the difference between the feature embeddings of the teacher and the student for the loss function L:

$$\sum_{x_i \in \mathcal{X}} \|f_T(x_i) - M(f_S(x_i))\|^2 \quad (3)$$

where $\|.\|$ represents the Euclidean norm. M is a mapping function that takes the feature embeddings of the student as input and aligns the embeddings of the student with the feature embeddings of the teacher.

As in [17,40], many researchers have proposed related knowledge distillation methods [1,30,34,51,54,57] based on Eq. 1. Notably, these methods transfer all outputs of the teacher network to the student network. As such, these methods can only be used in conventional knowledge distillation frameworks, and are therefore not directly applicable with specialization tasks.

3.2 Renormalized Knowledge Distillation

We introduce Renormalized Knowledge Distillation (RKD) as a mechanism to transfer only part of the outputs of the teacher to a specialized student network. Unlike conventional knowledge distillation approaches, RKD loss leverages a mapping function ϕ to select a subset of the logits from the teacher and normalizes this subset via the softmax function. Once the renormalized teacher logits are generated, the loss transfers the logit information solely for the classes of interest from the teacher to the student.

Formally, we define RKD loss as follows:

$$\hat{z_T} = \phi(z_T) \quad (4)$$

$$\mathcal{L}_{RKD} = \alpha \sum_{x_i \in \mathcal{X}} \mathcal{KL}(softmax(\frac{\hat{z_T}}{\tau}), softmax(\frac{z_S}{\tau})) \quad (5)$$

where $\hat{z_T}$ is the output of the mapping function ϕ for the teachers logits, ϕ is the mapping function that identifies the subset of the teacher logits corresponding with the specialized classes of interest, and $\mathcal{KL}$ is the KL divergence loss minimizing the difference between the teacher and student logits. By applying the mapping function, RKD loss can select any number of the logits from the teacher and transfer specific knowledge associated with the logits from the classes of interest regardless of the logit dimensions of the teacher and the student networks. Notice that RKD generalizes as conventional knowledge distillation loss, so that when the teacher and student logit dimensions are equal, Eq. 5 reduces to Eq. 2 and Eq. 1.

3.3 Regularization Loss for Feature Embeddings

RKD loss transfers the part of logits associated with classes of interest for subclass specialization from the teacher to the student, but it does not directly leverage the feature embeddings learned from a network to enhance classification performance for model training. It is well-known [10,14,24,26,46] that DNNs encode feature embeddings hierarchically. Accordingly, researchers have applied these hierarchically-generated features to a large variety of different problems [34,42,51,53,57] to achieve competitive performance. Inspired by previous works [34,51,57], we similarly utilize feature embeddings to enhance classification performance for the student by maximizing inter-class *angular* distance while simultaneously minimizing intra-class feature *spatial* variance.

To maximize the inter-class embedding distance given feature embeddings from the penultimate layer, we adopt Orthogonal Projection Loss (OPL) [37] to enforce class-wise orthogonality in the feature embeddings. OPL is defined as follows:

$$s = \Big(\sum_{\substack{i,j \in B \\ y_i = y_j}} \langle f(\boldsymbol{x}_i), f(\boldsymbol{x}_j) \rangle \Big) \; / \; \Big(\sum_{\substack{i,j \in B \\ y_i = y_j}} 1 \Big) \tag{6}$$

$$d = \Big(\sum_{\substack{i,k \in B \\ y_i \neq y_k}} \langle f(\boldsymbol{x}_i), f(\boldsymbol{x}_k) \rangle \Big) \; / \; \Big(\sum_{\substack{i,k \in B \\ y_i \neq y_k}} 1 \Big) \tag{7}$$

$$\mathcal{L}_{OPL} = (1 - s) + |d| \tag{8}$$

where $f(\boldsymbol{x})$ is a feature embedding given an input, $\langle .,. \rangle$ is a similarity measurement function that takes two input vectors and computes the cosine value between the vectors, and B is a mini-batch size. Minimizing OPL equates to enforcing orthogonality between embeddings of data points in different classes. This result is achieved by calculating the similarity of input pairs with the same class and the dissimilarity of input pairs with distinct classes in an image batch. OPL thus effectively increases the angular distance of inter-class embeddings in the feature representation space for model training.

Although OPL can render class-wise orthogonality in the feature embeddings, it nevertheless only considers the cosine similarity and dissimilarity between input pairs without regard for minimizing the spatial extent of intra-class embeddings within a single class. To help achieve this end, we introduce Intra-Class Variance (ICV) loss to enforce intra-class variance minimization for all feature embeddings with the same target class by calculating the variance of the outputs of the penultimate layer for each target class. In our experiments, we demonstrate that ICV loss can be used in tandem with OPL to further improve the effectiveness of learned latent embeddings.

Formally, we formulate the ICV loss using the following equations:

$$\boldsymbol{f_{var}} = \psi(f_S(\boldsymbol{X}_i)) \quad i = 1 \dots C \tag{9}$$

$$\mathcal{L}_{ICV} = \sum_{i=1}^{C} \|\boldsymbol{f_{var}}\|_F \tag{10}$$

where f_S is an approximate feature extractor function via student neural networks, $f_S(\boldsymbol{X}_i)$ is an $N \times D$ feature embedding matrix with the same target class, ψ is a function that calculates variance for every embedding, $\boldsymbol{f_{var}}$ is a $1 \times D$ vector, and $\|.\|_F$ is a normalization function that computes the Frobenius norm for the embedding vectors. To calculate gradients for weight updating, we need to ensure that ICV loss is differentiable. ICV is comprised of conventional variance and $L2$ norm for calculations for matrices, indicating that ICV is made up of differentiable functions, and therefore differentiable. Thus, if one defines the output of the network forward pass: $o_i = \mathcal{L}_{ICV}(f_S(\boldsymbol{X}_i))$, then the gradients required for backpropagation of the network can be obtained via $g_i = \frac{\partial \mathcal{L}_{ICV}}{\partial w}$, where w denotes a network parameter.

3.4 Training with Losses

Our final student model is trained using a combination of losses, including, including RKD and ICV losses, a task-specific loss, and OPL. The task-specific loss can be a conventional cross-entropy loss, say, in the case of classification problems, or a different, bespoke loss for function for different problem domains. Our total loss for the student network is defined:

$$\mathcal{L}_{total} = \mathcal{L}_{task} + \alpha\mathcal{L}_{RKD} + \beta\mathcal{L}_{OPL} + \gamma\mathcal{L}_{ICV} \tag{11}$$

where $\mathcal{L}_{task}$ is the task-specific loss, α is a hyperparameter for the RKD loss, β is a hyperparameter for OPL, and γ is a hyperparameter for the ICV loss. All of these hyperparameters can help to regularize the $\mathcal{L}_{task}$ loss and thereby improve model performance. We utilize a heuristically-chosen small positive value for all the hyperparameters, since $\mathcal{L}_{task}$ is still the most crucial guideline for student network training.

4 Experimental Results

We empirically evaluate CSKD on image classification and transfer learning specialization tasks. In addition, we conduct ablation studies that assess the effect of different numbers of subclasses for specialization performance, in addition to testing different hyperparameter values for each of the loss functions appearing in (11). For fair comparisons, we follow the same experimental settings as reported in relevant baseline KD research in our experiments [1,34]. We then choose the student networks in accordance with standard knowledge distillation teacher-student comparison practices [1,51].

Table 1. The table shows top-1 accuracy comparisons for distinct subclasses of the entire class for WRN. $\mathcal{L}_{\mathcal{CE}}$ represents the cross-entropy loss in the experiment. The hyperparameter of each loss term (e.g., $\mathcal{L}_{\mathcal{RKD}}$, $\mathcal{L}_{\mathcal{OPL}}$, and $\mathcal{L}_{\mathcal{ICV}}$) is 0.1.

Model: (Teacher, Student)	(WRN-40-2, WRN-16-1)			
Benchmark	CIFAR10	CIFAR100		
(Class Frac., Subclass Num.)	(50%, 5)	(5%, 5)	(10%, 10)	(50%, 50)
Teacher ($\mathcal{L}_{\mathcal{CE}}$)	94.98%	81.60%	77.5%	74.56%
Student ($\mathcal{L}_{\mathcal{CE}}$)	93.68%	93.20%	89.20%	72.82%
Student ($\mathcal{L}_{\mathcal{CE}} + \mathcal{L}_{\mathcal{RKD}}$)	93.82%	93.59%	89.60%	73.24%
Student ($\mathcal{L}_{\mathcal{CE}} + \mathcal{L}_{\mathcal{RKD}} + \mathcal{L}_{\mathcal{OPL}}$)	94.42%	93.99%	89.80%	73.32%
Student ($\mathcal{L}_{\mathcal{CE}} + \mathcal{L}_{\mathcal{RKD}} + \mathcal{L}_{\mathcal{OPL}} + \mathcal{L}_{\mathcal{ICV}}$)	**94.66%**	**95.19%**	**90.70%**	**74.24%**

4.1 Ablation Study Setup

We first conduct ablation studies with respect to the number of subclasses used for specialization, in addition to the hyperparameter values used in our loss function. Our experiments encompass several different benchmark architectures, including standard residual networks (Resnet) [16] and wide residual networks (WRN) [58] across the CIFAR10, CIFAR100, and Tiny ImageNet [23] benchmark datasets. We consider these ablation studies to better understand the effects of the proposed loss functions and to test the robustness of our method for different degrees of task specialization. For all ablation experiments, we train the teacher network on the normative training sets (e.g., all 10 or 100 classes for CIFAR10 and CIFAR100, respectively) and test the trained network on the images associated with specializeed classes (e.g., 5, 10, and 50 classes). Additionally, we train and test the student networks only on the identified subclasses.

We follow the same experimental settings as [34] to conduct our ablation experiments. For CIFAR10 and CIFAR100, we pad zeros to input images to have 40 × 40 images and randomly crop 32 × 32 cropped images. Additionally, we use random horizontal flipping for data augmentation. We utilize SGD with batch size 128, momentum 0.9, and weight decay 5 × 10-4 to train networks for 200 epochs. We also apply learning rate decay starting with 0.1 by multiplying by 0.2 at 60, 120, and 160 epochs. Finally, we use WRN-40-2 and WRN-16-1 for a teacher and a student network. For Tiny ImageNet, we resize input images to 256 × 256 and randomly crop the input images to generate 224 × 224 cropped images for network training. We also use random color jittering and horizontal flipping for data augmentation. We train the teacher and student network for 300 epochs and apply adjustable learning rates from 0.1 to small values by multiplying by 0.2 at 60, 120, 160, 200, and 250 epochs. In addition, we replicate these same experimental conditions for Resnet34 and Resnet18 network architectures as a teacher and a student network, respectively.

Table 2. The table shows top-1 accuracy comparisons for distinct subclasses of the entire class for Resnet. $\mathcal{L}_{\mathcal{CE}}$ represents the cross-entropy loss in the experiment. The hyperparameter of each loss term (e.g., $\mathcal{L}_{\mathcal{RKD}}$, $\mathcal{L}_{\mathcal{OPL}}$, and $\mathcal{L}_{\mathcal{ICV}}$) is 0.1.

Model: (Teacher, Student)	(Resnet34, Resnet18)		
Benchmark	Tiny ImageNet		
(Class Frac., Subclass Num.)	(2.5%, 5)	(5%, 10)	(25%, 50)
Teacher ($\mathcal{L}_{\mathcal{CE}}$)	56.00%	59.20%	60.08%
Student ($\mathcal{L}_{\mathcal{CE}}$)	73.60%	75.19%	67.03%
Student ($\mathcal{L}_{\mathcal{CE}} + \mathcal{L}_{\mathcal{RKD}}$)	76.80%	75.59%	67.99%
Student ($\mathcal{L}_{\mathcal{CE}} + \mathcal{L}_{\mathcal{RKD}} + \mathcal{L}_{\mathcal{OPL}}$)	**79.20%**	77.59%	**68.47%**
Student ($\mathcal{L}_{\mathcal{CE}} + \mathcal{L}_{\mathcal{RKD}} + \mathcal{L}_{\mathcal{OPL}} + \mathcal{L}_{\mathcal{ICV}}$)	78.40%	**77.99%**	68.15%

4.2 Ablation Study Experiment: Subclass Numbers for the Student

Table 3. Top-1 accuracy comparisons for a variety of hyperparameters of every penalty loss (e.g., $\mathcal{L}_{\mathcal{RKD}}$, $\mathcal{L}_{\mathcal{OPL}}$, and $\mathcal{L}_{\mathcal{ICV}}$) for WRN. All experimental results are conducted by 5 subclasses. $\mathcal{L}_{\mathcal{CE}}$ represents the cross-entropy loss in the experiment.

Model: (Teacher, Student)	(WRN-40-2, WRN-16-1)					
Benchmark	CIFAR10			CIFAR100		
Teacher ($\mathcal{L}_{\mathcal{CE}}$)	94.98%	94.98%	94.98%	81.60%	81.60%	81.60%
Student ($\mathcal{L}_{\mathcal{CE}}$)	93.68%	93.68%	93.68%	93.20%	93.20%	93.20%
Hyperparameters	0.1	0.15	0.2	0.1	0.15	0.2
Student ($\mathcal{L}_{\mathcal{CE}} + \mathcal{L}_{\mathcal{RKD}}$)	93.82%	94.14%	94.00%	93.59%	94.00%	94.19%
Student ($\mathcal{L}_{\mathcal{CE}} + \mathcal{L}_{\mathcal{RKD}} + \mathcal{L}_{\mathcal{OPL}}$)	94.42%	94.30%	94.46%	93.99%	94.39%	94.79%
Student ($\mathcal{L}_{\mathcal{CE}} + \mathcal{L}_{\mathcal{RKD}} + \mathcal{L}_{\mathcal{OPL}} + \mathcal{L}_{\mathcal{ICV}}$)	**94.66%**	94.36%	94.62%	**95.19%**	94.99%	**95.19%**

We test our method using randomly selected subclasses of varying sizes (e.g., 5, 10, 50) to better understand the effects of different subclasses for top-1 classification accuracy for task specialization on CIFAR10, CIFAR100, and Tiny ImageNet. We list empirical evaluation results in Table 1 and Table 2. As we can see, most student networks trained using CE achieve higher top-1 accuracy for small subclasses (e.g., 5, 10, and 50) compared with the teacher networks. For

CIFAR10 and CIFAR100 with 50 subclasses, the teacher has better top-1 accuracy than the top-1 accuracy of the students. We observe that the effectiveness of class specialization through knowledge distillation is sensitive to the relative number of subclasses (i.e., in proportion to the total number of teacher classes). Notably, the student performance is generally comparable to that of the teacher when the relative number of subclasses is large (e.g., 50%); however, in the case of a considerably small relative number of subclasses (e.g., 5–10%), the student performance is often dramatically better. We believe that these experimental results indicate the strong *data efficiency* potential encapsulated by the CSKD framework. Despite an extreme scarcity of "specialized" training data, it is nevertheless possible to successfully train a student network that substantially outperforms a fully-trained teacher on class specialized tasks (in one case we train a student network on only 2.5% of the Tiny ImageNet training data, however the student model renders a 40% relative improvement over the teacher).

Table 4. Top-1 accuracy comparisons for a variety of hyperparameters of every penalty loss (e.g., $\mathcal{L}_{\mathcal{RKD}}$, $\mathcal{L}_{\mathcal{OPL}}$, and $\mathcal{L}_{\mathcal{ICV}}$) for Resnet. All experimental results are conducted by 5 subclasses. $\mathcal{L}_{\mathcal{CE}}$ represents the cross-entropy loss in the experiment.

Model: (Teacher, Student)	(Resnet34, Resnet18)		
Benchmark	Tiny ImageNet		
Teacher ($\mathcal{L}_{\mathcal{CE}}$)	56.00%	56.00%	56.00%
Student ($\mathcal{L}_{\mathcal{CE}}$)	73.60%	73.60%	73.60%
Hyperparameters	0.1	0.15	0.2
Student ($\mathcal{L}_{\mathcal{CE}} + \mathcal{L}_{\mathcal{RKD}}$)	76.80%	76.80%	75.20%
Student ($\mathcal{L}_{\mathcal{CE}} + \mathcal{L}_{\mathcal{RKD}} + \mathcal{L}_{\mathcal{OPL}}$)	**79.20%**	79.20%	76.00%
Student ($\mathcal{L}_{\mathcal{CE}} + \mathcal{L}_{\mathcal{RKD}} + \mathcal{L}_{\mathcal{OPL}} + \mathcal{L}_{\mathcal{ICV}}$)	78.40%	76.80%	78.40%

4.3 Ablation Study Experiment: Hyperparameters for Loss Terms

We report the effects of varying hyperparameter values for the CIFAR10, CIFAR100, and Tiny ImageNet datasets in Table 3 and Table 4. All experimental results are conducted on five target subclasses. Setting the hyperparameter of all three loss terms (e.g., $\mathcal{L}_{\mathcal{RKD}}$, $\mathcal{L}_{\mathcal{OPL}}$, and $\mathcal{L}_{\mathcal{ICV}}$) to 0.1 achieves consistently competitive top-1 classification accuracy compared with other hyperparameter values. Specifically, the student model with all three hyperparameter values set to 0.1 achieves classification accuracy of 94.66% and 95.19% for CIFAR10 and CIFAR100, respectively. In addition, we see that most of the students utilizing all three loss functions in tandem have the highest top-1 classification accuracy compared with the student with only one or two losses. These results help validate our hypothesis that RKD, OPL, and ICV can effectively be leveraged in tandem to enhance class specialized KD performance by striking an agreeable balance between data-efficient knowledge distillation and model regularization.

In order to provide a qualitative analysis of our loss functions, we visualize the feature embeddings of the penultimate student layer by t-SNE [28] in Fig. 2 and Fig. 3. In particular, in both figures, (b) illustrates the latent structure inculcated to the student model through RKD loss; and (d) shows the effect of dense class embeddings induced by ICV.

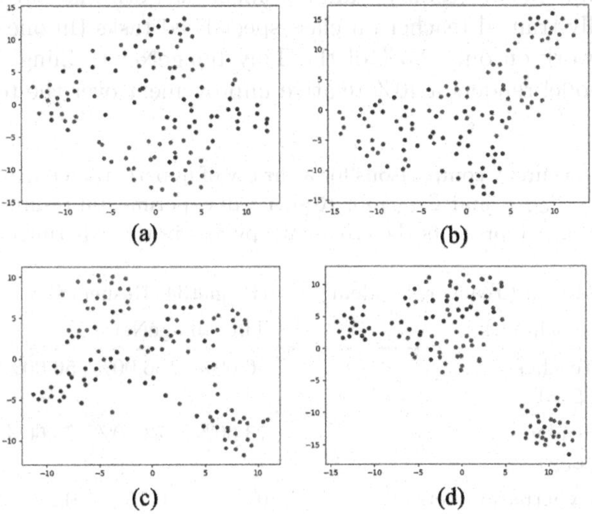

Fig. 2. Feature embedding t-SNE visualization of the penultimate layer for Tiny ImageNet with 5 target subclasses. Each plot represents: (a) $\mathcal{L}_{\mathcal{CE}}$, (b) $\mathcal{L}_{\mathcal{CE}} + \mathcal{L}_{\mathcal{RKD}}$, (c) $\mathcal{L}_{\mathcal{CE}} + \mathcal{L}_{\mathcal{RKD}} + \mathcal{L}_{\mathcal{OPL}}$, and (d) $\mathcal{L}_{\mathcal{CE}} + \mathcal{L}_{\mathcal{RKD}} + \mathcal{L}_{\mathcal{OPL}} + \mathcal{L}_{\mathcal{ICV}}$

4.4 Classification Performance Comparisons

We evaluate our proposed method for image classification with specialized image classes using four distinct benchmarks (CIFAR10, CIFAR100, Tiny ImageNet, and CUB200 [52]). In order to assess specialized image classification performance on our student networks, we randomly select five subclasses from the entire set of classes for each of the benchmark datasets. For the CIFAR10, CIFAR100, and Tiny ImageNet, datasets, the teacher network is trained on the *entire* training dataset (i.e., all classes) and tested on five selected subclasses. By contrast, the student networks are trained and tested using only the five selected subclasses. Because the CUB200 benchmark dataset consists of only 11,788 images for 200 different bird species, we acquire a pre-trained Resnet34 model on the ImageNet benchmark [41] and fine-tune this model on the CUB200 benchmark for the teacher network. For the student, we train the student network (e.g.,

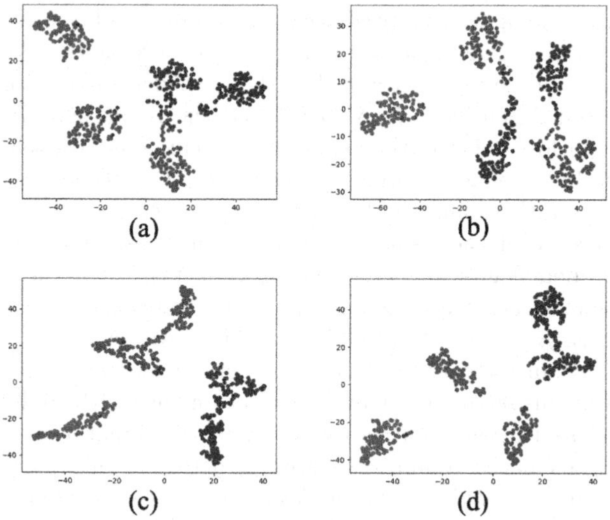

Fig. 3. Feature embedding t-SNE visualization of the penultimate layer for CIFAR100 with 5 target subclasses. Each plot represents: (a) $\mathcal{L}_{\mathcal{CE}}$, (b) $\mathcal{L}_{\mathcal{CE}} + \mathcal{L}_{\mathcal{RKD}}$, (c) $\mathcal{L}_{\mathcal{CE}} + \mathcal{L}_{\mathcal{RKD}} + \mathcal{L}_{\mathcal{OPL}}$, and (d) $\mathcal{L}_{\mathcal{CE}} + \mathcal{L}_{\mathcal{RKD}} + \mathcal{L}_{\mathcal{OPL}} + \mathcal{L}_{\mathcal{ICV}}$

Table 5. Top-1 classification accuracy (%) for image classification on CIFAR10, CIFAR100, Tiny ImageNet, and CUB200. We select hyperparameter 0.1 to the three losses in our proposed framework.

Class fraction (%)	CIFAR10	CIFAR100	Tiny ImageNet	CUB200
	50	5	2.5	2.5
Teacher	94.98%	81.60%	56.00%	63.96%
Student ($\mathcal{L}_{\mathcal{CE}}$)	93.68%	93.20%	73.60%	51.40%
Student (FitNet) [40]	93.86%	93.59%	75.20%	57.04%
Student (Attention) [57]	94.54%	94.40%	78.40%	55.63%
Student (VID) [1]	93.82%	93.79%	78.40%	59.15%
Student (ReKD) [34]	94.06%	92.79%	78.40%	59.15%
Student ($\mathcal{L}_{\mathcal{CE}} + \mathcal{L}_{\mathcal{RKD}}$) (Ours)	93.82%	93.59%	76.80%	58.45%
Student ($\mathcal{L}_{\mathcal{CE}} + \mathcal{L}_{\mathcal{RKD}} + \mathcal{L}_{\mathcal{OPL}}$) (Ours)	94.42%	93.99%	**79.20%**	57.74%
Student ($\mathcal{L}_{\mathcal{CE}} + \mathcal{L}_{\mathcal{RKD}} + \mathcal{L}_{\mathcal{OPL}}$ + $\mathcal{L}_{\mathcal{ICV}}$) (Ours)	**94.66%**	**95.19%**	78.40%	**61.97%**

Resnet18) on the images associated with five random subclasses *from scratch* and test the trained student network on the same subclasses. We evaluate our proposed losses and several state-of-the-art knowledge distillation approaches (e.g., FitNet [40], Attention [57], VID [1], and the Relational Knowledge Distillation (ReKD) loss [34]) on the same four benchmarks. We select these four approaches as the baseline techniques since these KD methods transfer the feature embeddings rather than the logits from the teacher to the student. These approaches thus avoid the issue of dimensionality misalignment between the teacher and student logits for class specialization tasks. For the hyperparameters of these approaches, we set $\lambda_{Attention}$ to 50. Additionally, we set λ_{ReKD-D} to 25and λ_{ReKD-D} to 50 in the ReKD loss. Then, we set λ_{FitNet} and λ_{VID} to 0.1. We use the same setting mentioned in Sect. 4.1 for model training and use the same experiment setting for Tiny ImageNet for the CUB200 dataset. Table 5 shows top-1 classification accuracy on the four benchmarks using five target subclasses. As we can see, a number of student networks outperform the teacher networks on *well-calibrated specialized classification tasks*, particularly when the class fraction percentage (i.e., number of subclasses divided by the total number of classes) is relatively low and the class representation is balanced. From these results, CSKD either achieves competitive classification performance or outperforms existing state-of-the-art knowledge distillation approaches for the tested class specialization tasks. Moreover, in each experiment, CSKD demonstrates significant performance improvements over a "naive" student network trained solely using $\mathcal{L}_{\mathcal{CE}}$. In the case of the challenging CUB200 dataset, despite the absence of pre-trained features, the more compact CSKD model utilizing our total loss function ($\mathcal{L}_{\mathcal{CE}} + \mathcal{L}_{\mathcal{RKD}} + \mathcal{L}_{\mathcal{OPL}} + \mathcal{L}_{\mathcal{ICV}}$) yields over 20% relative performance improvement over a naive student, and performs only marginally worse than the larger, pre-trained teacher model for the class specialization task. These results empirically validate the essential thrust of our CSKD approach. In place of greedily transferring the knowledge of feature embeddings of intermediate layers from the teacher to the student, our proposed approach instead (1) only transfers specific teacher logits to the student network and (2) regularizes the student network by concurrently enforcing orthogonality in between the classes and minimizing intra-class variance.

5 Conclusion

In this paper, we presented the CSKD framework combining RKD and ICV loss functions to render a compact and performant student network for the class specialization problem setting. The proposed RKD loss improves the efficiency of KD for class specialized tasks by transferring only the relevant portion of the teacher output of the teacher to a specialized student network. Additionally, ICV loss enforces spatially dense feature embeddings by minimizing class-wise variance. CSKD consistently outperforms other knowledge distillation approaches for specialized student networks.

Acknowledgments. We thank all the paper reviewers who provided constructive and knowledgeable feedback on our work for making our manuscript publishable.

References

1. Ahn, S., Hu, S.X., Damianou, A., Lawrence, N.D., Dai, Z.: Variational information distillation for knowledge transfer. In: Proceedings of the IEEE/CVF Conference on Computer Vision and Pattern Recognition, pp. 9163–9171 (2019)
2. Ba, J., Caruana, R.: Do deep nets really need to be deep? In: Advances in Neural Information Processing Systems, vol. 27 (2014)
3. Bargoti, S., Underwood, J.: Deep fruit detection in orchards. In: 2017 IEEE International Conference on Robotics and Automation (ICRA), pp. 3626–3633. IEEE (2017)
4. Bjerge, K., Nielsen, J.B., Sepstrup, M.V., Helsing-Nielsen, F., Høye, T.T.: An automated light trap to monitor moths (lepidoptera) using computer vision-based tracking and deep learning. Sensors **21**(2), 343 (2021)
5. Chan, T.H., Jia, K., Gao, S., Lu, J., Zeng, Z., Ma, Y.: PCANet: a simple deep learning baseline for image classification? IEEE Trans. Image Process. **24**(12), 5017–5032 (2015)
6. Choi, Y., Choi, J., El-Khamy, M., Lee, J.: Data-free network quantization with adversarial knowledge distillation. In: Proceedings of the IEEE/CVF Conference on Computer Vision and Pattern Recognition Workshops, pp. 710–711 (2020)
7. Cinbis, R.G., Verbeek, J., Schmid, C.: Weakly supervised object localization with multi-fold multiple instance learning. IEEE Trans. Pattern Anal. Mach. Intell. **39**, 189–203 (2016)
8. Dias, P.A., Tabb, A., Medeiros, H.: Apple flower detection using deep convolutional networks. Comput. Ind. **99**, 17–28 (2018)
9. Dietterich, T.G., et al.: Ensemble learning. In: The Handbook of Brain Theory and Neural Networks, vol. 2, no. 1, pp. 110–125 (2002)
10. Farabet, C., Couprie, C., Najman, L., LeCun, Y.: Learning hierarchical features for scene labeling. IEEE Trans. Pattern Anal. Mach. Intell. **35**(8), 1915–1929 (2012)
11. Frankle, J., Carbin, M.: The lottery ticket hypothesis: finding sparse, trainable neural networks. arXiv preprint arXiv:1803.03635 (2018)
12. Gabbay, F., Shomron, G.: Compression of neural networks for specialized tasks via value locality. Mathematics **9**, 2612 (2021)
13. Ghorbani, A., et al.: Deep learning interpretation of echocardiograms. NPJ Digit. Med. **3**(1), 1–10 (2020)
14. Guo, Y., Liu, Y., Oerlemans, A., Lao, S., Wu, S., Lew, M.S.: Deep learning for visual understanding: a review. Neurocomputing **187**, 27–48 (2016)
15. Han, S., Pool, J., Tran, J., Dally, W.: Learning both weights and connections for efficient neural network. In: Advances in Neural Information Processing Systems, vol. 28 (2015)
16. He, K., Zhang, X., Ren, S., Sun, J.: Deep residual learning for image recognition. In: Proceedings of the IEEE Conference on Computer Vision and Pattern Recognition, pp. 770–778 (2016)
17. Hinton, G., Vinyals, O., Dean, J., et al.: Distilling the knowledge in a neural network. arXiv preprint arXiv:1503.02531 (2015)
18. Huang, Z., Wang, N.: Like what you like: knowledge distill via neuron selectivity transfer. arXiv preprint arXiv:1707.01219 (2017)

19. Iandola, F.N., Han, S., Moskewicz, M.W., Ashraf, K., Dally, W.J., Keutzer, K.: Squeezenet: alexnet-level accuracy with 50x fewer parameters and < 0.5 mb model size. arXiv preprint arXiv:1602.07360 (2016)
20. Kang, D., Emmons, J., Abuzaid, F., Bailis, P., Zaharia, M.: Noscope: optimizing neural network queries over video at scale. arXiv preprint arXiv:1703.02529 (2017)
21. Kao, W.C., Xie, H.X., Lin, C.Y., Cheng, W.H.: Specific expert learning: enriching ensemble diversity via knowledge distillation. IEEE Trans. Cybern. (2021)
22. Kosaian, J., Phanishayee, A., Philipose, M., Dey, D., Vinayak, R.: Boosting the throughput and accelerator utilization of specialized CNN inference beyond increasing batch size. In: International Conference on Machine Learning, pp. 5731–5741. PMLR (2021)
23. Le, Y., Yang, X.: Tiny imagenet visual recognition challenge. CS 231N **7**, 3 (2015)
24. LeCun, Y., Bengio, Y., Hinton, G.: Deep learning. Nature **521**(7553), 436–444 (2015)
25. Lee, K., Shrivastava, A., Kacorri, H.: Hand-priming in object localization for assistive egocentric vision. In: Proceedings of the IEEE/CVF Winter Conference on Applications of Computer Vision, pp. 3422–3432 (2020)
26. Lee, S.H., Chan, C.S., Mayo, S.J., Remagnino, P.: How deep learning extracts and learns leaf features for plant classification. Pattern Recogn. **71**, 1–13 (2017)
27. Luo, J.H., Wu, J., Lin, W.: Thinet: a filter level pruning method for deep neural network compression. In: Proceedings of the IEEE International Conference on Computer Vision, pp. 5058–5066 (2017)
28. Van der Maaten, L., Hinton, G.: Visualizing data using t-SNE. J. Mach. Learn. Res. **9**(11) (2008)
29. Malinin, A., Mlodozeniec, B., Gales, M.: Ensemble distribution distillation. arXiv preprint arXiv:1905.00076 (2019)
30. Mirzadeh, S.I., Farajtabar, M., Li, A., Levine, N., Matsukawa, A., Ghasemzadeh, H.: Improved knowledge distillation via teacher assistant. In: Proceedings of the AAAI Conference on Artificial Intelligence, vol. 34, pp. 5191–5198 (2020)
31. Morgado, P., Vasconcelos, N.: Nettailor: tuning the architecture, not just the weights. In: Proceedings of the IEEE/CVF Conference on Computer Vision and Pattern Recognition, pp. 3044–3054 (2019)
32. Müller, R., Kornblith, S., Hinton, G.: Subclass distillation. arXiv preprint arXiv:2002.03936 (2020)
33. Papandreou, G., Zhu, T., Chen, L.C., Gidaris, S., Tompson, J., Murphy, K.: Personlab: person pose estimation and instance segmentation with a bottom-up, part-based, geometric embedding model. In: Proceedings of the European Conference on Computer Vision (ECCV), pp. 269–286 (2018)
34. Park, W., Kim, D., Lu, Y., Cho, M.: Relational knowledge distillation. In: Proceedings of the IEEE/CVF Conference on Computer Vision and Pattern Recognition, pp. 3967–3976 (2019)
35. Perez, L., Wang, J.: The effectiveness of data augmentation in image classification using deep learning. arXiv preprint arXiv:1712.04621 (2017)
36. Polikar, R.: Ensemble learning. In: Zhang, C., Ma, Y. (eds.) Ensemble Machine Learning, pp. 1–34. Springer, Boston (2012). https://doi.org/10.1007/978-1-4419-9326-7_1
37. Ranasinghe, K., Naseer, M., Hayat, M., Khan, S., Khan, F.S.: Orthogonal projection loss. In: Proceedings of the IEEE/CVF International Conference on Computer Vision, pp. 12333–12343 (2021)

38. Rastegari, M., Ordonez, V., Redmon, J., Farhadi, A.: XNOR-Net: ImageNet classification using binary convolutional neural networks. In: Leibe, B., Matas, J., Sebe, N., Welling, M. (eds.) ECCV 2016. LNCS, vol. 9908, pp. 525–542. Springer, Cham (2016). https://doi.org/10.1007/978-3-319-46493-0_32
39. Ravoor, P.C., Sudarshan, T.: Deep learning methods for multi-species animal re-identification and tracking-a survey. Comput. Sci. Rev. **38**, 100289 (2020)
40. Romero, A., Ballas, N., Kahou, S.E., Chassang, A., Gatta, C., Bengio, Y.: Fitnets: hints for thin deep nets. arXiv preprint arXiv:1412.6550 (2014)
41. Russakovsky, O., et al.: Imagenet large scale visual recognition challenge. Int. J. Comput. Vision **115**(3), 211–252 (2015)
42. Salehi, M., Sadjadi, N., Baselizadeh, S., Rohban, M.H., Rabiee, H.R.: Multiresolution knowledge distillation for anomaly detection. In: Proceedings of the IEEE/CVF Conference on Computer Vision and Pattern Recognition, pp. 14902–14912 (2021)
43. Shen, C., Wang, X., Song, J., Sun, L., Song, M.: Amalgamating knowledge towards comprehensive classification. In: Proceedings of the AAAI Conference on Artificial Intelligence, vol. 33, pp. 3068–3075 (2019)
44. Shen, C., Xue, M., Wang, X., Song, J., Sun, L., Song, M.: Customizing student networks from heterogeneous teachers via adaptive knowledge amalgamation. In: Proceedings of the IEEE/CVF International Conference on Computer Vision, pp. 3504–3513 (2019)
45. Shen, H., Han, S., Philipose, M., Krishnamurthy, A.: Fast video classification via adaptive cascading of deep models. In: Proceedings of the IEEE Conference on Computer Vision and Pattern Recognition, pp. 3646–3654 (2017)
46. Shrestha, A., Mahmood, A.: Review of deep learning algorithms and architectures. IEEE Access **7**, 53040–53065 (2019)
47. Singh, K.K., Lee, Y.J.: Hide-and-seek: forcing a network to be meticulous for weakly-supervised object and action localization. In: 2017 IEEE International Conference on Computer Vision (ICCV), pp. 3544–3553. IEEE (2017)
48. Syafrudin, M., Alfian, G., Fitriyani, N.L., Rhee, J.: Performance analysis of IoT-based sensor, big data processing, and machine learning model for real-time monitoring system in automotive manufacturing. Sensors **18**(9), 2946 (2018)
49. Teh, E.W., Rochan, M., Wang, Y.: Attention networks for weakly supervised object localization. In: BMVC, pp. 1–11 (2016)
50. Tsai, Y.-H., Zhong, G., Yang, M.-H.: Semantic co-segmentation in videos. In: Leibe, B., Matas, J., Sebe, N., Welling, M. (eds.) ECCV 2016. LNCS, vol. 9908, pp. 760–775. Springer, Cham (2016). https://doi.org/10.1007/978-3-319-46493-0_46
51. Tung, F., Mori, G.: Similarity-preserving knowledge distillation. In: Proceedings of the IEEE/CVF International Conference on Computer Vision, pp. 1365–1374 (2019)
52. Wah, C., Branson, S., Welinder, P., Perona, P., Belongie, S.: The Caltech-UCSD birds-200-2011 dataset (2011)
53. Wang, Y., Zhou, W., Jiang, T., Bai, X., Xu, Y.: Intra-class feature variation distillation for semantic segmentation. In: Vedaldi, A., Bischof, H., Brox, T., Frahm, J.-M. (eds.) ECCV 2020. LNCS, vol. 12352, pp. 346–362. Springer, Cham (2020). https://doi.org/10.1007/978-3-030-58571-6_21
54. Yim, J., Joo, D., Bae, J., Kim, J.: A gift from knowledge distillation: fast optimization, network minimization and transfer learning. In: Proceedings of the IEEE Conference on Computer Vision and Pattern Recognition, pp. 4133–4141 (2017)

55. Yu, X., Liu, T., Wang, X., Tao, D.: On compressing deep models by low rank and sparse decomposition. In: Proceedings of the IEEE Conference on Computer Vision and Pattern Recognition, pp. 7370–7379 (2017)
56. Yun, S., Park, J., Lee, K., Shin, J.: Regularizing class-wise predictions via self-knowledge distillation. In: Proceedings of the IEEE/CVF Conference on Computer Vision and Pattern Recognition, pp. 13876–13885 (2020)
57. Zagoruyko, S., Komodakis, N.: Paying more attention to attention: Improving the performance of convolutional neural networks via attention transfer. arXiv preprint arXiv:1612.03928 (2016)
58. Zagoruyko, S., Komodakis, N.: Wide residual networks. arXiv preprint arXiv:1605.07146 (2016)
59. Zaras, A., Passalis, N., Tefas, A.: Improving knowledge distillation using unified ensembles of specialized teachers. Pattern Recogn. Lett. **146**, 215–221 (2021)
60. Zhang, J., et al.: Fully automated echocardiogram interpretation in clinical practice: feasibility and diagnostic accuracy. Circulation **138**(16), 1623–1635 (2018)
61. Zhang, L., Bao, C., Ma, K.: Self-distillation: towards efficient and compact neural networks. IEEE Trans. Pattern Anal. Mach. Intell. **44**(8), 4388–4403 (2021)
62. Zheng, Z., Peng, X.: Self-guidance: improve deep neural network generalization via knowledge distillation. In: Proceedings of the IEEE/CVF Winter Conference on Applications of Computer Vision, pp. 3203–3212 (2022)
63. Zhong, Z., Li, J., Luo, Z., Chapman, M.: Spectral-spatial residual network for hyperspectral image classification: a 3-D deep learning framework. IEEE Trans. Geosci. Remote Sens. **56**(2), 847–858 (2017)
64. Zhu, M., Gupta, S.: To prune, or not to prune: exploring the efficacy of pruning for model compression. arXiv preprint arXiv:1710.01878 (2017)
65. Zhu, Y., Wang, Y.: Student customized knowledge distillation: bridging the gap between student and teacher. In: Proceedings of the IEEE/CVF International Conference on Computer Vision, pp. 5057–5066 (2021)

Gestalt-Guided Image Understanding for Few-Shot Learning

Kun Song, Yuchen Wu, Jiansheng Chen, Tianyu Hu, and Huimin Ma(✉)

University of Science and Technology Beijing, Beijing, China
{songkun,yuchen.wu}@xs.ustb.edu.cn, {jschen,Tianyu,mhmpub}@ustb.edu.cn

Abstract. Due to the scarcity of available data, deep learning does not perform well on few-shot learning tasks. However, human can quickly learn the feature of a new category from very few samples. Nevertheless, previous work has rarely considered how to mimic human cognitive behavior and apply it to few-shot learning. This paper introduces Gestalt psychology to few-shot learning and proposes Gestalt-Guided Image Understanding, a plug-and-play method called **GGIU**. Referring to the principle of totality and the law of closure in Gestalt psychology, we design Totality-Guided Image Understanding and Closure-Guided Image Understanding to extract image features. After that, a feature estimation module is used to estimate the accurate features of images. Extensive experiments demonstrate that our method can improve the performance of existing models effectively and flexibly without retraining or fine-tuning. Our code is released on https://github.com/skingorz/GGIU.

1 Introduction

In recent years, deep learning has shown surprising performance in various fields. Nevertheless, deep learning often relies on large amounts of training data. More and more pre-trained models are based on large-scale data. For example, CLIP [1] is trained on 400 million image-text pairs. However, a large amount of data come with extra costs in deep learning procedures, such as collection, annotation, and training. In addition, many kinds of data, such as medical image data, requires specialized knowledge to annotate. Data for some rare scenes are hard to obtain, such as car accidents. Therefore, there is a growing interest in training a better model using fewer data. Motivated by this, Few-Shot Learning (FSL) [2,3] is proposed to solve the problem of learning from small amounts of data.

The most significant obstacle to few-shot learning is the lack of data. In order to address this obstacle, existing few-shot learning approaches mainly employ metric learning, such as PN [4], and meta-learning, such as MAML [5]. Regardless of the technique, the ultimate goal is to extract more robust features for novel classes. Previous researches mainly focus on two aspects: designing a more robust feature extractor to represent the image feature better, such as meta-baseline [6], and using more dense features, such as DeepEMD [7]. However, few people consider how to mimic human learning patterns to enhance the effectiveness of few-shot learning.

L. Wang et al. (Eds.): ACCV 2022, LNCS 13842, pp. 409–424, 2023.
https://doi.org/10.1007/978-3-031-26284-5_25

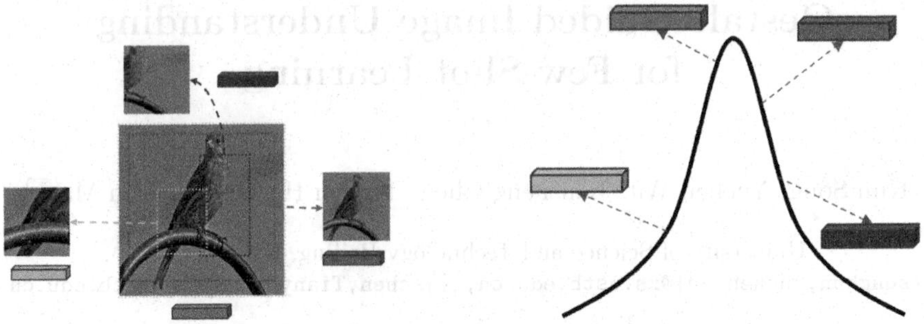

Fig. 1. Describe an image using a multivariate Gaussian distribution.

Modern psychology has extensively studied the mechanisms of human cognition. Gestalt psychology is one of them. On the one hand, Gestalt psychology states that conscious experience must be considered globally, which is the principle of totality of gestaltism. For example, as shown in the left of Fig. 1, the whole picture shows a bird standing on a tree branch. Its feature can represent the image. Meanwhile, given any patch of the image, human can easily determine that it is a part of a bird. These can be explained by the law of closure in Gestalt psychology: When parts of a whole picture are missing, our perception fills in the visual gap. Therefore, the feature of patches can also represent the image.

Motivated by Gestalt psychology, we imitate human learning patterns to redesign the image understanding process and apply it to few-shot learning. In this paper, we innovatively describe the image as a data distribution to represent the principle of totality and the law of closure in Gestalt psychology. We assume that image can be represented by a corresponding multivariate Gaussian distribution. As shown in the right of Fig. 1, the feature of the image can be considered as a sample of the potential distribution. Likewise, the feature of the largest patch, the image itself, is also a sample of the distribution. Then we design a feature estimation module with reference to Kalman filter to estimate the features of images accurately.

The main contributions of this paper are:

1. We introduce Gestalt psychology into the process of image understanding and propose a plug-and-play method without retraining or fine-tuning, called **GGIU**.
2. We innovatively propose to use multivariate Gaussian distribution to describe the image and design a feature estimation module with reference to Kalman filter to estimate image feature accurately.
3. We conduct extensive experiments to demonstrate the applicability of our method to a variety of different few-shot classification tasks. The experiment results demonstrate the robustness and scalability of our method.

2 Related Work

2.1 Few-Shot Learning

Most of the existing methods for few-shot learning is based on the meta-learning [8] framework. The motivation of meta-learning is learning to learn [9]. The network is trained on a set of meta-tasks during the training process to gain the ability to adapt itself to different tasks. The meta-learning methods for few-shot learning are mainly divided into two categories, optimization-based and metric-based.

Koch [10] apply metric learning to few-shot learning for the first time. They proposed to apply Siamese Neural Network to one-shot learning. A pair of convolutional neural networks with shared weights is used to extract the embedding features of each class separately. When inferring the category of unknown data, the unlabeled data and the training set samples are paired. The Manhattan Distance between unlabeled data and training data is calculated as the similarity, and the category with the highest similarity is used as the prediction of the samples. Matching Network [3] first conducts experiments on *mini*ImageNet for few-shot learning. It proposes an attention module that uses cosine distance to determine the similarity between the target object and each category and uses the similarity for the final classification. Prototypical Network [4] proposed the concept of category prototypes. Prototypical Network takes a class's prototype to be the mean of its support set in the embedding space. The similarity between the unknown data and each category's prototypes are measured, and the most similar category is selected as the final classification result. Satorras [11] uses graphical convolutional networks to transfer information between support and query sets and extended prototypical networks and matching networks to non-euclidean spaces to assist few-shot learning. DeepEmd [7] adopt the Earth Mover's Distance (EMD) as a metric to compute a structural distance between dense image representations to determine image relevance. COSOC [12] extracts image foregrounds using contrast learning to optimize the category prototypes. Yang [13] takes the perspective of distribution estimation to rectify the category prototype. SNE [14] encodes the latent distribution transferring from the already-known classes to the novel classes by label propagation and self-supervised learning. CSS [15] propose conditional self-supervised learning with a 3-stage training pipeline. CSEI [16] proposed an Erasing-then-Inpainting method to augment the data while training, which needs to retrain the model. AA [17] expands the novel data by adding extra "related base" data to few novel ones and fine-tunes the model.

2.2 Gestalt Psychology

Gestalt psychology is a psychology school that emerged in Austria and Germany in the early twentieth century. Gestalt principles, proximity, similarity, figure-ground, continuity, closure, and connection describe how humans perceive visuals in connection with different objects and environments. In this section, we mainly introduce the principle of totality and the law of closure.

Fig. 2. Although we see incomplete figures, our brain can easily complete them and regard them as rectangles (left) and circles (right).

Principle of Totality. The principle of totality points out that conscious experience must be considered globally. Wertheimer [18] described holism as fundamental to Gestalt psychology. Kohler [19] thinks: "In psychology, instead of being the sum of parts existing independently, wholes give their parts specific functions or properties that can only be defined in relation to the whole in question." Thus, the maxim that the whole is more than the sum of its parts is not a precise description of the Gestaltist [18].

Law of Closure. Gestalt psychologists held the view that humans often consider objects as complete rather than focusing on the gaps they may have. [20] For example, a circle has a good Gestalt in terms of completeness. However, we may also consider an incomplete circle as a complete one. This tendency to complete shapes and figures is called closure. [21] The law of closure states that even incomplete objects, such as forms, characters, and pictures, are seen as complete by people. In particular, when part of the entire picture is missing, our perception fills in the visual gaps. For example, as shown in Fig. 2, despite the incomplete shape, we still perceive a rectangle and a circle. If the law of closure did not exist, the image would depict different lines with different lengths, rotations, and curvatures. However, because of the law of closure, we perceptually combine the lines into whole shapes [22].

3 Method

3.1 Problem Definition

The dataset for few-shot learning consists of training set $\mathcal{D}_B$ and testing set $\mathcal{D}_V$ with no shared classes. We train a feature extractor $f_\theta(\cdot)$ on $\mathcal{D}_B$ containing lots of labeled data. During the evaluation, many N-way K-shot Q-query tasks $\mathcal{T} = \{(\mathcal{S}, \mathcal{Q})\}$ are constructed from $\mathcal{D}_V$. Each task contains a support set $\mathcal{S} = \{(x_i, y_i)\}_{i=1}^{K \times N}$ and a query set $\mathcal{Q}$. Firstly, N classes in $\mathcal{D}_V$ are randomly sampled for $\mathcal{S}$ and $\mathcal{Q}$. Then we calculate a classifier for N-way classification for each task based on $f_\theta(\cdot)$ and $\mathcal{S}$. At last, we calculate the feature of the query image $x \in \mathcal{Q}$ and classify it into one of the N class.

3.2 Metric-Based Few-Shot Learning Pipeline

Calculating a better representation for all classes in metric-based few-shot learning is critical. Under most circumstances, the features of all images in each category of the support set are calculated separately. The mean of features is used as the category representation, the prototype.

$$\boldsymbol{P_n} = \frac{1}{K} \sum_{(x_i, y_i) \in \mathcal{S}_n} f_\theta(x_i) \tag{1}$$

In Eq. 1, $\boldsymbol{P_n}$ represents the prototype of the n-th category and $\mathcal{S}_n$ represents the set of all images of the n-th class. When an image $x \in \mathcal{Q}$ and a distance function $d(,)$ are given, the feature extractor $f_\theta(\cdot)$ is used to calculate the feature of x. Then we calculate the distance between $f_\theta(x)$ and the prototypes in the embedding space. After that, a softmax function (Eq. 2) is used to calculate a distribution for classification.

$$p_\theta(y = n|x) = \frac{exp(-d(f_\theta(x), \boldsymbol{P_n}))}{\sum_{i=1}^{N} exp(-d(f_\theta(x), \boldsymbol{P_i}))} \tag{2}$$

Finally, the model is optimized by minimizing the negative log-likelihood $L(\theta) = -\log p_\theta(y = n|x)$, i.e.,

$$NLL(\theta) = d(f_\theta(x), \boldsymbol{P_n}) + \log \sum_{i=1}^{k} exp(-d(f_\theta(x), \boldsymbol{P_i})) \tag{3}$$

3.3 Gestalt Guide Image Understanding

Previous works mainly calculate prototypes with the features of images. Nevertheless, due to the limit of data volume, the image feature does not represent the information of the class well. In this section, inspired by Gestalt psychology, we reconceptualize images in terms of the principle of totality and the law of closure separately.

Given one image I, many patches of different sizes can be cropped from it. As shown in the right of Fig. 1, we assume that all patches follow a potential multivariate Gaussian distribution. All the patches can be regarded as samples of this distribution. Likewise, the largest patch of this image, the image itself, is also a sample of the distribution. Therefore, this potential Gaussian distribution can describe the image. The above can be expressed as follows: given a feature extractor $f_\theta(\cdot)$, for any patch p cropped from image I, its feature follows a multivariate Gaussian distribution, i.e., $f_\theta(p) \in \mathbb{R}^D$ and $f_\theta(p) \sim N_I(\boldsymbol{\mu_I}, \boldsymbol{\Sigma_I^2})$. Next, $\boldsymbol{\mu_I}$ can represent the feature of I. Finally, we estimate the image feature by the principle of totality and the law of closure.

Totality-Guided Image Understanding. The existing image understanding processes in few-shot learning are most from the totality of the image. The image can be considered as a sample of the potential multivariate Gaussian distribution, i.e., $f_\theta(I) \sim N_I(\boldsymbol{\mu_I}, \boldsymbol{\Sigma_I^2})$. Therefore, $N_I(\boldsymbol{\mu_I}, \boldsymbol{\Sigma_I^2})$ can be estimated by $f_\theta(I)$ (Eq. 4). $\boldsymbol{\mu_t}$ represents the estimate guided by the principle of totality.

$$\boldsymbol{\mu_t} = \hat{\boldsymbol{\mu_I}} = f_\theta(I) \tag{4}$$

Closure-Guided Image Understanding. Guided by the law of closure, we randomly crop patches from images as the sample of the potential distribution of image. For any patch $p \in I$, $f_\theta(p) \in \mathbb{R}^D$, $f_\theta(p) \sim N_I(\boldsymbol{\mu_I}, \boldsymbol{\Sigma_I^2})$. The joint probability density function of the feature of the i-th patch is shown as Eq. 5.

$$p_{N_I}(f_\theta(p_i)) = \frac{1}{(2\pi)^{\frac{D}{2}}|\boldsymbol{\Sigma}|^{-\frac{1}{2}}} \exp\left(-\frac{1}{2}\left(f_\theta(p_i) - \boldsymbol{\mu_I}\right)^T \boldsymbol{\Sigma}^{-1}\left(f_\theta(p_i) - \boldsymbol{\mu_I}\right)\right) \tag{5}$$

The log-likelihood function is:

$$\begin{aligned} \ell_{N_I}(\boldsymbol{\mu_I}, \boldsymbol{\Sigma_I}; p_1, \ldots, p_M) = &-\frac{MD}{2} log(2\pi) - \frac{M}{2} log(|\boldsymbol{\Sigma_I}|) \\ &-\frac{1}{2}\sum_{j=1}^{M}(p_i - \boldsymbol{\mu_I})^T \boldsymbol{\Sigma_I}^{-1}(p_i - \boldsymbol{\mu_I}) \end{aligned} \tag{6}$$

Solve the following maximization problem

$$\max_{\boldsymbol{\mu_I}, \boldsymbol{\Sigma_I}} \ell_{N_I}(\boldsymbol{\mu_I}, \boldsymbol{\Sigma_I}; p_1, \ldots, p_M) \tag{7}$$

Then we have

$$\boldsymbol{\mu_c} = \hat{\boldsymbol{\mu_I}} = \frac{1}{M}\sum_{i=1}^{M} f_\theta(p_i) \tag{8}$$

$\boldsymbol{\mu_c}$ represents the estimation guided by the law of closure.

3.4 Feature Estimation

We regard the process of estimating image feature guided by the totality and closure as two different observers following multivariate Gaussian distribution: O_t and O_c. The observations of O_t, O_c are $\boldsymbol{\mu_t} \in \mathbb{R}^{D\times1}$, $\boldsymbol{\mu_c} \in \mathbb{R}^{D\times1}$ and their random errors are $\boldsymbol{e_t}$, $\boldsymbol{e_c}$ respectively, where $\boldsymbol{e_t} \in \mathbb{R}^{D\times1}$ and $\boldsymbol{e_c} \in \mathbb{R}^{D\times1}$. $\boldsymbol{e_t}$ and $\boldsymbol{e_c}$ follow multivariate Gaussian distribution, i.e., $\boldsymbol{e_t} \sim N(\boldsymbol{0}, \boldsymbol{\Sigma_t^2})$, $\boldsymbol{e_c} \sim N(\boldsymbol{0}, \boldsymbol{\Sigma_c^2})$, where $\boldsymbol{\Sigma_t} \in \mathbb{R}^{D\times D}$ and $\boldsymbol{\Sigma_c} \in \mathbb{R}^{D\times D}$. In this section, we use Kalman filter to estimate image features $\boldsymbol{f} \in \mathbb{R}^{D\times1}$.

For O_t and O_c, we have

$$\boldsymbol{f} = \boldsymbol{\mu}_t + \boldsymbol{e}_t \tag{9}$$

$$\boldsymbol{\mu}_c = \boldsymbol{f} + \boldsymbol{e}_c \tag{10}$$

The prior estimates of $\boldsymbol{f}$ under O_t and O_c are

$$\hat{\boldsymbol{f}_t^-} = \boldsymbol{\mu}_t \tag{11}$$

$$\hat{\boldsymbol{f}_c^-} = \boldsymbol{\mu}_c \tag{12}$$

The feature $\boldsymbol{f}$ can be estimate by the prior estimates under O_t and O_c, we have

$$\hat{\boldsymbol{f}} = \hat{\boldsymbol{f}_c^-} + \boldsymbol{\lambda}(\hat{\boldsymbol{f}_t^-} - \hat{\boldsymbol{f}_c^-}) \tag{13}$$

i.e.,

$$\hat{\boldsymbol{f}} = \boldsymbol{\mu}_c + \boldsymbol{\lambda}(\boldsymbol{\mu}_t - \boldsymbol{\mu}_c) \tag{14}$$

$\boldsymbol{\lambda} = diag(\lambda_1, \lambda_2, \ldots, \lambda_D)$, where λ_i is a diagonal matrix, ranging from 0 to $\boldsymbol{I}$. The error between $\hat{\boldsymbol{f}}$ and $\boldsymbol{f}$ is

$$\boldsymbol{e} = \boldsymbol{f} - \hat{\boldsymbol{f}} \tag{15}$$

The error $\boldsymbol{e}$ follows a multivariate Gaussian distribution, i.e., $\boldsymbol{e} \sim N(\boldsymbol{0}, \boldsymbol{\Sigma}_e)$. where

$$\begin{aligned}\boldsymbol{\Sigma}_e &= E(\boldsymbol{e}\boldsymbol{e}^T) &(16)\\ &= E[(\boldsymbol{f} - \hat{\boldsymbol{f}})(\boldsymbol{f} - \hat{\boldsymbol{f}})^T] &(17)\\ &= E[(\boldsymbol{\lambda}\boldsymbol{e}^- - (\boldsymbol{I} - \boldsymbol{\lambda}\boldsymbol{e}_c))(\boldsymbol{\lambda}\boldsymbol{e}^- - (\boldsymbol{I} - \boldsymbol{\lambda}\boldsymbol{e}_c))^T] &(18)\end{aligned}$$

$\boldsymbol{e}^-$ represents the prior estimation of $\boldsymbol{e}$, Since $\boldsymbol{e}^-$ and $\boldsymbol{e}_c$ are independent of each other, we have $E(\boldsymbol{e}_c\boldsymbol{e}^-) = E(\boldsymbol{e}^-)E(\boldsymbol{e}_c) = 0$. Therefore, we have

$$\begin{aligned}\boldsymbol{\Sigma}_e &= \boldsymbol{\lambda}E(\boldsymbol{e}^-\boldsymbol{e}^{-T})\boldsymbol{\lambda}^T + (\boldsymbol{I} - \boldsymbol{\lambda})E(\boldsymbol{e}_c\boldsymbol{e}_c^T)(\boldsymbol{I} - \boldsymbol{\lambda})^T &(19)\\ &= \boldsymbol{\lambda}\boldsymbol{\Sigma}^-\boldsymbol{\lambda}^T + (\boldsymbol{I} - \boldsymbol{\lambda})\boldsymbol{\Sigma}_c(\boldsymbol{I} - \boldsymbol{\lambda})^T &(20)\end{aligned}$$

In order to estimate $\boldsymbol{f}$ accurately, we have to minimize $\boldsymbol{e}$, i.e.,

$$\min_{\lambda}\ tr(\boldsymbol{\Sigma}_e) = \min_{\lambda}\ [tr(\boldsymbol{\lambda}\boldsymbol{\Sigma}^-\boldsymbol{\lambda}^T) - 2tr(\boldsymbol{\lambda}\boldsymbol{\Sigma}_c) + tr(\boldsymbol{\lambda}\boldsymbol{\Sigma}_c\boldsymbol{\lambda}^T)] \tag{21}$$

We need to solve this equation:

$$\frac{\partial \boldsymbol{\Sigma}_e}{\partial \boldsymbol{\lambda}} = 0 \tag{22}$$

We have

$$\boldsymbol{\lambda} = (\boldsymbol{\Sigma}^-\boldsymbol{\Sigma}_c^{-1} + \boldsymbol{I})^{-1} \tag{23}$$

where

$$\boldsymbol{\Sigma}^- = E(\boldsymbol{e}^-\boldsymbol{e}^{-T}) \tag{24}$$
$$= E[(\boldsymbol{f} - \hat{\boldsymbol{f}})(\boldsymbol{f} - \hat{\boldsymbol{f}})^T] \tag{25}$$
$$= E[(\boldsymbol{\mu}_t + \boldsymbol{e}_t - \boldsymbol{\mu}_t)(\boldsymbol{\mu}_t + \boldsymbol{e}_t - \boldsymbol{\mu}_t)^T] \tag{26}$$
$$= \boldsymbol{\Sigma}_t \tag{27}$$

Therefore

$$\boldsymbol{\lambda} = (\boldsymbol{\Sigma}_t\boldsymbol{\Sigma}_c^{-1} + \boldsymbol{I})^{-1} \tag{28}$$

Although the error covariance matrix $\boldsymbol{\Sigma}_t$ and $\boldsymbol{\Sigma}_c$ of the observers O_t and O_c cannot be calculated, the relationship between $\boldsymbol{\lambda}$ and the number of patches can still be estimated, which can assist us in choosing the parameter. When the number of patches is large enough, the Closure-Guided image understanding can estimate the image features accurately. At this time, $\boldsymbol{\Sigma}_c$ is close to $\mathbf{0}$. According to Eq. 28, $\boldsymbol{\lambda}$ is close to $\mathbf{0}$. As the number of patches decreases, $\boldsymbol{\Sigma}_c$ gradually increases, and $\boldsymbol{\lambda}$ is close to $\boldsymbol{I}$.

3.5 The Overview of Our Approach

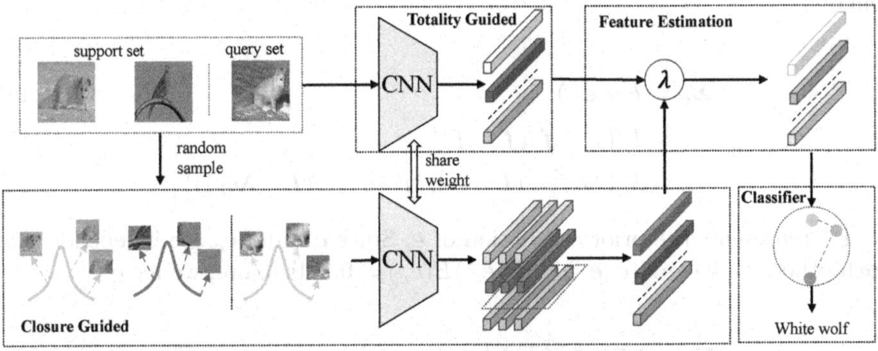

Fig. 3. There are two branches in our method. Totality-Guided module extracts the feature of the whole image. Guided by Closure-Guided module, image features are estimated from incomplete images. After that, we use feature estimation module to fuse the feature calculated by the Totality-Guided and Closure-Guided module. Finally, we classify the query image according to the image feature.

Our pipeline on 2-way 1-shot task is illustrated in Fig. 3. Given a few-shot learning task, firstly, we feed the image x into the feature extractor $f_\theta(\cdot)$ to extract the features. The prototype guided by the principle of totality $\boldsymbol{P}_n^t$ can be calculated by Eq. 29. Query features estimated by the principle of totality are the whole image feature extracted by the feature extractor (Eq. 4).

$$\boldsymbol{P}_n^t = \frac{1}{K} \sum_{(x_i,y_i)\in\mathcal{S}_n} f_\theta(x_i) \tag{29}$$

Meanwhile, guided by the law of closure, we randomly crop M patches from each image and feed them into a feature extractor with shared weights to calculate feature $f_\theta(p)$. For the convenience of calculating the categories prototypes, as shown in Eq. 30, we use all patches in the same category to calculate the prototype guided by the law of closure. Query features estimated by the law of closure can be calculated by Eq. 8.

$$\boldsymbol{P}_n^c = \frac{1}{K \times M} \sum_{(x_i,y_i)\in\mathcal{S}_n} \sum_{p_i\in x_i} f_\theta(p_i) \tag{30}$$

After calculating the category prototypes $\boldsymbol{P}_n^t$ and $\boldsymbol{P}_n^c$, they are fed into the feature estimate module to calculate the category prototype $\boldsymbol{P}_n$ (Eq. 31). Query features can be re-estimated by Eq. 14. Then the distances between the query feature and the category prototypes are calculated and the query set is classified according to Eq. 2.

$$\boldsymbol{P}_n = \boldsymbol{\lambda}\boldsymbol{P}_n^t + (\boldsymbol{I} - \boldsymbol{\lambda})\boldsymbol{P}_n^c \tag{31}$$

4 Experiment

4.1 Datasets

We test our method on *mini*ImageNet [3] and Caltech-UCSD Birds 200-2011 (CUB200) [23], which are widely used in few-shot learning.

***mini*ImageNet** is a subset of ILSVRC-2012 [24]. It contains 60,000 images in 100 categories, with 600 images in each category. Among them, 64 classes are used as the training set, 16 classes as the validation set, and 20 as the testing set.

Table 1. Results of the performance of different methods on *mini*ImageNet before and after adding GGIU The reported accuracy is 95% confidence interval.

Method	5-way 1-shot (%)	5-way 5-shot (%)
PN	61.59 ± 0.54	76.75 ± 0.46
PN+GGIU	64.34 ± 0.53 (↑ **2.75**)	79.49 ± 0.41 (↑ **2.74**)
CC	63.11 ± 0.74	80.43 ± 0.31
CC+GGIU	65.72 ± 0.77 (↑ **2.61**)	82.55 ± 0.29 (↑ **2.12**)
CL	63.74 ± 0.59	79.33 ± 0.31
CL+GGIU	65.50 ± 0.45 (↑ **1.76**)	80.76 ± 0.39 (↑ **1.43**)
CLIP	88.21 ± 0.33	97.47 ± 0.08
CLIP+GGIU	89.31 ± 0.33 (↑ **1.10**)	97.71 ± 0.06 (↑ **0.24**)

CUB200 contains 11788 images of birds in 200 species, which is widely used for fine-grained classification. Following the previous work [25], we split the categories into 130, 20, 50 for training, validation and testing.

4.2 Implementation Details

Since we propose a test-time feature estimation approach, we need to reproduce the performance of existing methods to validate our approach's effectiveness. Therefore, following Luo [12], we reproduce PN [4], CC [26], and CL [27]. The backbone we use in this paper is ResNet-12 [28], which is widely used in few-shot learning. We implement our method using PyTorch and test on an NVIDIA 3090 GPU. Since the authors do not provide a configuration file for the CUB200 dataset, we use the same configuration file as *mini*ImageNet. In the test phase, we randomly sampled five groups of test tasks, and each group of tasks contained 2000 episodes. Then five patches from each image are randomly cropped to rectify for prototypes and features with $\boldsymbol{\lambda} = diag(0.5, 0.5, \ldots, 0.5)$. The size of the patches is range from 0.08 to 1.0 of that of the original image.

4.3 Experiment Results

Results on In-Domain Data. Table 1 shows our performance on *mini*Image Net. Our method can effectively improve the performance based on existing methods for different tasks. We improve the performance on PN, CC, and CL by 2.75%, 2.61%, and 1.76%, separately.

We also test the performance on CLIP [1] to explore the performance of our method on the model pre-trained on large-scale data. We use the ViT-B/32 model published by OpenAI as a feature extractor and use PN for classification. Surprising, on such a high baseline, our method can still improve the 1-shot task by 1.10% and the 5-shot by 0.23%. The experiment results also illustrate that even with such large-scale training data, **GGIU** can still estimate more accurate image features.

Similarly, to test the effectiveness of our method on fine-grained classification, we also test the performance of our method on CUB200. As shown in Table 2, our method also significantly improves the performance based on existing methods of fine-grained classification.

Table 2. Results of the performance of different methods on CUB200 before and after adding GGIU. The reported accuracy is 95% confidence interval.

Method	5-way 1-shot (%)	5-way 5-shot (%)
PN	76.13 ± 0.21	88.06 ± 0.09
PN+GGIU	78.79 ± 0.24 (↑ **2.66**)	89.69 ± 0.17 (↑ **1.63**)
CC	70.57 ± 0.35	86.65 ± 0.16
CC+GGIU	72.60 ± 0.29 (↑ **2.03**)	87.90 ± 0.27 (↑ **1.25**)
CL	72.34 ± 0.48	85.93 ± 0.25
CL+GGIU	73.64 ± 0.46 (↑ **1.30**)	87.17 ± 0.25 (↑ **1.24**)

Table 3. Results of the performance reported by other methods with the performance of ours on *mini*ImageNet. The reported accuracy is 95% confidence interval. * represents the results reported by the original paper and † represents the results that we implement.

Method	Backbone	5-way 1-shot (%)	5-way 5-shot (%)
PN* [4]	Conv-4	49.42 ± 0.78	$\mathbf{68.20 \pm 0.66}$
PN†	Conv-4	50.15 ± 0.44	65.19 ± 0.51
DC* [13]	Conv-4	$\mathbf{54.62 \pm 0.64}$	–
Spot and Learn* [29]	Conv-4	51.03 ± 0.78	67.96 ± 0.71
PN+GGIU†	Conv-4	52.55 ± 0.52	67.36 ± 0.55
PN†	ResNet-12	61.59 ± 0.54	76.75 ± 0.46
CC* [26]	ResNet-12	55.45 ± 0.89	70.13 ± 0.68
CC†	ResNet-12	63.11 ± 0.74	80.43 ± 0.31
PN+TRAML* [30]	ResNet-12	60.31 ± 0.48	77.94 ± 0.57
PN+CL* [27]	ResNet-12	59.54 ± 0.47	74.46 ± 0.52
PN+CL†	ResNet-12	63.74 ± 0.59	79.33 ± 0.31
DC*	ResNet-18	61.50 ± 0.47	–
AA* [17]	ResNet-18	58.84 ± 0.77	80.35 ± 0.73
PN+GGIU†	ResNet-12	64.34 ± 0.53	79.49 ± 0.41
CC+GGIU†	ResNet-12	$\mathbf{65.72 \pm 0.77}$	$\mathbf{82.55 \pm 0.29}$
PN+CL+GGIU†	ResNet-12	65.50 ± 0.45	80.76 ± 0.39
CLIP†	ViT-B/32	88.21 ± 0.33	97.47 ± 0.08
CLIP+GGIU†	ViT-B/32	$\mathbf{89.31 \pm 0.33}$	$\mathbf{97.71 \pm 0.06}$

In addition, we compare the performance of our method with existing methods in Table 3.

Results on Cross-Domain Data. To validate the effectiveness and robustness of our approach, we conduct experiments on cross-domain tasks. We test the cross-domain performance on *mini*ImageNet and CUB200: Table 4 shows the model trained on *mini*ImageNet and tested on CUB200; Table 5 shows the

Table 4. Results of the performance of different methods trained on *mini*ImageNet and tested on CUB200 before and after adding GGIU. The reported accuracy is 95% confidence interval.

Method	5-way 1-shot (%)	5-way 5-shot (%)
PN	40.47 ± 0.21	56.14 ± 0.20
PN+GGIU	42.61 ± 0.49 (↑ **2.14**)	58.95 ± 0.49 (↑ **2.81**)
CC	43.56 ± 0.47	61.51 ± 0.39
CC+GGIU	45.88 ± 0.48 (↑ **2.32**)	64.77 ± 0.27 (↑ **3.26**)
CL	38.65 ± 0.44	52.36 ± 0.35
CL+GGIU	39.87 ± 0.31 (↑ **1.22**)	53.74 ± 0.21 (↑ **1.38**)

Table 5. Results of the performance of different methods trained on CUB200 and tested on miniImageNet. The reported accuracy is 95% confidence interval.

Method	5-way 1-shot (%)	5-way 5-shot (%)
PN	40.24 ± 0.35	55.47 ± 0.47
PN+GGIU	43.17 ± 0.57 (↑ **2.93**)	58.12 ± 0.49 (↑ **2.65**)
CC	43.54 ± 0.52	60.40 ± 0.39
CC+GGIU	44.27 ± 0.42 (↑ **0.73**)	60.94 ± 0.42 (↑ **0.54**)
CL	44.47 ± 0.56	61.84 ± 0.44
CL+GGIU	46.42 ± 0.68 (↑ **1.95**)	63.89 ± 0.45 (↑ **2.05**)

results of the model trained on CUB200 and tested on *mini*ImageNet. It can be seen that our method has adequate performance improvement on the cross-domain task of few-shot learning.

4.4 Result Analysis

Ablation Study. This section performs ablation experiments on *mini*ImageNet to explore the performance impact of feature rectification on the support set and query set.

As shown in Table 6, compared with only using **GGIU** to estimate query features, the performance of estimation support features is higher. For the 5-way 5-shot task, when **GGIU** is used to estimate query features, the performance is improved more. This can be explained as follows: The more support set samples, the more accurate the representation of the category prototype is. At this time, the accuracy of query features is the bottleneck of classification performance. Similarly, when the support set samples are few, the category prototypes calculated by the support set are inaccurate. At this moment, accurate category prototypes can significantly improve classification performance.

The Influence of the Fusion Parameters In our method, the fusion parameter $\boldsymbol{\lambda} = diag(\lambda_1, \lambda_2, \ldots, \lambda_D)$ is essential. This section explores the influence of $\boldsymbol{\lambda}$. We conduct three sets of experiments on PN for the different number of patches, 1, 5, and 10, respectively. As shown in Fig. 4, when $\lambda_i = 1$, image features $\hat{\boldsymbol{f}}$ are solely estimated by totality-guided image understanding (Eq. 14). The performance at this point is the baseline performance. As λ_i decreases, the influence of closure-guided image understanding increases, and the estimated image feature $\hat{\boldsymbol{f}}$ can represent the image better. So the performance gradually improves until an equilibrium point is reached. After the highest performance, continuing to decrease λ_i leads to a gradual decrease in model performance. It is worth noting that the performance will improve with a large enough number of patches even if $\lambda_i = 0$. It can be concluded that when the number of patches is large enough, closure-guided image understanding can well estimate image features. It can also be seen that the value of λ corresponding to the highest

Table 6. The results of ablation experiments. support with ✓ means adding GGIU while calculating prototypes; query with ✓ means adding GGIU while computing the feature of query images.

Support	Query	5-way 1-shot (%)	5-way 5-shot (%)
		61.59	76.75
✓		62.80	77.01
	✓	62.50	77.90
✓	✓	64.34	79.49

performance decreases as the number of patches increases. The more patches, the more accurate the estimation guided by the law of closure. What's more, when the number of patches is 1, 5, and 10, the best λ equals 0.7, 0.5, and 0.4, respectively. The above results are consistent with the deduction in Sect. 3.4.

The Influence of the Number of Patches. The number of patches is a very important hyper-parameter, and the inference efficiency might be low if this number is too big. This section explores how the number of patches influences the performance. As shown in Fig. 5, on *mini*ImageNet, we perform a 5-way 1-shot experiment on PN with $\lambda_i = 0.5$. Suppose only one patch is cropped to estimate the closure feature, in which case, it can be seen that it will also improve the performance substantially. However, as the number increases, the rate of increase in accuracy gradually slows down, which demonstrates that too many patches might lead to a marginal effect on the correction of the distribution, especially when the patches almost cover the whole image. Therefore, too many patches do not significantly improve the model performance.

The Relationship Between Intra-class Variations and $\boldsymbol{\lambda}$. In order to analyse the relationship between the optimal $\boldsymbol{\lambda}$ and intra-class variations, we

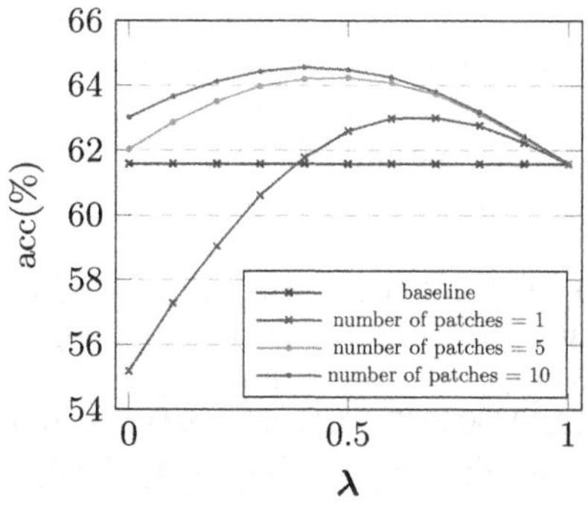

Fig. 4. Relationship between $\boldsymbol{\lambda}$ and performance

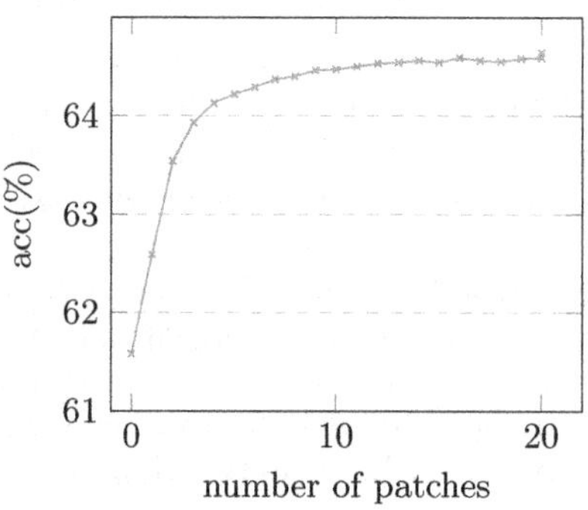

Fig. 5. The influence of the number of patches on performance

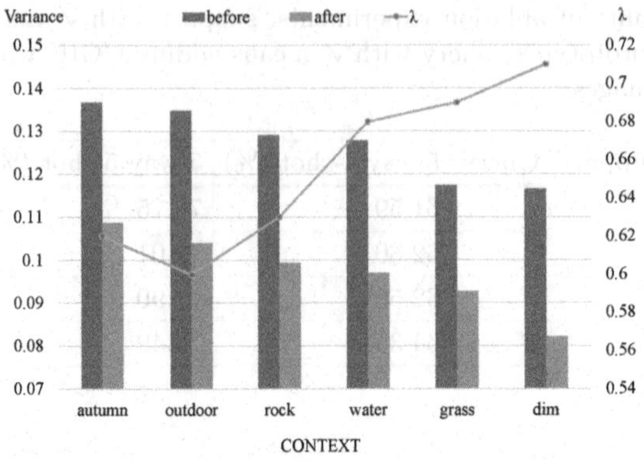

Fig. 6. The relationship between intra-class variations and $\boldsymbol{\lambda}$

conducted experiments on the NICO [31], which contains images labeled with category and context. We searched for the optimal $\boldsymbol{\lambda}$ in different contexts and calculated the intra-class variance in each context before and after using our method. As shown in Fig. 6, with the decrease of the intra-class variance, the optimal $\boldsymbol{\lambda}$ shows an increasing trend. Moreover, our method can significantly reduce the intra-class variance.

5 Conclusion

In this paper, we reformulate image features from the perspective of multivariate Gaussian distributions. We introduce Gestalt psychology into the process of image understanding to estimate more accurate image features. The Gestalt-guided image understanding consists of two modules: Totality-guided image understanding and Closure-guidied image understanding. Then we fed the features obtained from the above two modules into the feature estimation module and estimate image features accurately. We conduct many experiments on *mini*ImageNet and CUB200 for coarse-grained, fine-grained, and cross-domain few-shot image classification. The results demonstrate the effectiveness of **GGIU**. Moreover, **GGIU** even improved the performance based on CLIP. Finally, we analyze the influence of different hyper-parameters, and the results accord with our theoretical analysis.

Acknowledgement. This work was supported by the National Natural Science Foundation of China (No. U20B2062 and No. 62172036), the Fundamental Research Funds for the Central Universities (No. FRF-TP-20-064A1Z), the key Laboratory of Opto-Electronic Information Processing, CAS (No. JGA202004027), and the R&D Program of CAAC Key Laboratory of Flight Techniques and Flight Safety (No. FZ2021ZZ05).

References

1. Radford, A., et al.: Learning transferable visual models from natural language supervision. In: Proceedings of the 38th International Conference on Machine Learning, ICML 2021, 18–24 July 2021, Virtual Event. Proceedings of Machine Learning Research, vol. 139, pp. 8748–8763 (2021)
2. Fei-Fei, L., Fergus, R., Perona, P.: One-shot learning of object categories. IEEE Trans. Pattern Anal. Mach. Intell. **28**(4), 594–611 (2006). https://doi.org/10.1109/TPAMI.2006.79
3. Vinyals, O., Blundell, C., Lillicrap, T., Kavukcuoglu, K., Wierstra, D.: Matching networks for one shot learning. In: Advances in Neural Information Processing Systems: Annual Conference on Neural Information Processing Systems 2016, 5–10 December 2016, Barcelona, Spain, vol. 29, pp. 3630–3638 (2016)
4. Snell, J., Swersky, K., Zemel, R.S.: Prototypical networks for few-shot learning. In: Advances in Neural Information Processing Systems: Annual Conference on Neural Information Processing Systems 2017, 30 December 2017, Long Beach, CA, USA, vol. 30, pp. 4077–4087 (2017)
5. Finn, C., Abbeel, P., Levine, S.: Model-agnostic meta-learning for fast adaptation of deep networks. In: Proceedings of the 34th International Conference on Machine Learning, ICML 2017, Sydney, NSW, Australia, 6–11 August 2017. Proceedings of Machine Learning Research, vol. 70, pp. 1126–1135 (2017)
6. Chen, Y., Liu, Z., Xu, H., Darrell, T., Wang, X.: Meta-baseline: exploring simple meta-learning for few-shot learning. In: 2021 IEEE/CVF International Conference on Computer Vision, ICCV 2021, Montreal, QC, Canada, 10–17 October 2021, pp. 9042–9051 (2021). https://doi.org/10.1109/ICCV48922.2021.00893
7. Zhang, C., Cai, Y., Lin, G., Shen, C.: DeepEMD: few-shot image classification with differentiable earth mover's distance and structured classifiers. In: 2020 IEEE/CVF Conference on Computer Vision and Pattern Recognition, CVPR 2020, Seattle, WA, USA, 13–19 June 2020, pp. 12200–12210 (2020). https://doi.org/10.1109/CVPR42600.2020.01222
8. Hospedales, T., Antoniou, A., Micaelli, P., Storkey, A.: Meta-learning in neural networks: a survey. IEEE Trans. Pattern Anal. Mach. Intell. **44**(9), 5149–5169 (2021)
9. Thrun, S., Pratt, L.: Learning to learn: introduction and overview. In: Thrun, S., Pratt, L. (eds.) Learning to Learn, pp. 3–17. Springer, Cham (1998). https://doi.org/10.1007/978-1-4615-5529-2_1
10. Koch, G., Zemel, R., Salakhutdinov, R., et al.: Siamese neural networks for one-shot image recognition. In: ICML Deep Learning Workshop, vol. 2, Lille (2015)
11. Satorras, V.G., Estrach, J.B.: Few-shot learning with graph neural networks. In: 6th International Conference on Learning Representations, ICLR 2018, Vancouver, BC, Canada, 30 April–3 May 2018, Conference Track Proceedings (2018)
12. Luo, X., et al.: Rectifying the shortcut learning of background for few-shot learning. Adv. Neural. Inf. Process. Syst. **34**, 13073–13085 (2021)
13. Yang, S., Liu, L., Xu, M.: Free lunch for few-shot learning: distribution calibration. In: 9th International Conference on Learning Representations, ICLR 2021, Virtual Event, Austria, 3–7 May 2021 (2021)
14. Tang, X., Teng, Z., Zhang, B., Fan, J.: Self-supervised network evolution for few-shot classification. In: Zhou, Z. (ed.) Proceedings of the Thirtieth International Joint Conference on Artificial Intelligence, IJCAI 2021, Virtual Event/Montreal, Canada, 19–27 August 2021, pp. 3045–3051 (2021). https://doi.org/10.24963/ijcai.2021/419

15. An, Y., Xue, H., Zhao, X., Zhang, L.: Conditional self-supervised learning for few-shot classification. In: Zhou, Z. (ed.) Proceedings of the Thirtieth International Joint Conference on Artificial Intelligence, IJCAI 2021, Virtual Event/Montreal, Canada, 19–27 August 2021, pp. 2140–2146 (2021). https://doi.org/10.24963/ijcai.2021/295
16. Li, J., Wang, Z., Hu, X.: Learning intact features by erasing-inpainting for few-shot classification. In: Proceedings of the AAAI Conference on Artificial Intelligence, vol. 35, pp. 8401–8409 (2021)
17. Afrasiyabi, A., Lalonde, J.-F., Gagné, C.: Associative alignment for few-shot image classification. In: Vedaldi, A., Bischof, H., Brox, T., Frahm, J.-M. (eds.) ECCV 2020. LNCS, vol. 12350, pp. 18–35. Springer, Cham (2020). https://doi.org/10.1007/978-3-030-58558-7_2
18. Wagemans, J., et al.: A century of gestalt psychology in visual perception: II. conceptual and theoretical foundations. Psychol. Bull. **138**(6), 1218 (2012)
19. Henle, M.: The selected papers of Wolfgang Köhler. Philos. Phenomenol. Res. **33**(2), 270–271 (1972)
20. Hamlyn, D.W.: The Psychology of Perception: A Philosophical Examination of Gestalt Theory and Derivative Theories of Perception. Routledge, New York (2017)
21. Brennan, J.F., Houde, K.A.: History and Systems of Psychology. Cambridge University Press, Cambridge (2017)
22. Stevenson, H.: Emergence: The Gestalt Approach to Change. Unleashing Executive and Organizational Potential (2012). Retrieved July 2012
23. Wah, C., Branson, S., Welinder, P., Perona, P., Belongie, S.: The Caltech-UCSD Birds-200-2011 dataset (2011)
24. Russakovsky, O., et al.: ImageNet large scale visual recognition challenge. Int. J. Comput. Vision **115**(3), 211–252 (2015). https://doi.org/10.1007/s11263-015-0816-y
25. Li, W., et al.: LibFewShot: a comprehensive library for few-shot learning. arXiv preprint arXiv:2109.04898 (2021)
26. Gidaris, S., Komodakis, N.: Dynamic few-shot visual learning without forgetting. In: 2018 IEEE Conference on Computer Vision and Pattern Recognition, CVPR 2018, Salt Lake City, UT, USA, 18–22 June 2018, pp. 4367–4375 (2018). https://doi.org/10.1109/CVPR.2018.00459
27. Luo, X., Chen, Y., Wen, L., Pan, L., Xu, Z.: Boosting few-shot classification with view-learnable contrastive learning. In: 2021 IEEE International Conference on Multimedia and Expo, ICME 2021, Shenzhen, China, July 2021, pp. 1–6 (2021). https://doi.org/10.1109/ICME51207.2021.9428444
28. He, K., Zhang, X., Ren, S., Sun, J.: Deep residual learning for image recognition. In: 2016 IEEE Conference on Computer Vision and Pattern Recognition, CVPR 2016, Las Vegas, NV, USA, 27–30 June 2016, pp. 770–778 (2016). https://doi.org/10.1109/CVPR.2016.90
29. Chu, W.H., Li, Y.J., Chang, J.C., Wang, Y.C.F.: Spot and learn: a maximum-entropy patch sampler for few-shot image classification. In: Proceedings of the IEEE/CVF Conference on Computer Vision and Pattern Recognition, pp. 6251–6260 (2019)
30. Li, A., Huang, W., Lan, X., Feng, J., Li, Z., Wang, L.: Boosting few-shot learning with adaptive margin loss. In: 2020 IEEE/CVF Conference on Computer Vision and Pattern Recognition, CVPR 2020, Seattle, WA, USA, 13–19 June 2020, pp. 12573–12581 (2020). https://doi.org/10.1109/CVPR42600.2020.01259
31. Zhang, X., Zhou, L., Xu, R., Cui, P., Shen, Z., Liu, H.: Nico++: towards better benchmarking for domain generalization. arXiv preprint arXiv:2204.08040 (2022)

Depth Estimation via Sparse Radar Prior and Driving Scene Semantics

Ke Zheng[1], Shuguang Li[1(✉)], Kongjian Qin[2(✉)], Zhenxu Li[1], Yang Zhao[1], Zhinan Peng[1], and Hong Cheng[1]

[1] University of Electronic Science and Technology of China, Chengdu, Sichuan, China
lsg042@163.com

[2] China Automotive Technology and Research Center Co. Ltd., Tianjin, China

Abstract. Depth estimation is an essential module for the perception system of autonomous driving. The state-of-the-art methods introduce LiDAR to improve the performance of monocular depth estimation, but it faces the challenges of weather durability and high hardware cost. Unlike existing LiDAR and image-based methods, a two-stage network is proposed to integrate highly sparse radar data in this paper, in which sparse pre-mapping module and feature fusion module are proposed for radar feature extraction and feature fusion respectively. Considering the highly structured driving scenario, we introduce semantic information of the scenario to further improve the loss function, thus making the network more focused on the target region. Finally, we propose a novel depth dataset construction strategy by integrating binary mask-based filtering and interpolation methods based on the nuScenes dataset. And the effectiveness of our proposed method has been demonstrated through extensive experiments, which outperform existing methods in all metrics.

Keywords: Depth estimation · Multi-sensor · Autonomous driving · Driving scene characteristic

1 Introduction

The autonomous driving perception systems using a monocular camera can detect the target accurately, but it is difficult to confirm the location and scale of the target, due to the missing depth information. Recently, the use of convolution neural network is demonstrated to have significantly improved the accuracy of depth estimation [13,16–18,35]. All of the above work describes depth estimation as a regression problem. Accurate regression of depth values remains a challenge in this field. Thus, [3,11] were significant to transform the regression problem into the classification problem. Researchers [9,16] have attempted to improve the accuracy of depth estimation by solving multiple tasks in a joint manner.

K. Zheng and S. Li—Co-first authors.

L. Wang et al. (Eds.): ACCV 2022, LNCS 13842, pp. 425–441, 2023.
https://doi.org/10.1007/978-3-031-26284-5_26

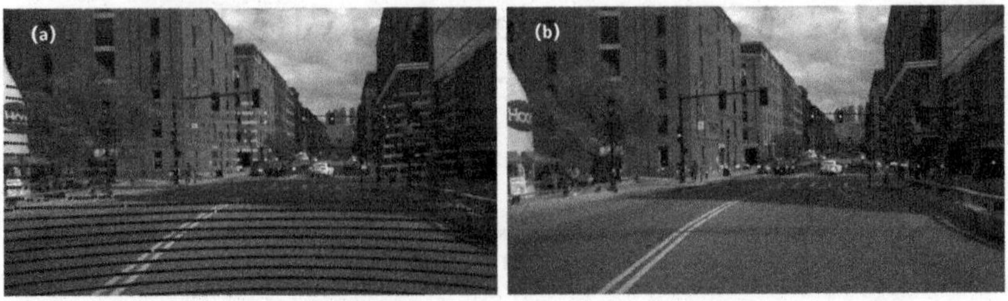

Fig. 1. Projection of the LiDAR and radar data points onto the image plane according to the calibration relationship shows their high degree of sparsity. (a) is the LiDAR projection point, (b) is the radar projection point.

However, monocular depth estimation is known as an ill-posed problem [10]. If pixels with known depth values are present in the image, the difficulty of the monocular depth estimation task is significantly reduced. A common method is the introduction of LiDAR data, i.e. the depth completion task [6,15,29]. Although LiDAR provides denser depth observations, the resolution is highly dependent on weather and is more expensive to acquire, the introduction of LiDAR also poses a considerable challenge to the overall system in real-time. The radar, on the other hand, is an all-weather sensor that performs well in bad weather and has a wide effective detection range. And radar has already been widely used in automotive industry applications such as Adaptive Cruise Control and Automatic Emergency Braking, which makes radar even more attractive to be applied for the depth estimation module. Therefore, we introduce radar measurements as prior knowledge to achieve high accuracy of depth estimation.

Conceivably, fusing radar and vision for depth estimation are great challenges for researchers in the field, due to the following technical barriers. 1) The only large dataset containing radar data is the nuScenes [2] dataset, but this dataset only provides the raw point cloud per frame. If only one frame of data is used as ground truth, only about 0.2% of pixels per image exist for supervision, as shown in Fig. 1(a), which would not be conducive to the model learning object contour information and detail information. 2) Compared to LiDAR, the radar point cloud is much more sparse, providing only about 0.003% of the prior depth value for each image, as shown in Fig. 1(b). Therefore, the feature of radar needs to be extracted in a way that is more suitable for dealing with sparse point clouds.

To address the above issues, we propose a new LiDAR data processing scheme to construct a denser and less noisy ground truth depth based on multiple frames of LiDAR data. The radar depth network (RDNet) is proposed as a two-stage network. Specially, considering the different input and purpose of the two stages, the sparse-coarse stage adopts a dual encoder-single decoder structure. Due to the high sparsity of the input radar data, the sparse pre-mapping module (SPM) is used to initially extract sparse feature. The dense depth map obtained in this stage is used as input to the coarse-fine stage, that single encoder-decoder structure and uses the feature fusion module (FFM), which introduces channel attention mechanism to fuse the features of the two stages. In addition, existing depth estimation networks do not take into account the characteristics of the driving scene. And they treat each pixel equally, while accurate depth values are often used in obstacle avoidance and 3D object detection, where the accuracy of the depth value of the target is required to be extremely high. Therefore, we further improve the loss function to make the network more focused on the corresponding region of the target. Our main contributions are as follows:

- We propose a denser and less noisy ground truth depth generation method, which solves the problem that the network can not learn details and edge information from sparse data.
- A new network structure better suited for collecting valid information from sparse radar data is proposed, including the sparse pre-mapping module and the feature fusion module.
- Improved loss function based on prior knowledge of the driving scenario, resulting in more accurate depth estimation of the target area.
- Various experiments have been carried out on the proposed new dataset based on nuScenes dataset to verify the effectiveness of the method compared to the state-of-the-art methods.

2 Related Work

2.1 Monocular Depth Estimation

Monocular depth estimation is currently addressed by training on large datasets with an attempt to acquire the ability to perform depth estimation. Early methods focused on depth estimation using hand-crafted features [25,27]. Recently, the main methods can be summarized as: 1) Introduction of attention mechanism [5,20,34]. For example, Li et al. [20] used channel attention mechanism in their model to extract the distinguishing features. 2) Dispersing the continuous depth into a certain number of bins, thus the regression problem is transformed into the classification problem [1,11]. Fu et al. [11] used spacing-increasing discretization strategy to discretize given depth in log domain. Bhat [1] used an adaptive discretization strategy to compute bin lengths for each image. 3) Jointly solving for multiple tasks [9,30]. [9] used generic architecture to jointly solve the three tasks of depth estimation, surface normal estimation and semantic segmentation. Wang et al. [30] proposed to decompose the image into several semantic segments, predicting a normalized depth map for each segment. 4) Optimizing coarse depth maps using CRF [19,31]. For example, Li et al. [19] introduced Hierarchical Conditional Random Fields to refine depth estimation from the super-pixel level to the pixel level.

We introduce highly sparse radar point cloud on top of the monocular depth estimation task, aiming to provide prior points and thus reduce the overall task difficulty. In addition, unlike existing methods that jointly solve multiple tasks, we introduce semantic information to make the overall depth estimation more relevant to the driving scenario requirements without increasing the computational effort of the network.

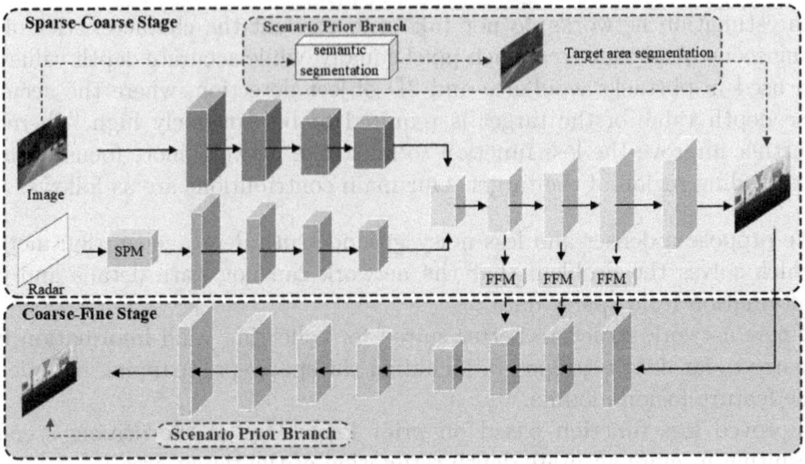

Fig. 2. Overview of proposed network architecture. It consists of two stages, the sparse-coarse stage and the coarse-fine stage.

2.2 Depth Completion

Depth completion task has additional characteristics compared to monocular depth estimation, such as the depth values of sparse points should be maintained as much as possible, and the transition between sparse points and their neighborhoods should be smooth.

These methods can be roughly divided into two catalogs: 1) After the network has predicted the coarse depth map, it is optimized by local neighborhoods. Cheng et al. [7] proposed convolutional spatial propagation network, which was propagated by recurrent convolution. Xu et al. [33] proposed to learn adaptive offsets in the network. 2) Using images to guide the recovery of depth maps. Tang et al. [26] calculated weights after extracting features from an image, and multiplied the weights when encoding sparse depth inputs. There were other methods that make use of feature representations in 3D space [4], surface normal [24,32] and so on.

The difference between radar and camera-based depth estimation and lidar and camera-based depth completion lies in the adequacy of the known depth information. In depth completion task, the image can be used as a guide to reconstructing the dense depth from the sparse input. However, because the input from radar is so sparse, it is more appropriate to provide prior depth information for the image using radar.

3 Proposed Method

In this section, we first introduce the overall architecture of RDNet, as well as the sparse pre-mapping module (SPM) for processing sparse radar data and the feature fusion module (FFM) for two stage feature fusion. The loss function used to train the network is then described, with further improvement of the loss function based on the driving scene semantics. Finally, the specific construction of the ground truth depth is presented.

3.1 Network Architecture

As shown in Fig. 2, we design an end-to-end depth estimation network framework based on radar and image. As estimating an accurate depth map directly from highly sparse radar and image via a single-stage network is a relatively difficult task, we design a two-stage network with the sparse-coarse stage and the coarse-fine stage respectively. The sparse-coarse stage takes the image and sparse radar data as input to predict a dense but coarse depth map. In this stage, in order to make full use of the feature of radar data for the effective fusion of image and radar, a dual encoder-single decoder architecture is used. That is, the image and radar are fused after extracting features separately, and then the final depth map is predicted by a decoder. Specially, the image encoder is constructed using ResNet-34 [14], which is pretrained on ImageNet [8]. In depth encoder, given the highly sparse nature of the radar data, we propose sparse pre-mapping module to extract the initial feature, and then use residual blocks to extract further feature. The decoder consists of four up-projection blocks [23], followed by a 3×3 convolution that maps the output to a depth map, and finally the depth map is restored to its initial resolution using bilinear upsampling. The coarse-fine stage uses the depth map of the previous prediction as input, which uses single encoder-decoder structure and feature fusion module to fuse the features of the two stages for obtaining a more accurate prediction. And as the driving scenario is relatively structured, we introduce the scenario prior branch to further enhance the network effect.

3.2 Sparse Pre-mapping Module

Considering the highly sparse nature of the radar input, standard convolution in sparse data processing would result in poor performance. And there are inconsistencies between LiDAR and radar data, even with effective noise filtering, there is still variability between them. To address the problem, the sparse pre-mapping module is set up before the residual block in the depth branch to enable mapping between data and the extraction of sparse feature. First, we briefly recall the sparsity-invariant convolution in [28], which takes a sparse feature map x and a binary mask m as input. It can be formulated as:

$$f(u,v) = \frac{\sum_{i,j=-k}^{k} m_{u+i,v+j} w_{i,j} x_{u+i,v+j}}{\sum_{i,j=-k}^{k} m_{u+i,v+j} + \delta} + b \tag{1}$$

where δ denotes a very small number to avoid the problem of dividing by zeros when there is no observed depth in the convolution region.

As shown in Fig. 3(a), our sparse pre-mapping module obtains a denser feature map by stacking sparsity-invariant convolutions. And in order to complete

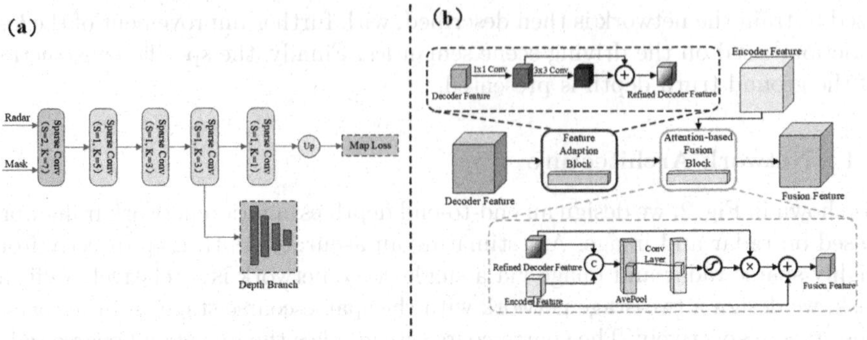

Fig. 3. (a) is the sparse pre-mapping module, which inputs are sparse radar depth map and binary mask, the whole module consists of several sparsity-invariant convolutions. (b) is the feature fusion module, the module takes features of the same resolution from the sparse-coarse stage and the coarse-fine stage as input and introduces channel attention mechanism that integrates the two stages to enhance feature representation.

the mapping between the data, the module output is bilinearly upsampled to the initial resolution and then supervision is applied. To further extract the feature of radar, the output of the fourth convolution is fed into the depth branch and further semantic features are extracted using the residual block.

3.3 Feature Fusion Module

To make the coarse-fine stage feature contain richer information and thus predict a more accurate final depth map, we use the strategy of decoder-encoder fusion in [26] to fuse feature from the sparse-coarse stage into the coarse-fine stage. And the feature representation is further enhanced by the introduction of channel attention mechanism. Specially, to adapt the decoder feature in the sparse-coarse stage to the encoder feature in the coarse-fine stage, we further refine the decoder feature and reduce its channel number using a convolution block with the residual connection. The structure is shown in Fig. 3(b), it can be formulated as:

$$Y_i = ReLU(Conv(F_i) + f_1(Conv(F_i))) \tag{2}$$

where F_i and Y_i denote the feature at layer i of the decoder and the corresponding output features and $f_1(\cdot)$ denotes the learnable feature refinement mapping.

After obtaining the refined feature Y_i and concatenating it with the encoder feature of the coarse-fine stage, We apply the channel attention mechanism to it, which adaptively adjusts the feature values of each channel, selectively enhancing feature containing useful information and suppressing useless feature through global information. Specially, global average pooling is used to obtain global context information, and then the attention vector is computed to enhance feature representation. It can be formulated as:

$$y = \sigma(f_2(Global[Y_i, F_i^s])) \odot Y_i + F_i^s \tag{3}$$

where *Global* represents the global pooling operation, F_i^s denotes the feature of current stage, $f_2(\cdot)$ denotes the learnable weight mapping, $\odot$ is element-wise product operator. $[\cdot, \cdot]$ denotes concatenate operation and σ denotes sigmoid function.

3.4 Loss Function

Differences in loss functions can have an impact on final depth estimation performance. We use $L1$ loss to calculate the loss between the ground truth depth and the predicted depth. Since the ground truth does not exist in every pixel, only the loss of valid pixels in the ground truth is computed, denoted as:

$$L_{depth} = \frac{1}{m} \sum_{(p,q) \in S} |d_{p,q} - \hat{d}_{p,q}| \tag{4}$$

where d and $\hat{d}$ denote the ground truth depth map and the predicted depth map respectively. S denotes the set of valid depths of d and m is the number of valid depths.

Further, we add edge-aware smoothness constraint [12] to encourage depth locally smooth. Since depth discontinuities usually occur at junctions, the image gradients are used for weighting, L_{smooth} is defined as:

$$L_{smooth} = e^{-|\partial_x(I)|}|\partial_x(\hat{d})| + e^{-|\partial_y(I)|}|\partial_y(\hat{d})| \tag{5}$$

where ∂_x, ∂_y denote the gradients along the x and y directions respectively, I denotes the input image.

In the training process, since we adopt the supervision to the depth prediction of the two stages and to the mapping result in the first stage, the overall loss function is represented as a weighted function:

$$L_{total} = \lambda_1(L_{coarse} + \lambda_2 L_{map} + \lambda_3 L_{smooth}) + (1 - \lambda_1)F_{final} \tag{6}$$

where λ_1, λ_2, λ_3 are hyperparameters. In addition, L_{coarse}, L_{map} and L_{final} denote the loss of the sparse-coarse stage, the loss of the sparse pre-mapping module and the loss of the coarse-fine stage respectively.

4 Scenario Semantics_Based Depth Estimation

Unlike other scenarios where the driving scene is relatively structured and has much prior knowledge that can be exploited. Much of the existing work has focused on the design of the network to improve its overall performance. However, the characteristics and requirements of the driving scenario have not been fully considered. Especially, in driving scenes, we usually want to get as accurate a depth value as possible for target areas, which include vehicles, pedestrians, etc.,

Fig. 4. Semantic segmentation result. (a) is the input image, (b) is the semantic segmentation result (In this scene, the blue and red areas are the target area we have defined). (Color figure online)

while we do not need to get very accurate depth values for not target areas. In summary, we want the depth estimation network to focus more on the target areas of the scene, rather than treating all areas equally.

Using this idea as a starting point, we analyze the results of the depth estimation. First, we introduce a semantic segmentation network, here using SegFormer, and the segmentation results are shown in Fig. 4. After obtaining the semantic labels for each scene, the pixels within the scene are classified into target/not target areas, and the RMSE/MAE of the two areas are calculated and the results are shown in Table 1. From the results, it can be seen that the errors in the target areas are even much larger than in the not target areas. We further analyze the number of pixels in the target and not target area and find that in most scenes the number of pixels in the target area is about 1/10 of the number of pixels in the not target area. We believe this can be seen as a sample imbalance. As the samples corresponding to not target area dominate the loss function, the model also tends to favor not target area during training, resulting in poorer performance in depth estimation for the target area.

Table 1. Errors in target area versus not target area.

Area	RMSE	MAE
Target area	8.379	4.929
Not target area	5.591	2.463

We make an improvement to Eq. 4 by introducing semantic segmentation results into the training process of the network, weighting the depth loss of different areas separately, thus making the network focus more on the target area. The improved depth loss is calculated as:

$$L_{depth} = \frac{1}{m}(\omega \sum_{(p,q)\in S_{TAR}} |d_{p,q} - \hat{d}_{p,q}| + (2-\omega) \sum_{(m,n)\in S_{NTAR}} |d_{m,n} - \hat{d}_{m,n}|) \quad (7)$$

where S_{TAR} and S_{NTAR} denotes the set of valid depths of target area and not target area, and m is the number of valid depths. ω is a weighting factor to balance the loss in the two areas.

Since we only improve the loss function without altering the network structure, we do not introduce any additional computational effort in the inference process of the network. This is in contrast to current networks that improve the depth estimation network into a multi-task network [30,31]. Moreover, our purpose is also different from existing networks. We are aiming to improve the accuracy of target area estimation, whereas they mostly aim to improve the representation of the backbone.

5 Dataset Generation

In contrast to the ground truth depth in the KITTI dataset [28], which is supervised for approximately 30% of the pixels, the nuScenes dataset provides only the raw LiDAR point cloud, which after projection is supervised for only approximately 0.2% of the pixels per frame. As the ground truth depth is too sparse and the evaluation metrics are only calculated at the pixels where supervision is present, the evaluation metrics do not give a full picture of the depth estimation and the model does not learn the details of the scene. Therefore, we propose a new LiDAR data processing scheme by integrating the binary mask-based filtering and interpolation method to construct a dense and less noisy ground truth on the basis of multi-frame LiDAR data.

Let L and m_L denote the aggregated multi-frame LiDAR data of size $H \times W$ and their corresponding sparse mask, respectively. The filtered LiDAR data can be represented as:

$$L' = g(L, m_L) \tag{8}$$

where g is the filtering operation.

Given that there are many unobserved points in the LiDAR data, the conventional filtering algorithms fail to consider the sparsity pattern of the data, which will alter the observation points intended to be the true values for training. We propose an observation point invariant filtering technique to alleviate the issue. Our proposed operation first masks non-observed points in the LiDAR data using a sparse mask m_L, and then finds the mean value of the depth of the observed points in region S of size $n \times m$. The outliers are obtained by testing whether the difference between the depth of the observation and the mean value is greater than threshold value, function g is defined as:

$$g = \begin{cases} L(p,q) & |L(p,q) - ave(p,q)| < \varepsilon \\ 0 & \text{otherwise} \end{cases} \tag{9}$$

and

$$ave(p,q) = \frac{1}{M} \sum_{(x,y) \in S} L(x,y) m_L(x,y) \tag{10}$$

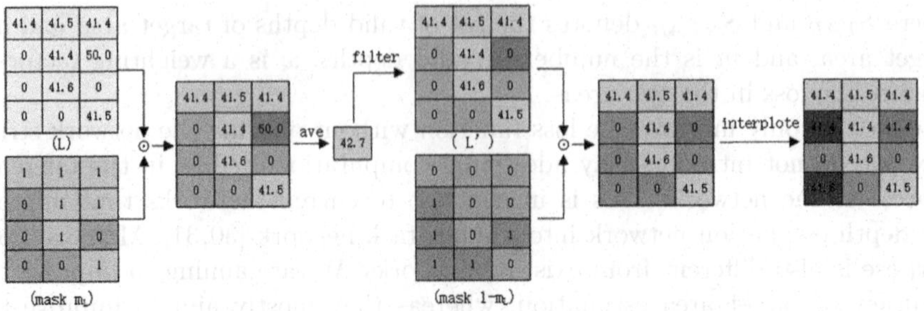

Fig. 5. Illustration of the entire process of LiDAR data processing.

where, M is the number of observed pixels, $ave(p,q)$ is the mean value at the current position (p,q), ε is the threshold value and S is a filter window of size $n \times m$.

Once the outliers are removed, the filtered LiDAR data L' and the sparse mask $1 - m_L$ are used for interpolation to obtain denser ground truth depth L_i, which can be formulated as:

$$L_i = f(L', 1 - m_L) \tag{11}$$

where f is the proposed interpolation operation.

Specifically, we use the sparse mask $1 - m_L$ to mask the observed points in the LiDAR data L' and interpolate the non-observed positions. Let T_x and T_y be the step in the x and y directions, the masked data points are traversed with the pre-determined step size, and find the nearest neighbors within a window of size (a,b) centered on the currently traversed data points (p,q). Since it is observed that the depth values vary much more in the y direction than in the x direction in the driving scenario, we set $a > b$. Function f is defined as:

$$f = \begin{cases} Nearest(p,q) & m_L(p,q) = 0 \\ L'(p,q) & m_L(p,q) \neq 0 \end{cases} \tag{12}$$

where $Nearest(p,q)$ denotes the search for nearest neighbor observations in a window centered on (p,q).

As we use a fixed step in the interpolation step, we sample the interpolated points in order to break the regularity of the interpolated data. The entire process of LiDAR data processing is illustrated in Fig. 5 and is represented as:

$$L_f = Sample(f(g(L, m_L), 1 - m_L)) \tag{13}$$

6 Experiments

6.1 Implementation Detail

We use the nuScenes dataset [2] to validate the above model. The dataset consists of 1000 scenes, each of which lasts 20 s long, with 40 keyframes. Each frame has a resolution of 1600×900, which we crop to the size of 1600×704 for training and testing. And nuScenes dataset contains driving scenarios in various conditions, which makes it more difficult to perform depth estimation on this dataset. We use 850 scenes and divide them into 810 scenes for training and 40 scenes for evaluation.

In order to train the proposed model, we perform the data augmentation with the color and horizontal flip transformation. We use Pytorch to deploy the network and train on an NVIDIA GeForce GTX TITAN X with 12G memory. In all experiments, batch size is set to 4. We use Adam optimizer with learning rate 0.0005, and the learning rate is reduced to half every 5 epochs. The weights in loss function are set to $\lambda_1 = 0.5, \lambda_2 = 0.3, \lambda_3 = 0.001$.

The standard metrics utilized for monocular depth estimation and depth completion work [1,31] are adopted for comparison. Let d and $\hat{d}$ be the predicted depth map and ground truth depth map, respectively, the evaluation metrics are:

Root Mean Square Error (RMSE): $\sqrt{\frac{1}{n}\sum_p^n (d_p - \hat{d}_p)^2}$
Mean Absolute Error (MAE): $\frac{1}{n}\sum_p^n |d_p - \hat{d}_p|$
Mean Absolute Relative Error (REL): $\frac{1}{n}\sum_p^n \frac{|d_p - \hat{d}_p|}{d_p}$
Threshold accuracy (δ_i): $max\left\{\frac{\hat{d}_p}{d_p}, \frac{d_p}{\hat{d}_p}\right\} < t, t = 1.25, 1.25^2, 1.25^3$.

6.2 Comparing Performance

To demonstrate the effectiveness of the proposed method, we train the models under the same conditions and compare our method with the image and LiDAR-based depth estimation methods proposed by Ma et al. [22] and Hu et al. [15] and the image and radar-based methods proposed by Lin et al. [21]. Those results are outlined in Table 2. Our method exhibits better performance with improvement in all evaluation metrics. In particular, compared to RadarNet [21], our RDNet reduces the RMSE by 0.518 m and the MAE by 0.338 m. It is also evident from the experiments that most LiDAR and image-based methods are not well adapted to radar, whereas our model has a unique capability to extract useful features from sparse data. The qualitative results shown in Fig. 6 imply that our method restores more accurate detail information than other methods, with fewer regions of incorrect depth estimation at image edges and object boundaries.

Table 2. Comparisons to advanced methods on nuScenes. Best results are in **bold** font.

Method	Error ↓			Acc ↑		
	RMSE	MAE	REL	δ_1	δ_2	δ_3
Hu et al. [15]	6.882	3.630	0.187	0.779	0.916	0.963
Lin et al. [21]	5.889	2.640	0.118	0.874	0.950	0.976
Ma et al. [22]	7.195	3.430	0.164	0.809	0.916	0.959
Ours	**5.371**	**2.302**	**0.103**	**0.897**	**0.960**	**0.980**

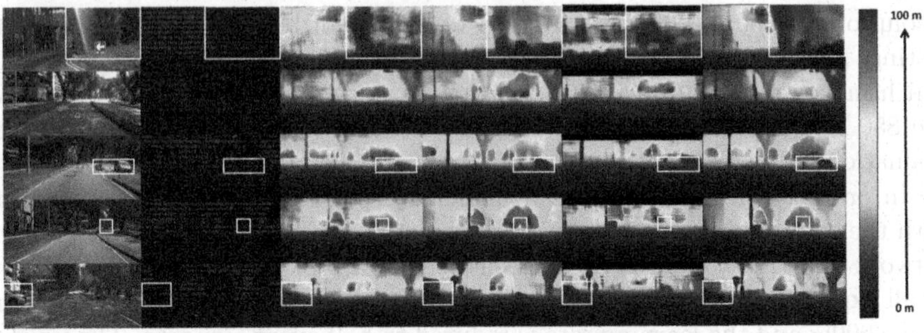

(a) Input image (b) Ground truth (c) Hu et al. [12] (d) Lin et al. [35] (e) Ma et al. [34] (f) Ours

Fig. 6. Qualitative results on nuScene. The proposed method obtains clearer details.

6.3 Ablation Studies

We conduct a series of ablation studies to verify the effectiveness of the various components proposed in our approach, including the late fusion structure, sparse pre-mapping module and feature fusion module. At the same time, we verify the effect of different loss functions, the role of adding radar data to the model, and analyze the setting of hyperparameters.

Image Branch Encoder: We use different baselines in the image encoder to extract the image feature, here only ResNet-34 and ResNet-18 are tested, as seen in Table 3, ResNet-34 achieves better performance, which we believe is due to the expanded network allowing better extraction of image feature. So we use ResNet-34 as the backbone in the subsequent comparison experiments.

Fusion Method: In this experiment, we compare the performance of the late fusion structure with the early fusion structure. The late fusion structure extracts advanced features from image and radar respectively through a series of blocks (final resolution is 1/32 of the input), then concatenates and fuses them. In the early fusion structure, the image and radar feature is extracted separately to 1/4 of the resolution of the input, then concatenate and extract advanced feature. As given in Table 3, the late fusion structure outperforms the early fusion structure. We believe that this is due to the fact that radar data is highly sparse and if the early fusion structure is used, most of the locations in the feature map are invalid at an early stage, resulting in the model failing to extract valid information.

Modules: We gradually add the sparse pre-mapping module (SPM) and feature fusion module (FFM) to the baseline to investigate the effectiveness of our proposed network. As can be seen in Table 3, the addition of the two modules improves the RMSE of the baseline by 0.535 m and the MAE by 0.374 m. To further verify the effectiveness of the sparse pre-mapping module in different fusion modes, we also deploy the module in the early fusion structure. This also means that for sparse radar data, sparsity-invariant convolution is more suitable for the initial extraction of its features.

Introduction of Radar: In order to verify the validation of radar optimizing effect on monocular depth estimation, we compare the model with only the image as input to our full model. The results are in agreement with our assumption that the introduction of radar does reduce the error in depth estimation compared to using only the image as input, implying the role of known depths at very few image locations in improving the performance of monocular depth estimation.

Table 3. Comparisons of performance under different model structures. In this table, EF for early fusion structure, LF for late fusion structure, SPM for sparse pre-mapping module, FFM for feature fusion module, R-18 (34) for the baseline of ResNet-18 (34), Only Image means the input is an only image.

Variant	RMSE	MAE	REL	δ_1
LF (R-18)	6.027	2.761	0.126	0.862
LF (R-34)	5.906	2.676	0.121	0.870
EF	6.725	3.018	0.125	0.855
EF+SPM	6.510	2.900	0.125	0.861
LF+SPM	5.838	2.639	0.123	0.873
LF+FFM	5.815	2.627	0.118	0.874
LF+SPM+FFM	**5.371**	**2.302**	**0.103**	**0.897**
Only image	5.877	2.637	0.117	0.877

Loss Function: In Table 4, we compare the effect of four conventional loss functions on the model, i.e., $L1$, $L2$, $LogL1$ and $BerHu$ loss, from which it can be seen that the best prediction results are obtained with $L1$ loss.

Table 4. Comparisons of performance under different loss functions.

Variant	RMSE	MAE	REL	δ_1
$LogL1$	5.854	2.560	0.110	0.886
$L2$	5.857	2.922	0.137	0.842
$BerHu$	6.263	3.345	0.172	0.807
$\boldsymbol{L1}$	**5.371**	**2.302**	**0.103**	**0.894**

Hyperparameters: We conduct a series of experiments for the parameter ω in Eq. 7 and the results obtained are shown in Fig. 7. Increasing the parameter ω can be interpreted as increasing the weight of the target area to make the network more inclined towards the target area. From the experimental results, it can be seen that the RMSE and MAE of the target area gradually decrease as ω increases. In contrast, when reducing the parameter ω within a certain range, for example, $\omega = 1$ to $\omega = 0$, RMSE and MAE for the not target area vary within 0.2 m. This is because the model already favors the not target area and continuing to increase the weight of the not target area on a small scale does not have a significant impact on the depth estimation results. When the parameter ω is set to 1.6, the error in the target area reaches its minimum value and we believe that a good balance between the target area and not target area is achieved. At a more extreme, when setting $\omega = 2.0$, i.e. training the network based on the loss of the target area only, the error increases instead, which we believe is due to the fact that there is some connection between the target area and not target area, which complements each other. The accuracy of its own depth estimation also suffers when the loss of the other side is completely ignored.

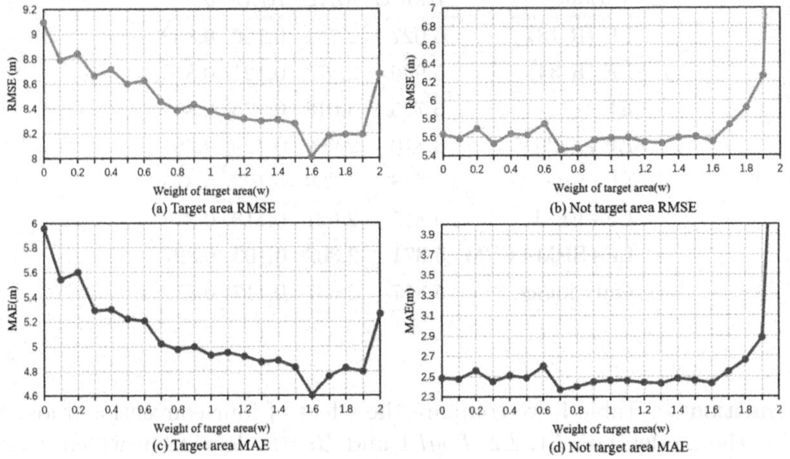

Fig. 7. Parameter experiments with ω. Roughly as the weight of the target area increases, the error in the target area decreases. When not target area is completely ignored, the target area error also increases, indicating that the depth estimates of the two regions are correlated.

7 Conclusion

In this paper, we propose a two-stage depth estimation model more suitable for autonomous driving by introducing radar and driving scenario semantics. We also propose a method for generating dense depth data to ensure that the model can learn better. The validity of our model has been demonstrated through extensive experiments, and its depth estimation results outperform existing methods. We have made some attempts at fusing radar and image for depth estimation and have demonstrated that highly sparse radar point cloud can indeed provide prior information that can improve overall depth estimation performance. In future work, our group will continue to investigate more efficient methods of encoding sparse radar data and methods of fusing radar and image. Also, lightweight radar and image-based networks for deployment in autonomous driving systems are the next steps in our work.

Acknowledgement. This research was funded by the Key R&D Projects of Science & Technology Department of Sichuan Province of China under Grant 2021YFG0070.

References

1. Bhat, S.F., Alhashim, I., Wonka, P.: AdaBins: depth estimation using adaptive bins. In: Proceedings of the IEEE/CVF Conference on Computer Vision and Pattern Recognition, pp. 4009–4018 (2021)
2. Caesar, H., et al.: nuScenes: a multimodal dataset for autonomous driving. In: Proceedings of the IEEE/CVF Conference on Computer Vision and Pattern Recognition, pp. 11621–11631 (2020)
3. Cao, Y., Wu, Z., Shen, C.: Estimating depth from monocular images as classification using deep fully convolutional residual networks. IEEE Trans. Circ. Syst. Video Technol. **28**(11), 3174–3182 (2017)
4. Chen, Y., Yang, B., Liang, M., Urtasun, R.: Learning joint 2D-3D representations for depth completion. In: Proceedings of the IEEE/CVF International Conference on Computer Vision, pp. 10023–10032 (2019)
5. Chen, Y., Zhao, H., Hu, Z., Peng, J.: Attention-based context aggregation network for monocular depth estimation. Int. J. Mach. Learn. Cybern. **12**(6), 1583–1596 (2021). https://doi.org/10.1007/s13042-020-01251-y
6. Cheng, X., Wang, P., Guan, C., Yang, R.: CSPN++: learning context and resource aware convolutional spatial propagation networks for depth completion. In: Proceedings of the AAAI Conference on Artificial Intelligence, vol. 34, pp. 10615–10622 (2020)
7. Cheng, X., Wang, P., Yang, R.: Depth estimation via affinity learned with convolutional spatial propagation network. In: Ferrari, V., Hebert, M., Sminchisescu, C., Weiss, Y. (eds.) ECCV 2018. LNCS, vol. 11220, pp. 108–125. Springer, Cham (2018). https://doi.org/10.1007/978-3-030-01270-0_7
8. Deng, J., Dong, W., Socher, R., Li, L.J., Li, K., Fei-Fei, L.: ImageNet: a large-scale hierarchical image database. In: 2009 IEEE Conference on Computer Vision and Pattern Recognition, pp. 248–255. IEEE (2009)
9. Eigen, D., Fergus, R.: Predicting depth, surface normals and semantic labels with a common multi-scale convolutional architecture. In: Proceedings of the IEEE International Conference on Computer Vision, pp. 2650–2658 (2015)
10. Eigen, D., Puhrsch, C., Fergus, R.: Depth map prediction from a single image using a multi-scale deep network. In: Advances in Neural Information Processing Systems, vol. 27 (2014)
11. Fu, H., Gong, M., Wang, C., Batmanghelich, K., Tao, D.: Deep ordinal regression network for monocular depth estimation. In: Proceedings of the IEEE Conference on Computer Vision and Pattern Recognition, pp. 2002–2011 (2018)
12. Godard, C., Mac Aodha, O., Brostow, G.J.: Unsupervised monocular depth estimation with left-right consistency. In: Proceedings of the IEEE Conference on Computer Vision and Pattern Recognition, pp. 270–279 (2017)
13. Gurram, A., Urfalioglu, O., Halfaoui, I., Bouzaraa, F., López, A.M.: Monocular depth estimation by learning from heterogeneous datasets. In: 2018 IEEE Intelligent Vehicles Symposium (IV), pp. 2176–2181. IEEE (2018)

14. He, K., Zhang, X., Ren, S., Sun, J.: Deep residual learning for image recognition. In: Proceedings of the IEEE Conference on Computer Vision and Pattern Recognition, pp. 770–778 (2016)
15. Hu, M., Wang, S., Li, B., Ning, S., Fan, L., Gong, X.: PENet: towards precise and efficient image guided depth completion. In: 2021 IEEE International Conference on Robotics and Automation (ICRA), pp. 13656–13662. IEEE (2021)
16. Jiao, J., Cao, Y., Song, Y., Lau, R.: Look deeper into depth: monocular depth estimation with semantic booster and attention-driven loss. In: Ferrari, V., Hebert, M., Sminchisescu, C., Weiss, Y. (eds.) ECCV 2018. LNCS, vol. 11219, pp. 55–71. Springer, Cham (2018). https://doi.org/10.1007/978-3-030-01267-0_4
17. Laina, I., Rupprecht, C., Belagiannis, V., Tombari, F., Navab, N.: Deeper depth prediction with fully convolutional residual networks. In: 2016 Fourth International Conference on 3D Vision (3DV), pp. 239–248. IEEE (2016)
18. Lee, J.H., Han, M.K., Ko, D.W., Suh, I.H.: From big to small: multi-scale local planar guidance for monocular depth estimation. arXiv preprint arXiv:1907.10326 (2019)
19. Li, B., Shen, C., Dai, Y., Van Den Hengel, A., He, M.: Depth and surface normal estimation from monocular images using regression on deep features and hierarchical CRFs. In: Proceedings of the IEEE Conference on Computer Vision and Pattern Recognition, pp. 1119–1127 (2015)
20. Li, R., Xian, K., Shen, C., Cao, Z., Lu, H., Hang, L.: Deep attention-based classification network for robust depth prediction. In: Jawahar, C.V., Li, H., Mori, G., Schindler, K. (eds.) ACCV 2018. LNCS, vol. 11364, pp. 663–678. Springer, Cham (2019). https://doi.org/10.1007/978-3-030-20870-7_41
21. Lin, J.T., Dai, D., Van Gool, L.: Depth estimation from monocular images and sparse radar data. In: 2020 IEEE/RSJ International Conference on Intelligent Robots and Systems (IROS), pp. 10233–10240. IEEE (2020)
22. Ma, F., Cavalheiro, G.V., Karaman, S.: Self-supervised sparse-to-dense: self-supervised depth completion from LiDAR and monocular camera. In: 2019 International Conference on Robotics and Automation (ICRA). pp. 3288–3295. IEEE (2019)
23. Ma, F., Karaman, S.: Sparse-to-dense: depth prediction from sparse depth samples and a single image. In: 2018 IEEE International Conference on Robotics and Automation (ICRA), pp. 4796–4803. IEEE (2018)
24. Qiu, J., et al.: DeepLiDAR: deep surface normal guided depth prediction for outdoor scene from sparse LiDAR data and single color image. In: Proceedings of the IEEE/CVF Conference on Computer Vision and Pattern Recognition, pp. 3313–3322 (2019)
25. Saxena, A., Sun, M., Ng, A.Y.: Make3D: learning 3D scene structure from a single still image. IEEE Trans. Pattern Anal. Mach. Intell. **31**(5), 824–840 (2008)
26. Tang, J., Tian, F.P., Feng, W., Li, J., Tan, P.: Learning guided convolutional network for depth completion. IEEE Trans. Image Process. **30**, 1116–1129 (2020)
27. Torralba, A., Oliva, A.: Depth estimation from image structure. IEEE Trans. Pattern Anal. Mach. Intell. **24**(9), 1226–1238 (2002)
28. Uhrig, J., Schneider, N., Schneider, L., Franke, U., Brox, T., Geiger, A.: Sparsity invariant CNNs. In: 2017 International Conference on 3D Vision (3DV), pp. 11–20. IEEE (2017)
29. Van Gansbeke, W., Neven, D., De Brabandere, B., Van Gool, L.: Sparse and noisy LiDAR completion with RGB guidance and uncertainty. In: 2019 16th International Conference on Machine Vision Applications (MVA), pp. 1–6. IEEE (2019)

30. Wang, L., Zhang, J., Wang, O., Lin, Z., Lu, H.: SDC-Depth: semantic divide-and-conquer network for monocular depth estimation. In: Proceedings of the IEEE/CVF Conference on Computer Vision and Pattern Recognition, pp. 541–550 (2020)
31. Wang, P., Shen, X., Lin, Z., Cohen, S., Price, B., Yuille, A.L.: Towards unified depth and semantic prediction from a single image. In: Proceedings of the IEEE Conference on Computer Vision and Pattern Recognition, pp. 2800–2809 (2015)
32. Xu, Y., Zhu, X., Shi, J., Zhang, G., Bao, H., Li, H.: Depth completion from sparse LiDAR data with depth-normal constraints. In: Proceedings of the IEEE/CVF International Conference on Computer Vision, pp. 2811–2820 (2019)
33. Xu, Z., Yin, H., Yao, J.: Deformable spatial propagation networks for depth completion. In: 2020 IEEE International Conference on Image Processing (ICIP), pp. 913–917. IEEE (2020)
34. Ye, X., Chen, S., Xu, R.: DPNet: detail-preserving network for high quality monocular depth estimation. Pattern Recogn. **109**, 107578 (2021)
35. Yin, W., Liu, Y., Shen, C., Yan, Y.: Enforcing geometric constraints of virtual normal for depth prediction. In: Proceedings of the IEEE/CVF International Conference on Computer Vision, pp. 5684–5693 (2019)

Flare Transformer: Solar Flare Prediction Using Magnetograms and Sunspot Physical Features

Kanta Kaneda[1(✉)], Yuiga Wada[1], Tsumugi Iida[1], Naoto Nishizuka[2], Yûki Kubo[2], and Komei Sugiura[1]

[1] Keio University, Tokyo, Japan
{k.kaneda,yuiga,tiida,komei.sugiura}@keio.jp
[2] National Institute of Information and Communications Technology, Tokyo, Japan
{nishizuka.naoto,kubo}@nict.go.jp

Abstract. The prediction of solar flares is essential for reducing the potential damage to social infrastructures that are vital to society. However, predicting solar flares accurately is a very challenging task. Existing methods predict flares using either physical features or images, but the main bottleneck is that they sometimes incorrectly predict a class that is smaller than the actual solar flare. In this paper, we propose the Flare Transformer, a solar flare prediction model that handles both images and physical features through the Magnetogram Module and the Sunspot Feature Module. The transformer attention mechanism is introduced to model the temporal relationships between input features. We also introduce a new differentiable loss function to balance the two major metrics of the Gandin–Murphy–Gerrity score and Brier skill score. We validate our model on a publicly available dataset. The results show that the Flare Transformer outperformed the baseline methods in terms of the Gandin–Murphy–Gerrity score and true skill statistic, and achieved better performance than those given by human experts.

Keywords: Solar flare prediction · Time-series forecasting · Transformer

1 Introduction

X-ray emissions, high energy particles, and coronal mass ejections released by solar flares disrupts GPS communication, causes radio blackouts, and poses health hazards to astronauts and flight crews [2]. It is estimated that the economic loss from a Carrington-class flare will be approximately US $163 billion in North America [22]. Therefore, the prediction of solar flares is essential for reducing the potential damage to our society. However, accurate predictions of solar flares remain a challenging task.

Given this background, this paper focuses on predicting the class of the largest solar flare that will occur within 24 h. Figure 1 shows an overview of our method. The inputs are a time-series of full-disk line-of-sight magnetograms

L. Wang et al. (Eds.): ACCV 2022, LNCS 13842, pp. 442–457, 2023.
https://doi.org/10.1007/978-3-031-26284-5_27

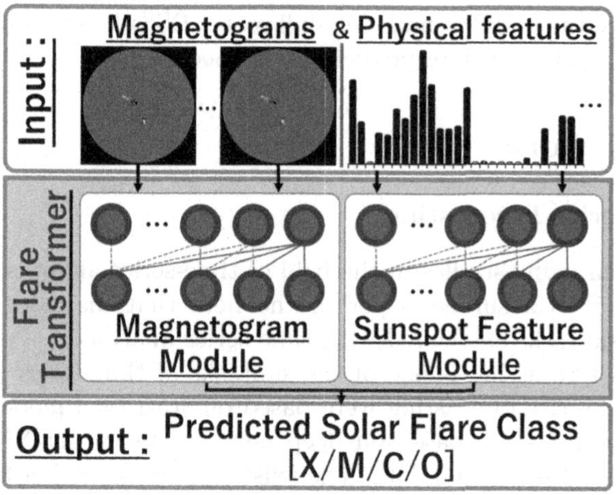

Fig. 1. Overview of the flare transformer.

taken by Helioseismic and Magnetic Imager (HMI) [24] onboard Solar Dynamic Observatory (SDO) [20] and region-level physical features extracted from active regions detected in the sunspot images. Note that a magnetogram is a grayscale solar image shown in Fig. 1, and details on physical features can be found in [6]. The output is the predicted solar flare class (see Table 1 for details).

Even for human experts, it is very challenging to predict solar flares. For example, their performance during the period 2000–2015 resulted in GMGS = 0.48 (Gandin–Murphy–Gerrity score) and $\mathrm{TSS}_{\geq \mathrm{M}} = 0.50$ (true skill statistics) [10] (these metrics would return a score of 1 for a perfect forecast). The main bottleneck of existing methods is that they sometimes incorrectly predict a class of flare that is smaller than that which actually occurs. For example, the DeFN [17] model incorrectly predicted 89% of X-class flares as M-class.

In this paper, we propose the Flare Transformer (FT), which handles time-series of images and physical features to produce accurate and reliable solar flare predictions. Our code is available at this URL[1]. Our model differs from existing methods in that it handles both line-of-sight magnetograms and physical features with the Magnetogram Module (MM) and the Sunspot Feature Module (SFM) (see Sect. 4.3). These modules allow the proposed FT model to capture useful features for solar flare prediction. We also introduce a transformer attention mechanism [27] to model the temporal relationships between input features. The main contributions of this paper are summarized as follows:

- We propose the FT, a solar flare prediction model that handles both line-of-sight magnetograms and physical features through the MM and SFM (see Sect. 4.3).
- We introduce the transformer attention mechanism [27] to model temporal relationships between input features.

[1] https://github.com/keio-smilab21/flare_transformer.

– We introduce the GMGS and Brier skill score (BSS) losses to balance the two major metrics of solar flare prediction (see Sect. 4.4).

2 Related Work

2.1 Time-Series Forecasting

There have been many studies in the field of time-series analysis (e.g., [29,30]). For instance, [5] is a survey paper in the field that discusses the methods, datasets, and subtasks associated with time-series analysis.

Time-series forecasting is one of the subtasks of time-series analysis. Early studies on time-series forecasting were based on statistical models (e.g., [3,9]). More recent methods are based on deep neural networks (DNNs), which can model complicated sequential data. The existing methods using DNNs typically use recurrent neural networks (RNNs) (e.g., [18,21,23,28]) or attention mechanisms (e.g., [13,29,30]). In terms of RNN based approaches, DeepAR [23] models future probabilistic distributions using an autoregressive RNN model, whereas DPT-DRNN [18] is an enhanced RNN model with a pre-training method that uses an autoencoder, allowing the PM2.5 concentration to be predicted from environmental monitoring data.

Inspired by the success of attention mechanism in other fields (e.g., natural language processing, computer vision), many models using a transformer [27] has been proposed for time-series forecasting. However, time-series data tends to have larger sequence length L than inputs in other fields, resulting in high computational complexity. Therefore, many existing studies have proposed methods to reduce the computational complexity. For example, LogSparse Transformer [13] propose the LogSparse attention, which reduce the computational complexity to $\mathcal{O}(L \log L)$. Adversarial Sparse Transformer [29] implements a sparse attention layer by using a sparse normalization transformation, α-entmax. Informer [30] is a transformer-based model for long-sequence time-series forecasting, which introduce a ProbSparse self-attention mechanism and generative style decoder.

2.2 Solar Flare Prediction

There have also been many studies on solar flare prediction (e.g., [11,17,19]). For instance, [7] comprehensively summarizes the methods and evaluation metrics for solar flare prediction tasks.

There are several standard datasets for solar flare prediction tasks. Nishizuka *et al.* published a dataset[2] consisting of physical features for sunspots extracted from images taken by the SDO [20] and Geostationary Operational Environmental Satellite (GOES). The dataset covers the period from June 2010 to December 2015 [16]. Angryk *et al.* published a dataset consisting of physical features of active regions extracted from Spaceweather HMI Active Region Patch series, which covers the period from May 2010 to December 2018 [1].

[2] Available at https://wdc.nict.go.jp/IONO/wdc/solarflare/index.html.

Table 1. The correspondence of flare class and X-ray intensity.

Flare class	Range of X-ray intensity [W/m^2]
X	$p_t > 10^{-4}$
M	$10^{-5} < p_t \leq 10^{-4}$
C	$10^{-6} < p_t \leq 10^{-5}$
O	$p_t \leq 10^{-6}$

Many methods have been proposed for solar flare prediction tasks. For example, Park *et al.* proposed a convolutional neural network based model to forecast solar flare occurrence using solar full-disk magnetograms [19]. DeFN [17] is a residual feed-forward network model which calculates the probability of solar flares occurring in the subsequent 24 h period using sunspot physical features. Tang *et al.* compared the performance of existing solar flare prediction models, and reported that the DeFN outperformed all other methods [4,8,25] for the 2010–2015 dataset [25]. The advantage of the DeFN model is that the physical features can be analyzed to search for those that are most effective for solar flare prediction. DeFN-R [15] is an extension of DeFN. While the DeFN is optimized for deterministic prediction, DeFN-R is optimized for a probability forecast based on the observation event rate.

3 Problem Statement

In this paper, we focus on the task of predicting the class of the largest solar flare that will occur within 24 h from time t, $\boldsymbol{y}_t$, which is defined as follows:

$$\boldsymbol{y}_t = \text{flareclass}(\max\{p_{t+1}, p_{t+2}, \ldots, p_{t+24}\}), \tag{1}$$

where p_t and flareclass$(\cdot)$ denote the maximum X-ray intensity within an hour of time t and the 1-of-K representation for classes X, M, C, and O, respectively. Table 1 defines each flare class in terms of p_t.

The input and output of this task are defined as follows:

- **Input:** Line-of-sight magnetograms and physical features. The physical features are extracted from solar images taken by the Atmospheric Imaging Assembly (AIA) [12] aboard the SDO [20].
- **Output:** A four-dimensional vector that denotes the predicted probabilities for each solar flare class.

In this task, it is desirable that $p(\hat{\boldsymbol{y}}_t)$ be as close as possible to $\boldsymbol{y}_t$, where $p(\hat{\boldsymbol{y}}_t)$ denotes the output of the model. However, because solar flare prediction is a class-imbalanced problem, it is unfavorable to output trivial predictions, such as predicting all flares as O-class. Thus, to avoid such predictions, it is desirable to output $p(\hat{\boldsymbol{y}}_t)$ that maximizes metrics such as GMGS [6] and BSS [15], which are standard metrics in the field of solar flare prediction.

In this paper, we do not model the solar flare prediction task as a regression problem for the following two reasons:

- The above classification task is the standard setting in the field of solar flare prediction (e.g., [17,19]).
- Predictions by human experts are given in the above classification setting, not as a regression task.

Therefore, it is reasonable to handle this task as a classification problem. Even though we do not explicitly consider the task as a regression problem, our model can be applied to regression tasks. We also assume that solar flare prediction is based on a full-disk image and not on a region-level image. We define line-of-sight magnetogram as a solar image taken by SDO/HMI [24]. We evaluate the model by GMGS, TSS [10] and BSS (see Sect. 5.2). The GMGS is a multi-class evaluation metric that considers class imbalance using the GMGS score matrix [6]. The BSS is a metric to evaluate the reliability of the forecast and is used not only in the field of solar flare prediction, but also in other fields such as weather forecasting [15].

4 Proposed Method

4.1 Novelty

The proposed method is unique in that the model handles both line-of-sight magnetograms and physical features through the Magnetogram Module (MM) and the Sunspot Feature Module (SFM) for solar flare prediction. We use the transformer attention mechanism [27] in the modules to model the temporal relationships between input features. Solar flare prediction is a time series forecasting task, and a certain degree of correlation is expected between time series features. Therefore, it is considered that the transformer, which performs correlation calculations in the self-attention and cross-attention mechanisms, is suitable for this task. The main differences between our method and existing methods (e.g., [17,19]) are as follows:

- While existing methods take only physical features [17] or images [19] as input, our model handles both features through the MM and the SFM. These modules are explained in Sect. 4.3.
- We introduce the GMGS and BSS losses to balance the two major metrics in solar flare prediction. These losses are explained in Sect. 4.4.

4.2 Input

The input $\boldsymbol{x}$ is defined as follows:

$$\boldsymbol{x} = (V_{t-k+1:t}, F_{t-k+1:t}) \ , \tag{2}$$

$$V_{t-k+1:t} = (\boldsymbol{v}_{t-k+1}, \boldsymbol{v}_{t-k+2}, ..., \boldsymbol{v}_t) \ , \tag{3}$$

$$F_{t-k+1:t} = (\boldsymbol{f}_{t-k+1}, \boldsymbol{f}_{t-k+2}, ..., \boldsymbol{f}_t) \ , \tag{4}$$

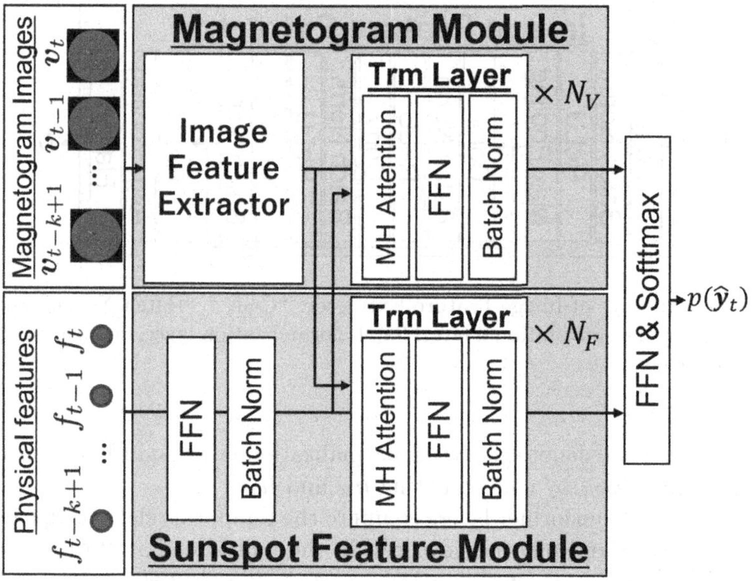

Fig. 2. Proposed method framework. Flare Transformer consists of Magnetogram Module and Sunspot Feature Module.

where $\boldsymbol{v}_t \in \mathbb{R}^{512\times512}$ and $\boldsymbol{f}_t \in \mathbb{R}^{90}$ denote the line-of-sight magnetogram and physical features at time t, respectively. We extract $\boldsymbol{f}_t$ by the method described in [17], and obtain $\boldsymbol{v}_t$ by resizing the line-of-sight magnetograms to 512×512 pixels.

4.3 Flare Transformer

Figure 2 shows the structure of our method. In the figure, MH Attention, Trm Layer, and FFN denote the multi-head attention, transformer layer, and feed-forward network, respectively. The proposed method consists of two main modules: MM and SFM. The difference between MM and SFM is the query taken in the source-target attention. In the MM, magnetogram features are taken as the query, while in the SFM, physical features are taken as the query. The MM first encodes the images $V_{t-k+1:t}$ as follows:

$$\boldsymbol{h}_V = f_{\mathrm{FE}}\left(V_{t-k+1:t}\right), \tag{5}$$

where f_{FE} denotes the Image Feature Extractor, which consists of multiple convolutional layers, max pooling layers, average pooling layers, and batch normalization layers, as shown in Fig. 3. The SFM also encodes the physical features $F_{t-k+1:t}$ as follows:

$$\boldsymbol{h}_F = f_{\mathrm{BN}}(f_{\mathrm{FFN}}(F_{t-k+1:t})), \tag{6}$$

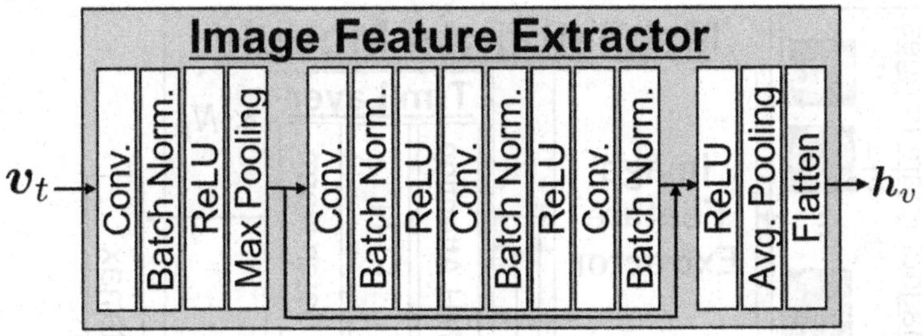

Fig. 3. The structure of Image Feature Extractor. "Conv," "Batch Norm," and "Avg pooling" denote the convolutional layer, batch normalization layer, and average pooling layer, respectively.

where f_{BN} and f_{FFN} denote the batch normalization layer and FFN, respectively. Then, we obtain $\boldsymbol{h}_{VF}$ by concatenating $\boldsymbol{h}_V$ and $\boldsymbol{h}_F$.

Next, the N_V transformer layers compute the temporal relationships between time-series images and physical features. In the multi-head attention block, $\boldsymbol{h}_V$ and $\boldsymbol{h}_{VF}$ are divided into $\boldsymbol{h}_V^{(i)} \in \mathbb{R}^{k\times d}$ and ${\boldsymbol{h}_{VF}}^{(i)} \in \mathbb{R}^{k\times 2d}$ $(i = 1, \ldots, N_{\mathrm{head}})$, where $d = H/N_{\mathrm{head}}$. Here, H and N_{head} denote the hidden layer size and number of heads, respectively.

The query $Q^{(i)} \in \mathbb{R}^{k\times d}$, key $K^{(i)} \in \mathbb{R}^{k\times 2d}$, and value $V^{(i)} \in \mathbb{R}^{k\times 2d}$ are computed for the i-th head as follows:

$$Q^{(i)} = W_q^{(i)} \boldsymbol{h}_V^{(i)}, \tag{7}$$

$$K^{(i)} = W_k^{(i)} \boldsymbol{h}_{VF}^{(i)}, \tag{8}$$

$$V^{(i)} = W_v^{(i)} \boldsymbol{h}_{VF}^{(i)}, \tag{9}$$

where $W_q^{(i)}$, $W_k^{(i)}$, and $W_v^{(i)}$ denote the weight matrices for $Q^{(i)}$, $K^{(i)}$, and $V^{(i)}$, respectively. The output of the transformer layer $\boldsymbol{h}_{\mathrm{trm}}$ is computed as follows:

$$\boldsymbol{h}_{\mathrm{trm}} = f_{\mathrm{BN}}(f_{\mathrm{FFN}}(\boldsymbol{h}_{\mathrm{mha}})), \tag{10}$$

$$\boldsymbol{h}_{\mathrm{mha}} = \left[\boldsymbol{f}_{\mathrm{attn}}^{(1)}; \boldsymbol{f}_{\mathrm{attn}}^{(2)}; \cdots; \boldsymbol{f}_{\mathrm{attn}}^{(N_{\mathrm{head}})}\right], \tag{11}$$

$$\boldsymbol{f}_{\mathrm{attn}}^{(i)} = \mathrm{softmax}\left(\frac{Q^{(i)} K^{(i)\top}}{\sqrt{d}}\right) V^{(i)}. \tag{12}$$

The output of the MM $\boldsymbol{h}_{\mathrm{MM}}$ is obtained by N_V transformer layers. Similarly, the output of the SFM $\boldsymbol{h}_{\mathrm{SFM}}$ is obtained by N_F transformer layers.

Finally, the predicted flare class y_t^* is obtained as follows:

$$y_t^* = \mathrm{argmax}_i(p(\hat{y}_{ti})), \tag{13}$$

$$p(\hat{\boldsymbol{y}}_t) = \mathrm{softmax}(f_{\mathrm{FFN}}(\boldsymbol{h}_{\mathrm{MM}}; \boldsymbol{h}_{\mathrm{SFM}})), \tag{14}$$

where $p(\hat{y}_{ti})$ denotes the predicted probability of i-th class.

4.4 Loss Function

In our model, we introduce GMGS and BSS losses to balance the GMGS [6] and $\mathrm{BSS}_{\geq \mathrm{M}}$ [15].

First, we define the GMGS loss for the following reason. The loss functions used in existing methods are not effective in improving the GMGS because they adjust the balance between classes using weights that are irrelevant to GMGS. Unlike existing methods, we can effectively improve the GMGS by using the score matrix of GMGS as weights. The GMGS loss $\mathcal{L}_{\mathrm{GMGS}}$ is defined as follows:

$$\mathcal{L}_{\mathrm{GMGS}} = -\frac{1}{NI}\sum_{n=1}^{N} s_{i^*j^*}\sum_{i=1}^{I} y'_{ni}\log(p(\hat{y}_{ni}))\,, \tag{15}$$

$$i^* = \mathrm{argmax}_i(y_{ni})\,, \tag{16}$$

$$j^* = \mathrm{argmax}_j(p(\hat{y}_{nj}))\,, \tag{17}$$

where N, I, $p(\hat{y}_{ni})$, y_{ni}, y'_{ni} and $s_{i^*j^*}$ denote the number of samples, number of classes, predicted probability of the i-th class for the n-th sample, label of the i-th class for the n-th sample, the label of the i-th class for label smoothed $\boldsymbol{y}_n$, and element (i^*, j^*) from the score matrix for GMGS [6], respectively.

Second, we define the BSS loss to improve the reliability of the forecast. We propose to use the BSS directly for the BSS loss because it is differentiable. The BSS loss $\mathcal{L}_{\mathrm{BSS}}$ is defined as follows:

$$\mathcal{L}_{\mathrm{BSS}} = -\frac{1}{NI}\sum_{n=1}^{N}\sum_{i=1}^{I}(p(\hat{y}_{ni}) - y_{ni})^2\,. \tag{18}$$

Overall, we use the following loss function:

$$\mathcal{L} = \mathcal{L}_{\mathrm{CE}} + \lambda_{\mathrm{GMGS}}\mathcal{L}_{\mathrm{GMGS}} + \lambda_{\mathrm{BSS}}\mathcal{L}_{\mathrm{BSS}}\,, \tag{19}$$

where $\mathcal{L}_{\mathrm{CE}}$ denotes the cross entropy loss between $\boldsymbol{y}_n$ and $\hat{\boldsymbol{y}}_n$, and λ_{GMGS} and λ_{BSS} denote the loss weights.

5 Experiments

5.1 Experimental Setup

In the experiments, we evaluated our method on a dataset that was collected by the following procedure. First, we downloaded hourly line-of-sight magnetograms from the SDO [20] web archives[3]. Next, we used an online physical feature database that is available at this URL[4].

Because the physical features in the dataset are region-level features, the following processes were performed to make the data suitable for input to our model.

[3] https://sdo.gsfc.nasa.gov/data/.

[4] https://wdc.nict.go.jp/IONO/wdc/solarflare/index.html.

Table 2. The number of samples included in each divided set.

Training set		Test set	
Year	Samples	Year	Samples
2010–2013	29247	2014	8127
2010–2014	37374	2015	8155
2010–2015	45529	2016	7795
2010–2016	53324	2017	7991

Table 3. Parameter settings and structures of FT.

Optimizer	Adam ($\beta_1 = 0.9$, $\beta_2 = 0.999$)
Learning rate	7.0×10^{-7}
Batch size	32
MM	$(H_{\mathrm{MM}}, A_{\mathrm{MM}}) = (128, 4)$
SFM	$(H_{\mathrm{SFM}}, A_{\mathrm{SFM}}) = (128, 4)$
Trm layer	$(N_V, N_F) = (1, 2)$
Loss weights	$(\lambda_{\mathrm{GMGS}}, \lambda_{\mathrm{BSS}}) = (0.01, 10)$

- If multiple sunspots were observed at a given time, we selected physical features from a randomly selected sunspot.
- If no sunspots were observed at a given time, we set the values of all physical features to zero.

The dataset contains 61315 samples, covering the period from June 2010 to December 2017. A sample consists of a line-of-sight magnetogram and 90 types of physical features [17]. The numbers of samples with ground truth labels of X, M, C, and O are 492, 4745, 19736, and 36342, respectively. The numbers of samples are imbalanced between classes because X-class and M-class solar flares are extremely unlikely events compared with other classes of flares. For example, only 2.9% of solar flares in 2017 were of X class.

In this study, we divided the training and test sets based on time-series cross-validation, which is a standard method used in time-series forecasting tasks [26]. With time-series cross-validation, the corresponding training set consists only of observations that occurred prior to the observations that form the test set. Table 2 presents the numbers of samples included in each divided set. The training and test sets were used for parameter training and evaluation, respectively.

The experimental setup is summarized in Table 3, where H_{MM} and A_{MM} denote the hidden size and number of attention heads in the MM, respectively, and H_{SFM} and A_{SFM} denote those in the SFM. Our model has 3.65 million parameters. The proposed model was trained on an RTX 2080 Ti with 11 GB of GPU memory and an Intel Core i9 processor. It took approximately 90 min to train our model. The inference time was approximately 65 ms.

5.2 Evaluation Metric

We evaluate the model by GMGS [6], TSS [10] and BSS [15]. The GMGS is defined as follows:

$$\mathrm{GMGS} = \mathrm{tr}(S^{\top} \cdot P)\,, \tag{20}$$

where S and P denote the I-rank scoring matrix with an element s_{ij} and I-categorical contingency table with an element p_{ij}, respectively. The GMGS is used as an important metric in recent studies for solar flare prediction [10]. The elements s_{ij} for the symmetric score matrix S is defined as follows:

$$s_{ii} = \frac{1}{I-1}\left[\sum_{k=1}^{i-1} a_k^{-1} + \sum_{k=i}^{I-1} a_k\right] \quad (1 \le i \le I)\,, \tag{21}$$

$$s_{ij} = \frac{1}{I-1}\left[\sum_{k=1}^{i-1} a_k^{-1} + \sum_{k=i}^{j-1}(-1) + \sum_{k=j}^{I-1} a_k\right] \quad (1 \le i \le j \le I)\,, \tag{22}$$

$$a_i = \frac{1-\sum_{k=1}^{i} p_k}{\sum_{k=1}^{i} p_k} \quad (1 \le i \le I)\,, \tag{23}$$

$$p_i = \sum_{j=1}^{I} p_{ij} \quad (1 \le j \le I)\,. \tag{24}$$

The TSS is defined as follows:

$$\mathrm{TSS} = \frac{\mathrm{TP}}{\mathrm{TP}+\mathrm{FN}} - \frac{\mathrm{FP}}{\mathrm{FP}+\mathrm{TN}}, \tag{25}$$

where TP, FP, FN, TN denote the number of true positive, false positive, false negative, and true negative samples for a contingency table, respectively.

The BSS is a standard metric for solar flare prediction that evaluates the reliability of the forecast [15]. The BSS is defined as follows:

$$\mathrm{BSS} = \frac{\mathrm{BS}-\mathrm{BS}_c}{0-\mathrm{BS}_c}, \tag{26}$$

$$\mathrm{BS} = \sum_{n=1}^{N}\sum_{i=1}^{I}(p(\hat{y}_{ni}) - y_{ni})^2\,, \tag{27}$$

$$\mathrm{BS}_c = \sum_{n=1}^{N}\sum_{i=1}^{I}(f - y_{ni})^2\,, \tag{28}$$

where N, I, y_{ni}, $p(\hat{y}_{ni})$, and f denote the number of samples, the number of classes, the label of the i-th class for the n-th sample, the predicted probability of the i-th class for the n-th sample, and climatological event rate, respectively.

Table 4. Quantitative comparison. The best scores are in bold.

Method	GMGS↑	$TSS_{\geq M}$↑	$BSS_{\geq M}$↑
DeFN [17]	0.375 ± 0.141	0.413 ± 0.150	-0.022 ± 0.782
DeFN-R [15]	0.302 ± 0.055	0.279 ± 0.162	0.036 ± 0.982
Ours (FT)	$\mathbf{0.503 \pm 0.059}$	$\mathbf{0.530 \pm 0.112}$	$\mathbf{0.082 \pm 0.974}$
Human [10,14]	0.48	0.50	0.16

5.3 Experimental Results

We conducted experiments based on time-series cross validation. Table 4 shows the quantitative results of the baselines and proposed method. The average and standard deviations of the scores are reported. The DeFN [17] and DeFN-R [15] models, which only take physical features as input, were used as the baseline methods. The results given by DeFN and DeFN-R were reproduced by ourselves. We evaluated the models by GMGS [6], $TSS_{\geq M}$ [10] and $BSS_{\geq M}$ [15]. "≥M" indicates that the model was evaluated after the output had been categorized as either "≥M" or "<M". We used GMGS and $BSS_{\geq M}$ as the primary metrics. GMGS is a metric for multi-categorical forecasts and satisfies equitability [10], whereas BSS is a standard metric for solar flare prediction that evaluates the reliability of the forecast [15]. The approximated version of GMGS is shown for the 2016 test set because no X-class flares occurred in 2016.

Table 4 indicates that the GMGS of the DeFN, DeFN-R, and FT methods are 0.375, 0.302 and 0.503 points, respectively. Therefore, the FT outperformed DeFN by 0.128 points in terms of GMGS. Table 4 also presents the performance of human experts. Kubo *et al.* reported that GMGS and $TSS_{\geq M}$ for daily forecasting operations by human experts were 0.48 and 0.50, respectively, for the period 2000–2015 [10]. Our method outperform human experts in terms of GMGS and $TSS_{\geq M}$, which indicates that FT is very promising.

Figure 4 shows the qualitative results. Figure 4(a) shows line-of-sight magnetograms from 21:00–23:00 on September 5, 2017 and standardized physical features. Note that, of the 90 features, only the first 20 positive values of $\boldsymbol{f}_t$ are displayed because of space limitation. The prediction is $y_t^* =$ "X". An X-class solar flare occurred at 12:02 on September 6, 2017, which is within 24 h of 23:00 on September 5, 2017. Therefore, the model was able to predict the correct maximum solar flare class. Similarly, a sample that was correctly predicted as an M–class solar flare is shown in Fig. 4(b). Figure 4(c) shows a failure case. The prediction is $y_t^* =$ "M" with $p(y_t^* =$ "M"$) = 0.47$ and $p(y_t^* =$ "X"$) = 0.40$. Because an X-class solar flare occurred within 24 h of time t, the prediction was incorrect. However, the predicted probability indicates that this was a marginal prediction. The above results indicate that our model gives better predictions than can be achieved by human experts in terms of GMGS and TSS.

5.4 Ablation Studies

We set the following ablation conditions:

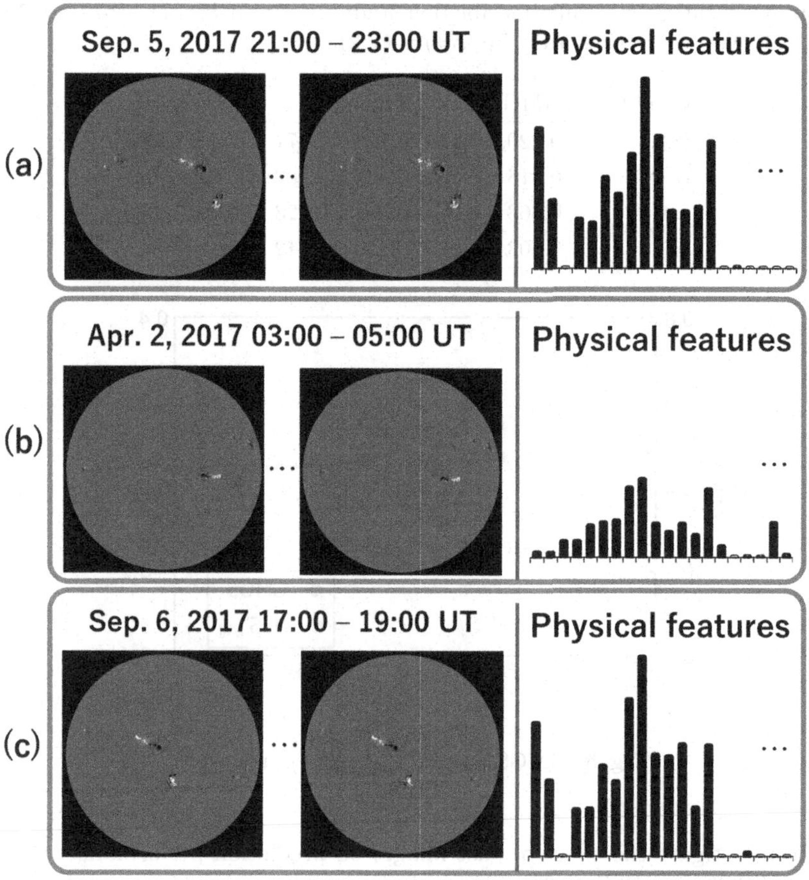

Fig. 4. Three samples of qualitative results. The model predicted the correct flare class for samples (a) and (b), while the model predicted the incorrect flare class for sample (c). (a) X-class solar flare. (b) M-class solar flare. (c) X-class solar flare but was predicted as M-class solar flare.

(a) We removed $F_{t-k+1:t}$ to investigate the performance when $V_{t-k+1:t}$ is used as the input.
(b) We changed N_F to investigate the performance when N_F is reduced.
(c) We changed N_V to investigate the performance when N_V is increased.

Table 5 presents the quantitative results for the ablation studies. The GMGS decreased by 0.283 points under condition (a). This result indicates that the introduction of the SFM to handle both image and physical features was beneficial to the performance. Under conditions (b) and (c), the scores of each metric fluctuated slightly. We selected condition (d) as the proposed method because the scores were well balanced. The above results indicate that handling both magnetograms and physical features as input was beneficial to performance.

Table 5. Quantitative results for ablation studies. (a) w/o $F_{t-k+1:t}$ (b) $(N_V, N_F) = (1,1)$ (c) $(N_V, N_F) = (2,2)$ (d) $(N_V, N_F) = (1,2)$.

Conditions	GMGS↑	$\text{TSS}_{\geq \text{M}}$↑	$\text{BSS}_{\geq \text{M}}$↑
(a)	0.220 ± 0.116	0.198 ± 0.371	-1.77 ± 0.225
(b)	0.516 ± 0.089	0.485 ± 0.082	0.052 ± 1.05
(c)	0.563 ± 0.070	0.551 ± 0.123	0.011 ± 0.965
(d)	0.503 ± 0.059	0.530 ± 0.112	0.082 ± 0.974

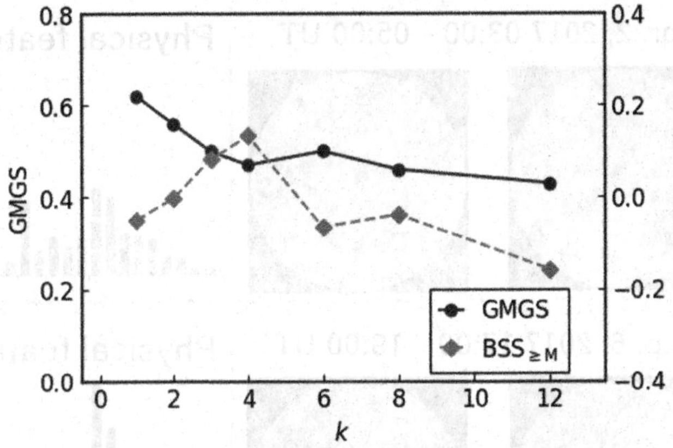

Fig. 5. GMGS and $\text{BSS}_{\geq \text{M}}$ plotted against k.

To investigate the effect of past images on prediction performance, we evaluated GMGS and $\text{BSS}_{\geq \text{M}}$ for various k. For example, $k = 4$ means that $\boldsymbol{x}_{t-3:t}$ was used as the model input. The results in Fig. 5 shows that the highest values of GMGS and $\text{BSS}_{\geq \text{M}}$ occur at $k = 1$ and $k = 4$, respectively. This indicates that $k = 1$ is sufficient to maximize GMGS. However, it also indicates that we need to consider an appropriate k in order to balance it with $\text{BSS}_{\geq \text{M}}$.

5.5 Error Analysis

Table 6 presents the confusion matrix for our method using 2017 test set. For the X-class, there were 20, 77, 16, and 7878 true positive (TP), false positive (FP), false negative (FN), and true negative (TN) samples, respectively.

Table 7 categorizes the failed cases. We define the influence of failure cases on GMGS as follows:

$$\text{GMGS}_{\text{Influence}} = \frac{c_{ij}(s_{ii} - s_{ij})}{N}, \tag{29}$$

where c_{ij} and s_{ij} denote the element (i, j) for the confusion matrix and for the GMGS score matrix [6], respectively. Table 7 indicates that the bottleneck is

Table 6. Confusion matrix for 2017 test set.

		Predicted flare class			
		O	C	M	X
Observed flare class	O	7269	210	34	12
	C	84	150	29	22
	M	18	50	34	43
	X	1	0	15	20

Table 7. The error analysis by $\mathrm{GMGS}_{\mathrm{Influence}}$ for 2017 test set.

Observed class	Predicted class	$\mathrm{GMGS}_{\mathrm{Influence}}$
X	M	0.1335
M	C	0.0885
C	O	0.0578
M	O	0.0442
X	O	0.0114

the misprediction of X-class for M-class flares. This suggests that methods to alleviate this bottleneck will effectively enhance the prediction performance.

6 Conclusion

In this paper, we proposed the Flare Transformer (FT), a method for predicting the maximum solar flare class that will occur within 24 h. The following contributions of this study can be emphasized:

- We proposed the FT method, which handles both line-of-sight magnetograms and physical features through the MM and SFM.
- We introduced transformer attention mechanism [27] to model temporal relationships between input features.
- We introduced the GMGS and BSS losses to balance the two major metrics in solar flare prediction.
- We have demonstrated that our model gives better predictions than can be achieved by human experts in terms of GMGS and TSS.

Acknowledgement. This work was partially supported by JSPS KAKENHI Grant Number 20H04269 and NEDO.

References

1. Angryk, R.A., et al.: Multivariate time series dataset for space weather data analytics. Sci. Data **7**(1), 1–13 (2020)

2. Bhattacharjee, S., Alshehhi, R., Dhuri, D., et al.: Supervised convolutional neural networks for classification of flaring and nonflaring active regions using line-of-sight magnetograms. Astrophys. J. **898**(2), 98 (2020)
3. Box, G., Jenkins, G., Reinsel, G., Ljung, G.: Time Series Analysis: Forecasting and Control. Wiley, Hoboken (2015)
4. Cinto, T., Gradvohl, S., Coelho, P., Silva, A.: A framework for designing and evaluating solar flare forecasting systems. Mon. Not. R. Astron. Soc. **495**(3), 3332–3349 (2020)
5. Esling, P., Agon, C.: Time-series data mining. ACM Comput. Surv. **45**(1), 1–34 (2012)
6. Gandin, L., Murphy, A.: Equitable skill scores for categorical forecasts. Mon. Weather Rev. **120**(2), 361–370 (1992)
7. Georgoulis, M., Bloomfield, S., et al.: The flare likelihood and region eruption forecasting (FLARECAST) project: flare forecasting in the big data & machine learning era. J. Space Weather Space Clim. **11**, A39 (2021)
8. Huang, X., Wang, H., Xu, L., Liu, J., Li, R., Dai, X.: Deep learning based solar flare forecasting model. I. Results for line-of-sight magnetograms. Astrophys. J. **856**(1), 7 (2018)
9. Hyndman, R., Koehler, A., et al.: Forecasting with Exponential Smoothing: The State Space Approach. Springer, Heidelberg (2008). https://doi.org/10.1007/978-3-540-71918-2
10. Kubo, Y., Den, M., Ishii, M.: Verification of operational solar flare forecast: case of regional warning center Japan. J. Space Weather Space Clim. **7**, A20 (2017)
11. Kusano, K., Iju, T., Bamba, Y., Inoue, S.: A physics-based method that can predict imminent large solar flares. Science **369**(6503), 587–591 (2020)
12. Lemen, J., et al.: The atmospheric imaging assembly (AIA) on the solar dynamics observatory (SDO). In: Chamberlin, P., Pesnell, W.D., Thompson, B. (eds.) The Solar Dynamics Observatory, pp. 17–40. Springer, New York (2011). https://doi.org/10.1007/978-1-4614-3673-7_3
13. Li, S., et al.: Enhancing the locality and breaking the memory bottleneck of transformer on time series forecasting. In: NeurIPS, vol. 32, pp. 5243–5253 (2019)
14. Murray, S., Bingham, S., Sharpe, M., Jackson, D.: Flare forecasting at the met office space weather operations centre. Space Weather **15**(4), 577–588 (2017)
15. Nishizuka, N., Kubo, Y., Sugiura, K., Den, M., Ishii, M.: Reliable probability forecast of solar flares: deep flare net-reliable (DeFN-R). Astrophys. J. **899**(2), 150 (2020)
16. Nishizuka, N., Sugiura, K., Kubo, Y., Den, M., et al.: Solar flare prediction model with three machine-learning algorithms using ultraviolet brightening and vector magnetograms. Astrophys. J. **835**(2), 156 (2017)
17. Nishizuka, N., Sugiura, K., Kubo, Y., Den, M., Ishii, M.: Deep flare net (DeFN) model for solar flare prediction. Astrophys. J. **858**(2), 113 (2018)
18. Ong, B.T., Sugiura, K., Zettsu, K.: Dynamically pre-trained deep recurrent neural networks using environmental monitoring data for predicting PM2.5. Neural Comput. Appl. **27**(6), 1553–1566 (2016)
19. Park, E., Moon, Y.-J., Shin, S., Yi, K., Lim, D., et al.: Application of the deep convolutional neural network to the forecast of solar flare occurrence using full-disk solar magnetograms. Astrophys. J. **869**(2), 91 (2018)
20. Pesnell, W., Thompson, B., Chamberlin, P.: The Solar Dynamics Observatory (SDO). In: Chamberlin, P., Pesnell, W.D., Thompson, B. (eds.) The Solar Dynamics Observatory, pp. 3–15. Springer, New York (2011). https://doi.org/10.1007/978-1-4614-3673-7_2

21. Rangapuram, S., Seeger, M., Gasthaus, J., Stella, L., et al.: Deep state space models for time series forecasting. In: NeurIPS, vol. 31, pp. 7785–7794 (2018)
22. Re, S.: Solar storm; how to calculate insured/reinsured losses? Space Weather Workshop (2016)
23. Salinas, D., Flunkert, V., Gasthaus, J., Januschowski, T.: DeepAR: probabilistic forecasting with autoregressive recurrent networks. Int. J. Forecast. **36**(3), 1181–1191 (2020)
24. Scherrer, P., et al.: The helioseismic and magnetic imager (HMI) investigation for the solar dynamics observatory (SDO). Sol. Phys. **275**(1), 207–227 (2012)
25. Tang, R., et al.: Solar flare prediction based on the fusion of multiple deep-learning models. Astrophys. J. Suppl. Ser. **257**(2), 50 (2021)
26. Tashman, L.: Out-of-sample tests of forecasting accuracy: an analysis and review. Int. J. Forecast. **16**(4), 437–450 (2000)
27. Vaswani, A., et al.: Attention is all you need. In: NeurIPS, vol. 30, pp. 5998–6008 (2017)
28. Wen, R., Torkkola, K., Narayanaswamy, B., Madeka, D.: A Multi-horizon Quantile Recurrent Forecaster. arXiv preprint arXiv:1711.11053 (2017)
29. Wu, S., Xiao, X., Ding, Q., Zhao, P., et al.: Adversarial sparse transformer for time series forecasting. In: NeurIPS, vol. 33, pp. 17105–17115 (2020)
30. Zhou, H., Zhang, S., Peng, J., Zhang, S., et al.: Informer: beyond efficient transformer for long sequence time-series forecasting. In: AAAI, pp. 11106–11115 (2021)

Neural Residual Flow Fields for Efficient Video Representations

Daniel Rho[1], Junwoo Cho[1], Jong Hwan Ko[1,2(✉)], and Eunbyung Park[1,2(✉)]

[1] Department of Artificial Intelligence, Sungkyunkwan University, Seoul, South Korea
{daniel231,jwcho000,jhko,epark}@skku.edu

[2] Department of Electrical and Computer Engineering, Sungkyunkwan University, Seoul, South Korea

Abstract. Neural fields have emerged as a powerful paradigm for representing various signals, including videos. However, research on improving the parameter efficiency of neural fields is still in its early stages. Even though neural fields that map coordinates to colors can be used to encode video signals, this scheme does not exploit the spatial and temporal redundancy of video signals. Inspired by standard video compression algorithms, we propose a neural field architecture for representing and compressing videos that deliberately removes data redundancy through the use of motion information across video frames. Maintaining motion information, which is typically smoother and less complex than color signals, requires a far fewer number of parameters. Furthermore, reusing color values through motion information further improves the network parameter efficiency. In addition, we suggest using more than one reference frame for video frame reconstruction and separate networks, one for optical flows and the other for residuals. Experimental results have shown that the proposed method outperforms the baseline methods by a significant margin.

1 Introduction

Neural fields [1–6] (also known as implicit neural representations or coordinate-based neural representations) are an emerging approach for representing various signals. Signals can be reconstructed using dense sampling once a neural network has been trained to map coordinates to corresponding signal values. Unlike other representation techniques that store discretely sampled data, neural fields use continuous coordinates as inputs, allowing them to represent signals at any resolution and at any arbitrary coordinate. It can accurately express both low and high frequencies of signals thanks to recent breakthroughs in input features [1,2,5]. This innovative representation approach has shown considerable promise in a number of areas, including computer graphics [5,7], physical simulations [2,8], and generative models [9,10], to name a few.

Supplementary Information The online version contains supplementary material available at https://doi.org/10.1007/978-3-031-26284-5_28.

L. Wang et al. (Eds.): ACCV 2022, LNCS 13842, pp. 458–474, 2023.
https://doi.org/10.1007/978-3-031-26284-5_28

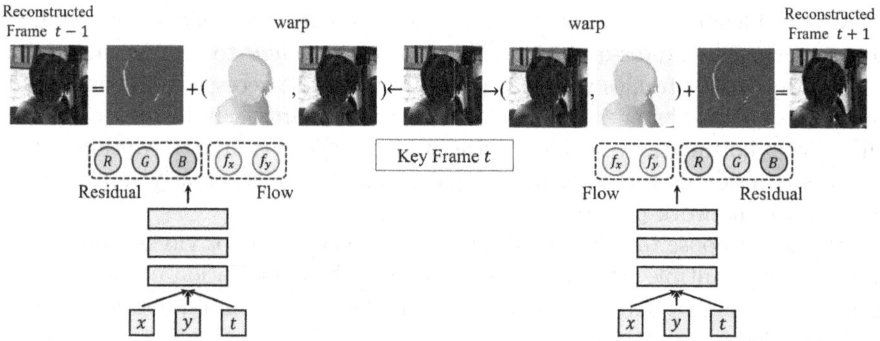

Fig. 1. Neural residual flow fields (NRFF) with a single reference frame. The gray-scale image on each side shows residuals, and the blue image in each parenthesis are optical flows. x, y, t are spatial and temporal coordinates. (Color figure online)

Although it has recently received widespread attention, its parameter efficiency has not been thoroughly investigated. In recent studies, neural fields require a large number of parameters to accurately represent signals [2,4,5]. Without enhancing parameter efficiency, transaction costs of neural fields are high because signals are stored as network parameters. This hinders us from utilizing it in many practical applications that may benefit from it.

In this paper, we study how to effectively represent videos using this new representation approach. A naive approach, equivalent to SIREN [2], would be to use a neural field as a function of spatial and temporal coordinates, $(r, g, b) = f_\theta(t, x, y)$, with continuous coordinates as inputs and three color channels as outputs. However, this approach does not exploit the spatial and temporal redundancy of video signals, and our goal is to improve the parameter efficiency by explicitly removing the redundancy.

We propose *Neural Residual Flow Fields (NRFF)*, a novel neural field scheme for video representation that leverages optical flows and residuals instead of raw colors. Our proposed scheme was inspired by standard video compression algorithms [11–13] that use motion information to deduplicate signals presented across frames. Optical flows allow us to reuse color values from other reference frames, which often preserve fine details. If those delicate patterns are on the surface of subjects inside video frames, reusing color values relieves the burden of learning similar patterns across frames and improves network parameter efficiency, rather than wasting network capacity on storing redundant raw signals for the entire frames.

Many video compression methods use block-wise motion estimations, where each motion vector is applied to all pixels within each block, significantly reducing the total number of required motion vectors. This is based on prior knowledge that smooth, low-frequency motion vector fields are often sufficient for describing movement between video frames. This gave us the idea that substituting optical flows for raw colors may greatly reduce the number of parameters.

However, video frames cannot be completely reconstructed solely by optical flows due to occlusion and dis-occlusion. Thus, we use *residuals* to recover the original signals from video frames with precision. It would not necessitate a large neural network since a substantial number of color values are likely to be reused from the motion information. To summarize, we train the networks to capture optical flows and residuals rather than raw color signals, which greatly improves the efficiency of network parameters.

We also propose to use more than one reference frame for video frame generation. Using multiple reference frames enables the exploitation of visible information over many reference frames, each of which may contain distinct exposed and occluded information.

In addition, we also suggest splitting the network into two subnetworks: one for optical flows and the other for residuals. Separating optical flows and residuals, assuming they have different dynamics, would improve the quality, and the experiment results support this.

Experimental results show that the proposed method significantly outperforms the baseline method, which relies on raw colors. Given similar sizes, *NRFF* reconstructs video frames more clearly and sharply. Quantitative results on the MPI Sintel [14] and UVG [15] datasets reveal that our method significantly outperformed its counterpart in terms of standard image reconstruction metrics (PSNR: 31.2 to 37.4, SSIM: 0.82 to 0.95). Although the proposed method is an initial attempt to improve the parameter efficiency of neural fields for videos, it also performs favorably with H.264 [11], a standard video compression algorithm. With the multi-reference frames method, *NRFF* matches the performance of H.264 using small group of pictures (GOP) on some videos without any model compression techniques such as pruning and entropy coding (Fig. 6).

In summary, our contribution is threefold.

- We show that using optical flows and residuals as output instead of colors can significantly improve video quality.
- We propose to use multiple reference frames for frame reconstruction, and this improves video quality without increasing the network size.
- We demonstrate that parameter efficiency can be improved by using separate neural fields-one for optical flows and the other for residuals-in addition to using a shared network for each group of pictures.

2 Related Works

Neural Fields. Neural fields map spatial and temporal coordinates to certain physical quantities [16]. Since a wide variety of tasks can be represented as fields, this approach has recently gained popularity and been used in several tasks, such as image representation [1,4,17], audio representation [2], 3D shape [7,18], and novel view synthesis [5,19–22]. Thanks to recent innovations [1,2,5,17,23], it can faithfully reconstruct even high frequency signals. Neural fields have also been applied to signals having both spatial and temporal dimensions, including video representation and novel view synthesis in 4D space [24–28]. As a new way of representing data, several attempts have been made to compress various signals, such as images [29–31] and videos [26,32]. However, the compression performance of neural field-based methods is currently far behind standard state-of-the-art compression algorithms.

Learning Based Video Compression. There have been several data-driven attempts to utilize neural networks for efficient video representation. Convolutional neural networks (CNNs) and auto-encoder architectures have been used to compress video signals [33,34]. These works train encoder and decoder networks on large-scale datasets and test them on unseen videos to achieve high video compression rates, assuming decoder networks are already shared and only core video information needs to be stored or sent. DVC [34] has achieved compression rates comparable to or slightly better than standard video compression algorithms, e.g., H.264 and H.265. However, these approaches are inherently vulnerable to the biases of training datasets. SIREN [2] is an attempt to represent various forms of signals, including videos, through neural fields, however, it does not consider its parameter efficiency. NeRV [26] is a variant of neural fields that achieves video compression performance comparable to that of the standard video compression algorithm, H.264. However, NeRV gives up two degrees of freedom and uses only time coordinates t as inputs for efficient rendering. Our proposed method, *NRFF*, explicitly removes the redundancy by combining residuals and flows with the help of key frames. IPF [32] is a concurrent work using optical flows and residuals for video compression. Ours and IPF are different in a number of ways. While IPF employs a separate network for each frame, we use a shared network for each group of pictures and take advantage of temporal redundancy between frames. We also propose to use more than one reference frame, whereas IPF only proposes to use one.

Optical Flow Estimation. Optical flow has been a core component of various computer vision tasks. Since the work of Horn and Schunck [35], many improvements have been proposed to make optical flow more accurate [35]. Recently, employing neural networks to improve optical flow estimation [36–38] rather than traditional algorithm-based methods [35] has been successful. To achieve more robust learning based optical flow methods, the ground truth optical flows have been collected by using an animated film and computer graphics [14,36]. Several strategies have been proposed to improve estimating performance by using occlusion masks [39] or transformer-based operations [40,41].

3 Method

Figure 1 illustrates an overview of optical flow-based neural fields for video representation. Our proposed neural fields generate optical flows for a given spatial and temporal coordinates (Sect. 3.1) to warp reference frames. In addition to optical flows, the residuals are also generated by neural fields, and these generated residuals are added to the warped frame to complete the reconstruction (Sect. 3.2). To improve video quality, we use more than one reference frame for video frame reconstruction. For each GOP, we use two neural networks: one for optical flows and the other for residuals.

3.1 Dense Optical Flow Estimation

Standard video compression methods use block-wise motion information to improve compression efficiency [11,13]. However, because of this algorithmic nature, results often contain block-shaped distortions or artifacts, which necessitate deblocking filters. We propose using dense, pixel-wise motion vectors through neural fields. It is known that neural networks can efficiently represent smooth signals [1,42]. Thus, we sample motion vectors densely by injecting dense coordinates and using small-size neural networks as neural fields.

3.2 Image Warping and Completion

We use generated flow fields to warp reference frames to predict target video frames. Let I be an image function that takes spatial and temporal coordinates (x, y, t) as inputs and produces corresponding colors (r, g, b) as outputs. A reference frame and a warped image are denoted as $\hat{I}_{ref}$ and $\tilde{I}$, respectively. The reference frame can be the key frame or a neighboring frame. Then, a warped image at time t, $\tilde{I}(x, y, t)$ can be written as

$$(\Delta x^t, \Delta y^t) = F_{\text{flow}}(x, y, t; \theta), \tag{1}$$

$$\tilde{I}(x, y, t) = \text{Interp}(\hat{I}_{ref}(t), x + \Delta x^t, y + \Delta y^t). \tag{2}$$

$F_{\text{flow}}(t)$, which is parameterized by θ, estimates optical flows between two frames (the video frame to be reconstructed and the reference frame). $\text{Interp}(\cdot)$ warps a video frame by a bicubic interpolation.

Reconstructing a video frame by simply warping the source video frame is likely to contain artifacts. Artifacts can be caused by a variety of factors, including imprecise optical flow estimation, occlusions and disocclusions between video frames, and accumulated errors from the recursive frame generation process. To alleviate the issues, we use residuals along with the optical flows.

The final equation for image completion can be written as

$$\hat{I}(x, y, t) = F_{\text{res}}(x, y, t; \psi) + \tilde{I}(x, y, t), \tag{3}$$

where F_{res} denotes the residual estimator with its own parameter ψ.

3.3 Key Frames

Inspired by the standard video compression algorithms, we store a key frame per GOP as a standalone frame so that other frames directly or indirectly depend on the key frame. Regarding key frames, there are two important considerations: the image quality of key frames and the location of key frames.

With a fixed total size, there is a trade-off between the key frame quality and the network size. Given a high-quality key frame, a network can exploit the high-frequency details of the key frame. However, since high-quality key frames require a larger memory size, the network size must be smaller in order to maintain the total size. Having a small-sized network tends to have difficulties covering long, dynamic frames. In contrast, a relatively low-quality key frame relegates expressing fine-details to the network. A large network handles long dynamic frames relatively easily, however, because the reference frames lack fine-details, the network must learn to compromise between learning fine-details and learning the optical flows of long and dynamic frames. We empirically found that the optimal ratio of the network size and the key frame quality (or the size) is related to the total number of frames in the GOP. The larger the GOP is, the larger the network size should be, and vice versa. The detailed experimental results can be found in the supplementary material.

Among possible positions, we chose the middle frame as the key frame to minimize the errors caused by the distance from the key frame. Experimental results showed that selecting the middle frame as the key frame is better in terms of reconstruction quality than selecting the first or last frame in the GOP.

We could use neural fields for key frame compression. However, existing neural field-based image compression methods usually underperform or are at most similar to standard image compression algorithms, such as JPEG, in terms of compression efficiency. Therefore, for keyframe compression, we employ the standard H.264 video compression technique. We still need more sophisticated training algorithms, input feature preprocessing techniques, and new network architectures to completely replace H.264 with neural fields in our proposed method.

3.4 End-to-End Training

Our proposed approach was designed to be differentiable throughout the entire process in order to build an end-to-end framework for video representation learning. We used MSE (mean squared error) loss to minimize the reconstruction errors. Let B be a mini-batch of frame indices excluding the key frame, then the loss function can be written as

$$L(\theta, \phi, \psi) = \frac{1}{Z} \sum_{t \in B} \sum_{x} \sum_{y} ||I(x, y, t) - \hat{I}(x, y, t)||^2. \tag{4}$$

$I(\cdot, \cdot, \cdot)$ is a ground truth color, and $\hat{I}(\cdot, \cdot, \cdot)$ is a reconstructed color. We multiplied the equation with the inverse of the constant Z to get the average loss over both time and space. Note that we excluded the key frame during the training process. The key frame will only be included during the evaluation.

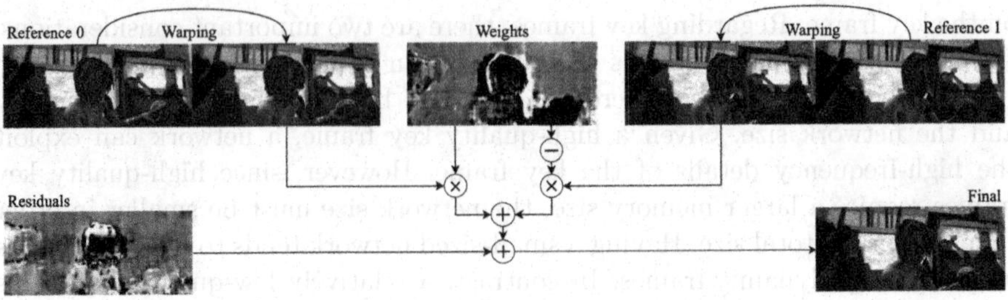

Fig. 2. Neural residual flow fields with multi-reference frames. ⊗ and ⊕ denote element-wise multiplication and addition, respectively. ⊖ in the figure denotes one minus inputs, that is, in the case of weights, one minus weights will be outputted.

3.5 Multi-reference Frames

To further improve the video quality, we propose to use more than one reference frame to reconstruct a video frame. For each video frame, the video representation network learns to warp the nearest two key frames, each of which is unique per GOP, to reconstruct a video frame. The key frame from the GOP, where a particular video frame to be reconstructed is located, is always referred to for frame reconstruction. A frame, preceding the key frame of the GOP in temporal order, uses the key frame from the nearest preceding GOP as the other reference. Likewise, a frame after the key frame will use the key frame of the nearest posterior GOP as another reference frame.

Overall architecture is presented in Fig. 2. First, two reference frames are warped by the generated optical flows. To combine information from multiple frames, we also need learned pixel-wise weights (from zero to one) that selectively aggregate two warped frames. The network would weigh more on closer pixels among two reference frames to aggregate warped images. The flow network now generates optical flow outputs for each reference frame and additional mixing weights. Lastly, the residuals are added to fully reconstruct a video frame.

3.6 Network Split

Simply replacing the output of neural fields from colors with optical flows and residuals means that optical flows and residuals are generated from the shared parameters. If the patterns of optical flows and residuals are similar, then sharing parameters may help to improve parameter efficiency. Otherwise, sharing parameters could degrade the quality. We analyze two different structures in the experimental section.

As proposed in IPF, we can also split in the temporal dimension. That is, we can use each network for each video frame. However, unlike the relationship between optical flows and residuals, neighboring video frames are usually highly correlated. To support our claim that separating the network for each video frame prevents the use of temporal redundancy and, thus, is less parameter efficient, we compare our method with IPF in the experimental section.

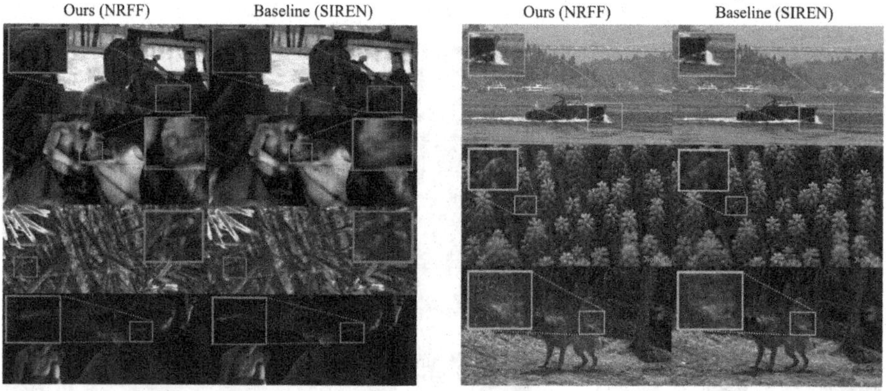

Fig. 3. The qualitative results between the proposed method and the baseline method under similar sizes (best viewed in color electronically). The left column examples are from the MPI Sintel dataset, and the right column examples are from the UVG dataset.

4 Experiments

We tested our approach on both synthetic and real-world datasets (MPI Sintel [14], UVG dataset [15]). We used the color-based neural field, SIREN [2], as the baseline. In addition to SIREN, we also compared our approach with the standard video compression algorithm, H.264 [11].

4.1 Dataset

The MPI Sintel [14] dataset was originally designed to evaluate the optical flow performance and it provides challenging and natural video sequences based on an open source animated film. We selected the MPI Sintel dataset for two reasons. First of all, it provides the ground truth optical flow, which allows us to evaluate the accuracy of the learned flows. Flow estimation is a core part of our algorithm, therefore, it is desirable to understand how flow estimation affects the final reconstruction. Second, it has been a good testbed for challenging scenarios, such as long range motion, motion blur, multi-frame analysis, and non-rigid motion. We tested our method on the entire 23 videos supported by MPI Sintel.

We also evaluated our method using seven 1080p videos from the Ultra Video Group (UVG) video collection [15], in order to compare it to other neural field-based video compression methods [26,32]. The dataset includes a variety of videos, ranging from nearly static scenes to fast-moving scenes.

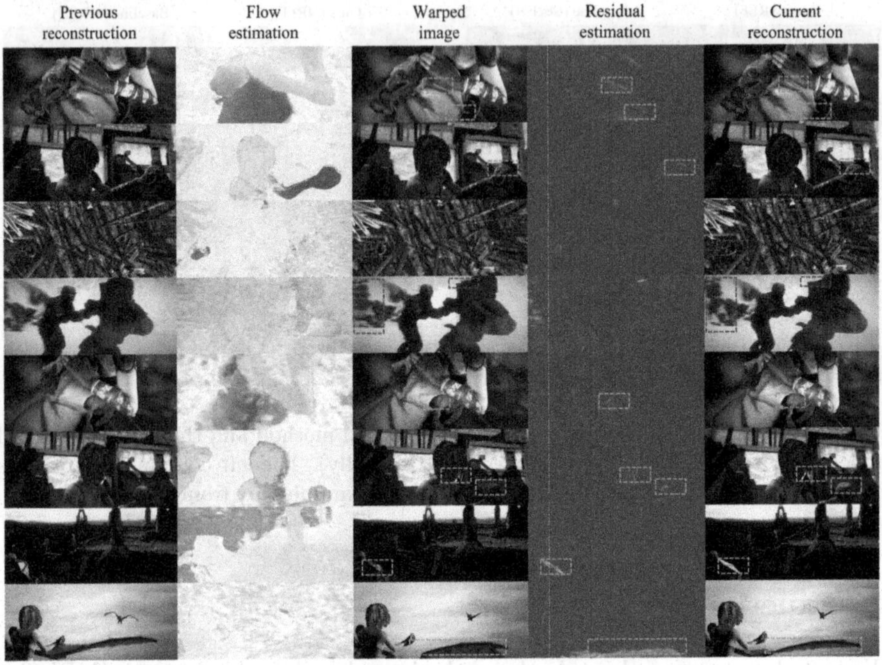

Fig. 4. The qualitative results on the MPI Sintel dataset.

4.2 Single Reference Frame Experiments

Experimental Setup. First, we used the SIREN backbone model and only changed the output linear layer to evaluate the effect of replacing colors with optical flows and residuals. For frame reconstruction, we applied optical flows recursively; that is, starting from the key frame, the reconstructed frame is used as the reference frame of the next frame. We used 16-bit precision weights for both ours and the baseline (SIREN) to halve the network size without quality degradation.

Our proposed method and the baseline are evaluated using the MPI Sintel video dataset. The resolution of all videos was reduced by half, resulting in 218×512 pixels. We set the size of the GOP to seven, so that each key frame covers six neighboring frames. In the case of a video containing 28 frames, for example, we trained four models for the video. We trained each model for total 30K iterations, and training the whole batch of frames at once was considered a single iteration. We used the Adam optimizer with a learning rate of 0.0005.

Results. As shown in Fig. 3, replacing colors with optical flows and residuals significantly improves video quality for similar model sizes. Our method effectively preserves fine details such as a person's beard, water drops, and bees, whereas the baseline method generates blurry outputs.

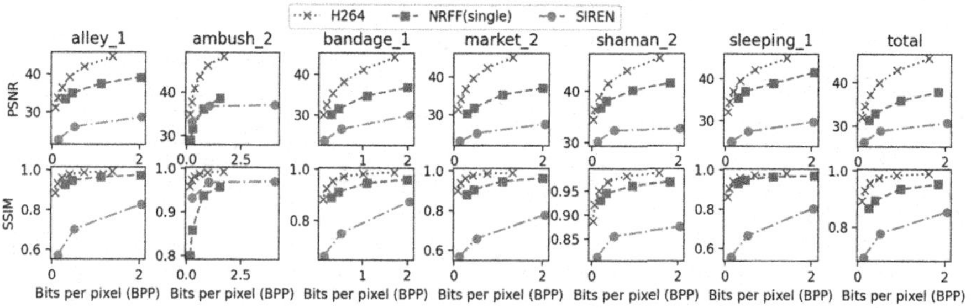

Fig. 5. The quantitative results of single reference frame experiments on the MPI Sintel dataset.

Figure 4 shows the qualitative results of single reference frame experiments in detail, including intermediate steps to complete video frames. The parts that need to be corrected by residuals are highlighted in Fig. 4. As we expected, the optical flow estimation is not necessarily accurate enough to reconstruct a video frame correctly. In fact, the final optical flow estimation includes many artifacts, particularly in the background regions. The residuals can successfully compensate for those artifacts created by inaccurate optical flows.

Due to the page limits, we presented the results of six videos and the average of all 23 videos in Fig. 5. We used two commonly used metrics in image reconstruction tasks; peak signal noise ratio (PSNR) and structural similarity index measure (SSIM). On average, our method enhanced PSNR from 30 to 37 and SSIM from 0.85 to 0.95 at around 2 bits per pixel. For example, in *alley_1*, we obtained 39.09 PSNR and 0.972 SSIM, as opposed to the baseline performance of 28.71 PSNR and 0.827 SSIM under similar model sizes. In *sleeping_1*, we got 41.56 PSNR and 0.970 SSIM, while the baseline only reached 29.78 PSNR and 0.806 SSIM. As illustrated in Fig. 3, this resulted in significant increases in visual quality.

There are a few exceptions where we did not gain much improvement. One example is demonstrated in the fourth row of Fig. 4. The scene is foggy and blurry, and the video shows a lot of camera movement. These factors, we believe, are the causes of the slightly lower quality of our approach. We hypothesize that, given a fixed number of parameters, inaccurate optical flow estimation may not be sufficiently compensated by residuals in some scenes. We also reported the performance of the H.264 video compression method [11] for your information.

4.3 Multi-reference Frames and Network Splitting Experiments

Experimental Setup. This section analyzes the effect of using multi-reference frames and separating networks on video quality. We replaced the activation function of the neural networks from sinusoidal activation [2] to swish function ($f(x) = x \times \mathsf{sigmoid}(\beta x)$) and used positional encoding [5]. Furthermore, we reused hidden layers to improve parameter efficiency. We used H.264 for key frame compression, and the middle frame of the GOP was chosen as the

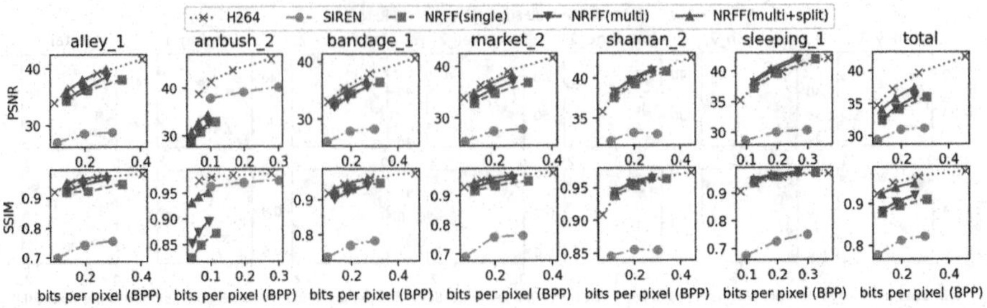

Fig. 6. The quantitative results of multi-reference frames experiments on the MPI Sintel dataset. The last column is the average performance on all 23 videos.

key frame. To modulate the total size for video representation, we controlled the quality factor in H.264. As for the network size, each network width was automatically set so that the total size of the network is proportional to the corresponding key frame size.

We ran experiments using full-resolution MPI Sintel videos. We set the size of the GOP to five for every method, including H.264. We also applied 16 bit precision in this experiment to both the baseline (SIREN) and our proposed methods for a fair comparison. We trained each model for 5K epochs, and the batch size was set to one. Since the single reference NRFF becomes unstable when the batch size is one, the whole batch of the GOP was used as the minibatch for the single reference NRFF and the number of epochs was therefore increased to 25K. As for the initial learning rate, we used optimal learning rate for each method: 1e-3 for SIREN and the single reference NRFF, and 1e-2 for the multiple reference NRFF. The NRFF network size was set to be equal to one-fourth of the key frame size. For example, a network with a key frame size of 10k bytes was set to have approximately 1,250 parameters of 16-bit precision.

We also evaluated each method using videos from the UVG dataset (the first 100 frames of each). In this experiment, we unbridled the GOP size limitation for SIREN and H.264. That is, we use the automatically chosen GOP size in the case of H.264, which is 100, and we use the same size for SIREN. For our method, the GOP size was set at 15. Since we empirically found that large GOP size requires an increased ratio of the network size and the key frame, we set the ratio to be approximately 0.5.

Results. Figure 6 and Fig. 7 show the quantitative results on the MPI Sintel and UVG datasets, respectively. The supplementary materials provide all of the experiment results on the MPI Sintel. First of all, the results indicate that using optical flows can enhance video quality for both synthetic and non-synthetic scenes, including static and dynamic scenes. Second, referencing more than one

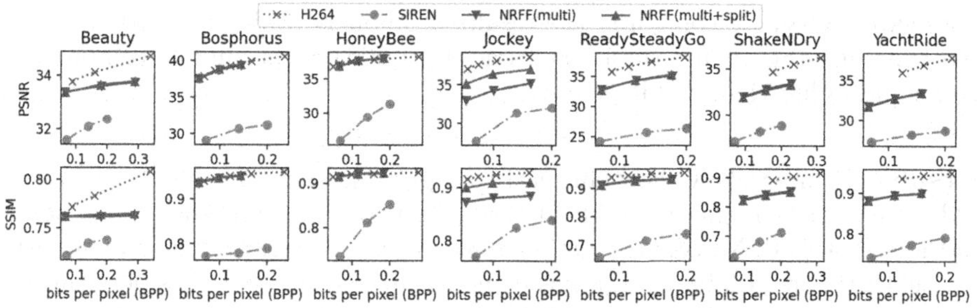

Fig. 7. The quantitative results of multi-reference frames experiments on the UVG dataset.

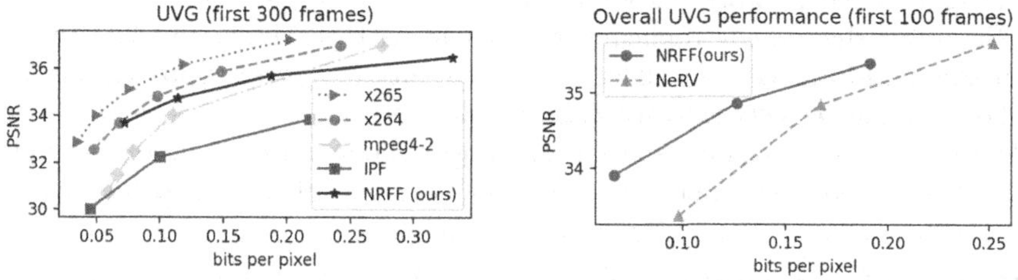

(a) comparison with IPF

(b) comparison with NeRV

Fig. 8. Overall compression performance on the UVG videos. All results in Fig. 8a were copied from IPF [32] except NRFF (ours).

frame enhances video quality without increasing the network size. Lastly, dividing the network into two subnetworks never degrades video quality and can significantly improve quality for some videos. This demonstrates that our assumption that optical flows and residuals have different dynamics is valid. Even without network compression methods, our method performs on par with or better than the H.264 algorithm on some videos, such as *shaman_2* and *sleeping_1* from MPI Sintel, and Bosphrous and HoneyBee from UVG. The optical flow-based method might worsen the video quality in some videos, such as *ambush_2*, which contains fast and large movements. However, the overall performance can be significantly improved by using optical flows instead of raw colors, as shown in Fig. 7, and the rightmost column of Fig. 6

4.4 Comparison with Other Neural-Field-Based Video Representation

Experimental Setup. We compared our method (multi-reference model with two subnetworks) to other neural field-based methods (IPF [32], NeRV) using the UVG dataset. Since these two methods are evaluated in different experimental settings, we separately compared them with ours. First, to compare ours with IPF, we used the same experimental settings as in IPF (first 300 frames with the size of the group of pictures (GOP) of five) to train our models. To compare ours with NeRV, we compared the results of publicly available NeRV codes in our experimental settings (first 100 frames for 1,500 epochs, and no compression techniques other than 16-bit precision).

Results. Figure 8 shows how efficiently our proposed neural fields can express a video compared to other neural field-based methods, including the current state-of-the-art method, NeRV. Our method outperforms a concurrent flow-based neural fields, IPF by a large margin, as demonstrated in Fig. 8a. This performance gap is significant considering that IPF has adopted additional compression techniques, such as quantization and entropy encoding, to improve compression performance, while ours does not. We conjecture that this performance gain pertains to using a shared network per GOP.

Figure 8b shows that replacing raw colors with residuals and flows results in better performance, especially in the low bits per pixel (bpp) region, even without enhancing network structure. Please note that NeRV relies on a neural field architecture that limits spatial sampling and only allows temporal sampling to improve compression performance, while ours permits spatial sampling.

4.5 Spatial and Temporal Interpolation

The neural fields have several advantages that no other video compression method has, one of which being the ability to sample values from arbitrary spatial and temporal coordinates, even at an unobserved point during encoding. To show this advantage of neural fields, we ran two experiments: spatial and temporal interpolation. For spatial interpolation, we first trained a neural network to represent a low-resolution video (with a resolution of (480, 270)). After that, without any post-processing methods, we simply upscaled the resolution four times in both height and width by sampling values in a much more dense grid. This is possible because neural fields take continuous coordinates as inputs. For temporal interpolation, we trained a neural network as in the main experiment with the fixed reference frames as for the multi-reference frame experiments. And then, the intermediate frame (for example, a frame in between the 5th and 6th frame) was sampled simply by injecting the corresponding temporal coordinates.

As shown in Fig. 9, simple dense sampling results in smooth interpolation in both time and space, even without any modifications and extra training techniques. Spatial interpolation of neural fields results in much smoother outputs than bilinear interpolation. For a fast-moving scene, interpolating two adjacent frames results in a blurry frame. However, our proposed method manages to represent the intermediate frame much more clearly. In addition to the fact that NRFF inherits the good properties of neural fields, the proposed method can offer new opportunities to improve the parameter efficiency of neural fields in other domains, such as NeRF [5] and light-field imaging, to name a few.

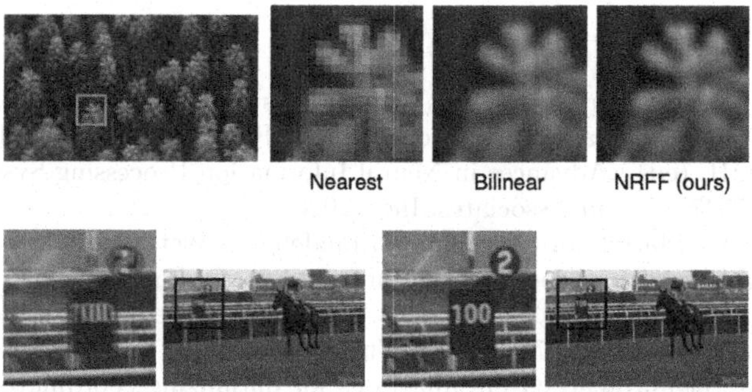

Fig. 9. The spatial and temporal interpolation. The first row shows the results of each method's spatial super resolution, upscaling from (480, 270) to (1920, 1080). The second row shows the results of temporal interpolation, which samples an intermediate frame that was not seen during video encoding.

5 Conclusions and Discussion

We present a way to exploit neural fields for efficient video representations. The video quality was greatly enhanced by explicitly leveraging reference frames through optical flows. The proposed approach, *Neural Residual Flow Fields (NRFF)* maintains smooth and less complex signals, which allows us to achieve more compact representations while maintaining quality.

Although the results were promising, there is still room for improvement. We observed that neural fields (or implicit neural representations) require a large number of parameters and long training iterations to capture high-frequency details. We believe that resolving this problem would significantly enhance encoding speed and make neural fields more accessible in various cases.

Note that we achieved fairly good compression rates without any model compression techniques, except for 16-bit precision weights. Incorporating unaccommodated network compression techniques into our proposed method would improve the performance much further. Weight pruning, entropy coding, and knowledge distillation could all be promising directions to investigate.

There are a handful of promising research directions that can further make this emerging representation more attractive. We believe we have only scratched the surface. A better understanding of the efficiency of neural network parameters in general might help answer a fundamental question about how a neural network preserves information in its parameters.

Acknowledgement. This research was supported by the Ministry of Science and ICT (MSIT) of Korea, under the National Research Foundation (NRF) grant (2021R1F1A1061259, 2022R1F1A1064184), Institute of Information and Communication Technology Planning Evaluation (IITP) grants for the AI Graduate School program (IITP- 2019-0-00421), the ITRC (Information Technology Research Center) support program (IITP-2021-0-02052), the ICT Creative Consilience program (IITP-2020-0-01821), and the Artificial Intelligence Innovation Hub program (IITP-2021-0-02068).

References

1. Tancik, M., et al.: Fourier features let networks learn high frequency functions in low dimensional domains. In: Larochelle, H., Ranzato, M., Hadsell, R., Balcan, M.F., Lin, H. (eds.) Advances in Neural Information Processing Systems, vol. 33, pp. 7537–7547. Curran Associates, Inc. (2020)
2. Sitzmann, V., Martel, J., Bergman, A., Lindell, D., Wetzstein, G.: Implicit neural representations with periodic activation functions. In: Larochelle, H., Ranzato, M., Hadsell, R., Balcan, M.F., Lin, H. (eds.) Advances in Neural Information Processing Systems, vol. 33, pp. 7462–7473. Curran Associates, Inc. (2020)
3. Tancik, M., et al.: Learned initializations for optimizing coordinate-based neural representations. In: Proceedings of the IEEE/CVF Conference on Computer Vision and Pattern Recognition (CVPR) (2021)
4. Martel, J.N.P., Lindell, D.B., Lin, C.Z., Chan, E.R., Monteiro, M., Wetzstein, G.: Acorn: adaptive coordinate networks for neural scene representation. ACM Trans. Graph. **40** (2021)
5. Mildenhall, B., Srinivasan, P.P., Tancik, M., Barron, J.T., Ramamoorthi, R., Ng, R.: NERF: representing scenes as neural radiance fields for view synthesis. In: Proceedings of the European Conference on Computer Vision (ECCV) (2020)
6. Cho, J., Nam, S., Rho, D., Ko, J.H., Park, E.: Streamable neural fields. In: Proceedings of the European Conference on Computer Vision (ECCV) (2022)
7. Mescheder, L., Oechsle, M., Niemeyer, M., Nowozin, S., Geiger, A.: Occupancy networks: learning 3D reconstruction in function space. In: Proceedings of the IEEE/CVF Conference on Computer Vision and Pattern Recognition (CVPR) (2019)
8. Raissi, M., Perdikaris, P., Karniadakis, G.E.: Physics-informed neural networks: a deep learning framework for solving forward and inverse problems involving nonlinear partial differential equations. J. Comput. Phys. **378**, 686–707 (2019)
9. Skorokhodov, I., Ignatyev, S., Elhoseiny, M.: Adversarial generation of continuous images. In: Proceedings of the IEEE/CVF Conference on Computer Vision and Pattern Recognition (CVPR) (2021)
10. Chen, Y., Liu, S., Wang, X.: Learning continuous image representation with local implicit image function. In: Proceedings of the IEEE/CVF Conference on Computer Vision and Pattern Recognition (CVPR) (2021)
11. Wiegand, T., Sullivan, G., Bjontegaard, G., Luthra, A.: Overview of the H. 264/AVC video coding standard. IEEE Trans. Circ. Syst. Video Technol. **13**, 560–576 (2003)
12. Le Gall, D.: MPEG: a video compression standard for multimedia applications. Commun. ACM **34**, 46–58 (1991)
13. Sullivan, G.J., Ohm, J.R., Han, W.J., Wiegand, T.: Overview of the high efficiency video coding (HEVC) standard. IEEE Trans. Circuits Syst. Video Technol. **22**, 1649–1668 (2012)
14. Butler, D.J., Wulff, J., Stanley, G.B., Black, M.: A naturalistic open source movie for optical flow evaluation. In: Proceedings of the European Conference on Computer Vision (ECCV) (2012)
15. Mercat, A., Viitanen, M., Vanne, J.: UVG dataset: 50/120fps 4K sequences for video codec analysis and development. In: Proceedings of the 11th ACM Multimedia Systems Conference, MMSys 2020, pp. 297–302. Association for Computing Machinery, New York (2020)

16. Xie, Y., et al.: Neural fields in visual computing and beyond. In: Computer Graphics Forum (2022)
17. Hertz, A., Perel, O., Giryes, R., Sorkine-Hornung, O., Cohen-Or, D.: SAPE: spatially-adaptive progressive encoding for neural optimization. In: Advances in Neural Information Processing Systems (2021)
18. Atzmon, M., Lipman, Y.: SAL: sign agnostic learning of shapes from raw data. In: Proceedings of the IEEE/CVF Conference on Computer Vision and Pattern Recognition (CVPR) (2020)
19. Martin-Brualla, R., Radwan, N., Sajjadi, M.S.M., Barron, J.T., Dosovitskiy, A., Duckworth, D.: NeRF in the wild: neural radiance fields for unconstrained photo collections. In: Proceedings of the IEEE/CVF Conference on Computer Vision and Pattern Recognition (CVPR) (2021)
20. Zhang, K., Riegler, G., Snavely, N., Koltun, V.: NERF++: analyzing and improving neural radiance fields. arXiv preprint arXiv:2010.07492 (2020)
21. Yariv, L., et al.: Multiview neural surface reconstruction by disentangling geometry and appearance. In: Advances in Neural Information Processing Systems (2020)
22. Yu, A., Ye, V., Tancik, M., Kanazawa, A.: pixelNeRF: neural radiance fields from one or few images. In: Proceedings of the IEEE/CVF Conference on Computer Vision and Pattern Recognition (CVPR) (2021)
23. Barron, J.T., Mildenhall, B., Tancik, M., Hedman, P., Martin-Brualla, R., Srinivasan, P.P.: Mip-NeRF: a multiscale representation for anti-aliasing neural radiance fields. In: Proceedings of the IEEE/CVF International Conference on Computer Vision (ICCV) (2021)
24. Li, Z., Niklaus, S., Snavely, N., Wang, O.: Neural scene flow fields for space-time view synthesis of dynamic scenes. In: Proceedings of the IEEE/CVF Conference on Computer Vision and Pattern Recognition (CVPR) (2021)
25. Gao, C., Saraf, A., Kopf, J., Huang, J.B.: Dynamic view synthesis from dynamic monocular video. In: Proceedings of the IEEE/CVF International Conference on Computer Vision (ICCV) (2021)
26. Chen, H., He, B., Wang, H., Ren, Y., Lim, S.N., Shrivastava, A.: NeRV: neural representations for videos. In: Advances in Neural Information Processing Systems (2021)
27. Peng, S., et al.: Neural body: implicit neural representations with structured latent codes for novel view synthesis of dynamic humans. In: Proceedings of the IEEE/CVF Conference on Computer Vision and Pattern Recognition (CVPR) (2021)
28. Chen, Z., et al.: VideoINR: learning video implicit neural representation for continuous space-time super-resolution. In: Proceedings of the IEEE/CVF Conference on Computer Vision and Pattern Recognition (CVPR) (2022)
29. Dupont, E., Golinski, A., Alizadeh, M., Teh, Y.W., Doucet, A.: COIN: compression with implicit neural representations. In: Neural Compression: From Information Theory to Applications - Workshop @ ICLR 2021 (2021)
30. Strümpler, Y., Postels, J., Yang, R., Van Gool, L., Tombari, F.: Implicit neural representations for image compression. arXiv preprint arXiv:2112.04267 (2021)
31. Dupont, E., Loya, H., Alizadeh, M., Goliński, A., Teh, Y.W., Doucet, A.: COIN++: data agnostic neural compression. arXiv preprint arXiv:2201.12904 (2022)
32. Zhang, Y., van Rozendaal, T., Brehmer, J., Nagel, M., Cohen, T.: Implicit neural video compression. In: ICLR Workshop on Deep Generative Models for Highly Structured Data (2022)

33. Rippel, O., Nair, S., Lew, C., Branson, S., Anderson, A.G., Bourdev, L.: Learned video compression. In: Proceedings of the IEEE/CVF International Conference on Computer Vision (ICCV) (2019)
34. Lu, G., Ouyang, W., Xu, D., Zhang, X., Cai, C., Gao, Z.: DVC: an end-to-end deep video compression framework. In: Proceedings of the IEEE/CVF Conference on Computer Vision and Pattern Recognition (CVPR) (2019)
35. Horn, B.K., Schunck, B.G.: Determining optical flow. Artif. Intell. **17**, 185–203 (1981)
36. Dosovitskiy, A., et al.: Flownet: learning optical flow with convolutional networks. In: Proceedings of the IEEE International Conference on Computer Vision (ICCV) (2015)
37. Ilg, E., Mayer, N., Saikia, T., Keuper, M., Dosovitskiy, A., Brox, T.: Flownet 2.0: evolution of optical flow estimation with deep networks. In: Proceedings of the IEEE/CVF Conference on Computer Vision and Pattern Recognition (CVPR) (2017)
38. Hur, J., Roth, S.: Iterative residual refinement for joint optical flow and occlusion estimation. In: Proceedings of the IEEE/CVF Conference on Computer Vision and Pattern Recognition (CVPR) (2019)
39. Zhao, S., Sheng, Y., Dong, Y., Chang, E.I.C., Xu, Y.: Maskflownet: asymmetric feature matching with learnable occlusion mask. In: Proceedings of the IEEE/CVF Conference on Computer Vision and Pattern Recognition (CVPR) (2020)
40. Teed, Z., Deng, J.: RAFT: recurrent all-pairs field transforms for optical flow. In: Proceedings of the European Conference on Computer Vision (ECCV) (2020)
41. Jiang, S., Campbell, D., Lu, Y., Li, H., Hartley, R.: Learning to estimate hidden motions with global motion aggregation. In: Proceedings of the IEEE/CVF International Conference on Computer Vision (ICCV) (2021)
42. Rahaman, N., et al.: On the spectral bias of neural networks. In: International Conference on Machine Learning (ICML), pp. 5301–5310 (2019)

Visual Explanation Generation Based on Lambda Attention Branch Networks

Tsumugi Iida[1(✉)], Takumi Komatsu[1], Kanta Kaneda[1], Tsubasa Hirakawa[2], Takayoshi Yamashita[2], Hironobu Fujiyoshi[2], and Komei Sugiura[1]

[1] Keio University, Tokyo, Japan
{tiida,tak3k_1999,k.kaneda,komei.sugiura}@keio.jp
[2] Chubu University, Kasugai, Japan
{hirakawa,takayoshi,fujiyoshi}@isc.chubu.ac.jp

Abstract. Explanation generation for transformers enhances accountability for their predictions. However, there have been few studies on generating visual explanations for the transformers that use multidimensional context, such as LambdaNetworks. In this paper, we propose the Lambda Attention Branch Networks, which attend to important regions in detail and generate easily interpretable visual explanations. We also propose the Patch Insertion-Deletion score, an extension of the Insertion-Deletion score, as an effective evaluation metric for images with sparse important regions. Experimental results on two public datasets indicate that the proposed method successfully generates visual explanations.

Keywords: Lambda networks · Transformer · Attention

1 Introduction

Visual explanations for deep neural networks are important in terms of enhancing accountability about biomedical image processing and providing scientific insight to experts. Specifically, in the task of predicting solar flares, for which the theoretical background remains unclear, visual explanations using magnetograms can provide scientists with insights into underlying solar activities.

In this paper, we focus on the task of visualizing important regions in an image as a visual explanation of the model's decisions. In this task, pixels that contributed to the model's prediction should be attended.

Explanation generation for convolutional neural networks has been studied intensively in recent years [28,34]. On the other hand, there have been few studies on generating visual explanations for transformers, especially those based on Lambda [3]. In addition, standard metrics for visual explanations (e.g. the Insertion-Deletion score [24]) are sometimes inappropriate for images with sparse important regions.

Supplementary Information The online version contains supplementary material available at https://doi.org/10.1007/978-3-031-26284-5_29.

L. Wang et al. (Eds.): ACCV 2022, LNCS 13842, pp. 475–490, 2023.
https://doi.org/10.1007/978-3-031-26284-5_29

Given this background, we propose the Lambda Attention Branch Networks (LABN), which generates interpretable visual explanations for Lambda-based transformers. In addition, we introduce the loss used in saliency guided training [12] to reduce the importance of regions irrelevant to the prediction. Figure 1 shows an overview of our method. It is composed of three modules: the Lambda Feature Extractor (LFE), Lambda Attention Branch (LAB), and Lambda Perception Branch (LPB). Attention in the original Lambda layer [3] is sometimes not clear as visual explanation. On the one hand, we obtain a clear visual explanation by introducing a branch structure dedicated to visual explanation generation.

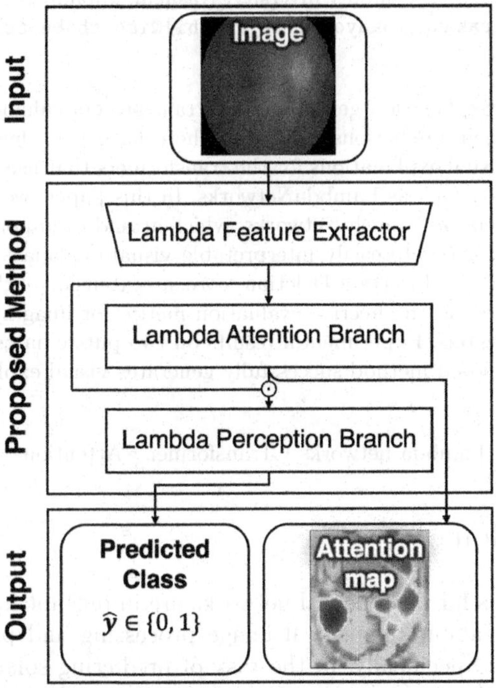

Fig. 1. Overview of our method.

We also propose the Patch Insertion-Deletion (PID) score, an extension of the Insertion-Deletion score, as an effective evaluation metric for images with sparse important regions. Unlike the Insertion-Deletion score, the PID score evaluates visual explanations in a patch-wise manner.

The main contributions of this study are as follows:

- We propose the LABN, which has a parallel branching structure to obtain clear visual explanations than those provided by attention in the Lambda Layer.
- We propose the PID score, which is an extension of the Insertion-Deletion score, as an effective evaluation metric for images with sparse important regions.
- We introduce the loss used in saliency guided training to improve the quality of the visual explanations.

2 Related Work

There have been many studies in the field of explanation generation [4,5,10, 22,24,28,34]. [6] is a comprehensive survey paper in this field that categorizes the methods according to their approach. [14] summarizes the characteristics of vision transformer models for various tasks. In the field of visual explanation generation, standard image classification datasets are used (e.g., ImageNet [7], CIFAR10, CIFAR100 IDRiD [25]).

Explanation generation methods can be classified into backpropagation methods, perturbation methods, and other methods. Backpropagation methods generate explanations by focusing on the gradient during backpropagation; Integrated Gradients [31], SmoothGrad [29], FullGrad [30], CAM [37], Grad-CAM [28], PatternNet [15], and LRP [4,5] are typical backpropagation methods. For example, [31] is a method that satisfies the two axioms of Sensitivity and Implementation Invariance. It integrates the gradient to generate an explanation. The authors of [29] pointed out that gradient-based explanations are often noisy and proposed an averaging method to reduce the noise. Reference [12] reduces noise using saliency guided training, which brings the gradients of less important regions closer to zero. Reference [30] theoretically proved that the two axioms that the explanation should satisfy do not hold simultaneously. It then proposed a generation method that can balance both axioms.

By contrast, perturbation methods generate explanations from changes in the output when the input is perturbed; LIME [26], RISE [24] and Shapley Sampling [16] are typical perturbation methods. For example, [24] proposed a method to generate an explanation from the relationship between masked image and output.

The Attention Branch Network [10] (ABN), IA-CNN [36] and IA-RED^2 [22] are categorized as other methods. The authors of [22] argued that attention in the transformer layer is not always appropriate as an explanation. Reference [13] further showed that attention in the transformer layer can be controlled without changing the prediction. ABN, which uses a branch structure, has been extended as Multi-ABN [17], ABEN [21], and PonNet [18]. The authors of [36] proposed a method to generate the explanations for each key point by connecting parallel branching structures to the CNN.

Sanity Check [2], ROAR [11], and [9] are representative studies that evaluate visual explanations. Reference [2] evaluated an explanation by comparing the explanations generated by trained and randomized models. In [11], images are re-trained without the important regions and the difference in accuracy is calculated. Re-training can eliminate the effect of out-of-distribution data. Reference [9] evaluated robustness using the distance between explanations with and without samples in the training set.

The proposed method differs from other visual explanation generation methods (e.g., Attention Rollout [1]) in that it generates explanations using a branch structure rather than attention in the Lambda layer. The proposed method is also different from ABN in that it generates visual explanations and delete pixels simultaneously.

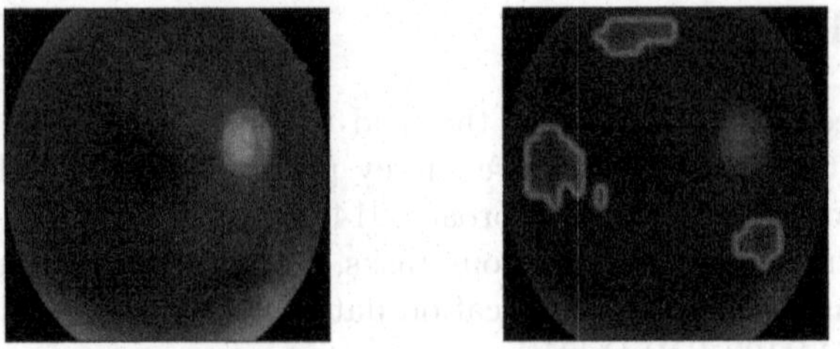

Fig. 2. Example data of visual explanation generation.

3 Problem Statement

In this paper, we focus on the task of visualizing important regions in an image as a visual explanation of the model's decisions. In particular, we focus on visual explanations of the Lambda-based transformer [3]. In this task, the pixels that contributed to the model's prediction should be attended. For example, Fig. 2 shows an example image from a standard dataset, Indian Diabetic Retinopathy Image Dataset (IDRiD) [25]. The left and right figures show the input image and the visual explanation, respectively.

The input and output of this task are defined as follows:

Input: Image $\boldsymbol{x} \in \mathbb{R}^{c_1 \times w_1 \times h_1}$
Output: Predicted probability for each class $\hat{\boldsymbol{y}} \in \mathbb{R}^{C}$

where C, c_1, w_1, and h_1 denote the number of classes, the number of channels, width and height of the input image, respectively. Additionally, the importance of each pixel is obtained as an attention map $\boldsymbol{\alpha} \in \mathbb{R}^{w_1 \times h_1}$, which is used as a visual explanation.

The Insertion-Deletion [24] and PID scores are used as an evaluation metrics (see Sect. 4 for details). Using the PID score, we can evaluate the match between the attention map and the region that contributed to the model's decision. In this paper, we assume that the model is based on a Lambda-based transformer. We also assume that the attention maps are not class specific. Furthermore, we focus on specific domains, such as medical care.

4 Proposed Method

Our method is inspired by transformer-based methods that capture the interactions among pixels, such as the Lambda ResNet [3]. Lambda ResNet can capture the relationship of the entire image with less computation than a typical self-attention mechanism used in simple vision transformers [8]. Since it does not assume patch partitioning, it is highly compatible with CNNs. Our method is also inspired by explanatory visualization methods, such as the Attention Branch Network [10], in the aspect of using a parallel branch structure. Attention Branch Network is a model that introduces a parallel branch structure to

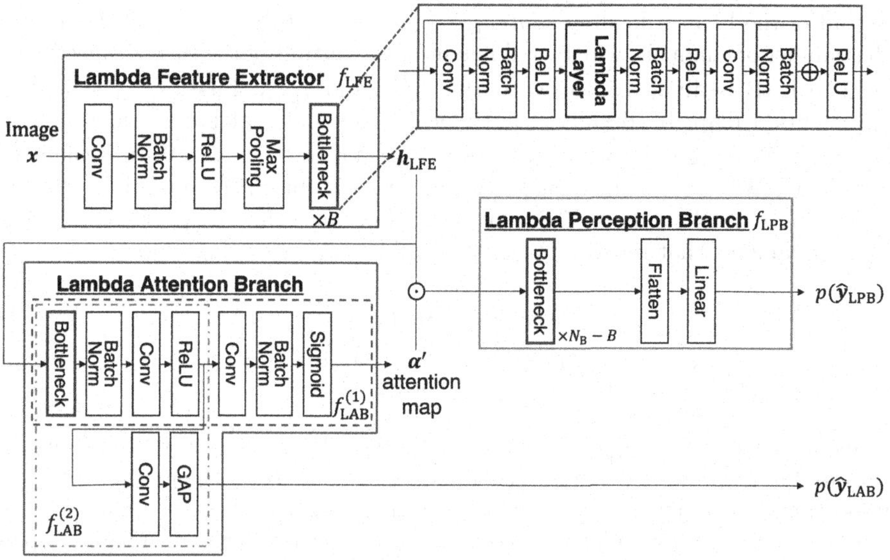

Fig. 3. Framework of LABN: our method consists of a Lambda Feature Extractor, Lambda Attention Branch and Lambda Perception Branch. Each module is explained in Sect. 4.1. "Conv" and "GAP" denote the convolution layer and global average pooling, respectively.

obtain an attention map. We use the parallel attention mechanism to obtain a clear visual explanation. This explanation is used to highlight important pixels for input.

The novelties of our approach are as follows:

- We introduce a structure that obtains an attention map as a parallel branch, which provides a clear explanation than a serial Lambda Layer.
- We propose the PID score, which is an extension of Insertion-Deletion score [24], as an effective explanation evaluation method for images with a large area of unimportant regions.

4.1 Structure

Figure 3 shows the structure of our method. It is composed of three modules: the LFE, LAB, and LPB. We assume that the backbone network contains N_B bottleneck layers. First, we divide the backbone network into the LFE and LPB modules at the B-th bottleneck layer. Next, we introduce the LAB, which is placed in parallel between the LFE and LPB.

The input of the LFE is image $\boldsymbol{x}$. The LFE contains B bottleneck layers and a batch normalization layer to extract features from $\boldsymbol{x}$. The bottleneck layer consists of a Lambda layer and multiple convolutional layers, batch normalization layers, and ReLU activation functions. The Lambda layer is described later. The output of the LFE is denoted as $\boldsymbol{h}_{\text{LFE}} \in \mathbb{R}^{c_2 \times w_2 \times h_2}$, where c_2, w_2 and h_2 denote the number of channels, width, and height of the output of the LFE, respectively.

The LAB is divided into two parts, $f^{(1)}_{\text{LAB}}$ and $f^{(2)}_{\text{LAB}}$. First, $f^{(1)}_{\text{LAB}}$ generates an attention map. It contains a bottleneck layer and a global average pooling layer. The input and output of $f^{(1)}_{\text{LAB}}$ are $\boldsymbol{h}_{\text{LFE}}$ and $\tilde{\boldsymbol{\alpha}} \in \mathbb{R}^{w_2 \times h_2}$. We upscale $\tilde{\boldsymbol{\alpha}}$ to obtain $\boldsymbol{\alpha} \in \mathbb{R}^{w_1 \times h_1}$ for the visual explanation. We obtain the final $\boldsymbol{\alpha}' \in \mathbb{R}^{w_2 \times h_2}$ by setting 0 to the values of $\tilde{\boldsymbol{\alpha}}$ that are below the value of θ_{α}, where θ_{α} is a hyperparameter that represents the threshold of the attention map.

$$\tilde{\boldsymbol{\alpha}} = f^{(1)}_{\text{LAB}}\left(\boldsymbol{h}_{\text{LFE}}\right), \tag{1}$$

$$\boldsymbol{\alpha}'_{ij} = \begin{cases} \tilde{\boldsymbol{\alpha}}_{ij} \ (\theta_{\alpha} < \tilde{\boldsymbol{\alpha}}_{ij}), \\ 0 \quad \text{(otherwise)}, \end{cases} \tag{2}$$

The reason for setting $\boldsymbol{\alpha}'_{ij} = 0$ for values less than θ_{α} is to use only the important regions for prediction. Because the input of LPB is $\boldsymbol{\alpha}' \odot \boldsymbol{h}_{\text{LFE}}$, the value for importance less than θ_{α} is 0. In other words, regions of low importance are masked to 0. Therefore, regions essential for prediction will be given higher alpha than other regions. This prevents the importance of regions that contribute to the model's predictions from decreasing.

The input and output of $f^{(2)}_{\text{LAB}}$ are $\boldsymbol{h}_{\text{LFE}}$ and $p(\hat{\boldsymbol{y}}_{\text{LAB}})$, respectively. Using $p(\hat{\boldsymbol{y}}_{\text{LAB}})$ in the loss function, we can train LAB directly for classification tasks. As a result, we can generate attention maps associated strongly with the classification result.

Next, f_{LPB} performs classification based on the outputs of the LFE and LAB. It contains $(N_{\text{B}} - B)$ bottleneck layers, flatten layers, and fully connected layers. The input and output of LPB are $\boldsymbol{\alpha}' \odot \boldsymbol{h}_{\text{LFE}}$ and $p(\hat{\boldsymbol{y}}_{\text{LPB}})$, respectively, where $\odot$ represents the Hadamard product.

4.2 Lambda Layer

We use the Lambda layer proposed in [3]. Self-attention [32] models relationships in a sequence by computing inner products. However, self-attention is computationally expensive and difficult to implement in images with long sequences. Unlike [8], Lambda Networks [3] reduced computational complexity by performing dimensionality reduction followed by inner product calculation. This allows the relationship between pixels to be modeled without having to split them into patches.

Figure 4 shows the structure of the Lambda layer. $\boldsymbol{h} \in \mathbb{R}^{c_3 \times w_3 \times h_3}$ denotes the input of the Lambda layer, where c_3, w_3 and h_3 denote the number of channels, width, and height of the input of the Lambda layer, respectively.

First, we generate the queries, keys and values as follows:

$$Q = V = \text{Conv}(\boldsymbol{h}), \tag{3}$$

$$K = \text{softmax}(\text{Conv}(\boldsymbol{h})). \tag{4}$$

Next, we compute the content lambda $\boldsymbol{\lambda}_\mathrm{c}$ and position lambdas $\boldsymbol{\lambda}_\mathrm{p}$:

$$\boldsymbol{\lambda}_\mathrm{c} = K^\top V, \tag{5}$$
$$\boldsymbol{\lambda}_\mathrm{p} = \mathrm{Conv}(V). \tag{6}$$

The output of the Lambda Layer $\boldsymbol{h}_\mathrm{L} \in \mathbb{R}^{c_3 \times w_3 \times h_3}$ is computed as follows:

$$\boldsymbol{h}_\mathrm{L} = (\boldsymbol{\lambda}_\mathrm{c} + \boldsymbol{\lambda}_\mathrm{p})^\top Q. \tag{7}$$

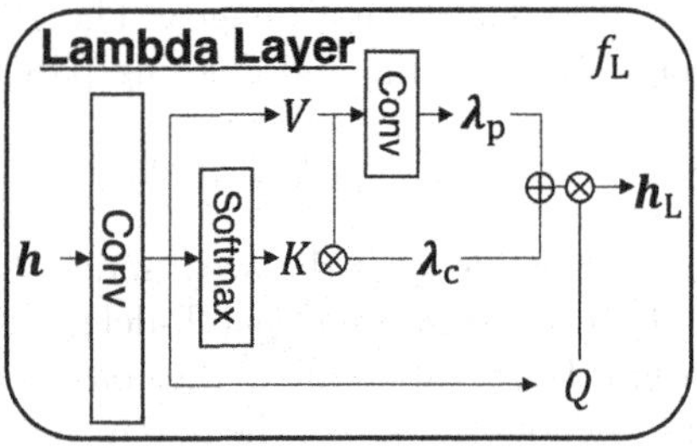

Fig. 4. Structure of Lambda layer.

We use the following loss function:

$$\mathcal{L} = \mathcal{L}_\mathrm{LPB} + \lambda_1 \mathcal{L}_\mathrm{LAB} + \lambda_2 \mathcal{L}_\mathrm{KL}, \tag{8}$$
$$\mathcal{L}_\mathrm{LPB} = \mathrm{CE}(\hat{\boldsymbol{y}}^{(\boldsymbol{x})}_\mathrm{LPB}, \boldsymbol{y}), \tag{9}$$
$$\mathcal{L}_\mathrm{LAB} = \mathrm{CE}(\hat{\boldsymbol{y}}_\mathrm{LAB}, \boldsymbol{y}), \tag{10}$$
$$\mathcal{L}_\mathrm{KL} = D_\mathrm{KL}\left(\hat{\boldsymbol{y}}^{(\boldsymbol{x})}_\mathrm{LPB} || \hat{\boldsymbol{y}}^{(\tilde{\boldsymbol{x}})}_\mathrm{LPB}\right), \tag{11}$$

where $\boldsymbol{y}$, CE, D_KL and λ_1, λ_2 denote the ground truth label, cross-entropy loss function, Kullback-Leibler divergence and weights, respectively. In addition, $\tilde{\boldsymbol{x}}$ and $\hat{\boldsymbol{y}}^{(\boldsymbol{z})}_\mathrm{LPB}$ denote the image masked by the bias image and the output of LPB when $\boldsymbol{z}$ is input, respectively. $\tilde{\boldsymbol{x}}$ and the bias image $\boldsymbol{b}$ are computed as follows:

$$\tilde{\boldsymbol{x}}_{ij} = \begin{cases} \boldsymbol{x}_{ij} \ (\theta_b < \tilde{\boldsymbol{\alpha}}_{ij}), \\ \boldsymbol{b}_{ij} \ (\text{otherwise}), \end{cases} \tag{12}$$
$$\boldsymbol{b}_{ij} = \frac{1}{N} \sum_{k=0}^{N} \boldsymbol{x}^{(k)}_{ij}, \tag{13}$$

where θ_b and $\boldsymbol{x}^{(k)}$ denote the hyperparameter of the mask ratio and the k-th sample of the training sets. Note that it is often impossible to define an appropriate bias image on generic datasets (e.g., ImageNet and VOC).

4.3 PID Score

The Insertion-Deletion (ID) score is a standard evaluation metric for visual explanations [24]. It measures the change in the probability of a predicted class when pixels are inserted according to the importance given by a method. However, the ID score often overestimates coarse explanations. This is inappropriate for problems with sparse important regions.

Therefore, we propose the PID score, an extension of the ID score, as an effective evaluation metric for images with sparse important regions. The PID score uses the maximum importance in the patch. This increases the influence of the details in the fine-grained explanation and allows for proper evaluation as well as for normal images. The PID score can appropriately evaluate such images by inserting and deleting patches. The PID score is defined as follows:

$$\mathrm{PID} = \mathrm{AUC}(\text{patch-insertion}) - \mathrm{AUC}(\text{patch-deletion}), \tag{14}$$

where AUC denotes the area under the curve.

Patch-insertion and patch-deletion curves are obtained by the following procedure: First, we divide $\boldsymbol{x}$ into patches (submatrices) $\boldsymbol{p}_{ij} \in \mathbb{R}^{c_1 \times m^2}$, where m, i, and j denote the size of the patch, vertical indices, and horizontal indices, respectively. When $m = 1$ and $\boldsymbol{b}_{ij} = 0$ for any i and j, the PID score is the same as the ID score.

Next, we apply max-pooling to attention map $\boldsymbol{\alpha}$ to create the attention map $\boldsymbol{\alpha}_{\mathrm{p}} \in \mathbb{R}^{m^2}$ for each patch. The elements of $\boldsymbol{\alpha}_{\mathrm{p}}$ are denoted as $\boldsymbol{\alpha}_{i_1 j_1}, \boldsymbol{\alpha}_{i_2 j_2}, \cdots, \boldsymbol{\alpha}_{i_m j_m}$ in ascending order. We define A_n as follows:

$$A_n = \{(i_k, j_k) | k \leqq n\}, \tag{15}$$

where n is the number of patches inserted or deleted.

Then, the inputs of patch-insertion $\boldsymbol{i}_n$ and patch-deletion $\boldsymbol{d}_n$ are represented using A_n as follows:

$$(\boldsymbol{i}_n, \boldsymbol{d}_n) = \begin{cases} (\boldsymbol{p}_{ij}, \boldsymbol{b}_{ij}) \ (i, j) \in A_n, \\ (\boldsymbol{b}_{ij}, \boldsymbol{p}_{ij}) \ (\text{otherwise}). \end{cases} \tag{16}$$

Finally, patch-insertion and patch-deletion curves are obtained by plotting n with $\boldsymbol{y}_c^{(\mathrm{ins,n})}$ and $\boldsymbol{y}_c^{(\mathrm{del},n)}$, respectively. Here, $\boldsymbol{y}^{(\mathrm{ins,n})}$, $\boldsymbol{y}^{(\mathrm{del},n)}$, and c respectively denote the outputs when $\boldsymbol{i}_n$ is input, outputs when $\boldsymbol{d}_n$ is input, and the class to which $\boldsymbol{x}$ belongs.

5 Experiments

5.1 Experimental Setup

IDRiD Dataset. The Indian Diabetic Retinopathy Image Dataset (IDRiD) [25] and DeFN magnetograms dataset [20] were used for the experimental evaluation. The IDRiD is a dataset for detecting diabetic retinopathy from retinal fundus images. It was annotated by medical experts. The images were classified into separate groups ranging from 0 (No apparent) to 4 (Severe) according to the International Clinical Diabetic Retinopa [35]. The IDRiD contained 516 samples. Among these samples, 168 were negative and 348 were positive. The training, validation, and test sets consisted of 330, 83, and 103 samples, respectively. For the IDRiD, we assigned binary label to five lesions of retinopathy grades 0–4, by converting grade 0 to negative and grades 1–4 to positive. The size of each image were 4288×2848. The input images were resized to 224×224. We used the IDRiD because it is a standard dataset for visual explanation generation tasks [19]. In addition, we added a bias image to the training set as a negative class.

Table 1. Parameter settings.

		IDRiD	DeFN magnetograms
Optimizer		AdamW	AdamW
Learning rate	LAB, Linear	1.0×10^{-3}	1.0×10^{-3}
	LFE, LPB	1.0×10^{-4}	1.0×10^{-4}
Weight decay		0.09	0.09
Batch size		32	32
Loss weights	Negative	2	1
	Positive	1	1
θ_a		0.75	0.5
θ_b		0.2	0.2
N_B		16	1
B		7	0

DeFN Magnetograms Dataset. The DeFN magnetograms dataset contained the hourly solar images taken by the Helioseismic and Magnetic Imager [27]. We collected magnetograms from the Solar Dynamic Observatory[1] [23] web archives. Because the mechanism underlying solar flares remains unclear, it is important to generate visual explanations that give insight into the related theory. The label of each magnetogram was the maximum solar flare class that occurred within 24 h. We assigned binary labels for the four solar flare classes of O, C, M, and X, by converting O and C classes to "$<$ M" and converting M and X to "$\geq$ M." The input images were resized to 512×512. The DeFN magnetograms dataset contained 61,315 samples, covering the period from June 2010 to December 2017. Among these samples, 56,078 images were labeled as "$<$ M" and 5,237 images were labeled as "$\geq$ M." The size of each image was 1024×1024. The training, validation, and test sets consisted of 45,530 samples from 2010–2015, 7,795 samples from 2016, and 7,990 samples from 2017, respectively. Similar to the IDRiD, we added a bias image to the training set as the "$<$ M" class. The training, validation, and test sets were used for parameter training, hyperparameter validation, and evaluation, respectively. The images were standardized for both datasets.

Hyperparameter Settings. Table 1 shows the hyperparameter settings of the proposed method. Here, the loss weights denote the weight of each class in the loss function. When $N_B = 1$ and $N_B = 16$, the model had 8,490 and 22 million parameters, respectively. The parameters were trained on an RTX 2080 with 11 GB of GPU memory and an Intel Core i9 processor. It took approximately 1 day and 40 min to train the model on the DeFN magnetograms dataset and IDRiD, respectively. The inference time was approximately 0.1 s. Warmup and cosine-decay were used to schedule the learning rate. We stopped the training when the loss on the validation set did not improve for six consecutive epochs.

[1] https://sdo.gsfc.nasa.gov/data/.

Table 2. Quantitative results on the IDRiD (upper table) and DeFN magnetograms dataset (lower table).

Method		RISE [24]	Lambda attention [33]	Ours (LABN)
ID↑		0.319 ± 0.015	-0.101 ± 0.074	$\mathbf{0.431 \pm 0.213}$
PID↑	$m = 2$	0.179 ± 0.080	-0.105 ± 0.073	$\mathbf{0.458 \pm 0.198}$
	$m = 4$	0.130 ± 0.045	-0.116 ± 0.081	$\mathbf{0.473 \pm 0.178}$
	$m = 8$	0.136 ± 0.050	-0.123 ± 0.078	$\mathbf{0.470 \pm 0.178}$
	$m = 16$	0.101 ± 0.033	-0.093 ± 0.054	$\mathbf{0.455 \pm 0.181}$
Method		RISE [24]	Lambda attention [33]	Ours (LABN)
ID↑		0.235 ± 0.145	0.374 ± 0.080	$\mathbf{0.506 \pm 0.170}$
PID↑	$m = 2$	0.261 ± 0.217	0.414 ± 0.129	$\mathbf{0.748 \pm 0.102}$
	$m = 4$	0.296 ± 0.199	0.403 ± 0.138	$\mathbf{0.755 \pm 0.100}$
	$m = 8$	0.379 ± 0.172	0.378 ± 0.162	$\mathbf{0.757 \pm 0.094}$
	$m = 16$	0.461 ± 0.164	0.291 ± 0.216	$\mathbf{0.756 \pm 0.096}$

Table 3. Confusion matrix.

	IDRiD	DeFN magnetograms
TP	56	161
TN	25	7566
FP	9	243
FN	13	20

5.2 Quantitative Results

We used RISE [24] and Lambda attention [33] as the baseline methods. We obtained Lambda attention by computing the average of $\boldsymbol{\lambda}_c^\top Q$ in the Lambda layer in the direction of channel dimension. Because there is no established explanation generation method for Lambda Networks, we constructed a baseline method based on standard explanation methods for transformers (e.g., attention rollout [33]). As a result, we selected the optimal $\boldsymbol{\lambda}_c^\top Q$ and named it Lambda attention. RISE is also a standard method that can be applied to general models.

We used the ID and PID scores as the primary evaluation metrics. The ID score is a standard evaluation method for explanation generation. We used the PID score because the IDRiD and DeFN magnetograms dataset contain sparse images.

Table 2 shows the quantitative results. The upper and lower tables show the results on the IDRiD and DeFN magnetograms dataset, respectively. We conducted the experiment four times for each method, and the average and standard deviations of the scores are reported. For the IDRiD and DeFN magnetograms dataset, only the positive and "$\geq$ M" data were used to calculate the score, respectively. This is because the negative and "$<$ M" data do not contain regions that are appropriate for explanation.

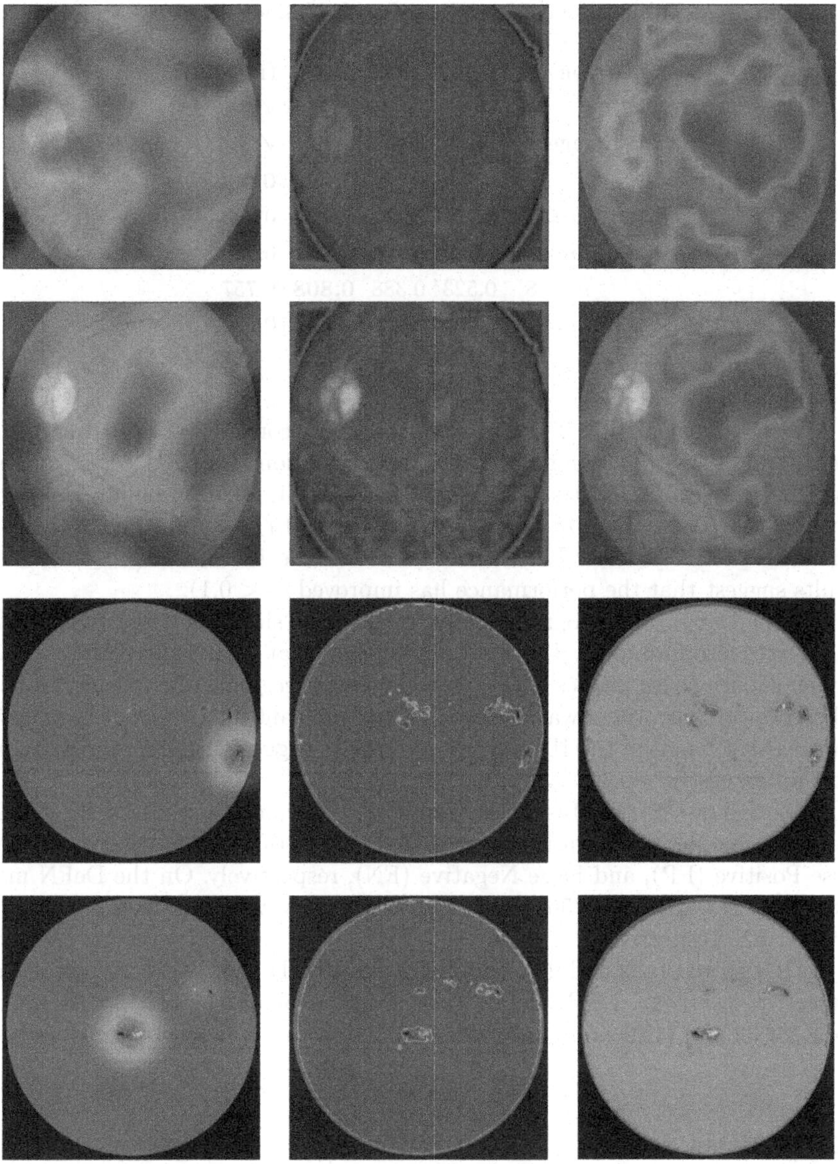

(a) RISE [24] (b) Lambda attention [33] (c) Ours

Fig. 5. Qualitative results. The top two and bottom two rows show the results on the IDRiD, DeFN magnetograms dataset, respectively. RISE focuses on areas that are too large, and Lambda attention focuses on the background. By contrast, the proposed method does not focus on inappropriate areas.

Table 4. Quantitative results of ablation studies.

Condition		(i)	(ii)	(iii)	(iv) Ours
$\mathcal{L}_{KL}$				✓	✓
Bias image			✓		✓
ID↑		0.044	0.124	0.460	**0.506**
PID↑	$m = 2$	0.311	0.446	**0.774**	0.748
	$m = 4$	0.489	0.405	**0.792**	0.755
	$m = 8$	0.523	0.388	**0.808**	0.757
	$m = 16$	0.556	0.382	**0.807**	0.756

For the IDRiD, Table 2 shows that the PID score at $m = 16$ was 0.101, −0.093, and 0.455 points for RISE, Lambda attention and LABN, respectively. The PID score of LABN was better than that of RISE by 0.354 points. Similarly, the ID and PID scores at $m = 2, 4, 8$ improved by 0.112 and 0.279, 0.343, 0.334 points, respectively, when LABN was used. For the PID score ($m \geq 2$), these results suggest that the performance has improved ($p < 0.1$).

For the DeFN magnetograms dataset, the table shows that the ID and PID scores were also improved. These results indicate that LABN successfully generates explanations for images with sparse important regions. In particular, for the PID score ($m \geq 2$), there was a statistically significant improvement ($p < 0.05$). On the other hand, in the ID score, these results suggest that there is no significant difference ($p > 0.1$).

Table 3 shows the confusion matrix of our method. On the IDRiD, 56, 25, 9 and 13 samples were classified as True Positive (TP), True Negative (TN), False Positive (FP), and False Negative (FN), respectively. On the DeFN magnetograms dataset, 161, 7566, 243, and 20 samples were classified as TP, TN, FP, and FN, respectively.

LABN failed on 22 and 263 samples in the IDRiD and DeFN magnetograms dataset, respectively. LABN failed to generate explanations with a PID score < 0.2 for 32 and 1029 samples in the IDRiD and DeFN magnetograms dataset, respectively.

5.3 Ablation Studies

We set the following ablation conditions:

1. w/o $\mathcal{L}_{KL}$
   We removed $\mathcal{L}_{KL}$ to investigate the effect of saliency guided training on explanation generation performance.
2. w/o bias image
   We eliminated the bias image from the training set to investigate the effect on explanation generation performance.

Table 4 shows the quantitative results for the ablation studies. Under condition (ii), both ID and PID scores decreased significantly. Note that condition (iii) led to higher performance in terms of PID. These results indicate that $\mathcal{L}_{\mathrm{KL}}$ contributed more to the model performance.

5.4 Qualitative Results

The top two and bottom two rows of Fig. 5 show the qualitative results for the IDRiD and DeFN magnetograms dataset samples, respectively. The first and second rows in the left column of the image illustrate that RISE attended large areas and did not focus on the important regions. In the first and second rows in the middle column of image, most of the area attended by Lambda attention are background, which is inappropriate. By contrast, LABN successfully attended the appropriate regions, which contributed to the accuracy, as shown in the first and second rows in the right column of image.

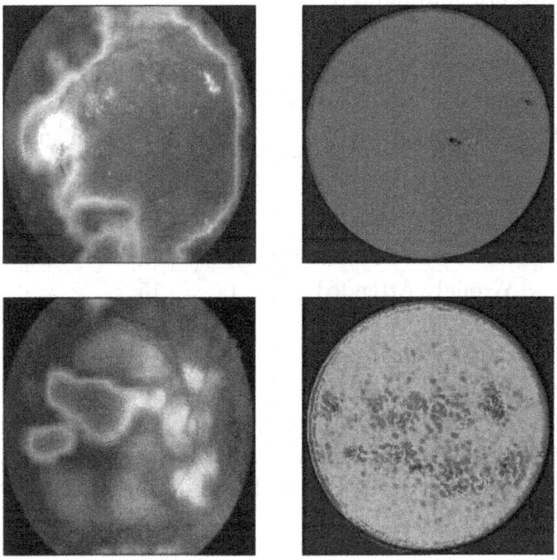

Fig. 6. Failure examples. Left and right figures are images from the IDRiD and DeFN magnetograms dataset, respectively.

The third and forth rows in the left and middle columns of images demonstrate that RISE and Lambda attention attended unrelated regions such as the circumference and background. By contrast, the third and forth rows in the right column of image show that LABN appropriately attended the active sunspots, which are important for solar flare prediction.

Figure 6 shows examples of failures (examples with a PID score less than 0.2), for the IDRiD and DeFN magnetograms dataset, respectively. In the left figure, a wide range of areas including the optic disk are highlighted. In the right figure, important regions were not appropriately highlighted, reducing the PID score.

For the IDRiD and DeFN magnetograms dataset, 32 and 100 samples, respectively were classed as failures. Note that we randomly selected 100 failure samples from the DeFN magnetograms dataset.

- IP (Incorrect Prediction)
  IP refers to a case in which the model predicts the incorrect class. The bottom left of Fig. 6 shows an example of an IP.
- OA (Over-Attended)
  OA refers to a case in which most of the image is highlighted. The top left of Fig. 6 shows an example of an OA image. In this example, the model paid attention to the entire image.
- IA (Insufficiently Attended)
  IA refers to a case in which the attended area is insufficient. The top right of Fig. 6 shows an example of an IA image. In this example, $||\boldsymbol{\alpha}||$ is small and the sunspots are not highlighted.

Table 5. Error analysis

Error ID	Description	IDRiD	DeFN magnetograms
IP	Incorrect Prediction	12	58
OA	Over-Attended	16	3
IA	Insufficiently Attended	3	24
WA	Wrongly Attended	1	15

- WA (Wrongly Attended)
  WA refers to a case in which the pixels that do not contribute to the accuracy are given attention. The bottom right of Fig. 6 shows an example of a WA image. The PID score at $m = 16$ is 0.1, which indicates that non-important regions are attended.

5.5 Error Analysis

Table 5 shows that the bottleneck of the IDRiD and DeFN magnetograms dataset are OA and IP errors, respectively. Therefore, we expect that improving the model accuracy and regularizing $\boldsymbol{\alpha}$ each dataset will be an effective approach to generating appropriate visual explanations.

6 Conclusions

In this paper, we focused on the task of generating an attention map as a visual explanation for a model's decisions. The main contributions of this paper are as follows:

- We introduced a parallel branch structure to obtain explanations that are clearer than those obtained using Lambda attention.
- We proposed the PID score as an effective evaluation measure for explaining images with sparse important regions.
- We introduced the loss used in saliency guided training [12] to reduce the importance of irrelevant regions.
- LABN outperformed the baseline methods in terms of the ID and PID scores.

Acknowledgement. This work was partially supported by JSPS KAKENHI Grant Number 20H04269 and NEDO.

References

1. Abnar, S., Zuidema, W.: Quantifying attention flow in transformers. arXiv preprint arXiv:2005.00928 (2020)
2. Adebayo, J., Gilmer, J., Muelly, M., Goodfellow, I., Hardt, M., Kim, B.: Sanity checks for saliency maps. In: NeurIPS, vol. 31 (2018)
3. Bello, I.: LambdaNetworks: modeling long-range interactions without attention. In: ICLR (2021)
4. Binder, A., Montavon, G., Lapuschkin, S., Müller, K.-R., Samek, W.: Layer-wise relevance propagation for neural networks with local renormalization layers. In: Villa, A.E.P., Masulli, P., Pons Rivero, A.J. (eds.) ICANN 2016. LNCS, vol. 9887, pp. 63–71. Springer, Cham (2016). https://doi.org/10.1007/978-3-319-44781-0_8
5. Chefer, H., Gur, S., Wolf, L.: Transformer interpretability beyond attention visualization. In: CVPR, pp. 782–791 (2021)
6. Das, A., Rad, P.: Opportunities and challenges in explainable artificial intelligence (XAI): a survey. arXiv preprint arXiv:2006.11371 (2020)
7. Deng, J., Dong, W., Socher, R., Li, L.J., Li, K., Fei-Fei, L.: ImageNet: a large-scale hierarchical image database. In: CVPR, pp. 248–255 (2009)
8. Dosovitskiy, A., Beyer, L., et al.: An image is worth 16×16 words: transformers for image recognition at scale. In: ICLR (2021)
9. Fel, T., Vigouroux, D., Cadène, R., Serre, T.: How good is your explanation? Algorithmic stability measures to assess the quality of explanations for deep neural networks. In: WACV, pp. 720–730 (2022)
10. Fukui, H., Hirakawa, T., et al.: Attention branch network: learning of attention mechanism for visual explanation. In: CVPR, pp. 10705–10714 (2019)
11. Hooker, S., Erhan, D., Kindermans, P.J., Kim, B.: A benchmark for interpretability methods in deep neural networks. In: NeurIPS, vol. 32 (2019)
12. Ismail, A.A., Corrada Bravo, H., Feizi, S.: Improving deep learning interpretability by saliency guided training. In: NeurIPS (2021)
13. Jain, S., Wallace, B.: Attention is not explanation. In: NAACL, pp. 3543–3556 (2019)
14. Khan, S., Naseer, M., Hayat, M., Zamir, S.W., et al.: Transformers in vision: a survey. arXiv preprint arXiv:2101.01169 (2021)
15. Li, H., Ellis, J., Zhang, L., Chang, S.F.: PatternNet: visual pattern mining with deep neural network. In: ICMR, pp. 291–299 (2018)
16. Lundberg, S., Lee, S.I.: A unified approach to interpreting model predictions. In: NeurIPS, pp. 4765–4774 (2017)

17. Magassouba, A., Sugiura, K., et al.: A multimodal classifier generative adversarial network for carry and place tasks from ambiguous language instructions. RA-L **3**(4), 3113–3120 (2018)
18. Magassouba, A., Sugiura, K., et al.: Predicting and attending to damaging collisions for placing everyday objects in photo-realistic simulations. Adv. Robot. **35**(12), 787–799 (2021)
19. Mitsuhara, M., Fukui, H., Sakashita, Y., et al.: Embedding human knowledge into deep neural network via attention map. In: VISAPP (2021)
20. Nishizuka, N., Sugiura, K., et al.: Deep flare net (DeFN) model for solar flare prediction. Astrophys. J. **858**(2), 113 (8 pp) (2018)
21. Ogura, T., Magassouba, A., Sugiura, K., et al.: Alleviating the burden of labeling: sentence generation by attention branch encoder-decoder network. RA-L **5**(4), 5945–5952 (2020)
22. Pan, B., Panda, R., Jiang, Y., et al.: IA-RED^2: interpretability-aware redundancy reduction for vision transformers. In: NeurIPS (2021)
23. Pesnell, W., Thompson, B., Chamberlin, P.: The solar dynamics observatory (SDO). Sol. Phys. **275**(1–2), 3–15 (2012). https://doi.org/10.1007/s11207-011-9841-3
24. Petsiuk, V., Das, A., Saenko, K.: RISE: randomized input sampling for explanation of black-box models. In: BMVC, p. 151 (13 pp) (2018)
25. Porwal, P., et al.: IDRiD: diabetic retinopathy - segmentation and grading challenge. Med. Image Anal. **59**, 101561 (2020)
26. Ribeiro, M., Singh, S., et al.: "Why should i trust you?": explaining the predictions of any classifier. In: KDD, pp. 1135–1144 (2016)
27. Scherrer, P., Schou, J., Bush, R., et al.: The helioseismic and magnetic imager (HMI) investigation for the solar dynamics observatory (SDO). Sol. Phys. **275**, 207–227 (2012). https://doi.org/10.1007/s11207-011-9834-2
28. Selvaraju, R., et al.: Grad-CAM: visual explanations from deep networks via gradient-based localization. In: ICCV, pp. 618–626 (2017)
29. Smilkov, D., Thorat, N., Kim, B., Viégas, F.B., Wattenberg, M.: SmoothGrad: removing noise by adding noise. arXiv preprint arXiv:1706.03825 (2017)
30. Srinivas, S., Fleuret, F.: Full-gradient representation for neural network visualization. In: Advances in Neural Information Processing Systems, vol. 32 (2019)
31. Sundararajan, M., Taly, A., Yan, Q.: Axiomatic attribution for deep networks. In: ICML, vol. 70, pp. 3319–3328 (2017)
32. Vaswani, A., Shazeer, N., Parmar, N., Uszkoreit, J., Jones, L., et al.: Attention is all you need. In: NeurIPS, pp. 6000–6010 (2017)
33. Vig, J.: A multiscale visualization of attention in the transformer model. In: ACL, pp. 37–42 (2019)
34. Wang, H., Wang, Z., Du, M., et al.: Score-CAM: score-weighted visual explanations for convolutional neural networks. In: CVPR, pp. 24–25 (2020)
35. Wu, L., et al.: Classification of diabetic retinopathy and diabetic macular edema. World J. Diabetes **4**(6), 290–294 (2013)
36. Zhang, Z., Chen, Y., Li, H., Zhang, Q.: IA-CNN: a generalised interpretable convolutional neural network with attention mechanism. In: IJCNN, pp. 1–8 (2021)
37. Zhou, B., Khosla, A., Lapedriza, A., Oliva, A., et al.: Learning deep features for discriminative localization. In: CVPR, pp. 2921–2929 (2016)

Shape Prior is Not All You Need: Discovering Balance Between Texture and Shape Bias in CNN

Hyunhee Chung and Kyung Ho Park(✉)

SOCAR AI Research, Seoul, Korea
{esther,kp}@socar.kr

Abstract. As Convolutional Neural Network (CNN) trained under ImageNet is known to be biased in image texture rather than object shapes, recent works proposed that elevating shape awareness of the CNNs makes them similar to human visual recognition. However, beyond the ImageNet-trained CNN, how can we make CNNs similar to human vision in the wild? In this paper, we present a series of analyses to answer this question. First, we propose AdaBA, a novel method of quantitatively illustrating CNN's shape and texture bias by resolving several limits of the prior method. With the proposed AdaBA, we focused on fine-tuned CNN's bias landscape which previous studies have not dealt with. We discover that fine-tuned CNNs are also biased to texture, but their bias strengths differ along with the downstream dataset; thus, we presume a data distribution is a root cause of texture bias exists. To tackle this root cause, we propose a granular labeling scheme, a simple but effective solution that redesigns the label space to pursue a balance between texture and shape biases. We empirically examine that the proposed scheme escalates CNN's classification and OOD detection performance. We expect key findings and proposed methods in the study to elevate understanding of the CNN and yield an effective solution to mitigate this texture bias.

1 Introduction

Discovering what Convolutional Neural Network (CNN) learned has become an important but challenging problem in modern computer vision studies [1,5,8,16]. Recent studies presented that CNNs, especially those trained under the ImageNet [3] dataset have texture bias that prioritizes image textures rather than object's shapes. This finding conflicts with human visual perceptions as humans utilize shape information to understand images [7]. The aforementioned texture bias of CNN is known to be a critical challenge due to the following

H. Chung and K. H. Park—Contributed equally to this work.

Supplementary Information The online version contains supplementary material available at https://doi.org/10.1007/978-3-031-26284-5_30.

L. Wang et al. (Eds.): ACCV 2022, LNCS 13842, pp. 491–506, 2023.
https://doi.org/10.1007/978-3-031-26284-5_30

reasons. First, CNN's texture bias might become a vulnerability from a security perspective as an adversary would attack the model by transforming the image's textures to mislead its understanding [14]. Second, CNN's texture bias might exhibit its inductive bias being distinct from the human visual system, which indicates insufficient robustness to be deployed in the real world. Under the insufficient robustness of CNN, it risks creating fatal damage to the humans who interact with the model (i.e., medical imaging [20,22]).

While prior analyses presented monumental findings and solutions regarding CNN's texture bias, we figure out several improvement avenues. First, prior studies primarily focused on analyzing the CNN's dynamics in ImageNet and its derived ones (i.e., Cue-conflict dataset [7]). While fine-tuning has become a de facto technique in modern computer vision tasks (i.e., image recognition [11,18,24], object detection [2,21]), there were no bias analyses on the fine-tuned CNNs. To this end, we postulate several questions regarding the fine-tuned CNN'S dynamics: Do CNNs fine-tuned on various downstream datasets show texture bias, just as the case in ImageNet? If so, do fine-tuned CNNs show similar bias strength regardless of the downstream dataset?

Second, we urge that the frozen label space assumption should be released for a more practical solution against CNN's texture bias. We denote a frozen label space as the assumption that does not change the labeling scheme but uses the given dataset as it was originally labeled. As Hermann et al. [14] once proposed, a data distribution (as well as label space) of conventional datasets (i.e., ImageNet) is a root cause of CNN's texture bias [14]. Nevertheless, we analyze that previously-proposed solutions have not tackled this root cause but primarily focused on additional actions given a trained model [7,25]. Suppose the practitioners establish a labeled dataset from unlabeled samples before model training. What if we can mitigate the texture bias when the practitioners design a labeling scheme? What if we can resolve texture bias before model training? If the practitioners can resolve the texture bias by simply changing the labeling scheme, we expect it to become a powerful solution in the real world.

To accomplish these improvement avenues, our study proposes a series of analyses that scrutinize answers to the aforementioned questions. The contributions of our study are as follows. First, we seek a quantitative tool to analyze CNN's bias for an accurate understanding of its dynamics. While [15] once presented a solid baseline of this quantitative tool, we discovered several limits. Therefore, as an advanced version of the baseline, we propose a novel bias analysis method denoted Adaptive Bias Analysis (AdaBA), and empirically examine that our method coherently exhibits the same result as the previously-proposed method while it improves its limits. Second, we further analyze the dynamics of fine-tuned CNNs, which prior studies have not actively scrutinized. We analyze that fine-tuned CNNs are also biased to textures, but the strength of texture bias differs in downstream datasets. Thus, we presume a root cause of CNN's texture preference is indeed a data distribution, just as proposed in a recent discovery [14]. Third, we propose a novel viewpoint (problem setup) to mitigate CNN's texture bias by redesigning the labeling scheme before model training. We propose a Granular labeling scheme, a novel label space design to acquire a balance between texture and shape bias. Upon the synthetically-created datasets, we experimentally examine a CNN trained under the proposed granular labeling scheme (which embraces the balance between texture and shape bias) is advantageous in two tasks: classification and out-of-distribution (OOD) detection. Lastly, we further analyze how the representation acquired under the granular labeling scheme differs from the others through measuring representation similarity with Centered Kernel Alignment (CKA) [10,17].

2 Related Works

As the original motivation of CNN's design stems from neuroscience [9,27], it has been regarded to recognize the image based on shape information, just as human perception [4,23]. However, Geirhos et al. empirically validates that CNNs have texture bias, especially when they are trained under ImageNet. To reduce this texture bias, Geirhos et al. propose a manual injection of share awareness by style-transferring ImageNet dataset [3]. Hermann et al. (which shares the most similar motivation with our study) unveiled the origins of texture bias and the reason why CNNs are inherently biased to texture information in the ImageNet dataset; The data distribution of ImageNet causes the model to classify labels by texture characteristics, not shape information. Moreover, this study further claimed that simply elevating shape awareness does not always contribute to the best performance; thus, a careful approach to bias mitigation should be considered. Beyond the aforementioned analyses on ImageNet-trained CNNs, Islam et al. designed a method of quantitatively analyzing CNN's shape and texture biases [15].

3 Adaptive Bias Analysis (AdaBA)

3.1 Baseline and Its Improvement Avenues

Before performing analyses, we strongly necessitate an effective, quantitative tool to understand the CNN's texture and shape bias in a more precise manner. While Islam et al. presented a solid baseline approach, we scrutinized several improvement avenues. The detailed limits of the baseline is described below. For descriptions on the baseline, please refer to the original publication [15], and we also provided brief illustrations in the supplementary materials.

Dependence on Heuristically-Defined Texture Patterns. The baseline requires heuristically-defined texture patterns to create shape and texture pairs. We analyze it induces the proposed method to exhibit CNN's bias focused on this heuristically-chosen texture, not the original texture information included in the original sample. If we desire to claim that 'CNN trained on dataset X is biased to textures', defining the shape and texture pairs should not rely on heuristically-chosen texture patterns. Instead, we presume an improved approach should be capable of establishing shape and texture pairs without any human interventions.

Training Additional Style Transfer Model. Furthermore, the baseline also requires establishing an additional style transfer method to create texture and shape pairs. Then, it implies that we should additionally train an auxiliary model to utilize the baseline. We further expect this point as a risky factor of the baseline. What if the style transfer model is not qualified enough to provide texture-invariant samples?

Misled Interpretation on Bias Score Derived from Softmax Function. Last but not least, we figured out the case where the final bias scores are not consistently sustained with the calculated mutual information which implicitly describes the strength of bias. Referring to the baseline's score calculation procedures, it applies a softmax operation to the set of mutual information from texture pair, shape pair, and residuals. For example, suppose the case where the set of mutual information is [−0.9959, 0.8498, 1] for texture, shape, and residuals, respectively. As mutual information exhibits similar implications to the correlation coefficient, the aforementioned set shows a strong texture bias rather than shape. But, when we apply a softmax operation to this set, a mutual information score for texture becomes 0.22 while this score for shape becomes 0.77, exhibiting high shape bias. We analyze the softmax operation as a risk of reversing the original mutual information on texture and shape information, which might cause a misinterpretation of CNN's bias.

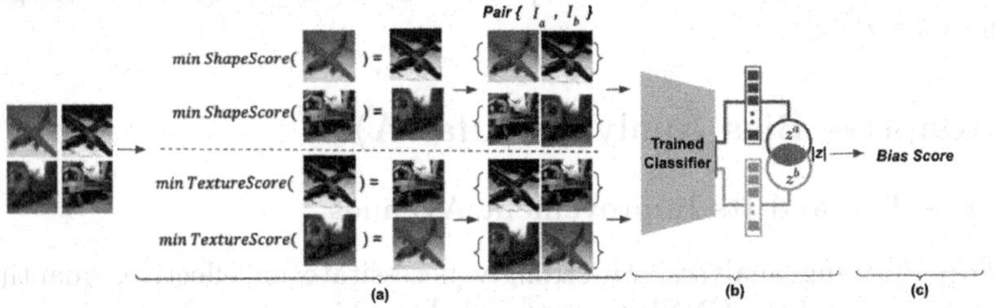

Fig. 1. Illustration of AdaBA. (a) defining texture and shape pair with shape and texture score respectively, (b) extracting mutual information from defined pair, (c) we use raw mutual information with absolute value. Note that mutual information of (b) defined in Islam et al. [15]

3.2 Methodology

Considering the baseline's improvement avenues, we design a novel approach to analyze CNN's bias in a quantitative manner. We denote our approach as the Adaptive Bias Analysis (AdaBA), as the proposed method can yield shape and texture bias scores adaptive to the dataset where the CNN is trained. We visualized an overall architecture of AdaBA in Fig. 1. The proposed AdaBA resolves the aforementioned drawbacks by changing baseline approach's two components:

1) Texture and shape pair generation without auxiliary components, 2) Redesigning bias score by eliminating softmax operation. The detailed descriptions are described below.

Texture and Shape Pair Generation Without Auxiliary Components. To resolve the first and second drawbacks of the baseline, AdaBA creates shape and texture pairs leveraging the theoretical definition of texture and shape proposed in style transfer studies. Referring to the design philosophy of the baseline, the texture pair includes two samples that have similar textures and different shapes. For a shape pair, vice versa. The pair generation procedures of AdaBA are as follows. First, it extracts the feature map from CNN on a given image sampled from the target dataset. We denote F^a as a feature map of the image I_a. Second, it generates two matrices of shape matrix and gram matrix, which are proposed in style-transfer studies [6]. A shape matrix can be defined as F_{ij}, and the texture(gram) matrix is justified as $\sum_k F^a_{ik} F^b_{jk}$ where i, j indicates width and heights at the feature map. Note that gram matrix implicit the meaning of texture information in a given image [6]. Third, for every sample in the dataset, it calculates the shape and gram matrix. It then selects a single sample as an anchor and calculates both shape score and texture score with the other samples following the equations provided in 1 and 2, respectively (i.e., measures the shape and texture scores from a pair consists of one anchor sample and one the other sample). The shape score exhibits a euclidean distance of shape matrices over the euclidean distance between gram matrices. The large shape score means a large shape difference over texture difference; thus, it implies two samples in a pair have dissimilar shape characteristics over the texture information. For the texture score, vice versa. Note that the texture score is a reciprocal of the shape score. Lastly, given an anchor sample, we establish a shape pair by selecting another sample that records the lowest shape score. For a texture sample, the texture pair is established with another sample with the lowest texture score. Throughout these procedures, key benefits of AdaBA's pair generation are particularly vivid. It does not require heuristically-chosen texture patterns or an additional style-transferring model, while it establishes shape and texture pairs based on a solid theoretical definition of texture and shape.

$$Shape\ Score = \frac{Euclidean(F^a_{ij}, F^b_{ij})}{Euclidean(\sum_k F^a_{ik} F^b_{jk}, \sum_k F^a_{ik} F^b_{jk})} \quad (1)$$

$$Texture\ Score = \frac{Euclidean(\sum_k F^a_{ik} F^b_{jk}, \sum_k F^a_{ik} F^b_{jk})}{Euclidean(F^a_{ij}, F^b_{ij})} \quad (2)$$

Redesigning Bias Score by Eliminating Softmax Operation. AdaBA employs the absolute value of calculated mutual information on shape and textures instead of softmax operation. As we discovered the use of softmax operation risks misleading CNN's bias, we designed AdaBA to eliminate its use. By eliminating the use of softmax operation on MIs, we presume the risk of bias misinterpretation decreases compared to the baseline approach. Given mutual information values on texture and shape, our study used an absolute value of them as the final texture and shape bias score. We presume the mutual information values resulting from the CNN include the original representation of the biases without much noise; thus, retrieving an absolute value of each mutual information better describes the CNN's biases.

3.3 Validation on AdaBA

To prove a usefulness of the proposed AdaBA, we establish two questions for examining AdaBA's validity: 1) whether the AdaBA generates shape and texture pairs well, 2) whether the AdaBA yields bias analysis results on ImageNet-trained CNNs consistent with previously-proven discoveries. Upon the proven validity of AdaBA, we further scrutinize the behaviors of fine-tuned CNNs, which is absent in prior works.

Does AdaBA Generate the Pair Well? First, as our AdaBA aims to alternate the baseline's pair generation procedure, we hereby examine whether the established shape and texture pairs well consistent with the original purpose; shape pairs should include samples that share a similar shape and dissimilar texture, while texture pairs include samples that have dissimilar shapes and similar texture. Given a concatenated dataset that includes both original CIFAR-10 and Stylized CIFAR-10, we sampled several shapes and texture pairs (which are established through AdaBA) and visualized them in Fig. 2. Following the established pairs, we qualitatively evaluate that the generated shape and texture pair satisfy their original purpose; therefore, the proposed AdaBA effectively creates those pairs without heuristically-defined patterns and auxiliary style-transfer model.

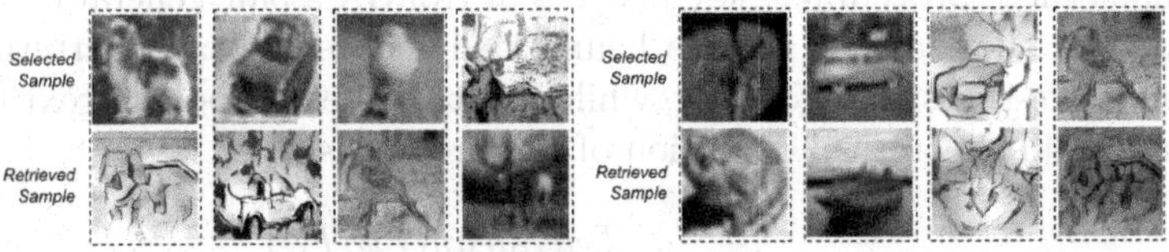

(a) Retrieved shape Pairs (b) Retrieved texture Pairs

Fig. 2. Generated shape and texture pairs with AdaBA. We observe that generated shape and texture pairs fully satisfy the original purpose of pair generation.

Does AdaBA Yield Consistent Results with Prior Discoveries? Furthermore, our study also aims to evaluate whether the proposed AdaBA effectively describes CNN's shape and texture bias. We actively referred to their evaluation logic of the baseline study, which is a state-of-the-art. The baseline study evaluated their approach's performance by comparing the resulted shape and texture bias scores with prior studies' discoveries. For example, various studies suggested that ImageNet-trained CNNs are biased toward image textures rather than shapes. The baseline approach also yields a higher texture bias score than the shape score; thus, the prior studies and the baseline approach are saying the same proposition, so this study urges that their approach is correct. Please refer to the original publication [15] for more detailed explanations. Upon this evaluation logic, we examine whether the AdaBA yields bias analysis results consistent with the following discoveries [15]: *1) ImageNet-trained CNN is biased to the texture, 2) Style-transferring reduces this texture bias.* For the dataset in these examinations, we alternatively used TinyImageNet (a subsample of ImageNet) as analyses on the original ImageNet required large computation overheads.

To implement a CNN trained under TinyImageNet, we trained the ResNet-50 model with the TinyImageNet dataset. With these trained CNNs, we compared the bias analysis results at AdaBA and the baseline to prove whether our approach yields similar analysis results consistent with the previous well-proven discoveries. Note that we also showed the baseline method's analysis results as it is known to show similar results to previous discoveries. The results are shown in Table 1.

Table 1. Bias analysis on TinyImageNet

Dataset	AdaBA			Baseline		
	Shape	Texture	$\frac{Shape}{Texture}$	Shape	Texture	$\frac{Shape}{Texture}$
Original	0.2202	0.3756	0.5863	518	1530	0.3386
Stylized	0.3842	0.3295	1.1660	755	1293	0.5839

From the results, we discovered that AdaBA consistently accomplishes similar results to the previous discoveries. First, we figure out that both AdaBA and baseline method yields texture bias of TinyImageNet-trained CNN (which is a proxy of ImageNet-trained one), where this texture bias has been proved in previous works. Second, we also discovered both approaches yield that the CNNs trained under the style-transferred samples exhibit a reduced texture bias rather than the one trained with original samples. The AdaBA and baseline method exhibits an increased $\frac{Shape}{Texture}$ value, where the larger value implies enhanced shape bias and reduced texture bias. Throughout these results, we conclude that the proposed AdaBA exhibits bias analysis results consistent with the baselines while it improves the baseline's limits simultaneously. We acknowledge that the proposed AdaBA should be examined in various datasets or problem settings but skipped in this work as our study primarily focuses on the proposing solution to mitigate texture bias, not analyzing the bias.

3.4 Bias Analysis on Fine-tuned CNNs with AdaBA

Upon the theoretical and experimental justification of our AdaBA, we extend our focus toward the fine-tuned CNNs. As no analyses exist on the fine-tuned CNNs' biases, we hereby scrutinize them. We postulate two questions regarding fine-tuned CNN's bias as follows: *1) What bias do CNNs fine-tuned in various benchmark datasets expose?* Do they show texture bias just as it showed in ImageNet? *2) Does style transferring indeed mitigate texture bias in the fine-tuned CNNs, just as it does in ImageNet?* While prior study [7] suggested that style-transferring contributes to mitigating texture bias, is this method also valid in the fine-tuning regime? For implementation details, we employ four conventionally-utilized benchmark datasets: CIFAR-10, CIFAR-100, TinyImageNet, and Stanford Cars. Given these datasets, we fine-tuned the ResNet-50 classifier from the ImageNet-trained weights and retrieved shape and texture scores from our AdaBA and the baseline method. To answer the first question, we examine whether the fine-tuned CNNs are also implicit texture bias, as in ImageNet-trained CNN. Moreover, for the second question, we style-transfer CIFAR-10 with AdaIN and fine-tune the CNN on this style-transferred dataset. We then compare its texture and shape bias with the one fine-tuned in the original CIFAR-10 dataset. Throughout these setups, the experiment results are shown in Tables 2 and 3, and key findings are illustrated as follows.

Table 2. Bias analysis on Fine-tuned CNNs. AdaBA results in similar trends to the baseline method. Note that MI means mutual information.

Dataset	AdaBA				Baseline			
	ShapeMI	TextureMI	Shape	Texture	ShapeMI	TextureMI	Shape	Texture
CIFAR-10	−0.2170	0.7271	0.2170	0.7271	−0.0994	−0.1039	1026	1022
CIFAR-100	−0.1631	0.2174	0.1631	0.2174	−0.4479	0.4659	586	1462
TinyImageNet	−0.2202	0.3756	0.2202	0.3756	−0.4731	0.5674	534	1514
Stanford-Cars	−0.0045	0.7894	0.0045	0.7894	0.4339	−0.5646	1496	552

Fine-Tuned CNNs Exhibit Texture Bias, but Their Strengths Depend on the Downstream Dataset. We scrutinize that fine-tuned CNNs also bear texture bias just as the one trained in ImageNet; thus, the analysis proposed in Geirhos et al. [7] is also valid in the fine-tuning scenario. Furthermore, we hypothesize a recent study of Hermann et al. [14] on the origins of CNN's texture preferences also supports this result, implying that both ImageNet and widely-used benchmark datasets' data distribution become a root cause of texture bias. As Hermann et al. [14] once noted, we also suspect that the model architecture is not a big concern, but the data distribution or label space design provokes this bias.

Table 3. Bias analysis on CIFAR-10

Dataset	AdaBA (Ours)			Baseline		
	Shape	Texture	$\frac{Shape}{Texture}$	Shape	Texture	$\frac{Shape}{Texture}$
Original	0.2170	0.7217	0.2984	180	1259	0.1430
Stylized	0.2882	0.1576	1.8286	330	516	0.6395

Style-Transferring Do Mitigate Fine-Tuned CNN's Texture Bias. Following Table 3, we observe that style-transferring the downstream dataset contributes to mitigating the texture bias of the fine-tuned CNN. Just as the prior study pointed out, we presume style-transferring samples unify the texture information in the dataset, thus enhancing the awareness of shape information to the CNN. As a comparative strength of shape over the texture increases after style transfer, we result that the method proposed in [7] is also valid under the fine-tuning regime.

4 Mitigating Texture Bias by Redesigning Label Space

4.1 Granular Labeling Scheme

Objective. Upon the takeaway that data distribution is one promising cause of CNN's texture bias, then, how can we figure out the optimal balance between texture and shape bias? Prior works primarily focused on mitigating texture bias under the frozen label space, which implies that a training set was regarded as a sanctuary. But, our study ideates that a practical solution should release this assumption and changes the data distribution by redesigning the label space. Accordingly, we suggest a novel labeling scheme denoted as a Granular labeling scheme where the samples belonging to each label share the shape and texture characteristics simultaneously, while conventional label spaces are established based on the human practitioner's needs.

(a) Unlabeled (b) Shape-biased (c) Texture-biased (d) Granular (Ours)

Fig. 3. Various labeling schemes that we utilized in the study. Given unlabeled samples shown in (a), the machine learning practitioners can create various labeling schemes: shape-biased, texture-biased, and granular schemes.

Dataset. As we aim to examine the effectiveness of various labeling schemes, we necessitate a set of unlabeled samples that can be annotated under various labeling schemes. To the best of our knowledge, we could not figure out publicized datasets fulfilling this requirement; thus, we synthetically created the dataset. Given CIFAR-10, first, we style-transferred the original CIFAR-10 where its samples bear similar shapes and dissimilar textures from the original CIFAR-10. As we showed in Fig. 3, our study concatenated them with the original CIFAR-10 and regarded it as a set of unlabeled samples; thus, the concatenated samples have ten shapes (airplane, automobile, $\cdots$), and two textures (Natural and Artistic). Given these concatenated samples, we created different training and test sets based on various labeling schemes. We denote a concatenated dataset consisting of original CIFAR-10 with three styles (Mosaic Realism, Rococo, and Neoplasticism) as Concatenated Set 1, 2, and 3, respectively.

Labeling Scheme. We postulated three labeling schemes that the machine learning practitioners would utilize: Shape-biased, Texture-biased, and Granular labeling scheme (which is our proposition). Suppose the practitioners have a set of unlabeled samples as shown in Fig. 3. (a), where there exists two shape characteristics (airplane, truck) and two texture characteristics (natural, artistic). Under the shape-biased scheme shown in Fig. 3. (b), the practitioners can establish two labels of airplane and truck, and samples within each label share similar shape characteristics but different textures (i.e., samples in label 1 have a similar airplane-like object, but their texture varies). Conversely, under the texture-biased scheme shown in Fig. 3. (c), samples at each label share similar texture characteristics but dissimilar shapes (i.e., samples in label 1 share natural texture but have different objects). Lastly, the granular labeling scheme (a novel labeling scheme proposed in our study) lets the samples at each label be differentiated in both shape and texture characteristics; thus, the number of labels increases compared to the prior ones. Referring to Fig. 3. (d), the samples in label 1 have different shape and texture characteristics from the other labels. Based on these schemes, we examined whether the proposed granular labeling scheme conveys better representation power to the model for various computer vision tasks.

Classification Setting and Evaluation. Our study posits two classification tasks: shape classification and texture classification. Suppose we solve a binary classification between truck and airplane labels given a concatenated dataset (which simultaneously includes Natural and Shape texture) for shape classification. In this case, the practitioners conventionally design a labeling scheme where the samples at each class share similar objects and different textures. The samples in the truck label share look-a-like truck objects, but their textures vary from Natural to Artistic ones. Following the defined labeling schemes in Sect. 4.1 (b), we can say the practitioners followed the shape-biased labeling scheme, and the model would solve binary classification between truck and airplane labels. As the granular scheme divides each label more finite manner, the model shall solve a 4-class classification with the following labels: (truck, natural), (truck, artistic), (airplane, natural), and (airplane, artistic). For a proper comparative evaluation, we concatenated the prediction results under the granular scheme based on the shape property. For example, we concatenated (truck, natural) and (truck, artistic) prediction results as a truck label, and the airplane label consists of prediction results of (airplane, natural) and (airplane, artistic). Therefore, the model under the granular scheme solves the 4-class classification, but the machine learning practitioners can practically acquire binary labels by concatenating prediction results fit to their classification objective. Last but not least, we utilized Accuracy and F1-score as evaluation metrics.

4.2 Bias Analysis on Concatenated Sets

In this section, we aimed to determine whether the strength of texture changes depending on the labeling scheme. Moreover, we further examine whether our granular scheme achieves a different bias level compared to the other schemes. Given three concatenated sets, we performed bias analysis with the proposed

AdaBA and measured the strength difference between texture and shape bias (denoted as *Diff*). Following the results of Table 4, we scrutinized that different labeling schemes exhibit different bias landscapes of the CNN; thus, redesigning the label space would presumably influence a balance between texture and shape biases. We further discovered that our proposed granular scheme achieves a balanced point between shape-biased and texture-biased schemes' one-side bias. This implies representation with conventional labeling schemes (i.e., shape-biased or texture-biased) overly biased to one side, but the proposed granular scheme can contribute to the balanced landscape of CNN's bias in texture and shape.

Table 4. Bias analysis on CNNs trained under various labeling schemes. Note that Diff implies an absolute value of difference between **Shape** and **Texture** scores.

Labeling scheme	Training sets								
	C-Set1			C-Set2			C-Set3		
	Shape	Texture	Diff	Shape	Texture	Diff	Shape	Texture	Diff
Granular	0.5588	0.2785	**0.2803**	0.2356	0.6270	**0.3914**	0.6227	0.8230	**0.2003**
Texture-biased	0.3526	0.4641	0.1115	0.0584	0.7532	0.6948	0.5330	0.6587	0.1258
Shape-biased	0.1698	0.9615	0.7917	0.9959	0.8498	0.1461	0.0461	0.9460	0.9000

5 Is Granular Labeling Scheme Advantageous in Classification Performance?

5.1 Setup

We first and foremost examined whether the proposed granular labeling scheme contributes to better performance at two classification tasks: shape classification and texture classification. The shape classification is a 10-class classification where the samples at each label include similar object shapes and different textures of natural and artistic. For the shape classification, we primarily compared the proposed granular scheme's performance with a shape-biased scheme. Note that the CNN trained under the granular scheme solves 20-class classification while the shape-biased scheme lets the model solve 10-class classification. On the other hand, the texture classification is a binary classification with two labels: natural and artistic. We validate the effectiveness of the granular scheme with the texture-biased scheme. While the granular scheme shares the same setting with the one at shape classification, the texture-biased scheme takes binary label space: the samples at each class share similar textures and different shapes. We followed the evaluation procedures described in Sect. 4.1 for a proper comparative study. The experiment results are described in Table 5.

5.2 Analysis

We discovered the proposed granular labeling scheme contributed to precise classification performances compared to other paradigms. We hypothesize conveying both shape and texture characteristics contributed to more qualified representation, and this improved representation supported a significant classification performance at both tasks. Throughout the experiment results, we reconfirmed a common notion: the samples within a single label shall share similar characteristics, and the more minimized variance contributes to the better representation power for classification. Accordingly, we figured out that a simple label space change can improve classification performance; thus, this finding can be a useful guideline for machine learning practitioners.

Table 5. Shape and texture classification results under various labeling schemes. Denote that C-set means concatenated sets.

Labeling scheme	Training sets					
	C-Set1		C-Set2		C-Set3	
Shape classification						
	Accuracy	**F1-score**	**Accuracy**	**F1-score**	**Accuracy**	**F1-score**
Shape-biased	0.7751	0.8625	0.7851	0.8679	0.6990	0.8153
Granular	**0.7835**	**0.8884**	**0.7858**	**0.8769**	**0.7337**	**0.8403**
Texture classification						
	Accuracy	**F1-score**	**Accuracy**	**F1-score**	**Accuracy**	**F1-score**
Texture-biased	0.9611	0.9796	0.9621	0.9804	0.9837	0.9916
Granular	**0.9857**	**0.9905**	**0.9900**	**0.9938**	**0.9933**	**0.9950**

6 Does Granular Labeling Scheme Contribute to Better OOD Detection?

6.1 Setup

Furthermore, we validate whether the proposed granular labeling scheme contributes to better OOD detection performance. Among previous OOD detection methods [12,13,19], we employed an approach proposed in Vaze et al. [26] due to its supreme performance in various benchmark datasets. As we trained the CNN with the dataset stemming from CIFAR-10, we utilized two OOD datasets that do not share the same semantics with the training set: CIFAR-100 and SVHN. Furthermore, we synthetically created additional OOD samples for a more precise experiment: a stylized OOD dataset. The stylized OOD dataset includes samples that have been style-transferred with the AdaIn under the same procedure of creating the training set. Note that we created style-transferred OOD samples using the same style type used in the training set. (i.e., If the training set includes stylized samples in Rococo type, the OOD dataset also includes stylized OOD samples with Rococo type) We established the OOD detectors based on three CNNs trained under different labeling schemes and examined which labeling scheme contributes to better OOD detection performance. Following prior OOD detection studies, we also employed Area Under ROC curve

(AUROC) for an evaluation metric to comprehensively evaluate the OOD detection performance under various threshold levels. Note that we utilized the same implementation settings with the one elaborated on Sect. 5. We described experiment results in Table 6.

6.2 Analogy

We figured out a model trained under the granular labeling scheme was not always superficial in detecting OOD samples. While the granular scheme accomplished precise OOD detection performance in the Stylized OOD dataset, the shape-biased scheme achieved better performance in most original OOD datasets. We presume an underlying reason for this result also lies in the overfitted representations under the granular labeling schemes. As the granular labeling scheme acquires overly optimized representations to the training samples, it weakens the general understanding of various samples, including the ones that exist at different distributions. Based on the results of the OOD detection, we expect the proposed granular scheme is vulnerable in vision tasks which include samples at different distributions. Still, we acknowledge our analogy is at an empirical level; thus, a more in-depth analysis of this phenomenon is highly required in the follow-up studies.

Table 6. The OOD detection performances under various labeling schemes. Denote that C-set means concatenated sets.

OOD set	Labeling scheme	Original OOD			Style-transfered OOD		
		C-Set1	C-Set2	C-Set3	C-Set1	C-Set2	C-Set3
CIFAR-100	Shape-biased	**0.7787**	**0.8068**	0.7023	0.7154	0.7391	0.7153
	Texture-biased	0.4604	0.5564	0.6598	0.7524	0.7317	0.6102
	Ours	0.6745	0.7651	**0.7942**	**0.8636**	**0.8696**	**0.7527**
SVHN	Shape-biased	**0.7734**	**0.7981**	0.6876	0.7756	0.5583	0.7458
	Texture-biased	0.4355	0.4575	0.4764	0.7503	0.5446	0.5841
	Ours	0.7187	0.7587	**0.6943**	**0.8892**	**0.5858**	**0.7746**

7 What Representation Do Various Schemes Acquire?

7.1 Setup

While we discover that the proposed granular scheme acquires a balance between texture and shape information, the follow-up question arises: *How does the learned representation look like?* To excavate an answer, we compared the representation similarities among CNNs trained under various labeling schemes with CKA [17]. We analyzed layer-wise representation similarity within a single model, that implies a similarity among convolution layers in a single CNN. We aim to analyze the knowledge capacity of a trained model. Suppose particular layers in a single model acquires representations similar to the other layers. This case implies that the model failed to learn various characteristics of the training samples; thus, it limits the representation power of a given sample. Conversely, if layers within the same model bear lower similarities to each other, it shows that the model can illustrate various characteristics of a given sample; thus, the model considers a wide range of patterns exists to solve a classification task. Upon the aforementioned setups, we visualized layer-wise representation similarities within each model Fig. 4. Note that both x and y-axis in the figure imply the convolution layers index at ResNet-50.

7.2 Analogy

Following the results in Fig. 4, we figured out the CNN trained under the granular scheme has a large capacity of knowledge as its layer-wise representation similarity is comparatively lower than the others. While the CNNs trained under shape-biased and texture-biased schemes bear many similar representations within their layers, the model trained under the proposed scheme has smaller similar representations. We analyze this smaller similarity among layers let the model scrutinize various patterns of a given sample, and this larger knowledge capacity contributed to the precise classification performance. We figured out the effectiveness of the proposed granular labeling scheme comes from the quality of representation. The representation trained under the proposed scheme has a larger knowledge capacity, and it acquires a presumably qualified contextual understanding of a given data at high-level layers of the neural networks. For more in-depth analyses, we additionally revealed that a CNN trained under the granular labeling scheme exhibits distinct high-level representations from the other schemes; thus, this distinct representation particularly contributes to better performances. We skipped the description in this paper due to page limits, please refer the supplementary materials for detailed analyses.

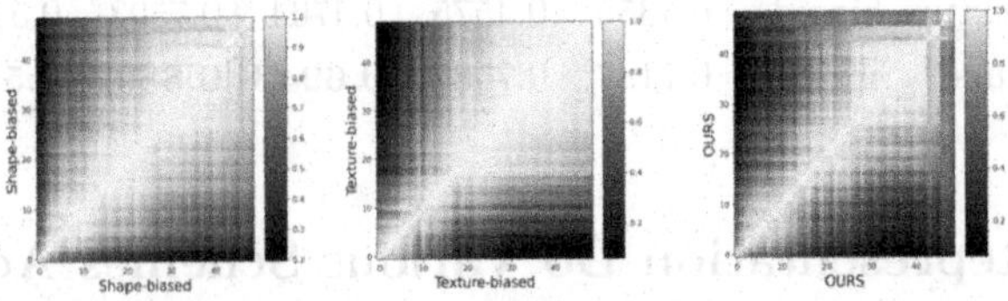

(a) Shape-biased (b) Texture-biased (c) Granular (Ours)

Fig. 4. Layer-wise representation similarities among layers within the same model.

8 Conclusion

Throughout the study, we present a series of analyses that scrutinize CNN's texture and shape bias. First, we propose AdaBA, a novel bias analysis method that sustains the baseline method's performance as well as more lightweight procedures. Upon the AdaBA, we explore how the fine-tuned CNNs expose a biased landscape in various downstream datasets and result in a data distribution is a root cause of texture bias of fine-tuned CNNs, as well as ImageNet-trained CNNs [14]. To this end, we suggest a granular labeling scheme that can mitigate the CNN's texture preference by simply redesigning the label space. We empirically examine that the granular labeling scheme exhibits a balanced bias between texture and shape, and it yields escalated performances on classification and OOD detection. Lastly, we analyze that the granular labeling scheme acquires more qualified representation power, describing a fruitful illustration of a given sample.

Acknowledgement. This work is supported by the Korea Agency for Infrastructure Technology Advancement grant funded by the Ministry of Land, Infrastructure, and Transport (Grant RS-2022-0014579).

References

1. Adebayo, J., Gilmer, J., Muelly, M., Goodfellow, I., Hardt, M., Kim, B.: Sanity checks for saliency maps. In: Advances in Neural Information Processing Systems, vol. 31 (2018)
2. Cheng, B., Wei, Y., Shi, H., Feris, R., Xiong, J., Huang, T.: Revisiting RCNN: on awakening the classification power of faster RCNN. In: Ferrari, V., Hebert, M., Sminchisescu, C., Weiss, Y. (eds.) ECCV 2018. LNCS, vol. 11219, pp. 473–490. Springer, Cham (2018). https://doi.org/10.1007/978-3-030-01267-0_28
3. Deng, J., Dong, W., Socher, R., Li, L.J., Li, K., Fei-Fei, L.: ImageNet: a large-scale hierarchical image database. In: 2009 IEEE Conference on Computer Vision and Pattern Recognition, pp. 248–255. IEEE (2009)
4. Dodge, S., Karam, L.: A study and comparison of human and deep learning recognition performance under visual distortions. In: 2017 26th International Conference on Computer Communication and Networks (ICCCN), pp. 1–7. IEEE (2017)
5. Doshi-Velez, F., Kim, B.: Towards a rigorous science of interpretable machine learning. arXiv preprint arXiv:1702.08608 (2017)
6. Gatys, L.A., Ecker, A.S., Bethge, M.: Image style transfer using convolutional neural networks. In: Proceedings of the IEEE Conference on Computer Vision and Pattern Recognition, pp. 2414–2423 (2016)
7. Geirhos, R., Rubisch, P., Michaelis, C., Bethge, M., Wichmann, F.A., Brendel, W.: ImageNet-trained CNNs are biased towards texture; increasing shape bias improves accuracy and robustness. arXiv preprint arXiv:1811.12231 (2018)
8. Ghorbani, A., Wexler, J., Zou, J.Y., Kim, B.: Towards automatic concept-based explanations. In: Advances in Neural Information Processing Systems, vol. 32 (2019)
9. Gu, J., et al.: Recent advances in convolutional neural networks. Pattern Recogn. **77**, 354–377 (2018)
10. Guo, C., Wu, D.: Canonical correlation analysis (CCA) based multi-view learning: an overview. arXiv preprint arXiv:1907.01693 (2019)
11. He, K., Zhang, X., Ren, S., Sun, J.: Deep residual learning for image recognition. In: Proceedings of the IEEE Conference on Computer Vision and Pattern Recognition, pp. 770–778 (2016)

12. Hendrycks, D., Gimpel, K.: A baseline for detecting misclassified and out-of-distribution examples in neural networks. arXiv preprint arXiv:1610.02136 (2016)
13. Hendrycks, D., Mazeika, M., Dietterich, T.: Deep anomaly detection with outlier exposure. arXiv preprint arXiv:1812.04606 (2018)
14. Hermann, K., Chen, T., Kornblith, S.: The origins and prevalence of texture bias in convolutional neural networks. Adv. Neural. Inf. Process. Syst. **33**, 19000–19015 (2020)
15. Islam, M.A., et al.: Shape or texture: understanding discriminative features in CNNs. arXiv preprint arXiv:2101.11604 (2021)
16. Kim, B., Wattenberg, M., Gilmer, J., Cai, C., Wexler, J., Viegas, F., et al.: Interpretability beyond feature attribution: quantitative testing with concept activation vectors (TCAV). In: International Conference on Machine Learning, pp. 2668–2677. PMLR (2018)
17. Kornblith, S., Norouzi, M., Lee, H., Hinton, G.: Similarity of neural network representations revisited. In: International Conference on Machine Learning, pp. 3519–3529. PMLR (2019)
18. Krizhevsky, A., Sutskever, I., Hinton, G.E.: ImageNet classification with deep convolutional neural networks. In: Advances in Neural Information Processing Systems, vol. 25 (2012)
19. Lee, K., Lee, H., Lee, K., Shin, J.: Training confidence-calibrated classifiers for detecting out-of-distribution samples. arXiv preprint arXiv:1711.09325 (2017)
20. Li, J., et al.: A systematic collection of medical image datasets for deep learning. arXiv preprint arXiv:2106.12864 (2021)
21. Papageorgiou, C., Poggio, T.: A trainable system for object detection. Int. J. Comput. Vision **38**(1), 15–33 (2000)
22. Shen, D., Wu, G., Suk, H.I.: Deep learning in medical image analysis. Annu. Rev. Biomed. Eng. **19**, 221–248 (2017)
23. Simonyan, K., Zisserman, A.: Very deep convolutional networks for large-scale image recognition. arXiv preprint arXiv:1409.1556 (2014)
24. Sun, Q.S., Zeng, S.G., Liu, Y., Heng, P.A., Xia, D.S.: A new method of feature fusion and its application in image recognition. Pattern Recogn. **38**(12), 2437–2448 (2005)
25. Tuli, S., Dasgupta, I., Grant, E., Griffiths, T.L.: Are convolutional neural networks or transformers more like human vision? arXiv preprint arXiv:2105.07197 (2021)
26. Vaze, S., Han, K., Vedaldi, A., Zisserman, A.: Open-set recognition: a good closed-set classifier is all you need. arXiv preprint arXiv:2110.06207 (2021)
27. Wu, Z., Shen, C., Van Den Hengel, A.: Wider or deeper: revisiting the ResNet model for visual recognition. Pattern Recogn. **90**, 119–133 (2019)

What Role Does Data Augmentation Play in Knowledge Distillation?

Wei Li[(✉)], Shitong Shao, Weiyan Liu, Ziming Qiu, Zhihao Zhu, and Wei Huan

School of Instrument Science and Engineering, Southeast University, Nanjing 210096, Jiangsu, China
{li-wei,shaoshitong,liuweiyan,qiuziming,zhuzhihao, huan-wei}@seu.edu.cn

Abstract. Knowledge distillation is an effective way to transfer knowledge from a large model to a small model, which can significantly improve the performance of the small model. In recent years, some contrastive learning-based knowledge distillation methods (i.e., SSKD and HSAKD) have achieved excellent performance by utilizing data augmentation. However, the worth of data augmentation has always been overlooked by researchers in knowledge distillation, and no work analyzes its role in particular detail. To fix this gap, we analyze the effect of data augmentation on knowledge distillation from a multi-sided perspective. In particular, we demonstrate the following properties of data augmentation: **(a)** data augmentation can effectively help knowledge distillation work even if the teacher model does not have the information about augmented samples, and our proposed diverse and rich **J**oint **D**ata **A**ugmentation (JDA) is more valid than single *rotating* in knowledge distillation; **(b)** using diverse and rich augmented samples to assist the teacher model in training can improve its performance, but not the performance of the student model; **(c)** the student model can achieve excellent performance when the proportion of augmented samples is within a suitable range; **(d)** data augmentation enables knowledge distillation to work better in a few-shot scenario; **(e)** data augmentation is seamlessly compatible with some knowledge distillation methods and can potentially further improve their performance. Enlightened by the above analysis, we propose a method named **C**osine **C**onfidence **D**istillation (CCD) to transfer the augmented samples' knowledge more reasonably. And CCD achieves better performance than the latest SOTA HSAKD with fewer storage requirements on CIFAR-100 and ImageNet-1k. *Our code is released at* https://github.com/liwei-group/CCD.

1 Introduction

With the vigorous development of deep learning, numerous excellent models (e.g., ResNet [14], ShuffleNet [40], ViT [7]) have been proposed. In this trend, the evaluation metrics of some image upstream and downstream tasks have been greatly improved [11, 20,21,26,34]. For instance, ResNet50 has achieved 77.15% on the ImageNet-1k [27]

Supplementary Information The online version contains supplementary material available at https://doi.org/10.1007/978-3-031-26284-5_31.

L. Wang et al. (Eds.): ACCV 2022, LNCS 13842, pp. 507–525, 2023.
https://doi.org/10.1007/978-3-031-26284-5_31

classification task in 2015, and VIT-H/14 has achieved 88.55% accuracy on the same task in 2020. However, ResNet50 has only 25.5 million parameters, compared to 632 million parameters of VIT-H/14. The massive storage requirements of the large models render them to deploy in real-time applications challengingly. For the purpose of developing efficient models, knowledge distillation [15], as an effective technique, has been widely used in model compression [1]. To be specific, knowledge distillation aims to transfer knowledge from a pre-trained teacher network with big-scale parameters to a lightweight student network. This significant training technique generally enables the student model to outperform traditional training techniques by a large margin. Commonly, the cases where the teacher model is fixed and not fixed are referred to as offline knowledge distillation [32,37,42] and online knowledge distillation [2,10,41], respectively. And online knowledge distillation can improve the performance of student models more effectively than offline knowledge distillation but requires more computational and storage costs [9]. To make the design choices that other researchers can better apply, we choose offline knowledge distillation as the standard for the study in this work.

Contrastive learning aims to encode the correlations between a sample pair $(\mathcal{X}_i, \mathcal{X}_j)$. Specifically, if $\mathcal{X}_i$ and $\mathcal{X}_j$ are similar, contrastive learning makes the distance between them close; otherwise, makes the distance between them as far as possible. In recent years, contrastive learning has been considered as an effective solution in the self-supervised domain. The popular contrastive learning methods, such as MOCO [13] and SimCLR [3], have been widely recognized and applied by related researchers. It is worth noting that, contrastive learning has also been applied in knowledge distillation as a novel way to transfer knowledge. The knowledge distillation methods, SSKD [35] and HSAKD [4], based on contrastive learning and self-supervised representational learning, utilize the same data augmentation (i.e., rotations $\{0°, 90°, 180°, 270°\}$). And the above methods are state-of-the-art (SOTA) in 2020 and 2021, respectively. However, the phenomenon that data augmentation changes the information of the training samples is not discussed in their works. In addition, because SSKD and HSAKD use *rotating* in training to duplicate the samples, their steps n_{step} (defined in Eq. 1, where n_{iter}, n_{bs} and n_{epoch} refer to the number of batch sizes in an epoch, batch size, and the number of epochs, respectively) were four times higher than vanilla Knowledge Distillation (vanilla KD) in their comparative experiments. Based on the above analysis, we know that using data augmentation in SSKD and HSAKD changes the information of the training samples and n_{step}. So it is unclear that whether the performance improvement of SSKD and HSAKD is brought by data augmentation. Due to these reasons, we urgently need to analyze the role that data augmentation plays in knowledge distillation.

$$n_{step} = n_{iter} \times n_{bs} \times n_{epoch} \tag{1}$$

Some works such as [8,33] and [6] have noted the role of data augmentation in knowledge distillation. They find that the knowledge distillation approaches, including vanilla KD and CRD [31], can also improve the performance of the student model to a certain extent by means of only data augmentation. However, the above work only discusses the role of data augmentation in knowledge distillation from a one-sided perspective. So they do not consider the impact of multiple factors, including different n_{step}, the diversity of data augmentation, the proportion of augmented samples in the all training samples, and the few-shot scenario, on the performance of student models. In order to fill this gap, we further evaluate the effect of data augmentation on knowledge distillation under different factors, and the main conclusions we find in our extensive experiments can be summarized as follows:

- Data Augmentation is effective in knowledge distillation, even though the teacher model has no information about the augmented sample. The effectiveness of self-supervised methods (i.e., SSKD and HSAKD) can be attributed to *rotating*, to a certain extent.
- By increasing n_{step}, knowledge distillation both with and w/o (i.e., without) data augmentation will improve the performance of the student model. In addition, the knowledge distillation with diverse and rich data augmentation is more valid than single rotational knowledge distillation.
- Transferring knowledge only from the augmented samples doesn't necessarily work, but transferring knowledge from both the original and augmented samples can effectively make the augmented samples work.
- Using diverse and rich augmented samples to assist the teacher model in training can improve its performance, but not the performance of the student model.
- The student model can achieve excellent performance when the proportion of augmented samples is within a suitable range. But too many augmented samples will lead to a drop in performance.
- Data augmentation enables knowledge distillation to work better in few-shot scenarios.
- Data augmentation is seamlessly compatible with some knowledge distillation methods and can potentially further improve their performance.

Inspired by these conclusions, we propose a method named **C**osine **C**onfidence **D**istillation (CCD) to transfer the probabilistic knowledge of augmented samples more reasonably. And CCD achieves better performance compared to the latest SOTA HSAKD with fewer storage requirements on CIFAR-100 and ImageNet-1k.

2 Related Work

Knowledge Distillation with Self-Supervision. Recently, knowledge distillation methods (i.e., SSKD and HSAKD) with self-supervision have achieved state-of-the-art on both CIFAR-100 and ImageNet-1k. Among them, SSKD and HSAKD adopt the idea of contrastive learning directly and indirectly, respectively. Specifically, SSKD tends to compute the sample-based metric matrices for the teacher and student models separately and align them to achieve self-supervision and knowledge distillation. And HSAKD generates bivariate distribution labels based on augmented and semantic categories, then minimizes the loss between logits of the teacher and student models. The core idea of the above two methods is to transfer the information learned by the teacher model through self-supervision to the student model, which needs to be implemented through data augmentation. Therefore, it is clear that data augmentation has become an essential part of self-supervised knowledge distillation [19,28,30].

Data Augmentation. In the field of computer vision, data augmentation is a simple and effective way to improve model performance [5,12,16,18,36]. For instance, data augmentation rules such as *rotating*, shear, and contrast are widely applied in visual tasks (e.g., object classification and object detection) and avoid the overfitting problem of the model to a certain extent [22,29]. If data augmentation really works, SSKD and HSAKD only use *rotating* to augment data, which obviously lacks variety. So for our study, We turn our attention to other more diverse data augmentation. AutoAugment [5], an effective and popular data augmentation method, adopts 16 commonly used data augmentation rules as its sub-policies. These sub-policies efficiently augment the dataset by searching for their optimal hyperparameters through reinforcement learning (RL). However, for knowledge distillation, we do not know what individualized samples are adapted for a particular pair of teacher and student models. And finding the optimal conversion probability again requires a lot of training costs. Therefore, we will apply 14 data augmentation (i.e., the sub-policies in AutoAugment) to convert the original samples in random order and with the same probability in our study.

In a word, inspired by SSKD and HSAKD applying data augmentation for their self-supervised methods, we employ a wide and abundant variety of data augmentation rules to conduct our research.

3 Contributions

For all experimental results shown in this Sect. 3.1 to this Sect. 3.5, we conduct evaluations on the standard CIFAR-10 [17] benchmark across the ResNet56-ResNet20 [14] pair and the standard CIFAR-100 [17] benchmark across the WRN-40-2-WRN-16-2 [38] pair. Note that the top-1 test accuracy of teachers ResNet-56 and WRN-40-2 are 93.56% and 76.44%, respectively. And all black horizontal lines in figures of this paper represent the test accuracy of the teacher model. All "$\times$**number**" in this paper represents a multiple of the increase about n_{step} compared to the benchmark, and more detailed benchmark settings can be found in Appendix A. Besides, we utilize the standard training settings following [4,35] on CIFAR-100 and following [23] on CIFAR-10. *All teacher models do not utilize the augmented samples for representation learning, unless otherwise specified in this paper.* To get plausible results, we report the mean test accuracy with 3 runs. Note that to introduce our research more logically, we will present our contributions following the form of progressive exploration.

3.1 Inspired by SSKD and HSAKD

By regarding the similarity between self-supervised samples as the transferring knowledge, SSKD has achieved excellent performance. But when investigating previous work, we find that SSKD is sensitive in certain scenes. For example, the validation accuracy of ResNet-18 (student) trained on ImageNet-1k with ResNet-34 (teacher), obtained by the original work, is 71.62% [35]. Still, the validation accuracy got by work [23] is 70.09%. There is a 1.53% point difference between the above two results, which is relatively large under the same hyperparameter configuration. Meanwhile, the validation accuracy for vanilla KD got by work [23] is 71.23% and got by us is 71.16%. Intuitively, SSKD is inferior to vanilla KD under certain circumstances, which drives us to rethink the effectiveness of SSKD.

Table 1. These two tables show the experimental results of decoupling of SSKD and HSAKD on CIFAR-10 and CIFAR-100, respectively. Where "Baseline" stands for training using only vanilla KD. In addition, "(number)" refer to the increased validation accuracy compared to the baseline.

CIFAR-10

n_{step}	Methods	Options	Acc.(%)
×2	Baseline	$\mathcal{L}_{ce} + \mathcal{L}_{kd}$	93.24
×2	SSKD	$\mathcal{L}_{SSKD}$	$92.56_{(-0.68)}$
×2	SSKD	$\mathcal{L}_{SSKD} - \mathcal{L}_{ss}$	$92.72_{(-0.52)}$
×2	HSAKD	$\mathcal{L}_{HSAKD}$	$93.22_{(-0.02)}$
×2	HSAKD	$\mathcal{L}_{HSAKD} - \mathcal{L}_{kl\_q}$	$92.68_{(-0.56)}$
×4	Baseline	$\mathcal{L}_{ce} + \mathcal{L}_{kd}$	93.47
×4	SSKD	$\mathcal{L}_{SSKD}$	$92.73_{(-0.74)}$
×4	SSKD	$\mathcal{L}_{SSKD} - \mathcal{L}_{ss}$	$92.73_{(-0.74)}$
×4	HSAKD	$\mathcal{L}_{HSAKD}$	$93.46_{(-0.01)}$
×4	HSAKD	$\mathcal{L}_{HSAKD} - \mathcal{L}_{kl\_q}$	$92.88_{(-0.59)}$

CIFAR-100

n_{step}	Methods	Options	Acc.(%)
×2	Baseline	$\mathcal{L}_{ce} + \mathcal{L}_{kd}$	74.72
×2	SSKD	$\mathcal{L}_{SSKD}$	$75.51_{(+0.79)}$
×2	SSKD	$\mathcal{L}_{SSKD} - \mathcal{L}_{ss}$	$75.31_{(+0.59)}$
×2	HSAKD	$\mathcal{L}_{HSAKD}$	$76.73_{(+2.01)}$
×2	HSAKD	$\mathcal{L}_{HSAKD} - \mathcal{L}_{kl\_q}$	$75.55_{(+0.83)}$
×4	Baseline	$\mathcal{L}_{ce} + \mathcal{L}_{kd}$	74.87
×4	SSKD	$\mathcal{L}_{SSKD}$	$76.16_{(+1.29)}$
×4	SSKD	$\mathcal{L}_{SSKD} - \mathcal{L}_{ss}$	$76.31_{(+1.44)}$
×4	HSAKD	$\mathcal{L}_{HSAKD}$	$77.20_{(+2.33)}$
×4	HSAKD	$\mathcal{L}_{HSAKD} - \mathcal{L}_{kl\_q}$	$76.05_{(+1.18)}$

A natural direction to tackle this problem is to decouple SSKD and HSAKD. To this end, we first give the standard cross-entropy (CE) loss of original samples in Eq. 2:

$$\begin{aligned} \mathbf{p}^S(\mathbf{x};\tau) &= \mathbf{softmax}\left(f^S(\mathbf{x})/\tau\right), \\ \mathcal{L}_{ce} &= \mathbb{E}_{\mathbf{x}\in\mathcal{X}}\,\mathbf{CE}\left(\mathbf{p}^S(\mathbf{x};1),\mathbf{y}\right), \end{aligned} \tag{2}$$

where τ, $f^S(\cdot)$, $\mathbf{CE}(\cdot,\cdot)$, $\mathcal{X}$ and $\mathbf{y}$ refer to the temperature hyperparameter, the student backbone network, the cross entropy loss function, the original sample set and the hard label about $\mathbf{x}$, respectively. Then we denote the vanilla KD loss of original samples as Eq. 3.

$$\begin{aligned} \mathbf{p}^T(\mathbf{x};\tau) &= \mathbf{softmax}\left(f^T(\mathbf{x})/\tau\right), \\ \mathcal{L}_{kd} &= \tau^2\mathbb{E}_{\mathbf{x}\in\mathcal{X}}\,\mathbf{KL}\left(\mathbf{p}^T(\mathbf{x};\tau)\,\|\,\mathbf{p}^S(\mathbf{x};\tau)\right), \end{aligned} \tag{3}$$

where $f^T(\cdot)$ denotes the teacher backbone network and $\mathbf{KL}(\cdot\|\cdot)$ denotes the Kullback-Leibler divergence. Thereby, the vanilla KD loss of augmented samples can be defined as follows:

$$\mathcal{L}_T = \tau^2\mathbb{E}_{\widetilde{\mathbf{x}}\in\widetilde{\mathcal{X}}}\,\mathbf{KL}\left(\mathbf{p}^T(\widetilde{\mathbf{x}};\tau)\,\|\,\mathbf{p}^S(\widetilde{\mathbf{x}};\tau)\right), \tag{4}$$

where $\widetilde{\mathcal{X}}$ stands for the augmented sample set. Hence, we give the corresponding Eqs. 5 for SSKD and HSAKD:

$$\begin{aligned} \mathcal{L}_{SSKD} &= \lambda_1 * \mathcal{L}_{ce} + \lambda_2 * \mathcal{L}_{kd} + \lambda_3 * \mathcal{L}_{ss} + \lambda_4 * \mathcal{L}_T, \\ \mathcal{L}_{HSAKD} &= \mathcal{L}_{ce} + \mathcal{L}_{kl\_q} + \mathcal{L}_{kl\_p}. \end{aligned} \tag{5}$$

In $\mathcal{L}_{SSKD}$, λ_1, λ_2, λ_3 and λ_4 are the balancing weights. And the sub-task loss function $\mathcal{L}_{ss}$ concurs with L_{ss} in [35]. In $\mathcal{L}_{HSAKD}$, $\mathcal{L}_{kl\_q}$ (feature-based) and $\mathcal{L}_{kl\_p}$ (response-based) are the loss functions as mentioned in the work of [4][1]. Although $\mathcal{L}_{SSKD}$ and $\mathcal{L}_{HSAKD}$ are not similar in form, by Eq. 6 we can rewrite the form of $\mathcal{L}_{kl\_p}$.

[1] For the sake of simplicity, $\mathcal{L}_{kl\_q}$ and $\mathcal{L}_{kl\_p}$ here have an additional process of calculating mathematical expectations compared to the original paper.

$$
\begin{aligned}
\mathcal{L}_{kl\_p} &= \tau^2 \mathbb{E}_{\mathbf{x}\in\mathcal{X}} \frac{1}{M} \sum_{j=1}^{M} \mathbf{KL}\left(\mathbf{p}^T\left(t_j\left(\mathbf{x}\right);\tau\right) \| \mathbf{p}^S\left(t_j\left(\mathbf{x}\right);\tau\right)\right), \\
&= \tau^2 \mathbb{E}_{\mathbf{x}\in\mathcal{X}} \frac{1}{M} \mathbf{KL}\left(\mathbf{p}^T\left(\mathbf{x};\tau\right) \| \mathbf{p}^S\left(\mathbf{x};\tau\right)\right) + \tau^2 \mathbb{E}_{\widetilde{\mathbf{x}}\in\widetilde{\mathcal{X}}} \frac{M-1}{M} \mathbf{KL}\left(\mathbf{p}^T\left(\widetilde{\mathbf{x}};\tau\right) \| \mathbf{p}^S\left(\widetilde{\mathbf{x}};\tau\right)\right), \\
&= \frac{1}{M}\mathcal{L}_{kd} + \frac{M-1}{M}\mathcal{L}_T,
\end{aligned}
\tag{6}
$$

where M and $\{t_j\left(\cdot\right)\}_{j=1}^{M}$ refer to the number of rotation operators (i.e., rotations $\{0°, 90°, 180°, 270°\}$) and a set of data augmentation operators, respectively. According to the code and original paper provided by the author, we can default M to 4. Furthermore, we find that λ_2/λ_4 is $1/3$ in SSKD-related codes. Then an obvious conclusion is that $\mathcal{L}_{kl\_p} \propto \lambda_2 * \mathcal{L}_{kd} + \lambda_4 * \mathcal{L}_T$.

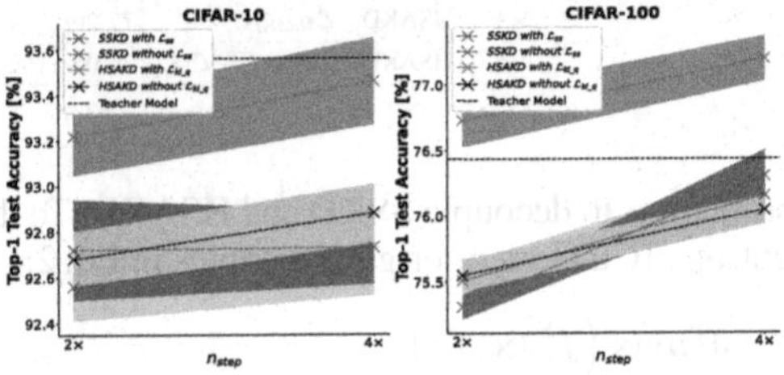

Fig. 1. This line graph is used to clearly demonstrate the roles $\mathcal{L}_{kl\_q}$ and $\mathcal{L}_{ss}$ play in knowledge distillation, and the shaded area in this figure indicates the standard deviation. Of note is that, $\mathcal{L}_{ss}$ does not play a positive role. In contrast, $\mathcal{L}_{kl\_q}$ can significantly improve the performance of the student model.

The above analysis shows that the essential difference between SSKD and HSAKD is that SSKD transfers the correlation information between samples after the global average pooling (GAP) layer ($\mathcal{L}_{ss}$). In contrast, HSAKD transfers the self-supervised augmented distribution information of the outputs of middle layers ($\mathcal{L}_{kl\_q}$).

For the purpose of verifying the role of $\mathcal{L}_{ss}$, $\mathcal{L}_{kl\_q}$ and *rotating*, we conduct the decoupling experiment on the two benchmarks mentioned above. For a fair comparison, we quadruple the baseline's n_{epoch} to ensure its n_{step} being the same as the other methods. Then, we get the results shown in Table 1 and Fig. 1. The results displayed in Fig. 1 show that ***it is effective to transfer the self-supervised augmented distribution information of middle layers' outputs***. On the contrary, transferring the correlation information of GAP's output is ineffective. Meanwhile, the analysis in Table 1 shows that SSKD without $\mathcal{L}_{ss}$ and HSAKD without $\mathcal{L}_{kl\_q}$ (i.e., $\lambda_1 * \mathcal{L}_{ce} + \lambda_2 * \mathcal{L}_{kd} + \lambda_4 * \mathcal{L}_T$ and $\mathcal{L}_{ce} + \mathcal{L}_{kl\_p}$) have a slight decrease in performance compared to the baseline on CIFAR-10. Still, there is a manifest improvement in the performance on CIFAR-100 compared with the baseline. Intuitively, the common part of SSKD and HSAKD is the traditional CE loss on the original training samples and the vanilla KD loss on both original and rotated samples. Just relying on this common part, the student model has a significant performance gain on CIFAR-100 compared with baseline. Thus we can illustrate that ***rotating is effective in knowledge distillation***.

From the results presented in Table 1, we find that adding the rotated samples into CIFAR-10 will cause performance damage during the training process. We argue that this phenomenon is due to the rotation operator itself. In follow-up experiments, we demonstrate that utilizing more diverse data augmentation operators is equally effective on CIFAR-10.

3.2 The Role of Data Augmentation

Although in Sect. 3.1 we have bespoken that *rotating* is beneficial to knowledge distillation. However, only utilizing *rotating* as data augmentation in training lacks diversity, so it cannot demonstrate that other data augmentation is also effective. Similarly, in the work [33], only CutMix [36] and Mixup [39] are discussed. To ensure the diversity of data augmentation, we propose the **J**oint **D**ata **A**ugmentation (JDA), which is composed of cascaded sub-policies. Assuming that we define N various sub-policies $\{\mathbf{sp}_i(\cdot)\}_{i=1}^{N}$. And we also define a Bernoulli operator $g(f;q)$ as shown in Eq. 7.

$$g(f;q) = \begin{cases} f(\cdot) & ,w.p.\ q \quad , \\ \mathbf{identity}(\cdot) & ,w.p.\ 1-q \quad , \end{cases} \tag{7}$$

$\mathbf{identity}(\cdot)$ refers to the identity transformation, i.e. $\mathbf{identity}(x) = x$. Then for an original sample $\mathbf{x} \in \mathcal{X}$, we can denote its augmented sample $\widetilde{\mathbf{x}}$ in Eq. 8.

$$\widetilde{\mathbf{x}} = g(\mathbf{sp}_N;q) \circ g(\mathbf{sp}_{N-1};q) \cdots g(\mathbf{sp}_2;q) \circ g(\mathbf{sp}_1;q)(\mathbf{x}), \tag{8}$$

where $\circ$ denotes composition, and note that all sub-policies in Eq. 8 have the same probability of occurrence. This approach can ensure that all sub-policies have similar effects on the original sample and provide the convenience for related experiments in this paper. Moreover, JDA not only is easy to be set up but also guarantees a huge difference compared to the optimal hyperparameters searched by AutoAugment. JDA also eliminates the possibility that the data augmentation work is a result of AutoAugment's hyperparameter search. In particulat, as demonstrated in Appendix B, JDA is more effective than AutoAugment because it can perform richer and more varied transformation on one single image. In more detail, q is set to 0.5 by default unless otherwise specified in our experiments. We consider choosing 14 sub-policies in AutoAugment for JDA, and the detailed hyperparameter settings for 14 sub-policies can be apparent from Appendix A. Furthermore, we introduce the mini-batch component of model training to explain how our data augmentation works. For the original sample set $\mathcal{X}$ (i.e., the original mini-batch), our proposed JDA transforms all elements in it in turn and composes a new augmented sample set $\widetilde{\mathcal{X}}$. Then the new mini-batch composed of $\mathcal{X}$ and $\widetilde{\mathcal{X}}$ serves as the real input of the model.

Table 2. Left: *CIFAR-10.* **Right:** *CIFAR-100.* **KD+JDA:** *vanilla KD+joint data augmentation.* The numbers and numerical subscripts in the table represent the test accuracy and standard deviation, respectively.

SSKD	HSAKD	KD+JDA	SSKD	HSAKD	KD+JDA
$92.73_{(\pm 0.16)}$	$93.46_{(\pm 0.19)}$	$93.51_{(\pm 0.14)}$	$76.17_{(\pm 0.17)}$	$77.20_{(\pm 0.17)}$	$76.86_{(\pm 0.16)}$

In SSKD and HSAKD, hard labels for classification, which are provided by augmented samples, are not applied to supervise the student model in training. SSKD considers it is unnecessary for the student model to correctly identify these labels in

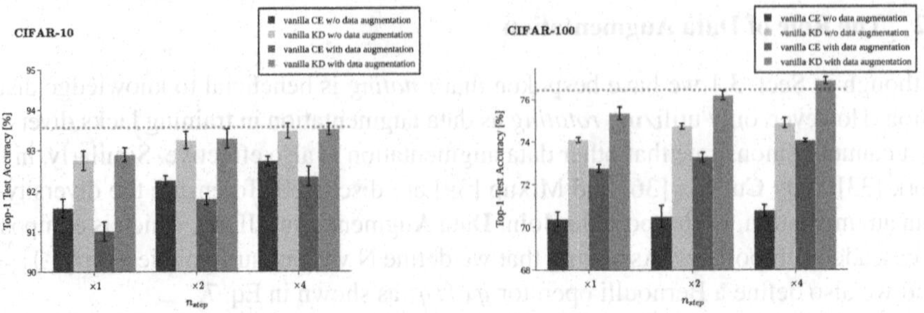

Fig. 2. This two figures show the comparative experimental results on CIFAR-10 and CIFAR-100. Among them, blue •, orange •, green • and red • represent four different loss combinations of $\{\mathcal{L}_{ce}\}$, $\{\mathcal{L}_{ce}, \mathcal{L}_{kd}\}$, $\{\mathcal{L}_{ce}, \mathcal{L}_{A}\}$ and $\{\mathcal{L}_{ce}, \mathcal{L}_{A}, \mathcal{L}_{kd}, \mathcal{L}_{T}\}$, respectively. And error bars in figures indicate standard deviation. (Color figure online)

knowledge distillation, and HSAKD inherits this behavior from SSKD. Although the knowledge distillation methods mentioned above do not attach importance to the hard labels provided by augmented samples, we will utilize them and manifest them can work (Sect. 3.3). We give the CE loss with respect to the augmented samples in the following Eq. 9.

$$\mathcal{L}_{A} = \mathbb{E}_{\widetilde{\mathbf{x}} \in \widetilde{\mathcal{X}}} \, \mathbf{CE}\left(\mathbf{p}^{S}\left(\widetilde{\mathbf{x}}; 1\right), \widetilde{\mathbf{y}}\right), \tag{9}$$

where $\widetilde{\mathbf{y}}$ stands for the hard label about $\widetilde{\mathbf{x}}$. So in our proposed data-augmented knowledge distillation, the overall loss can be expressed as $\mathcal{L}_{oa}$ (in Eq. 10, where $|\mathcal{X}|$ and $|\widetilde{\mathcal{X}}|$ refer to the number of elements in the original sample set and the augmented sample set, respectively). At the same time, this method can be denoted as KD+JDA.

$$\mathcal{L}_{oa} = \frac{|\mathcal{X}|}{|\mathcal{X}| + |\widetilde{\mathcal{X}}|}\mathcal{L}_{ce} + \frac{|\widetilde{\mathcal{X}}|}{|\mathcal{X}| + |\widetilde{\mathcal{X}}|}\mathcal{L}_{A} + \frac{|\mathcal{X}|}{|\mathcal{X}| + |\widetilde{\mathcal{X}}|}\mathcal{L}_{kd} + \frac{|\widetilde{\mathcal{X}}|}{|\mathcal{X}| + |\widetilde{\mathcal{X}}|}\mathcal{L}_{T}. \tag{10}$$

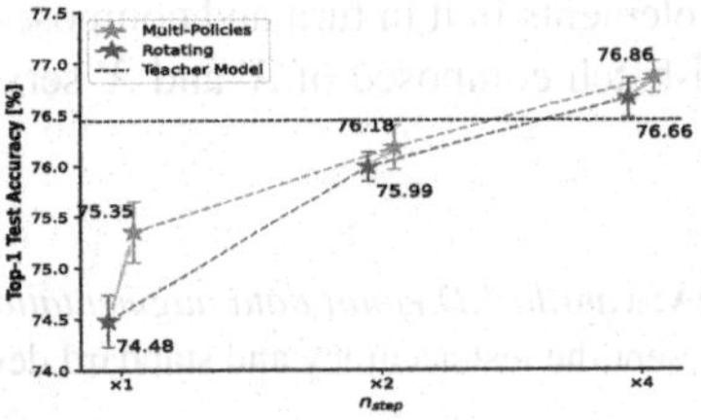

Fig. 3. In this figure, the purple arrows represent the changes in test accuracy from single *rotating* to JDA. So this scattergraph shows that our proposed JDA outperforms single *rotating*. (Color figure online)

We should verify the validity of $\mathcal{L}_{oa}$ through rigorous experiments that n_{step} is the same for all comparative methods. Therefore, we set up three various n_{step} (i.e., ×1, ×2, ×4) in our experiments. Then we compare four different methods, and finally show the results in Fig. 2.

From Fig. 2, we can find that $\mathcal{L}_{oa}$ achieves the best performance on both CIFAR-10 and CIFAR-100. Especially on CIFAR-100, when n_{step} required for all methods is ×4, we can see that its test accuracy has exceeded SSKD and is only slightly lower than HSAKD by observing Table 2. So we can infer that **data augmentation is useful in knowledge distillation, and the magnitude of its improvement depends on the nature of the dataset itself.**

Table 3. The data are obtained under the premise that the teacher model uses only the original samples in training. And ✓ and × represent whether they use the corresponding loss or not, respectively.

CIFAR-10 • Acc. of teacher model:93.56%						CIFAR-100 • Acc. of teacher model:76.44%					
$\mathcal{L}_{ce}$	$\mathcal{L}_{kd}$	$\mathcal{L}_A$	$\mathcal{L}_T$	Acc.(%)	Std.(%)	$\mathcal{L}_{ce}$	$\mathcal{L}_{kd}$	$\mathcal{L}_A$	$\mathcal{L}_T$	Acc.(%)	Std.(%)
✓	×	✓	×	91.79	0.13	✓	×	✓	×	73.27	0.22
✓	✓	✓	✓	93.28	0.24	✓	✓	✓	✓	76.18	0.22
✓	✓	×	×	92.70	0.21	✓	✓	×	×	74.66	0.12
✓	×	×	✓	91.88	0.17	✓	×	×	✓	74.57	0.16
✓	×	×	×	92.14	0.08	✓	×	×	×	71.17	0.12
✓	✓	×	✓	93.15	0.11	✓	✓	×	✓	75.32	0.16

Intuitively, the green rectangle is lower than blue rectangle on CIFAR-10, while the conclusion is converse on CIFAR-100. This illustrates that **adding data augmentation to the single vanilla CE loss during the training phase may harm the performance of the student model.**

Our proposed JDA composed of cascade sub-policies can indeed better transfer the "dark knowledge" from the teacher model to the student model. Unfortunately, both $\mathcal{L}_{SSKD} - \mathcal{L}_{ss}$ and $\mathcal{L}_{HSAKD} - \mathcal{L}_{kl\_q}$ lack $\mathcal{L}_A$ compared to $\mathcal{L}_{oa}$, so we cannot conclude that diverse and abundant data sub-policies are more effective than single *rotating* by directly comparing the relevant experimental results. In order to thoroughly verify the conclusion mentioned above under the premise of a fair comparison, we first add $\mathcal{L}_A$ for utilizing single *rotating* in training, and then conduct additional experiments on the CIFAR-100 benchmark and draw the results in Fig. 3. We can find that diverse and rich sub-policies have improved the test accuracy to some extent compared with single *rotating* at different n_{step}. In this way, we can conclude that **diverse and rich data augmentation is more valid than single *rotating*.**

3.3 Decoupling the Overall Loss

Section 3.2 has demonstrated that $\mathcal{L}_{oa}$ in knowledge distillation is excellent. And this also effectively shows that data augmentation is quite helpful for the performance improvement of the student model. Therefore, another question that needs to be answered urgently is thrown: Which part of $\mathcal{L}_{oa}$ plays a positive role? Is it $\mathcal{L}_{ce}$, $\mathcal{L}_{kd}$, $\mathcal{L}_A$, or $\mathcal{L}_T$?

Following the above analysis, we argue that decoupling the overall loss is the necessary work. For the experiments in this subsection, the setting of n_{step} we uniformly adopted is ×2, and the real input of the student model and the teacher model is the same as that of Sect. 3.2. Finally the decoupling experimental results can be found in Table 3. By analyzing the results in Table 3, we can find that some conclusions drawn on CIFAR-10 and CIFAR-100 are not consistent. In instance, simply adding $\mathcal{L}_T$ to $\mathcal{L}_{ce}$ hurts performance on the CIFAR-10 benchmark, but improves it on the CIFAR-100 benchmark. Of course, this similar conclusion also appears in the previous Section. So we only discuss the common conclusions on these two benchmarks. Intuitively, we conclude that $\mathcal{L}_T$ plays a weaker role than $\mathcal{L}_{kd}$ in knowledge distillation by comparing the results of two different loss combinations (i.e., $\{\mathcal{L}_{ce}, \mathcal{L}_{kd}\}$ and $\{\mathcal{L}_{ce}, \mathcal{L}_{T}\}$). On the other hand, when we add the combination of $\mathcal{L}_A$ and $\mathcal{L}_T$ to the combination of $\mathcal{L}_{ce}$ and $\mathcal{L}_{kd}$. The results are always some improvement on both CIFAR-10 and CIFAR-100. To sum up, **transferring knowledge only from the augmented samples does not necessarily work, but when the real input contains original samples in training, the augmented samples can effectively play an auxiliary advantage in improving the performance of the student model**. So we infer that **the augmented samples play an auxiliary advantage and the original samples play a key advantage in knowledge transfer**. In addition, by comparing combinations $\{\mathcal{L}_{ce}, \mathcal{L}_{kd}, \mathcal{L}_A, \mathcal{L}_T\}$ and $\{\mathcal{L}_{ce}, \mathcal{L}_{kd}, \mathcal{L}_T\}$ in Table 3, and even Fig. 3 and Table 1, we can find that the traditional CE loss of the augmented samples (i.e., $\mathcal{L}_A$) has an apparent positive effect. Therefore, our opinion is different from that in the SSKD paper [35]. We argue that the traditional CE loss of the augmented samples is also an essential part of knowledge distillation based on data augmentation. However, the teacher model does not use both original and augmented samples for the above experiments in training. As a result, it's natural to trust that transferring knowledge using only augmented samples is less effective than using only original samples, which might be because the teacher model lacks relevant knowledge. Considering this problem, we let the teacher model learn the information of the augmented samples and conduct the decoupling experiment again, and finally present the results in Table 4.

Table 4. The data are obtained under the premise that the teacher model uses both the original samples and the augmented samples in training. And ✓ and × represent whether they use the corresponding loss or not, respectively. "(number)" in this table refers to the increased validation accuracy compared to that in Table 3.

CIFAR-10 • Acc. of teacher model: $94.16_{(+0.60)}$%						CIFAR-100 • Acc. of teacher model: $77.81_{(+1.37)}$%					
$\mathcal{L}_{ce}$	$\mathcal{L}_{kd}$	$\mathcal{L}_A$	$\mathcal{L}_T$	Acc.(%)	Std.(%)	$\mathcal{L}_{ce}$	$\mathcal{L}_{kd}$	$\mathcal{L}_A$	$\mathcal{L}_T$	Acc.(%)	Std.(%)
✓	×	✓	×	$91.79_{(+0.00)}$	$0.13_{(+0.00)}$	✓	×	✓	×	$73.27_{(+0.00)}$	$0.22_{(+0.00)}$
✓	✓	✓	✓	$92.94_{(-0.33)}$	$0.15_{(-0.10)}$	✓	✓	✓	✓	$75.81_{(-0.37)}$	$0.27_{(+0.06)}$
✓	✓	×	×	$93.25_{(+0.54)}$	$0.23_{(+0.01)}$	✓	✓	×	×	$75.69_{(+1.03)}$	$0.09_{(-0.04)}$
✓	×	×	✓	$91.77_{(-0.11)}$	$0.15_{(-0.02)}$	✓	×	×	✓	$74.23_{(-0.34)}$	$0.27_{(+0.10)}$
✓	×	×	×	$92.14_{(+0.00)}$	$0.08_{(+0.00)}$	✓	×	×	×	$71.17_{(+0.00)}$	$0.12_{(+0.00)}$
✓	✓	×	✓	$92.99_{(-0.16)}$	$0.15_{(+0.03)}$	✓	✓	×	✓	$75.59_{(+0.27)}$	$0.12_{(-0.04)}$

Comparing Table 3 and Table 4, it is surprising that when we utilize the teacher model to transfer knowledge with the augmented sample information, the performance of the student model has not been improved except that a loss combination $\{\mathcal{L}_{ce}, \mathcal{L}_{kd}\}$ is applied. It can be clearly inferred that **in the training stage of the teacher model, combing the original with augmented samples can effectively improve the performance of the teacher model compared with only using the original samples. But this is not available for the student model**. Meanwhile, when we observe Fig. 6 in Appendix C, we can easily find that the augmented sample information is rather incorrect when the teacher model only uses the original samples in training. In contrast, the augmented sample information produced by the teacher model trained with both the original and augmented samples is relatively correct. So this means that **for augmented samples, "dark knowledge", which leads to misclassification, also plays a significant role in knowledge distillation**.

In particular, the teacher model transfers relatively correct information hurting the performance of the student model, and transfers relatively incorrect information improving the performance of the student model. This fact is contrary to our experience. By carefully observing Fig. 6 (CIFAR-10) in Appendix C, we can find that the visualization of the original samples is changed after the teacher model has been trained with the additional augmented samples. So we argue that the reason causes the above conclusion is that **the teacher model trains both the original and augmented samples, which will damage the original samples' reasonable information**.

3.4 The Probability of Data Augmentation

This subsection will analyze the data augmentation based on the value of p. We know that p refers to the probability of each sub-policy being executed. Specifically, the larger p is, the greater the number of executed sub-policies in the augmented samples is. Generally, executing data augmentation with a high probability is not the best choice. We believe that this conclusion typically applies to JDA. To fully explore the effects of different p values, we set p to be 0.0, 0.2, 0.4, 0.6, 0.8 and 1.0 for experiments, respectively. Finally, we show the results in Fig. 4 **(a) and (b)**.

As displayed in Fig. 4 **(a) and (b)**, we can clearly determine that the curve has a peak in each subfigure and illustrate the regularity of both curves. On a deeper level, this phenomenon implies that JDA's benefits are only available when p is inside a specified range. So we can draw a conclusion that **p is sensitive in training, and p with a reasonable setting can effectively achieve knowledge transfer**. Furthermore, although these experimental results guarantee that each mini-batch has at least half of the original samples, the student model performs poorly on both benchmarks when p is greater than 0.5. Therefore, we can infer that **the ratio of the augmented samples should not be set too large when using the augmented samples to assist knowledge distillation. Otherwise, the student model cannot obtain excellent performance**.

3.5 Few-Shot Analysis

In the real world, many datasets usually do not have a large amount of labeled data. Therefore, the study of few-shot learning becomes extremely important for solving this problem. For SSKD and HSAKD, they verified that their methods are robust by simulating a few-shot scenario that has only a small amount of labeled training samples. But since we have proved that data augmentation can strongly improve the student model's performance, it is reasonable for us to deduce that data augmentation may have contributed to the performance improvement of both SSKD and HSAKD in a few-shot scenario. In order that we can evaluate the role of data augmentation in a few-shot scenario, we follow the training setting in SSKD and HSAKD and randomly retain 25%, 50%, 75%, and 100% training samples in CIFAR-100. Particularly, we compare the performance of vanilla KD with and w/o JDA by training $\times 2$, and the experimental results can be found in Fig. 4 **(c) and (d)**.

Figure 4 **(c) and (d)** illustrate that vanilla KD with data augmentation further strengthens the generalization ability of models when the labeled data are insufficient. Specifically, we can discover that when fewer training samples are retained, the student model with JDA can outperform the student model w/o JDA better. It means that **the potential of knowledge distillation with data augmentation in few-shot learning is enormous.**

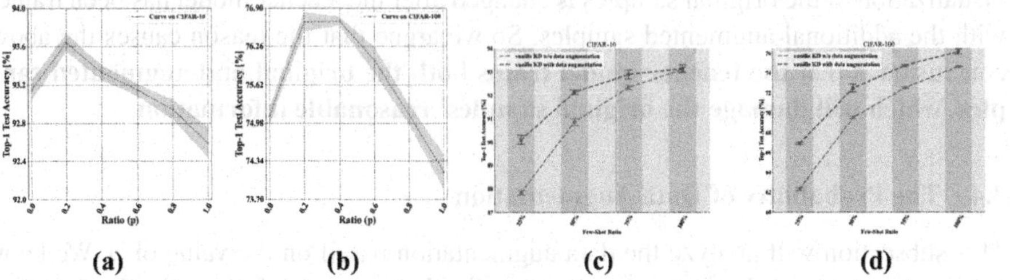

Fig. 4. (a) and (b): These two figures show the performance of the student model for different p values on CIFAR-10 and CIFAR-100, and the shaded area indicates the standard deviation. In addition, all experiments are performed with n_{step} set to $\times 2$. **(c) and (d):** These two figures demonstrate that our proposed joint data augmentation can work greatly in few-shot scenarios. Moreover, the error bars in figures refer to standard deviation.

3.6 Wide Comparison

In theory, JDA, as a data-focused method, can be perfectly combined with other methods that focus on "what to distill". To make sure data augmentation is robust in various teacher-student pairs and can be seamlessly compatible with some knowledge distillation methods, we conduct more extensive experiments on CIFAR-100. The hyperparameter settings of the experiments are the same as the CIFAR-100 benchmark mentioned in Sect. 3. And the results are shown in Table 5. By comparing whether to add JDA to vanilla KD, SPKD [32], and CRD [31][2], we can intuitively find that JDA effectively improves the performance of all methods. In addition, as can be seen in Appendix B, comparing the latest SOTA HSAKD and JDA, HSAKD requires additional computational modules, and more complex feature-based distillation, and spending two times n_{step} to get the same results. The above analysis fully demonstrates that **data augmentation is broadly effective and can be easily combined with other knowledge distillation methods.**

3.7 Cosine Confidence Knowledge Distillation

Inspired by Sect. 3.4, the strength of the augmented sample must be in a suitable range to exert a positive effect. If the model can adaptively assign appropriate weight to each

[2] The reason JDA is not added to SSKD and HSAKD is that these methods themselves use *rotating* as their data augmentation. If we are to force the inclusion of JDA, it will destroy the original character of these approaches.

Table 5. Top-1 test accuracy (%) comparison of different methods across various teacher-student pairs on CIFAR-100. The results of SSKD and HSAKD are copied from [4], and the remaining results are obtained by our run. "($\pm number$)" in this table refers to the standard deviation, and the red number in the upper left corner of the accuracy symbolizes the ranking of closely-related methods. In particular, the rightmost column represents each method's n_{step} in our experiments.

Teacher Student	WRN-40-2 WRN-16-2	WRN-40-2 WRN-40-1	ResNet56 ResNet20	ResNet32×4 ResNet8×4	VGG13 MobileNetV2	n_{step}
Teacher Student	76.44 $^{11}73.57_{(\pm 0.23)}$	76.44 $^{11}71.95_{(\pm 0.59)}$	73.44 $^{11}69.62_{(\pm 0.26)}$	79.63 $^{11}72.95_{(\pm 0.24)}$	74.64 $^{11}73.51_{(\pm 0.26)}$	 ×2
SSKD	$^{7}76.16_{(\pm 0.17)}$	$^{7}75.84_{(\pm 0.04)}$	$^{10}70.80_{(\pm 0.02)}$	$^{7}75.83_{(\pm 0.29)}$	$^{7}76.21_{(\pm 0.16)}$	×4
HSAKD	$^{2}77.20_{(\pm 0.17)}$	$^{1}77.00_{(\pm 0.21)}$	$^{4}72.58_{(\pm 0.33)}$	$^{2}77.26_{(\pm 0.14)}$	$^{5}77.45_{(\pm 0.21)}$	×4
KD	$^{10}74.36_{(\pm 0.11)}$	$^{10}73.21_{(\pm 0.10)}$	$^{9}71.68_{(\pm 0.30)}$	$^{10}72.34_{(\pm 0.12)}$	$^{10}75.94_{(\pm 0.21)}$	×2
KD+JDA	$^{5}76.80_{(\pm 0.13)}$	$^{4}76.18_{(\pm 0.18)}$	$^{6}72.37_{(\pm 0.28)}$	$^{6}76.50_{(\pm 0.22)}$	$^{3}77.64_{(\pm 0.23)}$	×2
SPKD	$^{9}74.84_{(\pm 0.38)}$	$^{9}73.51_{(\pm 0.17)}$	$^{7}72.11_{(\pm 0.10)}$	$^{9}72.77_{(\pm 0.25)}$	$^{9}76.13_{(\pm 0.25)}$	×2
SPKD+JDA	$^{6}76.58_{(\pm 0.31)}$	$^{4}76.18_{(\pm 0.26)}$	$^{3}72.73_{(\pm 0.11)}$	$^{5}76.64_{(\pm 0.36)}$	$^{6}77.33_{(\pm 0.14)}$	×2
CRD	$^{8}74.88_{(\pm 0.16)}$	$^{8}74.43_{(\pm 0.16)}$	$^{8}71.94_{(\pm 0.20)}$	$^{8}73.58_{(\pm 0.20)}$	$^{8}76.14_{(\pm 0.17)}$	×2
CRD+JDA	$^{4}76.84_{(\pm 0.23)}$	$^{3}76.27_{(\pm 0.16)}$	$^{5}72.38_{(\pm 0.08)}$	$^{4}77.12_{(\pm 0.11)}$	$^{4}77.61_{(\pm 0.06)}$	×2
CCD(ours)+JDA	$^{3}77.16_{(\pm 0.14)}$	$^{6}76.07_{(\pm 0.10)}$	$^{2}72.82_{(\pm 0.16)}$	$^{3}77.16_{(\pm 0.18)}$	$^{2}77.71_{(\pm 0.12)}$	×2
CCD(ours)+JDA	$^{1}77.34_{(\pm 0.12)}$	$^{2}76.78_{(\pm 0.11)}$	$^{1}73.24_{(\pm 0.08)}$	$^{1}77.59_{(\pm 0.18)}$	$^{1}78.11_{(\pm 0.10)}$	×4

Table 6. Top-1 accuracy (%) and Top-5 accuracy (%) comparison on ImageNet-1k. We follow the experimental setting in [4,31,35] and mark the highest Top-1 validation accuracy by **bold black**.

Teacher	Student	Acc.	Teacher	Student	KD	AT [37]	CC [25]	SPKD	RKD [24]	CRD	SSKD	HSAKD	DKD [42]	KD+JDA	CCD(ours)+JDA
ResNet-34	ResNet-18	Top-1	73.31	69.75	70.66	70.70	69.96	70.62	71.34	71.38	71.62	72.16	71.70	72.16	**72.22**
		Top-5	91.42	89.07	89.88	90.00	89.17	89.80	90.37	90.49	90.67	90.85	90.41	90.99	90.86
		n_{step}	–	×1	×1	×1	×1	×1	×1	×1	×4	×4	×1	×4	×4

augmented sample, the knowledge imparted by the teacher model can be more reasonable. As a result, here we propose a method called **C**osine **C**onfidence **D**istillation (CCD) to help transfer the knowledge of the augmented samples. First, as denoted in Eq. 11, we need to calculate the confidence of the teacher model with the augmented samples, which is measured by the cosine distance.

$$d = \mathbf{cosine}(\widetilde{\mathbf{x}}, \mathbf{x}) = \frac{\left\langle f^T\left(\widetilde{\mathbf{x}}\right), f^T\left(\mathbf{x}\right)\right\rangle}{\|f^T\left(\widetilde{\mathbf{x}}\right)\|_2 \cdot \|f^T\left(\mathbf{x}\right)\|_2}. \tag{11}$$

Of particular note is that d provides a way of quantitatively presenting the strength of the augmented samples, the basis of which is clearly shown in Fig. 5. For the given augmented samples, the strength of their data augmentation is negatively correlated with their cosine confidence weight. Thus, the cosine distance is reasonable to measure whether the augmented samples have a high confidence level to facilitate distillation. Due to $d \in [-1, 1]$, utilizing it directly as weight makes the expectation of KL loss close to zero and model optimization difficult. We multiply $d+1$ as a weight $\in [0, 2]$ by $\mathbf{KL}\left(\mathbf{p}^T\left(\widetilde{\mathbf{x}}; \tau\right) \| \mathbf{p}^S\left(\widetilde{\mathbf{x}}; \tau\right)\right)$. This means that the stronger an augmented sample is, the greater the distance between the original sample and the augmented sample is, and the smaller d is. Thus, less knowledge is transferred from the teacher model to the student model. So we denote new $\mathcal{L}_{\hat{\mathcal{T}}}$ in Eq. 12 instead of $\mathcal{L}_{\mathcal{T}}$. And the new overall loss is shown in Eq. 13.

$$\mathcal{L}_{\hat{\mathcal{T}}} = \tau^2 \mathbb{E}_{(\mathbf{x},\widetilde{\mathbf{x}})\sim(\mathcal{X},\widetilde{\mathcal{X}})} \left(\mathbf{cosine}\left(\widetilde{\mathbf{x}},\mathbf{x}\right)+1\right) * \mathbf{KL}\left(\mathbf{p}^T\left(\widetilde{\mathbf{x}};\tau\right) \| \mathbf{p}^S\left(\widetilde{\mathbf{x}};\tau\right)\right). \quad (12)$$

$$\mathcal{L}_{\hat{oa}} = \frac{|\mathcal{X}|}{|\mathcal{X}|+|\widetilde{\mathcal{X}}|}\mathcal{L}_{ce} + \frac{|\widetilde{\mathcal{X}}|}{|\mathcal{X}|+|\widetilde{\mathcal{X}}|}\mathcal{L}_{A} + \frac{|\mathcal{X}|}{|\mathcal{X}|+|\widetilde{\mathcal{X}}|}\mathcal{L}_{kd} + \frac{|\widetilde{\mathcal{X}}|}{|\mathcal{X}|+|\widetilde{\mathcal{X}}|}\mathcal{L}_{\hat{\mathcal{T}}}. \quad (13)$$

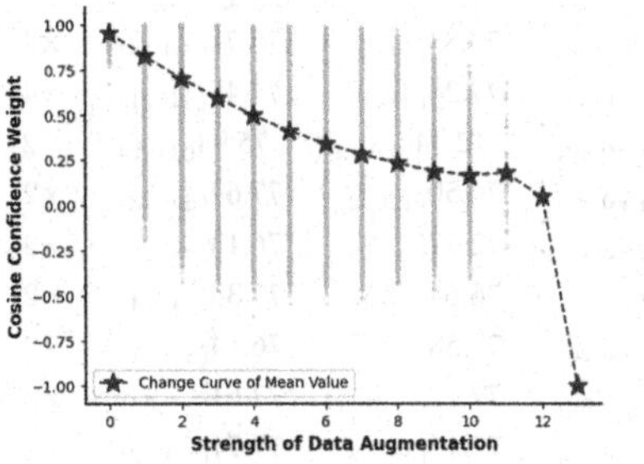

Fig. 5. The horizontal axis displays the number of transformations the augmented sample has undergone, and the vertical axis shows the cosine distance between the original sample and the associated augmented sample on CIFAR-100.

The apparent here to note in Eq. 12 is that $\widetilde{\mathbf{x}}$ in sample pair $(\mathbf{x},\widetilde{\mathbf{x}})$ is transformed by $\mathbf{x}$. The comparative experimental results of CCD are also shown in Table 5. In addition, comparative results on the ImageNet-1k [27] benchmark can be found in Table 6. In particular, CCD is modified from the vanilla KD and exceeds the performance of KD+JDA. For n_{step} being ×2, we can observe that CCD outperforms KD+JDA, SPKD+JDA, and CRD+JDA on all teacher-student pairs except WRN-40-2-WRN-40-1 in CIFAR-100, fully indicating that CCD is an excellent method. For n_{step} is ×4, CCD surpasses HSAKD on all teacher-student pairs except WRN-40-2-WRN-40-1 in CIFAR-100. In the ImageNet-1k benchmark, we also achieve a SOTA distillation result. It should be emphasized that CCD achieves almost the same performance when the training time of HSAKD is twice that of CCD. Hence, CCD is more outstanding than HSAKD under all-around consideration.

4 Conclusion

In this paper, we conduct a multi-angle analysis of the role that data augmentation plays in knowledge distillation. Then we conclude that data augmentation can effectively improve the performance of knowledge distillation, and so forth (more detailed conclusions are shown at the end of Sect. 1). Furthermore, inspired by Sect. 3.4, we propose an excellent method named CCD to transfer knowledge of the augmented samples and the performance of CCD is better than that of the latest SOTA HSAKD. In future, our work will focus on "what kind of augmented samples should be used for distillation" or "how to better utilize the information of augmented samples", other than "what to distill".

Acknowledgement. This work was supported in part by the Aeronautical Science Foundation of China under Grant 20200058069001 and in part by the Fundamental Research Funds for the Central Universities under Grant 2242021R41094.

A Hyperparameter Settings

All experiments performed in this paper followed the settings in Table 7. In particular, we ensure that their batch sizes are the same for experiments with different n_{step}. Furthermore, for our proposed JDA, the magnitudes of the corresponding sub-policies are shown in Table 8. Most of these settings are obtained with modifications based on AutoAugment.

Table 7. This table shows the hyperparameter settings when n_{step} is $\times 1$. At the same time, we achieve the setting where n_{step} is $\times 2$ and $\times 4$ by increasing epoch. Crucially, we use the same batch size of the real input for all methods including JDA and *rotating*.

Datasets	Learning rate	Optimizer	Weight decay	Batch size	Epoch	Scheduled epoch	Gamma
CIFAR-10	0.1	SGD	1e–4	128	182	[91, 136]	0.1
CIFAR-100	0.05	SGD	1e–4	256	240	[150, 180, 210]	0.1
ImageNet-1k	0.1	SGD	1e–4	1024	100	[30, 60, 90]	0.1

Table 8. This table shows the 14 sub-policies and their hyperparameter settings used in our experiments. Part of this table is copied from [5]. And the execution order of the sub-policies can be found in our released codes.

Operation name	Description	Magnitude in CIFAR-10	Magnitude in CIFAR-100
ShearX	Shear the image along the horizontal axis with rate *magnitude*	0.24	0.15
ShearY	Shear the image along the vertical axis with rate *magnitude*	0.24	0
TranslateX	Translate the image in the horizontal direction by *magnitude* number of pixels	$\frac{45}{331}$	$\frac{15}{331}$
TranslateY	Translate the image in the vertical direction by *magnitude* number of pixels	$\frac{45}{331}$	$\frac{120}{331}$
Rotate	Rotate the image *magnitude* degrees	6	9
AutoContrast	Maximize the the image contrast, by making the darkest pixel black and lightest pixel white	–	–
Invert	Invert the pixels of the image	–	–
Equalize	Equalize the image histogram	–	–
Solarize	Invert all pixels above a threshold value of *magnitude*	204.8	153.6
Posterize	Reduce the number of bits for each pixel to *magnitude* bits	0	8
Brightness	Adjust the brightness of the image. A *magnitude* = 0 gives a black image, whereas *magnitude* = 1 gives the original image	0.54	0.27
Contrast	Control the contrast of the image. A *magnitude* = 0 gives a gray image, whereas *magnitude* = 1 gives the original image	0.63	0.27
Color	Adjust the color balance of the image, in a manner similar to the controls on a colour TV set. A *magnitude* = 0 gives a black & white image, whereas *magnitude*=1 gives the original image	0.27	0.36
Sharpness	Adjust the sharpness of the image. A *magnitude* = 0 gives a blurred image, whereas *magnitude* = 1 gives the original image	0.81	0.45

B Additional Method Comparisons

First, we present a series of computational cost comparisons of SOTA algorithms for knowledge distillation in Table 9. Second, we compare the difference in performance between JDA and AutoAugment on knowledge distillation in Table 10. The results show that both JDA and CCD achieve the best performance in their respective comparisons.

Table 9. GFLOPs: *Giga Floating-point Operations Per Second.* We utilize facebook's open-source project fvcore to calculate GFLOPs. For operators that fvcore does not support statistics, we count their totals in NUO. **NUO:** *The Number of Unsupported Operators.* **TP:** *ThroughPut (images/s).* We calculated the throughput of all methods from start to finish under an NVIDIA RTX 3080 Ti. Meanwhile, all methods are executed 5000 times to reduce interference. This table presents the comparison results of related knowledge distillation methods on other vital indicators. In general, response-based methods are more portable and reproducible than feature-based methods. Methods that do not use additional modules are more lightweight in training than methods that use additional modules. JDA+CCD does not require additional modules and is very close to the original KD regarding GFLOPs, NUO and TP. Therefore, we can conclude that our proposed JDA+CCD is lightweight.

Methods	Additional modules	Location of distillation	GFLOPs↓	NUO↓	TP↑
vanilla KD	No	Response-based	0.4313672	36	16244.3
SPKD	No	Feature-based	0.4314327	50	16192.9
CRD	Yes	Feature-based	0.4355945	115	6726.8
SSKD	Yes	Feature-based	0.4314327	68	14303.8
HSAKD	Yes	Feature-based	1.0127411	93	7679.0
CCD+JDA (ours)	No	Response-based	0.4313672	49	15192.6

Table 10. Performance comparison of JDA and AutoAugment on offline knowledge distillation. All experiments in this table use the same hyperparameter settings. As a result, we find that JDA beats AutoAugment on four teacher-student pairs.

Teacher	WRN-40-2	WRN-40-2	ResNet56	ResNet32×4	VGG13	n_{step}
Student	WRN-16-2	WRN-40-1	ResNet20	ResNet8×4	MobileNetV2	
KD+JDA	76.80%	76.18%	72.37%	76.50%	77.64%	×2
KD+AutoAugment	76.34%	75.68%	71.97%	75.73%	77.64%	×2

C Additional Visualization

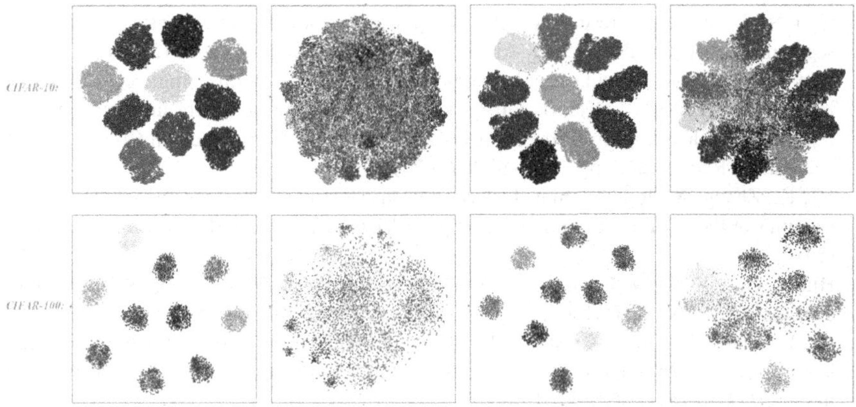

Fig. 6. The figure contains T-SNE visualizations of the output of the teacher model's GAP for eight different scenarios. The four columns from left to right refer to the four cases of $(\mathcal{X}, \mathcal{X})$, $\left(\widetilde{\mathcal{X}}, \mathcal{X}\right)$, $\left(\mathcal{X}, \widetilde{\mathcal{X}} + \mathcal{X}\right)$ and $\left(\widetilde{\mathcal{X}}, \mathcal{X} + \widetilde{\mathcal{X}}\right)$, where (A, B) stands for the teacher model trained with B. Then, we adopt T-SNE to visualize A.

References

1. Beyer, L., Zhai, X., Royer, A., Markeeva, L., Anil, R., Kolesnikov, A.: Knowledge distillation: a good teacher is patient and consistent. In: Proceedings of the IEEE/CVF Conference on Computer Vision and Pattern Recognition, pp. 10925–10934 (2022)
2. Chen, D., Mei, J.P., Wang, C., Feng, Y., Chen, C.: Online knowledge distillation with diverse peers. In: Proceedings of the AAAI Conference on Artificial Intelligence, vol. 34, pp. 3430–3437 (2020)
3. Chen, T., Kornblith, S., Norouzi, M., Hinton, G.: A simple framework for contrastive learning of visual representations. In: International Conference on Machine Learning, pp. 1597–1607. PMLR (2020)
4. Yang, C., An, Z., Cai, L., Xu, Y.: Hierarchical self-supervised augmented knowledge distillation. In: Proceedings of the Thirtieth International Joint Conference on Artificial Intelligence (IJCAI), pp. 1217–1223 (2021)
5. Cubuk, E.D., Zoph, B., Shlens, J., Le, Q.V.: Randaugment: practical automated data augmentation with a reduced search space. In: Proceedings of the IEEE/CVF Conference on Computer Vision and Pattern Recognition Workshops, pp. 702–703 (2020)
6. Das, D., Massa, H., Kulkarni, A., Rekatsinas, T.: An empirical analysis of the impact of data augmentation on distillation (2020)
7. Dosovitskiy, A., et al.: An image is worth 16×16 words: transformers for image recognition at scale. In: International Conference on Learning Representations (2020)
8. Fu, J., et al.: Role-wise data augmentation for knowledge distillation. arXiv preprint arXiv:2004.08861 (2020)
9. Gou, J., Yu, B., Maybank, S.J., Tao, D.: Knowledge distillation: a survey. Int. J. Comput. Vision **129**(6), 1789–1819 (2021)

10. Guo, Q., et al.: Online knowledge distillation via collaborative learning. In: Proceedings of the IEEE/CVF Conference on Computer Vision and Pattern Recognition, pp. 11020–11029 (2020)
11. Guo, S.: Dpn: Detail-preserving network with high resolution representation for efficient segmentation of retinal vessels. J. Ambient Intell. Hum. Comput., 1–14 (2021)
12. Han, J., et al.: You only cut once: boosting data augmentation with a single cut (2022)
13. He, K., Fan, H., Wu, Y., Xie, S., Girshick, R.: Momentum contrast for unsupervised visual representation learning. In: Proceedings of the IEEE/CVF Conference on Computer Vision and Pattern Recognition, pp. 9729–9738 (2020)
14. He, K., Zhang, X., Ren, S., Sun, J.: Deep residual learning for image recognition. In: Proceedings of the IEEE Conference on Computer Vision and Pattern Recognition, pp. 770–778 (2016)
15. Hinton, G., Vinyals, O., Dean, J.: Distilling the knowledge in a neural network (2015). https://doi.org/10.48550/ARXIV.1503.02531, https://arxiv.org/abs/1503.02531
16. Ho, D., Liang, E., Chen, X., Stoica, I., Abbeel, P.: Population based augmentation: efficient learning of augmentation policy schedules. In: International Conference on Machine Learning, pp. 2731–2741. PMLR (2019)
17. Krizhevsky, A., Hinton, G., et al.: Learning multiple layers of features from tiny images (2009)
18. Lim, S., Kim, I., Kim, T., Kim, C., Kim, S.: Fast autoaugment. Adv. Neural Inf. Process. Syst. **32**, 1–11 (2019)
19. Liu, S., Tian, Y., Chen, T., Shen, L.: Don't be so dense: sparse-to-sparse gan training without sacrificing performance. Int. J. Comput. Vision **20**(X) (2022)
20. Liu, S., et al.: Paint transformer: feed forward neural painting with stroke prediction. In: Proceedings of the IEEE/CVF International Conference on Computer Vision (ICCV), pp. 6598–6607 (2021)
21. Liu, Z., Farrell, J., Wandell, B.A.: Isetauto: detecting vehicles with depth and radiance information. IEEE Access **9**, 41799–41808 (2021)
22. Liu, Z., Mao, H., Wu, C.Y., Feichtenhofer, C., Darrell, T., Xie, S.: A convnet for the 2020s. In: Proceedings of the IEEE/CVF Conference on Computer Vision and Pattern Recognition, pp. 11976–11986 (2022)
23. Matsubara, Y.: torchdistill: a modular, configuration-driven framework for knowledge distillation. In: Kerautret, B., Colom, M., Krähenbühl, A., Lopresti, D., Monasse, P., Talbot, H. (eds.) RRPR 2021. LNCS, vol. 12636, pp. 24–44. Springer, Cham (2021). https://doi.org/10.1007/978-3-030-76423-4_3
24. Park, W., Kim, D., Lu, Y., Cho, M.: Relational knowledge distillation. In: Proceedings of the IEEE/CVF Conference on Computer Vision and Pattern Recognition (CVPR) (2019)
25. Peng, B., et al.: Correlation congruence for knowledge distillation. In: Proceedings of the IEEE/CVF International Conference on Computer Vision, pp. 5007–5016 (2019)
26. Razavi, M., Alikhani, H., Janfaza, V., Sadeghi, B., Alikhani, E.: An automatic system to monitor the physical distance and face mask wearing of construction workers in covid-19 pandemic. SN Comput. Sci. **3**(1), 1–8 (2022)
27. Russakovsky, O.: Imagenet large scale visual recognition challenge. Int. J. Comput. Vision **115**(3), 211–252 (2015)
28. Sharma, S.: Game theory for adversarial attacks and defenses. arXiv preprint arXiv:2110.06166 (2021)
29. Singh, B., Najibi, M., Davis, L.S.: Sniper: efficient multi-scale training. In: Bengio, S., Wallach, H., Larochelle, H., Grauman, K., Cesa-Bianchi, N., Garnett, R. (eds.) Advances in Neural Information Processing Systems, vol. 31. Curran Associates, Inc. (2018). https://proceedings.neurips.cc/paper/2018/file/166cee72e93a992007a89b39eb29628b-Paper.pdf

30. Sun, C., Shrivastava, A., Singh, S., Gupta, A.: Revisiting unreasonable effectiveness of data in deep learning era. In: Proceedings of the IEEE International Conference on Computer Vision (ICCV) (2017)
31. Tian, Y., Krishnan, D., Isola, P.: Contrastive representation distillation. In: International Conference on Learning Representations (2019)
32. Tung, F., Mori, G.: Similarity-preserving knowledge distillation. In: Proceedings of the IEEE/CVF International Conference on Computer Vision, pp. 1365–1374 (2019)
33. Wang, H., Lohit, S., Jones, M., Fu, Y.: Knowledge distillation thrives on data augmentation. arXiv preprint arXiv:2012.02909 (2020)
34. Wieczorek, M., Rychalska, B., Dąbrowski, J.: On the unreasonable effectiveness of centroids in image retrieval. In: Mantoro, T., Lee, M., Ayu, M.A., Wong, K.W., Hidayanto, A.N. (eds.) ICONIP 2021. LNCS, vol. 13111, pp. 212–223. Springer, Cham (2021). https://doi.org/10.1007/978-3-030-92273-3_18
35. Xu, G., Liu, Z., Li, X., Loy, C.C.: Knowledge distillation meets self-supervision. In: Vedaldi, A., Bischof, H., Brox, T., Frahm, J.-M. (eds.) ECCV 2020. LNCS, vol. 12354, pp. 588–604. Springer, Cham (2020). https://doi.org/10.1007/978-3-030-58545-7_34
36. Yun, S., Han, D., Oh, S.J., Chun, S., Choe, J., Yoo, Y.: Cutmix: regularization strategy to train strong classifiers with localizable features. In: Proceedings of the IEEE/CVF International Conference on Computer Vision, pp. 6023–6032 (2019)
37. Zagoruyko, S., Komodakis, N.: Paying more attention to attention: improving the performance of convolutional neural networks via attention transfer. In: International Conference on Learning Representations (ICLR) (2016)
38. Zagoruyko, S., Komodakis, N.: Wide residual networks. In: BMVC (2016)
39. Zhang, H., Cisse, M., Dauphin, Y.N., Lopez-Paz, D.: mixup: beyond empirical risk minimization. In: International Conference on Learning Representations (2018)
40. Zhang, X., Zhou, X., Lin, M., Sun, J.: Shufflenet: an extremely efficient convolutional neural network for mobile devices. In: Proceedings of the IEEE Conference on Computer Vision and Pattern Recognition, pp. 6848–6856 (2018)
41. Zhang, Y., Xiang, T., Hospedales, T.M., Lu, H.: Deep mutual learning. In: Proceedings of the IEEE Conference on Computer Vision and Pattern Recognition (CVPR) (2018)
42. Zhao, B., Cui, Q., Song, R., Qiu, Y., Liang, J.: Decoupled knowledge distillation. In: Proceedings of the IEEE/CVF Conference on Computer Vision and Pattern Recognition (CVPR), pp. 11953–11962 (2022)

TCVM: Temporal Contrasting Video Montage Framework for Self-supervised Video Representation Learning

Fengrui Tian[1], Jiawei Fan[2], Xie Yu[2], Shaoyi Du[1(✉)], Meina Song[2], and Yu Zhao[3]

[1] Xi'an Jiaotong University, Xi'an, China
tianfr@stu.xjtu.edu.cn, dushaoyi@gmail.com
[2] Beijing University of Posts and Telecommunications, Beijing, China
{jwfan,yuxie130,mnsong}@bupt.edu.cn
[3] Harbin Institute of Technology, Harbin, China

Abstract. Extracting appropriate temporal differences and ignoring irrelevant backgrounds are two important perspectives on preserving sufficient motion information in video representation, such as driver behavior monitoring and driver fatigue detection. In this paper, we propose a unified contrastive learning framework called Temporal Contrasting Video Montage (TCVM) to learn action-specific motion patterns, which can be implemented in a plug-and-play way. On the one hand, Temporal Contrasting (TC) module is designed to guarantee appropriate temporal difference between frames. It utilizes high-level feature space to capture raveled temporal information. On the other hand, Video Montage (VM) module is devised for alleviating the effect from video background. It demonstrates similar temporal motion variances in different positive samples by implicitly mixing up the backgrounds of different videos. Experimental results show that our TCVM reaches promising performances on both large action recognition dataset (i.e. Something-Somethingv2) and small datasets (i.e. UCF101 and HMDB51).

1 Introduction

Video representation learning is a fundamental task in computer vision, which promotes the performance of related downstream tasks, e.g., action recognition [12,31,44], video retrieval [46], and temporal action detection [32]. Especially, in auto-piloting [4] or co-piloting [30] areas, video representation learning is significant for driver fatigue detection and driver abnormal behavior detection. Unlike image-related tasks, video representation learning needs neural networks to capture both spatial and temporal features, which makes the problem more complicated and challenging. Moreover, compared with image labeling, video

F. Tian and J. Fan—Equal Contribution.
This work was done when Fengrui Tian and Jiawei Fan were interns at Megvii Research.

Supplementary Information The online version contains supplementary material available at https://doi.org/10.1007/978-3-031-26284-5_32.

L. Wang et al. (Eds.): ACCV 2022, LNCS 13842, pp. 526–542, 2023.
https://doi.org/10.1007/978-3-031-26284-5_32

labeling is relatively cumbersome and time-consuming work. Many researchers have turned to self-supervised learning methods recently.

There are several mainstream directions in self-supervised video representation learning, including pretext task designing and contrastive learning. Some studies focus on designing video-related pretext tasks, such as solving video jigsaw puzzles [1,24,27], predicting video clip order [46] and predicting video speed [3,23,42]. Besides, contrastive representation learning shows great potential in computer vision tasks. Some researcher turn to design effective contrastive learning frameworks, e.g. MoCo [20], BYOL [16], SimCLR [7]. VideoMoCo [34] and Feichtenhofer *et al.* [13] try to apply these contrastive learning methods to video representation learning.

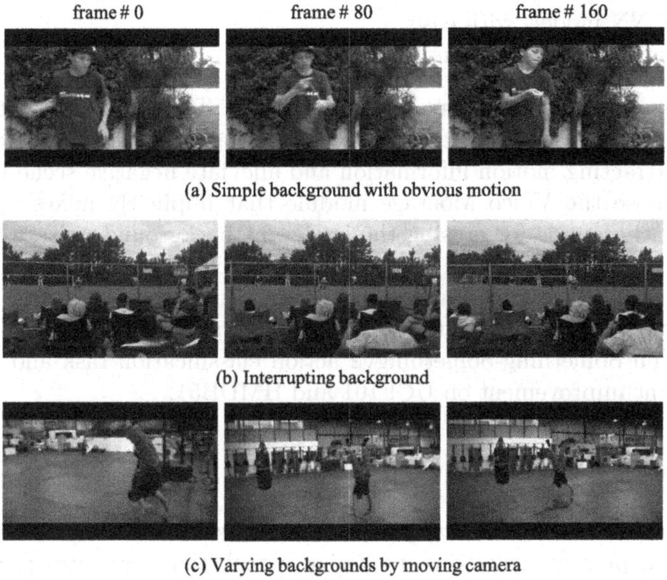

Fig. 1. The ability of extracting motion information in video stream is severely constrained by the complex combinations of views, backgrounds, and motions.

Although previous studies have gained great achievements, there still exist two main problems in video contrastive learning. **i) Extracting motion information.** On the one hand, in a specific video clip, frames with high similarity easily contributes to similar representations. On the other hand, some videos contains interrupting background and camera movements (shown in Fig. 1(b) and Fig. 1(c)), which affects the localization to action subject. The above aspects make it difficult to extract motion variances. **ii) Unavoidable scene-bias.** Actions and backgrounds are mutually relevant. For example, playing football often happens in playground, and swimming pool obviously implies swimming. The apparently distinguishable backgrounds provide a shortcut for discriminating positive and negative samples, which drastically affects the performance of contrastive learning.

In order to tackle the above problems and capture more action-specific motion patterns in video representation learning, we propose a unified framework called Temporal Contrasting Video Montage (TCVM), which contains two modules Temporal Contrasting (TC) and Video Montage (VM). The TC module utilizes high-level feature differences among neighboring frames to model temporal information, and simultaneously retains necessary spatial features in videos, which does not need to pre-process dataset in advance or to add extra datas in other modalities. The VM module mixes up all background information of different videos within a batch to decrease the differences of background bias between positive and negative samples, which improves the performance of contrastive learning. Overall, the proposed TCVM framework can be efficiently inserted into 2D or 3D CNN models with ease.

The contributions are summarized as follows.

- We propose a plug-and-play framework called Temporal Contrasting Video Montage (TCVM) for video contrastive learning, which can improve the ability of extracting motion information and alleviate negative scene bias.
- We propose the Video Montage module that implicitly mixes up different video backgrounds for erasing the irrelevant background noises. To demonstrate motion features, we present the Temporal Contrasting module that models frame-wise foreground variances in videos.
- Experimental results show that the proposed method outperforms significant margin on Something-Somethingv2 action classification task and achieves a significant improvement on UCF101 and HMDB51.

2 Related Works

Video Representation Learning from Pretext Tasks. Learning video representations from pretext tasks aims to design a task that generates pseudo video labels. There are some methods that extend successful pretext tasks from image domain to video domain, such as solving video jigsaw puzzle [1,24,27], identifying video clips rotation [25] and video colorization [41]. However, these methods cannot capture strong spatio-temporal video representations. By the nature of temporal consistency in videos, researchers design different pretext tasks such as identifying video clip order [14,40], sorting video frames [29], and predicting video clip order [46]. Besides, many recent studies focus on predicting video speeds [3,23,42] or relative playback speed of the same video [6]. However, the representations learned by these methods are similar to the designed tasks and often irrelevant to downstream tasks [33].

Video Contrastive Learning. Contrastive learning-based methods aim to build representations by maximizing the similarity of the same instance with different views (positive pairs) and minimizing similarity of different instances (negative pairs). Recently, contrastive representation learning shows great potential in computer vision tasks. MoCo [20] presents a contrastive learning framework that uses a momentum encoder to build dynamic negative pairs. To explore the ability in video representation learning, some methods research on designing basic contrastive learning frameworks. [34] and [13] try to extend successful image contrastive learning methods to videos. CVRL [36] indicates the importance of temporal consistency in different views of videos. TCLR [10] tries to model video representation by local-local and global-local contrastive pairs. Due to the unique temporal features in video domain compared to images, Some researchers start to design contrastive pairs for learning more temporal representations. For example, DPC [17] and Mem-DPC [18] try to learn temporal representations by contrasting the predicted frame features.

Motion Learning in Videos. The motion information in video plays an important role in video representation learning. But existing methods often pay insufficient attention to temporal features. Ding *et al.* [11] presents a self-supervised contrastive learning method that merges different foregrounds and backgrounds in videos. Choi *et al.* [9] propose to mask actors with a human detector and further present a novel adversarial loss. On the other hand, many researchers focus on multi-view video contrastive learning methods [2,35,39]. For example, contrastive learning with the optical flow is widely researched in recent years and achieves impressive results [19,45]. Besides, Huang *et al.* [22] presents a video representation learning that naturally uses different types of frames in compressed video streams to decouple the motion and context information. The above-mentioned methods try to learn motions by prepossessing the input videos or adding another modality. Our paper focuses on modeling motion features extracted from the encoders.

3 Method

Given only RGB video clips, the proposed method builds the video representations in a contrastive learning way. Like all contrastive learning methods, our approach has an encoder that maps the input videos to latent representations in a unit sphere, and supervises it by the contrastive loss to identify the representations from the same or different video clips. Unlike other video contrastive learning methods, the proposed method employs modeling the high-level video representations and adopts recognizing similar actions in different videos for learning motion representations.

3.1 Overview of Temporal Contrasting Video Montage Framework

The overview of the proposed method is shown in Fig. 2. First, the Video Montage module distributes clips with the same motion patterns to different videos. In this way, it generates a video with several clips that share similar motion

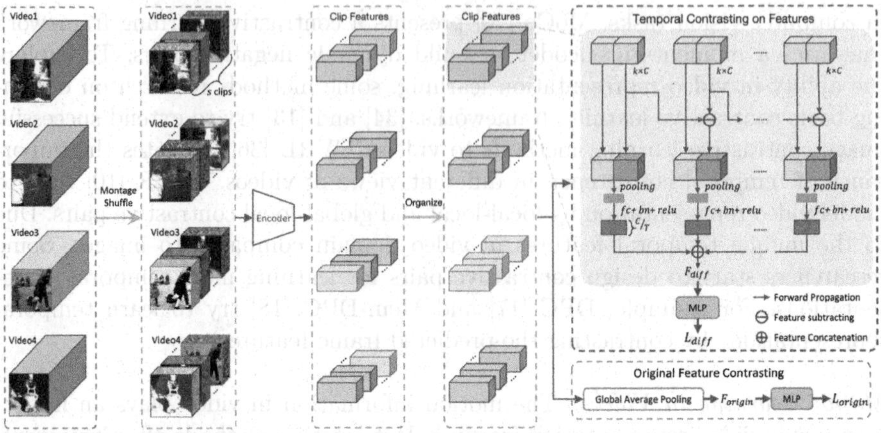

Fig. 2. The overview of the proposed method. First, the input videos are re-organized for sharing similar motion clips in different videos by the proposed Video Montage framework. Second, the generated videos are fed into the backbone encoder to extract frame-wise features of each clip. Third, the clip features are introduced into Temporal Contrasting module and Original Feature Contrasting module for video contrastive learning.

representation with other clips in other videos. Second, the generated videos are introduced to the backbone encoder to extract the frame-wise spatio-temporal features, before the representations are re-organized. Third, to model the motion information in videos, the extracted features are introduced into the Temporal Contrasting (TC) module for erasing background bias in each clip and building the temporal representations. The video representations from TC module are supervised by L_{diff}. On the other hand, to ensure that the model learns necessary scene features in videos, Original Feature Contrasting (OFC) module projects the extracted features to a lower dimension and supervises it by the contrastive loss L_{origin}. The goal of contrastive learning is to pull together positive samples and push away negative samples. The loss function is defined as the following equations.

$$L_{origin} = \sum_{j \in P_o(i)} -log \frac{exp(cos(z_i, z_j)/\tau)}{\sum_{k \in A_o} exp(cos(z_i, z_k)/\tau)}, \tag{1}$$

$$L_{diff} = \sum_{j \in P_d(i)} -log \frac{exp(cos(z_i, z_j)/\tau)}{\sum_{k \in A_d(i)} exp(cos(z_i, z_k)/\tau)}, \tag{2}$$

where z_i denotes the ith clip representation and $P_o(i)$ and $P_d(i)$ contain all positive sample representations of ith clip in OFC and TC module, respectively. $A_o(i)$ and $A_d(i)$ include the representations of negative samples of ith clip in OFC and TC module, separately. Cosine similarity is applied to calculate the distance between different samples. Following [26], we use the summation over positive samples located outside of the *log*. τ indicates a temperature parameter to control the smoothness of the distance. In this module, the final loss L is defined as Eq. (3).

$$L = \alpha L_{origin} + \beta L_{diff}, \tag{3}$$

where α and β are all set to 0.5 in this study. The model is trained in an end-to-end manner.

3.2 Video Montage Module

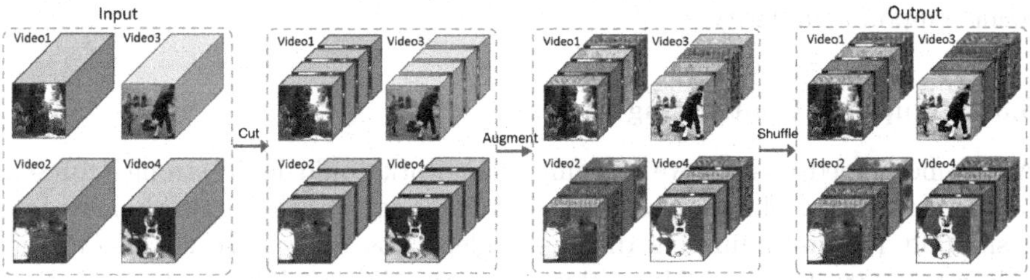

Fig. 3. Video Montage module. First, the input video is segmented into s clips uniformly. Next, each clip is added augmentation independently. Last, the clips are shuffle cross all batches.

The Philosophy of Video Montage. Video Montage is especially designed for video contrastive learning, which aims to alleviate negative scene-bias. Different from adding static frames, Video Montage mixes up the background information in positive-negative sample sets within a batch, which can decrease the dissimilarity resulting from background. Video Montage mixes up the background information by model itself, rather than introducing additional operation, in which the background information is fused after random shuffling.

Video Montage Operation. To imitate recognizing the same action in different videos by human eyes, the VM framework is proposed as shown in Fig. 3. There are three steps in VM framework: cut, augment, and shuffle.

Cut. First, s clips with the same motion are created by segmenting videos uniformly. For example, the input video consists of 16 frames and suppose $s = 2$. The input video is cut into 2 clips uniformly with 8 frames in each clip. In other words, the first clip consists of the front 1–8 frames, the second clip consists of the next 8–16 frames.

Augment. Each clip is augmented independently as shown in Fig. 3. The clips from the same video are considered as positive pairs. Adding independent augmentation could create dissimilar positive clip pairs and hence boosts the network to learn the action representation with less background bias. In this way, the same motion clips with different background information are created as positive pairs.

Shuffle. The same motion clips are distributed in current mini-batch. The new s clips in the same batch are concatenated in the temporal dimension to generate the new videos. The junction of different clips is considered as an augmentation at temporal dimension. It forces the network to separate two adjacent clips by inner temporal similarity.

3.3 Temporal Contrasting Module

Above-mentioned contrastive learning framework could learn strong representations of videos. However, as described in Sect. 1, the network may take a shortcut for discriminating different video clips because scenes in different videos vary greatly. To address this issue, the TC module is proposed. Suppose $X = \{x_1, x_2, ..., x_T\}$ as the feature tensor of a video clip extracted from the encoder. $x_i \in \mathbb{R}^{C \times H \times W}$ is the ith frame feature. T is the number of video frames. In TC module, the feature of ith frame x_i is used to model the differences with its next frame.

$$d_i = \begin{cases} x_i^j - x_{i+1}^j & j \leq k \times C, \\ x_i^j & j > k \times C, \end{cases} \tag{4}$$

where d_i denotes the feature differences at the ith frame. $k \in [0, 1]$ controls the proportion of channels that calculates the current frame feature differences with the next frame. Then, the feature d_i is transformed by one spatial pooling layer and one fully connected layer with the batch normalization and the ReLU function for encoding the ith frame features. The difference information between i and $i+1$ frames represents the changing information cross temporal dimension. Besides, the original feature represents scene information in videos. The proportion of the channels that model differences in each frame is important in TC, and it is set to $k = 0.5$ in this study. We will discuss the k values in ablation study.

Another challenge is how to fuse the feature difference information for the final video representation. Temporal pooling operation is widely used in recent self-supervised video representation learning methods [1,14,34]. This operation may not be appropriate for fusing the feature difference information. First, the feature difference information is well-ordered in temporal dimension. Videos are generated from the first frame to the last frame. The feature difference information between two ordered frames represents the changing information in the video. However, the pooling operation is irrelevant in time. If we flip frame-wise features in temporal dimension, the final video representation ought to be different. Unfortunately, the representation is the same with non-flipped features by using the temporal pooling. Second, pooling operation would destroy the difference information between two neighboring frames. Since the difference information is calculated by the subtraction of two neighboring frames, the addition among temporal dimension equals to adding the first and the last frame features and omits the information in other frames. To model the difference information at every frame, one fully connected layer f is used to project the ith frame difference information d_i from C dimensions to C/T dimensions. The final difference feature F_{diff} is conducted by the following equation.

$$F_{diff} = concat\{f(d_1), f(d_2), ..., f(d_T)\}. \tag{5}$$

The dimensions of the final difference features equals to C. In this way, the final representation contains the motion difference information among all temporal frames.

The network extracts the features of the generated videos without any down-sampling operation at the temporal dimension. Then the frame-wise features are cut to k segments uniformly representing the feature of s video clips. Clips' features from the same videos are positive pairs. Besides, the features from different videos are negative pairs. Last, the network will learn from the features of each clip by using the TC module and the OFC module.

4 Results

4.1 Implementation Details

Backbone Selections. To have an apple-to-apple comparison, we follow the common practices [13,36] and choose the Slow path in SlowFast [12] as the R3D50 backbone. We also use the TSM50 [31] model as a strong backbone to test the potential of the proposed method. To use our framework, all the down-sampling operations in temporal dimension are removed.

Self-supervised Pre-training. We conduct experiments on Kinetics-400 [5] (K400) dataset. K400 is a large scale action recognition dataset. It contains about 240k videos in training dataset and 20k validation videos. We use the training videos without any labels for pre-training our models. 32 frames sampled from each video with the temporal stride 2 are segmented into $s = 4$ clips in our experiments. Each clip consists of 8 frames. The image size in each frame is set as 112×112. We follow [8] and use the random grayscale, random color jitter, random gaussian blur, random horizontal flip for augmentation. Temperature parameter τ is set to 0.07 following [20]. The model is trained for 200 epochs. SGD is used as our optimizer with the momentum of 0.9. Batch size is set to 4 per GPU and we use 8 NVIDIA 2080ti GPUs in self-supervised pre-training. The learning rate is set as 0.1 decaying as $0.1\times$ at 50, 100, and 150 epochs.

Table 1. Action recognition results on SSv2. Epoch denotes the pretraining epoch. ◊ denotes the MoCo result from our implementation. † denotes the results from 10 × 3 view evaluation following the common practice [12,13,36]. ‡ denotes the results from 2 × 3 view evaluation following [31].

Method	Year	Pretrain	Network	Input size	Params	Epoch	Top1	Top5
Modist [45]	2021	K400	R3D50	16 × 224 × 224	31.8M × 2	600	54.9	
RSPNet [6]	2021	K400	R3D18	16 × 224 × 224	33.2M	50	44.0	
RSPNet [6]	2021	K400	S3D-G	16 × 224 × 224	11.6M	50	55.0	
MoCo [13]	2021	K400	R3D50	8 × 224 × 224	31.8M × 2	200	54.4	
BYOL [13]	2021	K400	R3D50	8 × 224 × 224	31.8M × 2	200	55.8	
SwAV [13]	2021	K400	R3D50	8 × 224 × 224	31.8M × 2	200	51.7	
SimCLR [13]	2021	K400	R3D50	8 × 224 × 224	31.8M × 2	200	52	
MoCo◊ [8]	2020	K400	R3D50	8 × 224 × 224	31.8M × 2	200	54.6	84.3
Random	2022		R3D50	8 × 224 × 224	31.8M		54.2	82.9
Supervised	2022	K400	R3D50	8 × 224 × 224	31.8M		55.9	84.0
Ours	2022	K400	R3D50	8 × 224 × 224	31.8M	200	55.8	84.6
Ours†	2022	K400	R3D50	8 × 224 × 224	31.8M	200	**59.3**	**87.2**
Random	2022		TSM50	8 × 224 × 224	24.3M		56.7	84.0
Random‡	2022		TSM50	8 × 224 × 224	24.3M		57.6	84.8
Supervised	2022	K400	TSM50	8 × 224 × 224	24.3M		58.5	85.2
Supervised‡	2022	K400	TSM50	8 × 224 × 224	24.3M		60.8	86.5
Ours	2022	K400	TSM50	8 × 224 × 224	24.3M	200	58.5	85.2
Ours‡	2022	K400	TSM50	8 × 224 × 224	24.3M	200	**60.7**	**86.6**

Downstream Action Classification. We found an interesting phenomenon that different activity datasets are related to motion information at different levels [21]. As shown in Table 3b, It remains a relatively high action recognition accuracy on K400 dataset by randomly selecting one frame in each video to train and test the network. In contrast, the action recognition performance decreases dramatically on Something-Somethingv2 (SSv2) with the same training strategy. Figure 4 could explain these experimental results. Some classes in K400 dataset are more related to their static scene information. In contrast to SSv2 dataset, videos share similar appearances and backgrounds. Temporal information among different frames plays a more important role in the action classification task. Considering this observation, it is more appropriate to use SSv2 rather than K400 for evaluating the video representation ability of the proposed method.

Something-Something v2 [15] (SSv2) is a large and challenging video classification dataset. We fine-tune the pre-trained model on SSv2 dataset with a learning rate of 0.005. Following [46], the models are loaded with the weights from the pre-trained model except for TC module and the final fully-connected

Table 2. Action retrieval results on UCF101 and HMDB51. ◊ denotes the MoCo result from our implementation.

Method	Year	Pretraining Dataset	Backbone	Top1	Top5	Top10
Huang [22]	2021	UCF101	C3D	41	41.7	57.4
PacePred [42]	2021	K400	C3D	31.9	49.7	59.2
PRP [47]	2020	UCF101	R3D	22.8	38.5	46.7
Mem-DPC [18]	2020	K400	R3D18	20.2	40.4	52.4
TCLR [10]	2021	K400	R3D18	56.2		
SpeedNet [3]	2020	K400	S3D-G	13.0	28.1	37.5
CoCLR [19]	2020	UCF101	S3D	53.3	69.4	76.6
CSJ [24]	2021	K400	R3D34	21.5	40.5	53.2
MoCo◊ [20]	2020	K400	R3D50	45.0	61.4	70.0
Ours	2022	K400	R3D50	54.2	70.2	78.2
Ours	2022	K400	TSM50	**55.5**	**71.5**	**79.1**

(a)Action retrieval results on UCF101.

Method	Year	Pretraining Dataset	Backbone	Top1	Top5	Top10
Huang [22]	2021	UCF101	C3D	16.8	37.2	50.0
PacePred [42]	2021	K400	C3D	12.5	32.2	45.4
PRP [47]	2020	UCF101	R3D	8.2	25.3	36.2
Mem-DPC [18]	2020	K400	R3D18	7.7	25.7	40.6
BE [43]	2021	UCF101	R3D34	11.9	31.3	
TCLR [10]	2021	K400	R3D18	22.8		
CoCLR [19]	2020	UCF101	S3D	23.2	43.2	53.5
MoCo◊ [20]	2020	K400	R3D50	20.7	41.3	55.5
Ours	2022	K400	R3D50	25.4	47.5	60.1
Ours	2022	K400	TSM50	**25.9**	**49.8**	**64.1**

(b) Action retrieval results on HMDB51.

layer with randomly initialized. The video is sampled to 8 frames with 224×224 resolutions following [12].

Video Retrieval. The proposed method is also tested on UCF101 [38] and HMDB51 [28] datasets for video retrieval task. Because these datasets are relative small and easy to overfit, results may be quite different without some specific tricks in fine-tuning stage. In contrast, video retrieval task only uses the representations from the encoder, which could reveal the representation ability of the encoder more fairly by using different self-supervised methods. We only report

Table 3. Ablation studies on TCVM framework. D^2 denotes that we use feature subtraction again on the previous have-subtracted features. C, A and S separately denote *cut, argument* and *shuffle* operation.

Method			Dataset		
TC	VM	OFC	UCF101	HMDB51	SSv2
✓			28.2	14.0	53.9
✓	✓		31.5	13.8	55.0
✓	✓	✓	56.1	25.6	55.8

(a) Ablation study on the proposed TC, VM, OFC.

Dataset	Frames	Top1
K400	1	59.22
K400	8	72.67
SSv2	1	18.42
SSv2	8	60.60

(b)Ablation study on temporal information of K400/SSv2.

k	D^2	UCF101	HMDB51	SSv2
0.25		52.8	23.2	56.4
1		56.4	28.0	55.8
0.5	✓	55.1	26.9	55.7
0.5		**57.3**	**25.9**	**58.6**

(c) Ablation study on TC module.

Operation	UCF101	HMDB51	SSv2
C-A	41.6	17.4	43.3
A-C-S	27.2	12.7	52.3
C-S-A	49.0	19.9	43.4
C-A-S	56.1	25.6	55.8

(d) Ablation study on VM operations.

LR Annealing		Video Sampler		# Frames		Dataset		
CosineLR	StepLR	TSN sampler	I3D sampler	16	32	UCF101	HMDB51	SSv2
✓			✓		✓	55.0	21.2	56.3
	✓	✓			✓	51.6	21.2	56.2
	✓		✓	✓		50.6	19.3	56.0
	✓		✓		✓	56.1	25.6	55.8

(e) Ablation study on the proposed LR annealing, video sampler, #Frames.

their video retrieval results in this paper. After training on K400 dataset, the model is fixed as a feature encoder to test video retrieval tasks.

4.2 Compared with State-of-the-Art Methods

Table 1 shows the results of the proposed method on SSv2. Since few self-supervised learning methods focus on SSv2, we try our best to investigate all the methods and make comparisons as fair as possible. We also implement MoCo [8] by ourselves. We follow SlowFast [12] and TSM [31]'s evaluation settings to test backbone R3D50 and TSM's accuracy, which is 10×3 views and 2×3 views, respectively. 1-view test results are also provided in Table 1, which means that one video is only sampled into one clip for test. Our method's results are comparative with k400 supervised results in both TSM50 and R3D50 backbone, and the top-5 accuracy even outperforms k400 supervised model. As for TSM, top-1 accuracy can be improved from 57.6% to 60.6% by using the proposed method, which is only 0.1% lower than k400 supervised result. It is noticed that MoCo, BYOL, SwAV and SimCLR [13] uses teacher-student Network architecture. In contrast, the proposed method outperforms MoCo, SwAV, and SimCLR and reaches similar results with BYOL but only uses half of parameters in the pre-training step.

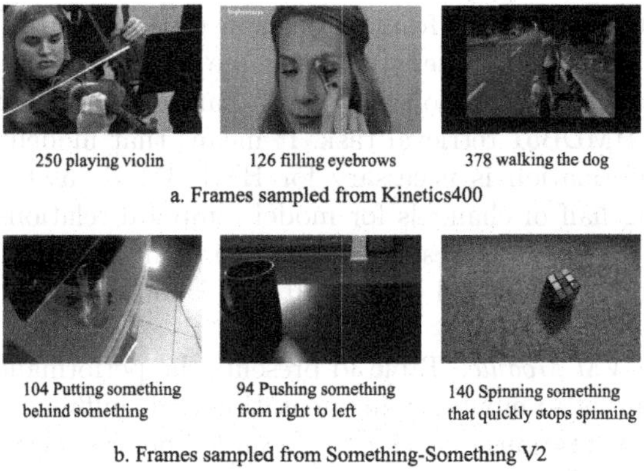

Fig. 4. The visualization of some frames from K400 and SSv2 datasets.

Table 2a and Table 2b present the video retrieval results on UCF101 and HMDB51 datasets. A series of state-of-the-art methods [3,10,18–20,22,24,42,43, 47] are also listed in the tables. For UCF101 retrieval, our method reaches 54.2% Top1 accuracy, 70.2% Top5 accuracy, and 78.2% Top 10 accuracy, respectively. Furthermore, our method achieves a higher performance by using TSM model. It is noticed that CSJ [24] presents a video jigsaw method to reason about the continuity in videos. Our method makes a great improvement compared with CSJ. Our method outperforms other methods significantly. For HMDB51 retrieval, the proposed method achieves 25.4% Top1, 47.5% Top5, and 60.1% Top10 accuracy by using R3D50 backbone and reaches 25.9% top1, 49.8% Top5, and 64.1% Top10 accuracy by using TSM model.

4.3 Ablation Study

Ablations on the Entire Framework. To demonstrate the combinations of the proposed TC and VM modules, Table 3a presents the ablation studies on the entire module. Without the proposed VM module, the performance of self-supervised video representation learning drops from 55.0% to 53.9% on the SSv2 action classification task. It also leads to a bad performance on UCF101 and HMDB51 video retrieval tasks. Interestingly, by adding the OFC module, it leaves the apparent performance gap on UCF101 and HMDB51 video retrieval task. But on SSv2 action classification task, the performance only raises from 55.0% to 55.8%. One of the reasons is that fine-tuning on the downstream task may learn non-linear relationship from the pre-trained models. It closes the gap between different pretrained models.

Ablations on the TC Module. We conduct ablation study of proposed TC module, as shown in Table 3c. We first test the different proportions of k. $k = 0.25$ denotes that the network uses quarter of channels to model temporal differences among features. $k = 1$ means that all the channels are used for modeling relationships with neighboring frames. Besides, we also try to model high level temporal information. D^2 denotes that the frame subtraction operation performs again on the previous subtraction features. It can be seen that D^2 shows better performance on HMDB51 retrieval task. It means that modeling more complicated motion information is necessary for HMDB51 dataset. Compared with $k = 1$, only using half of channels for model temporal relationship with neighboring channels may bring necessary scene motion information. There is a clear performance gap between $k = 1$ and $k = 0.5$ on UCF101.

Ablations on the VM Module. Table 3d presents the performance of two important operations in video montage: augmentation and shuffling. "aug before cut" denotes that augmentations of each clip are added before video shuffling. In this way, clips from the same videos shares the same augmentation. "aug after shuffle" denotes the augmentation is added after the pseudo videos are generated. In this way, clips in the same pseudo videos have the same augmentation. It can be seen that adding augmentation to each videos independently boosts the video understanding performances in each dataset. "wo shuffle" means that the shuffling operation in Fig. 3 is removed. The results show that the performance raises about 15%, 17%, 12% on UCF101 retrieval, HMDB51 retrieval and SSv2 classification tasks respectively by adding the shuffling performance.

Ablations on the Training Strategy. To make a fair comparison, some other training strategies [8,34] are also tested. In Table 3e, the "cosinelr" means we use the cosine learning rate decaying strategies following the common practices [8]. The "TSN sampler" means during the self-supervised training, the input videos are sampled by tsn sampler in [31]. And the "I3D sampler" denotes the input video sampling strategy that is used in the proposed method. The "16 frames" denotes the frame number of input videos in self-supervised training step drops from 32 to 16, which means each clip consists of 4 frames instead of 8.

4.4 Visualization Analysis

To further verify the effectiveness of the proposed method, class activation maps [37] (CAM) results are also presented on UCF101 compared with MoCo, as shown in Fig. 5. (a) shows that our method could localize motion areas more precisely in static background and discriminate action. (b) shows MoCo could easily be distracted by complex surroundings while our model still focuses on actors. (c) implies our model could correctly capture actors' motion clues even when the background is quickly changing.

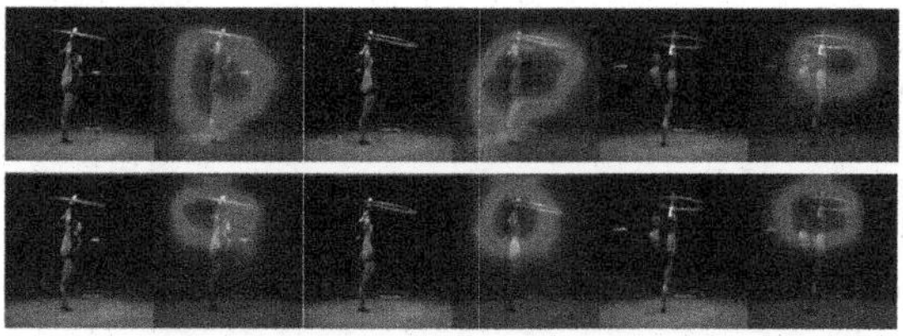

(a) Simple background with obvious motion

(b) Interrupting background

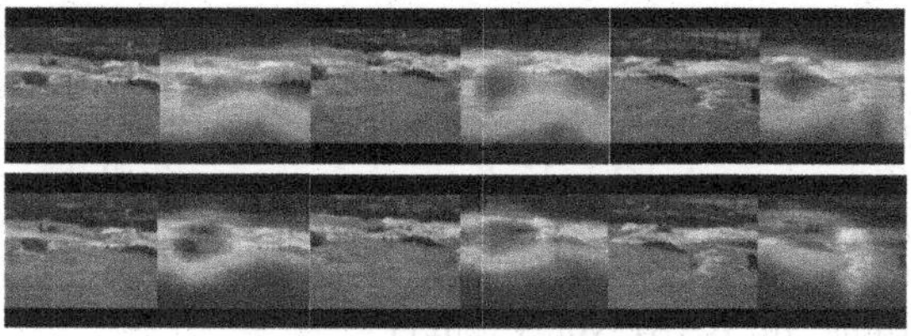

(c) Varying backgrounds by moving camera

Fig. 5. CAM results on the UCF101 dataset. The first and second raws denote the visualization results of MoCo and the proposed method, respectively.

5 Conclusions

In this paper, we propose the Temporal Contrasting Video Montage framework for self-supervised video representation learning. First, the input videos are processed by the VM module to generate video clips. Second, video clips are introduced to the encoder to extract features. Third, the proposed TC module encodes the high-level features by the frame-wise feature differences. Last, the features from the TC module are optimized in a contrastive learning strategy. Experimental results present that our methods outperforms other state-of-the-art methods for UCF101 and HMDB51 retrieval tasks and reaches the similar accuracy with supervised counterparts on the large scale action recognition dataset SSv2. The proposed method relies on a large scale of unlabelled videos for pretraining, which may limit its range of applications. In the future, we will try to train the proposed method on other datasets.

Acknowledgements. This work was supported by the National Key Research and Development Program of China under Grant No. 2020AAA0108100, the National Natural Science Foundation of China under Grant No. 61971343, 62088102 and 62073257, and the Key Research and Development Program of Shaanxi Province of China under Grant No. 2022GY-076.

References

1. Ahsan, U., Madhok, R., Essa, I.: Video Jigsaw: unsupervised learning of spatiotemporal context for video action recognition. In: 2019 IEEE Winter Conference on Applications of Computer Vision (WACV), pp. 179–189 (2019). https://doi.org/10.1109/WACV.2019.00025
2. Alwassel, H., Mahajan, D., Korbar, B., Torresani, L., Ghanem, B., Tran, D.: Self-supervised learning by cross-modal audio-video clustering. In: NeurIPS (2020)
3. Benaim, S., et al.: SpeedNet: learning the speediness in videos. In: CVPR, pp. 9922–9931 (2020)
4. Biondi, F.N., Alvarez, I.J., Jeong, K.A.: Human-vehicle cooperation in automated driving: a multidisciplinary review and appraisal. Int. J. Hum.-Comput. Interact. **35**, 932–946 (2019)
5. Carreira, J., Zisserman, A.: Quo Vadis, action recognition? A new model and the kinetics dataset. In: CVPR, pp. 6299–6308 (2017)
6. Chen, P., et al.: RSPNet: relative speed perception for unsupervised video representation learning. In: AAAI, vol. 1 (2021)
7. Chen, T., Kornblith, S., Norouzi, M., Hinton, G.: A simple framework for contrastive learning of visual representations. In: ICML, pp. 1597–1607. PMLR (2020)
8. Chen, X., Fan, H., Girshick, R., He, K.: Improved baselines with momentum contrastive learning. arXiv preprint arXiv:2003.04297 (2020)
9. Choi, J., Gao, C., Messou, J.C., Huang, J.B.: Why can't i dance in the mall? Learning to mitigate scene bias in action recognition. arXiv preprint arXiv:1912.05534 (2019)
10. Dave, I., Gupta, R., Rizve, M.N., Shah, M.: TCLR: temporal contrastive learning for video representation. arXiv preprint arXiv:2101.07974 (2021)
11. Ding, S., et al.: Motion-aware self-supervised video representation learning via foreground-background merging. arXiv preprint arXiv:2109.15130 (2021)
12. Feichtenhofer, C., Fan, H., Malik, J., He, K.: Slowfast networks for video recognition. In: ICCV, pp. 6202–6211 (2019)
13. Feichtenhofer, C., Fan, H., Xiong, B., Girshick, R., He, K.: A large-scale study on unsupervised spatiotemporal representation learning. In: Proceedings of the IEEE/CVF Conference on Computer Vision and Pattern Recognition, pp. 3299–3309 (2021)
14. Fernando, B., Bilen, H., Gavves, E., Gould, S.: Self-supervised video representation learning with odd-one-out networks. In: CVPR, pp. 3636–3645 (2017)
15. Goyal, R., et al.: The "something something" video database for learning and evaluating visual common sense. In: Proceedings of the IEEE International Conference on Computer Vision, pp. 5842–5850 (2017)
16. Grill, J.B., et al.: Bootstrap your own latent: a new approach to self-supervised learning. arXiv preprint arXiv:2006.07733 (2020)
17. Han, T., Xie, W., Zisserman, A.: Video representation learning by dense predictive coding. In: ICCV Workshops (2019)

18. Han, T., Xie, W., Zisserman, A.: Memory-augmented dense predictive coding for video representation learning. In: Vedaldi, A., Bischof, H., Brox, T., Frahm, J.-M. (eds.) ECCV 2020. LNCS, vol. 12348, pp. 312–329. Springer, Cham (2020). https://doi.org/10.1007/978-3-030-58580-8_19
19. Han, T., Xie, W., Zisserman, A.: Self-supervised co-training for video representation learning. In: NeurIPS (2020)
20. He, K., Fan, H., Wu, Y., Xie, S., Girshick, R.: Momentum contrast for unsupervised visual representation learning. In: CVPR, pp. 9729–9738 (2020)
21. Huang, D.A., et al.: What makes a video a video: analyzing temporal information in video understanding models and datasets. In: CVPR, pp. 7366–7375 (2018)
22. Huang, L., Liu, Y., Wang, B., Pan, P., Xu, Y., Jin, R.: Self-supervised video representation learning by context and motion decoupling. In: CVPR, pp. 13886–13895 (2021)
23. Huang, Z., Zhang, S., Jiang, J., Tang, M., Jin, R., Ang, M.H.: Self-supervised motion learning from static images. In: CVPR, pp. 1276–1285 (2021)
24. Huo, Y., et al.: Self-supervised video representation learning with constrained spatiotemporal jigsaw. In: IJCAI, pp. 751–757 (2021). https://doi.org/10.24963/ijcai.2021/104
25. Jing, L., Yang, X., Liu, J., Tian, Y.: Self-supervised spatiotemporal feature learning via video rotation prediction. arXiv preprint arXiv:1811.11387 (2018)
26. Khosla, P., et al.: Supervised contrastive learning. In: NeurIPS, vol. 33, pp. 18661–18673 (2020)
27. Kim, D., Cho, D., Kweon, I.S.: Self-supervised video representation learning with space-time cubic puzzles. In: AAAI, vol. 33, pp. 8545–8552 (2019). https://doi.org/10.1609/aaai.v33i01.33018545
28. Kuehne, H., Jhuang, H., Garrote, E., Poggio, T., Serre, T.: HMDB: a large video database for human motion recognition. In: ICCV, pp. 2556–2563. IEEE (2011)
29. Lee, H.Y., Huang, J.B., Singh, M., Yang, M.H.: Unsupervised representation learning by sorting sequences. In: ICCV, pp. 667–676 (2017)
30. Li, Y., et al.: MPC-based switched driving model for human vehicle copiloting considering human factors. Transp. Res. Part C Emerg. Technol. **115**, 102612 (2020). https://doi.org/10.1016/j.trc.2020.102612. https://www.sciencedirect.com/science/article/pii/S0968090X18308179
31. Lin, J., Gan, C., Han, S.: TSM: temporal shift module for efficient video understanding. In: ICCV, pp. 7083–7093 (2019)
32. Lin, T., Zhao, X., Su, H., Wang, C., Yang, M.: BSN: boundary sensitive network for temporal action proposal generation. In: Ferrari, V., Hebert, M., Sminchisescu, C., Weiss, Y. (eds.) ECCV 2018. LNCS, vol. 11208, pp. 3–21. Springer, Cham (2018). https://doi.org/10.1007/978-3-030-01225-0_1
33. Misra, I., van der Maaten, L.: Self-supervised learning of pretext-invariant representations. In: CVPR, pp. 6707–6717 (2020)
34. Pan, T., Song, Y., Yang, T., Jiang, W., Liu, W.: VideoMoCo: contrastive video representation learning with temporally adversarial examples. In: CVPR, pp. 11205–11214 (2021)
35. Patrick, M., et al.: Space-time crop & attend: improving cross-modal video representation learning. In: ICCV (2021)
36. Qian, R., et al.: Spatiotemporal contrastive video representation learning. In: Proceedings of the IEEE/CVF Conference on Computer Vision and Pattern Recognition, pp. 6964–6974 (2021)

37. Selvaraju, R.R., Cogswell, M., Das, A., Vedantam, R., Parikh, D., Batra, D.: Grad-CAM: visual explanations from deep networks via gradient-based localization. In: ICCV, pp. 618–626 (2017)
38. Soomro, K., Zamir, A.R., Shah, M.: UCF101: a dataset of 101 human actions classes from videos in the wild. arXiv preprint arXiv:1212.0402 (2012)
39. Sun, C., Myers, A., Vondrick, C., Murphy, K., Schmid, C.: VideoBERT: a joint model for video and language representation learning. In: ICCV, pp. 7464–7473 (2019)
40. Suzuki, T., Itazuri, T., Hara, K., Kataoka, H.: Learning spatiotemporal 3D convolution with video order self-supervision. In: Leal-Taixé, L., Roth, S. (eds.) ECCV 2018. LNCS, vol. 11130, pp. 590–598. Springer, Cham (2019). https://doi.org/10.1007/978-3-030-11012-3_45
41. Vondrick, C., Shrivastava, A., Fathi, A., Guadarrama, S., Murphy, K.: Tracking emerges by colorizing videos. In: Ferrari, V., Hebert, M., Sminchisescu, C., Weiss, Y. (eds.) ECCV 2018. LNCS, vol. 11217, pp. 402–419. Springer, Cham (2018). https://doi.org/10.1007/978-3-030-01261-8_24
42. Wang, J., Jiao, J., Liu, Y.-H.: Self-supervised video representation learning by pace prediction. In: Vedaldi, A., Bischof, H., Brox, T., Frahm, J.-M. (eds.) ECCV 2020. LNCS, vol. 12362, pp. 504–521. Springer, Cham (2020). https://doi.org/10.1007/978-3-030-58520-4_30
43. Wang, J., et al.: Removing the background by adding the background: towards background robust self-supervised video representation learning. In: CVPR, pp. 11804–11813 (2021)
44. Wang, L., et al.: Temporal segment networks for action recognition in videos. IEEE Trans. Pattern Anal. Mach. Intell. **41**(11), 2740–2755 (2018)
45. Xiao, F., Tighe, J., Modolo, D.: MoDist: motion distillation for self-supervised video representation learning. arXiv preprint arXiv:2106.09703 (2021)
46. Xu, D., Xiao, J., Zhao, Z., Shao, J., Xie, D., Zhuang, Y.: Self-supervised spatiotemporal learning via video clip order prediction. In: CVPR, pp. 10334–10343 (2019)
47. Yao, Y., Liu, C., Luo, D., Zhou, Y., Ye, Q.: Video playback rate perception for self-supervised spatio-temporal representation learning. In: CVPR, pp. 6548–6557 (2020)

SCFNet: A Spatial-Channel Features Network Based on Heterocentric Sample Loss for Visible-Infrared Person Re-identification

Peng Su, Rui Liu(✉), Jing Dong, Pengfei Yi, and Dongsheng Zhou

Key Laboratory of Advanced Design and Intelligent Computing Ministry of Education, School of Software Engineering, Dalian University, Dalian, China
liurui@dlu.edu.cn

Abstract. Cross-modality person re-identification between visible and infrared images has become a research hotspot in the image retrieval field due to its potential application scenarios. Existing research usually designs loss functions around samples or sample centers, mainly focusing on reducing cross-modality discrepancy and intra-modality variations. However, the sample-based loss function is susceptible to outliers, and the center-based loss function is not compact enough between features. To address the above issues, we propose a novel loss function called Heterocentric Sample Loss. It optimizes both the sample features and the center of the sample features in the batch. In addition, we also propose a network structure combining spatial and channel features and a random channel enhancement method, which improves feature discrimination and robustness to color changes. Finally, we conduct extensive experiments on the SYSU-MM01 and RegDB datasets to demonstrate the superiority of the proposed method.

1 Introduction

Person re-identification (ReID) is a popular research direction in the image retrieval field, which is a cross-device pedestrian retrieval technology. Person ReID faces many challenges due to factors such as lighting changes, viewpoint changes, and pose changes. Most existing methods [1–4] focus on matching person images with visible images. However, when at night, it can be hard for visible cameras to capture clear pictures of pedestrians due to insufficient light. Nowadays, many new outdoor monitors have integrated infrared image capture devices. Therefore, how to use both visible and infrared images for cross-modality person re-identification (VI-ReID) has become a very significant research issue.

Compared with the single-modality person ReID task, VI-ReID is more challenging with a higher variation in data distribution. Two classic methods have been investigated to solve the VI-ReID problem. The first method is based on modal conversion [5,6], which eliminates modal discrepancy by converting the images of two modalities to each other. However, the operation of modal conversion is relatively complicated. And inevitably, some key information is lost and

L. Wang et al. (Eds.): ACCV 2022, LNCS 13842, pp. 543–559, 2023.
https://doi.org/10.1007/978-3-031-26284-5_33

some noise is introduced in the conversion process. Another method is based on representation learning [7,8]. This method often maps visible and infrared images into a unified feature space and then learns a discriminative feature representation for each person. Among them, local features as a common feature representation method have been widely used in person ReID. Inspired by previous work MPANet [9], we propose a Spatial-Channel Features Network (SCFNet) structure based on local feature representation in this paper. SCFNet extracts local features in simultaneously spatial and channel dimensions, enhancing the feature representation capability. Besides the above two methods, metric learning [10,11] is often used in VI-ReID as a key technique. In particular, triplet loss and its variants [10,12] are the most dominant metric learning methods. However, most of the existing triplet losses are designed around samples or sample centers. The loss designed around the sample is vulnerable to the influence of anomalous samples, and the loss designed around the center is not tight enough. To solve this problem, an improved heterocentric sample loss is proposed in this paper. It can tolerate the outliers between samples and also takes into account the modal differences. More importantly, it can make the distribution of sample features relatively more compact.

In addition, we also propose a local random channel enhancement method for VI-ReID. By randomly replacing pixel values in local areas of a color image, the model is made more robust to color changes. We combine it with global random grayscale to improve the performance of the model with only a small increase in computational effort.

The summary of our main contributions is as follows:

- A SCFNet network structure is proposed that allows local features to be extracted from the spatial and channel levels. In addition, a local random channel enhancement is used to further increase the identification and robustness of the features.
- A heterocentric sample loss is proposed, which optimizes the structure of the feature space from both sample and center perspectives, enhancing intra-class tightness and inter-class separability.
- Experiments on two benchmark datasets show that our proposed method obtain good performance in the VI-ReID tasks. In particular, 69.37% of mAP and 72.96% of Rank1 accuracy are obtained on SYSU-MM01.

2 Related Work

VI-ReID was first proposed by WU et al. [13] in 2017. They contributed a standard dataset SYSU-MM01 and designed a set of evaluation criteria for this problem. Recently, the research on VI-ReID has been divided into two main types: modality conversion and representation learning. In addition, metric learning, as a key algorithm used in these two methods, is often studied separately. This paragraph will introduce the related work from these three aspects.

Modality Conversion Based Methods. This method usually interconverts images of two modalities by GAN to reduce the modal difference. Wang et al.

[5] proposed AlignGAN to transform visible images into infrared images, and they used pixel and feature alignment to alleviate intra- and inter-modal variation. In the second year, they also used GAN to generate cross-modal paired images in [14] to solve the VI-ReID problem by performing set-level and instance-level feature alignment. Choi et al. [15] used GAN to generate cross-modality images with different morphologies to learn decoupled representations and automatically decouple identity discriminators and identity irrelevant factors from infrared images. Hao et al. [16] designed a module to distinguish images for better performance in cross-modality matching using the idea that generators and discriminators are mutually constrained in GAN. The method based on modality conversion reduces inter-modal differences to a certain extent, but inevitably it also introduces some noise. In addition, methods based on modality conversion are relatively complex and usually require more training time.

Representation Learning Based Methods. The method of representation learning aims to use a feature extraction structure to map the features of both modalities into the unified feature space. Ye et al. [7] proposed an AGW baseline network, which used Resnet50 as a feature extraction network and added a non-local attention module to make the model focus more on global information. In the same year, they also proposed a DDAG [8] network to extract contextual features by intra-modal weighted aggregation and inter-modal attention. Liu et al. [17] incorporated feature skip connections in the middle layers into the latter layers to improve the discrimination and robustness of the person feature. Huang et al. [18] captured the modality-invariant relationship between different character parts based on appearance features, as a supplement to the modality-shared appearance features. Zhang et al. [19] proposed a multi-scale part-aware cascade framework that aggregates local and global features of multiple granularities in a cascaded manner to enrich and enhance the semantic information of features. Representation learning approaches often achieve good results through well-designed feature extraction structures. And, the representation of person features has evolved from the initial global features to richer features, such as local features, multi-scale features, etc.

Metric Learning. Metric learning aims to learn the degree of similarity between person images. As a key algorithm in VI-ReID, metric learning has been used in both modal transformation methods and representation learning methods. In VI-ReID, triplet loss and its variants are the most used metric learning methods. Ye et al. [10] proposed the BDTR loss, which dealt with both cross-modal variation and intra-modal variation to ensure feature discriminability. Li et al. [12] proposed a strategy of batch sampling all triples for the imbalance problem of modal optimization in the optimization process of triplet loss. Liu et al. [20] proposed Heterocentric Triplet Loss (HcTri) to improve intra-modality similarity by constraining the center-to-center distance of two heterogeneous intra-class modalities. Wu et al. [9] introduced a Center Cluster Loss (CC) to study the relationship between identities. The CC loss establishes the relationship between class centers on the one hand, and clusters the samples toward the center on the other hand. Inspired by the work of [9] and [20], we design a heterocentric sample loss that combines samples and sample centers. This loss makes the sample distribution more compact while optimizing the intra-class similarity and inter-class difference, which further enhances the identification and robustness of features.

3 Methodology

3.1 Network Architecture

We chose the MPANet as our baseline model. It has 3 main characteristics, Modality Alleviation Module (MAM) eliminates the modality-related information, Pattern Alignment Module (PAM) discovers nuances in images to increase the recognizability of features, and Modal Learning (ML) mitigates modal differences with mutual mean learning. We chose this baseline for two main reasons. First, it provides a framework to learn modal invariant features of the person. Secondly, the performance of this network is very superior.

The architecture of SCFNet is shown in Fig. 1. First, we preprocess the visible images using local random channel enhancement (CA) and random grayscale (RG). Then the infrared image and the processed visible image are put into a backbone network consisting of Resnet50 and MAM to extract features. The 3D features map obtained are represented $X \in \mathbb{R}^{C \times H \times W}$. In MPANet, the authors use the PAM to extract nuances features from multiple patterns in the channel dimension. But in the spatial dimension, some detailed features of a person should also be included. Therefore, we introduce a spatial feature extraction module to extract local information about the person from the space dimension.

To extract the spatial feature, a simple method referring to [20] and [21] is to cut the feature map horizontally and extract each small piece of feature as a representation of local features. As shown in Fig. 1, we divide the obtained 3D feature map X in the horizontal direction, evenly into p parts. The features of each small segment are denoted as $X_i \in \mathbb{R}^{C \times \frac{H}{p} \times W}, i \in \{1, 2, \ldots, p\}$. Then, each local feature is reshaped into a d-dimensional vector using pooling and convolution operations. The formula is expressed as $S_i \in \mathbb{R}^d, i \in \{1, 2, \ldots, p\}$, where

$$S_i = Reshape(Conv(pool(X_i))) \tag{1}$$

Finally, all local features are concatenated as the final spatial feature representation $S \in \mathbb{R}^{D_1}$, where $D_1 = p \times d$.

For channel features, we directly use the PAM in MPANet. However, to decrease the number of parameters during computation, we downsampled the features by adding a 1×1 convolutional layer at the end. Finally, the channel features obtained after the reshaping operation are represented as $C \in \mathbb{R}^{D_2}$. For the convenience of calculation, the value of D_2 we set is equal to D_1.

Since local features are susceptible to factors like occlusion and pose variations, we also incorporate global features. The global feature representation $G \in \mathbb{R}^{D_3}$ is obtained by directly using pooling and reshaping operations on the 3D feature X. Finally, we join S, C, and G as the final feature representation in the inference stage.

3.2 Loss Function

Triplet loss serves an essential role in VI-ReID. Traditional triplet losses are designed by constraining the samples, but the heterocentric triplet loss [20] replaces the samples with the center of each class. It is formulated as follows:

$$L_{hc\_tri} = \sum_{i=1}^{P}[\rho + ||c_v^i - c_t^i||_2 - \min_{\substack{n\in\{v,t\}\\ j\neq i}} ||c_v^i - c_n^j||_2]_+ \\ + \sum_{i=1}^{P}[\rho + ||c_t^i - c_v^i||_2 - \min_{\substack{n\in\{v,t\}\\ j\neq i}} ||c_t^i - c_n^j||_2]_+ \quad (2)$$

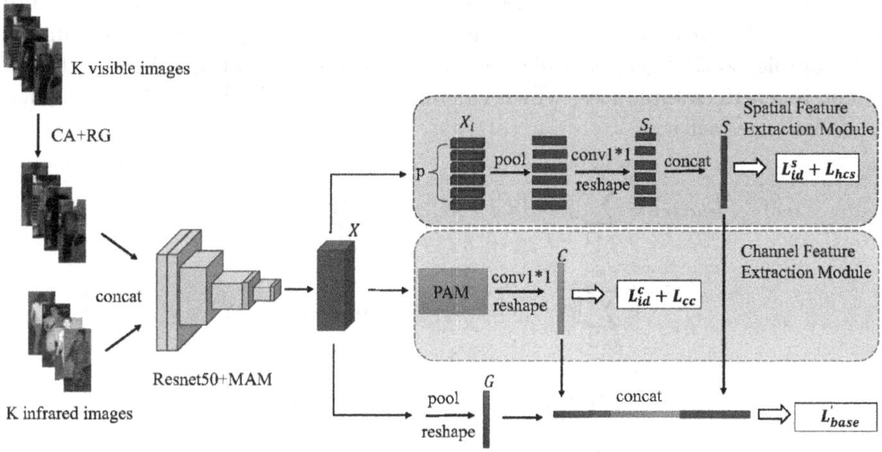

Fig. 1. The architecture of the proposed SCFNet.

where ρ is the margin parameter for heterocentric triplet loss, $[x]_+ = max(x, 0)$ denotes the standard hinge loss, and $||x_a - x_b||_2$ denotes the Euclidean distance of two feature vectors x_a and x_b. P denotes the total number of different classes in a mini-batch. c_v^i and c_t^i denote the central features of the visible and infrared class i in a mini-batch, respectively. They are obtained by averaging the samples of all classes i in the respective modalities, which are calculated as follows:

$$c_v^i = \frac{1}{K}\sum_{j=1}^{K} v_j^i \tag{3}$$

$$c_t^i = \frac{1}{K}\sum_{j=1}^{K} t_j^i \tag{4}$$

where v_j^i and t_j^i denote the feature representation of the jth visible image and the jth infrared image of class i respectively.

The heterocentric triplet loss uses class centers instead of samples, relaxing the tight constraint in traditional triplet loss and alleviating the effects caused by anomalous samples, allowing the network to converge better. However, optimizing only the central features means that the change for each sample feature is small. If the mini-batch is large, then the constraints assigned to each sample are small. This will cause the network to have a slow convergence rate. Therefore, we propose to optimize the central features while also applying constraints to each sample, so that each sample is closer to its central features. We refer to this as heterocentric sample loss, which calculation formula is shown in Eq. 5. where δ is a balance factor.

$$\begin{aligned} L_{hc\_tri} = &\sum_{i=1}^{P}[\rho + ||c_v^i - c_t^i||_2 - \min_{\substack{n\in\{v,t\}\\ j\neq i}} ||c_v^i - c_n^j||_2]_+ \\ &+ \sum_{i=1}^{P}[\rho + ||c_t^i - c_v^i||_2 - \min_{\substack{n\in\{v,t\}\\ j\neq i}} ||c_t^i - c_n^j||_2]_+ \\ &+ \delta\sum_{j=1}^{K}[||v_j^i - c_v^i||_2 + ||t_j^i - c_t^i||_2] \end{aligned} \tag{5}$$

The heterocentric sample loss is based on the heterocentric triplet loss and adds the constraint that the Euclidean distance between the sample feature and the central feature should be as close as possible. In this way, the samples of the same class will be more compact in the feature space. The difference between the heterocentric triplet loss and the heterocentric sample loss is shown in Fig. 2(a) and (b).

Heterocentric sample loss reduces both cross-modality variation and intermodal variation in a simple way. It has two main advantages: (1) It is not a strong constraint, so it can be robust to abnormal samples. (2) By constraining the relationship between samples and centers, each sample is brought near the center

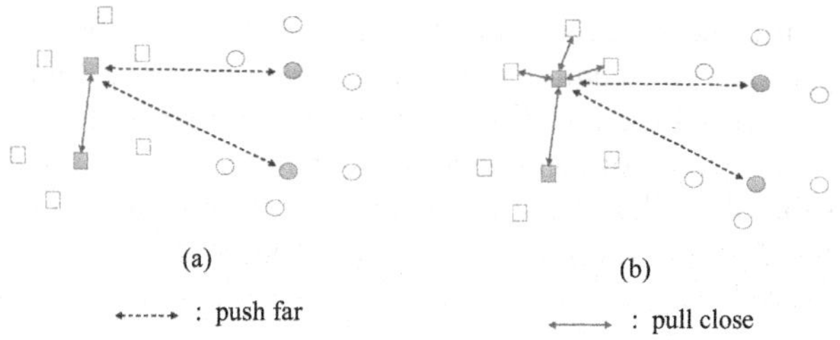

Fig. 2. Comparison of Heterocentric Triplet Loss (a) and Heterocentric Sample Loss (b). Different colors represent different modalities, and different shapes represent different classes. Graphs with dashed borders and no color fills represent sample characteristics. A color-filled graph represents the central feature computed from the sample features of the corresponding color.

of its class, thus making the final features more compact and discriminative, and also speeding up the convergence rate.

Besides triplet loss, identity loss is also often used in the VI-ReID. Given an image y, p_i denotes the class predicted by the network, q_i denotes the true label, and the identity loss is denoted as:

$$L_{id} = -\sum_{1}^{n} q_i \log p_i \tag{6}$$

Let the total loss of the baseline model MPANet be denoted as L_{base}, from which the central cluster loss is removed and denoted as $L_{base}^{'}$. We use the identity loss L_{id}^{s} and heterocentric sample loss L_{hcs} for spatial features, the central cluster loss L_{cc} and identity loss L_{id}^{c} for channel features, and $L_{bass}^{'}$ for total features (as shown in Fig. 1). Finally, the total loss of SCFNet is defined as:

$$Loss = L_{bass}^{'} + L_{hcs} + L_{id}^{s} + L_{cc} + L_{id}^{c} \tag{7}$$

3.3 Local Random Channel Enhancement

There is naturally a large difference in appearance between infrared and visible images, which is the most significant cause of modal differences. We observe that the infrared images and the images extracted from each channel of visible images separately are similar to the grayscale images in appearance. Therefore, we propose a local random channel enhancement method to alleviate the influence of modal differences.

The specific operation of local random channel enhancement is as follows. Given a visible image, we randomly select a region in the visible light image and randomly replace the pixel value of the region with one of the pixel values of the R, G, B channel or the gray value. Figure 3 shows the effect before and after the transformation.

Local random channel enhancement makes visible and infrared images more similar in appearance by changing the pixel values of the images, while this random enhancement also forces the network to be less sensitive to color changes and facilitates its learning of color-independent features. In our experiments, we use a combination of random channel enhancement and random grayscale. With only a slight increase in computational effort, the model's performance can be effectively improved.

4 Experiments

4.1 Experimental Setting

Datasets. We conducted experiments on two public VI-ReID datasets, SYSU-MM01 [5] and RegDB [22]. For a fair comparison, we report the results of Rank-k and mAP on each dataset.

SYSU-MM01 is the first and the largest public dataset proposed for VI-ReID. It was captured by 6 cameras. The training set has 22258 visible pictures and 11909 infrared pictures and the testing set has 3803 infrared pictures and 301 visible pictures. We followed the evaluation protocol proposed in [5] for the SYSU-MM01 and mainly reported the results for single and multi-shot settings in all and indoor search modes.

The RegDB dataset has 412 person IDs with 10 visible and 10 infrared images for each person. 206 person IDs are randomly selected for training in the training phase and the remaining 206 person IDs are used for testing. For the RegDB dataset, we used the two most common search settings: "Infrared to Visible" and "Visible to Infrared".

Implementation Details. This paper implements an improved model based on Pytorch. All evaluation indicators are trained on the Sitonholy SCM artificial

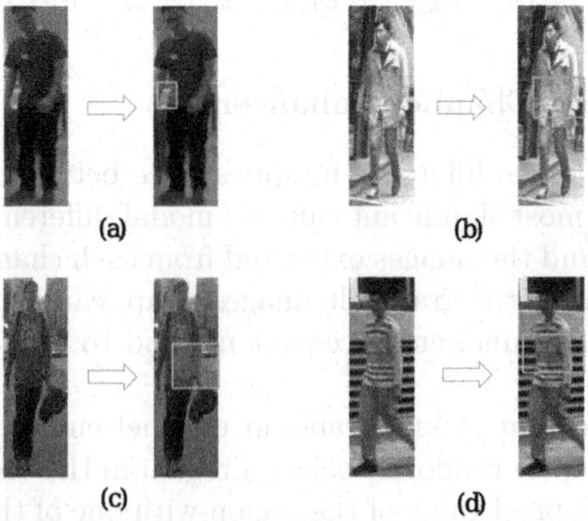

Fig. 3. Local Random channel enhancement. a, b, c, and d correspond to the 4 conversions, respectively.

intelligence cloud platform, using a single Tesla P100-SXM2 graphics card on Ubuntu operating system. Our network uses the Adam optimizer with training epochs of 140 generations and an initial learning rate of 3.5×10^{-4}, at the 80th and 120th generations time decay. In the spatial feature extraction module, the number of blocks that the feature map is divided is 6, and each local feature is represented as a 512-dimensional vector. Therefore, the final dimension of the spatial feature is $D_1 = 3072$. In the channel feature extraction module, the dimension D_2 of the downsampled channel feature is also set to 3072. The final dimension $D_3 = 2048$ of the global feature. For the loss of heterocentric samples, after experimental comparison, we set the marginal ρ to 0.9 and δ to 0.1 on the SYSU-MM01 dataset and set the marginal ρ to 0.3 and δ to 1 on the RegDB dataset. The probabilities used for local random channel enhancement and global grayscale are 0.8 and 0.3, respectively.

4.2 Ablation Experiment

Effectiveness of the Components. To validate the effectiveness of each module of SCFNet, we conducted ablation experiments on two public datasets. The SYSU-MM01 dataset tests the single-shot settings in all search mode, and the RegDB dataset tests the settings from visible to infrared. The results of the ablation experiments are shown in Table 1 and Table 2. Among them, "CL" indicates that the channel feature extraction module is used, "SL" indicates that the spatial feature extraction module is used, "CA" indicates that the local random channel enhancement is used, and "HCS" indicates that the heterocentric sample loss is used. For the RegDB dataset, the spatial feature extraction module is not used and the Center Cluster Loss of the channel feature extraction module is replaced by the heterocentric sample loss. This is because RegDB is a small dataset, using the channel feature extraction module is enough to extract effective features, and using a too complex structure will increase the difficulty of network convergence.

From the comparison between version 1 and version 2 in Table 1, it can be found that using the spatial feature extraction module can improve Rank1 and mAP by 1.22 and 0.88, respectively, compared to the situation of not using it. This illustrates the effectiveness of the method to mine person feature information from the spatial dimension. From version 2 and version 3 in Table 1, and

Table 1. Performance evaluation of each component on the SYSU-MM01 dataset.

Version	CL	SL	CA	HCS	Rank1	mAP
Version1	✓				68.14	65.17
Version2	✓	✓			69.36	66.05
Version3	✓	✓	✓		71.46	67.36
Version4	✓	✓		✓	71.60	67.80
Version5	✓	✓	✓	✓	72.96	69.37

version 1 and version 2 in Table 2, it can be seen that using local random channel enhancement can enhance the model performance. This indicates that the use of local random channel enhancement can reduce the sensitivity of the model to color. From the comparison of version 2 and version 4 in Table 1, and version 1 and version 3 in Table 2, it can be seen that the use of heterocentric sample loss can improve the mAP by 1.75 and the Rank1 by 2.24 on SYSU-MM01 and improve the mAP by 0.53 and the Rank1 by 1.41 on RegDB. This is because the heterocentric sample loss reduces intra-class separability and makes the model feature distribution more compact. So it is easier to distinguish different classes. We can see from Table 1 version 5 and Table 2 version 4, the best results were achieved by using these modules together. This illustrates that these modules are complementary in terms of performance and combining them allows us to achieve the highest performance of our model.

Discussion on δ of Heterocentric Sample Loss. The hyperparameter δ is a balancing factor whose value can greatly affect the performance of the model. So we conduct several experiments to determine the value of δ in Eq. 5. The experimental results are shown in Fig. 4.

On the SYSU-MM01 dataset, the network is very sensitive to changes in δ. We changed δ from 0 to 0.3, as can be seen from Fig. 4(a), when δ is less than 0.1, the Rank1 and mAP of the network tend to rise, and when δ is greater than 0.1, the performance of the network begins to gradually decline, especially when $\delta = 0.3$, its Rank1 and mAP drop to around 60% and 55%, respectively. Therefore, on this dataset, δ is taken as 0.1. On the RegDB dataset, the change of the network to δ is quite different from SYSU-MM01. When δ takes 0, the loss function degenerates into a heterocentric loss, and in our network structure, Rank1 and mAP are only about 61% and 56%. When δ takes a small value of 0.1, the performance of the network increases rapidly, which also shows the effectiveness of the loss of heterocentric samples. When the value of δ is between 0 and 1, the performance of the network gradually increases. When the value of δ is between 1 and 3, the Rank1 and mAP of the network only change slightly.

Table 2. Performance evaluation of each component on the RegDB dataset.

Version	CL	CA	HCS	Rank1	mAP
Version1	✓			83.11	80.58
Version2	✓	✓		84.91	81.61
Version3	✓		✓	84.52	81.11
Version4	✓	✓	✓	86.33	82.10

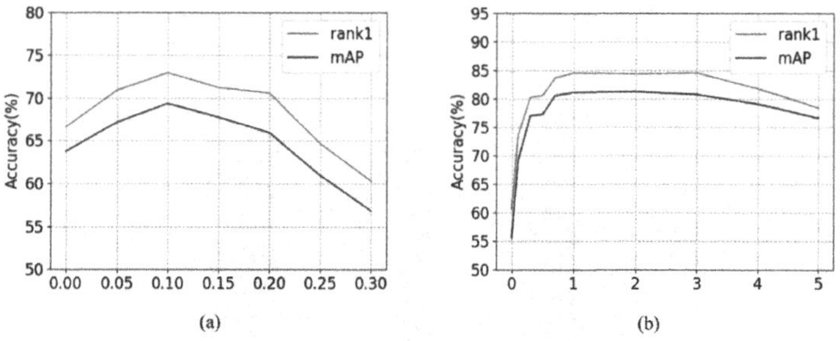

Fig. 4. Evaluating the weight parameter δ in the loss of heterocentric samples. (a) shows the results in SYSU-MM01, and (b) shows the results in RegDB.

When the value of δ is greater than 3, the Rank1 and mAP of the network begin to have a downward trend. So, on this dataset, we set the value of δ to 1.

For this phenomenon, we speculate that this is due to the different data distributions in these two datasets. In the RegDB dataset, there is a one-to-one correspondence between the poses in the infrared image and the visible image for the same person, so its intra-class discrepancy is smaller than that of SYSU-MM01. The δ control item in Eq. 5 is to shorten the distance between the sample and the center and reduce the intra-class discrepancy. Therefore, the RegDB dataset is less sensitive to δ changes than the SYSU-MM01 dataset.

Discussion on the Number P of Spatial Feature Divisions. The number of spatial features that are divided determines the size of each local feature and also affects the representability of each local feature. We design some experiments to compare the number of segmented spatial features. Experiments are carried out with the number of divisions being 4, 6, and 8 respectively in SYSU-MM01 dataset. The experimental results are shown in Fig. 5. It can be seen that when the number of divisions is 6, the effect is the best, with Rank1 and mAP being 69.36% and 66.05%, respectively. When the number of divisions is 4, the granularity of the obtained local features is not fine enough, and it is difficult to capture more detailed features. The experimental results drop again when the number of divisions is 8. This is because features with too small a granularity are easily affected by noise, so it is hard for the network to capture effective local information. After comparison, the number of divisions of 6 is a reasonable value, so in the experiment, we take p = 6.

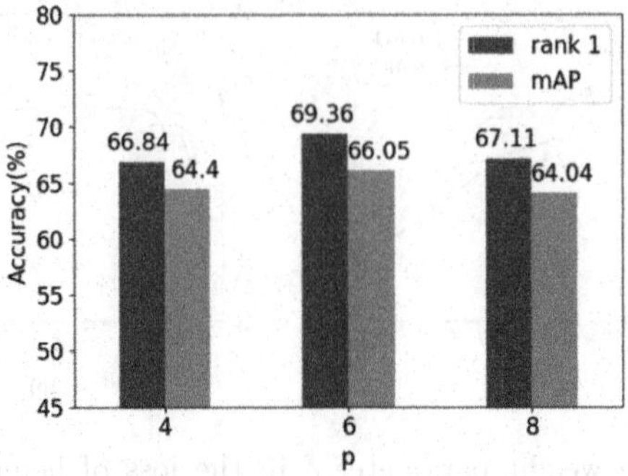

Fig. 5. Influence of the number p of spatially local features

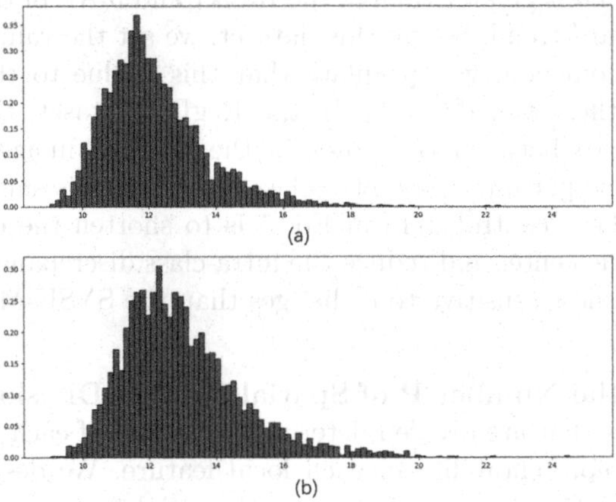

Fig. 6. The histogram of sample-to-center distance.

Visualization of Sample-to-Center Distance. To demonstrate the performance of heterocentric sample loss more precisely, we train two models on the spatial branch of SCFNet using heterocentric triplet loss and heterocentric sample loss respectively. Then, the data from all test sets are fed into the network and the distance to the center is calculated for each category. These distances are plotted as histograms as shown in Fig. 6, where Fig. 6(a) is using heterocentric sample loss and Fig. 6(b) is using heterocentric triplet loss. We calculated their means and variances, where means for (a) are 12.16 with a variance of 2.34 and means for (b) are 12.96 with a variance of 3.17. This suggests that heterocentric sample loss can make the intra-class features more compact.

Table 3. Comparison of SYSU-MM01 dataset with state-of-the-art methods in all search and indoor search modes. 1st and 2nd best results are marked in red and blue, respectively.

Method	Single-shot				Multi-shot			
	R1	R10	R20	mAP	R1	R10	R20	mAP
All Search								
Zero-Pad [13]	14.80	54.12	71.33	15.95	19.13	61.4	78.41	10.89
cmGAN [23]	26.97	67.51	80.56	27.80	31.49	72.74	85.01	22.27
D^2RL [6]	28.90	70.60	82.40	29.20	–	–	–	–
HI-CMD [15]	34.94	77.58	–	35.94	–	–	–	–
HPILN [24]	41.36	84.78	94.51	42.95	47.56	88.13	95.98	36.08
AlignGAN [5]	42.40	85.00	93.70	40.70	51.50	89.40	95.70	33.90
AGW [7]	47.50	–	–	47.65	–	–	–	–
DDAG [8]	54.75	90.39	95.81	53.02	–	–	–	–
DGTL [25]	57.34	–	–	55.13	–	–	–	–
cm-SSFT [26]	61.60	89.20	93.90	63.20	63.40	91.20	95.70	62.00
HcTri [20]	61.68	93.10	97.17	57.51	–	–	–	–
MSA [28]	63.13	–	–	59.22	–	–	–	–
SFANet [29]	65.74	92.98	97.05	60.83	–	–	–	–
MPANet [9]	70.58	96.21	98.80	68.24	75.58	97.91	99.43	62.91
Ours	72.96	96.67	98.82	69.37	78.50	97.67	99.19	63.99
Indoor search								
Zero-Pad [13]	20.58	68.38	85.79	26.92	24.43	75.86	91.32	18.64
cmGAN [23]	31.63	77.23	89.18	42.19	37.00	80.94	92.11	32.76
HPILN [24]	45.77	91.82	98.46	56.52	53.50	93.71	98.93	47.48
AlignGAN [5]	45.90	87.60	94.40	54.30	57.10	92.70	97.40	45.30
AGW [7]	54.17	–	–	62.97	–	–	–	–
DDAG [8]	61.02	94.06	98.41	67.98	–	–	–	–
DGTL [25]	63.11	–	–	69.20	–	–	–	–
cm-SSFT [26]	70.50	94.90	97.70	72.60	73.00	96.30	99.10	72.40
HcTri [20]	63.41	91.69	95.28	68.17	–	–	–	–
MSA [28]	67.18	–	–	72.74	–	–	–	–
SFANet [29]	71.60	96.60	99.45	80.05	–	–	–	–
MPANet [9]	76.74	98.21	99.57	80.95	84.22	99.66	99.96	75.11
Ours	77.36	97.76	99.34	80.87	85.73	99.30	99.88	76.07

4.3 Comparison with the State-of-the-art Methods

In this section, we compare our approach with some SOTA methods on SYSU-MM01 and RegDB. The results are shown in Table 3 and Table 4, respectively. The compared methods include Zero-Pad [13], D^2RL [6], cmGAN [23], HI-CMD [15], HPILN [24], AlignGAN [5], DDAG [8], AGW [7], DGTL [25], cm-SSFT [26], MPMN [27], MSA [28], SFANet [29], HAT [30], HcTri [20], MPANet [9]. The results of the comparison method are directly taken from the original article, where "–" means that the corresponding result is not reported in the corresponding article.

As can be seen from Table 3, on the SYSU-MM01 dataset, our method improves the Rank1 and mAP of both single and multiple searches in all search mode compared to the highest-performing MPANet. In the indoor search mode,

Table 4. Comparison with state-of-the-art methods on the REGDB dataset with different query settings. 1st and 2nd best results are marked in red and blue, respectively.

Method	R1	R10	R20	mAP
Visible to infrared				
Zero-Pad [13]	17.75	34.21	44.35	18.90
D^2RL [6]	43.40	–	–	44.10
AlignGAN [5]	57.90	–	–	53.60
DDAG [8]	69.34	86.19	91.49	63.46
AGW [7]	70.05	–	–	66.37
HAT [30]	71.83	87.16	92.16	67.56
cm-SSFT [26]	72.30	–	–	72.90
SFANet [29]	76.31	91.02	94.27	68.00
MPANet [9]	83.70	–	–	80.90
DGTL [25]	83.92	–	–	73.78
MPMN [27]	86.56	96.68	98.28	82.91
HcTri [20]	91.05	97.16	98.57	83.28
Ours	85.79	99.80	100.00	81.91
Infrared to visible				
Zero-Pad [13]	16.63	34.68	44.25	17.82
AlignGAN [5]	56.30	–	–	53.40
DDAG [8]	68.06	85.15	90.31	61.80
HAT [30]	70.02	86.45	91.61	66.30
SFANet [29]	70.15	85.24	89.27	63.77
DGTL [25]	81.59	–	–	71.65
MPANet [9]	82.80	–	–	80.70
MPMN [27]	84.62	95.51	97.33	79.49
HcTri [20]	89.30	96.41	98.16	81.46
Ours	86.33	99.41	99.80	82.10

the mAP of only single search is slightly lower than MPANet by 0.08, and the remaining Rank1 and mAP are 1.03 higher than MPANet on average.

Table 4 shows the comparison results on the RegDB dataset. It can be seen that the performance of our method reaches a high level compared to current mainstream methods. Among them, the mAP, Rank10, and Rank20 of our method reach the highest level under the search from infrared to visible. HcTri reaches the highest Rank1 and mAP on RegDB. This is because the HcTri is more focused on extracting spatial features. Meanwhile, in the RegDB dataset, there are many image pairs with one-to-one correspondence of spatial locations. Therefore HcTri network is more advantageous. In the SYSU-MM01 dataset, which is more in line with realistic scenarios, the results of our method are much higher than those of HcTri.

5 Conclusion

In this paper, we propose a novel SCFNet for VI-ReID tasks. It mines the feature information of the person in spatial and channel dimensions. To motivate the model to learn color-independent features, we use a random channel enhancement method. Also, a heterocentric sample loss optimization network training process is introduced to make the person feature more compact and distinguishable. Many experiments were conducted on SYSU-MM01 and RegDB to demonstrate the effectiveness of our proposed method. In the future, we will also explore the generalizability of the method and its performance in single-modality person ReID.

Acknowledgements. This work was supported by the Project of NSFC (Grant No. U1908214, 61906032), Special Project of Central Government Guiding Local Science and Technology Development (Grant No. 2021JH6/10500140), the Program for Innovative Research Team in University of Liaoning Province (LT2020015), the Support Plan for Key Field Innovation Team of Dalian(2021RT06), the Science and Technology Innovation Fund of Dalian (Grant No. 2020JJ25CY001), the Support Plan for Leading Innovation Team of Dalian University (Grant No. XLJ202010), the Fundamental Research Funds for the Central Universities (Grant No. DUT21TD107), Dalian University Scientific Research Platform Project (No. 202101YB03).

References

1. Sun, Y., Zheng, L., Li, Y., Yang, Y., Tian, Q., Wang, S.: Learning part-based convolutional features for person re-identification. IEEE Trans. Pattern Anal. Mach. Intell. **43**(3), 902–917 (2019). https://doi.org/10.1109/TPAMI.2019.2938523
2. Wang, G., Yuan, Y., Chen, X., Li, J., Zhou, X.: Learning discriminative features with multiple granularities for person re-identification. In: Proceedings of the 26th ACM International Conference on Multimedia, pp. 274–282 (2018). https://doi.org/10.1145/3240508.3240552

3. Xia, B.N., Gong, Y., Zhang, Y., Poellabauer, C.: Second-order non-local attention networks for person re-identification. In: Proceedings of the IEEE/CVF International Conference on Computer Vision, pp. 3760–3769 (2019). https://doi.org/10.1109/ICCV.2019.00386
4. Zheng, L., Huang, Y., Lu, H., Yang, Y.: Pose-invariant embedding for deep person re-identification. IEEE Trans. Image Process. **28**(9), 4500–4509 (2019). https://doi.org/10.1109/TIP.2019.2910414
5. Wang, G., Zhang, T., Cheng, J., Liu, S., Yang, Y., Hou, Z.: RGB-infrared cross-modality person re-identification via joint pixel and feature alignment. In: Proceedings of the IEEE/CVF International Conference on Computer Vision, pp. 3623–3632 (2019). https://doi.org/10.1109/ICCV.2019.00372
6. Wang, Z., Wang, Z., Zheng, Y., Chuang, Y.Y., Satoh, S.: Learning to reduce dual-level discrepancy for infrared-visible person re-identification. In: Proceedings of the IEEE/CVF Conference on Computer Vision and Pattern Recognition, pp. 618–626 (2019). https://doi.org/10.1109/CVPR.2019.00071
7. Ye, M., Shen, J., Lin, G., Xiang, T., Shao, L., Hoi, S.C.: Deep learning for person re-identification: a survey and outlook. IEEE Trans. Pattern Anal. Mach. Intell. **44**(6), 2872–2893 (2021). https://doi.org/10.1109/TPAMI.2021.3054775
8. Ye, M., Shen, J., J. Crandall, D., Shao, L., Luo, J.: Dynamic dual-attentive aggregation learning for visible-infrared person re-identification. In: Vedaldi, A., Bischof, H., Brox, T., Frahm, J.-M. (eds.) ECCV 2020. LNCS, vol. 12362, pp. 229–247. Springer, Cham (2020). https://doi.org/10.1007/978-3-030-58520-4_14
9. Wu, Q., et al.: Discover cross-modality nuances for visible-infrared person re-identification. In: Proceedings of the IEEE/CVF Conference on Computer Vision and Pattern Recognition, pp. 4330–4339 (2021). https://doi.org/10.1109/CVPR46437.2021.00431
10. Ye, M., Lan, X., Wang, Z., Yuen, P.C.: Bi-directional center-constrained top-ranking for visible thermal person re-identification. IEEE Trans. Inf. Forensics Secur. **15**, 407–419 (2019). https://doi.org/10.1109/TIFS.2019.2921454
11. Zhu, Y., Yang, Z., Wang, L., Zhao, S., Hu, X., Tao, D.: Hetero-center loss for cross-modality person re-identification. Neurocomputing **386**, 97–109 (2020). https://doi.org/10.1016/j.neucom.2019.12.100
12. Li, W., Qi, K., Chen, W., Zhou, Y.: Unified batch all triplet loss for visible-infrared person re-identification. In: 2021 International Joint Conference on Neural Networks (IJCNN), pp. 1–8. IEEE (2021). https://doi.org/10.1109/IJCNN52387.2021.9533325
13. Wu, A., Zheng, W.S., Yu, H.X., Gong, S., Lai, J.: RGB-infrared cross-modality person re-identification. In: Proceedings of the IEEE International Conference on Computer Vision, pp. 5380–5389 (2017). https://doi.org/10.1109/ICCV.2017.575
14. Wang, G.A., et al.: Cross-modality paired-images generation for RGB-infrared person re-identification. In: Proceedings of the AAAI Conference on Artificial Intelligence, vol. 34, pp. 12144–12151 (2020). https://doi.org/10.1609/aaai.v34i07.6894
15. Choi, S., Lee, S., Kim, Y., Kim, T., Kim, C.: Hi-CMD: hierarchical cross-modality disentanglement for visible-infrared person re-identification. In: Proceedings of the IEEE/CVF Conference on Computer Vision and Pattern Recognition, pp. 10257–10266 (2020). https://doi.org/10.1109/CVPR42600.2020.01027
16. Hao, Y., Li, J., Wang, N., Gao, X.: Modality adversarial neural network for visible-thermal person re-identification. Pattern Recogn. **107**, 107533 (2020). https://doi.org/10.1016/j.patcog.2020.107533

17. Liu, H., Cheng, J., Wang, W., Su, Y., Bai, H.: Enhancing the discriminative feature learning for visible-thermal cross-modality person re-identification. Neurocomputing **398**, 11–19 (2020). https://doi.org/10.1016/j.neucom.2020.01.089
18. Huang, N., Liu, J., Zhang, Q., Han, J.: Exploring modality-shared appearance features and modality-invariant relation features for cross-modality person re-identification. arXiv preprint arXiv:2104.11539 (2021). https://doi.org/10.48550/arXiv.2104.11539
19. Zhang, C., Liu, H., Guo, W., Ye, M.: Multi-scale cascading network with compact feature learning for RGB-infrared person re-identification. In: 2020 25th International Conference on Pattern Recognition (ICPR), pp. 8679–8686. IEEE (2021). https://doi.org/10.1109/ICPR48806.2021.9412576
20. Liu, H., Tan, X., Zhou, X.: Parameter sharing exploration and hetero-center triplet loss for visible-thermal person re-identification. IEEE Trans. Multimedia **23**, 4414–4425 (2020). https://doi.org/10.1109/TMM.2020.3042080
21. Sun, Y., Zheng, L., Yang, Y., Tian, Q., Wang, S.: Beyond part models: person retrieval with refined part pooling (and a strong convolutional baseline). In: Proceedings of the European Conference on Computer Vision (ECCV), pp. 480–496 (2018). https://doi.org/10.48550/arXiv.1711.09349
22. Nguyen, D.T., Hong, H.G., Kim, K.W., Park, K.R.: Person recognition system based on a combination of body images from visible light and thermal cameras. Sensors **17**(3), 605 (2017). https://doi.org/10.3390/s17030605
23. Dai, P., Ji, R., Wang, H., Wu, Q., Huang, Y.: Cross-modality person re-identification with generative adversarial training. In: IJCAI, vol. 1, p. 6 (2018). https://doi.org/10.24963/ijcai.2018/94
24. Zhao, Y.B., Lin, J.W., Xuan, Q., Xi, X.: HPILN: a feature learning framework for cross-modality person re-identification. IET Image Proc. **13**(14), 2897–2904 (2019). https://doi.org/10.1049/iet-ipr.2019.0699
25. Liu, H., Chai, Y., Tan, X., Li, D., Zhou, X.: Strong but simple baseline with dual-granularity triplet loss for visible-thermal person re-identification. IEEE Signal Process. Lett. **28**, 653–657 (2021). https://doi.org/10.1109/LSP.2021.3065903
26. Lu, Y., et al.: Cross-modality person re-identification with shared-specific feature transfer. In: Proceedings of the IEEE/CVF Conference on Computer Vision and Pattern Recognition, pp. 13379–13389 (2020). https://doi.org/10.1109/CVPR42600.2020.01339
27. Wang, P., et al.: Deep multi-patch matching network for visible thermal person re-identification. IEEE Trans. Multimedia **23**, 1474–1488 (2020). https://doi.org/10.1109/TMM.2020.2999180
28. Miao, Z., Liu, H., Shi, W., Xu, W., Ye, H.: Modality-aware style adaptation for RGB-infrared person re-identification. In: IJCAI, pp. 916–922 (2021). https://doi.org/10.24963/ijcai.2021/127
29. Liu, H., Ma, S., Xia, D., Li, S.: SFANet: a spectrum-aware feature augmentation network for visible-infrared person reidentification. IEEE Trans. Neural Netw. Learn. Syst. (2021). https://doi.org/10.1109/TNNLS.2021.3105702
30. Ye, M., Shen, J., Shao, L.: Visible-infrared person re-identification via homogeneous augmented tri-modal learning. IEEE Trans. Inf. Forensics Secur. **16**, 728–739 (2020). https://doi.org/10.1109/TIFS.2020.3001665

Multi-modal Segment Assemblage Network for Ad Video Editing with Importance-Coherence Reward

Yunlong Tang[1,2], Siting Xu[1], Teng Wang[1,2], Qin Lin[2], Qinglin Lu[2], and Feng Zheng[1(✉)]

[1] Southern University of Science and Technology, Shenzhen, China
{tangyl2019,xust2019,wangt2020}@mail.sustech.edu.cn, f.zheng@ieee.org
[2] Tencent Inc., Shenzhen, China
{angelqlin,qinglinlu}@tencent.com

Abstract. Advertisement video editing aims to automatically edit advertising videos into shorter videos while retaining coherent content and crucial information conveyed by advertisers. It mainly contains two stages: video segmentation and segment assemblage. The existing method performs well at video segmentation stages but suffers from the problems of dependencies on extra cumbersome models and poor performance at the segment assemblage stage. To address these problems, we propose M-SAN (Multi-modal Segment Assemblage Network) which can perform efficient and coherent segment assemblage task end-to-end. It utilizes multi-modal representation extracted from the segments and follows the Encoder-Decoder Ptr-Net framework with the Attention mechanism. Importance-coherence reward is designed for training M-SAN. We experiment on the Ads-1k dataset with 1000+ videos under rich ad scenarios collected from advertisers. To evaluate the methods, we propose a unified metric, Imp-Coh@Time, which comprehensively assesses the importance, coherence, and duration of the outputs at the same time. Experimental results show that our method achieves better performance than random selection and the previous method on the metric. Ablation experiments further verify that multi-modal representation and importance-coherence reward significantly improve the performance. Ads-1k dataset is available at: https://github.com/yunlong10/Ads-1k.

Keywords: Ad video editing · Segment assemblage · Advertisement dataset · Multi-modal · Video segmentation · Video summarization

1 Introduction

With the boom of the online video industry, video advertising has become popular with advertisers. However, different online video platforms have different requirements for the content and duration of ad videos. It is time-consuming

---

Supplementary Information The online version contains supplementary material available at https://doi.org/10.1007/978-3-031-26284-5_34.

L. Wang et al. (Eds.): ACCV 2022, LNCS 13842, pp. 560–576, 2023.
https://doi.org/10.1007/978-3-031-26284-5_34

and laborious for advertisers to edit their ad videos into a variety of duration tailored to the diverse requirements, during which they have to consider which part is important and whether the result is coherent. Therefore, it is of great importance to automatically edit the ad videos to meet the requirements of duration, and the edited videos should be coherent and retain informative content.

Ad video editing is a task aiming to edit an ad video into its shorter version to meet the duration requirements, ensuring coherence and avoiding losing important ad-related information. Video segmentation and segment assemblage are the two main stages in ad video editing task [20], as Fig. 1 shows. An ad video will be cut into several segments with a small duration during the video segmentation stage. At the segment assemblage stage, the output will be produced by selecting and assembling a subset of the input segments of the source ad video. The key to video segmentation is to preserve the local semantic integrity of each video segment. For instance, a complete sentence of a speech or caption in the source video should not be split into two video segments. Existing method [20] has achieved this by aligning shots, subtitles, and sentences to form the segments. However, at the segment assemblage stage, the only pioneer work [20] suffers the following problems: (1) To calculate the individual importance of each segment and the coherence between segments, extra models are required to perform video classification [14,21] and text coherence prediction [7], which is inefficient during inference. (2) Without globally modeling the context of videos, graph-based search adopted by [20] produces results with incoherent segments or irrelevant details.

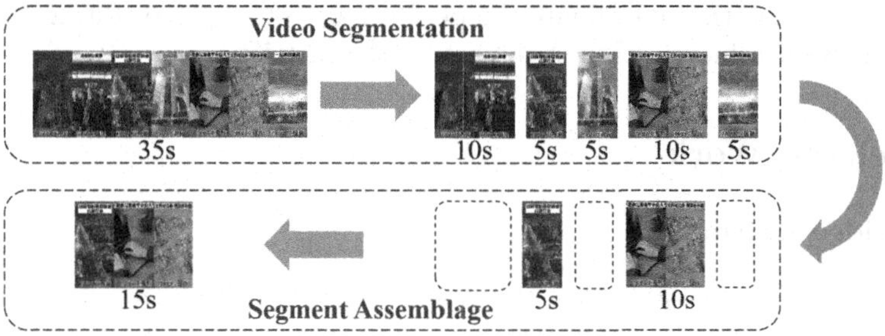

Fig. 1. The two stages of ad video editing: video segmentation and segment assemblage.

To tackle these problems, we propose an end-to-end Multi-modal Segment Assemblage Network (M-SAN) for accurate and efficient segment assemblage. It is free of extra cumbersome models during inference and strikes a better balance between importance and coherence. Specifically, we obtain segments at the video segmentation stage by boundary detection and alignment. Different from daily life videos, ad videos usually have sufficient multi-modal content like speech and caption, which contain abundant video semantics [13]. Therefore, pretrained unimodal models are applied to respectively extract the representation of shots, audios, and sentences, which are concatenated together yielding a multi-modal representation of segments. During the assemblage stage, we adopt a pointer network with RNN-based decoder to improve the temporal dependency between selected segments. Importance-coherence reward is designed for training M-SAN with Policy Gradient [35]. Importance reward measures the amount of important ad-related information contained in the output. Coherence reward measures the text coherence between every two adjacent selected segments, which is computed as the mean of PPL (perplexity) [30] of sentences generated by concatenating the two texts extracted from adjacent selected segments.

To evaluate our methods, we propose the new metric, Imp-Coh@Time, which takes the importance, coherence, and duration of the outputs into consideration at the same time instead of evaluating importance or coherence respectively. We experiment on the Ads-1k dataset with 1000+ ad videos collected from advertisers. Experimental results show that our method achieves better performance than random selection and the previous method [20] on the metric. Ablation experiments further verify that multi-modal representation and importance-coherence reward significantly improve the performance.

Our work mainly focuses on segment assemblage in ad video editing, and its main contributions can be summarized as follows:

- We propose M-SAN to perform segment assemblage efficiently and improve the result of ad video editing, without relying on an extra model when inference.
- We propose importance-coherence reward and train M-SAN with policy gradient to achieve a better trade-off between importance and coherence.
- We collect the dataset Ads-1k with 1000+ ad videos and propose Imp-Coh@Time metrics to evaluate the performance of ad video editing methods. Our M-SAN achieves state of the art on the metrics.

2 Related Work

2.1 Video Editing

There are three main categories of automated video editing [20]. They're video summarization, video highlight detection, and task-specific automated editing.

Video Summarization. The most relevant task to ad video editing is video summarization. It is a process that extracts meaningful shots or frames from video by analyzing structures of the video and time-space redundancy in an automatic or semi-automatic way. To perform video summarization, a load of work focuses on supervised learning based on frames [8,11,23,34], shots [9,16,39], and clips [19]. Other than these works, DSN [40] is the first proposed unsupervised video summarization model training with diversity-representativeness reward by policy gradient. Without utilizing annotation, the method reached fully unsupervised. Our rewards design mainly refers to [40].

Video Highlight Detection. Learning how to extract an important segment from videos is the main motivation we focus on video highlight detection. [31] proposed frameworks that exploit users' previously created history. Edited videos created by users are utilized in [38] to achieve highlight detection in an unsupervised way since human-edited videos tend to show more interesting or important scenes. [36] presents an idea that shorter videos tend to be more likely to be selected as highlights. Combining the above ideas, we exploit the ability of MSVM [20] which extracts potential selling points in segments that tend to be of higher importance when performing assemblage.

Task-Specific Automated Editing. Videos can be presented in various forms such as movies, advertisements, sports videos, etc. To extract a short video from a long video also exists in specific scenarios of movies [2,15,37]. [15] pointed out that a movie has the trait that its computational cost is high. Also, sports videos have been explored to extract highlights [25,26] for sports having the characteristic that they are high excitement. Advertisements are rich in content and they vary in duration among different platforms [20].

2.2 Text Coherence Prediction and Evaluation

Text information is extracted in the stage of video segmentation. When assembling segments, texts that are concatenated act as a reference to coherence. [4] proposed narrative incoherence detection, denoted semantic discrepancy exists causes incoherence. In [20], next sentence prediction (NSP) [7] is exploited to assess coherence. Although our work is inspired by their thoughts, we use perplexity as the coherence reward and metric to evaluate sentence coherence.

2.3 Neural Combinatorial Optimization

Combinatorial optimization problem is a problem that gets extremum in discrete states. Common problems like Knapsack Problem (KP), Travelling Salesman Problem (TSP), and Vehicular Routing Problem (VRP) belong to combinatorial optimization problem. Pointer Network (Ptr-Net) is proposed in [33] and performs better than heuristic algorithm in solving TSP. Later, Ptr-Net has been exploited with reinforcement learning [3,6,12,18,29]. In [10], Ptr-Net is used to solve the length-inconsistency problem in video summarization. In our network, Ptr-Net is used to model the video context and select the tokens from the input sequence as output.

3 Method

3.1 Problem Formulation

Given a set of N segments of ad video $S = \{s_i\}_{1\leq i\leq M}$, our goal is to select a subset $A = \{a_i\}_{1\leq i\leq N} \subseteq S$ which can be combined into the output video so that it can take the most chance to retain the important information and be coherent as well as meeting the requirements of duration. We denote the ad importance of the segment a_i as $imp(a_i)$, the coherence as $coh(a_i, a_j)$ and the duration as

$dur(a_i)$. Overall the task of segment assemblage can be regarded as a constrained combinatorial optimization problem and is defined formally as follows:

$$\max_{A \subseteq S} \sum_{a_i \in A} imp(a_i) + \sum_{a_i \prec a_j} coh(a_i, a_j),$$
$$s.t.\ T_{min} \leq \tau(A) \leq T_{max},\ and\ \forall a_i, a_j \in A, a_i \neq a_j, \tag{1}$$

where

- T_{min} and T_{max} are the lower and upper bound of requirement duration,
- $\tau(A) = \sum_{a_i \in A} dur(a_i)$,
- $a_i \prec a_j$ is defined as $(a_i, a_j \in A) \wedge (i < j) \wedge (\forall a_k \in A,\ k < i \vee k > j)$.

This is an NP-hard problem that can not be solved in polynomial time. Instead of utilizing graph modeling and optimization [20] to search for an optimal solution, we adopt a neural network with pointer [33] that follows the framework of neural combinatorial optimization to optimize the objective directly.

3.2 Architecture

The architecture of M-SAN is shown as Fig. 2. It incorporates a multi-modal video segmentation module [20] (MVSM), multi-modal representation extraction module (MREM) and assemblage module (AM). To preserve the local semantic integrity of each segment, we adopt MVSM to obtain video segments with reasonable boundaries. With ASR and OCR, MVSM also captures the texts from each segment. Given the segments and the corresponding texts, MREM extracts the segment-level representations of shots, audios, and texts, which are further jointed into the multi-modal representations. AM utilizes these representations to model the context of video and make decisions by a pointer network [33].

Video Segmentation Module. At the video segmentation stage, we first apply MVSM [20] to obtain the segments of each input video. MVSM splits a video into the video track and audio track and extracts shots $\{v_i\}$, audios $\{\alpha_i\}$, ASR and OCR results $\Omega_i = \{\omega_i\}$ (a sentence with words ω_i) from video to generate the boundaries of the content in each modality. The boundaries of audio space and textual space are first merged to form the joint space, followed by merging the boundaries of visual space and joint space to yield the final segments set $\{s_i\}$. The segment $s_i = (\{v_{p_0}, ..., v_{q_0}\}, \{\alpha_{p_1}, ..., \alpha_{q_1}\}, \Omega_i)$ preserves the integrity of local atomic semantic, where $p_{(\cdot)} < q_{(\cdot)}$.

Multi-modal Representation Extraction Module. MREM integrates three kinds of representation extractors: pre-trained Swin-Transformer [24], Vggish [14] and BERT [7] models. The Swin-Transformer extracts the visual representations $\{\tilde{v}_{p_0}, ..., \tilde{v}_{q_0}\}$ from shots $\{v_{p_0}, ..., v_{q_0}\}$ in each segment $\{s_i\}$, and a segment-level visual representation is computed as the mean of single shot-level representaton. Vggish and BERT model extract audio representations $\{\tilde{\alpha}_{p_1}, ..., \tilde{\alpha}_{q_1}\}$ and

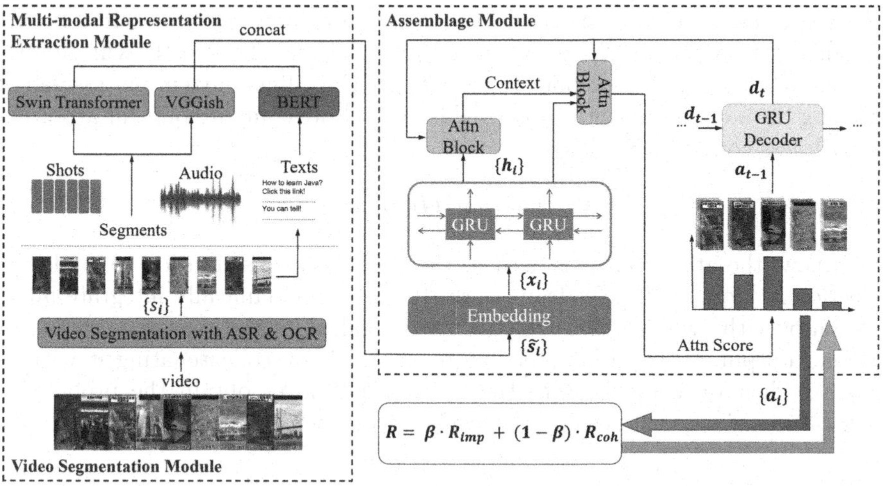

Fig. 2. The Architecture of M-SAN

text representations $\tilde{\Omega}_i$, respectively. Similarly, a segment-level audio representation $\bar{\alpha}_i$ is given by the mean of $\{\tilde{\alpha}_{p_1}, ..., \tilde{\alpha}_{q_1}\}$. Since each segment contains at most one sentence, it is $\tilde{\Omega}_i$ that is the segment-level text representation. These three modalities will be jointed by concatenating directly to yield the final multi-modal representation of segment $\tilde{s}_i = [\bar{v}_i, \bar{\alpha}_i, \tilde{\Omega}_i]^T$.

Assemblage Module. The output of an Encoder-Decoder Ptr-Net [33] is produced by iteratively copying an input item that is chosen by the pointer [27], which is quite suitable for segment assemblage task. Therefore, our assemblage module (AM) follows this framework.

The encoder integrates a linear embedding layer and a bi-directional GRU. To enhance interactions between modalities, the linear embedding layer will perform preliminary a fusion of the three modalities and produce embedding token x_i corresponding to segment s_i. To enhance the interaction between segments and model the context of the whole video, a Bi-GRU is adopted to further embed the tokens:

$$H = GRU_e(X), \tag{2}$$

where $X = [x_1, ..., x_M]$, $H = [h_1, ..., h_M]$, and the hidden state h_i is the context embedding for segment s_i.

Given the output of encoder H and X, the GRU [5] decoder with Attention mechanism [1] predicts the probability distribution of segment to be selected from S at every time-step t to get the result A:

$$p_\theta(A|S) = \prod_{t=1}^{N} p_\theta(a_t|a_{1:t}, S) = \prod_{t=1}^{N} p_\theta(a_t|a_{1:t-1}, H, X) = \prod_{t=1}^{N} p_\theta(a_t|a_{t-1}, d_t), \tag{3}$$

where θ is the learnable parameter, and d_t is the hidden state computed by the decoder at time-step t. With d_t as the query vector, the decode will glimpse [32] the whole output of encoder H to compute the bilinear attention. Instead of utilizing the additive attention adopted in [10,12], we compute bilinear attention μ_t with less computational cost:

$$\mu_t = Softmax(H^T W_{att_1} d_t) \ . \tag{4}$$

Until now, the attention μ_t is used as the probability distribution to guide the selection in most of Ptr-Net framework [10,12]. To dynamically integrate information over the whole video [28], we further calculate the context vector c_t of encoder output and update query vector d_t to $\tilde{d}_t$ by concatenating it with c_t (Eq. 5). Then we compute attention a second time to obtain the probability distribution of segment selection at current time-step t (Eq. 6).

$$c_t = H\mu_t = \sum_{m=1}^{M} \mu_t^{(m)} h_m, \quad \tilde{d}_t = \begin{bmatrix} d_t \\ c_t \end{bmatrix}, \tag{5}$$

$$\tilde{\mu}_t = Softmax(MH^T W_{att_2} \tilde{d}_t), \tag{6}$$

$$p_\theta(a_t | a_{t-1}, d_t) = \tilde{\mu}_t, \tag{7}$$

where M can mask the position i corresponding to selected segment a_i to $-\infty$. The segment at this position will be not selected at later time-steps, since $Softmax(\cdot)$ modifies the corresponding probability to 0. Finally, the segment selected at time-step t will be sampled from the distribution:

$$a_t \sim p_\theta(a_t | a_{t-1}, d_t) \ . \tag{8}$$

If the sum of duration of selected segments $\tau(A)$ exceeds the tolerable duration limit $[T_{min}, T_{max}]$, the current and following segments selected will be replaced by [EOS] token [17].

3.3 Reward Design

Importance Reward. To extract important parts from original ad videos, rewards related to the importance of selected segments should be fed back to the network during training. We design importance reward R_{imp}:

$$R_{imp} = \frac{1}{|A|} \sum_{a_i \in A} imp(a_i), \tag{9}$$

$$imp(a_i) = \frac{1}{|L_{a_i}|} \sum_{\ell \in L_{a_i}} w_l \cdot \ell, \tag{10}$$

where A is the set of selected segments, L_{a_i} stands for the total number of labels of the selected segment, $\ell \in \{1, 2, 3, 4\}$ is narrative techniques label hierarchy, and w_l is the weight of one label. According to Eq. 9, the importance of a single segment is the weighted average of its annotated labels' weight. The labels listed in the supplementary are divided into four groups with four levels ranging from 1 to 4 according to their ad-relevance. We compute the importance reward of one output as the mean of the importance of single segments.

Coherence Reward. Importance reward mainly focuses on local visual dymantics within the single segment, neglecting temporal relationships between adjacent segments. We introduce a linguistic coherence reward to improve the fluency of caption descriptions of two segments.

Specifically, the extracted texts from adjacent segments are combined in pairs while retaining the original order, that is, preserving the same order of the original precedence. Then we compute the perplexity [30] (PPL) for each combined sentence by GPT-2 pre-trained on 5.4M advertising texts:

$$\begin{aligned} PPL(\Omega_1, \Omega_2) &= p(\omega_1^{(1)}, \omega_2^{(1)}, ..., \omega_m^{(1)}, \omega_1^{(2)}, \omega_2^{(2)}, ..., \omega_n^{(2)})^{-1/(m+n)} \\ &= \sqrt[m+n]{\prod_{i=1}^{m+n} \frac{1}{p(\omega_i | \omega_1, \omega_2, ..., \omega_{i-1})}}, \end{aligned} \tag{11}$$

where $\Omega_i = (\omega_1^{(i)}, ..., \omega_m^{(i)})$ is one sentence with m words. PPL reflects the *incoherence* of a sentence. We also maintain a PPL map to store the PPL of sentences in pairs as Fig. 3 shown.

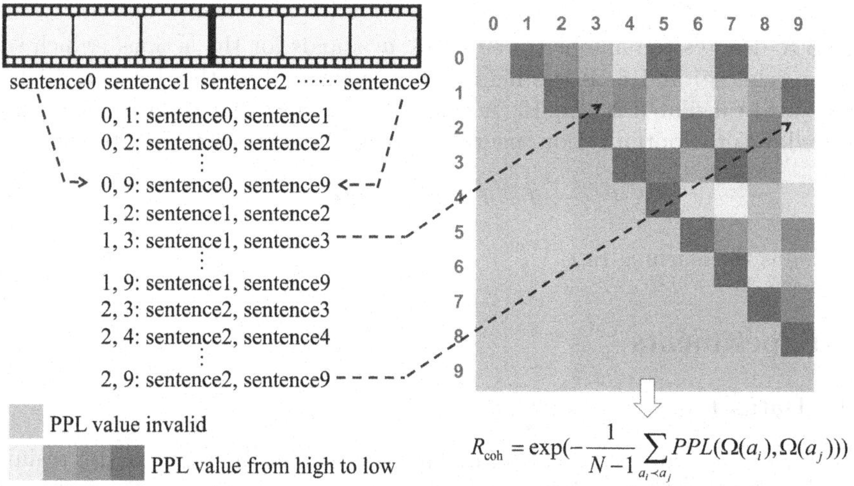

Fig. 3. PPL map and coherence reward

The grey parts represent invalid PPL values, which means sentence pairs that violate the original orders have no valid PPL value. The rest means valid PPL values vary from high to low as the color of red deepens. Since the smaller the PPL, the better the text coherence, the coherence reward is computed with a transfer function:

$$R_{coh} = exp(-\frac{1}{N-1} \sum_{a_i \prec a_j} PPL(\Omega(a_i), \Omega(a_j))), \tag{12}$$

where $N = |A|$ is the total number of selected segments, $\Omega(a_i)$ is the sentence recognized from segment a_i. The $exp(\cdot)$ ensures the R_{coh} at the same order of magnitude with R_{imp}.

Importance-Coherence Reward. To balance the importance and coherence of the selection, we make R_{imp} and R_{coh} complement each other and jointly guide the learning of M-SAN:

$$R = \beta \cdot R_{imp} + (1-\beta) \cdot R_{coh}, \tag{13}$$

where the coefficient of reward β is a hyperparameter.

3.4 Training

Policy gradient (a.k.a. REINFORCE algorithm [35]) is adopted during training:

$$\nabla_\theta J(\theta) \approx \frac{1}{K} \sum_{k=1}^{K} \sum_{t=1}^{N} (R(A^{(k)}) - b^{(k)}) \nabla_\theta \log p_\theta(a_t^{(k)} | a_{t-1}^{(k)}, d_t^{(k)}), \tag{14}$$

where K denotes the number of episodes, a_t stands for the actions (which segment to choose) and d_t is the hidden state estimated by the decoder. $R(A)$ is the reward calculated by Eq. 13. A baseline value b so that the variance can be reduced. For the optimization, neural network parameter θ is updated as:

$$\theta = \theta - \eta \nabla_\theta(-J(\theta)), \tag{15}$$

where η is the learning rate.

4 Experiments

4.1 Dataset

In [20], there are only 50 ad videos used in experiments. To obtain better training and evaluation, we collect 1000+ ad videos from the advertisers to form the Ads-1k dataset. There are 942 ad videos for training and 99 for evaluation in total. However, the annotation methods of the training set and test set are somehow different. Instead of preparing the ground-truth for each data, we annotate each video with multi-labels shown in the supplementary.

Table 1. Dataset statistics. N_{seg} and N_{label} are respectively the average number of segments and labels of each video. D_{seg} and D_{video} are the average duration of a segment and a video, respectively.

Dataset	N_{seg}	$D_{seg}(s)$	$D_{video}(s)$	N_{label}
Training Set	13.90	2.77	34.60	30.18
Test Set	18.81	1.88	34.21	35.77
Overall	14.37	2.68	35.17	30.71

We counted the average segment number for videos, the average length of segments in seconds, the average duration of each video in seconds, and the average label number for each video for the training set, test set, and the whole dataset respectively. The results are shown as Table 1. Besides, the number of annotated segment pairs and the proportion are counted. The number of *coherent*, *incoherent*, and *uncertain* pairs are 6988, 9551, and 2971, occupying 36%, 49%, and 15%, respectively.

4.2 Metric

Imp@T. We define the score of ad importance given the target duration T as follows:

$$Imp@T = \frac{1}{|A|} \sum_{a_i \in A} imp(a_i) \cdot \mathbb{I}[c_1 \cdot T \leq \tau(A) \leq c_2 \cdot T], \tag{16}$$

where A is the set of selected segments, $imp(a_i)$ is defined by Eq. 10. $\mathbb{I}(\cdot)$ is the indicator function. c_1 and c_2 are two constant that produced a interval based on given target duration T. We set $c_1 = 0.8$ and $c_2 = 1.2$, since a post-processing of 0.8× slow down 1.2× or fast forward can resize the result close to the target T without distortion in practice. Take $T = 10$ as example, the interval will be $[8, 12]$, which means if the duration of result $\tau(A) = \sum_{a_i \in A} dur(a_i) \in [8, 12]$ then this result is valid and gain the score.

Coh@T. The coherence score given the target duration T is defined as follows:

$$Coh@T = \frac{1}{|A| - 1} \sum_{a_i \prec a_j} coh(a_i, a_j) \cdot \mathbb{I}[c_1 \cdot T \leq \tau(A) \leq c_2 \cdot T], \tag{17}$$

where the $coh(a_i, a_j)$ is the coherence small score between the text of segment i and the text of segment j. With the annotation for coherence on test set, we can score the results produced by our models. When scoring, for each combination of consecutive two segments i and j in output, if it is in the *coherent* set, the $coh(a_i, a_j)$ will be 1. If it is in the *incoherent* set, the $coh(a_i, a_j)$ will be 0. Otherwise, it is in the *uncertain* set, the $coh(a_i, a_j)$ will be 0.5.

Imp-Coh@T. The overall score is defined as follows:

$$ImpCoh@T = \frac{Imp@T}{|A|-1} \cdot \sum_{a_i \prec a_j} coh(a_i, a_j), \tag{18}$$

where Imp@T is defined in Eq. 16. The score reflects the ability of trade-off among importance, coherence and total duration.

4.3 Implementation Details

Baselines. Besides SAM (Segments Assemblage Module) proposed in [20], we also adopt two random methods to perform the segment assemblage task, given the segments produced by MVSM [20].

- Given the set of input segments $S = \{s_i\}_{1 \leq i \leq M}$ and the target time T, *Random* will first produce a random integer $1 \leq r \leq M$. Then it will randomly pick up r segments from S to get the result $A = \{a_i\}_{1 \leq i \leq r}$, regardless of the target time T.
- Given $S = \{s_i\}_{1 \leq i \leq M}$ and T, *Random-Cut* randomly picks up segments from S and add to A until $c_1 \cdot T \leq \tau(A) \leq c_2 \cdot T$, ensuring satisfying the requirement of duration.
- Given $S = \{s_i\}_{1 \leq i \leq M}$ and T, the *SAM* [20] will utilize an extra model to perform video classification or named entity recognition to obtain some labels for each segment and compute an importance score for each segment. It also utilizes an extra BERT [7] to perform next sentence prediction (NSP) to compute a coherence score for each pair of segments. Then the segments $\{s_i\}_{1 \leq i \leq M}$ will be modeled as a graph with $|S|$ nodes and $|S|(|S|-1)$ edges, where the weight of nodes are the importance score and the weight of edges are the coherence score. DFS with pruning is then adopted to search on the graph to collect a set that maximizes the sum of importance scores and coherence scores.

Parameters and Training Details. The Swin-Transformer [24] we used is the Large version with an output size of 1536. The 5 shots will be extracted to generate a visual representation every second of the video. The segment-level representation is computed as the mean of. The output size of BERT [7] and Vggish [14] are 768 and 128 respectively, and the sampling rate of Vggish is 5 every second to align with visual information. The dimension of $\tilde{s}_i$ is 2432. The linear embedding layer performs a projection from 2432 to 768. We optimize the sum reward $R = 0.5 \cdot R_{imp} + 0.5 \cdot R_{coh}$, where the R_{imp} and R_{coh} are given in Eqs. 9 and 12 with $\beta = 0.5$. The w_l in R_{imp} for all ℓ are 0.25. The optimizer we used is Adam [22]. More details can be found in supplementary.

4.4 Performance Comparison

We evaluated our M-SAN on the test set of Ads-1k with the three baselines mentioned above. The results are provided by Table 2. It shows that our M-SAN is state of the art on segment assemblage task given target duration $T = 10$

and $T = 15$. There is a significant improvement from Random to Random-Cut. Therefore, simply sticking to the time limit can improve performance by leaps and bounds. From Random-Cut to SAM, the improvement at $T = 10$ is also obvious, while the difference between them at $T = 15$ is not. This is probably because a longer duration budget forces Random-Cut to select more segments. The segment pairs as result have a greater chance of being coherent pairs.

Although SAM performs better than Random-Cut, its ability to trading-off between importance and coherence is still weak. M-SAN addresses the problem by trained with importance-coherence reward and achieves a better performance.

Table 2. Performance comparison results. Our M-SAN is state of the art on segment assemblage task.

	Imp-Coh@10			Imp-Coh@15		
	Imp	Coh	Overall	Imp	Coh	Overall
Random	10.68	12.57	7.92	17.04	20.49	12.71
Random-Cut	55.35	73.75	41.77	60.39	76.56	47.25
SAM [20]	72.54	86.55	65.97	63.97	78.58	58.09
M-SAN(ours)	**80.29**	**92.16**	**74.19**	**77.00**	**90.83**	**70.28**

4.5 Ablation Studies

Ablation of Modalities. We explore the effect of the representation from different modalities. Table 3 shows that incorporating the text or audio representation can both improve the overall score while incorporating the former have much effect. After leveraging all representations from three modalities, the overall scores increase significantly, which has 8.7 and 3.75 gains compared with utilizing visual information only on Imp-Coh@10 and Imp-Coh@15, respectively. Even though adding text representation to video-audio dual modalities hurts the Imp@15, other scores increase obviously, which demonstrates the significance of multi-modal representation.

Ablation of Reward and Glimpse. To verify the effectiveness of reward and glimpse (two-stage attention calculation), we design the following ablation experiments with target duration $T = 10$ and $T = 15$. The results in the Table 4 show that M-SAN gains a higher score on all metrics than the one without glimpse. Therefore, dynamically integrating information over the whole video by Glimpse can improve the performance. Similarly, we perform ablation on the rewards: importance-coherence reward (M-SAN), importance reward only, and coherence reward only. Results in Table 4 show that M-SAN trained with

Table 3. Ablation study on modalities

Modalities			Imp-Coh@10			Imp-Coh@15		
V	T	A	Imp	Coh	Overall	Imp	Coh	Overall
✓			78.43	89.59	65.49	74.48	87.41	66.53
✓	✓		78.57	89.73	71.04	76.22	89.85	68.56
✓		✓	79.49	91.49	72.72	**77.08**	90.36	69.91
✓	✓	✓	**80.29**	**92.16**	**74.19**	77.00	**90.83**	**70.28**

importance-reward only gained relatively low scores comparing the other two kinds of rewards. The one trained with coherence-reward only gained a higher score than the one with importance-reward. M-SAN trained with importance-coherence reward prominently performs better than the other two.

Table 4. Ablation study on glimpse and rewards.

	Imp-Coh@10			Imp-Coh@15		
	Imp	Coh	Overall	Imp	Coh	Overall
M-SAN	**80.29**	**92.16**	**74.19**	**77.00**	**90.83**	**70.28**
w/o glimpse	79.80	91.50	72.97	75.98	90.58	70.12
coh-rwd only	79.37	91.02	72.28	76.90	90.29	69.75
imp-rwd only	67.82	71.62	49.87	67.42	79.89	54.10

Analysis of Reward Ratio. To further figure out which reward ratio brings the optimal results, we experiment on $T = 10$ and $T = 15$, assigning 0.0/0.3/0.5/0.7 to β. Importance score, coherence score, and overall score results are shown in the line chart Fig. 4. The results at the $T = 10$ and $T = 15$ are presented by lines painted in blue and orange respectively. Given target duration $T = 10$ s, all three scores reach a peak at $\beta = 0.5$. Given target duration $T = 15$ s, the importance score and overall score reach a peak at $\beta = 0.3$, which is slightly higher than the score at $\beta = 0.5$. Coherence score reaches the highest point at $\beta = 0.5$. On the whole, the performance is relatively good at the condition of β=0.5.

4.6 Qualitative Analysis

One result is shown as Fig. 5. The source video is about a collectible game app with real mobile phones as the completion rewards. Its original duration is 36 s, and the target duration is 15 s. SAM [20] generates videos with too many foreshadowing parts, which exceeds the duration limitation. M-SAN produces a 15 s result and tends to select latter segments in source videos, which is reasonable

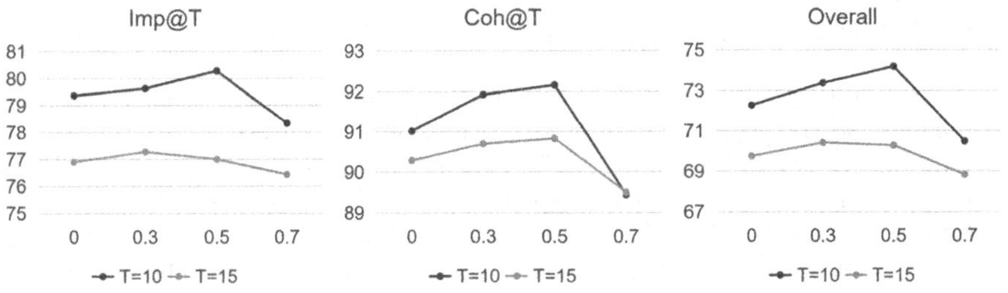

Fig. 4. Comparison results with $\beta = 0.0/0.3/0.5/0.7$. (Color figure online)

because key points usually appear in the latter part of the ad with the front part doing foreshadowing. The result of M-SAN first demonstrates using the app to get a new phone and then shows a scene where a new phone is packed, which emphasizes the rewards of completing the game. This verifies the M-SAN focuses on more informative segments and does better than SAM in duration control.

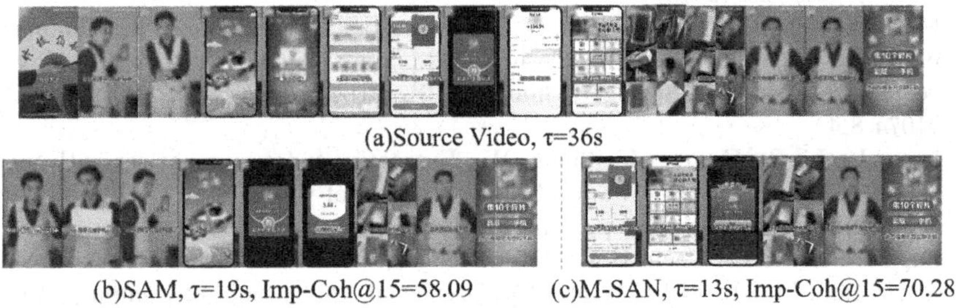

Fig. 5. Visualization. Source video and videos assembled by SAM and M-SAN given target duration $T = 15$ s. τ is the actual duration of result.

5 Conclusion

The two main stages of ad video editing are video segmentation and segment assemblage. Existing methods perform poorly at the segment assemblage stage. To improve the performance of segment assemblage, we proposed M-SAN to perform segment assemblage end-to-end. We also proposed importance-coherence reward based on the characteristics of ad and train M-SAN with policy gradient. We collected an ad video dataset with 1000+ ad videos and proposed Imp-Coh@Time metrics. Experimental results show the effectiveness of M-SAN and verify that multi-modal representation and importance-coherence reward bring a significant performance boost.

Acknowledgment. This work is supported by the National Natural Science Foundation of China under Grant No. 61972188 and 62122035.

References

1. Bahdanau, D., Cho, K., Bengio, Y.: Neural machine translation by jointly learning to align and translate. arXiv preprint arXiv:1409.0473 (2014)
2. Bain, M., Nagrani, A., Brown, A., Zisserman, A.: Condensed movies: story based retrieval with contextual embeddings. In: Proceedings of the Asian Conference on Computer Vision (2020)
3. Bello, I., Pham, H., Le, Q.V., Norouzi, M., Bengio, S.: Neural combinatorial optimization with reinforcement learning. arXiv preprint arXiv:1611.09940 (2016)
4. Cai, D., Zhang, Y., Huang, Y., Lam, W., Dolan, B.: Narrative incoherence detection. arXiv preprint arXiv:2012.11157 (2020)
5. Cho, K., et al.: Learning phrase representations using RNN encoder-decoder for statistical machine translation. arXiv preprint arXiv:1406.1078 (2014)
6. Deudon, M., Cournut, P., Lacoste, A., Adulyasak, Y., Rousseau, L.-M.: Learning heuristics for the TSP by policy gradient. In: van Hoeve, W.-J. (ed.) CPAIOR 2018. LNCS, vol. 10848, pp. 170–181. Springer, Cham (2018). https://doi.org/10.1007/978-3-319-93031-2_12
7. Devlin, J., Chang, M.W., Lee, K., Toutanova, K.: BERT: pre-training of deep bidirectional transformers for language understanding. arXiv preprint arXiv:1810.04805 (2018)
8. Fajtl, J., Sokeh, H.S., Argyriou, V., Monekosso, D., Remagnino, P.: Summarizing videos with attention. In: Carneiro, G., You, S. (eds.) ACCV 2018. LNCS, vol. 11367, pp. 39–54. Springer, Cham (2019). https://doi.org/10.1007/978-3-030-21074-8_4
9. Feng, L., Li, Z., Kuang, Z., Zhang, W.: Extractive video summarizer with memory augmented neural networks. In: Proceedings of the 26th ACM International Conference on Multimedia, pp. 976–983 (2018)
10. Fu, T.J., Tai, S.H., Chen, H.T.: Attentive and adversarial learning for video summarization. In: 2019 IEEE Winter Conference on Applications of Computer Vision (WACV), pp. 1579–1587. IEEE (2019)
11. Ghauri, J.A., Hakimov, S., Ewerth, R.: Supervised video summarization via multiple feature sets with parallel attention. In: 2021 IEEE International Conference on Multimedia and Expo (ICME), pp. 1–6s. IEEE (2021)
12. Gong, Y., et al.: Exact-K recommendation via maximal clique optimization. In: Proceedings of the 25th ACM SIGKDD International Conference on Knowledge Discovery & Data Mining, pp. 617–626 (2019)
13. Guo, D., Zeng, Z.: Multi-modal representation learning for video advertisement content structuring. In: Proceedings of the 29th ACM International Conference on Multimedia, pp. 4770–4774 (2021)
14. Hershey, S., et al.: CNN architectures for large-scale audio classification. In: 2017 IEEE International Conference on Acoustics, Speech and Signal Processing (ICASSP), pp. 131–135. IEEE (2017)
15. Huang, Q., Xiong, Y., Xiong, Y., Zhang, Y., Lin, D.: From trailers to storylines: an efficient way to learn from movies. arXiv preprint arXiv:1806.05341 (2018)
16. Ji, Z., Xiong, K., Pang, Y., Li, X.: Video summarization with attention-based encoder-decoder networks. IEEE Trans. Circuits Syst. Video Technol. **30**(6), 1709–1717 (2019)
17. Kikuchi, Y., Neubig, G., Sasano, R., Takamura, H., Okumura, M.: Controlling output length in neural encoder-decoders. arXiv preprint arXiv:1609.09552 (2016)

18. Kool, W., Van Hoof, H., Welling, M.: Attention, learn to solve routing problems! arXiv preprint arXiv:1803.08475 (2018)
19. Koutras, P., Maragos, P.: SUSiNet: see, understand and summarize it. In: Proceedings of the IEEE/CVF Conference on Computer Vision and Pattern Recognition Workshops (2019)
20. Lin, Q., Pang, N., Hong, Z.: Automated multi-modal video editing for ads video. In: Proceedings of the 29th ACM International Conference on Multimedia, pp. 4823–4827 (2021)
21. Lin, R., Xiao, J., Fan, J.: NeXtVLAD: an efficient neural network to aggregate frame-level features for large-scale video classification. In: Leal-Taixé, L., Roth, S. (eds.) ECCV 2018. LNCS, vol. 11132, pp. 206–218. Springer, Cham (2019). https://doi.org/10.1007/978-3-030-11018-5_19
22. Liu, M., Zhang, W., Orabona, F., Yang, T.: Adam: a stochastic method with adaptive variance reduction. arXiv preprint arXiv:2011.11985 (2020)
23. Liu, Y.-T., Li, Y.-J., Wang, Y.-C.F.: Transforming multi-concept attention into video summarization. In: Ishikawa, H., Liu, C.-L., Pajdla, T., Shi, J. (eds.) ACCV 2020. LNCS, vol. 12626, pp. 498–513. Springer, Cham (2021). https://doi.org/10.1007/978-3-030-69541-5_30
24. Liu, Z., et al.: Swin transformer: hierarchical vision transformer using shifted windows. In: Proceedings of the IEEE/CVF International Conference on Computer Vision, pp. 10012–10022 (2021)
25. Merler, M., et al.: The excitement of sports: automatic highlights using audio/visual cues. In: Proceedings of the IEEE Conference on Computer Vision and Pattern Recognition Workshops, pp. 2520–2523 (2018)
26. Merler, M., et al.: Automatic curation of golf highlights using multimodal excitement features. In: 2017 IEEE Conference on Computer Vision and Pattern Recognition Workshops (CVPRW), pp. 57–65. IEEE (2017)
27. Messaoud, S., et al.: DeepQAMVS: query-aware hierarchical pointer networks for multi-video summarization. In: Proceedings of the 44th International ACM SIGIR Conference on Research and Development in Information Retrieval, pp. 1389–1399 (2021)
28. Mnih, V., Heess, N., Graves, A., et al.: Recurrent models of visual attention. In: Advances in Neural Information Processing Systems, vol. 27 (2014)
29. Nazari, M., Oroojlooy, A., Snyder, L.V., Takác, M.: Deep reinforcement learning for solving the vehicle routing problem. arXiv preprint arXiv:1802.04240 (2018)
30. Radford, A., Wu, J., Child, R., Luan, D., Amodei, D., Sutskever, I., et al.: Language models are unsupervised multitask learners. OpenAI Blog **1**(8), 9 (2019)
31. Rochan, M., Krishna Reddy, M.K., Ye, L., Wang, Y.: Adaptive video highlight detection by learning from user history. In: Vedaldi, A., Bischof, H., Brox, T., Frahm, J.-M. (eds.) ECCV 2020. LNCS, vol. 12366, pp. 261–278. Springer, Cham (2020). https://doi.org/10.1007/978-3-030-58589-1_16
32. Vinyals, O., Bengio, S., Kudlur, M.: Order matters: sequence to sequence for sets. arXiv preprint arXiv:1511.06391 (2015)
33. Vinyals, O., Fortunato, M., Jaitly, N.: Pointer networks. In: Advances in Neural Information Processing Systems, vol. 28 (2015)
34. Wang, J., Wang, W., Wang, Z., Wang, L., Feng, D., Tan, T.: Stacked memory network for video summarization. In: Proceedings of the 27th ACM International Conference on Multimedia, pp. 836–844 (2019)
35. Williams, R.J.: Simple statistical gradient-following algorithms for connectionist reinforcement learning. Mach. Learn. **8**(3), 229–256 (1992)

36. Xiong, B., Kalantidis, Y., Ghadiyaram, D., Grauman, K.: Less is more: learning highlight detection from video duration. In: Proceedings of the IEEE/CVF Conference on Computer Vision and Pattern Recognition, pp. 1258–1267 (2019)
37. Xiong, Y., Huang, Q., Guo, L., Zhou, H., Zhou, B., Lin, D.: A graph-based framework to bridge movies and synopses. In: Proceedings of the IEEE/CVF International Conference on Computer Vision, pp. 4592–4601 (2019)
38. Yang, H., Wang, B., Lin, S., Wipf, D., Guo, M., Guo, B.: Unsupervised extraction of video highlights via robust recurrent auto-encoders. In: Proceedings of the IEEE International Conference on Computer Vision, pp. 4633–4641 (2015)
39. Zhang, K., Grauman, K., Sha, F.: Retrospective encoders for video summarization. In: Ferrari, V., Hebert, M., Sminchisescu, C., Weiss, Y. (eds.) ECCV 2018. LNCS, vol. 11212, pp. 391–408. Springer, Cham (2018). https://doi.org/10.1007/978-3-030-01237-3_24
40. Zhou, K., Qiao, Y., Xiang, T.: Deep reinforcement learning for unsupervised video summarization with diversity-representativeness reward. In: Proceedings of the AAAI Conference on Artificial Intelligence, vol. 32 (2018)

CCLSL: Combination of Contrastive Learning and Supervised Learning for Handwritten Mathematical Expression Recognition

Qiqiang Lin[1], Xiaonan Huang[1], Ning Bi[1,2], Ching Y. Suen[3], and Jun Tan[1,2(✉)]

[1] School of Mathematics and Computational Science, Sun Yat-Sen University, Guangzhou 510275, People's Republic of China
{linqq9,wangchy53,mcsbn,mcstj}@mail2.sysu.edu.cn

[2] Guangdong Province Key Laboratory of Computational Science, Sun Yat-Sen University, Guangzhou 510275, People's Republic of China

[3] Centre for Pattern Recognition and Machine Intelligence, Concordia University, Montreal, QC H3G 1M8, Canada
suen@encs.concordia.ca

Abstract. Handwritten Mathematical Expressions differ considerably from ordinary linear handwritten texts, due to their two-dimentional structures plus many special symbols and characters. Hence, HMER(Handwritten Mathematical Expression Recognition) is a lot more challenging compared with normal handwriting recognition. At present, the mainstream offline recognition systems are generally built on deep learning methods, but these methods can hardly cope with HEMR due to the lack of training data. In this paper, we propose an encoder-decoder method combining contrastive learning and supervised learning(CCLSL), whose encoder is trained to learn semantic-invariant features between printed and handwritten characters effectively. CCLSL improves the robustness of the model in handwritten styles. Extensive experiments on CROHME benchmark show that without data enhancement, our model achieves an expression accuracy of 58.07% on CROHME2014, 55.88% on CROHME2016 and 59.63% on CROHME2019, which is much better than all previous state-of-the-art methods. Furthermore, our ensemble model added a boost of 2.5% to 3.4% to the accuracy, achieving the state-of-the-art performance on public CROHME datasets for the first time.

Keywords: Handwritten · Semantic invariant · Contrastive learning

1 Introduction

Handwritten mathematical expression recognition (HMER) is more challenging than other handwritten forms such as handwritten digits and words [1–4] because handwritten mathematical expressions (HMEs) do not only have different

L. Wang et al. (Eds.): ACCV 2022, LNCS 13842, pp. 577–592, 2023.
https://doi.org/10.1007/978-3-031-26284-5_35

writing styles but also include a large number of mathematical symbols, complex two-dimensional structures and the limitation of small trainable datasets.

The traditional grammar-based HMER model is divided into three steps [5–7]: symbol segmentation, symbol recognition, and structural analysis. However, it does not bring satisfication in the recognition of handwritten mathematical formulas with complex two-dimensional structures.

With continuous improvement of computing capability, deep learning has attracted more and more attention and reaped fruitful results in search technology, machine learning, machine translation, natural language processing, multimedia learning, recommendation and personalization technology, and other related fields. Ha et al. [8] firstly applied neural networks to recognize individual characters and symbols, and then Ramadhan et al. [9] established the convolutional neural network model to recognize mathematical formula symbols. Hai et al. [10] proposed a combination of convolutional neural networks and Long Short-Term Memory (LSTM) to effectively identify online and offline handwritten characters. However, these methods can only recognize single characters. Bahdanau et al. [11] proposed an Encoder-Decoder architecture framework based on attention mechanism, which made a significant breakthrough in machine translation. And then the Encoder-Decoder framework has gradually been applied to the field of mathematical expression recognition(MER).

Zhang et al. [4] firstly applied the Encoder-Decoder framework in the field of mathematical expression recognition with a proposal of an end-to-end offline recognition model, referred to as "watch, attend and parse (WAP)". Different from the previous models, WAP uses the attention mechanism to automatically segment symbols, so that the input HMEs images are modeled with the output of one-dimensional character sequences in LaTeX format. The original WAP model employs a fully convolutional networks(FCN) encoder and a recurrent neural network decoder using gated recurrent units(GRU) equipped with an attention mechanism as the parser to generate LaTeX sequences. Subsequently, Zhang et al. [12] further improved the WAP model, using DenseNet [14] network as the encoder, and proposed multi-scale attention model which solve the problem of mathematical symbol recognition well.

Truong et al. [15] proposed a weakly supervised learning method based on WAP model, which assisted the encoder to extract more useful high-level features by adding a symbol classifier near the encoder. In terms of the improvement of decoder, "Bidirectionally Trained TRansformer (BTTR)" [16] replaces the GRU network decoder by a bi-directional Transformer decoder, and alleviates lack of coverage by employing positional encodings. An Attention aggregation based bi-directional Mutual Learning Network (ABM) is proposed by Biao et al. [17] to better learn complementary context information. Truong et al. [18] put forward a relation-based sequence representation, which reduced the ambiguity caused by the use of "_" and "{ }" and enhanced the recognition of offline handwritten mathematical expressions (HMEs) by reconstructing structural relations. Zhang et al. [19] proposed a tree-structured decoder to improve the decoding ability of dealing with complicated mathematical expressions.

Moreover, to improve the robustness of the recognizer with respect to writing styles, Wu et al. [20] proposed a novel paired adversarial learning method to extract semantic-invariant features. Le et al. [21] proposed the model of dual loss attention, which has two losses including decoder loss and context matching loss, in order to detect the semantic invariant features of the encoder from handwritten and printed mathematical expressions and improve the performance of LaTeX grammar for the encoder. However, there is a big difference in distribution between printed and handwritten MEs, so these methods cannot learn semantic-invariant features effectively. Therefore, we propose to explore how to effectively make full use of easily generated printed mathematical expressions (PMEs) to improve the recognition accuracy of HEMR model. Inspired by contrastive learning [22–25,32], this paper proposes a new method based on BTTR model: a combination of self-supervised contrastive learning and supervised learning to enable the encoder to learn the semantic-invariant features between printed and handwritten. The contributions of this paper are listed below:

- With reference to self-supervised contrastive learning, we apply contrastive learning to feature extraction in printed and handwritten, so that the encoder can learn semantic-invariant features between the two different forms, and the extracted features are learned as similar as possible.
- Considering the case of shared encoder, large batchsize and large-scale datasets are required for contrastive learning, otherwise it is difficult to gain ideal results. In this paper, hybrid contrastive learning and supervised learning methods are employed to ensure the correct updating of parameters in the training process, and also guarantee the encoder learns semantic-invariant relationship between PMEs and HMEs.
- Extensive experiments on various CROHME benchmarks show that our method on both a single model and an ensemble model outperform state-of-the-art results.

2 Related Work

Contrastive Learning. In recent years, contrastive learning has set off a wave of interest in the field of computer vision (CV). Models based on the idea of contrastive learning, such as MoCo [23], SimCLR [22], MoCov2 [24], SimSiam [25], emerge one after another. As a self-supervised representation learning method, contrastive learning has outperformed supervised learning in some tasks of CV. The main idea of contrastive learning is to narrow the distance between positive samples and expand the distance between negative samples. A pair of positive samples is usually obtained by two different random transformations of the same image. A typical model among them is SimCLR as shown in Fig. 1. SimCLR employs ResNet as the base encoder f(.) and add nonlinear projection head g(.), mapping representation to the space where the contrastive loss is applied. Inspired by these papers, we apply a contrastive learning architecture to handwritten mathematical expression recognition(HMER).

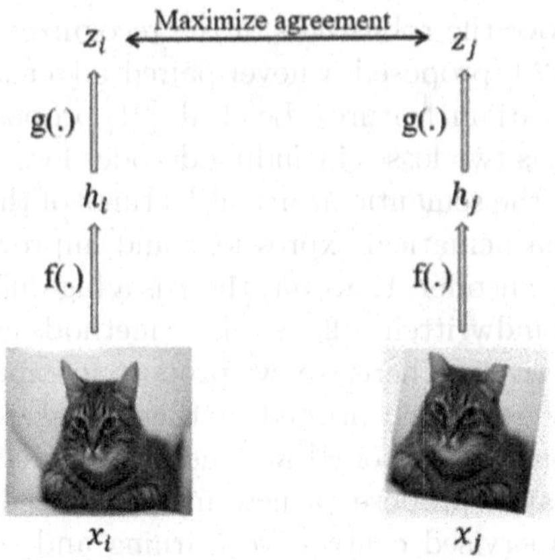

Fig. 1. The structure of SimCLR

BTTR. BTTR [16] uses DenseNet [12] as the encoder and transformer decoder as the decoder, which can perform both left-to-right (L2R) and right-to-left (R2L) decoding. The training phase is achieved by generating two target sequences (L2R and R2L) from the target LATEX sequence, and computing the training loss for the same batch. Approximate joint search [27] is used during inference to improve recognition performance. This article improves on BTTR and uses it as the baseline (Fig. 2).

3 Method

In this paper, we propose a method combining self-supervised contrast learning and supervised learning(CCLSL). HMEs come from CROHME training set, and we use Python and LATEX provided by CROHME training set to generate images of printed mathematical expressions (PMEs). On a batchsize, we assume that there are N pairs of paired samples, each of which contains HMEs image and PMEs image of the same size and the same label, denoted as $x^{pair} = (x^h, x^p)$,and the corresponding label is marked Y^{pair}.

Our handwritten expression recognition system is shown in Fig. 3. CCLSL contains two parts, one is the encoder-decoder model based on supervised learning; the other part is the self-supervised contrastive learning model, which adds a projection head g(.) to the encoder for maximizing the similarity between corresponding printed and handwritten pixel features in the space where contrastive loss is applied. We define parameters of encoder block and decoder block as θ_e and θ_d respectively. Encoder-decoder parameters and projection head parameters are defined as θ and θ_g. In the given training set $D = (x^{pair}, Y^{pair})$, θ is updated by maximizing probability of prediction while (θ_e, θ_g) is updated by maximizing the similarity of the corresponding pixel features of each printed and handwritten MEs in the contrastive space.

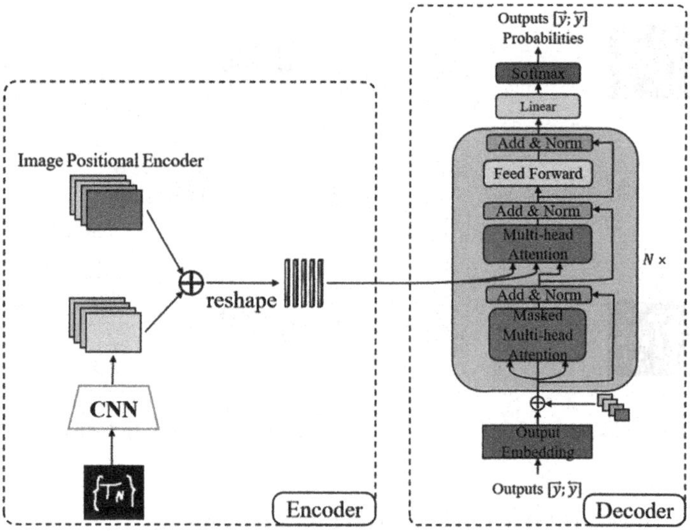

Fig. 2. The architecture of BTTR model. L2R and R2L sequences $[\overrightarrow{y};\overleftarrow{y}]$ are concatenated through the batch dimension as the input to the decoder.

3.1 Encoder

In the encoder block, CNN network is used as the feature extractor of HMEs image, which is composed of DenseNet [14] and a 1×1 convolutional layer. The role of the last convolution layer is to adjust the size of image features to the embedded dimensions d model for subsequent processing. Image pair $x^{pair} = [x^h; x^p]$ is processed through CNN networks to obtain feature map $h^{pair} = [h^h; h^p]$, which is added to 2-D image positional encodings $E^{pair} = [E^P; E^p]$ to obtain feature map with positional information, and then is flattened to 1-D feature map $f^{pair} = [f^h; f^p]$, where $x^{pair} \in R^{2\times H\times W\times 1}$, $h^{pair} \in R^{2\times H'\times W'\times d}$, $E^{pair} \in R^{2\times H'\times W'\times d}$, $f^{pair} \in R^{2\times H'W'\times d}$.

3.2 Transformer Decoder

In the decoder part of this article, we use the standard transformer decoder [26], it consists of N Transformer Decoder Layers, each layer contains three parts: Multi-Head Attention, Masked Multi-Head Attention, Feed-Forward Network.

Multi-head Attention. Multi-head attention is concatenated from single-head attention. For a given Q, K, V, we compute the head in the projected subspace

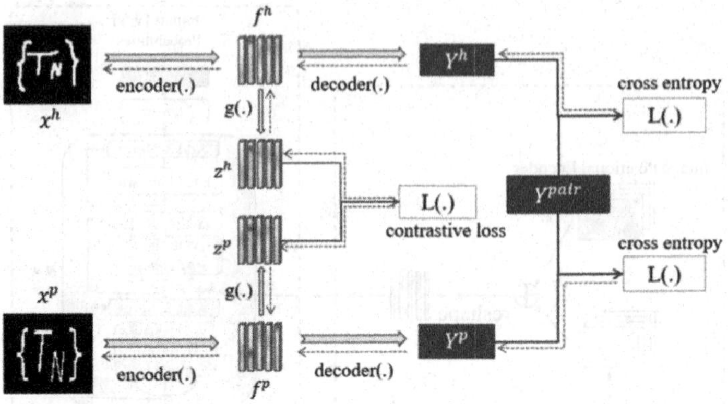

Fig. 3. A hybrid system of self supervised contrastive learning and supervised learning

by utilizing the scaled dot-product attention module.

$$H_i = \frac{(QW_i^Q)(KW_i^K)^T}{\sqrt{d_{model}}}(VW_i^V) . \tag{1}$$

where $W_i^Q \in R^{d_{model} \times d_q}$, $W_i^K \in R^{d_{model} \times d_k}$, $W_i^V \in R^{d_{model} \times d_v}$, represent the projection matrix.

After that, the h heads are concatenated and projected through the projection matrix $W^O \in R^{hd_v \times d_{model}}$ to get the new feature vector:

$$multihead = [H_1; H_2; ...; H_h]W^O . \tag{2}$$

Masked Multi-head Attention. In the process of decoding, the information of future moment cannot be obtained at the current moment, so it is necessary to use the mask technique to cover the information of the future moment during training process.

Feed-Forward Network. FFN is a fully connected network including two linear transformations and a nonlinear function, where the nonlinear function generally adopts the relu activation function.

3.3 Supervised Training

Referring to BTTR [16], we apply a bidirectional training strategy for supervised learning. First, two specific symbols "SOS" and "EOS" were introduced into the dictionary to indicate the beginning and end of a sequence. For a given paired label $Y^{pair} = \{y_1, y_2, ..., y_T\}$ it is represented as $\overrightarrow{Y^{pair}} = \{\text{"}SOS\text{"}, y_1, y_2, ..., y_T, \text{"}EOS\text{"}\}$ from left to right (L2R) and $\overleftarrow{Y^{pair}} =$

$\{``EOS", y_T, y_{T-1}, ..., y_1, ``SOS"\}$ from right to left, where y_i represents mathematical symbols and T is the length of the LaTeX sequence symbols. Considering that the transformer does not actually care about the order of the input symbols, we can use a single transformer decoder for bidirectional language modeling. We use cross-entropy as the objective function, conditioned on the image x^{pair} and the encoder-decoder parameter θ, to maximize the probability of the predicted symbols of the bidirectional target LaTeX sequence.

$$\mathcal{L}_{CE}\left(Y^{pair}|x^{pair}\right) = \frac{1}{2}\left(\mathcal{L}\left(Y^{pair}|x^{p}\right) + \mathcal{L}\left(Y^{pair}|x^{h}\right)\right), \tag{3}$$

$$\mathcal{L}\left(Y^{pair}|x\right) = \frac{1}{2T}\left(\sum_{j=1}^{T} log\, p\left(\overrightarrow{y_j} \mid \overrightarrow{y_{<j}}, x\right) + \sum_{j=1}^{T} log\, p\left(\overleftarrow{y_j} \mid \overleftarrow{y_{<j}}, x\right)\right). \tag{4}$$

3.4 Contrastive Training

In order to enable the encoder to effectively learn the same semantic-invariance features between printed and handwritten MEs and improve the robustness of this model in writing style, self-supervised contrastive learning is introduced to supervised learning. Inspired by the idea of SimCLR [22], we add a projection head after the encoder to map the representation to the space where the contrastive loss is applied. The projection head g(.) will consist of an MLP with one hidden layer:

$$z^p = g(f^p) = \sigma\left(W^g \sigma\left(f^p\right)\right). \tag{5}$$

$$z^h = g(f^h) = \sigma\left(W^g \sigma\left(f^h\right)\right). \tag{6}$$

where σ stands for ReLU nonlinearity, $\theta_g = W^g \in R^{d \times d}$.

As for $z^p = [z_1^p, z_2^p, ..., z_{H'W'}^p]$ and $z^h = [z_1^h, z_2^h, ..., z_{H'W'}^h]$, A contrastive loss function [31] enables the corresponding positional features of z^p and z^h as close as possible:

$$\mathcal{L}_{CL}\left(x^{pair}\right) = \mathcal{L}_{CL}\left(x^p, x^h\right), \tag{7}$$

$$\mathcal{L}_{CL}\left(x^p, x^h\right) = \sum_{i=1}^{H'W'}\left(\mathcal{L}_{NCE}\left(z_i^p, z_i^h; z^p \bigcup z^h\right) + \mathcal{L}_{NCE}\left(z_i^h, z_i^p; z^p \bigcup z^h\right)\right). \tag{8}$$

where $\mathcal{L}_{NCE}(.)$ is a contrastive loss function, called InfoNCE [32]:

$$\mathcal{L}_{NCE}\left(u, v_+; U\right) = -log\frac{exp(u^T \cdot v_+/\tau)}{\sum_{v \in U \backslash u} exp(u^T \cdot v/\tau)}. \tag{9}$$

where u, v, v_+ are l_2 normalized.

3.5 Combination of Contrastive Learning and Supervised Learning

According to SimCLR, under the condition of not introducing memory bank, the pre-training effect can only be acceptable if the batch size is large enough. However, the available HMEs training data are insufficient to support mass training. Therefore, a new method is proposed in this paper: Skip the pre-training stage of contrastive learning, and directly carry out the combination of self-supervised contrastive learning and supervised learning. The specific operation is to minimize the hybrid loss function:

$$\mathcal{L}_{hybrid} = \mathcal{L}_{CE}\left(Y^{pair}|x^{pair}\right) + \lambda\mathcal{L}_{CL}\left(x^{pair}\right) . \tag{10}$$

where λ is a hyperparameter that controls the tradeoff between decoder loss $\mathcal{L}_{CE}$ and contrastive loss $\mathcal{L}_{CL}$.

In the inference phase, after discarding the projection head g(.), our model is capable of recognizing both HMEs and PMEs. Similar to BTTR, the decoder employs approximate joint search [27] to improve decoding performance.

4 Experiments

4.1 Experimental Setup and Results

Table 1. Performance of the BTTR as baseline system on CROHME 2014, CROHME 2016 and CROHME 2019.

Model	2014 ExpRate	2016 ExpRate	2019 ExpRate
BTTR [16]	53.96	52.31	52.96
Baseline	55.68	53.44	55.46

Experimental Setup. We evaluated our method on the CROHME Competition dataset [28–30]. The training set selected in this paper is CROHME2014 training set, which contains 8836 HMEs pictures in total. We use Matplotlib library to generate corresponding PMEs. CROHME2014 test set containing 986 images, CROHME2016 test set containing 1147 images, and CROHME2019 test set containing 1199 images are employed to test the performance of the model. And we employ Expression Rate (ExpRate) metrics to evaluate HMEs recognition systems. Adadelta algorithm with gradient shear is chosen to learn parameters with batch size set to 15, batch image size set to 440000 and max epochs set to 200. When the expression rate on the validation set does not increase after 30 epochs, the learning rate is set up to decrease. During training $\tau = 2$ is used to soften the output distribution. The model is trained on two NVIDIA 2080Ti GPUs with 11 $\times$ 2 GB memory.

Experimental Results. First of all, we rerun related work BTTR [16] as the baseline system. Table 1 shows the results of test on CRHOMR2014, CRHOMR2016 and CRHOMR2019. It is worth mentioning that the open source code reproduction results provided by BTTR are a bit better than their paper results, so we adopt the rerun model as the baseline model.

Table 2. Performance comparison of offline HMER systems on CROHME test sets. * Refers to the ensemble of recognition models utilizing multiple different initializations. In particular, all models listed in the table trained without any augmentation.

Model	2014 ExpRate	2016 ExpRate	2019 ExpRate
Single model			
WAP [13]	48.38	46.82	–
WS-WAP [15]	53.65	51.96	–
PAL-v2 [20]	48.88	49.61	–
Dual loss attention [21]	51.88	51.53	–
DenseWAP-TD [19]	49.1	48.5	51.4
ABM [17]	56.85	52.92	53.96
SAN [34]	56.2	53.6	53.5
BTTR [16]	53.96	52.31	52.96
Baseline	55.68	53.44	55.46
CCLSL(our)	**58.07**	**55.88**	**59.63**
Ensemble model			
WS-WAP* [15]	55.68	52.57	–
PAL-v2* [20]	54.87	57.89	–
DenseWAP-TD* [19]	54.00	52.10	54.60
BTTR* [16]	57.91	54.49	56.88
CCLSL* (our)	**60.61**	**58.32**	**62.97**

In Table 2, we compare our model with other offline HMER systems on the CROHME 2014/2016/2019 test sets respectively. To ensure fairness of comparison, none of the systems employ integration of multiple models. Results show that our model achieves an expression accuracy of 58.07% on CROHME2014, 55.88% on CROHME2016 and 59.63% on CROHME2019, with an improvement of 2.39%/2.44%/4.17% on CROHME2014/2016/2019 compared to the baseline model, and the recognition performance of our model on the three test sets are obviously better than the most advanced method. In addition, under the condition of $\lambda = 0.02$, two single models are retrained with different initializations, and the six single models trained by $\lambda = 0.01$, 0.02, 0.03, 0.04, 0.05, and 0.06 are simply integrated together. It can be seen from Table 2 that the recognition effect of the integrated model in CROHME test sets also achieves state-of-the-art at present.

Table 3. Performance of CCLSL under different hyperparameters λ.

Dataset		2014 ExpRate	2016 ExpRate	2019 ExpRate
Baseline		55.68	53.44	55.46
λ	0.01	57.76	55.18	58.21
	0.02	58.07	**55.88**	**59.63**
	0.03	57.76	54.84	58.38
	0.04	57.76	55.01	57.88
	0.05	57.05	55.01	59.13
	0.06	**58.17**	55.10	58.79
	0.07	55.23	54.92	57.46

4.2 Ablation Experiments

Ablation: Superparameter λ. We evaluate the performance of CCLSL under different superparameter λ shown in Table 3. We set 0.01, 0.02, 0.03, 0.04, 0.05, 0.06, 0.07 to λ for the experiment. By comparison, we find that the model achieves the best results in all of the three test sets when $\lambda = 0.02$. ExpRates in CROHME 2014/2016/2019 test sets are 58.07%, 55.88%, 59.63%, significantly improved compared with the baseline model, indicating that Combination of contrastive learning and supervised learning(CCLSL) can effectively improve model recognition performance.

Ablation: Different Training Methods. In order to further illustrate the effectiveness of our method, some ablation experiments are performed in Table 4. It can be seen from Table 4 that Pre and Mixed have limited performance improvement of the model. It is likely that the two data distributions are quite different, resulting in a far difference in the feature maps extracted by the encoder. The method proposed in this paper can enable the model, especially the encoder, to learn the semantic invariant features of PMEs and HMEs images.

4.3 Encoder Migration

At present, the classic model WAP [12] in the offline handwritten mathematical formula also uses the DenseNet as the encoder and the double-layers GRU network with the attention mechanism as the decoder. In this paper, the encoder trained by the CCLSL method will be migrated to the WAP model for retraining, while using WAP [12], WAP variant WS-WAP [15], ABM [17] as the baseline model.

Table 4. The performance of BTTR under different training methods. Pre refers to pre-training the BTTR model using PMEs images, and then fine-tuning them in HMEs images. Mixed refers to training models using both PMEs and HMEs images. CCLSL is the method proposed in this paper.

Dataset	Pre	Mixed	CCLSL	ExpRate
CROHME 2014	×	×	×	55.68
	✓	×	×	56.14
	×	✓	×	56.04
	×	×	✓	**58.07**
CROHME 2016	×	×	×	53.44
	✓	×	×	53.87
	×	✓	×	54.34
	×	×	✓	**55.88**
CROHME 2019	×	×	×	55.46
	✓	×	×	55.96
	×	✓	×	55.54
	×	×	✓	**59.63**

Table 5. Performance of encoder migrated to WAP. The V1 model freezes the encoder parameters of WAP and only trains the decoder. V2 uses the cross-entropy function as the loss function on the basis of V1 to fine-tune the encoder and decoder to make them more adaptable, and V3 uses label smoothing [33] as the loss function on the basis of V1 to fine-tune the encoder and decoder to improve the generalization ability of the model.

Model	2014 ExpRate	2016 ExpRate	2019 ExpRate
WAP [13]	48.38	46.82	–
WS-WAP [15]	53.65	51.96	–
ABM [17]	56.85	52.92	53.96
V1(our)	51.47	48.64	49.54
V2(our)	56.54	53.55	55.57
V3(our)	**59.69**	**54.92**	**57.88**

Table 5 shows that, by transferring the encoder DenseNet trained by CCLSL to the WAP model for fine-tuning, the recognition accuracy of the model is greatly improved, and the performance is also significantly better than other WAP model variants. This further verifies that the encoder trained with CCLSL can indeed learn semantically invariant features between print and handwriting, significantly improving the robustness of the model in terms of writing style.

4.4 Visualization

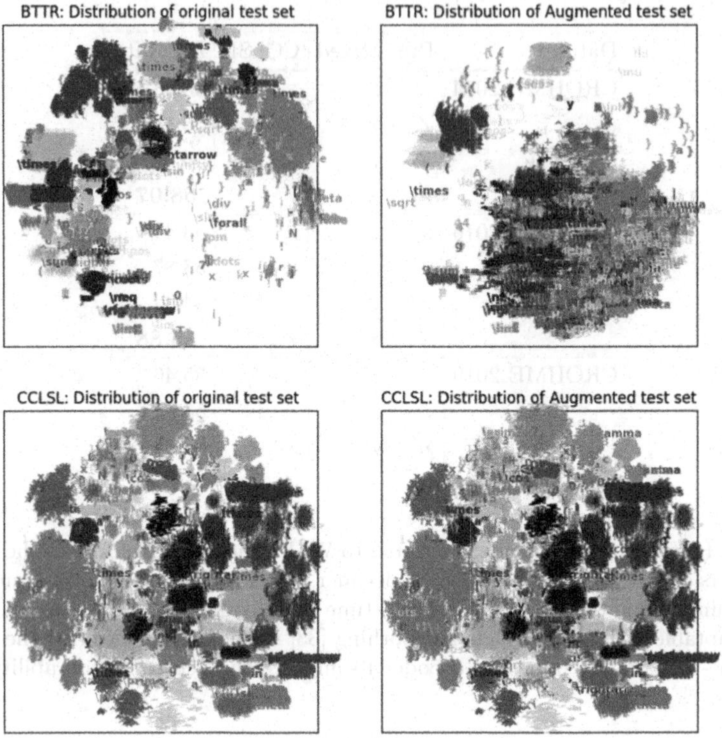

Fig. 4. Visualize the context vectors extracted by the BTTR and CCLSL on the test set and augmented test set.

To visualize the semantically invariant features learned by the proposed system, we show the context vectors for each symbol class in BTTR and CCLSL on the CROHME 2014 test set and the corresponding data augmentation set. The augmented dataset is generated using A general geometric augmentation tool [35] for text images. For visualization, we utilize t-SNE to map the data from high-dimensional to 2-dimensional. It can be observed from Fig. 4 that the character shape changes in the enhanced image, which leads to a large deviation of the context vector extracted by BTTR, and the system proposed in this paper can capture similar context vector representations and learn semantic invariance.

Finally, Fig. 5 illustrates the attention-based decoding process, where mintcream is the background, black is the font, and other color areas are the

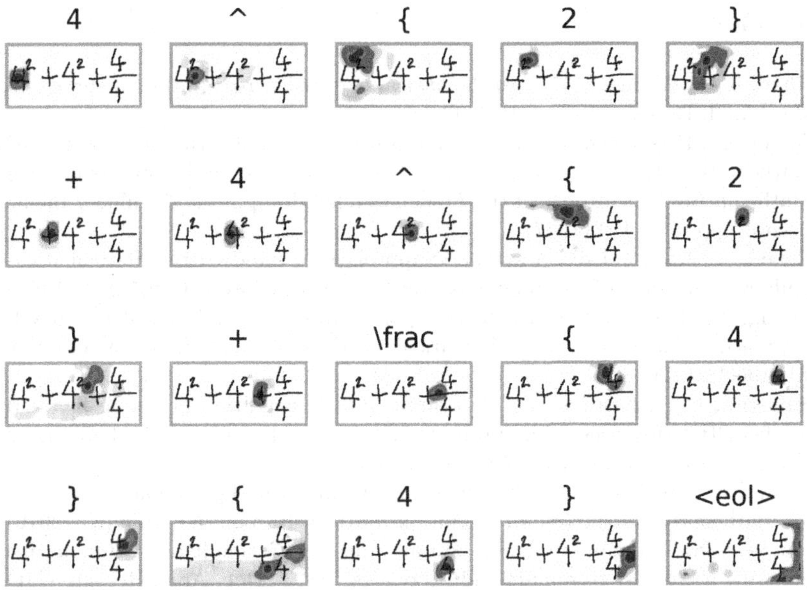

Fig. 5. Visualization of the attention process.

focus areas of attention. The darker the color, the higher the attention weight. It can be seen that the attention can not only capture the spatial position of each character, but also use the spatial structure information to assist the decoder in parsing symbols '{' and '}'.

5 Conclusion

In this paper, we have proposed a new method (CCLSL) to effectively recognize offline HMEs: a combination of self-supervised contrastive learning and supervised learning to enable the encoder to learn the semantic-invariant features between PMEs and HMEs, improving the model's robustness in writing style. Extensive experiments on various CROHME datasets show that our method on both single and integrated models achieved state-of-the-art performance.

Based on the research results of this paper, the future research direction is proposed: mining more latex expressions of mathematical formulas, generating PMEs pictures to participate in the training of the model. Using the method proposed in this paper to ensures that the encoder can learn the semantically invariant features of PMEs and HMEs, the decoder can learn more latex syntax and further improve the performance of the decoder.

Acknowledgments. This work was supported by the Guangdong Provincial Government of China through the Computational Science Innovative Research Team" program and Guangdong Province Key Laboratory of Computational Science at the Sun Yat-sen University, and the National Science Foundation of China (11971491, 11471012).

References

1. Chan, K.-F., Yeung, D.-Y.: Mathematical expression recognition: a survey. Int. J. Doc. Anal. Recogn. **3**(1), 3–15 (2000)
2. Anderson, R.H.: Syntax-directed recognition of hand-printed two-dimensional mathematics. In: Symposium on Interactive Systems for Experimental Applied Mathematics: Proceedings of the Association for Computing Machinery Inc., Symposium, pp. 436–459 (1967)
3. Belaid, A., Haton, J.-P.: A syntactic approach for handwritten mathematical formula recognition. IEEE Trans. Pattern Anal. Mach. Intell. **1**, 105–111 (1984)
4. Zhang, J., et al.: Watch, attend and parse: an end-to-end neural network based approach to handwritten mathematical expression recognition. Pattern Recogn. **71**, 196–206 (2017)
5. Simistira, F., Katsouros, V., Carayannis, G.: Recognition of online handwritten mathematical formulas using probabilistic svms and stochastic context free grammars. Pattern Recogn. Lett. **53**, 85–92 (2015)
6. Álvaro, F., Sánchez, J.-A., Benedí, J.-M.: An integrated grammar-based approach for mathematical expression recognition. Pattern Recogn. **51**, 135–147 (2016)
7. MacLean, S., Labahn, G.: A new approach for recognizing handwritten mathematics using relational grammars and fuzzy sets. Int. J. Doc. Anal. Recogn. (IJDAR) **16**(2), 139–163 (2013)
8. Ha, J., Haralick, R.M., Phillips, I.T.: Understanding mathematical expressions from document images. In: Proceedings of 3rd International Conference on Document Analysis and Recognition, vol. 2, pp. 956–959. IEEE (1995)
9. Ramadhan, I., Purnama, B., Al Faraby, S.: Convolutional neural networks applied to handwritten mathematical symbols classification. In: 2016 4th International Conference on Information and Communication Technology (ICoICT), pp. 1–4. IEEE (2016)
10. Dai, H., Le Duc, A., Nakagawa, M.: Combination of lstm and cnn for recognizing mathematical symbols. In: Proceedings of the 17th Information-Based Induction Sciences Workshop (2014)
11. Bahdanau, D., Cho, K., Bengio, Y.: Neural machine translation by jointly learning to align and translate. arXiv preprint arXiv:1409.0473 (2014)
12. Zhang, J., Du, J., Dai, L.: Multi-scale attention with dense encoder for handwritten mathematical expression recognition. In: 2018 24th International Conference on Pattern Recognition (ICPR), pp. 2245–2250. IEEE (2018)
13. Wang, J., Du, J., Zhang, J., Wang, Z.-R.: Multi-modal attention network for handwritten mathematical expression recognition. In: 2019 International Conference on Document Analysis and Recognition (ICDAR), pp. 1181–1186 (2019)
14. Huang, G., Liu, Z., Van Der Maaten, L., Weinberger, K.Q.: Densely connected convolutional networks. In: Proceedings of the IEEE Conference on Computer Vision and Pattern Recognition, pp. 4700–4708 (2017)
15. Truong, T.-N., Nguyen, C.T., Phan, K.M., Nakagawa, M.: Improvement of end-to-end offline handwritten mathematical expression recognition by weakly supervised learning. In: 2020 17th International Conference on Frontiers in Handwriting Recognition (ICFHR), pp. 181–186. IEEE (2020)
16. Zhao, W., Gao, L., Yan, Z., Peng, S., Du, L., Zhang, Z.: Handwritten mathematical expression recognition with bidirectionally trained transformer. In: Lladós, J., Lopresti, D., Uchida, S. (eds.) ICDAR 2021. LNCS, vol. 12822, pp. 570–584. Springer, Cham (2021). https://doi.org/10.1007/978-3-030-86331-9_37

17. Bian, X., Qin, B., Xin, X., Li, J., Su, X., Wang, Y.: Handwritten mathematical expression recognition via attention aggregation based bi-directional mutual learning. arXiv preprint arXiv:2112.03603 (2021)
18. Truong, T.-N., Ung, H.Q., Nguyen, H.T., Nguyen, C.T., Nakagawa, M.: Relation-based representation for handwritten mathematical expression recognition. In: Barney Smith, E.H., Pal, U. (eds.) ICDAR 2021. LNCS, vol. 12916, pp. 7–19. Springer, Cham (2021). https://doi.org/10.1007/978-3-030-86198-8_1
19. Zhang, J., Du, J., Yang, Y., Song, Y.-Z., Wei, S., Dai, L.: A tree-structured decoder for image-to-markup generation. In: International Conference on Machine Learning, pp. 11076–11085. PMLR (2020)
20. Wu, J.-W., Yin, F., Zhang, Y.-M., Zhang, X.-Y., Liu, C.-L.: Handwritten mathematical expression recognition via paired adversarial learning. Int. J. Comput. Vision, 1–16 (2020)
21. Le, A.D.: Recognizing handwritten mathematical expressions via paired dual loss attention network and printed mathematical expressions. In: Proceedings of the IEEE/CVF Conference on Computer Vision and Pattern Recognition Workshops, pp. 566–567 (2020)
22. Chen, T., Kornblith, S., Norouzi, M., Hinton, G.: A simple framework for contrastive learning of visual representations. In: International Conference on Machine Learning, pp. 1597–1607. PMLR (2020)
23. He, K., Fan, H., Wu, Y., Xie, S., Girshick, R.: Momentum contrast for unsupervised visual representation learning. In: Proceedings of the IEEE/CVF Conference on Computer Vision and Pattern Recognition, pp. 9729–9738 (2020)
24. Chen, X., Fan, H., Girshick, R., He, K.: Improved baselines with momentum contrastive learning. arXiv preprint arXiv:2003.04297 (2020)
25. Chen, X., He, K.: Exploring simple siamese representation learning. In: Proceedings of the IEEE/CVF Conference on Computer Vision and Pattern Recognition, pp. 15750–15758 (2021)
26. Vaswani, A., et al.: Attention is all you need. Adv. Neural Inf. Process. Syst. **30**, 5998–6008 (2017)
27. Liu, L., Utiyama, M., Finch, A., Sumita, E.: Agreement on target-bidirectional neural machine translation. In: Proceedings of the 2016 Conference of the North American Chapter of the Association for Computational Linguistics: Human Language Technologies, pp. 411–416 (2016)
28. Mouchere, H., Viard-Gaudin, C., Zanibbi, R., Garain, U.: ICFHR 2014 competition on recognition of on-line handwritten mathematical expressions. In: 2014 14th International Conference on Frontiers in Handwriting Recognition (CROHME 2014), pp. 791–796. IEEE (2014)
29. Mouchère, H., Viard-Gaudin, C., Zanibbi, R., Garain, U.: ICFHR 2016 crohme: competition on recognition of online handwritten mathematical expressions. In: 2016 15th International Conference on Frontiers in Handwriting Recognition (ICFHR), pp. 607–612. IEEE (2016)
30. Mahdavi, M., Zanibbi, R., Mouchere, H., Viard-Gaudin, C., Garain, U.: ICDAR 2019 CROHME+TFD: competition on recognition of handwritten mathematical expressions and typeset formula detection. In: 2019 International Conference on Document Analysis and Recognition (ICDAR), pp. 1533–1538. IEEE (2019)
31. Hadsell, R., Chopra, S., LeCun, Y.: Dimensionality reduction by learning an invariant mapping. In: 2006 IEEE Computer Society Conference on Computer Vision and Pattern Recognition (CVPR 2006), vol. 2, pp. 1735–1742. IEEE (2006)
32. Oord, A.V.D., Li, Y., Vinyals, O.: Representation learning with contrastive predictive coding. arXiv preprint arXiv:1807.03748 (2018)

33. Müller, R., Kornblith, S., Hinton, G.E.: When does label smoothing help? Adv. Neural Inf. Process. Syst. **32** (2019)
34. Yuan, Y., et al.: Syntax-aware network for handwritten mathematical expression recognition. In: Proceedings of the IEEE/CVF Conference on Computer Vision and Pattern Recognition, pp. 4553–4562 (2022)
35. Luo, C., Zhu, Y., Jin, L., Wang, Y.: Learn to augment: joint data augmentation and network optimization for text recognition. In: Proceedings of the IEEE/CVF Conference on Computer Vision and Pattern Recognition, pp. 13746–13755 (2020)

Dynamic Feature Aggregation for Efficient Video Object Detection

Yiming Cui(✉)

University of Florida, Gainesville, USA
cuiyiming@ufl.edu

Abstract. Video object detection is a fundamental yet challenging task in computer vision. One practical solution is to take advantage of temporal information from the video and apply feature aggregation to enhance the object features in each frame. Though effective, those existing methods always suffer from low inference speeds because they use a fixed number of frames for feature aggregation regardless of the input frame. Therefore, this paper aims to improve the inference speed of the current feature aggregation-based video object detectors while maintaining their performance. To achieve this goal, we propose a vanilla dynamic aggregation module that adaptively selects the frames for feature enhancement. Then, we extend the vanilla dynamic aggregation module to a more effective and reconfigurable deformable version. Finally, we introduce inplace distillation loss to improve the representations of objects aggregated with fewer frames. Extensive experimental results validate the effectiveness and efficiency of our proposed methods: On the ImageNet VID benchmark, integrated with our proposed methods, FGFA and SELSA can improve the inference speed by 31% and 76% respectively while getting comparable performance on accuracy. Codes are available at https://github.com/YimingCuiCuiCui/DFA.

Keywords: Video object detection · Dynamic feature aggregation

1 Introduction

Object detection is an essential task in computer vision which aims to localize and categorize objects of interest in a single or sequence of images [3,9,14,27,31,38,44]. With the excellent performance of deep learning-based computer vision methods on image object detection tasks [9,14,27,31], researchers have begun to extend image object detection to the more challenging video domain. Compared with still images, videos have the issues of feature degradation caused by camera jitters or fast motion that rarely happen in the image domains [3,45], which increase the difficulty of object detection in videos. Therefore, directly applying object detectors from image domains on a frame-by-frame basis for video analysis always produces poor performance. Existing works can be divided into two directions to solve the issues caused by video feature degradation.

Since the same object always reappears in multiple frames, videos can provide rich temporal information, which provides hints for video analysis. Therefore,

L. Wang et al. (Eds.): ACCV 2022, LNCS 13842, pp. 593–609, 2023.
https://doi.org/10.1007/978-3-031-26284-5_36

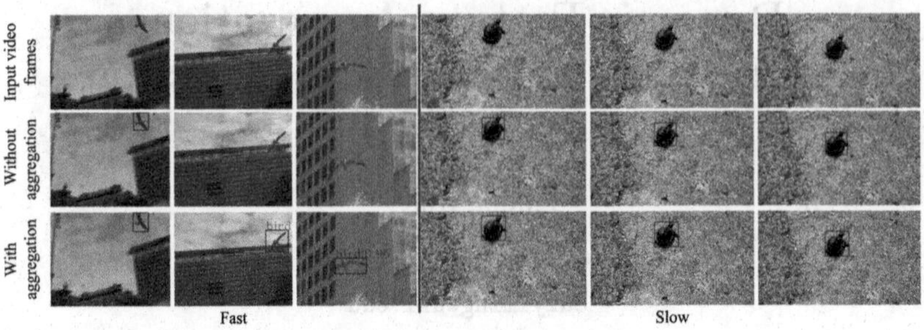

Fig. 1. Comparison of video object detection results with/without feature aggregation on objects with different motion speeds.

one direction for video object detection is to exploit this temporal information with post-processing pipelines [13,20,21] where still image object detection approaches are first applied on each frame. Then those detected objects are assembled across frames with temporal hints like motion estimation or object tracking. However, those approaches are not trained end-to-end, and the detection results on a single frame and across frames cannot be optimized jointly. Therefore, if the predictions in a single frame are inaccurate, they cannot be optimized and refined during the post-processing procedure.

Another direction for video object detection is aggregating features across multiple frames to eliminate feature degradation in videos. These methods assume that frames with poor features only account for a small ratio compared with the whole video sequences. By aggregating temporal features, the performance of video object detection can be boosted. These methods can be categorized as local, global, and combinations, depending on how to aggregate features. The first sub-direction methods [35,44,45] exploit the local temporal information in videos to enhance the target frame features in a short time range and ignore the global information. To address this issue, the second sub-direction methods [7,11,12,30] introduce attention modules to use global temporal information. However, these methods ignore local temporal information due to GPU memory limits or computational constraints. The third sub-direction methods [3,19] make a combination of local and global temporal information but always suffer from low inference speeds.

Though getting better performance than post-processing methods, those feature aggregation-based object detectors always have a lower inference speed. Therefore, besides focusing on the performance of video object detection, recent works [2,18,19,39,40] also design efficient frameworks to improve the inference speeds. However, these methods are designed for a specific framework and cannot be generalized to other video object detectors. What makes things worse, these approaches are always efficient during the inference process at the sacrifice of performance like accuracy or recall.

Unlike the existing efficient video object detectors, we aim to design a plug-and-play module that can be integrated into most existing methods to balance their inference speeds and performances. To achieve this, we first notice that the low inference speeds of the current feature aggregation-based object detectors are caused by their aggregation processes, which are proportional to the number of frames used for aggregation [3,5,44,45]. It is natural to think whether it is necessary to always use a fixed number of frames for feature aggregation. For objects with fast motion speeds[1], feature aggregation can improve the video object detection performance. As shown in Fig. 1(a), the flying bird cannot be correctly detected in multiple frames without feature aggregation. On the contrary, when the objects are with slow motion speeds, as shown in Fig. 1(b), original Faster R-CNN [27] without any feature aggregation can already detect the turtle in the current frame correctly. Therefore, using too many aggregation frames for videos with slow motion is unnecessary since the model with a few aggregation frames or even without aggregation can already perform well.

In this paper, we attempt to improve the efficiency of the current feature aggregation-based video object detectors in a simple yet effective way. We notice that there is no need to always use a fixed number of frames to aggregate features for video object detection regardless of the inputs. Therefore, we design modules to aggregate features dynamically. We first propose a vanilla dynamic feature aggregation strategy which can adaptively select frames for aggregation based on the inputs. Then, we extend the vanilla strategy to a deformable version which is more effective and reconfigurable. Finally, we introduce an inplace distillation loss to enhance the object feature representations when only a few frames are used for aggregation. Our contributions can be summarized as follows:

- To the best of our knowledge, we are the first to adaptively and dynamically aggregate features for video object detection to balance the model efficiency and performance.
- We design a vanilla dynamic feature aggregation (DFA) module and then extend it to a deformable version which can adaptively and reconfigurably enhance the object feature representations based on the input frame. Inplace distillation loss is introduced to improve the feature representations of those aggregated with fewer frames for better performance.
- Our proposed method is a plug-and-play module which can be integrated into most of the recent state-of-the-art video object detectors. Evaluated on the ImageNet VID benchmark, the performance of video object detection can be preserved with a much better inference speed when integrated with our proposed method.

2 Related Works

Still Image Object Detection. Image object detection task aims to localize and categorize objects of interest in a still image. Current deep learning-based models can be classified into two main directions: Two-stage object detector

[1] For better analysis, we use the same way as FGFA [44] to categorize objects in every single frame based on their motion speeds.

and one-stage object detector. Among them, R-CNN-based two-stage object detectors [1,9,14,23,27] first generate a fixed number of proposals with Region Proposal Network (RPN) [9] to localize and classify the object candidates coarsely. Then, they refine these proposals to output fine-grained predictions. To improve the inference speed of those models mentioned above, one-stage models [22,25,26,33] are introduced to predict the locations and categories of objects directly based on the extracted features from CNN without region proposals. For simplicity and generalization, our method is built based on Faster R-CNN [27], which is one of the state-of-the-art object detectors and can be easily extended to others.

Video Object Detection. Different from image object detection, the video object detection task must handle situations caused by motion to generate good predictions in each frame. Post-processing-based methods detect every frame separately and assemble those detected objects with various metrics like optical flow. Seq-NMS [13] assembles bounding boxes at different frames with the criteria of IoU threshold and re-ranks the linked bounding boxes. TCN [21] uses tubelet modules and applies a temporal convolutional network to embed temporal information to improve the detection across frames. Despite the simplicity, those methods are not trained end-to-end and perform poorly.

To solve the issues, feature aggregation-based methods [3,38,44,45] usually enhance the object representations using the temporal information to eliminate the feature degradation caused by motions. Among them, FGFA [44] first warps the feature maps from the local adjacency frames to the keyframe based on the flow motion and then aggregates those warped features to improve the object representations for the following detection network. SELSA [38] aggregates features in a global full-sequence level. In SELSA, proposals across space-time domains with similar semantics are linked, and their features are weight-averaged for aggregation to provide richer information to handle issues like motion blur and pose changes. MANet [35] jointly aggregates object features on both pixel-level and instance-level. The pixel-level aggregation is used to model detailed motion, while the instance-level calibration is introduced to capture global motion cues. MEGA [3] takes global and local information into account where global features are first aggregated into local features. Then these global-enhanced local features are fused into the key frame for better detection performance. TF-Blender [5] improves the feature aggregation process using the temporal relations between frames. TransVOD [16] introduces the Transformer to aggregate the spatial and temporal information in a multi-head self-attention mode.

Compared with post-processing-based methods, feature aggregation-based video object detectors usually perform better with a lower inference speed. In this paper, we mainly focus on feature aggregation-based methods.

Efficient Networks. Though deep learning-based methods perform better on multiple computer vision tasks, their complexities become higher, making them unsuitable for applications with constrained computational budgets but a short

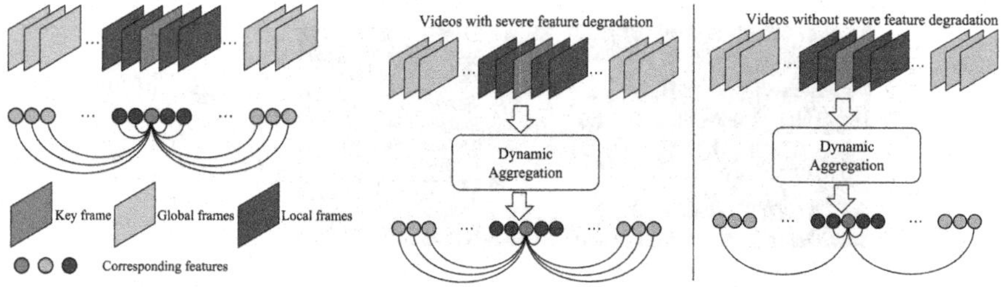

(a) Existing methods (b) Our method

Fig. 2. The framework of our proposed method, which uses dynamic aggregation strategies for video object detection tasks. (a) Current methods aggregate features with a fixed number of frames (e.g., 10) regardless of the input frames. (b) Our methods adaptively select frames for dynamic aggregation according to the input frame. For videos with severe feature degradation, 8 frames are selected to enhance the key frame representation. For videos with high qualities, 4 frames are used for aggregation for fast inference speed.

response time like mobile platforms. Therefore, recent works [17,19,29,32,39] begin to focus on how to speed up the detection process for real-time applications. Towards this goal, lightweight networks like Mobile-Net [17,29], Efficient-Net [32] and automated neural architecture search models [24,37] are introduced to take the place of heavy backbones like ResNet [15] to reduce the computation complexity for mobile applications. Besides replacing the backbones, LSTS [19] learns semantic-level spatial correspondences between neighboring frames to reduce the information redundancy in video frames to accelerate the detection process. CenterNet-HP [39] replaces two-stage detectors like Faster R-CNN [27] with one-stage model CenterNet [8] for real-time video object detection. Detection results from previous frames are propagated in the form of a heatmap to enhance the performance of the future frames. Other works [34,36] improve the detection speeds with the help of compressed video information. Though efficient during inference, those methods usually require carefully designed modules and make great changes to the existing video object detectors, making it infeasible to generalize to other methods. Also, these methods generally have a worse performance despite high inference speeds. On the contrary, our proposed approaches are plug-and-play modules which can be easily integrated into the existing detectors to balance their efficiency and performance.

Recently, dynamic networks have been introduced, which allow selective inference paths. Slimmable networks [41–43] are models trained executable at different widths, which can be adaptive to multiple computational resources and get even better performance compared with their counterparts trained individually. For object detection, dynamic proposals are introduced for efficient inferernce [6]. In this paper, we borrow the idea from slimmable networks to make our model adaptive to different input videos and able to adjust the numbers of frames for aggregation according to the input frame for video object detection.

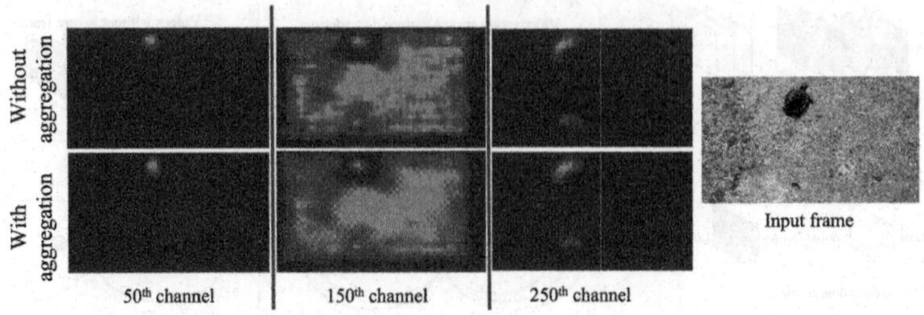

Fig. 3. Comparison of features maps with/without aggregation on objects with slow motion speeds.

3 Methodology

The key idea of our method is to replace the fixed number of frames with a dynamic size in the current feature aggregation-based video object detectors. Therefore, instead of using fixed frames, our model can adaptively choose the frames for aggregation according to the inputs, as shown in Fig. 2. Our proposed method is a plug-and-play module which can be easily integrated into most feature aggregation-based video object detectors. In the following sections, we first review the current feature aggregation-based video object detectors and analyze the inefficiency of their aggregation processes. Then, we propose a vanilla dynamic aggregation strategy to adapt the model to different input frames. Next, we extend the vanilla dynamic aggregation to a deformable version which is more effective and reconfigurable. Finally, inplace distillation loss is introduced to enhance the feature representations aggregated with fewer adjacent frames.

3.1 Preliminary

The canonical feature aggregation methods [3,7,35,38,44] generally work on a fixed number of frames k, which can be summarized as: Given the current frame $\boldsymbol{I}_i$ and its neighboring frames $\boldsymbol{I}_j \in \mathcal{N}(\boldsymbol{I}_i)$, their corresponding features $\boldsymbol{f}_j$ are weighted averaged in order to aggregate the temporal feature $\Delta \boldsymbol{f}_i$:

$$\Delta \boldsymbol{f}_i = \sum_{\boldsymbol{I}_j \in \mathcal{N}(\boldsymbol{I}_i)} (w_{ij} \times \boldsymbol{f}_j), \tag{1}$$

where w_{ij} denotes the weights for aggregation and the size of $\mathcal{N}(\boldsymbol{I}_i)$ is k. Then the aggregated feature $\Delta \boldsymbol{f}_i$ is fed into a task network $\mathcal{N}_{task}$ for object detection:

$$\boldsymbol{b}_i, \boldsymbol{c}_i = \mathcal{N}_{task}(\Delta \boldsymbol{f}_i), \tag{2}$$

where $\boldsymbol{b}_i, \boldsymbol{c}_i$ denote the predicted bounding boxes and their corresponding categories in the current frame $\boldsymbol{I}_i$. However, the performance and inference speed of these feature aggregation-based models are heavily influenced by k. For example, when k decreases from 31 to 3, the inference speed of FGFA [44] can increase from 5.8 FPS to 20.6 FPS (tested on a single Titan RTX GPU) while the mAP drops from 74.6% to 72.5%.

Therefore, we wonder whether it is necessary to use so many frames (e.g., 21 frames) for feature aggregation regardless of the input videos. Figure 3 compares the feature maps with and without feature aggregation on objects with slow motion speeds. We visualize the 50-th, 150-th and 250-th feature channels in Fig. 3. As shown in the figure, there is little difference between the feature maps with and without aggregation. Meanwhile, we also calculate the cosine similarity scores of the whole feature map with and without aggregation, which is 0.9819. Therefore, for objects with slow motion speeds, there is not too much improvement with feature aggregation.

3.2 Vanilla Dynamic Aggregation

In the current feature aggregation based video object detectors [10,35,38,44], as described in Eq. 1, w_{ij} is calculated as the cosine similarity between the neighboring feature $\boldsymbol{f}_j$ and $\boldsymbol{f}_i$, which is unrelated to k. Since k will not affect the aggregation process during inference time, it is possible to update the feature aggregation module in the current methods to a dynamic version, where an adaptive number of frames is applied to eliminate feature degradation to boost the performance of video object detection.

A simple idea to achieve this is to classify the current frames into multiple categories and determine the number of frames used for aggregation based on the categories. That is how our vanilla dynamic aggregation method comes out. In detail, we make the number of frames for feature aggregation dynamic based on the input frame: For frames where objects are with low motion speeds, fewer or even no neighboring frames are taken into account for feature aggregation. On the contrary, for those frames with severe feature degradation, more or even the whole neighboring frames are used to boost the detection performance.

To make the number of frames for feature aggregation dynamic, instead of a fixed number k, we use k_v frames for aggregation, which is determined by the current input frame $\boldsymbol{I}_i$. To achieve this, we first categorize the current frame $\boldsymbol{I}_i$ into θ categories based on the motion speeds of the objects in $\boldsymbol{I}_i$, where θ is a configurable parameter. Then, we use a function $\mathcal{S}_v(k, \delta)$ to determine k_v, where δ is an integer within the range of $(0, \theta]$ to represent the category of $\boldsymbol{I}_i$. Then, the formulation of $\mathcal{S}_v(k, \delta)$ can be represented as:

$$k_v = \mathcal{S}_v(k, \delta) = \left\lceil \delta \frac{k}{\theta} \right\rceil, \tag{3}$$

where $\lceil\rceil$ denotes the ceiling function. When θ is defined, k_v will have multiple discrete choices depending on δ. For example, when θ is chosen to be 3, $\delta = 1$ represents frames where objects are with slow motion speeds, and $k_v = \lceil \frac{k}{3} \rceil$ frames are used for feature aggregation for fast inference speed. Similarly, $\delta = 3$ means the current frame contains objects which move fast and we use $k_v = k$ frames to eliminate the feature degradation for a better performance. Given the number of frames k_v for the current frame $\boldsymbol{I}_i$, we slice a subset of the neighboring $\mathcal{N}_v(\boldsymbol{I}_i)$ from the whole range $\mathcal{N}(\boldsymbol{I}_i)$ using:

$$\mathcal{N}_v(\boldsymbol{I}_i) = \mathcal{G}(\mathcal{N}(\boldsymbol{I}_i), k_v), \tag{4}$$

where $\mathcal{G}(\cdot, k_v)$ is a sampling function to select k_v neighboring frames from k total neighboring frames. Therefore, Eq. 1 will be updated as:

$$\Delta \boldsymbol{f}_i^v = \sum_{\boldsymbol{I}_j \in \mathcal{N}_v(\boldsymbol{I}_i)} \left(w_{ij}^v \times \boldsymbol{f}_j\right), \tag{5}$$

where $\Delta \boldsymbol{f}_i^v, w_{ij}^v$ represent the aggregated features of frame $\boldsymbol{I}_i$ with dynamic neighborhood $\mathcal{N}_v(\boldsymbol{I}_i)$ and the corresponding weights.

During the training and inference processes, the category of the current frame $\boldsymbol{I}_i$, denoted as δ, is predicted based on the features of $\boldsymbol{I}_i$ and its neighboring frames $\mathcal{N}(\boldsymbol{I}_i)$. In detail, $\boldsymbol{f}_i$ is first concatenated with $\boldsymbol{f}_j$ and then fed into a mini-network $\mathcal{N}_{mot}^v$ to predict the category of $\boldsymbol{I}_i$, summarized as:

$$\delta = \mathcal{N}_{mot}^v(\texttt{cat}(\boldsymbol{f}_i, \boldsymbol{f}_j)), \qquad \forall \boldsymbol{I}_j \in \mathcal{N}(\boldsymbol{I}_i) \tag{6}$$

Following FGFA [44] and MEGA [3], we measure the motion speed of an object in a frame with motion IoU, denoted as s_m, using the averaged intersection-over-union (IoU) scores with its corresponding instances in the neighboring frames. Then, we divide each frame into θ classes based on s_m to generate the ground truth category δ^{gt}. For example, when θ is set to be 3, objects are classified into slow ($s_m > 0.9$), medium ($s_m \in [0.7, 0.9]$) and fast ($s_m < 0.7$) groups, respectively. Therefore, each frame is divided into 3 categories based on the motion speeds of the objects it contains. Cross entropy loss ($\mathcal{L}_{CE}$) between δ and δ^{gt} is calculated as the loss $\mathcal{L}_{mot}^v$ to optimize the network $\mathcal{N}_{mot}^v$, as:

$$\mathcal{L}_{mot}^v = \mathcal{L}_{\mathrm{CE}}\left(\delta, \delta^{gt}\right) \tag{7}$$

3.3 Deformable Dynamic Aggregation

With our proposed vanilla dynamic aggregation method, the current video object detectors' performance and inference speed can theoretically be well balanced. However, there are two issues.

The categories to determine k_v need to be predefined during the training process and are not reconfigurable at inference time. In other words, a well-trained model is not adaptive to multiple configurations. Take $\theta = 3$, which represents objects with slow, medium, and fast motion speeds, as an example. Given a frame with the motion IoU of $s_m = 0.75$, it is always categorized as medium motion speeds, which requires $\lceil \frac{2}{3} k \rceil$ frames for aggregation. If we would like to regard the frame as slow motion speeds to use fewer frames for aggregation when computational resources are limited, we need to modify the category ranges (for example, $s_m \in [0.5, 0.7]$ for the medium group) and train a new model again. It is inconvenient and unsuitable for real-world applications when θ is set to be very large or the category ranges are switched frequently.

Moreover, experiments show that the vanilla dynamic aggregation module does not perform well in detecting objects of small sizes. Most of the existing feature aggregation-based video object detectors use Faster R-CNN [27] as the baseline method without feature pyramid networks [23]. Therefore, more frames are required to enhance the feature representations of small objects during the aggregation process. However, in the vanilla dynamic aggregation module, the sizes of objects in the frames are not considered.

To solve the issues mentioned above, we extend the vanilla dynamic aggregation module to a deformable version, which is more effective and reconfigurable. Instead of classifying the input frame $\boldsymbol{I}_i$ into θ categories, we use a function σ to project the $s \in [0, 1]$ in the range of 0 and 1, where s is a score which takes both the motion IoU $s_m \in [0, 1]$ and size $s_s \in [0, 1]$ of objects in the current frame $\boldsymbol{I}_i$ into account. Therefore, Eq. 3 is updated to be:

$$k_d = \mathcal{S}_d(k, s) = \lceil \sigma(s)\, k \rceil = \lceil \sigma(s_m s_s)\, k \rceil \tag{8}$$

During the inference time, we can determine k_d by selecting σ in configure files. In real-world applications, when there are enough computational resources like applications on servers, we can choose σ, which casts $s \in [0, 1]$ in the range of 0 and 1. When there are not enough resources, like applications on cellphones or servers where partial machines are under maintenance, we can reload the configure file where a new σ projects s in a new range (e.g. $[0, 0.5]$) to use fewer frames for aggregation without the need of training a new model.

Similarly, given k_d for the current frame $\boldsymbol{I}_i$, the sampled neighboring frames $\mathcal{N}_d(\boldsymbol{I}_i)$ is represented as Eq. 9 and deformable dynamic aggregation process is summarized as Eq. 10.

$$\mathcal{N}_d(\boldsymbol{I}_i) = \mathcal{G}(\mathcal{N}(\boldsymbol{I}_i), k_d), \tag{9}$$

$$\Delta \boldsymbol{f}_i^d = \sum_{\boldsymbol{I}_j \in \mathcal{N}_d(\boldsymbol{I}_i)} (w_{ij}^d \times \boldsymbol{f}_j), \tag{10}$$

where Δf_i^d, w_{ij}^d represents the enhanced features of frame $\boldsymbol{I}_i$ with dynamic neighborhood $\mathcal{N}_d(\boldsymbol{I}_i)$ generated from the deformable dynamic aggregation module and the corresponding weights.

During the training and inference processes, we use mini-networks $\mathcal{N}_{mot}^d$ and $\mathcal{N}_{size}$ to estimate the averaged motion IoU s_m and size s_s of objects in the current frame I_i, respectively. Similar to Eq. 6, the above process can be represented as:

$$\begin{aligned} s_m &= \mathcal{N}_{mot}^d(\texttt{cat}(\boldsymbol{f}_i, \boldsymbol{f}_j)), \qquad \forall \boldsymbol{I}_j \in \mathcal{N}(\boldsymbol{I}_i) \\ s_s &= \mathcal{N}_{size}(\boldsymbol{f}_i) \end{aligned} \tag{11}$$

For each frame $\boldsymbol{I}_i$, we calculate the averaged bounding box area of the objects it contains as the ground truth s_s^{gt} for $\mathcal{N}_{size}$. Motion IoU ground truths s_m^{gt} are measured with the same pipeline as FGFA [44] and the vanilla dynamic aggregation module. Then, mean square error loss $\mathcal{L}_{\mathrm{MSE}}$ is applied to optimize $\mathcal{N}^d_{mot}$ and $\mathcal{N}_{size}$ as:

$$\begin{aligned}\mathcal{L}^d_{mot} &= \mathcal{L}_{\mathrm{MSE}}\left(s_m, s_m^{gt}\right)\\ \mathcal{L}_{size} &= \mathcal{L}_{\mathrm{MSE}}\left(s_s, s_s^{gt}\right)\end{aligned} \tag{12}$$

3.4 Inplace Distillation Loss

Our proposed method aims to balance the inference speed and performance of the existing feature aggregation-based video object detectors. Therefore, when it comes to the situation that k_v (k_d) is small, we would like features $\Delta \boldsymbol{f}_i^v (\Delta \boldsymbol{f}_i^d)$ aggregated with $k_v(k_d)$ frames similar to $\Delta \boldsymbol{f}_i$ aggregated with k frames. Here we borrow the idea from knowledge distillation that the performance of student models can be boosted when trained with the soft predictions of teacher models.

In our method, we treat the full model with k frames for aggregation as the teacher network and those with fewer frames $k_v(k_d) < k$ as the student models. We add an extra $\mathcal{L}_{dst}$ during the training process to ensure the features aggregated with fewer neighboring frames can perform similarly to those aggregated with the whole neighboring frames, so that the detection accuracy of objects with small/medium sizes can be improved. Here we use deformable dynamic aggregation as an example. We calculate the mean square loss ($\mathcal{L}_{\mathrm{MSE}}$) between $\Delta \boldsymbol{f}_i$ and $\Delta \boldsymbol{f}_i^d$ as:

$$\mathcal{L}_{dst} = \mathcal{L}_{\mathrm{MSE}}\left(\Delta \boldsymbol{f}_i, \Delta \boldsymbol{f}_i^d\right) \tag{13}$$

Inplace distillation loss is only applied during the training process; thus, it will not affect the inference speed.

4 Experiments

4.1 Experiment Setup

For mini-network $\mathcal{N}^v_{mot}$ and $\mathcal{N}^d_{mot}$, a one-layer convolutional layer is used to fuse the concatenated features. Then a global average pooling operation is applied to reduce the spatial and temporal resolutions. Next, the pooled feature is fed into a 2-layer MLP for classification ($\mathcal{N}^v_{mot}$) or regression ($\mathcal{N}^d_{mot}$). The object size estimation network $\mathcal{N}_{size}$ has the same architecture as $\mathcal{N}^d_{mot}$ except that the input is $\boldsymbol{f}_i$ rather than the concatenation of $\boldsymbol{f}_i$ and $\boldsymbol{f}_j$. For vanilla dynamic aggregation, θ is set to be 3 unless otherwise stated.

We evaluate our proposed methods on the ImageNet VID benchmark [28] as the recent state-of-the-art video object detection models [3,7,38,44]. Following the widely used protocols in [3,38,44], we train our model on a combination of ImageNet VID and DET datasets. We implement our method mainly based on mmtracking [4][2]. The whole network is trained on 8 T A100 GPUs. During the inference process, 30 neighboring frames are used for feature aggregation.

[2] There are around 2% mAP fluctuations in performance, and we take the mean after running 5 experiments.

Table 1. Performance comparison with the recent state-of-the-art video object detection approaches on ImageNet VID validation set.

	Methods	FPS	mAP	$AP_{0.5}$	$AP_{0.75}$	AP_s	AP_m	AP_l
ResNet-50	FGFA [44]	5.8	46.7	74.3	51.5	5.7	21.8	52.9
	FGFA + Vanilla DA	7.9	46.4	73.9	51.1	5.3	20.8	52.5
	FGFA + Deformable DA	7.6	46.6	74.1	51.2	6.5	21.6	52.7
	SELSA [38]	5.0	48.1	77.9	52.8	8.3	26.2	54.3
	SELSA + Vanilla DA	9.4	47.2	76.5	51.1	7.6	25.8	52.9
	SELSA + Deformable DA	8.8	47.9	77.5	52.4	8.7	26.1	53.5
	Temporal ROI Align [10]	1.5	48.1	79.0	52.1	7.0	26.2	54.1
	Temporal ROI Align + Vanilla DA	3.9	46.9	77.8	51.0	6.4	25.7	52.6
	Temporal ROI Align + Deformable DA	3.5	47.8	78.8	51.7	7.2	25.9	53.5
ResNet-101	FGFA	5.1	50.2	77.6	56.1	7.3	24.0	56.3
	FGFA + Vanilla DA	7.5	49.7	77.2	54.9	6.9	24.1	56.1
	FGFA + Deformable DA	7.1	50.1	77.5	55.8	7.9	23.8	56.1
	SELSA [38]	4.5	52.1	81.3	57.4	9.0	28.1	58.1
	SELSA + Vanilla DA	8.5	51.2	80.0	57.0	7.8	26.8	57.3
	SELSA + Deformable DA	8.0	52.0	81.0	56.8	9.1	27.9	57.8
	Temporal ROI Align [10]	1.2	51.3	82.4	56.1	10.4	28.7	56.9
	Temporal ROI Align + Vanilla DA	3.6	50.4	81.8	55.3	9.3	27.5	55.1
	Temporal ROI Align + Deformable DA	3.3	50.9	82.0	55.6	10.5	29.5	56.3

4.2 Main Results

In this section, we conduct experiments on vanilla, and deformable feature aggregation with the current video object detectors on the ImageNet VID benchmark [28]. We compare state-of-the-art feature aggregation-based video object detectors integrated with our proposed methods. The results are summarized in Table 1. For local aggregation methods like FGFA [44], our proposed dynamic aggregation can significantly improve the inference speeds while maintaining comparable performance like mAP and $AP_{0.5}$. We argue that this is because FGFA aggregates feature with local temporal neighboring frames, which share much redundant information. Therefore, removing those redundancies during the inference process will not affect the final predictions much, especially when the objects are at slow motion speeds.

For global aggregation methods like SELSA [38], and Temporal ROI Align [10], our proposed methods can still improve the inference speeds by a large margin yet at the sacrifice of performance like $AP_{0.75}$. We argue that this is because global aggregation methods select features with similar representations for aggregation, and removing several frames during aggregation may have a harmful effect on precisely localizing the bounding boxes, considering $AP_{0.75}$ drops more compared with $AP_{0.5}$. Meanwhile, we notice that compared with vanilla dynamic aggregation, the deformable version has much better performance (even better than the original model) on small object detection when taking the object sizes into account, which validates the effectiveness of the pro-

Fig. 4. Examples of FGFA [44] integrated with deformable dynamic aggregation on detecting objects with different motion speeds in the video.

posed modules. We argue that our methods can adaptively select the frames for aggregation, which provide adequate but not redundant information for object detection. Figure 4 shows several examples of video object detection results integrated with deformable dynamic aggregation. From the figure, our proposed methods can precisely predict the bounding boxes and categories of objects in each video frame.

4.3 Model Analysis

In this section, we conduct extensive ablation study experiments to analyze the structures and parameters of our proposed modules. By default, we use FGFA [44] with the backbone of ResNet-50 [15] as the model to conduct experiments unless otherwise stated. In this section, we mainly analyze the proposed deformable dynamic aggregation module.

Analysis of Sampling Function $\mathcal{G}$. We conduct experiments with deformable feature aggregation on the choices of sampling function as Table 3. "Nearest" and "Furthest" represent choosing the closest and furthest k_d frames for aggregation, while "Bin" means binning the k frames into k_d buckets and sample 1 frame from each bucket. For example, suppose the current frame is the 11th of a video with 21 frames and $k_d = 7$, Table 2 shows the comparison of selected frames with different sampling functions. Besides the three sampling functions mentioned above, we also compare with random sampling results. From Table 2, "Nearest" sampling has the best performance compared with the other methods, while "Bin" sampling has a comparable result. "Furthest" and "Random" sampling methods have a poor performance, and we argue that this is because there are

Table 2. Comparison of video object detection results with different sampling functions $\mathcal{G}$ on FGFA [44] with ResNet-50 [15] as the backbone.

Method	Nearest	Furthest	Bin
Selected frames	8, 9, 10, 11, 12, 13, 14	1, 2, 3, 11, 19, 20, 21	2, 5, 8, 11, 14, 17, 20

Table 3. Example of selected frames with different sampling function $\mathcal{G}$.

Method	mAP	$AP_{0.5}$	$AP_{0.75}$	AP_s	AP_m	AP_l
Nearest	47.0	74.5	51.2	6.5	21.6	53.3
Furthest	45.5	72.5	50.1	5.6	20.8	51.7
Random	45.9	73.7	50.6	5.8	21.0	52.3
Bin	46.8	74.4	51.0	6.4	21.4	53.2

Table 4. Comparison of video object detection results with different mapping functions σ on FGFA [44] with ResNet-50 [15] as the backbone.

Function	FPS	mAP	$AP_{0.5}$	$AP_{0.75}$	AP_s	AP_m	AP_l
Linear ($y = 1 - x$)	7.6	46.6	74.1	51.2	6.5	21.6	52.7
Sqrt ($y = 1 - \sqrt{x}$)	7.9	46.4	74.0	51.0	6.1	21.3	52.6
Quadratic ($y = 1 - x^2$)	7.4	46.7	74.3	51.4	6.7	21.5	52.7
Learnable ($y = \texttt{MLP}(x)$)	7.1	46.9	74.4	51.5	6.9	21.9	53.1

not enough effective and informative frames for aggregation when using these two strategies.

Analysis of Mapping Function σ. We analyze the choice of mapping function σ in our proposed deformable feature aggregation as Table 4. We compare four different mapping functions, namely, linear, square root, quadratic and learnable function by retraining the models with the corresponding σ. From Table 4, compared with linear function, when choosing learnable networks as mapping function, the performance is the best at the sacrifice of inference speed. Square root and quadratic functions can balance the inference speed and accuracy by mapping s into different distributions.

Comparison with Knowledge Distillation. We also conduct experiments to compare with knowledge distillation results. We notice that FGFA aggregated with 15 frames have a similar inference speed as our proposed methods. Therefore, we use an FGFA aggregated with 15 frames to distill the knowledge from an FGFA aggregated with 30 frames and compare the results with our proposed method in Table 5. From the table, the distillation-only method is not as good as our proposed methods despite similar inference speed. Also, the model will perform worse if trained without inplace distillation loss.

Table 5. Comparison between FGFA distilled from a model aggregated with more frames and the proposed method. † means models without inplace distillation loss.

Method	FPS	mAP	$AP_{0.5}$	$AP_{0.75}$	AP_s	AP_m	AP_l
FGFA (15 frames)	7.7	46.2	73.7	50.5	5.8	20.8	52.0
FGFA (15 frames) + Distill	7.7	46.4	73.9	50.7	5.9	21.0	52.3
FGFA (30 frames)	5.8	46.7	74.3	51.5	5.7	21.8	52.9
FGFA (30 frames)+ Ours†	7.6	46.5	73.9	50.9	5.6	20.9	52.5
FGFA (30 frames)+ Ours	7.6	46.6	74.1	51.2	6.5	21.6	52.7

Table 6. Comparison between FGFA integrated with/without our proposed methods on video object detection on objects with different motion speeds.

Method	FPS	mAP	AP_{50}	AP_{slow}	AP_{medium}	AP_{fast}
FGFA	5.8	46.8	74.3	83.8	72.2	50.5
FGFA + Ours	7.6	46.5	74.2	83.7	71.8	50.3

Analysis of Motion Speeds. Besides object sizes, we also compare experimental results to analyze the effects on object motion speeds. Following MEGA [3], we categorize objects into slow, medium, and fast groups and calculate their corresponding accuracy as Table 6. The table shows that the performance on objects with slow motion speeds drops a little when integrated with a deformable dynamic aggregation module. However, detection accuracy on objects with medium motion speeds decreases by 0.4%. We argue that this is because situations like occlusion or rare positions are always treated as objects with medium motion speeds, and a few frames are sampled for aggregation, which are not enough to handle those cases.

5 Conclusion

Existing feature aggregation-based video object detectors usually apply a fixed number of frames to enhance objects' representations and boost performance. Therefore, the performance and inference speed are heavily influenced by the number of frames used for aggregation. In this paper, we aim to perform dynamic aggregation to the current methods to balance the performance and inference speed. We first propose vanilla dynamic aggregation and then extend to a deformable version which can adaptively and reconfigurably select frames used for feature enhancement according to the input frames. Furthermore, we introduce the inplace distillation loss to boost the performance of frames not fully aggregated. Extensive experiments on the ImageNet VID benchmark validate the effectiveness and efficiency of our proposed methods. We hope our approaches can bring some ideas to the efficient video object detection field.

References

1. Cai, Z., Vasconcelos, N.: Cascade R-CNN: high quality object detection and instance segmentation. IEEE Trans. Pattern Anal. Mach. Intell. (2019). https://doi.org/10.1109/tpami.2019.2956516
2. Chen, K., et al.: Optimizing video object detection via a scale-time lattice. In: Proceedings of the IEEE Conference on Computer Vision and Pattern Recognition, pp. 7814–7823 (2018)
3. Chen, Y., Cao, Y., Hu, H., Wang, L.: Memory enhanced global-local aggregation for video object detection. In: Proceedings of the IEEE/CVF Conference on Computer Vision and Pattern Recognition, pp. 10337–10346 (2020)
4. MMT Contributors: MMTracking: OpenMMLab video perception toolbox and benchmark (2020). https://github.com/open-mmlab/mmtracking
5. Cui, Y., Yan, L., Cao, Z., Liu, D.: TF-blender: temporal feature blender for video object detection. In: Proceedings of the IEEE/CVF International Conference on Computer Vision, pp. 8138–8147 (2021)
6. Cui, Y., Yang, L., Liu, D.: Dynamic proposals for efficient object detection. arXiv preprint arXiv:2207.05252 (2022)
7. Deng, J., Pan, Y., Yao, T., Zhou, W., Li, H., Mei, T.: Relation distillation networks for video object detection. In: Proceedings of the IEEE/CVF International Conference on Computer Vision, pp. 7023–7032 (2019)
8. Duan, K., Bai, S., Xie, L., Qi, H., Huang, Q., Tian, Q.: Centernet: keypoint triplets for object detection. In: Proceedings of the IEEE/CVF International Conference on Computer Vision, pp. 6569–6578 (2019)
9. Girshick, R.: Fast R-CNN. In: Proceedings of the IEEE International Conference on Computer Vision (2015)
10. Gong, T., et al.: Temporal ROI align for video object recognition. In: Proceedings of the AAAI Conference on Artificial Intelligence, vol. 35, pp. 1442–1450 (2021)
11. Han, L., Wang, P., Yin, Z., Wang, F., Li, H.: Exploiting better feature aggregation for video object detection. In: Proceedings of the 28th ACM International Conference on Multimedia, pp. 1469–1477 (2020)
12. Han, L., Wang, P., Yin, Z., Wang, F., Li, H.: Class-aware feature aggregation network for video object detection. IEEE Trans. Circ. Syst. Video Technol. **32**(12), 8165–8178 (2021)
13. Han, W., et al.: Seq-NMS for video object detection. arXiv preprint arXiv:1602.08465 (2016)
14. He, K., Gkioxari, G., Dollár, P., Girshick, R.: Mask R-CNN. In: Proceedings of the IEEE International Conference on Computer Vision, pp. 2961–2969 (2017)
15. He, K., Zhang, X., Ren, S., Sun, J.: Deep residual learning for image recognition. In: Proceedings of the IEEE Conference on Computer Vision and Pattern Recognition, pp. 770–778 (2016)
16. He, L., et al.: End-to-end video object detection with spatial-temporal transformers. In: Proceedings of the 29th ACM International Conference on Multimedia, pp. 1507–1516 (2021)
17. Howard, A.G., et al.: Mobilenets: efficient convolutional neural networks for mobile vision applications. arXiv preprint arXiv:1704.04861 (2017)
18. Jiang, Z., Gao, P., Guo, C., Zhang, Q., Xiang, S., Pan, C.: Video object detection with locally-weighted deformable neighbors. In: Proceedings of the AAAI Conference on Artificial Intelligence, vol. 33, pp. 8529–8536 (2019)

19. Jiang, Z., et al.: Learning where to focus for efficient video object detection. In: Vedaldi, A., Bischof, H., Brox, T., Frahm, J.-M. (eds.) ECCV 2020. LNCS, vol. 12361, pp. 18–34. Springer, Cham (2020). https://doi.org/10.1007/978-3-030-58517-4_2
20. Kang, K., et al.: T-CNN: tubelets with convolutional neural networks for object detection from videos. IEEE Trans. Circuits Syst. Video Technol. **28**(10), 2896–2907 (2017)
21. Kang, K., Ouyang, W., Li, H., Wang, X.: Object detection from video tubelets with convolutional neural networks. In: 2016 IEEE Conference on Computer Vision and Pattern Recognition (CVPR) (2016). https://doi.org/10.1109/cvpr.2016.95
22. Law, H., Deng, J.: CornerNet: detecting objects as paired keypoints. In: Ferrari, V., Hebert, M., Sminchisescu, C., Weiss, Y. (eds.) Computer Vision – ECCV 2018. LNCS, vol. 11218, pp. 765–781. Springer, Cham (2018). https://doi.org/10.1007/978-3-030-01264-9_45
23. Lin, T.Y., Dollár, P., Girshick, R., He, K., Hariharan, B., Belongie, S.: Feature pyramid networks for object detection. In: Proceedings of the IEEE Conference on Computer Vision and Pattern Recognition, pp. 2117–2125 (2017)
24. Liu, C., et al.: Progressive neural architecture search. In: Proceedings of the European Conference on Computer Vision (ECCV), pp. 19–34 (2018)
25. Liu, W., et al.: SSD: single shot MultiBox detector. In: Leibe, B., Matas, J., Sebe, N., Welling, M. (eds.) ECCV 2016. LNCS, vol. 9905, pp. 21–37. Springer, Cham (2016). https://doi.org/10.1007/978-3-319-46448-0_2
26. Redmon, J., Divvala, S., Girshick, R., Farhadi, A.: You only look once: unified, real-time object detection. In: Proceedings of the IEEE Conference on Computer Vision and Pattern Recognition, pp. 779–788 (2016)
27. Ren, S., He, K., Girshick, R., Sun, J.: Faster R-CNN: towards real-time object detection with region proposal networks. In: Advances in Neural Information Processing Systems, pp. 91–99 (2015)
28. Russakovsky, O., et al.: Imagenet large scale visual recognition challenge. Int. J. Comput. Vision **115**(3), 211–252 (2015)
29. Sandler, M., Howard, A., Zhu, M., Zhmoginov, A., Chen, L.C.: Mobilenetv 2: inverted residuals and linear bottlenecks. In: Proceedings of the IEEE Conference on Computer Vision and Pattern Recognition, pp. 4510–4520 (2018)
30. Shvets, M., Liu, W., Berg, A.C.: Leveraging long-range temporal relationships between proposals for video object detection. In: Proceedings of the IEEE/CVF International Conference on Computer Vision, pp. 9756–9764 (2019)
31. Sun, P., et al.: Sparse R-CNN: end-to-end object detection with learnable proposals. In: Proceedings of the IEEE/CVF Conference on Computer Vision and Pattern Recognition, pp. 14454–14463 (2021)
32. Tan, M., Le, Q.: Efficientnet: rethinking model scaling for convolutional neural networks. In: International Conference on Machine Learning, pp. 6105–6114. PMLR (2019)
33. Tian, Z., Shen, C., Chen, H., He, T.: FCOS: fully convolutional one-stage object detection. In: Proceedings of International Conference on Computer Vision (ICCV) (2019)
34. Wang, S., Lu, H., Deng, Z.: Fast object detection in compressed video. In: Proceedings of the IEEE/CVF International Conference on Computer Vision, pp. 7104–7113 (2019)
35. Wang, S., Zhou, Y., Yan, J., Deng, Z.: Fully motion-aware network for video object detection. In: Proceedings of the European Conference on Computer Vision (ECCV), pp. 542–557 (2018)

36. Wang, X., Huang, Z., Liao, B., Huang, L., Gong, Y., Huang, C.: Real-time and accurate object detection in compressed video by long short-term feature aggregation. Comput. Vis. Image Underst. **206**, 103188 (2021)
37. Wu, B., et al.: FBNet: hardware-aware efficient convnet design via differentiable neural architecture search. In: Proceedings of the IEEE/CVF Conference on Computer Vision and Pattern Recognition, pp. 10734–10742 (2019)
38. Wu, H., Chen, Y., Wang, N., Zhang, Z.: Sequence level semantics aggregation for video object detection. In: Proceedings of the IEEE/CVF International Conference on Computer Vision, pp. 9217–9225 (2019)
39. Xu, Z., Hrustic, E., Vivet, D.: Centernet heatmap propagation for real-time video object detection. In: ECCV (2020)
40. Yao, C.-H., Fang, C., Shen, X., Wan, Y., Yang, M.-H.: Video object detection via object-level temporal aggregation. In: Vedaldi, A., Bischof, H., Brox, T., Frahm, J.-M. (eds.) ECCV 2020. LNCS, vol. 12359, pp. 160–177. Springer, Cham (2020). https://doi.org/10.1007/978-3-030-58568-6_10
41. Yu, J., Huang, T.: Autoslim: towards one-shot architecture search for channel numbers (2019)
42. Yu, J., Huang, T.: Universally slimmable networks and improved training techniques (2019)
43. Yu, J., Yang, L., Xu, N., Yang, J., Huang, T.: Slimmable neural networks (2018)
44. Zhu, X., Wang, Y., Dai, J., Yuan, L., Wei, Y.: Flow-guided feature aggregation for video object detection. In: Proceedings of the IEEE International Conference on Computer Vision, pp. 408–417 (2017)
45. Zhu, X., Xiong, Y., Dai, J., Yuan, L., Wei, Y.: Deep feature flow for video recognition. In: Proceedings of the IEEE Conference on Computer Vision and Pattern Recognition, pp. 2349–2358 (2017)

Computational Photography, Sensing, and Display

Robust Human Matting via Semantic Guidance

Xiangguang Chen[1], Ye Zhu[2], Yu Li[3(✉)], Bingtao Fu[1], Lei Sun[1], Ying Shan[2], and Shan Liu[1]

[1] Platform Technologies, Tencent Online Video, Shenzhen, China
seanxgchen@tencent.com
[2] ARC Lab, Tencent PCG, Shenzhen, China
samuelzhu@tencent.com
[3] International Digital Economy Academy (IDEA), Shenzhen, China
liyu@idea.edu.cn

Abstract. Automatic human matting is highly desired for many real applications. We investigate recent human matting methods and show that common bad cases happen when semantic human segmentation fails. This indicates that semantic understanding is crucial for robust human matting. From this, we develop a fast yet accurate human matting framework, named Semantic Guided Human Matting (**SGHM**). It builds on a semantic human segmentation network and introduces a light-weight matting module with only marginal computational cost. Unlike previous works, our framework is data efficient, which requires a small amount of matting ground-truth to learn to estimate high quality object mattes. Our experiments show that trained with merely 200 matting images, our method can generalize well to real-world datasets, and outperform recent methods on multiple benchmarks, while remaining efficient. Considering the unbearable labeling cost of matting data and widely available segmentation data, our method becomes a practical and effective solution for the task of human matting. Source code is available at https://github.com/cxgincsu/SemanticGuidedHumanMatting.

1 Introduction

Human matting aims to predict an alpha matte to extract human foreground from an input image or video, which has many important applications in visual processing. To achieve that, a green screen is often required for studio solutions. However, a green screen is not always available in many real scenarios, such as daily video conferencing and background replacement effects shot with mobile devices. Therefore, human matting methods without a green screen are highly desired. Many previous works use an additional trimap for matting, which indicates three kinds of regions in an image, namely foreground, background, and unknown. However, it requires careful manual annotation to obtain a trimap.

X. Chen and Y. Zhu—These authors contributed equally to this work.

L. Wang et al. (Eds.): ACCV 2022, LNCS 13842, pp. 613–628, 2023.
https://doi.org/10.1007/978-3-031-26284-5_37

Background matting approaches [1,2] are recently proposed which use a pre-recorded background image as a prior. Though decent results are obtained, it only can handle cases with a static background and a fixed camera pose.

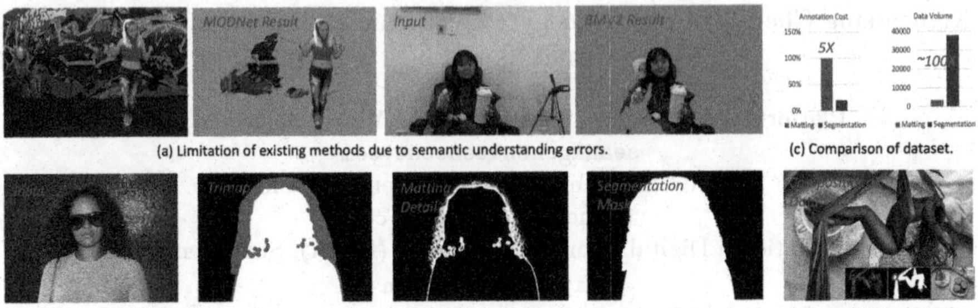

(a) Limitation of existing methods due to semantic understanding errors. (c) Comparison of dataset.

(b) The visualization of trimap used in traditional matting method, matting detail, and human segmentation mask. (d) Compositional matting data.

Fig. 1. Limitation of existing method and motivations of this work. (a) Common failure cases of latest works [1,3] happen when semantic understanding fails. (b) Traditional matting methods rely on the input of a trimap. Since matting details are located around the human mask boundaries [4], the coarse segmentation mask can also be leveraged as a prior in matting. (c) Currently, segmentation data is much easier to annotate, and the amount of publicly available data is much larger than that of matting data. (d) Compositing foreground with different backgrounds can enlarge the matting dataset size but it has a domain gap as it looks unreal [5].

Many recent works focus on developing methods towards automatic human matting. Some early attempts [6,7] try to generate pseudo trimap as a first step and predict a matte from the trimap. Due to limited training data, these methods cannot generalize well to real-world examples [2]. Another drawback of these methods is that they cannot run in real-time which is required for many applications, such as background replacement in live video conferencing. The recent work MODNet [3] proposes a fast and fully automatic portrait matting method. RVM [8] is another recent work which leverages temporal information in a video to improve robustness and stability.

In this work, we aim to develop a robust, accurate, and fast method for automatic human matting, which shares the same goal as MODNet [3] and RVM [8]. We investigate the failure cases of existing automatic methods on real-world examples and observe that these failure cases are often due to inaccurate semantic understanding. As shown in Fig. 1(a), parts of the background are wrongly predicted as foreground or part of human body are wrongly segmented. This indicates a weak semantic understanding ability of these state-of-the-art (SOTA) methods. In order to enhance their ability of semantic understanding while keeping fine-grained details of matting, we seek to utilize semantic segmentation task to guide matting process. There are three reasons behind this motivation. 1) Segmentation mask determines the overall accuracy of foreground and background predictions, and fine-grained structures only appear around the mask. This indicates that a semantic human mask can replace a trimap (Fig. 1(b)) and be used as a prior condition for matting [4]. 2) The labeling of high-quality matting requires skillful annotators and is very time-consuming. For that, the amount of available training data for matting is quite limited (at the order of hundreds and thousands) compare to segmentation task, which require only simple line drawings around boundaries. As a matter of fact, there are many human segmentation datasets at a scale that is two or more magnitude larger (Fig. 1(c)). A larger amount of data is of great significance to the generalization ability on real-world images. 3) Synthetic datasets created by compositing images (Fig. 1(d)) are also used in training matting models, but they have a clear limitation due to the drastic domain gap between synthetic and real-world images. This prevents the trained models from generalizing to real-world examples. The work [5] analyzes the domain gap issue systematically. Our approach does not suffer from this issue by using less of such data.

Based on the above analysis, We propose a multi-stage framework to predict semantic segmentation mask and matting alpha successively. A segmentation sub-network is first employed for the task of segmentation, and then it is reused to guide the matting process to focus on the surrounding area of the segmentation mask. To achieve real-time efficiency as well as better performance, we let the two tasks share the encoder part of the model, which has been proved superior to separated encoders in [5]. By this design, our matting module successfully handled many challenging cases. In summary, our network consists of a shared encoder, a segmentation decoder and a matting decoder, and the segmentation decoder feeds useful intermediate information to the matting decoder. In training, a two-stage pipeline is proposed. Firstly, the encoder and the segmentation decoder are trained with publicly available segmentation datasets. With these data, our segmentation sub-network is trained to predict robust human masks. Secondly, 269 matting images are employed to train the matting decoder. To comprehensively evaluate the performance of matting methods, we adopt 5 benchmarks to carry out qualitative and quantitative comparison. One of them is our self-collected dataset from complex scenarios, such as diverse background, multiple human, body accessories, and low light. Our method outperforms all other methods across all benchmarks.

We summarize our contribution as follows:

1. We develop a robust, accurate and efficient human matting framework, which utilizes shared encoder for both segmentation and matting. It gives our method the ability to use powerful semantic understanding to guide matting process meanwhile help to reduce computation.
2. The proposed framework can make fully use of coarse mask training data and reduce matting reliance on high-quality and large number of annotations. With only about 200 matting images, our method is able to produce high quality alpha details.
3. Extensive experiments show our method achieves the state-of-the-art results on multiple benchmarks.

2 Related Work

In this section, we review matting with auxiliary input and automatic matting, which are related to our work. We also review segmentation as segmentation provides the rough mask of human region.

Matting with Auxiliary Input. Early methods are mostly optimization or filter based which require an additional trimap as input [9–19]. Deep learning is introduced in trimap-based matting methods in [20–22] that use a deep network for trimap-based matting. These trimap-based methods are often general to different matting target objects but it requires the user to provide trimap annotations. Background mattings [1,2] are recently proposed to replace the trimap input with a pre-recorded background image as a prior condition. Although background matting can generate decent results on static background, it cannot be applied to camera moving circumstances. Recently proposed mask-guided method [4] achieves SOTA results once a coarse is provided. In their work, mask is generated from manual annotation or segmentation output, which greatly limits the convenience of use. Our goal is to incorporate the mask generation into the matting process, so as to realize fully automatic matting and still keeping real-time running.

Automatic Matting. Fully automatic matting without any additional input has been pursued [23–25]. Methods in [26,27] studies class agnostic matting but cannot generalize well. Some methods like [6,7,28–30] dedicate to human matting. In this direction, the latest MODNet [3] aims at fast portrait matting and RVM [8] is towards robust human matting using temporal information. For MODNet, it performs well in the portrait image, but easily fails in full body image. Recent work P3M-Net [31] proposes a dual decoder to do human matting, which is similar to us. But there are several significant differences: 1) P3M-Net use segmentation decoder to generate a pseudo trimap while our segmentation predicts real mask. P3M-Net predicts alpha details only on trimap unknown region. This setting tends to output false matting results when trimap is wrongly predicted. Our matting decoder treats mask as guidance and regresses alpha at the whole image. Under this setting, the matting decoder is given an opportunity to correct semantic errors. 2) Our segmentation decoder and matting decoder are trained at two separate stages. At the segmentation training stage, the segmentation decoder is strongly supervised by a large dataset. As a result, the segmentation decoder predicts more robust results than the weakly supervised result in P3M-Net. 3) Another advantage of our model is it is data-efficient in that we only use a very small amount of high-precision data to train the matting decoder.

Segmentation. Semantic segmentation assigns a semantic class label to every pixel in the scene. Its difference with matting is that it predicts a hard binary mask that belongs to either foreground or background and cannot generate fine details and transparent value as in matte. So directly applying segmentation mask to image and video composition will generate hard boundary at the foreground object, leaving noticeable artifacts when replacing the backgrounds. However, segmentation can provide strong semantic cues of the object location which facilitate our matting task. Many deep learning-based semantic segmentation are fully convolutional and some effective modules like Atrous Spatial Pyramid Pooling (ASPP) [32] are proposed. We follow them in our segmentation network design.

3 Method

Given a color image I, the matting task can be formulated as follows:

$$I = \alpha F + (1 - \alpha)B,\ \alpha \in [0, 1], \tag{1}$$

where F, B are foreground and background, and α is the alpha matte denoting where is foreground part located. For image matting problem, we should predict the alpha matte from the input color image, which is a hard and ill-posed task. As mentioned earlier, existing methods rely on additional auxiliary inputs like trimap or pre-captured background. Automatic method like RVM is not robust against semantic error. Based on this, we try to design a framework to better leverage the semantic prior from segmentation, but produce fine detail and transparent matte values. A straightforward way is to rely on a semantic segmentation mask and generate the matting results using a new matting network. This setup is developed and demonstrated in mask-guided (MG) matting [4]. The two-step setup treats segmentation and matting as two separate tasks and has a few drawbacks. First of all, the matting network only uses the predicted segmentation map and ignores the rich semantic features. Second, using a separate matting network will extract features again from image and introduce additional computation, which slows down the speed noticeably on high resolution.

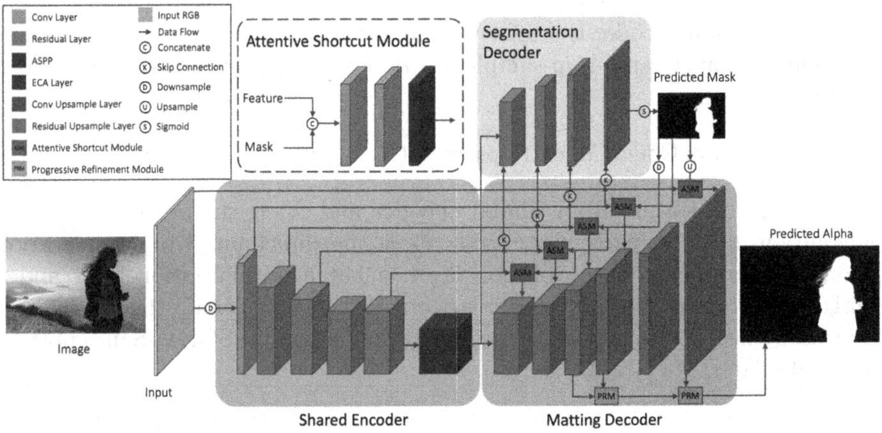

Fig. 2. The network structure of our SGHM. High-resolution image is first downsampled for the shared encoder, then the segmentation decoder is used to generate a coarse semantic mask prediction. We propose an Attentive Shortcut Module (ASM) to adaptively fuse shared features and masks. Finally, the matting decoder refines the unknown area of human margin and predicts the alpha matte.

Based on above analysis, we propose a new human matting method named Semantic Guided Human Matting (SGHM), which uses a segmentation network to guide human matting. Specifically, we share the encoder between segmentation and matting task. Thus, matting task can learn accurate semantic understanding from reusing the rich semantic features in encoder and focus on predicting alpha details in matting decoder.

As shown in Fig. 2, our SGHM consists of a shared encoder to extract image features, a segmentation decoder to predict image segmentation mask, and a matting decoder with Progressive Refinement Module (PRM) [4] to predict a high-resolution matting result. We propose to use an Attentive Shortcut Module (ASM) to combine the features from encoder and mask from segmentation decoder for matting decoder.

3.1 Shared Encoder

As mentioned above, we propose to improve matting results by using semantic human segmentation features. So we make the segmentation and matting tasks share an encoder. More specifically, we first train the encoder and segmentation decoder as segmentation model, and then fix the parameters of the encoder and train the matting decoder with segmentation features extracted from encoder. We adopt ResNet50 [33] as feature extraction backbone followed by a ASPP module [32] for shared encoder, which extracted features at $\frac{1}{4}, \frac{1}{8}, \frac{1}{16}, \frac{1}{32}, \frac{1}{64}$ scale for two decoders with an input image at $\frac{1}{4}$ scale, which can be denoted as F_0, F_1, F_2, F_3, F_4.

3.2 Segmentation Decoder

Our segmentation decoder is a light-weight and efficient module, which contains 4 convolution layers and 4 up-sample layers. For each convolution layer, it can be defined as:

$$X_i = \text{Conv}(\text{Concat}(\text{Upsample}(X_{i+1}), F_i), i = 3, 2, 1, 0, \tag{2}$$

where F_i is the feature from shared encoder and X_i is the output feature of convolution layer. In particular, $X_4 = F_4$ is the direct input of segmentation decoder. Following each convolution layer, a batch normalization layer and a ReLU layer are attached except the last one. Finally, we obtain the output segmentation mask S. We denote our segmentation branch as SGHM-S in reporting the results later.

3.3 Matting Decoder

Our matting decoder inputs the segmentation features and segmentation mask of different scales, outputs the matting results at $1, \frac{1}{4}, \frac{1}{8}$ scale. Firstly, we use ASM module to combine the features from encoder and segmentation mask. Then we sequentially process the features of different scales by several upsample blocks.

We predict matting results at $1, \frac{1}{4}, \frac{1}{8}$ scale by output modules. Finally, we adopt PRM module to produce the final high-resolution matting result based on the matting results at three output scales.

Attentive Shortcut Module. Our model proposed to use semantic segmentation to improve human matting by sharing encoder of segmentation and matting. In addition to features from shared encoder, we also feed the segmentation mask of different scales as input of matting decoder. For matting decoder, how to fuse the features and mask from segmentation is of vital importance. One direct way is to concatenate these two inputs for further processing. We propose to use ASM to fuse these two inputs. With the help of ASM, we can get more adaptive features for matting decoder. Specifically, the ASM contains two convolution layers, two SpectralNorm layers [34] and an efficient channel attention layer [35]. Channel attention can produce an adaptive feature by calculating a channel-wise weight vector corresponding to input feature.

Upsample Block. Upsample block process input features sequentially from $\frac{1}{64}$ scale to the original scale. First, it element-wisely adds the feature of the current scale and the feature of the previous scale upsampled by residual blocks from $\frac{1}{64}$ scale to $\frac{1}{2}$ scale. Then, for $\frac{1}{2}$ scale and 1 scale, we replace the residual blocks with a single transposed convolution layer with batch normalization and ReLU for efficiency.

Output Block. We predict matting result at $1, \frac{1}{4}, \frac{1}{8}$ scale. For each output scale, we attach a matting result prediction block after the upsample block. Each prediction block contains a convolution layer, batch normalization layer, ReLU and convolution layer sequentially.

Progressive Refinement Module. We adopt Progressive Refinement Module (PRM) [4] to further refine the output matting alphas from output blocks. PRM can selectively fuse the matting alphas from the previous scale and the current scale with a self-guidance mask, which can preserve the confident regions from the previous scale and focus on refining uncertain regions at the current scale. Specifically, the self-guidance mask of the current scale is generated from matting alpha obtained at the previous scale as follows:

$$g_l = \begin{cases} 0 & 0 < \alpha_{l-1} < 1, \\ 1 & \text{otherwise}, \end{cases} \tag{3}$$

where α_{l-1} is the matting alpha of previous scale. The α_{l-1} is upsampled to match the size of the raw matting output α_l' of the current scale. With the self-guidance mask g_l, the refined matting alpha of current scale can be calculated as following:

$$\alpha_l = \alpha_l' g_l + \alpha_{l-1}(1 - g_l). \tag{4}$$

Note that features in confident region predicted from the previous scale are preserved and the current scale only focuses on refining the uncertain region.

4 Training

We train our SGHM model in two stages. First, we train the segmentation network using widely available segmentation datasets. In this stage, the parameters of shared encoder and segmentation decoder are updated simultaneously. After segmentation net is trained, the shared encoder can extract powerful semantic features to provide information for the matting task. Next, we fix the shared encoder and segmentation decoder, and train the matting decoder only. The coarse mask output from segmentation net is also used as the input at this stage. During inference, the two decoders are executed successively. Segmentation mask is first predicted and fed to matting decoder to produce matting result.

4.1 Segmentation Training

We train the segmentation sub-network with about $47.2k$ paired images, which are from SPD [36] (about $2.5k$), Portrait Matting [7] (about $1.7k$), dataset released in [37] (about $5.2k$), human parsing dataset [38] (about $4.7k$), Privacy-Preserving dataset [31] (about $9.4k$) and green screen dataset from BMV2 [1] (about $23.7k$). We treat green screen data as segmentation mask since it provides more body posture diversity than alpha details. Note that we drop some image pairs by annotation checking, and collect about $35k$ background images from internet for random background composition.

We adopt Binary Cross Entropy (BCE) loss to train segmentation model. For data augmentation, we adopt random affine transformation, random horizontal flipping, random noise, random color jitters, random composite, and random crop to 320×320. We train our segmentation model on 8 NVIDIA Tesla A100 GPUs with a batch size of 10 for each GPU. We use Adam as optimizer, and the learning rate is initialized to $5e^{-4}$. The model is totally trained for 100 epochs with a cosine learning rate decay scheduler.

4.2 Matting Training

We train matting model on the foreground images of AIM [20] dataset except transparent object images. The total foreground images are 269, and we use MS COCO dataset as background images.

Following MG [4], we adopt l_1 regression loss, composition loss [20], Laplacian loss [39] for training matting model. We denote the ground-truth alpha with $\hat{\alpha}$ and the prediction alpha with α. Then the combined loss function can be formulated as:

$$L(\hat{\alpha}, \alpha) = L_{l_1}(\hat{\alpha}, \alpha) + L_{comp}(\hat{\alpha}, \alpha) + L_{lap}(\hat{\alpha}, \alpha). \tag{5}$$

We apply this combined loss on all output matting alphas at 1, $\frac{1}{4}$, $\frac{1}{8}$ scale with adaptive weights g_l calculated in Eq. 3 to force the training to more focused on the unknown region at each scale. Moreover, we set different weights for different scales to form the final loss function as follows:

$$L_{tot} = \sum_{l} \omega_l L(\hat{\alpha}_l \cdot g_l, \alpha_l \cdot g_l), \tag{6}$$

where ω_l is the loss weight of different scales. We set $\omega_{\frac{1}{8}} : \omega_{\frac{1}{4}} : \omega_1 = 1 : 2 : 3$ in our experiments.

We train our matting model with 100, 000 iterations on 4 NVIDIA Tesla A100 GPUs with a batch size of 8 for each GPU. We use Adam as optimizer, and the learning rate is initialized to $1e^{-3}$. We adopt the same data augmentation with training of segmentation, with a random crop of 1280×1280. We also adopt mask perturbation for augmentation. Note that we fix the parameters of segmentation model during training matting module, which can force the matting decoder to focus more on features to predict alpha details. If not fixed matting performance will drop as it will overfit to the small set of matting data.

5 Experiment

5.1 Benchmarks

To verify the effectiveness of the proposed method, we evaluate the performance on the following 5 benchmarks, including three real-world datasets and two composition datasets.

AIM [20]. We select 12 human images from AIM dataset for testing. Each foreground human image is composited to 20 backgrounds which are selected from top-240 of BG-20K [40] test set.

D646 [26]. Similar to AIM, 11 foreground images are composited with the last 220 backgrounds from BG-20K test set.

PPM-100 [3]. This dataset provides 100 finely annotated portrait images with various backgrounds. Images from PPM-100 are more realistic and natural than composition images.

P3M-500-NP [31]. We use the face kept images rather than face masked from P3M. The purpose is to avoid the unknown impact of face blur on evaluation. This benchmark has a great diversity of body postures.

RWCSM-289. To further verify our model generalization, we build a real-world complex scene matting dataset, denoted as RWCSM-289. It contains a variety of complex living and working scenarios. Its sources are hand-hold captured videos, online video meetings, TV shows, live videos, and Vlogs. Many of them come from youtube and are used by RVM [8]. It is worth noting that this dataset include motion and multi-person scenes, which is helpful to evaluate model robustness. The ground truth alpha is annotated by PhotoShop.

5.2 Quantitative Comparison

We compare our approach with the state-of-the-art automatic matting methods, including LFM [27], SHM [6], HATT [26], BSHM [41], MODNet [3], P3MNet [31], video matting method RVM [8] and mask-guided method MG [4]. We use inference size 512 for MODNet since it provides the best results on PPM-100. For RVM, We generate 10 frames video by repeating 10 times for every single image and take last frame result as evaluation target. For P3MNet the recommended testing resize strategy is used. For MG, we feed our segmentation result to its network as mask guidance. Both MG and our method keep the short size of images to 1280 when testing. We use mean absolute difference (MAD), mean squared error (MSE), spatial gradient (Grad) [42], and connectivity (Conn) [42] as alpha matting quality metrics. Note that MAD and MSE values are scaled by 10^3 and all metrics are calculated over the whole image.

Table 1. Quantitative results on real-world benchmarks. '↓': lower values are better.

Dataset	Method	MAD↓	MSE↓	Grad↓	Conn↓
PPM-100	LFM	15.80	9.40	-	-
	SHM	15.20	7.20	-	-
	HATT	13.70	6.70	-	-
	BSHM	11.40	6.30	-	-
	P3MNet	15.61	12.86	56.37	130.42
	MODNet	8.60	4.40	64.26	80.82
	RVM	10.95	6.53	63.13	105.19
	SGHM (ours)	**5.97**	**2.58**	**48.20**	**51.17**
P3M-500-NP	LFM	18.80	13.10	31.93	19.50
	SHM	12.20	9.30	20.30	17.09
	HATT	17.60	7.20	19.99	27.42
	P3MNet	6.50	3.50	**10.35**	12.51
	MODNet	12.82	7.41	16.02	20.23
	RVM	11.10	7.06	15.30	19.17
	SGHM (ours)	**6.49**	**3.11**	11.39	**10.16**
RWCSM-289	P3MNet	32.92	31.09	28.42	77.37
	MODNet	18.95	15.76	19.65	46.18
	RVM	14.36	11.25	15.68	28.52
	SGHM (ours)	**9.23**	**6.57**	**13.52**	**18.68**

Table 1 and Table 2 show the results of different matting methods evaluated on real-world and composition datasets. It shows that our method outperforms other methods across all real-world datasets in all metrics. Specifically, our method is ahead of compared method on PPM-100. On P3M-500-NP, we achieve the results (MAD 6.49, MSE 3.11) that are on par with the P3MNet (MAD 6.50, MSE 3.50) by only introducing face-masked P3M data into the segmentation stage. For complex scene data RWCSM-289 which covers more diversity of background, number of humans, body accessories, illumination, and image resolution, we significantly outperform P3MNet and MODNet, and are better than video-based approach RVM. On the composition datasets, SGHM still achieves the best results, showing consistently excellent performance of the proposed method.

5.3 Qualitative Comparison

This section shows qualitative comparisons on real-world benchmarks. We reveal alpha details in Fig. 3 and model robustness in Fig. 4. In Fig. 3 rows 1 to 4, we

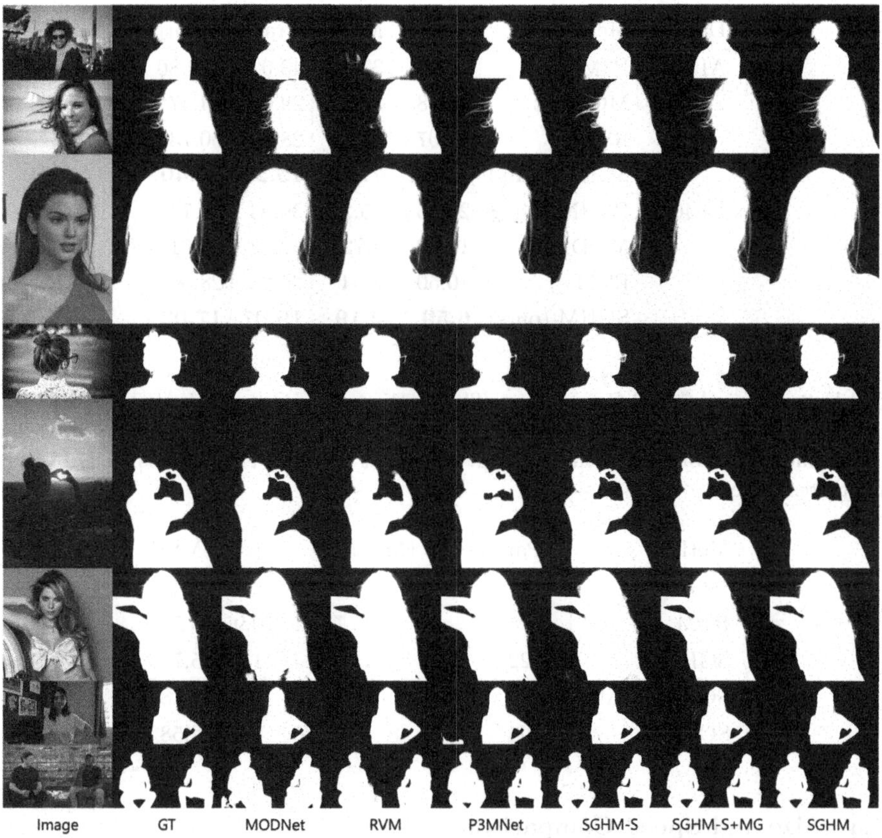

Fig. 3. Visual comparison of different methods on alpha details. SGHM-S denotes the segmentation results of our method. SGHM denotes the final matting results. SGHM-S+MG denotes using SGHM-S as extra input for MG. Our proposed method produces superior results from coarse to fine. Best viewed on monitor with zooming in for detail.

compare hair details and find ours predict fine-grained hair details comparable to mask-based method MG, which are more accurate than P3MNet, MODNet and RVM. Multiple body postures are displayed in rows 5 to 8. Other methods tend to get semantic errors (can be found in MODNet at row of 6, MG at row of 7, P3MNet at row 5) while SGHM produces more accurate alpha matte. It is worth noting that our method has the ability to correct semantic errors in the coarse masks (see row 1, 6 and 8 from SGHM-S to SGHM).

In Fig. 4, we select two SOTA methods MODNet and RVM for robustness comparison from four categories of videos. The extracted foreground is composited with a green background for visualization. Our method predicts much fewer semantic errors and demonstrates better robustness against semantic understanding errors than the other two methods.

Table 2. Quantitative results on composition benchmarks. '↓': lower values are better.

Dataset	Method	MAD↓	MSE↓	Grad↓	Conn↓
AIM	P3MNet	44.78	37.70	43.02	100.80
	MODNet	33.18	23.58	29.08	74.47
	RVM	27.07	17.54	28.84	60.73
	SGHM (ours)	**14.34**	**7.18**	**19.29**	**29.40**
D646	P3MNet	20.25	15.27	36.93	54.74
	MODNet	10.52	4.72	32.62	28.61
	RVM	10.50	4.94	35.24	28.60
	SGHM (ours)	**6.59**	**2.19**	**19.07**	**17.02**

Table 3. Size and Speed Comparison. The matting metrics are evaluated on PPM-100 dataset and the speed is evaluated on HD size on an NVIDIA A100 GPU. SGHM-S is the segmentation branch of our method. It runs at over 100 FPS as it is on 1/4 of full image resolution.

Method	#Parameters (M)	FPS	MAD	MSE
MODNet	6.49	20.76	8.60	4.40
RVM	3.75	71.81	10.95	6.53
SGHM-S	40.22	106.14	11.84	5.72
SGHM-S+MG	69.92	18.14	8.83	4.18
SGHM	43.94	34.76	5.97	2.58

5.4 Size and Speed Comparison

As mentioned in Sect. 3, MG uses a segmentation mask as extra input to its matting network. Unlike MG, we incorporate the mask generation stage into matting framework. Since MG uses an independent matting network, it introduces more parameters and its total parameter number is the combination of segmentation and matting networks. The speed is thus slowed down. Our SGHM shares the encoder with the segmentation, which causes marginal extra parameters (Table 3). SGHM also runs faster than MG on the same setting and can achieve 34 FPS on HD image (1920 × 1080) on NVIDIA A100. For the matting quality, our method achieves better performance. This shows we have both speed and accuracy advantages over MG. MODNet and RVM are also compared. Although they have fewer parameters, they both have limitation. MODNet runs slower on HD inference size (20.76 FPS) than 512 (81.01 FPS). RVM predicts unsatisfactory fine-grained alpha results across all benchmarks.

5.5 Ablation Studies

Role of Segmentation Task. We propose to introduce segmentation task to improve the performance and generalization of alpha matting in two ways. One is sharing encoder features and other is coarse mask guidance. Table 4 shows our ablation study results on PPM-100. The results lead to two conclusions: (1) Sharing semantic features is very beneficial to matting task, which helps to reduce MAD from 7.83 to 5.97. (2) Mask guidance plays an indispensable role in

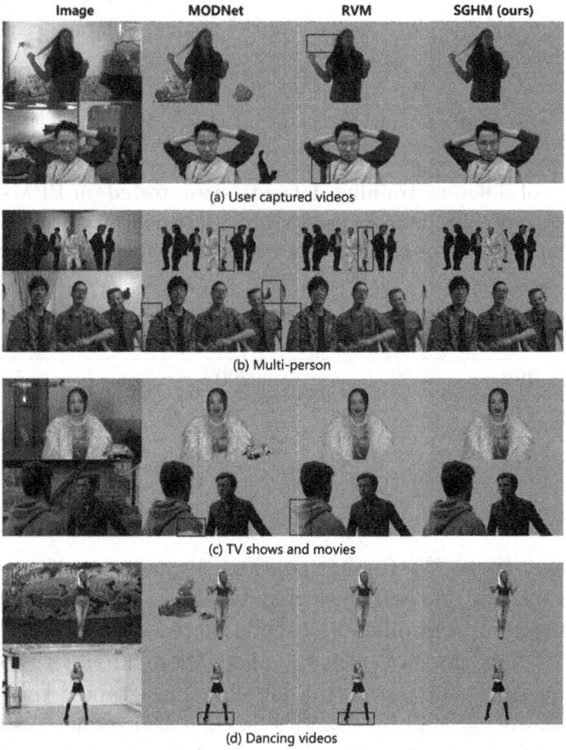

Fig. 4. Visual comparison of different methods on four categories of videos. Our method is more robust to semantic errors.

guiding matting process, as matting performance drops dramatically when mask guidance is removed. Since our matting model is trained on only hundreds of images, robust semantic features and good mask guidance are both helpful for improving model generalization.

Role of ASM. We propose ASM to combine semantic features and segmentation mask for matting decoder. As listed in the fourth and fifth rows of Table 4, model gets worse results without ASM. SpectralNorm and ECA layer are the two key components in ASM. In-depth analysis reveals that MAD drops from 5.97 to 7.11 when ECA layer is removed, while MAD is 6.33 when SpectralNorm is removed. This indicates ECA layer contributes more as it channel-wisely re-weight the features to adapt them for matting. Note that in the first row, we remove the mask input to only use the features from encoder and keep the same Conv layers with proposed ASM.

Table 4. Ablation study on different settings, tested on PPM-100.

ASM	Mask guidance	Sharing encoder weights	MAD	MSE
√			17.23	8.54
		√	10.46	5.07
√	√		7.83	4.04
	√	√	7.50	3.52
√	√	√	**5.97**	**2.58**

Table 5. Results of different training datasets sizes, tested on PPM-100. The LargeSeg dataset consists of 140k human masks which are collected from multiple publicly available datasets. The size of D646 is 362 which is selected from the Distinctions-646 training set.

	Segmentation dataset size	Matting dataset size	MAD	MSE
Baseline	40k	200+	5.97	2.58
+LargeSeg	170k	200+	**5.16**	**2.04**
+D646	40k	600+	5.71	2.45

Role of Dataset Size. We further conduct an experiment to verify the data efficiency of our method. As can be seen in Table 5, a larger segmentation dataset improves matting results significantly, while increasing the matting dataset size improves slightly. Note that it is easy to collect these human segmentation masks from publicly available datasets. But labeling fine-grained matting requires a much higher annotation skill level and it is time and money costing. This is an important and practical finding that we can efficiently improve matting performance by collecting more coarse human masks in an easy and fast way rather than paying for the high cost fine-detailed alpha annotating.

6 Conclusion

In this work, we investigate the major challenge in robust human matting and reveal that it is from the semantic understanding. Based on this, we propose a semantic guided human matting method. We introduce an additional matting decoder to the semantic segmentation network. By reusing the features from semantic segmentation encoder, the matting decoder is aware of global semantic information and also can generate fine matting details. With very small number of matting data, we can train a robust, accurate and real-time matting model which achieves top performance on multiple benchmark datasets. We believe that our proposed framework is a practical pipeline for matting application which does not rely on large number of high annotation cost matting data.

References

1. Lin, S., Ryabtsev, A., Sengupta, S., Curless, B., Seitz, S., Kemelmacher-Shlizerman, I.: Real-time high-resolution background matting. In: CVPR (2021)
2. Sengupta, S., Jayaram, V., Curless, B., Seitz, S., Kemelmacher-Shlizerman, I.: Background matting: the world is your green screen. In: CVPR (2020)
3. Ke, Z., Sun, J., Li, K., Yan, Q., Lau, R.W.: Modnet: real-time trimap-free portrait matting via objective decomposition. In: AAAI (2022)
4. Yu, Q., et al.: Mask guided matting via progressive refinement network. In: CVPR (2021)
5. Li, J., Zhang, J., Maybank, S.J., Tao, D.: Bridging composite and real: towards end-to-end deep image matting. Int. J. Comput. Vision **130**(2), 246–266 (2022)
6. Chen, Q., Ge, T., Xu, Y., Zhang, Z., Yang, X., Gai, K.: Semantic human matting. In: ACM MM (2018)
7. Shen, X., Tao, X., Gao, H., Zhou, C., Jia, J.: Deep automatic portrait matting. In: ECCV (2016)
8. Lin, S., Yang, L., Saleemi, I., Sengupta, S.: Robust high-resolution video matting with temporal guidance. In: WACV (2022)
9. Aksoy, Y., Ozan Aydin, T., Pollefeys, M.: Designing effective inter-pixel information flow for natural image matting. In: CVPR (2017)
10. Chen, Q., Li, D., Tang, C.K.: KNN matting. TPAMI **35**(9), 2175–2188 (2013)
11. Chuang, Y.Y., Curless, B., Salesin, D.H., Szeliski, R.: A bayesian approach to digital matting. In: CVPR (2001)
12. Gastal, E.S., Oliveira, M.M.: Shared sampling for real-time alpha matting. In: Computer Graphics Forum, vol. 29, pp. 575–584 (2010)
13. Levin, A., Lischinski, D., Weiss, Y.: A closed-form solution to natural image matting. TPAMI **30**(2), 228–242 (2007)
14. Levin, A., Rav-Acha, A., Lischinski, D.: Spectral matting. TPAMI **30**(10), 1699–1712 (2008)
15. Sun, J., Jia, J., Tang, C.K., Shum, H.Y.: Poisson matting. In: ToG, vol. 23, pp. 315–321 (2004)
16. Chen, T., Wang, Y., Schillings, V., Meinel, C.: Grayscale image matting and colorization. In: ACCV (2004)
17. Pham, V.Q., Takahashi, K., Naemura, T.: Real-time video matting based on bilayer segmentation. In: ACCV (2009)

18. Park, Y., Yoo, S.I.: A convex image segmentation: Extending graph cuts and closed-form matting. In: ACCV (2010)
19. Sindeev, M., Konushin, A., Rother, C.: Alpha-flow for video matting. In: ACCV (2012)
20. Xu, N., Price, B., Cohen, S., Huang, T.: Deep image matting. In: CVPR (2017)
21. Forte, M., Pitié, F.: F, B, alpha matting. CoRR abs/2003.07711 (2020)
22. Liu, Y., et al.: Tripartite information mining and integration for image matting. In: ICCV (2021)
23. Yang, S., Wang, B., Li, W., Lin, Y., He, C., et al.: Unified interactive image matting. arXiv preprint arXiv:2205.08324 (2022)
24. Dai, Y., Price, B., Zhang, H., Shen, C.: Boosting robustness of image matting with context assembling and strong data augmentation. In: CVPR, pp. 11707–11716 (2022)
25. Chen, G., et al.: PP-matting: high-accuracy natural image matting. arXiv preprint arXiv:2204.09433 (2022)
26. Qiao, Y., et al.: Attention-guided hierarchical structure aggregation for image matting. In: CVPR (2020)
27. Zhang, Y., et al.: A late fusion CNN for digital matting. In: CVPR (2019)
28. Zhu, B., Chen, Y., Wang, J., Liu, S., Zhang, B., Tang, M.: Fast deep matting for portrait animation on mobile phone. In: ACM MM (2017)
29. Sun, Y., Tang, C.K., Tai, Y.W.: Human instance matting via mutual guidance and multi-instance refinement. In: CVPR (2022)
30. Xing, Y., Li, Y., Wang, X., Zhu, Y., Chen, Q.: Composite photograph harmonization with complete background cues. In: ACM MM (2022)
31. Li, J., Ma, S., Zhang, J., Tao, D.: Privacy-preserving portrait matting. arXiv (2021)
32. Chen, L.C., Papandreou, G., Schroff, F., Adam, H.: Rethinking atrous convolution for semantic image segmentation. arXiv (2017)
33. He, K., Zhang, X., Ren, S., Sun, J.: Deep residual learning for image recognition. In: CVPR (2016)
34. Miyato, T., Kataoka, T., Koyama, M., Yoshida, Y.: Spectral normalization for generative adversarial networks. arXiv (2018)
35. Wang, Q., Wu, B., Zhu, P., Li, P., Zuo, W., Hu, Q.: ECA-Net: efficient channel attention for deep convolutional neural networks. In: CVPR. IEEE (2020)
36. supervise.ly: Supervisely person dataset. supervise.ly (2018)
37. Wu, Z., Huang, Y., Yu, Y., Wang, L., Tan, T.: Early hierarchical contexts learned by convolutional networks for image segmentation. In: ICPR. IEEE (2014)
38. Gong, K., Liang, X., Li, Y., Chen, Y., Yang, M., Lin, L.: Instance-level human parsing via part grouping network. In: ECCV (2018)
39. Hou, Q., Liu, F.: Context-aware image matting for simultaneous foreground and alpha estimation. In: ICCV (2019)
40. Li, J., Zhang, J., Maybank, S.J., Tao, D.: End-to-end animal image matting. arXiv (2020)
41. Liu, J., et al.: Boosting semantic human matting with coarse annotations. In: CVPR (2020)
42. Rhemann, C., Rother, C., Wang, J., Gelautz, M., Kohli, P., Rott, P.: A perceptually motivated online benchmark for image matting. In: CVPR. IEEE (2009)

RDRN: Recursively Defined Residual Network for Image Super-Resolution

Alexander Panaetov, Karim Elhadji Daou, Igor Samenko, Evgeny Tetin, and Ilya Ivanov(✉)

Huawei, Moscow Research Center, Munich, Russia
{panaetov.alexander1,karim.daou,samenko.igor, evgeny.tetin,ivanov.ilya1}@huawei.com

Abstract. Deep convolutional neural networks (CNNs) have obtained remarkable performance in single image super-resolution (SISR). However, very deep networks can suffer from training difficulty and hardly achieve further performance gain. There are two main trends to solve that problem: improving the network architecture for better propagation of features through large number of layers and designing an attention mechanism for selecting most informative features. Recent SISR solutions propose advanced attention and self-attention mechanisms. However, constructing a network to use an attention block in the most efficient way is a challenging problem. To address this issue, we propose a general recursively defined residual block (RDRB) for better feature extraction and propagation through network layers. Based on RDRB we designed recursively defined residual network (RDRN), a novel network architecture which utilizes attention blocks efficiently. Extensive experiments show that the proposed model achieves state-of-the-art results on several popular super-resolution benchmarks and outperforms previous methods by **up to 0.43 dB.**

Keywords: Super-resolution · Recursively defined residual network · Recursively defined residual block

1 Introduction

The main purpose of super-resolution (SR) is to reconstruct high-resolution image (HR) from given low-resolution counterpart (LR). SR is an ill-posed problem since the mapping between LR and HR images is ambiguous (one-to-many). Recovering missing details is a challenging task, especially for a high upscale factor. Despite being a difficult problem, SR plays an important role in various image processing tasks with applications in face recognition, medical imaging,

Supported by Huawei.

Supplementary Information The online version contains supplementary material available at https://doi.org/10.1007/978-3-031-26284-5_38.

L. Wang et al. (Eds.): ACCV 2022, LNCS 13842, pp. 629–645, 2023.
https://doi.org/10.1007/978-3-031-26284-5_38

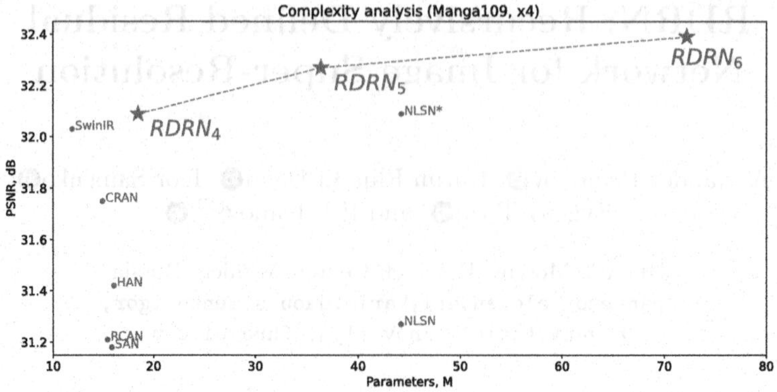

Fig. 1. Number of parameters and performance on Manga109 with upscale factor ×4 (BI model)

surveillance, digital zoom, etc. While many existing SR methods reconstruct HR image from several LR images, in this paper we focus on single image super-resolution (SISR).

In recent years, convolutional neural networks (CNNs) have achieved remarkable results in many computer vision tasks, including SISR. Deep CNNs have shown improvement over the traditional algorithms. Network depth in existing solutions has been significantly increased from three layers in SRCNN [5] to more than 400 in recent works [25,35]. However, very deep networks can suffer from training difficulties and hardly achieve any extra performance gain. A further increase in CNN depth does not lead to an improvement in quality and makes them unsuitable for various applications. The difficulty of training can be explained by the fact that network is not able to efficiently use information from intermediate layers. This issue can be partially solved using residual learning [8]. Combining features from different layers through skip connections is a fruitful idea in SISR. Additional connections along the network's depth could help to learn more powerful feature representations, making training more stable and accelerating convergence.

Another approach to address the training difficulty is related to the mechanism of attention. Recently, this direction has become very popular and profitable for SISR. The intuition behind attention is a simulation of the human vision system, which can focus on the most informative parts of an image and ignore the irrelevant information. Recent works show that attention can effectively reduce the width and depth of a network while maintaining comparable or better performance due to enhanced discriminative learning ability [35].

In this paper, we combine both approaches. The design of recursively defined residual block (RDRB) is shown in Fig. 2. It consists of two parts: basic block (Fig. 2-a) and recursive block (Fig. 2-b). We have found that enhanced spatial attention (ESA) introduced in [20] is very effective for the super-resolution task, and we take advantage of its benefits even more than in the original paper. We

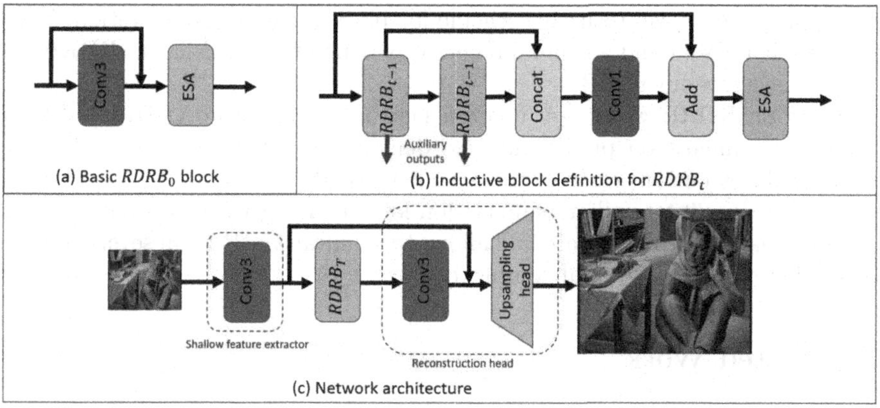

Fig. 2. Architecture of the proposed recursively defined residual block (RDRB) and network (RDRN). RDRB is defined in a recurrent manner. (a) We define basic block $RDRB_0$ as a convolution layer followed by ESA block from [20]. (b) Each subsequent block $RDRB_t$ can be defined using the architecture of previous block $RDRB_{t-1}$ according to the scheme. (c) Full network architecture

include ESA in the basic block, which is repeated in the final architecture multiple times. Compared to previous work, our RDRB contains more connections between intermediate layers. It combines hierarchical cues along the network depth to obtain richer feature representations. Experiments show that the effect of the proposed RDRB is more visible for lower upscale factors ($\times 2$, $\times 3$). For the upscale factor $\times 2$, the RDRB-based model outperforms all recent solutions without any bells and whistles. It can be explained by the recurrent nature of RDRB. Shallow features are propagated to all levels of RDRB via long skip connections from input. For lower upscale factors, they contain more relevant information, as fewer details will be missed compared to higher upscale factors. Finally, to further improve RDRB, especially for large upscale factors, we insert non-local sparse attention [24] into the block.

Based on RDRB we design recursively defined residual network (RDRN) as shown in Fig. 2-c. Following [19], we add batch normalization (BN) and apply adaptive deviation modulator (AdaDM) to the final model. For training, we use intermediate supervision (IS) loss, which improves convergence and allows to simultaneously train several models of different computational complexity without a large overhead.

The proposed RDRN shows superior performance on the most popular SR benchmarks. Our model produces better visual quality, recovers more details and outperforms current state-of-the-art solutions with a significant margin of up to 0.43 dB.

In summary, the main contributions of our paper are:

- We propose a novel recursive scheme for block architecture definition. Using that scheme we build a general recursively defined residual block (RDRB) for more accurate image SR.
- Based on RDRB we design a novel network architecture (RDRN). Extensive experiments on public datasets demonstrate that the proposed model outperforms current state-of-the-art SR methods.
- We introduce intermediate supervision loss. Training with IS helps to obtain additional performance gain and allows to simultaneously train several models of different computational complexity.

2 Related Work

Super-resolution algorithms can be categorized into two types: traditional and deep learning based methods. In this section, we will focus on the second category as the most successful in computer vision.

Dong et al. [5] proposed the first three-layer convolutional neural network (SRCNN) to learn the mapping from LR to HR directly. This pioneering work achieved superior performance against the previous traditional methods.

Following this work, many networks achieved better performance using deeper and wider architectures [12,13]. SRCNN used interpolated image as input, however, it is more efficient to upscale the feature maps at the end of the network. To address this issue, Shi et al. proposed ESPCN [27] with a sub-pixel layer, which is widely used in modern SISR networks. As in the case of the classification task, a further increase in depth and width of plain architecture leads to quality degradation. However, residual blocks [8] and dense connections [9] allow to train more powerful networks, and Lin et al. proposed a very deep and wide EDSR [17] based on a modified residual block. Dense connections were used in RDN [36] to utilize hierarchical features from all convolutional layers.

Following [17], most of the recent SISR works do not use batch normalization (BN) [11], as it harms network's performance. However, Liu et al. [19] showed that normalization layers reduce the standard deviation of feature pixels (the main reason for quality degradation) and proposed adaptive deviation modulator (AdaDM) to solve that issue. AdaDM can successfully enable BN layers, significantly improving performance and allowing to train larger models.

Attention mechanism is used as a simulation of the human vision system that focuses on the most informative parts of an image, ignores irrelevant information and enhances discriminative learning ability. Recently, attention has been successfully applied to the SISR problem, and attention-based methods have shown superiority over pure CNN solutions. Zhang et al. [35] proposed channel attention (CA) to adaptively rescale each feature channel-wise by modeling the interdependencies across feature channels. Such CA mechanism improves the representational ability of residual channel attention network (RCAN).

Liu et al. [20] designed lightweight and powerful enhanced spatial attention (ESA). Dai et al. [3] introduced second-order attention network (SAN) to adaptively refine features using second-order feature statistics. Niu et al. [25] presented a new holistic attention network (HAN), which consists of a layer attention module (LAM) and a channel-spatial attention module (CSAM) and models the holistic interdependencies among layers, channels, and positions. Zhang et al. [37] proposed a context reasoning attention network (CRAN) that can adaptively modulate the convolution kernel according to the global context enhanced by semantic reasoning.

Attention could be considered as an additional tool to improve the expressiveness and convergence of CNNs. An alternative is the self-attention mechanism [30], which was introduced for natural language processing models. Recent work has shown that even pure transformers can achieve SOTA results on computer vision tasks [7]. However, hybrid models that contain both convolution and self-attention show outstanding results as they integrate the advantages of both approaches. Following this new trend, Liang et al. [16] introduced the SwinIR architecture based on Swin Transformer [21]. SwinIR has improved upon all previous methods by a significant margin.

3 Proposed Method

In this section, we first present the overview of RDRN. Then we give a detailed description of the proposed RDRB. We proceed with outlining the technical part of training, including the loss function and implementation details. Finally, we provide arguments about the advantages of the proposed method and differences from other approaches.

3.1 Network Architecture

As shown in Fig. 2, RDRN can be divided into three parts: shallow feature extractor, RDRB deep feature extractor and reconstruction head. Let us denote I_{LR} and I_{SR} the low resolution input and the output of RDRN. Following [3,25,35], we use only one convolutional layer to extract the shallow features F_{SF} from LR input:

$$F_{SF} = H_{SF}(I_{LR}), \tag{1}$$

where H_{SF} denotes the convolution operation. The extracted shallow feature map F_{SF} is used as input to the deep feature extractor:

$$F_{DF}^{T} = H_{RDRB}^{T}(F_{SF}), \tag{2}$$

where T is the recursion depth and H_{RDRB}^{T} stands for RDRB feature extraction model, which will be introduced in the next subsection. RDRB is the core of the proposed network. Finally, the extracted deep features F_{DF} are combined with the shallow features to stabilize training, after which they are processed by the reconstruction module:

$$I_{SR}^{T} = H_{rec}(F_{SF}, F_{DF}^{T}), \tag{3}$$

where H_{rec} denotes the reconstruction head that consists of a convolutional layer and a sub-pixel layer [27]. The long skip connection propagates low-frequency information directly to the reconstruction module, which can help the deep feature extractor to focus on the extraction of high-frequency information [16].

There are several choices for loss function to optimize the model, such as ℓ_1, ℓ_2, perceptual, adversarial loss. We found that for the proposed method, ℓ_1 loss is the most suitable one, and we minimize the following loss function:

$$L^T(\Theta) = \frac{1}{m}\sum_{i=1}^{m} \left\| H^T_{RDRN}(I^i_{LR}) - I^i_{HR} \right\|_1 = \frac{1}{m}\sum_{i=1}^{m} \left\| I^{T,i}_{SR} - I^i_{HR} \right\|_1, \quad (4)$$

where H^T_{RDRN}, Θ, and m denote the function of the proposed RDRN, the set of learned parameters, and the number of training pairs, respectively. Following [18] we fine-tune the model using ℓ_2 loss. More details are provided in the experiments section.

3.2 Recursively Defined Residual Block (RDRB)

As discussed above, recursively defined residual block is the core of RDRN, and here we give the overall description of the proposed block architecture. The definition will be done in a recursive manner. First, we define basic block RDRB_0 as a convolutional layer followed by ESA block introduced in [20]:

$$F^0_{DF} = H^0_{RDRB}(F_{SF}) := ESA(conv_{3\times3}(F_{SF}) + F_{SF}). \quad (5)$$

We found that ESA mechanism is highly effective for the super-resolution task, and we take advantage of its benefits even more than in the original paper. We include ESA in the basic block, which is repeated in our architecture multiple times.

Finally, we use induction to define RDRB_t for any natural t:

$$F^t_{DF} = H^t_{RDRB}(F_{SF}), \quad (6)$$

$$F^{t,1}_{DF} = H^t_{RDRB}(F^t_{DF}), \quad (7)$$

$$H^t_{RDRB}(F_{SF}) := ESA\Big(F_{SF} + conv_{1\times1}\Big(concat(F^{t-1}_{DF}, F^{t-1,1}_{DF})\Big)\Big). \quad (8)$$

The scheme of building RDRB_0 and RDRB_t is depicted in Fig. 2. Same as in the basic block, ESA is included in RDRB_t. It's worth noticing that we don't share weights in any part ot the model. To avoid any misunderstanding about the definition of the proposed RDRB, we provide PyTorch implementation of the basic block and the recursive step in the supplementary.

3.3 Intermediate Supervision (IS)

As shown in Fig. 2, for $t > 0$ each RDRB_t contains two additional auxiliary outputs. Each of these intermediate outputs is paired with a reconstruction head and an additional loss. The final loss function is a weighted sum of the original loss and all intermediate losses (due to recursive definition of the block we have $2^{T+1} - 2$ additional loss terms):

$$L(\Theta) = w_0 L^T(\Theta) + \sum_{i=1}^{2^{T+1}-2} w_i L_i(\Theta), \tag{9}$$

where w_i denotes the weight and $L_i(\Theta)$ is the loss based on the intermediate output. The computational overhead for IS training is minimal, because the size and complexity of the added heads is much smaller than the size and complexity of RDRB. Using IS loss function gives two advantages. First, it allows to simultaneously train several models of different computational complexity using the output of intermediate reconstruction head. Second, as proved in the ablation study, training with IS enables us to achieve a performance gain.

3.4 Implementation Details

Here we specify the implementation details of RDRN. We use RDRB_5 in our final architecture. Following [19], we add batch normalization (BN) [11] and adaptive deviation modulator (AdaDM) to every 3×3 convolution. Finally, starting from recursion level 3 we add non-linear spatial attention block [24] after ESA. According to our experiments, both tricks improve the final score and allow us to outperform SwinIR [16].

Experiments show that for IS training it is better to zero out the weights for losses based on auxiliary outputs from RDRB_1 and RDRB_2 and set the remaining weights to one.

3.5 Discussion

In our research we aim to follow the best practices from recent work. The motivation is two-fold: to outperform current SoTA models and to design general method for architecture definitions which can be used by other researchers. We conducted a large number of experiments and devised the following recipe:

- Following [18], we use advanced two-stage training procedure.
- We apply batch normalization [11] and add AdaDM to the model as in [19].
- We have an attention mechanism both inside the basic block and between blocks.

We explored deformable convolution [38] for image SR. In all experiments it improves the final score but doubles the training time, and we decided to postpone this direction for future research.

The main novelty of our work is the proposed architecture of RDRB and RDRN. We present the concept of building a network that utilizes multiple connections between different blocks. It is not limited to one specific architecture: the basic block can be replaced with any other block and the merging of two blocks can vary as well. The proposed architecture significantly differs from the ones presented in prior works.

Difference from Residual Feature Aggregation Network (RFANet). RFANet [20] utilizes skip-connection only on block level while blocks are connected consequently. The intuition behind RDRN is that combining of hierarchical cues along the network depth allows to get richer feature representations. Compared to RFANet, our architecture is able to combine features at deeper layers. In ablation study we have comparison of residual-block (RB) from RFANet and $RDRB_0$ under the same conditions. We also show in Table 1 the benefits of our recursive way of building the network compared to plain connection of blocks in RFANet.

Difference from Non-local Sparse Network (NLSN). NLSN [24] is a recent state-of-the-art SR method. The model achieved remarkable results after it was updated with BN and AdaDM [19]. The original architecture of NLSN is plain and made of residual blocks and non-local sparse attention (NLSA) blocks. The main difference is again coming from more effective way of merging information from different layers. Our smaller $RDRN_4$ model has similar performance with NLSN and 72% less FLOPs (Table 6).

Difference from Deep Recursive Residual Network (DRRN). The architecture of the recursive block (RB) from [28] is defined by a recursive scheme, similarly to the proposed RDRB. The key differences between the two approaches are as follows. First, according to the definition of RB, the weights of the same sub-blocks are shared, while RDRB does not reuse weights. However, weights sharing can be an effective way to reduce the number of parameters, and we save this direction for future research. Second, DRRN contains sequences of RB blocks. In contrast, our model is based on one large RDRB. Our recursive definition helps to stack blocks together with additional skip connections, granting extra performance gain.

4 Experiments

In this section, we compare RDRN to the state-of-the-art algorithms on five benchmark datasets. We first give a detailed description of experiment settings. Then we analyze the contribution of the proposed neural network. Finally, we provide a quantitative and visual comparison with the recent state-of-the-art methods for the most popular degradation models.

4.1 Implementation and Trainings Details

Datasets and Metrics. We train our models on DF2K dataset, which combines DIV2K [29] and Flickr2K together as in [16,19,37]. For testing, we choose five

Table 1. Ablation study on the advantage of RDRB

Model/PSNR	Set5	Set14	B100	Urban100	Manga100
RFANet(RB)	32.65	29.00	27.86	27.11	31.73
RFANet(RDRB$_0$)	32.67 **+0.02**	29.02 **+0.02**	27.86 0.00	27.16 **+0.05**	31.81 **+0.08**
RDRN(RB)	32.73 **+0.08**	29.02 **+0.02**	27.86 0.00	27.14 **+0.03**	31.80 **+0.07**

standard datasets: Set5 [2], Set14 [32], B100 [22], Urban100 [10], and Manga109 [23]. We train and test RDRN for three degradation methods. Degraded data is obtained by bicubic interpolation (BI), blur-downscale (BD) and downscale-noise (DN) models from Matlab. We employ peak signal-to-noise ratio (PSNR) and structural similarity (SSIM) [31] to measure the quality of super-resolved images. All SR results are evaluated on Y channel after color space transformation from RGB to YCbCr. Metrics are calculated using Matlab.

Training Settings. The proposed network architecture is implemented in PyTorch framework, and models are trained from scratch. In our network, patch size is set to 64×64. We use Adam [14] optimizer with a batch size of 48 (12 per each GPU). The initial learning rate is set to 10^{-4}. After 7.5×10^5 iterations it is reduced by 2 times. Default values of β_1 and β_2 are used, which are 0.9 and 0.999, respectively, and we set $\epsilon = 10^{-8}$. During training we augment the images by randomly rotating 90°, 180°, 270° and horizontal flipping.

For all the results reported in the paper, we train the network 9×10^5 iterations with ℓ_1 loss function. After that we fine-tune the model 1.5×10^4 iterations with a smaller learning rate of 10^{-5} using MSE loss. Complete training procedure takes about two weeks on a server with 4 NVIDIA Tesla V100 GPUs.

4.2 Ablation Study

Impact of the Architecture. To highlight the advantages of the proposed architecture, we compare RDRN and RFANet [20]. For comparison we use vanilla RFANet with 32 RFA blocks and 64 channels. Each RFA block consists of 4 residual blocks (RB). First, we demonstrate that adding ESA to our basic block RDRB$_0$ is beneficial. For that purpose we train RFANet(RDRB$_0$), changing RB to RDRB$_0$. As shown in Table 1, our basic block gives a performance gain to RFANet. Second, we show that the proposed recurrent scheme of building a network is better than stacking RFA blocks. We train RDRN(RB) using RB instead of RDRB$_0$ to demonstrate that. Even when using the same basic block, RDRN(RB) still outperforms RFANet(RFA, RB). For fair comparison, we change the depth of both networks to keep the computation complexity for all models similar.

Impact of Intermediate Supervision (IS). To show the impact of using IS during training, we train the same network with and without IS. Table 2 demonstrates the effectiveness of the proposed loss function. Experiments show

Table 2. Ablation study on the advantage of IS

Model/PSNR	Set5	Set14	B100	Urban100	Manga100
$RDRN_4$	32.68	29.00	27.84	27.07	31.77
$RDRN_4$-IS	32.71	29.01	27.85	27.07	31.83
	+0.03	**+0.01**	**+0.01**	0.00	**+0.06**
$RDRN_5$	32.67	28.99	27.86	27.09	31.86
$RDRN_5$-IS	32.73	29.05	27.88	27.19	31.97
	+0.06	**+0.06**	**+0.02**	**+0.10**	**+0.11**

that IS provides a bigger gain for larger models. It corresponds with the intuition that it is harder to train a larger model using SGD, and IS helps to propagate gradients better. Using IS allows to simultaneously train several SISR models of different computational complexity.

4.3 Results with Bicubic (BI) Degradation Model

We compare the proposed algorithm with the following 11 state-of-the-art methods: SRCNN [4], RDN [36], RCAN [35], SAN [3], NLSN [24], DRLN [1], HAN [25], RFANet [20], CRAN [37], SwinIR [16] and NLSN* [19]. We take all results from the original papers. Following [3,16,17,25,35], we provide a self-ensemble model and denote it RDRN+.

Quantitative Results. Table 3 reports the quantitative comparison of ×2, ×3, ×4 SR. Compared to the existing methods, RDRN scores best in all scales of the reconstructed test sets. Even without self-ensemble, our model outperforms other solutions at upscale factors 2× and 3×. For 4× SR, RDRN achieves the best PSNR values, however, SSIM score of competitors on several datasets is higher. The effect of the proposed method is more significant for lower upscale factors. This can be explained by the RDRB design. Shallow features are propagated for all levels of RDRB using long skip connections from input. For lower upscale factors, shallow features contain more important information as less information will be missed compared to higher upscale factors.

Visual Results. We give visual comparison of various competing methods on Urban100 dataset for ×2 SR in Fig. 3. Our model obtains better visual quality and recovers a more detailed image. Most compared methods recover grid textures of buildings with blurring artifacts, while RDRN produces sharper images. The proposed method can maintain a regular structure in difficult cases where previous approaches fail. To further illustrate the analysis above, we show such cases for ×2 SR in Fig. 4. The recovered details are more faithful to the ground truth.

4.4 Results with Bicubic Blur-Downscale (BD) Degradation Model

Following [1,20,25,35–37], we provide results for blur-downscale degradation, where HR image is blurred by a 7×7 Gaussian kernel with standard deviation $\sigma = 1.6$ and then downscaled by bicubic interpolation with scaling factor ×3.

Table 3. Quantitative results with BI degradation model. The best and second best results are highlighted in **bold** and underlined

Methods	Scale	Set5		Set14		B100		Urban100		Manga109	
		PSNR	SSIM	PSNR	SSIM	PSNR	SSIM	PSNR	SSIM	PSNR	SSIM
Bicubic	×2	33.66	0.9299	30.24	0.8688	29.56	0.8431	26.88	0.8403	30.80	0.9339
SRCNN [4]	×2	36.66	0.9542	32.45	0.9067	31.36	0.8879	29.50	0.8946	35.60	0.9663
RDN [36]	×2	38.24	0.9614	34.01	0.9212	32.34	0.9017	32.89	0.9353	39.18	0.9780
RCAN [35]	×2	38.27	0.9614	34.12	0.9216	32.41	0.9027	33.34	0.9384	39.44	0.9786
SAN [3]	×2	38.31	0.9620	34.07	0.9213	32.42	0.9028	33.10	0.9370	39.32	0.9792
NLSN [24]	×2	38.34	0.9618	34.08	0.9231	32.43	0.9027	33.42	0.9394	39.59	0.9789
DRLN [1]	×2	38.27	0.9616	34.28	0.9231	32.44	0.9028	33.37	0.9390	39.58	0.9786
HAN [25]	×2	38.27	0.9614	34.16	0.9217	32.41	0.9027	33.35	0.9385	39.46	0.9785
RFANet [20]	×2	38.26	0.9615	34.16	0.9220	32.41	0.9026	33.33	0.9389	39.44	0.9783
CRAN [37]	×2	38.31	0.9617	34.22	0.9232	32.44	0.9029	33.43	0.9394	39.75	0.9793
SwinIR [16]	×2	38.42	0.9623	34.46	0.9250	32.53	0.9041	33.81	0.9427	39.92	0.9797
NLSN* [19]	×2	38.43	0.9622	34.40	0.9249	32.50	0.9036	33.78	0.9419	39.89	0.9798
RDRN(ours)	×2	38.54	0.9627	34.67	0.9261	32.53	0.9043	34.12	0.9442	40.35	0.9807
RDRN+ (ours)	×2	**38.59**	**0.9629**	**34.76**	**0.9265**	**32.56**	**0.9046**	**34.27**	**0.9452**	**40.48**	**0.9810**
Bicubic	×3	30.39	0.8682	27.55	0.7742	27.21	0.7385	24.46	0.7349	29.95	0.8556
SRCNN [4]	×3	32.75	0.9090	29.30	0.8215	28.41	0.7863	26.24	0.7989	30.48	0.9117
RDN [36]	×3	34.71	0.9296	30.57	0.8468	29.26	0.8093	28.80	0.8653	34.13	0.9484
RCAN [35]	×3	34.74	0.9299	30.65	0.8482	29.32	0.8111	29.09	0.8702	34.44	0.9499
SAN [3]	×3	34.75	0.9300	30.59	0.8476	29.33	0.8112	28.93	0.8671	34.30	0.9494
NLSN [24]	×3	34.85	0.9306	30.70	0.8485	29.34	0.8117	29.25	0.8726	34.57	0.9508
DRLN [1]	×3	34.78	0.9303	30.73	0.8488	29.36	0.8117	29.21	0.8722	34.71	0.9509
HAN [25]	×3	34.75	0.9299	30.67	0.8483	29.32	0.8110	29.10	0.8705	34.48	0.9500
RFANet [20]	×3	34.79	0.9300	30.67	0.8487	29.34	0.8115	29.15	0.8720	34.59	0.9506
CRAN [37]	×3	34.80	0.9304	30.73	0.8498	29.38	0.8124	29.33	0.8745	34.84	0.9515
SwinIR [16]	×3	34.97	0.9318	30.93	0.8534	29.46	0.8145	29.75	0.8826	35.12	0.9537
NLSN* [19]	×3	34.95	0.9316	30.86	0.8513	29.45	0.8141	29.77	0.8812	35.20	0.9534
RDRN(ours)	×3	35.04	0.9322	30.99	0.8530	29.50	0.8152	29.87	0.8830	35.44	0.9543
RDRN+ (ours)	×3	**35.10**	**0.9326**	**31.04**	**0.8539**	**29.53**	**0.8158**	**30.02**	**0.8848**	**35.58**	**0.9549**
Bicubic	×4	28.42	0.8104	26.00	0.7027	25.96	0.6675	23.14	0.6577	24.89	0.7866
SRCNN [4]	×4	30.48	0.8628	27.50	0.7513	26.90	0.7101	24.52	0.7221	27.58	0.8555
RDN [36]	×4	32.47	0.8990	28.81	0.7871	27.72	0.7419	26.61	0.8028	31.00	0.9151
RCAN [35]	×4	32.63	0.9002	28.87	0.7889	27.77	0.7436	26.82	0.8087	31.22	0.9173
SAN [3]	×4	32.64	0.9003	28.92	0.7888	27.78	0.7436	26.79	0.8068	31.18	0.9169
NLSN [24]	×4	32.59	0.9000	28.87	0.7891	27.78	0.7444	26.96	0.8109	31.27	0.9184
DRLN [1]	×4	32.63	0.9002	28.94	0.7900	27.83	0.7444	26.98	0.8119	31.54	0.9196
HAN [25]	×4	32.64	0.9002	28.90	0.7890	27.80	0.7442	26.85	0.8094	31.42	0.9177
RFANet [20]	×4	32.66	0.9004	28.88	0.7894	27.79	0.7442	26.92	0.8112	31.41	0.9187
CRAN [37]	×4	32.72	0.9012	29.01	0.7918	27.86	0.7460	27.13	0.8167	31.75	0.9219
SwinIR [16]	×4	32.92	0.9044	29.09	0.7950	27.92	0.7489	27.45	0.8254	32.03	0.9260
NLSN* [19]	×4	32.86	0.9025	29.11	0.7940	27.92	0.7481	27.49	0.8247	32.09	0.9251
RDRN(ours)	×4	32.94	0.9039	29.17	0.7951	27.96	0.7490	27.49	0.8241	32.27	0.9259
RDRN+ (ours)	×4	**33.00**	**0.9046**	**29.24**	**0.7961**	**28.01**	**0.7499**	**27.63**	**0.8266**	**32.47**	**0.9273**

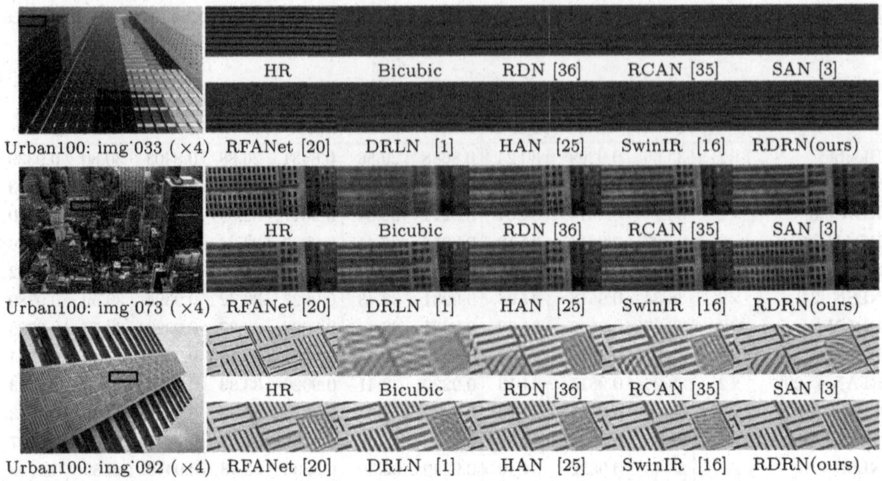

Fig. 3. Visual comparison for 4× SR with BI model on Urban100 dataset.

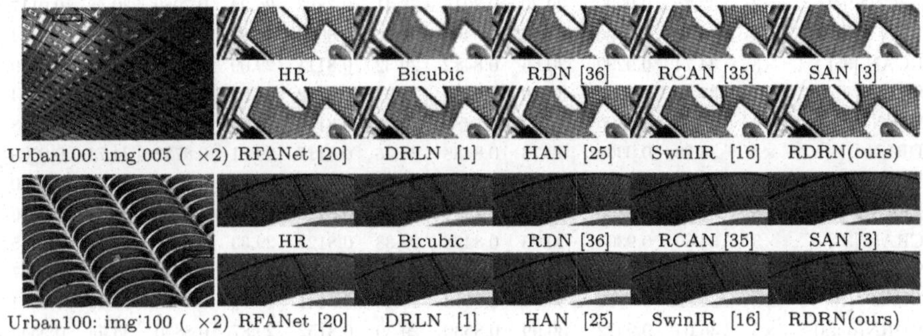

Fig. 4. Visual comparison for 2× SR with BI model on Urban100 dataset.

Quantitative Results. In Table 4 we compare the proposed RDRN model with the following super-resolution methods: SPMSR [26], SRCNN [4], FSRCNN [6], VDSR [12], IRCNN [33], SRMDNF [34], RDN [36], RCAN [35], SRFBN [15], SAN [3], HAN [25], FRANet [20] and CRAN [37]. As shown, our solution achieves consistently better performance than other methods even without self-ensemble (i.e. RDRN+).

Visual Results. We show visual comparison for ×3 SR with BD degradation in Fig. 5. The proposed model can recover grid textures and stripes even under heavy blur conditions. In the provided examples, RDRN reconstructs all stripes in the correct direction. In contrast, compared models have problems with the stripes direction and blurred areas.

Table 4. Quantitative results with BD degradation model. The best and second best results are highlighted in **bold** and underlined

Method	Scale	Set5		Set14		B100		Urban100		Manga109	
		PSNR	SSIM	PSNR	SSIM	PSNR	SSIM	PSNR	SSIM	PSNR	SSIM
Bicubic	×3	28.78	0.8308	26.38	0.7271	26.33	0.6918	23.52	0.6862	25.46	0.8149
SPMSR [26]	×3	32.21	0.9001	28.89	0.8105	28.13	0.7740	25.84	0.7856	29.64	0.9003
SRCNN [4]	×3	32.05	0.8944	28.80	0.8074	28.13	0.7736	25.70	0.7770	29.47	0.8924
FSRCNN [6]	×3	26.23	0.8124	24.44	0.7106	24.86	0.6832	22.04	0.6745	23.04	0.7927
VDSR [12]	×3	33.25	0.9150	29.46	0.8244	28.57	0.7893	26.61	0.8136	31.06	0.9234
IRCNN [33]	×3	33.38	0.9182	29.63	0.8281	28.65	0.7922	26.77	0.8154	31.15	0.9245
SRMDNF [34]	×3	34.01	0.9242	30.11	0.8364	28.98	0.8009	27.50	0.8370	32.97	0.9391
RDN [36]	×3	34.58	0.9280	30.53	0.8447	29.23	0.8079	28.46	0.8582	33.97	0.9465
RCAN [35]	×3	34.70	0.9288	30.63	0.8462	29.32	0.8093	28.81	0.8647	34.38	0.9483
SRFBN [15]	×3	34.66	0.9283	30.48	0.8439	29.21	0.8069	28.48	0.8581	34.07	0.9466
SAN [3]	×3	34.75	0.9290	30.68	0.8466	29.33	0.8101	28.83	0.8646	34.46	0.9487
HAN [25]	×3	34.76	0.9294	30.70	0.8475	29.34	0.8106	28.99	0.8676	34.56	0.9494
RFANet [20]	×3	34.77	0.9292	30.68	0.8473	29.34	0.8104	28.89	0.8661	34.49	0.9492
CRAN [37]	×3	34.90	0.9302	30.79	0.8485	29.40	0.8115	29.17	0.8706	34.97	0.9512
RDRN(ours)	×3	35.07	0.9317	31.07	0.8524	29.54	0.8152	29.72	0.8792	35.53	0.9538
RDRN+(ours)	×3	**35.12**	**0.9320**	**31.15**	**0.8533**	**29.57**	**0.8157**	**29.86**	**0.8812**	**35.66**	**0.9543**

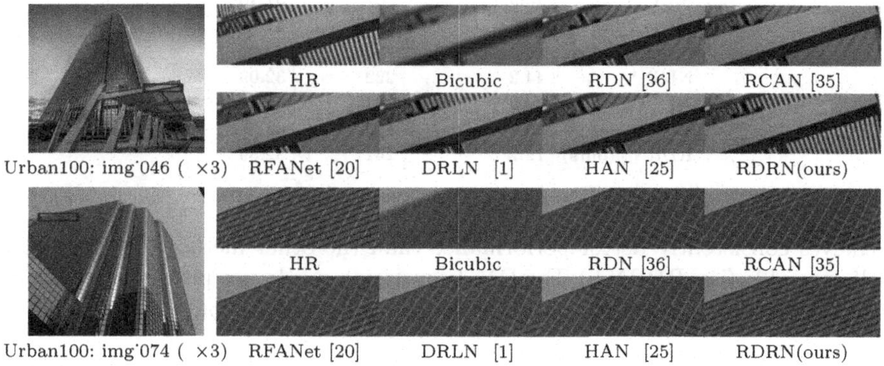

Fig. 5. Visual comparison for 3× SR with BD model on Urban100 dataset.

4.5 Results with Bicubic Downscale-Noise (DN) Degradation Model

We apply our method to super-resolve images with the downscale-noise (DN) degradation model, which is widely used in recent SISR papers [4,6,12,33,33, 36,37]. For DN degradation, HR image is first downscaled with scaling factor ×3, after which Gaussian noise with noise level 30 is added to it.

Quantitative Results. In Table 5 we compare the proposed RDRN model with the following super-resolution methods: SRCNN [4], FSRCNN [6], VDSR [12], IRCNN_G [33], IRCNN_C [33], RDN [36], CRAN [37]. As shown, our solution

Table 5. Quantitative results with DN degradation model. The best and second best results are highlighted in **bold** and underlined

Method	Scale	Set5		Set14		B100		Urban100		Manga109	
		PSNR	SSIM	PSNR	SSIM	PSNR	SSIM	PSNR	SSIM	PSNR	SSIM
Bicubic	×3	24.01	0.5369	22.87	0.4724	22.92	0.4449	21.63	0.4687	23.01	0.5381
SRCNN [4]	×3	25.01	0.6950	23.78	0.5898	23.76	0.5538	21.90	0.5737	23.75	0.7148
FSRCNN [6]	×3	24.18	0.6932	23.02	0.5856	23.41	0.5556	21.15	0.5682	22.39	0.7111
VDSR [12]	×3	25.20	0.7183	24.00	0.6112	24.00	0.5749	22.22	0.6096	24.20	0.7525
IRCNN_G [33]	×3	25.70	0.7379	24.45	0.6305	24.28	0.5900	22.90	0.6429	24.88	0.7765
IRCNN_C [33]	×3	27.48	0.7925	25.92	0.6932	25.55	0.6481	23.93	0.6950	26.07	0.8253
RDN [36]	×3	28.47	0.8151	26.60	0.7101	25.93	0.6573	24.92	0.7354	28.00	0.8591
CRAN [37]	×3	28.74	0.8235	26.77	0.7178	26.04	0.6647	25.43	0.7566	28.44	0.8692
RDRN(ours)	×3	28.81	0.8244	26.87	0.7201	26.11	0.6674	25.73	0.7654	28.80	0.8739
RDRN+(ours)	×3	**28.84**	**0.8251**	**26.90**	**0.7205**	**26.12**	**0.6678**	**25.82**	**0.7678**	**28.88**	**0.8750**

Table 6. Number of parameters, FLOPs (for $3 \times 64 \times 64$ input) and performance on Manga109 with upscale factor ×4 (BI model)

Model	Parameters, M	FLOPs, G	PSNR, dB
SwinIR	11.9	54	32.03
RCAN	15.6	65	31.21
SAN	15.9	67	31.18
HAN	16.9	67	31.42
$RDRN_4$ (ours)	18.5	62	32.09
NLSN*	44.2	222	32.09
NLSN	44.2	222	31.27
$RDRN_5$ (ours)	36.4	123	32.27
$RDRN_6$ (ours)	72.2	241	32.39

achieves consistently better performance than the other methods even without self-ensemble (i.e. RDRN+). For DN degradation we do not provide visual comparison, because the results of recent work is not available publicly.

4.6 Model Complexity Analyses

Figure 1 and Table 6 demonstrate comparison with recent SR works in terms of model size, FLOPs and performance on Manga109 dataset. Compared to NLSN, the number of parameters of our $RDRN_5$ is reduced by **18%.**

Recursion Depth Analysis. We show comparison of the proposed models with different recursion depth $T = 4, 5, 6$. The smaller network $RDRN_4$ outperforms most of the recent methods on Manga109 dataset with comparable complexity and number of parameters.

5 Conclusion

In this paper, we propose a recursively defined residual network (RDRN) for highly accurate image SR. Specifically, recursively defined residual block

(RDRB) allows us to build and train a large and powerful network. To stabilize the training and further improve the quality we apply intermediate supervision (IS) loss function. Training with IS allows the network to learn more informative features for more accurate reconstruction. RDRN achieves superior SISR results under different degradation models, such as bicubic interpolation (BI), blur-downscale (BD) and downscale-noise (DN). Extensive experiments demonstrate that the proposed model outperforms recent state-of-the-art solutions in terms of accuracy and visual quality. The proposed network architecture is general and could be applied to other low-level computer vision tasks. The architecture of the core RDRB block can be simply described by two schemes: basic block and the recursive block. We implement only simple ideas and believe that the proposed approach could be significantly improved either using manual or automatic search.

Acknowledgements. Authors would like to thank Ivan Mazurenko, Li Jieming and Liao Guiming from Huawei for fruitful discussions, guidance and support.

References

1. Anwar, S., Barnes, N.: Densely residual laplacian super-resolution. IEEE Trans. Pattern Anal. Mach. Intell. (TPAMI) (2020)
2. Bevilacqua, M., Roumy, A., Guillemot, C., Alberi-Morel, M.L.: Low-complexity single-image super-resolution based on nonnegative neighbor embedding. In: BMVC (2012)
3. Dai, T., Cai, J., Zhang, Y., Xia, S.T., Zhang, L.: Second-order attention network for single image super-resolution. In: CVPR (2019)
4. Dong, C., Loy, C.C., He, K., Tang, X.: Learning a deep convolutional network for image super-resolution. In: Fleet, D., Pajdla, T., Schiele, B., Tuytelaars, T. (eds.) ECCV 2014. LNCS, vol. 8692, pp. 184–199. Springer, Cham (2014). https://doi.org/10.1007/978-3-319-10593-2_13
5. Dong, C., Loy, C.C., He, K., Tang, X.: Image super-resolution using deep convolutional networks. TPAMI **38**(2), 295–307 (2015)
6. Dong, C., Loy, C.C., Tang, X.: Accelerating the super-resolution convolutional neural network. In: Leibe, B., Matas, J., Sebe, N., Welling, M. (eds.) ECCV 2016. LNCS, vol. 9906, pp. 391–407. Springer, Cham (2016). https://doi.org/10.1007/978-3-319-46475-6_25
7. Dosovitskiy, A., et al.: An image is worth 16x16 words: transformers for image recognition at scale. arXiv preprint arXiv:2010.11929 (2020)
8. He, K., Zhang, X., Ren, S., Sun, J.: Deep residual learning for image recognition. In: CVPR (2016)
9. Huang, G., Liu, Z., Van Der Maaten, L., Weinberger, K.Q.: Densely connected convolutional networks. In: CVPR (2017)
10. Huang, J.B., Singh, A., Ahuja, N.: Single image super-resolution from transformed self-exemplars. In: CVPR (2015)
11. Ioffe, S., Szegedy, C.: Batch normalization: accelerating deep network training by reducing internal covariate shift. arXiv preprint arXiv:1502.03167 (2015)
12. Kim, J., Kwon Lee, J., Mu Lee, K.: Accurate image super-resolution using very deep convolutional networks. In: CVPR (2016)

13. Kim, J., Kwon Lee, J., Mu Lee, K.: Deeply-recursive convolutional network for image super-resolution. In: CVPR (2016)
14. Kingma, D.P., Ba, J.: Adam: A method for stochastic optimization. arXiv preprint arXiv:1412.6980 (2014)
15. Li, Z., Yang, J., Liu, Z., Yang, X., Jeon, G., Wu, W.: Feedback network for image super-resolution. In: CVPR (2019)
16. Liang, J., Cao, J., Sun, G., Zhang, K., Gool, L.V., Timofte, R.: Swinir: image restoration using swin transformer. arXiv preprint arXiv:2108.10257 (2021)
17. Lim, B., Son, S., Kim, H., Nah, S., Mu Lee, K.: Enhanced deep residual networks for single image super-resolution. In: CVPR (2017)
18. Liu, J., Tang, J., Wu, G.: Residual feature distillation network for lightweight image super-resolution. In: European Conference on Computer Vision Workshops (2020)
19. Liu, J., Tang, J., Wu, G.: Adadm: enabling normalization for image super-resolution. arXiv preprint arXiv:2111.13905 (2021)
20. Liu, J., Zhang, W., Tang, Y., Tang, J., Wu, G.: Residual feature aggregation network for image super-resolution. In: CVPR, pp. 2356–2365. IEEE (2020)
21. Liu, Z., et al.: Swin transformer: hierarchical vision transformer using shifted windows. arXiv preprint arXiv:2103.14030 (2021)
22. Martin, D., Fowlkes, C., Tal, D., Malik, J.: A database of human segmented natural images and its application to evaluating segmentation algorithms and measuring ecological statistics. In: ICCV (2001)
23. Matsui, Y., et al.: Sketch-based manga retrieval using manga109 dataset. Multimed. Tools Appl. **76**(20), 21811–21838 (2017)
24. Mei, Y., Fan, Y., Zhou, Y.: Image super-resolution with non-local sparse attention. In: CVPR, pp. 3517–3526. Computer Vision Foundation/IEEE (2021)
25. Niu, B., et al.: Single image super-resolution via a holistic attention network. In: Vedaldi, A., Bischof, H., Brox, T., Frahm, J.-M. (eds.) ECCV 2020. LNCS, vol. 12357, pp. 191–207. Springer, Cham (2020). https://doi.org/10.1007/978-3-030-58610-2_12
26. Peleg, T., Elad, M.: A statistical prediction model based on sparse representations for single image super-resolution. TIP **23**(6), 2569–2582 (2014)
27. Shi, W., et al.: Real-time single image and video super-resolution using an efficient sub-pixel convolutional neural network. In: CVPR (2016)
28. Tai, Y., Yang, J., Liu, X.: Image super-resolution via deep recursive residual network. In: CVPR (2017)
29. Timofte, R., Agustsson, E., Van Gool, L., Yang, M.H., Zhang, L.: Ntire 2017 challenge on single image super-resolution: Methods and results. In: CVPRW (2017)
30. Vaswani, A., et al.: Attention is all you need. In: arXiv preprint arXiv:1706.03762 (2017)
31. Wang, Z., Bovik, A.C., Sheikh, H.R., Simoncelli, E.P.: Image quality assessment: from error visibility to structural similarity. IEEE Trans. Image Process. **13**(4), 600–612 (2004)
32. Zeyde, R., Elad, M., Protter, M.: On single image scale-up using sparse-representations. In: International Conference on Curves and Surfaces (2010)
33. Zhang, K., Zuo, W., Gu, S., Zhang, L.: Learning deep CNN denoiser prior for image restoration. In: CVPR (2017)
34. Zhang, K., Zuo, W., Zhang, L.: Learning a single convolutional super-resolution network for multiple degradations. In: CVPR (2018)
35. Zhang, Y., Li, K., Li, K., Wang, L., Zhong, B., Fu, Y.: Image super-resolution using very deep residual channel attention networks. In: ECCV (2018)

36. Zhang, Y., Tian, Y., Kong, Y., Zhong, B., Fu, Y.: Residual dense network for image super-resolution. In: CVPR (2018)
37. Zhang, Y., Wei, D., Qin, C., Wang, H., Pfister, H., Fu, Y.: Context reasoning attention network for image super-resolution. In: Proceedings of the IEEE/CVF International Conference on Computer Vision (ICCV), pp. 4278–4287 (October 2021)
38. Zhu, X., Hu, H., Lin, S., Dai, J.: Deformable convnets v2: More deformable, better results. In: CVPR (2019)

Super-Attention for Exemplar-Based Image Colorization

Hernan Carrillo[1(✉)], Michaël Clément[1], and Aurélie Bugeau[1,2]

[1] Univ. Bordeaux, CNRS, Bordeaux INP, LaBRI, 5800 Talence, France
{hernan.carrillo-lindado,michael.clement,aurelie.bugeau}@labri.fr
[2] Institut universitaire de France (IUF), Paris, France

Abstract. In image colorization, exemplar-based methods use a reference color image to guide the colorization of a target grayscale image. In this article, we present a deep learning framework for exemplar-based image colorization which relies on attention layers to capture robust correspondences between high-resolution deep features from pairs of images. To avoid the quadratic scaling problem from classic attention, we rely on a novel attention block computed from superpixel features, which we call super-attention. Super-attention blocks can learn to transfer semantically related color characteristics from a reference image at different scales of a deep network. Our experimental validations highlight the interest of this approach for exemplar-based colorization. We obtain promising results, achieving visually appealing colorization and outperforming state-of-the-art methods on different quantitative metrics.

Keywords: Attention mechanism · Colorization · Superpixel

1 Introduction

Colorization is the process of adding plausible color information to grayscale images. Ideally, the result must reach a visually pleasant image, avoiding possible artifacts or improper colors. Colorization has gained importance in various areas such as photo enhancement, the broadcasting industry, films post-production, and legacy content restoration. However, colorization is an inherently ill-posed problem, as multiple suitable colors might exist for a single grayscale pixel, making it a challenging task. This ambiguity on the color decision usually leads to random choices or undesirable averaging of colors. In order to overcome these issue, several colorization approaches have been proposed, and can be categorized into three types: automatic learning-based colorization, scribble-based colorization, and exemplar-based colorization.

Automatic learning-based methods [1–4] bring fairly good colors to grayscale images by leveraging on color priors learned from large-scale datasets. Nonetheless, this type of methods lacks from user's decision. In scribble-based methods [5–9], the user initially adds color scribbles in different image regions, later propagated to the whole image using similarities between neighboring pixels. The last group of methods [10–15] transfers color from one (or many) reference

L. Wang et al. (Eds.): ACCV 2022, LNCS 13842, pp. 646–662, 2023.
https://doi.org/10.1007/978-3-031-26284-5_39

color image to the input grayscale, thus biasing the final image's color to the user's preference. However, when semantic similarities do not exist between the reference image and input image, the efficacy of these exemplar-based methods highly decreases. Therefore, there is an interest in adding information from large-scale datasets and associate it with semantic information of a reference image to no longer depend on just naive pixel-wise matching. To deal with the aforementioned challenge, [16] proposed an attention mechanism for image colorization, mainly by calculating non-local similarities between different feature maps (input and reference images). However, attention mechanisms come with a complexity problem, namely a quadratic scaling problem, due to its non-local operation. This is why it has to be applied on features with low dimensions. On the other hand, low-resolution features lack of detailed information for calculating precise and robust pixel-wise similarities. For instance, low-resolution deep features mainly carry high-level semantic information related to a specific application (*i.e.*, segmentation, classification) that can be less relevant for high-resolution similarity calculation or matching purposes.

In this work, we propose a new exemplar-based colorization method that relies on similarity calculation between high-resolution deep features. These features contain rich low-level characteristics which are important in the colorization task. To overcome the complexity issue, we extend to the colorization task the super-attention block presented in [17] that performs non-local matching over high-resolution features based on superpixels. The main contributions made are:

- A new end-to-end deep learning architecture for exemplar-based colorization improving results over state-of-the-art methods.
- A multiscale attention mechanism based on superpixels features for reference-based colorization.
- A strategy for choosing relevant target/reference pairs for the training phase.

2 Related Work

Colorization with Deep Learning. Automatic learning-based colorization methods use large-scale datasets to learn a direct mapping between each of the grayscale pixels from the input image to a color value. In the earliest work [18], a grayscale image is fed to a neural network that predicts UV chrominance channels from the YUV luminance-chrominance color space by a regression loss. In [19] a grayscale image is given as input to a VGG network architecture, and for each of the pixels a histogram of hue and chroma is predicted, which serve as a guide to the final colorization result. Other deep learning architectures have been used, such as Generative Adversarial Networks (GANs). For instance, [4] proposes to couple semantic and perceptual information to colorize a grayscale image. This is done using adversarial learning and semantic hints from high-level classification features. A different approach is proposed by [20] by defining an autoregressive problem to predict the distribution of every pixel's colors conditioned on the previous pixel's color distribution and the input grayscale image, and addresses the problem using axial attention [21]. Overall, automatic learning-based methods reduce colorization time, in contrast to purely manual colorization; however, this type of methods lacks user-specific requirements.

Exemplar-Based Methods. Another family of colorization methods is the Exemplar-based methods, which requires a reference color image from which colors can be transferred to a target grayscale image. Early work uses matching between luminance and texture information, assuming that similar intensities should have similar colors [10]. Nonetheless, this approach focuses on matching global statistics while ignoring coherent spatial information, leading to unsatisfactory results. In [11], a variational formulation is proposed where each pixel is penalized for taking a chrominance value from a reduced set of possible candidates chosen from the reference image. In more recent approaches, He *et al.* [13] was the first deep learning approach for exemplar-based colorization. This is done by using the PatchMatch [22] algorithm to search semantic correspondences between reference and target images and leveraging the colors learned from the dataset. The authors of [23], inspired by the style transfer techniques, used AdaIN [24] to initially generate a colorized image to be refined by a second neural network. Finally, [13,25] use a deep learning framework that leverages on matching semantic correspondences between two images for transferring colors from the reference color image to the grayscale one. However, these methods mostly rely on pure semantic (low-level) features which leads to imprecise colorization results.

Attention Layers. Recently, attention mechanisms have been a hot topic in particular with transformer architectures [26]. These networks include (self-)attention layers as their main block. The attention layers learn to compute similarities called attention maps between input data or embedded sequences by means of a non-local matching operation [27,28]. Attention layers have been use for colorization in some recent works. For example, [20] presents a transformer architecture capable of predicting high-fidelity colors using the self-attention mechanism. Then, [16] extends the attention mechanism for searching similarities between two different feature maps (target and reference) applied to colorization application, achieving interesting results. Lately, [15] proposes an attention-based colorization framework that considers semantic characteristics and a color histogram of a reference image as priors to the final result. Finally, [29] implements the axial attention mechanism to guide the transfer of the color's characteristics from the reference image to the target image. Even though these methods provide promising colorization results, they may suffer from the quadratic complexity problem of attention layers, and therefore can only perform non-local matching at low resolution. In this work, by using attention layers at the superpixels level, we allow this matching to be done at full resolution.

3 Colorization Framework

Our objective is to colorize a target grayscale image T by taking into consideration the characteristics of a color reference image R. Let $T_L \in R^{H \times W \times 1}$

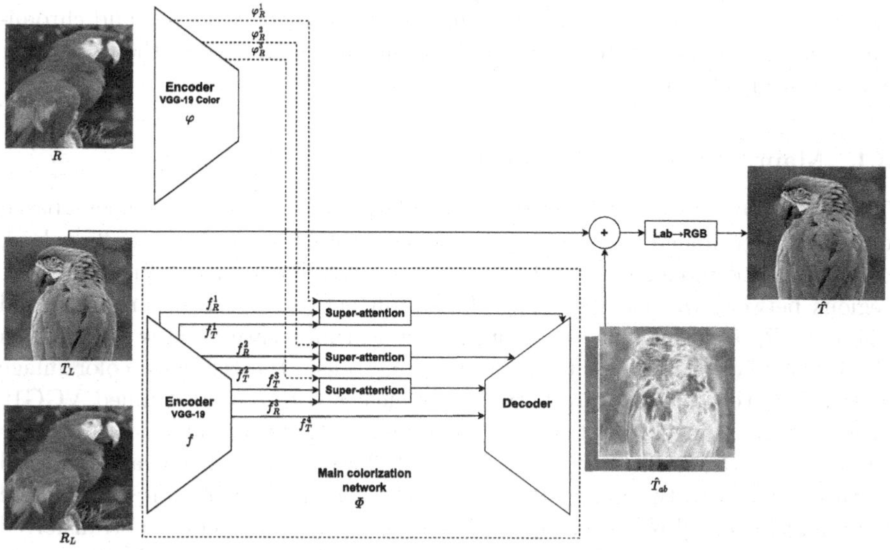

Fig. 1. Diagram of our proposal for exemplar-based image colorization. The framework consists of two main parts: 1) the color feature extractor φ extracts multi-level feature maps φ_R^ℓ from a color reference image R, and 2) the main colorization network Φ, which learns to map a luminance channel image T_L to its chrominance channels $\hat{T}_{ab}$ given a reference color image R. This colorization guidance is done by super-attention modules, which learn superpixel-based attention maps from the target and reference feature maps from distinct levels f_T^ℓ and f_R^ℓ respectively.

be the luminance component of the target, specifically the channel L from the color space Lab in CIELAB, and $R_{Lab} \in R^{H \times W \times 3}$ the color reference image in CIELAB color space. In this work, we choose to work with the luminance-chrominance CIELAB color, as it is perceptually more uniform than other color spaces [30].

In order to colorize the input image, we train a deep learning model Φ that learns to map a grayscale image to its chrominance channels (ab) given a reference color image:

$$\hat{T}_{ab} = \Phi(T_L \mid R). \tag{1}$$

Our proposed colorization framework is composed of two main parts: an external feature extractor φ for color images, and the main colorization network Φ, which relies on super-attention blocks applied at different levels (see Fig. 1). The main colorization network Φ is based on a classical Unet-like encoder-decoder architecture [31], with the addition of the super-attention blocks [17] which allows transferring color hints from the color reference image to the main colorization network. Next, this network predicts the target's chrominance channels $\hat{T_{ab}}$. And, as a final stage, target luminance and chrominance channels are concatenated into $\hat{T}_{Lab}$ and then converted to the RGB color space $\hat{T}$ using Kornia [32].

3.1 Main Colorization Network

The main colorization network Φ aims to colorize a target grayscale image based on a reference image when semantic-related content appears, or pulling back to the learned model when this relation is not present in certain objects or regions between the images. The colorization network receives target T_L and reference R_L as input images to obtain deep learning feature maps f_T^ℓ, f_R^ℓ from the ℓth level of the architecture. In the same sense, the reference color image R is fed to the color feature extractor, which is a frozen pre-trained VGG19 encoder that retrieves multiscale feature maps φ_R^ℓ. Specifically, feature maps are extracted from the first three levels of the encoders. Then, all extracted features are fed to the super-attention blocks, where a correlation is computed between target and reference encoded features. Next, the content is transferred from the reference features to the target by multiplying the similarity matrix and the color reference features. Then, the color features coming from the super-attention modules are transferred to the future prediction by concatenating them to the decoder features. Finally, the decoder predicts the two (ab) chrominance channels $\hat{T}_{ab}$ that are then concatenated to the target luminance channel T_L, then the prediction is converted from CIELAB color space to RGB color space to provide the final RGB image $\hat{T}$. However, this conversion between color spaces arises clipping problems on *RGB* values as in certain cases the combination of predicted Lab values can be outside the conversion range.

3.2 Super-Attention as Color Reference Prior

The super-attention block injects colors priors from a reference image R to the main colorization network Φ. This block relies on multi-level deep features to calculate robust correspondence matching between the target and reference images. Specifically, the super-attention block is divided into two parts: The super-features encoding layer and the super-features matching layer. The super-features encoding layer provides a compact representation of high-resolution deep features using superpixels. For the colorization application, we focus on features maps extracted at the first three levels of the architecture, as they provide a long range of high and low-level features that suit content and style transfer applications [17]. Figure 2 depicts the diagram of the super-attention block where f_T^ℓ, f_R^ℓ and φ_R^ℓ are feature maps from the encoder f and the encoder φ at level ℓ of T_L, R_L and R respectively. This encoding block leverages on superpixels decomposition of the target and reference grayscale images. Each of these decompositions contain N_T and N_R superpixels, respectively, with P_i pixels each, where i is the superpixel index. Finally, the encoding is done by means of a channel-wise *maxpooling* operation resulting in super-features F of size $C \times N$, where $N \ll H \times W$ (a typical superpixel segmentation yields $N \approx \sqrt{H \times W}$ elements).

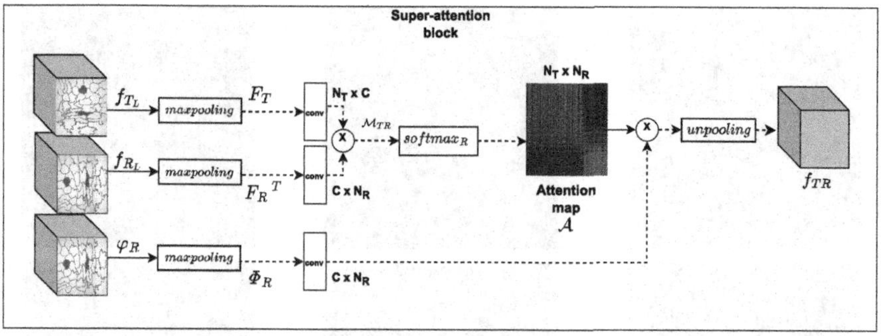

Fig. 2. Diagram of our super-attention block. This layer takes a reference luminance feature map f_R, reference color feature map φ_R and a target luminance feature map f_T, as an input, and learns an attention map at superpixel level by means of a robust matching between high-resolution encoded feature maps.

In summary, this encoding part makes possible operations such as correlation between high-resolution features in our colorization framework.

The super-features matching layer computes the correlation between encoded high-resolution deep-learning features. This layer is inspired by the classic attention mechanism [16] on images to achieve a robust matching between target and reference super-features. However, contrary to [17] our similarity matrix (attention map) is learned by the model. Figure 2 illustrates the process. This layer exploits the non-local similarities between the target F_T and reference F_R super-features, by computing the attention map at layer ℓ as:

$$\mathcal{A}^{\ell} = \text{softmax}({\mathcal{M}^{\ell}}_{TR}/\tau). \tag{2}$$

The softmax operation normalizes row-wise the input into probability distributions, proportionally to the number of target superpixels N_R. Then, the correlation matrix $\mathcal{M}_{TR}$ between target and reference super-features reads:

$$\mathcal{M}^{\ell}_{TR}(i,j) = \frac{(F^{\ell}_T(i) - \mu^{\ell}_T) \cdot (F^{\ell}_R(j) - \mu^{\ell}_R)}{\left\|F^{\ell}_T(i) - \mu^{\ell}_T\right\|_2 \left\|F^{\ell}_R(j) - \mu^{\ell}_R\right\|_2} \tag{3}$$

where μ_T, μ_R are the mean of each super-feature and i, j are the current superpixels from the target and reference respectively.

In terms of complexity, the super-feature encoding approach allows to overcome the quadratic complexity problem in the computation of standard attention maps. Indeed, instead of computing attention maps of quadratic size in the number of pixels, our super-attention maps are computed on a much smaller number of superpixels (*i.e.*, linear in the number of pixels). More details about this complexity reduction can be found in [17].

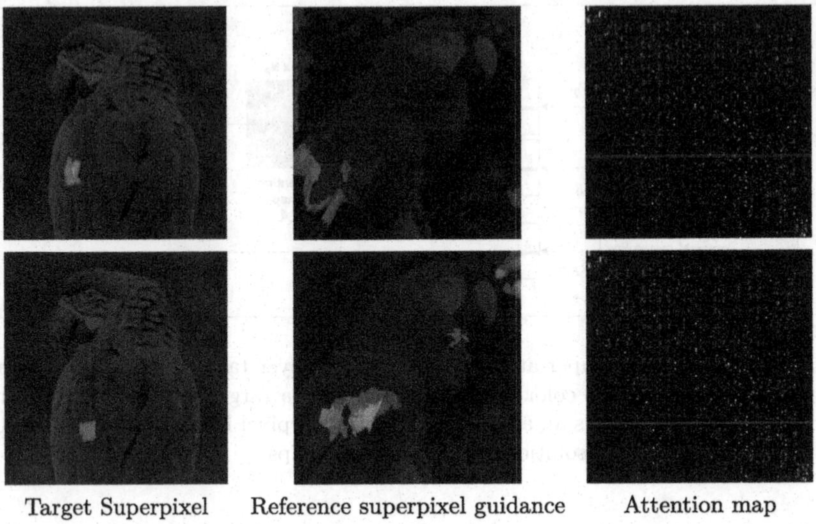

Fig. 3. Example of guidance maps from the super attention mechanism. First column highlights one superpixel in the target; second column, the reference in which the lightness of the superpixels are scaled according to the computed attention map; third column: the attention map with in red the row corresponding to the target superpixel and that is used to generate the second column. (Color figure online)

To illustrate the use of this block between target and reference images, Fig. 3 shows some examples of matching using the super-attention block at the first level of the architecture. Mainly, it shows that for one superpixel from the target feature maps, the learned attention map successfully looks for superpixels on the reference feature maps with similar characteristics to the reference image.

3.3 Training Losses

Designing the loss function in any deep learning model is one of the key parts of the training strategy. In the classical case of automatic colorization, one would like to predict $\hat{T}_{ab}$ by reconstructing the colors from the groundtruth image T_{ab}. But, this idea does not work within exemplar-based colorization as the predicted $\hat{T}_{ab}$ colors should take into account color's characteristics from a reference image $\hat{T}_{ab} = \phi(T_L \mid R)$. Then, the goal is to guarantee an accurate and well-grounded transfer of color characteristics to the target from the reference. In this work, we propose a coupled strategy of two loss terms, L1 smooth and LPIPS, to help reconstruct the final image.

L1 Smooth Loss. To close the gap between T_{ab} and $\hat{T}_{ab}$, and to restore the missing color information when reference similarities may not be found, we propose to use a reconstruction term based on Huber loss, also called L1 smooth. This loss is more robust to outliers than the L_2 loss [33], as well as avoiding averaging colors due to the multi-modal ambiguity problem on colorization [13,14,16]. Notice that T_{ab} and $\hat{T}_{ab}$ are assumed to be the flattened images into 1D vectors, then the L1 smooth is computed as follows:

$$L_{1_{smooth}} = \begin{cases} 0.5\left(T_{ab} - \hat{T}_{ab}\right)^2, & \text{if } \left|T_{ab} - \hat{T}_{ab}\right| < 1 \\ \left|T_{ab} - \hat{T}_{ab}\right| - 0.5, & \text{otherwise.} \end{cases} \tag{4}$$

VGG-Based LPIPS Loss. To encourage the perceptual similarity between the groundtruth target image T and the predicted one $\hat{T}$, we use the LPIPS loss [34]. Given a pretrained network, LPIPS is computed as a weighted L_2 distance between deep features of the predicted colorized target image $\hat{T}$ and the groundtruth T. In this work, feature maps f_T^ℓ and $\hat{f}_T^\ell$ are obtained from a pre-trained VGG network, and the loss term is computed as:

$$L_{LPIPS} = \sum_\ell \frac{1}{H^\ell W^\ell} \left\| \omega^\ell \odot \left(f_T^\ell - \hat{f}_T^\ell\right) \right\|_2^2 \tag{5}$$

where H^ℓ (resp. W^ℓ) is the height (resp. the width) of feature map f_T^ℓ at layer ℓ and ω^ℓ are weights for each feature. The features are unit-normalized in the channel dimension. Note that to compute this VGG-based LPIPS loss, both input images have to be in RGB color space and normalized on range $[-1, 1]$. As our initial prediction is in the Lab color space, to apply backpropagation it has to be converted to RGB in a differentiable way using Kornia [32].

Total Loss. Finally, these two loss terms, $L_{1_{smooth}}$ and L_{LPIPS}, are summed by means of different fixed weights which allow to balance the total loss. The joint total loss used on the training phase is then:

$$L_{total} = \lambda_1 L_{1_{smooth}} + \lambda_2 L_{LPIPS} \tag{6}$$

where λ_1, λ_2 are fixed weights for each of the individual losses.

Notice that some previous exemplar-based colorization methods proposed to add an additional histogram loss to favor color transfer [14,15,29]. As we do not want to enforce a complete transfer of all colors from the reference, but only the ones that are relevant with the source image, we have decided not to use it in our model. We provide in supplementary material additional experiments that show that adding this loss can decrease the quality of the results.

3.4 Implementation Details

For the training phase, we set the weights of two terms of the loss to $\lambda_1 = 10$, $\lambda_2 = 0.1$. These values were chosen empirically to obtain a good balance between the $L_{1_{smooth}}$ reconstruction term and the LPIPS semantic term. We train the network with a batch size of 8 and for 20 epochs. We use the Adam optimizer with a learning rate of 10^-5 and $\beta_1 = 0.9$, $\beta_2 = 0.99$. Finally, our model is trained with a single GPU NVIDIA RTX 2080 Ti and using PyTorch 1.10.0.

For the super-attention modules, superpixel segmentations are calculated on the fly using the SLIC algorithm [35]. Multiscale superpixels grid are calculated using downsampled versions of the grayscale target T_L and reference R_L images. These downsamplings are performed to match the size of the feature maps from the first three levels of the encoder f.

4 Dataset and References Selection

We train the proposed colorization network using the COCO dataset [36]. Unlike ImageNet, this dataset proposes a smaller quantity of images but with more complex scene structures, depicting a wider diversity of objects classes. Additionally, the dataset provides the segmentation of objects, which we rely on for our target/reference images pairs strategy. In detail, this dataset is composed of 100k images for training and 5k images for validation. For the training procedure, we resize the images to the size of 224×224 pixels.

Another key aspect of the training strategy in exemplar-based methods is to find a suitable semantic reference to the target image. To build the target-reference pairs of images, we took inspiration from [13] to design our ranking of reference images. There, they proposed a correspondence recommendation pipeline based on grayscale images. Here, our approach focuses on searching target-reference matches from a wide variety of segmented objects as well as natural scenes images. Our proposal ranks four images semantically related to each target image. First, to increase the variety of reference images within a category, we extract each meaningful object whose size is greater than 10% of the size of the current image. To retrieve the image objects, we use the segmentations provided by the dataset. Second, we compute semantically rich features from the fifth level of a pre-trained VGG-19 [37] encoder φ_T^5 and φ_R^5. Next, for each target, reference images are ranked based on the L_2 distance between these pre-computed features. Finally, during training, target-reference pairs of images are sampled using a uniform distribution with a weight of 0.25 by randomly choosing either itself (*i.e.*, the groundtruth target image is used as the reference) or the other top-4 semantically closest references.

Figure 4 shows examples of target-reference matching based on our proposal. The target images are presented in the first column, and the following columns represent its corresponding reference based on the nearest L_2 distance between feature maps. The second column (Top 1) shows the references most semantically-related to the target, while the last column (Top 4) shows the references the least semantically-related to the target.

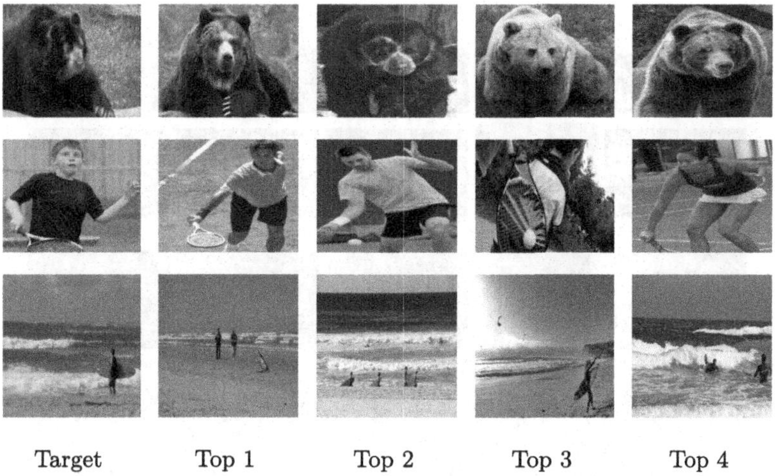

Fig. 4. Illustration of our reference selection method. The first column shows example target images, and the next four columns show the closest references in the dataset, in decreasing order of similarity.

5 Experimental Validations

In this section, we present quantitative and qualitative experimental results to illustrate the interest of our proposed approach. First, we propose an analysis of our method with an ablation study comparing different architectural choices and training strategies. Then, we compare our results to three state-of-the-art exemplar-based colorization approaches.

5.1 Analysis of the Method

We start by analyzing quantitatively and qualitatively certain variants of our proposed colorization framework. Within this ablation study, we analyze two variants. The first one is a baseline colorization model without using references. It uses the same generator architecture without any attention layer. The second variant is our framework using a standard attention layer in the bottleneck of the architecture instead of our super-attention blocks. Finally, the third model is our proposed exemplar-based colorization framework, which includes the use of references with super-attention blocks in the top three levels of the network.

Evaluation Metrics. To evaluate quantitatively the results of these different methods, we used three metrics. Two metrics that compare the result with the groundtruth color image, and a third metric that compare the prediction of colors with respect to the reference color image. The first one is the structural similarity (SSIM) metric [38], which analyzes the ability of the model to reconstruct the

Fig. 5. Colorization results obtained using different variants of our colorization framework. In the first two lines, the chosen color reference is different from the target to be colorized (but semantically similar). In the last two lines, the reference correspond to the actual color version of the target image. These results allow to assess the ability of our method to transfer relevant colors from different types of reference images. (Color figure online)

original image content. The second one is the learned perceptual image patch similarity (LPIPS) metric [34] which correlates better with the human perceptual similarity. These first two metrics (SSIM and LPIPS) evaluate the quality of the output colorization compared to the groundtruth. The third metric employed is the histogram intersection similarity (HIS) [39] which is computed between the predicted colorization and the reference image. The goal of this third metric is to evaluate if the colors from the reference have been correctly transferred to the prediction. However, unless the groundtruth and reference share the same color distribution, these metrics are inherently contradictory (*i.e.*, good HIS would lead to bad SSIM/LPIPS). In this work, we do not necessarily want to fully transfer the colors from the reference image. Instead, we view the reference as colors hints that can be used by our network to predict a more plausible colorization. Therefore, we consider the HIS between the groundtruth target images and the reference images to be optimal in this sense (*i.e.*, it would be the score obtained with perfect predictions). On the COCO validation set, we obtain this reference, HIS $= 0.542$, by picking the top 1 reference for each groundtruth. In the following, to better assess the quality of our results, we propose to report ΔHIS, that is the absolute difference between the average HIS obtained with our predictions and the reference HIS of 0.542.

Table 1. Quantitative analysis of our model. SSIM and LPIPS metrics are calculated with respect to the target groundtruth image. ΔHIS is the absolute difference between the average HIS from the reference and the colorized prediction.

Method	SSIM↑	LPIPS↓	ΔHIS↓
Ours without reference	0.920	0.164	–
Ours with standard attention	0.921	0.172	0.081
Ours	**0.925**	**0.160**	**0.054**

Results and Ablation Study. The results displayed in Table 1 are the averages of the metrics calculated using the evaluation set of 5000 pairs of target/reference images from the COCO validation set. From this Table, we can observe that our full colorization framework achieves the best SSIM and LPIPS scores in comparison with the other variants of the model, suggesting that the use of reference images with super-attention blocks helps in getting better colorization results. We can also notice that the full model achieves a better ΔHIS than the standard attention variant. This suggests that, instead of forcing a global transfer of colors from the references, our model is capable of picking specific and plausible colors from the reference images to generate better colorization results. Note that the ΔHIS is not reported for the first line, as this variant does not use any reference image.

In addition to this quantitative evaluation, in Fig. 5 we present a qualitative comparison of our method and its ablation variants. From the results of the first and second row, we can see that the method with standard attention proposes a more global transfer of colors, leading to the appearance of brighter colors related to the reference. However, this also causes unnatural colorization around the carriage in the first example (*i.e.*, green, orange) and on the hand of the baseball player in the second example. On the other hand, our method with super-attention block overcomes this issue and provides a more natural colorization, using specific colors from the reference. The last two lines show the effectiveness of our method when the reference is semantically identical to the target. Indeed, it shows that color characteristics from the reference image are passed with high fidelity, as compared to the variant without reference where it results in opaque colors.

In overall, our framework achieves a better visually pleasant colorization, and the transfer of color characteristics from the reference is done in specific areas of the target image, thanks to our super-attention blocks. When the color reference and the super-attention modules are not included in the framework, we observe a decrease of quality in the resulting colors (*i.e.*, averaging of colors and/or wrong colorization).

5.2 Comparison with Other Exemplar-Based Colorization Methods

To evaluate the performance of our exemplar-based colorization framework, we compare our results quantitatively and qualitatively with three other

Fig. 6. Comparison of our proposed method with different reference-based colorization methods: Deep Exemplar [13], Just Attention [15] and XCNET [29].

state-of-the-art exemplar-based image colorization approaches [13,15,29]. In order to fairly compare these methods, we run their available codes for the three approaches using the same experimental protocol and the same evaluation set as in the previous section.

A quantitative evaluation is proposed to compare the three state-of-the-art methods and our framework. For comparing the methods, we again use SSIM, LPIPS, and ΔHIS. As shown in Table 2 our colorization framework preserves stronger perceptual structural information from the original target image with respect to all three state-of-the-art methods. LPIPS score measures the perceptual similarity between colorized results and target groundtruth. Our method surpasses [13,15,29] significantly. Finally, our method achieves a smaller ΔHIS with respect to all compared state-of-the-art methods.

Figure 6 shows colorization results from [13,15,29] and our approach. For the first two images, our proposal produces a more visually pleasant colorization and

Table 2. Quantitative comparison with three state-of-the-art exemplar based-colorization methods. Ours correspond to our full model with references and super-attention blocks.

Method	SSIM ↑	LPIPS ↓	ΔHIS↓
XCNET [29]	0.867	0.270	0.139
Deep Exemplar [13]	0.894	0.241	0.127
Just Attention [15]	0.896	0.239	0.125
Ours	**0.925**	**0.160**	**0.054**

provides more natural colors than the other methods. In contrast, the results for the first two images of [15] and [29] shows a high amount of color bleeding, mainly on top of the baseball player and on the background, as their approach captured global color distribution from the reference image. For the fourth image, methods [15] and [29] failed to transfer the color information from the red jacket. Conversely, [13] and our approach did transfer the jacket's color. Next, in the results for the fifth image, we observe unnatural colors as the green water for methods [13] and [29]. In contrast, our method and [15] achieves a natural colorization by transferring colors from the reference. Finally, for the results of the last image, [13] encourages the transfer of the colors from the reference image better than the other methods; however, the final colorization results seem unnatural. For the same image, the colorization results on method [29] and ours seem to be the right balance between color transfer and naturalness from the learned colorization model.

6 Conclusion

In this paper, we have proposed a novel end-to-end exemplar-based colorization framework. Our framework uses a multiscale super-attention mechanism applied on high-resolution features. This method learns to transfer color characteristics from a reference color image, while reducing the computational complexity compared to a standard attention block. In this way, we coupled into one network both semantic similarities from a reference and the learned automatic colorization process from a large dataset. Our method outperforms quantitatively state-of-the-art methods, and achieves qualitatively competitive colorization results in comparison with the state-of-the-art methods.

The transfer of colors which serve as a colorization guidance in the exemplar-based methods plays a key relevance on the final results. In our method, we introduce semantically-related colors characteristics from the reference by means of the super-attention block. This block let us find correspondences at different levels of the architecture, resulting in rich low-level and high-level information. Yet we suspect that concatenating the retrieved features of our attention mechanism do not completely enforce a strong guidance from reference color and future work should concentrate on a scheme for finer transfer. One solution we will consider is adding segmentation masks as in [3] inside our model.

Finally, another future line of research aims at coupling scribbles [5,9,40] as user hints with our reference-based colorization framework.

Acknowledgements. This study has been carried out with financial support from the French Research Agency through the PostProdLEAP project (ANR-19-CE23-0027-01).

References

1. Zhang, R., Isola, P., Efros, A.A.: Colorful image colorization. In: Leibe, B., Matas, J., Sebe, N., Welling, M. (eds.) ECCV 2016. LNCS, vol. 9907, pp. 649–666. Springer, Cham (2016). https://doi.org/10.1007/978-3-319-46487-9_40
2. Deshpande, A., Lu, J., Yeh, M.C., Chong, M.J., Forsyth, D.: Learning diverse image colorization. In: Conference on Computer Vision and Pattern Recognition (2017)
3. Su, J.W., Chu, H.K., Huang, J.B.: Instance-aware image colorization. In: Conference on Computer Vision and Pattern Recognition (2020)
4. Vitoria, P., Raad, L., Ballester, C.: ChromaGAN: adversarial picture colorization with semantic class distribution. In: Winter Conference on Applications of Computer Vision (2020)
5. Levin, A., Lischinski, D., Weiss, Y.: Colorization using optimization. ACM Trans. Graph. **23** (2004)
6. Huang, Y.C., Tung, Y.S., Chen, J.C., Wang, S.W., Wu, J.L.: An adaptive edge detection based colorization algorithm and its applications. In: ACM International Conference on Multimedia (2005)
7. Qu, Y., Wong, T.T., Heng, P.A.: Manga colorization. In: International Conference on Computer Graphics and Interactive Techniques (2006)
8. Zhang, R., et al.: Real-time user-guided image colorization with learned deep priors. ACM Trans. Graph. **36**, 119 (2017)
9. Zhang, L., Li, C., Simo-Serra, E., Ji, Y., Wong, T.T., Liu, C.: User-guided line art flat filling with split filling mechanism. In: Conference on Computer Vision and Pattern Recognition (2021)
10. Welsh, T., Ashikhmin, M., Mueller, K.: Transferring color to greyscale images. ACM Trans. Graph. (2002)
11. Bugeau, A., Ta, V.T., Papadakis, N.: Variational exemplar-based image colorization. IEEE Trans. Image Process. **23**, 298–307 (2014)
12. Chia, A.Y.S., et al.: Semantic colorization with internet images. ACM Trans. Graph. **30**, 156 (2011)
13. He, M., Chen, D., Liao, J., Sander, P.V., Yuan, L.: Deep exemplar-based colorization. ACM Trans. Graph. **37**, 1–16 (2018)
14. Lu, P., Yu, J., Peng, X., Zhao, Z., Wang, X.: Gray2ColorNet: transfer more colors from reference image. In: ACM International Conference on Multimedia, pp. 3210–3218 (2020)
15. Yin, W., Lu, P., Zhao, Z., Peng, X.: Yes, "attention is all you need", for exemplar based colorization. In: ACM International Conference on Multimedia (2021)
16. Zhang, B., et al.: Deep exemplar-based video colorization. In: Conference on Computer Vision and Pattern Recognition (2019)
17. Carrillo, H., Clément, M., Bugeau, A.: Non-local matching of superpixel-based deep features for color transfer. In: International Conference on Computer Vision Theory and Applications (2022)
18. Cheng, Z., Yang, Q., Sheng, B.: Deep colorization. In: International Conference on Computer Vision (2015)

19. Larsson, G., Maire, M., Shakhnarovich, G.: Learning representations for automatic colorization. In: Leibe, B., Matas, J., Sebe, N., Welling, M. (eds.) ECCV 2016. LNCS, vol. 9908, pp. 577–593. Springer, Cham (2016). https://doi.org/10.1007/978-3-319-46493-0_35
20. Kumar, M., Weissenborn, D., Kalchbrenner, N.: Colorization transformer. In: International Conference on Learning Representations (2021)
21. Ho, J., Kalchbrenner, N., Weissenborn, D., Salimans, T.: Axial attention in multidimensional transformers. arXiv preprint arXiv:1912.12180 (2019)
22. Barnes, C., Shechtman, E., Finkelstein, A., Goldman, D.B.: PatchMatch: a randomized correspondence algorithm for structural image editing. ACM Trans. Graph. **28**, 24 (2009)
23. Xu, Z., Wang, T., Fang, F., Sheng, Y., Zhang, G.: Stylization-based architecture for fast deep exemplar colorization. In: Conference on Computer Vision and Pattern Recognition (2020)
24. Huang, X., Belongie, S.: Arbitrary style transfer in real-time with adaptive instance normalization. In: International Conference on Computer Vision (2017)
25. Iizuka, S., Simo-Serra, E.: DeepRemaster: temporal source-reference attention networks for comprehensive video enhancement. ACM Trans. Graph. **38**, 1–13 (2019)
26. Vaswani, A., et al.: Attention is all you need. In: Advances in Neural Information Processing Systems (2017)
27. Yu, J., Lin, Z., Yang, J., Shen, X., Lu, X., Huang, T.S.: Generative image inpainting with contextual attention. In: Conference on Computer Vision and Pattern Recognition (2018)
28. Wang, X., Girshick, R., Gupta, A., He, K.: Non-local neural networks. In: Conference on Computer Vision and Pattern Recognition (2018)
29. Blanch, M.G., Khalifeh, I., Smeaton, A., Connor, N.E., Mrak, M.: Attention-based stylisation for exemplar image colourisation. In: IEEE International Workshop on Multimedia Signal Processing (2021)
30. Connolly, C., Fleiss, T.: A study of efficiency and accuracy in the transformation from RGB to CIELAB color space. IEEE Trans. Image Process. **6**(7), 1046–1048 (1997)
31. Ronneberger, O., Fischer, P., Brox, T.: U-Net: convolutional networks for biomedical image segmentation. In: Navab, N., Hornegger, J., Wells, W.M., Frangi, A.F. (eds.) MICCAI 2015. LNCS, vol. 9351, pp. 234–241. Springer, Cham (2015). https://doi.org/10.1007/978-3-319-24574-4_28
32. Riba, E., Mishkin, D., Ponsa, D., Rublee, E., Bradski, G.: Kornia: an open source differentiable computer vision library for PyTorch. In: Winter Conference on Applications of Computer Vision, pp. 3674–3683 (2020)
33. Ren, S., He, K., Girshick, R., Sun, J.: Faster R-CNN: towards real-time object detection with region proposal networks. In: Advances in Neural Information Processing Systems (2015)
34. Zhang, R., Isola, P., Efros, A.A., Shechtman, E., Wang, O.: The unreasonable effectiveness of deep features as a perceptual metric. In: Conference on Computer Vision and Pattern Recognition (2018)
35. Achanta, R., Shaji, A., Smith, K., Lucchi, A., Fua, P., Süsstrunk, S.: SLIC superpixels compared to state-of-the-art superpixel methods. IEEE Trans. Pattern Anal. Mach. Intell. **34**, 2274–2282 (2012)
36. Lin, T.-Y., et al.: Microsoft COCO: common objects in context. In: Fleet, D., Pajdla, T., Schiele, B., Tuytelaars, T. (eds.) ECCV 2014. LNCS, vol. 8693, pp. 740–755. Springer, Cham (2014). https://doi.org/10.1007/978-3-319-10602-1_48

37. Simonyan, K., Zisserman, A.: Very deep convolutional networks for large-scale image recognition. In: International Conference on Learning Representations (2015)
38. Wang, Z., Bovik, A., Sheikh, H., Simoncelli, E.: Image quality assessment: from error visibility to structural similarity. IEEE Trans. Image Process. **13**(4), 600–612 (2004)
39. Isola, P., Zhu, J.Y., Zhou, T., Efros, A.A.: Image-to-image translation with conditional adversarial networks. In: Conference on Computer Vision and Pattern Recognition (2017)
40. Heu, J., Hyun, D.Y., Kim, C.S., Lee, S.U.: Image and video colorization based on prioritized source propagation. In: International Conference on Image Processing (2009)

Author Index

L. Wang et al. (Eds.): ACCV 2022, LNCS 13842, pp. 663–665, 2023.
https://doi.org/10.1007/978-3-031-26284-5

Printed in the United States
by Baker & Taylor Publisher Services